TIM CLARK (ED.)

Molecular Modeling Annual

CD-ROM AND PRINT ARCHIVE EDITION
JOURNAL OF MOLECULAR MODELING
VOLUME 2, 1996

 Springer

Aims and Scope and Call for Papers

The Journal of Molecular Modeling is the first fully electronic journal in chemistry. It is the perfect medium for presenting results of scientific research produced in the computer, offering the unique possibility of applying all the daily tools of computational chemistry to the world of scientific publishing to the benefit of both authors and readers.

The aim of the Editor and the Publisher is to produce a high quality chemistry journal that takes advantage of modern electronic communication techniques. The use of electronic media in every step of the publication process leads to unprecedentedly short publication times. Thus the journal offers many advantages important to the active scientist in the field:

- fully peer reviewed; fully citable (ISSN; CAS; ISI);
- no page charges; full colour;
- no extra costs for sophisticated presentation materials;
- widespread visibility of your research in the Internet.

Graphical abstracts of accepted papers are generally accessible on the WWW-site of the Springer Verlag (URL http://science.springer.de/ or hhtp://www.springer-ny.com/) or at the Computer-Chemie-Centrum Erlangen. (URL: http://www.organik.uni-erlangen.de) Complete papers are made accessible through servers to subscribers with a minimum availability time for individual papers of two years. The mandate to make a journal a permanent archive of science is fullfilled by the publication of the *Molecular Modeling Annual 1996 - CD-ROM and print archive edition of the Journal of Molecular Modeling Vol. 2.*

The Journal of Molecular Modeling publishes all quality science that passes the critical review of experts in the field and falls within the scope of the journal coverage, including:

- computer-aided molecular design
- visualisation, classification and handling of chemical data
- rational drug design, de novo ligand design and receptor modeling
- computationally enhanced desktop tools for the life sciences
- protein and peptide modeling
- molecular mechanics/dynamics simulation of polymers and biopolymers
- prediction of biological activities (QSAR) and physico-chemical properties (QSPR)
- genetic algorithms and neural nets
- catalyst modeling

The editorial office of the *Journal of Molecular Modeling* is located at the Computer-Chemie-Centrum of the Institut für Organische Chemie, Friedrich-Alexander-Universität Erlangen-Nürnberg. As the Editor-in-Chief I encourage all scientists to submit their contributions making use of full colour graphics and/or other sophisticated presentation techniques, such as videos, VRML-scenes, 3D graphics representation, program demos and to include supplemental material in their publications.

I look forward to welcoming you as a new author of this exciting journal.
Best wishes,

Dr. Tim Clark

Editor-in-Chief: Tim Clark
Journal of Molecular Modeling
Computer-Chemie-Centrum
Nägelsbachstraße 25
91052 Erlangen
Germany

Tel: +49-(0)9131-852948
Fax: +49-(0)9131-856565
jmolmod@organik.uni-erlangen.de

Publisher: © Springer-Verlag Berlin Heidelberg 1996
Springer-Verlag GmbH & Co. KG
Berlin, Germany

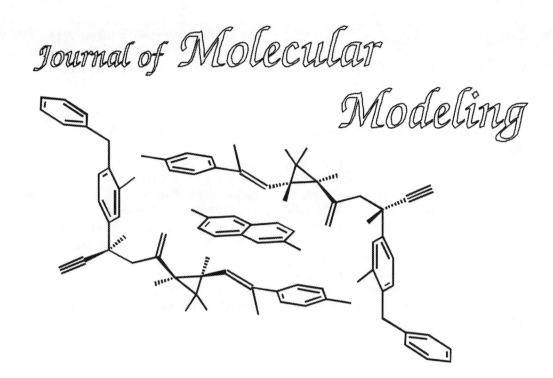

Journal of Molecular Modeling

Editor

Tim Clark, Computer-Chemie-Centrum Erlangen, Germany

International Editorial Board

Jürgen Brickmann, TH Darmstadt, Germany

Frank Blaney, Smithkline Beecham, United Kingdom

Dan Dolata, University of Ohio, USA

Dave Jackson, Oxford Molecular Ltd., United Kingdom

Peter Murray-Rust, Nottingham University, United Kingdom

Graham Richards, Oxford University, United Kingdom

Jean L. Rivail, Universite de Nancy, France

Henry Rzepa, Imperial College London, United Kingdom

James J.P. Stewart, Colorado, USA

Assistant Editor

Henryette Roth, Computer-Chemie-Centrum Erlangen, Germany

Springer

Copyright

Subscription information

Recommended annual subscription rates:

Online edition (ISSN 0948-5023)
personal subscription	DM 125.–
library or industrial subscription	DM 396.–

CD-ROM (ISSN 1430-8622)
Volume 2 1996 (ISBN 3-540-14567-2)	DM 148.–
reduced price for personal subscribers	DM 73.–
reduced price for library/industrial subscribers	DM 100.–

Online edition (ISSN 0948-5023) + **CD-ROM** (ISSN 1430-8622) **Volume 2 1996** (ISBN 3-540-14567-2)
personal subscription	DM 198.–
library or industrial subscription	DM 496.–

Print edition (ISSN 0949-183X) -
CD-ROM + Print Archive Edition
Volume 2 1996 (ISBN 3-540-14566-4)	DM 298.–

All prices are suggested retail prices plus 15% VAT in Germany. Customers in other EU countries without VAT Identification Number, please add local VAT. Prices exclude carriage charges. Prices are subject to change without notice.

Orders can be sent directly to:

Springer-Verlag
Hans-Martin Scheidt
Heidelberger Platz 3
D-14197 Berlin, Germany
Tel. (0) 30/82787-303 , FAX (0) 30/82787-301
E-mail: subscriptions@springer.de

Electronic subscription order forms are available at the Springer WWW-server
(http://science.springer.de/jmm/orderfo.htm)

Cancellations must be received by September 30 to take effect at the end of the same year.

Changes of address / e-mail address: Allow six weeks for all changes to become effective. All communications should include both old and new addresses (with Postal Codes).

Production

Springer-Verlag
Beate Gärtner
Journal Production Department II
Postfach 10 52 80
D-69042 Heidelberg, Germany
Tel (0) 6221/487-905, FAX (0) 6221/487-967
E-mail: gaertner@springer.de

Cover Design

Konzept & Design, Ilvesheim

Publisher

© Springer-Verlag Berlin Heidelberg 1996
Springer-Verlag GmbH & Co. KG
Berlin, Germany
Printed in Germany

SPIN 10532229
Printed on acid-free paper

Contents

Guidelines for Authors

General Guidelines for Manuscript Preparation and Submission

The *Journal of Molecular Modeling* covers all aspects of computational chemistry. Papers may be submitted either electronically or as conventional hardcopy, but accepted papers must be made available in electronic form. The following files can be used for the production of the Journal, other formats will be handled on an experimental basis:

AMI Pro 1.1 - 3.0
Microsoft Publisher 2.0
Microsoft Word
Microsoft Works 2.0 - 3.0
Rich Text Format (RTF)
Text-only (ASCII)
Ventura Publisher
Word for Macintosh
WordPerfect
Word Star

Papers should follow the guidelines issued by the *American Chemical Society* for the *Journal of Organic Chemistry* whereever no other specifications are given. Manuscripts must be prepared with a word processor using 10 point *Times Roman* on DIN A4 (21.0 x 29.7 cm). Pages must be numbered. Authors should consult recent issues of the Journal and *The ACS Style Guide*, American Chemical Society: Washington, DC, 1986, for format guidance. Any author who is not fully fluent in idiomatic English is urged to gain assistance with manuscript preparation from a fluent colleague, as manuscripts with grammar deficiencies are sometimes handicapped during the scientific review process.

Papers should be submitted directly via e-mail to jmolmod@organik.uni-erlangen.de or via anonymous FTP to iris2.organik.uni-erlangen.de or via an e-mail message describing how to download the PostScript version or common word-processor files of the paper from the author's local machine.

Alternatively four copies of hardcopy papers or the manuscript as 3 1/2" PC or Macintosh diskettes may be sent by conventional mail to:

Journal of Molecular Modeling
Computer-Chemie-Centrum
Universität Erlangen-Nürnberg
Nägelsbachstr. 25
D - 91052 Erlangen
Germany

Supplementary material will be published to accompany the papers and its use is strongly encouraged. The work should be reported so that it can be repeated using the information given in the paper and the supplementary material. Detailed Guidelines can be found in *J.Med.Chem.* **1988**, *31*, 2230.

Titles, Keywords, Authors and Abstracts

Titles are of great importance for current awareness and for information retrieval. Additionally, 3 to 5 keywords must be given, chosen carefully to provide information on the contents and to function as "points of entry" for retrieval purposes. Be consistent in author designation; supply given name, middle initial, and last name for complete identification. The corresponding author's mailing address, phone number, fax number, and e-mail address should be included.

All manuscripts must be accompanied by a graphical abstract, which should state briefly the purpose of the research (if this is not contained in the title), the principal results, and major conclusions. The abstract should be acompanied by up to 3 graphics. For a typical paper, an 80- to 200-word abstract is usually adequate. If the manuscript is accepted for publication, the abstract will be published at the journals WWW-page. Abstracts should therefore be submitted as separate file in ASCII or HTML format with the graphics as GIF, JPG or RGB-file.

Contents of Manuscripts

Papers will be accepted from academic, government, industrial and private research institutions and from software developers. Papers (article and notes) are accepted on the understanding that the work described is original and has not been published elsewhere and that the author has obtained any necessary authorization for publication of the material submitted. Authors are solely responsible for the factual accuracy of their contributions. The *Journal of Molecular Modeling* will publish a limited number of reviews on current problems of interest. These will either be solicited by the editor or should be submitted after consultation with him.

All sections of the paper must be presented in as concise a manner as possible consistent with clarity of expression. An introductory paragraph or statement should be given, placing the work in the appropriate context and clearly stating the purpose and objectives of the research. The background discussion should be brief and restricted to pertinent material. An extensive review of prior work is not appropriate and documentation of the literature should be selective rather than exhaustive, particularly if reviews can be cited.

The computational procedure and discussion sections should be clearly distinguished with a separate heading; other headings should be used sparingly. Tabulation of results is encouraged whenever it leads to a more effective presentation or economical use of space.

Structural Drawings

Structures and figures will be taken directly from the authors source **without** reduction during the layout process. Equations, schemes and blocks of structures are presented in the Journal either in one-column or two-column format. Schemes are numbered with Arabic numerals and should be provided with a descriptive footnote. Within schemes, structures should be numbered with boldface Arabic numerals. Each graphical file should contain only one scheme, equation or block of structures.

Graphical material must be supplied in the actual size that is appropriate for one- or two-column format (8.5 or 17.8 cm width, respectively). The Journal's editors would like to achieve as much uniformity as possible in the printed chemical structures. As such, the bond length, line thickness, and bold thickness of the structures should be *0.508 cm, 0.021 cm*, and *0.071 cm* respectively, and the *10-pt Times Roman* font should be used. All atom labels and text should be in the "plain" style (lightface) with the exception of compound numbers, which should be boldface.

For the users of ChemDraw 2.x and 3.x programs CambridgeSoft provides stationary pads and style sheets that include the requested standard measurements. The stationary pads/style sheets are included in the program or can be downloaded via World Wide Web from http://www.camsoft.com/suppfiles/jmolmod.html.

Authors using the ChemIntosh and ChemWindow programs will find the necessary parameters incorporated into the 'JOC style' to be found in version 3.0 of each, but have to change the font from *Helvetica* to *Times Roman*. Authors using other programs should select parameters consistent with the following specifications:

10-pt Times Roman font
fixed length 14.4 pt (0.508 cm, 0.2 in.)
bold width 2.0 pt (0.071 cm, 0.0278 in.)
line width 0.6 pt (0.021 cm, 0.0083 in.)
tolerance 3 pt (0.106 cm, 0.0417 in.)
margin width 1.6 pt (0.056 cm, 0.0222 in.)
hash spacing 2.5 pt (0.088 cm, 0.0345 in.)
bond spacing 18% of width
Set the page to 100%

Using the ruler or appropiate margin settings, create structure blocks, schemes and equations having maximum widths of 8.5 cm (one-column format) or 17.8 cm (two-column format). Embolden compound numbers, but not atom labels or captions.

The following format for chemical structure diagrams will be accepted:

MDL Mol-file (MOL)
ISIS-Draw Sketch file (SKC)
ChemWindow (CW3)
ChemDraw (CHM)
C-Design or ChemStructure
WordPerfect Graphic (WPG)
Windows™ Meta-file (WMF)
Encapsulated PostScript (EPS)

Other file formats will be handled on an experimental basis.

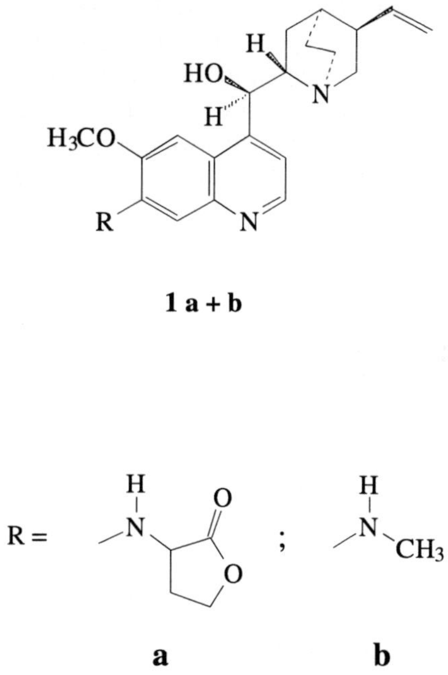

1 a + b

a   **b**

Figure 1. *Example for a structural drawing in one-column format.*

Tables

These should be numbered consecutively with Arabic numerals and must be provided separately from the text. Footnotes in tables should be given letter designations and be cited in the table by **non-superscript** letters in square brackets. The sequence of letters should proceed by line rather than by column. If a footnote is cited both in the text and in a table, insert a lettered footnote in the table to refer to the numbered footnote in the text. Each table should be provided with a descriptive heading, which, together with the individual column headings, should make the table, as nearly as possible, self-explanatory. In setting up tabulations, authors are requested to keep in mind the type area of the journal page (17.8 x 23.9 cm) and the column width (8.5 cm), and to make tables conform to the limitations of these dimensions. Arrangements that leave many columns partially filled or that contain much blank space should be avoided.

Abbreviations and chemical formulae can be used freely in headings and columns of tables, as long as they are provided as WMF files.

Figures

The importance of good quality illustrations cannot be overstressed. It is the responsibility of the author to submit the graphics ready for direct reproduction. Authors are encouraged to supply figures, both monochrome and colour, separately from the text, one file for each. Figures should be constructed in keeping with the column width, line width, and font size specified for the Structural Drawings. Colour illustrations should be submitted as a compressed graphics file format. The following format will be accepted, all other graphic file formats will be handled on an experimental base:

DS4 – Micrografx Designer 4.0
EPS – Encapsulated PostScript
GIF – Graphics Interchange Format
HPGL – Hewlett Packard Graphics Language
JPG – JPEG, JFIF Compliant
MAC – MacPaint
MGX – Micrografx Clipart
PCD – Kodak Photo CD
PIC – Lotus
PICT – Macintosh (Pixel Graphic)
PSD – Adobe™ PhotoShop 3.0,
RGB – Silicon Graphics image
TGA –Truevision Targa
TIFF – Tagged Image File Format
WMF –Windows Metafile

The files must be named in an unambiguous way so that the figure is to be identified. The figure number should **not** be typed in the graphic itself. If a figure is split into several parts (e.g. a, b, c), each part should be supplied as one file. The letters a, b, c must be typed in the graphic itself, with boldface small characters in 10pt s *Times Roman* on the left bottom side of the graphic.

The legends can be grouped together in one text file. They should be typed in plain (lightface) 10pts *Times Roman*.

No extra charge is imposed for colour graphics or other sophisticated presentation material.

Supplementary material

The electronic edition of this journal can accommodate almost any type of supplementary data, e.g. PDB and XYZ files, VRML-scenes, source code, program executables, video sequences, etc. The file format should be chosen to be as common as possible. For text files plain ASCII is the most suitable. A supplementary material available statement that describes the material should be placed at the end of the printed manuscript text. Consult a current issue of *J.Org.Chem.* for the proper wording of this statement. Captions or legends for figures, tables, etc., must appear directly on the figure. The CD-ROM edition of the Journal will include all the supplementary material to guarantee longevity.

Author's Manuscript Submission Checklist

- ❑ corresponding author's mailing address, phone number, fax number and **email** address
- ❑ complete set of files relevant to the publication. Text, tables, graphics etc. must be in separated files
- ❑ signed Copyright Transfer form (available via WWW)
- ❑ graphical abstract for the *Journal of Molecular Modeling* WWW-page
- ❑ structures and figures corresponding to the guideline
- ❑ a letter of permission from each individual whose unpublished work or private communication is cited
- ❑ supplementary material, if appropriate
- ❑ a properly-worded supplementary material available statement
- ❑ relevant data as PDB or XYZ-file
- ❑ description of data files (PDB, MPEG etc.)
- ❑ printed listing of submitted files indicating operating system and file format

Copyright Transfer

Papers must be accompanied by a signed copy (sent by conventional mail) of the copyright transfer form (available as PostScript and PDF file *via* WWW or from the editorial office).

Reprints

Authors will receive one full colour reprint free of charge. Additional offprints may be ordered, for prices please send email to em-helpdesk@springer.de.

Publication charges

There is no publication charge for contributors - including colour figures, 3D-data files, videos and supplementary materials

Journal of Molecular Modeling - Copyright Transfer Form

The undersigned author has submitted a manuscript entitled:

_____

_____

(the "Work") for publication in the *Journal of Molecular Modeling*, (the "Journal"), published by Springer-Verlag.

1. The author transfers to Springer-Verlag (the "Publisher") during the full term of copyright, the exclusive rights comprised in the copyright of the Work, including but not limited to the right to publish the Work, and the material contained therein throughout the world, in all languages, and in all media of expression now known or later developed, and to license or permit others to do so.
2. Notwithstand the above, the author retains the following: Proprietary rights other than copyright, such as patent rights. The right to make copies of all or part of the Work for the author's use in classroom teaching. The right to make copies of the Work for internal distribution within the institution that employs the author. The right to use figures and tables from the Work, and up to 250 words of text, for any purpose. The right to make oral presentations of material from the Work. The author agrees that all copies made under any of the above conditions will include a notice of copyright and a citation of the Journal.
3. In the case of a Work prepared under US Government contract, the US Government may reproduce, royalty-free, all or portions of the Work and may authorize others to do so, for official US Government purposes only, if the US government contract so requires.
4. If the Work was written as a work made for hire in the course of employment, the Work is owned by the company/ employer which must sign this agreement in the space provided below. In such case, the Publisher hereby licenses back to such employer the right to use the Work internally or for promotional purposes only.
5. The author represents that the Work is the author's original work. If the Work was prepared jointly, the author agrees to inform the co-authors of the terms of this Agreement and to obtain their permission to sign on their behalf. The Work is submitted only to this Journal, and has not been published before. (If excerpts from copyrighted works are included, the author will obtain written permission from the copyright owners and show credit to the sources in the Work.) The author also represents that, to the best of his or her knowledge, the Work contains no libelous or unlawful statements, does not infringe on the rights of others, or contain material or instructions that might cause harm or injury.

Check one:

☐ Author's own work

☐ US Government work

☐ Work made for hire for Employer

_____
Author's signature and date

_____
Typed or printed name

_____

_____
Institution or company (Employer)

J. Mol. Model. **1996**, *2*, A15

Editorial

A year ago, the *Journal of Molecular Modeling* was a promising newcomer among the host of chemistry journals and I described my pleasure at its level of acceptance in my editorial for the first print and CD-ROM edition. One year later, we can look back on a journal that has almost three times the number of pages and contributions as last year and that has won the (sometimes grudging) respect of many practitioners of molecular modeling. Interest in both the online and the hardcopy editions has been well up to expectations and the level of the published papers has been at least maintained, if not improved. We are particularly pleased that the World Association of Theoretically Oriented Chemists (WATOC) will offer discounted subscriptions to the *Journal of Molecular Modeling* to its members from the 1997 volume (volume 3) and are proud to be associated with one of the leading organisations devoted to the use of computational methods in chemistry.

This association helps to emphasise one of the goals of the Journal of Molecular Modeling. Molecular modeling is an area that combines contributions from many different disciplines, from theroretical physics to medicinal chemistry. Our goal is that the Journal should become a forum for everyone associated with the subject and that it should bring specialists from very different areas together in order to provoke synergistic contact between method developers, applications specialists and all the other disciplines involved. For this reason, the *Journal of Molecular Modeling* will continue to publish pure applications next to basic theory in the hope that authors and readers will begin to appreciate the breadth of the subject.

Our second year has also seen considerable technical development in the direction that we originally intended for the Journal. This year's papers include VRML scenes, video sequences and "hypertables" in which the molecular structures can be accessed by clicking the appropriate cell of the table. This sort of development necessarily means that there will be an increasing gap between the electronic and the hardcopy editions as the years go by. We are already approaching the situation, which we hope to realise in 1997, that a fully electronic version of a publication will be accompanied by a more conventional "print like" version, but that the two will only have the information they contain in common. This sort of presentation will become standard for electronic media as we begin to realise that browsing through an electronic journal is a very different process to flipping through the pages of a printed edition. Our goal must now be to find the electronic format that most appeals to readers without sacrificing the scientific quality of the Journal. I can never emphasise too strongly my conviction that our mission is to be an excellent scientific journal that happens to be published electronically, not an electronic journal that happens to contain science.

Perhaps the most positive aspect of the Journal for me is that it continues to be an exciting experiment that has won the enthusiasm of the members of the editorial team, everyone involved at Springer and many of our readers, who often feel a special affection for our common child. The continuing success of the Journal is due to many people. Henryette Roth continues to rule the editorial office with a firm hand and to supply the necessary organisational talent that is completely missing in my character. We are pleased to welcome Gudrun Schürer, who will take over some of the editorial chores. Finally, a very special thanks to everyone at Springer, Heidelberg. Their support and enthusiasm has been exceptional and they have always conveyed the impression that the *Journal of Molecular Modeling* lies close to their hearts. I often have the impression that I am possibly the most spoilt editor in all of chemistry.

So, welcome to the *Molecular Modeling Annual 1996*, which I hope you will enjoy enough to become a subscriber and regular author of the Journal of Molecular Modeling.

Tim Clark   Erlangen, November 1996

J. Mol. Model. **1996**, 2, 1 -15

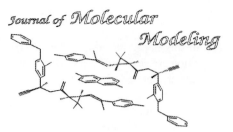

Theoretical Methods for the Representation of Solvent

Modesto Orozco*, Cristobal Alhambra and Xavier Barril

Departament de Bioquímica. Facultat de Química. Universitat de Barcelona. Martí i Franquès 1, Barcelona 08028. SPAIN (modesto@luz.bq.ub.es)

José M. López, María A. Busquets and Francisco J.Luque*

Departament de Farmàcia. Unitat Fisicoquímica. Facultat de Farmàcia. Universitat de Barcelona. Avgda Diagonal sn. Barcelona 08028. SPAIN.

Received: 21 July 1995 / Accepted: 17 December 1995 / Published: 5 March 1996

Abstract

A review of the theoretical approaches for the representation of the solvent effect on molecular structure and reactivity is presented. The main characteristics of the different methods available for the description of solvation phenomena are outlined. The advantages and shortcomings of the computational approaches are discussed. Comparison of the different methodologies might help a non-expert user select the most suitable method for the treatment of a particular system in solution.

Keywords: solvation, computer simulation, continuum models, discrete methods, quantum mechanics-molecular mechanics methods

Introduction

The complexity of chemical phenomena makes it necessary to use molecular models. A given model must incorporate all the relevant features of the process of interest, so that the results can be interpreted and contrasted with experimental evidence. Choice of the model is made by defining three basic elements: the material constituents or "building-blocks" of the system, the physical rules underlying the chemical process, and the mathematical formalism required to describe such a process according to the physics of the problem. The results provided by the model can be rationalized on the grounds of these elements. Comparison with experiment will show the weaknesses of the model, which must then be revised and improved. The final objective is the development of a theoretical model able to explain the chemical behaviour.

The evolution of theoretical chemistry clearly illustrates this scientific method. Thus, in the early eighties neither theory nor computational resources allowed the study of chemical processes in condensed phases. Reduced systems, in which the solvent was ignored, were used, while large improvements were made in the study of processes in the gas phase. Since then, we have witnessed the impressive increase in computer power, the optimization of computational algorithms available in standard computer programs, and the improvement in the accuracy of quantum chemical techniques. As a result, a very precise quantum mechanical (QM) description of chemical systems in the gas phase can now be achieved. On the other hand, this evolution has been accompanied by the development of sophisticated methods for the representation of solvent, which has provided the theoretical framework and the technical resources required to initiate the study of chemical processes in condensed phases.

* *To whom correspondence should be addressed*

J. Mol. Model. **1996**, *2*

In this paper the most recent methods for the study of solvation are reviewed. First, the nature of the solvent effect and the changes induced in the structure of the solute and in the thermodynamic and kinetic characteristics of chemical interactions are discussed. Second, the classical methods for the description of solvent effects are presented. Finally, the treatment of systems in solution by QM methods are examined. The strengths and weaknesses of the different methods and their range of applicability are discussed.

The nature of the solvent effect

From a rigorous theoretical point of view the concepts of solute and solvent are meaningless, since a dilute solution is an ensemble of molecules which should be treated at the same level, irrespective of their nature and population. The differentiation between solute and solvent obeys to practical considerations, because of the difficulty of treating correctly the bulk solution due to: i) the large number of solvent molecules needed to simulate a dilute solution, and ii) the high level of accuracy often required to describe the solute. In this context, the "solvent effect" can be interpreted as the change experienced by a chemical system (the solute) upon transfer from the gas phase to a dilute solution.

There are different ways in which the solute is influenced by solvation. In the following the solvent effect is examined according to the nature of the changes induced in the solute,

which may affect i) the molecular (nuclear and electronic) structure and ii) the thermodynamics and kinetics of chemical processes. The nature of these effects is quite different and, consequently, the theoretical method chosen to study a given process must be able to capture properly the nature of the factors involved in these effects.

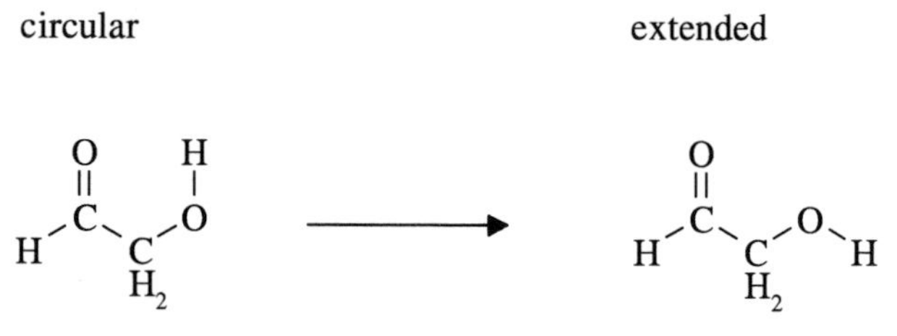

Changes in molecular structure

The solvent can introduce notable changes in the molecular structure, both in terms of nuclear and electronic distributions. Thus, changes in the nuclear configuration may arise from the tendency of polar solvents to stabilize structures with large charge separation (see below). The net effect can be a change in the conformational space of the solute, so that the relative population of the conformers having the largest polarity is increased. An example is the destabilization of the conformations with intramolecular hydrogen-bonds in polar solvents (Figure 1). Another illustrative case is the change in the equilibrium between isomers, as shown by the isomerization of formic acid (Figure 2). Nuclear changes involving the formation or breaking of covalent bonds are also largely dependent on the solvent. There are numerous examples reported in the literature about changes in the preferred species for tautomeric processes [1]. In general, a polar solvent displaces the tautomeric equilibrium so as to increase the population of the most polar tautomer. This effect can revert the stability in gas phase, even for apolar solvents [2]. In all these cases the solvent effect is mainly modulated by classical electrostatic interactions.

The influence of the solvent on the electronic distribution (for a given nuclear configuration) modulates the chemical reactivity. This is reflected in the change of different properties upon solvation, such as the enlargement in the dipole moment for neutral polar molecules [3], the change in the molecular electrostatic potential, the variation in the molecular volume [4], or the displacement in the spectroscopic properties [5]. The magnitude of the polarizing effect is surprisingly large, as noted by increases of 20-30% in dipole moments for neutral solutes in aqueous solution [3]. Furthermore, even in apolar solvents like chloroform the solvent polarization is not negligible, as indicated by increases of 8-10% in the dipole moments determined for neutral molecules from theoretical calculations (see below). In most cases the solvent effect is exerted through the electrostatic perturbation of the solute, but in some cases dispersion interactions are known to play a major effect [5].

circular   extended

$\Delta G^{vac} = +1.3$ kcal/mol

$\Delta \mu^{vac} = +1.5$ D

$\Delta G^{aq} = -3.0$ kcal/mol

$\Delta \mu^{aq} = +2.3$ D

Figure 1. *Conformational equilibrium of HCOCH$_2$OH in gas phase and in aqueous solution. The free energy in solution, ΔG_{aq}, was determined as:*
$\Delta G^{vac} + \Delta G_{hyd}$(extended) $- \Delta G_{hyd}$(circular), where ΔG_{vac} is the free energy in gas phase at 298 K and ΔG_{hyd} is the free energy of hydration determined from the AM1/MST method using the standard parameters (see text and ref. 45f).

Changes in thermodynamics and kinetics of chemical interactions

The best known effect of solvent is the modulation of chemical reactivity, even in the most apolar solvents [6]. This effect is especially relevant for polar solvents like water, where

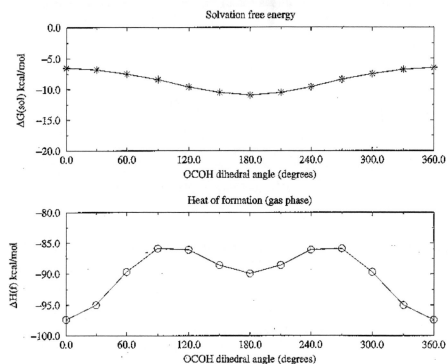

Figure 2. *Changes in the heat of formation in gas phase and the free energy of hydration upon rotation of the C-O bond of formic acid. Values of ΔH(f) and ΔG(sol) were determined at the AM1 and AM1/ MST levels, respectively.*

it can invert the sign of the free energy change for a chemical reaction in the gas phase. The solvent has three major effects on the chemical reactivity: i) the modulation of the intrinsic reactive characteristics of reactants; ii) the introduction of viscosity effects; and iii) the direct interaction of solvent molecules with reactants, products, and transition states, which may lead to a differential stabilization of these species.

The change in the intrinsic reactive properties is mainly related to the polarization of the solute charge distribution (see above). The inclusion of viscosity effects is expected to modulate greatly the dynamics of the molecular system. This effect is not dramatic for a small and mobile solvent, or for processes with high intrinsic (gas phase) energy barriers, but it can be relevant when the solvent molecules are large and low mobile, or when the energy barrier is small. Indeed, viscosity effects are also involved in processes controlled by diffusion.

The interaction of solvent molecules with reactants, products and transition states may greatly influence the chemical reaction. This effect is especially important for polar solvents like water, where specific interactions, i.e. hydrogen bonds, can be established with the reactive species leading to changes in their relative stabilities. All chemical interactions are susceptible to the solvent effect, but the largest influence may be expected to occur i) when the polarities of reactants, transition state or products are very different, and ii) when the number of specific solute-solvent interactions

(the number of hydrogen-bond donor and acceptors) changes during the interaction. There are a large number of processes that illustrate this effect Some well-known examples are discussed in this review, but more detailed explanations can be found elsewhere [see, for instance, ref. 7].

Ion-ion interactions are particularly influenced by the solvent. The shielding of electrostatic interactions in polar solvents can stabilize two species equally charged (with the same sign) and separated by a finite distance, but this situation would obviously be unstable in gas phase [8]. Close anion-cation interactions are extremely stable in the gas phase, but the configurations corresponding to a solvent-separated ion pair are favored in polar solvents [8]. This difference ultimately stems from the preferential stabilization of the isolated (or solvent-separated) ions by the solvent. An additional example is the ionization of acids and bases. A simple calculation (Figure 3) based on the experimental proton affinities of water and the hydroxyl anion [9] suggests that the free energy of ionization of water in gas phase at 298 K is around 218 kcal/mol. In contrast, the free energy in aqueous solution, as determined from the experimental pKa, is around 24 kcal/mol, and that determined from the thermodynamic cycle in Figure 3 using experimental free energies of hydration [10] is around 21 kcal/mol.

The effect of polar solvents on ion-molecule interactions has also been the subject of a large number of experimental and theoretical studies [7c,11]. A classical process is the SN^2 reaction, in which water molecules induce large changes in

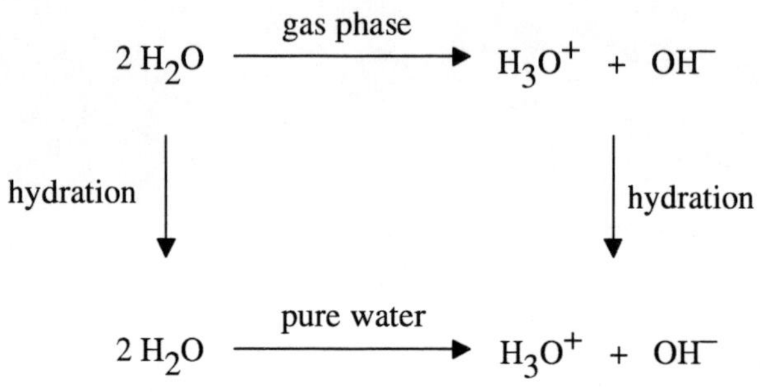

$$2\,H_2O \xrightarrow{\text{gas phase}} H_3O^+ + OH^-$$

hydration $\downarrow$ $\qquad\qquad\qquad$ hydration $\downarrow$

$$2\,H_2O \xrightarrow{\text{pure water}} H_3O^+ + OH^-$$

$\Delta G^{vac} \approx 218$ kcal/mol
(from experimental proton affinities)

$\Delta\Delta G^{sol} \approx -197$ kcal/mol
(from experimental free energies of hydration)

$\Delta G^{aq} \approx 21$ kcal/mol (from thermodynamic cycle)

$\Delta G^{aq} \approx 24$ kcal/mol (from experimental pKa)

Figure 3. *Thermodynamic cycle for the ionization of water in gas phase and in aqueous solution (see text for details).*

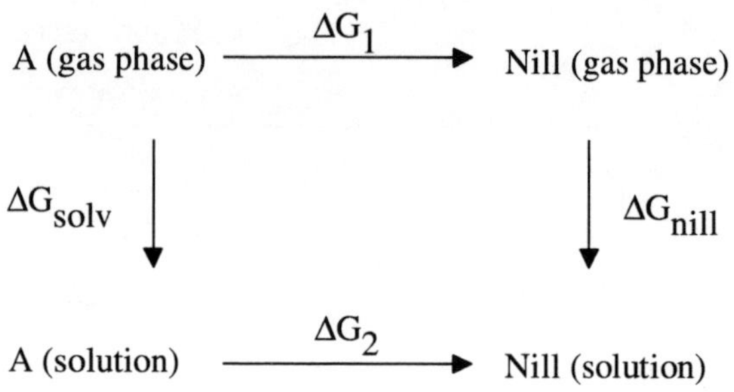

the kinetic and thermodynamic characteristics. A remarkable example is the nucleophilic attack of a hydroxyl group to a carbonyl center. This reaction is very exothermic and occurs without activation barrier in the gas phase (for instance,see ref. 11e,i and references therein), but it is clearly endothermic in aqueous solution, and the activation barrier is largely due to the preferential solvation of reactants with respect to transition state and products [11e]. Similar considerations apply

Table 1. *Differences in free energy of hydration ($\Delta\Delta G_{hyd}$) related to hydrogen-bond formation in aqueous solution. Geometries for hydrogen-bond structures were determined at the AM1 level in the gas phase. Values of ΔG_{hyd} were computed using the standard AM1-MST method. Errors in hydrogen-bond geometries are not expected to introduce dramatic changes in the evaluation of $\Delta\Delta G_{hyd}$. All the values are in kcal/mol. $\Delta\Delta G_{hyd}$ is defined as ΔG_{hyd}(bound)-ΔG_{hyd}(unbound).*

Donor	Acceptor	$\Delta\Delta G_{hyd}$
H_2O	H_2O	+2.4
H_2O	NH_3	+3.1
NH_3	NH_3	+1.9
NH_3	H_2O	+1.3
HF	H_2O	+3.2
HF	NH_3	+4.5

$$A \text{ (gas phase)} \xrightarrow{\Delta G_1} \text{Nill (gas phase)}$$

$\Delta G_{solv} \downarrow$ $\qquad\qquad\qquad$ $\downarrow \Delta G_{nill}$

$$A \text{ (solution)} \xrightarrow{\Delta G_2} \text{Nill (solution)}$$

$$\Delta G_{solv} = \Delta G_1 + \Delta G_{nill} - \Delta G_2 = \Delta G_1 - \Delta G_2$$

Figure 4. *Thermodynamic cycle used to compute the free energy of solvation.*

for the attack of other nucleophilic agents to carbonyl centers [see, for instance, ref. 11g,h].

In addition to ion-ion and ion-molecule interactions, other chemical reactions are highly susceptible to the solvent. In particular, attention has been paid to the solvent effect on Claisen rearrangements [7b,12], Diels-Alder reactions [13], benzoin condensation [14], reduction reactions [15], racemization [16] and aldolic condensations [17], among others [for details of solvent effects on chemical reactions, see ref. 7]. In all cases the solvent plays an important role in the thermodynamics and kinetics of each reaction.

Finally, non-bonded interactions, specially hydrogen-bonds, are significantly affected by polar solvents. Thus, the formation of hydrogen-bonded structures in aqueous solution is largely disfavored due to efficient competition of water molecules for the hydrogen-bond donor and acceptor groups. A prototypical example is the hydrogen-bond association of nucleic acid bases, which is exothermic in the gas phase, while the weak stacking interaction is preferred in aqueous solution [18]. The reason for this behavior stems from the differences in the free energy of hydration of the hydrogen-bonded complexes of nucleic bases with respect to the free bases.

A more general picture of the influence of polar solvent on hydrogen-bond interactions can be obtained from inspection of Table 1, where the solvent effect on the formation of several hydrogen-bond complexes was determined using theoretical methods (see below). The results show the large disturbing effect of water, as noted by the positive values of the differences in free energy of hydration for the formation of the hydrogen-bond complex, which can revert the sign of the free energy of association in the gas phase. The implications of these results on molecular recognition in aqueous solution, specially in biological systems, cannot be underestimated.

Theoretical calculation of solvent effects

As noted before, the solvent may influence the nuclear and electronic distributions of the solute, as well as the chemical reactivity and intermolecular interactions. In the last 20 years different methods have been developed to gain insight into these effects. All these methods can broadly be grouped into two main categories depending on the treatment of the solute: i) classical methods and ii) QM methods. In both cases, the solvent is described classically, either in terms of discrete particles (the microscopic level) or as a continuum medium (the macroscopic level).

The former methods treat the solute as a classical particle, whose interactions are determined by classical force-fields, which difficults a correct representation of the solvent-induced changes on solute properties. The QM methods account for solvent polarization, and accordingly for changes in structure and reactivity upon solvation. Unfortunately, they are computationally very expensive and this impedes their application to large systems. The (classical or QM) methods based on a microscopic description of the solvent determine the average representation of the system in solution from the ensemble of configurations collected from Monte Carlo (MC) or Molecular Dynamics (MD) simulations. Quite accurate estimates of the change in the free energy of reaction in solution when MD or MC simulations are coupled to statistical mechanical algorithms [19]. The expensiveness of these simulations is, nevertheless, a critical factor for large systems, or when the solute is represented quantum mechanically. In this context, the treatment of the solvent as a continuum medium is advantageous, since the degrees of freedom of the solvent are not longer considered.

Table 2 *Comparative characteristics of Monte Carlo and Molecular Dynamics for the study of solvated systems.*

Treating the solvent as a continuum medium, however, implies the lack of a detailed description of solute-solvent interactions. Nevertheless, precise results can be obtained through a well-balanced selection of the properties that characterize the solvent continuum model.

Classical methods

The classical methods determine the free energy of solvation (the reversible work needed to transfer a molecule from gas phase to solution) as the difference between the works required to annihilate the molecule in gas phase and in dilute solution (Figure 4). If the work involved in the annihilation of intramolecular interactions is the same in gas phase and in solution, i.e. the molecule has a similar electronic and nuclear configuration in the two phases, the thermodynamic cycle in Figure 4 can be simplified, and the free energy of solvation can be determined from the work required to annihilate the intermolecular interactions of the solute in solution. The annihilation is usually performed in two steps (electrostatic decoupling), where the steric interactions are annihilated after removing the electrostatic interactions .

The calculation of electrostatic and steric contributions to the free energy of solvation can be performed considering the solvent as a finite set of discrete particles (discrete methods) or as a continuum which reacts against the solute (continuum methods).

Classical discrete methods

The microscopic description of solute and solvent is encoded in the force-field [19,20], which has different energy terms for bonded (stretching, bending, torsion) and non-bonded (electrostatic, van der Waals) interactions. These terms adopt

Properties	Monte Carlo	Molecular Dynamics
Control of T, P	Easy	Difficult (weak coupling methods)
Conf. variables	Internals	Cartesians
Reduction of configurational space	Easy	Difficult (holonomic constrains)
Sampling for small solutes	Easy	Easy
Sampling for large solutes	Very difficult	Average difficulty
Study of time-dependent processes	Very difficult	Average difficulty
Setup of the simulation system	Average difficulty	Easy

very simple expressions, which facilitates the evaluation of the potential energy of the system. The averaged representation of the solute-solvent system is obtained from MD and MC techniques [21].

The sampling of the solute-solvent configurational space by MD is performed from the trajectories determined by integration of the equations of motion, which can be performed following Newton, Lagrange or Hamilton formalisms (for review, see ref. 21). MC samplings are obtained from the application of Boltzmann probabilistic rules to a very large set of randomly selected configurations. In principle, MD and MC samplings should be identical for infinite simulations if the system is in equilibrium. However, the use of MD and MC techniques present several differences for their application to computational simulations, a concise summary of them being given in Table 2.

In general, MD and MC have a similar efficiency in the sampling of the configurational space for small solutes. MD techniques offer the advantage that the time evolution of the system may be analyzed, which is often interesting to study time-correlation functions or transport properties. On the other hand, MC calculations allow complete control of the simulation system and an easy reduction of configurational variables, although caution must be taken to avoid an artefactual simplification of the configurational space. When the solute is large and flexible, MD techniques are prefered, since in these cases the sampling in internal coordinates used in standard MC simulations is inefficient. Strategies to increase the reliability of MC techniques in the sampling of conformational movements in large solutes have been discussed elsewhere [21f]. Both MD and MC techniques provide Boltzmann-averaged samplings, which can be used to determine the free energy of solvation. This can be done following two different strategies based on: i) linear free energy response (LFER) theory, or ii) statistical mechanical (SM) methods.

According to LFER (for details see ref. 22), the electrostatic component of the free energy of solvation is half the magnitude of the averaged solute-solvent electrostatic interaction energy (E_{sx} in eq.1). The steric components can be computed upon a proper scaling of the solute-solvent van der Waals interaction (E_{vw} in eq. 2) or by using parametric equations expressed in terms of the molecular volume, the solvent accessible surface or similar descriptors (ζ in eq. 2). In eq. 2 the brackets mean that the averages are done for Boltzmann samplings, and the scaling parameters (α, β) are adjustable variables fitted to reproduce dispersion-repulsion and cavitation contributions to the free energy of solvation. Parametrized LFER-based discrete methods provide good estimates of free energy of hydration, and its application in molecular modeling studies is promising [23].

$$\Delta G_{ele} = \frac{1}{2}\langle E_{sx}\rangle \qquad (1)$$

$$\Delta G_{ster} = \alpha\langle E_{vw}\rangle + \beta\langle\zeta\rangle \qquad (2)$$

Some of the most powerful techniques available for the study of systems in solution have emerged from the coupling of SM theory with MD and MC simulations. Among them, free energy perturbation (FEP) and thermodynamic integration (TI) are of special interest for the calculation of the free energy of solvation [24]. It is not our purpose to explain in detail these methods, but just to present briefly their essential characteristics. For a detailed explanation on these and related techniques, we address the reader to the original works in ref. 24, and to other recent reviews [19].

FEP and TI techniques compute the free energy of solvation using the thermodynamic cycle in Figure 4. The free energy of solvation is determined as the difference in the works involved in the annihilation of the solute in gas phase and in solution ($\Delta G1$ and $\Delta G2$ in Figure 4) through a reversible pathway. The requisite of reversibility for the A(solute)->B(dummy) interconversion implies that such a mutation has to be divided into smaller steps: A->A'->A''->...->B''->B'->B, in such a way that the change in every step is small enough as to make every (micro)process reversible. In practical simulations this is achieved through the use of a "coupling" parameter (λ), which controls the change between the Hamiltonians for the two states, A and B (eq. 3; $\lambda=0$: state A, $\lambda=1$: state B).

$$H_\lambda = (1-\lambda)H_A + \lambda H_B \qquad (3)$$

The use of the coupling parameter within the FEP framework allows the calculation of the free energy of solvation according to eq. 4. The equivalent expression obtained from TI theory is shown in eq. 5. The Boltzmann averages are obtained from MD or MC samplings, and $\Delta\lambda$ defines the number of steps (windows) involved in the annihilation process ($\lambda:0->1$).

$$\Delta G_{solv} = \sum_{\lambda=0}^{1-\Delta\lambda} RT\,ln\left\langle exp\left[-\left(E_{\lambda+\Delta\lambda} - E_\lambda\right)\Big/ RT\right]\right\rangle_\lambda \qquad (4)$$

$$\Delta G_{solv} = -\sum_{\lambda=0}^{1-\Delta\lambda}\left[\int_\lambda^{\lambda+\Delta\lambda}\left\langle\frac{\partial E_{\lambda'}}{\partial\lambda'}\right\rangle_{\lambda'} d\lambda'\right] \qquad (5)$$

Other related algorithms have been suggested. Among them, slow growth (SG, [25]), where a very small window ($\Delta\lambda=d\lambda$) is sampled only with a single configuration, the muticonfigurational thermodynamic integration (MCTI, [26]), which is based on a discontinous integration of ($\partial E/\partial\lambda$), and the finite difference thermodynamic integration (FDTI, [27]), which has shown an excellent performance in different studies of complex solutes [1f,27]. This latter algorithm, which combines TI and FEP methods, computes the free energy of solvation according to the expression given by eq. 6.

$$\Delta G_{solv} = -\int_{\lambda=0}^{\lambda=1}\left(\frac{\partial}{\partial\lambda}\left[-RT ln\left\langle exp^{-\left(E_{\lambda+\delta\lambda}-E_{\lambda}\right)}\middle/RT\right\rangle_{\lambda}\right]\right) \quad (6)$$

The SM methods have been successfully used in the study of solvation in different solvents, as well as in the calculation of transfer free energies between immiscible solvents [28]. A couple of recent studies have shown that the method can provide, without any specific ad hoc parametrization, free energies of solvation with an average error of 1 kcal/mol from experimental data for neutral polar solutes [29]. However, it should be noted that the goodness of the results is guaranteed only when: i) a reliable force-field is used, ii) polarization effects are not very important, and iii) MD or MC simulations are large enough to provide representative Boltzmann samplings at every step of the simulation.

Classical continuum methods

These methods treat the solvent as a continuum medium which reacts against the (unperturbed) solute charge distribution. Calculation of the electrostatic free energy of solvation can be done following the theory of polarizable solvents [22a] at different degrees of complexity. Extension of the method to incorporate other components of the free energy of solvation (dispersion-repulsion and cavitation) is not difficult.

The classical continuum models can be classified according to two main characteristics: i) the shape of the cavity that defines the solute/solvent interface, and ii) the description of the solute charge distribution and the solvent reaction field. With regard to the solute/solvent interface, a large number of cavities have been used. The simplest methods define cavities of regular shape, such as spheres, ellipsoids or cylinders, while the most accurate methods use cavities adapted to the actual molecular shape. Regarding the treatment of the solute charge distribution and the solvent reaction field, Tomasi and Persico in their recent and exhaustive review [30] categorized the different treatments in five formalisms: i) *multipole expansions*, ii) *apparent surface charge*, iii) *image charge*, iv) *finite difference*, and v) *finite elements*. It is not our purpose to review all the methods, but just to comment on the most popular ones. The reader is referred to ref. 30 and 31 for a more complete discussion.

Methods based on multipole expansions are probably the simplest ones. They typically use a regular cavity and the multipole expansion is truncated at different levels: monopole (Born model, eq. 7 [32]), dipole (Bell and Onsager , eqs. 8-9 [33]) or higher order terms [27]. In eqs. 7-9, ε is the dielectric contrant, q and μ are the charge and dipole moment, R is the radius of the cavity and α is the solute polarizability.

$$\Delta G_{ele} = -\frac{\varepsilon-1}{2\varepsilon}\left[\frac{q^2}{R}\right] \quad (7)$$

$$\Delta G_{ele} = -\frac{\varepsilon-1}{2\varepsilon+1}\left[\frac{\mu^2}{R^3}\right] \quad (8)$$

$$\Delta G_{ele} = -\frac{\varepsilon-1}{2\varepsilon+1}\left[\frac{\mu^2}{R^3}\left(1-\frac{\varepsilon-1}{2\varepsilon+1}\frac{2\alpha}{R^3}\right)^{-1}\right] \quad (9)$$

Born and Bell-Onsager models have been very popular because of their simplicity, even though the quantitative quality of the results may not very good when complex molecules are dealt with. These two expressions have been recently applied to account for long range interactions in MD and MC simulations, where cutoff techniques are used to evaluate non-bonded interactions [19-21].

More elaborated methods have been developed based on multipole expansions. Among them, the Generalized Born Model (GBM; eq.10) [34] treats the solute as a set of monopoles (centered at nuclei), each independently solvated. Empirical corrections are introduced to correct the overestimation of the free energy of solvation, which arises from the direct incremental application of the Born formalism (eqs. 11-13). GBM and other methods based on multipole expansion are available in the QM framework (see below).

$$\Delta G_{ele} = -\frac{1}{2}\left(1-\frac{1}{\varepsilon}\right)\sum_{i,j}\frac{q_i q_j}{f_{GB}} \quad (10)$$

$$f_{GB} = \left(r_{ij}+\alpha_{ij}^2 e^{-D}\right)^{0.5} \quad (11)$$

$$\alpha_{ij} = \left(\alpha_i\alpha_j\right)^{0.5} \quad (12)$$

$$D = {r_{ij}^2}\middle/{\left(2\alpha_{ij}\right)^2} \quad (13)$$

r_{ij}: distance between two atoms (center of charges)
α_i: Born radii of atom -i.

Methods based on the apparent surface charge are also widely used and different QM formalisms are available (see below). In these methods the reaction field is represented in terms of an imaginary surface charge spread over the cavity, and the electrostatic free energy of solvation can be represented with expressions analogous to eq. 14 [35], where σ is the apparent surface charge distribution, ρ is the solute charge distribution., Φ_ρ is the solute electrostatic potential on the cavity surface, and Φ_σ is the solvent reaction potential. At the classical level the polarization of the solute charge distribution by the solvent is typically ignored, i.e. $\rho = \rho(o)$.

$$\Delta G_{ele} = \frac{1}{2} \int_S \sigma \Phi_\rho dS = \frac{1}{2} \int \Phi_\sigma \rho dr \qquad (14)$$

The above expression, which arises from the application of Poisson and Laplace equations with the appropriate boundary conditions at the cavity surface, can be used to obtain fast estimates of the electrostatic free energy of solvation, as suggested in the eighties by Tomasi and coworkers [35]. Recent studies have confirmed the goodness of this strategy [36] for the calculation of free energies of solvation in a large series of prototypical solutes. In addition, new strategies have been suggested that, upon a detailed expansion of eq. 14, can capture at least partially the mutual solute<->solvent polarization effects [36b].

Finite difference methods (FDM) are very popular for the study of the free energy of solvation of large systems [37]. The electrostatic component is determined according to eq. 15. Here the internal electrostatic potential (Φ_i^{intra}) is determined from the unperturbed solute charge distribution, while the total electrostatic potential (Φ_i^{tot}) is determined by solving the Poisson equation, where the dielectric constant is considered to be a function of the distance (eq. 16; Φ is the total potential).

$$\Delta G_{ele} = \frac{1}{2} \sum_i q_i \left(\Phi_i^{tot} - \Phi_i^{intra} \right) \qquad (15)$$

$$\nabla\left[\varepsilon(r)\nabla\Phi(r)\right] = -4\pi\left(\rho_{int}(r) + \rho_{ext}(r)\right) \qquad (16)$$

The solution of the Poisson equation can be found at a linear or nonlinear level, depending on the complexity of the relationship between the external charge distribution (ρ_{ext}) and the electrostatic potential (Φ). In any case, the analytical solution is not feasible, it being necessary to use finite difference methods. This is accomplished by using finite grids to map the entire space, including solute and solvent. The electrostatic potential at each point is then determined as a function of the potentials at the nearest grid points. The process is repeated until convergence.

FDM provides good estimates of the free energy of solvation for small molecules [38] provided that the grid of points is dense enough. Otherwise, rather inaccurate results are obtained. Alternative strategies have been devised to alleviate this problem, such as the focusing procedure [39]. Furthermore, some of the FDM algorithms, like DelPhi [37c], have been extensively applied to the study of solvation in macromolecules. When large systems are considered, cau-

tion is needed due to the large magnitude of the solvation free energy, the difficulties in defining a priori the dielectric constant inside the macromolecule, and the use of a rigid set of charges for the solute charge distribution.

The most advanced classical continuum theories have widespread application in the study of biochemical systems, where efficient algorithms are necessary. They are also becoming very useful when coupled to MD and MC simulations, where they are used to simulate the environment effect beyond the cutoff used for the non-bonded interactions. However, the shortcomings of these methods have to be properly considered. In most cases non-electrostatic contributions to the free energy of solvation are neglected, and a detailed treatment of the solvent-induced polarization of the solute is difficult. Indeed, the description of the solute charge distribution and the definition of the cavity may not be accurate enough. Finally, specific solute-solvent interactions are not dealt with.

Quantum mechanical methods

The treatment of the solute-solvent system at the QM level is impossible with the current computational resources. The cost of this approach limits the description of the solvent to a few molecules within the supermolecule approach, mainly to analyze specific interactions with the solute. Thus, the general approach is to treat the solute at the QM level, while the solvent is represented classically using discrete or continuum representations. In the discrete models the free energy of solvation is computed from the work needed to annihilate the coupling interactions between the solute and the solvent Hamiltonian (see below). In the continuum models the free energy of solvation is determined as the work necessary to build up the solute in the solvent minus the work necessary to perturb nuclear and electron distributions upon transfer from gas phase to solution. It is worth noting that in both strategies the solvent-induced polarization of the solute is explicitly introduced.

Quantum discrete methods

These methods use a classical force field to represent the solvent interactions [40]. The total Hamiltonian of the solute-solvent system is decomposed into three terms (eq. 17): i) a QM Hamiltonian for the intramolecular solute interactions (QM); ii) a classical Hamiltonian for the solvent-solvent (intra and intermolecular) interactions (MM); and iii) a coupling term to account for the solute-solvent interactions (QM/MM).

$$\hat{H}^{ef} = \hat{H}^{MM} + \hat{H}^{QM} + \hat{H}^{QM/MM} \qquad (17)$$

Owing to the expense of QM calculations, most algorithms use semiempirical Hamiltonians for the solute [40,41], but recently methods based on density functional theory, and also at the *ab initio* level with small basis sets have been developed [42]. Furthermore, all the methods consider that the only change on the solute electron distribution results from electrostatic interactions, neglecting changes due to dispersion. Indeed, the solvent is usually assumed to be self-polarized, the specific solute->solvent polarization being, accordingly, tipically neglected at the molecular level. Finally, rigid models are used to represent simple solvents like water, avoiding the need of an intramolecular term in the force field of the solvent.

Assuming the previous considerations and the validity of an empirical model for the solvent intermolecular interactions, the Hamiltonian for the solute-solvent system can be expanded as shown in eq. 18, where a set of charges and van der Waals parameters is used to represent the solvent molecules. In eq.18 s denotes the solvent charge sites, m and i refer to the solute nuclei and electrons, N is the number of doubly occupied molecular orbitals, and E_{vw} is the solute-solvent van der Waals interaction energy as determined from standard force-fields. The average representation of the solute-solvent system is obtained through MC or MD simulations. Modified equations have been suggested by Warshel and coworkers for aqueous solution, where the electrostatic properties of water are represented by means of a polarizable dipole [40a-d,h] instead of a set of rigid charges.

$$\hat{H}^{ef} = \hat{H}^{MM} + \hat{H}^{QM} + \sum_{s=1}^{S} \sum_{i=1}^{2N} \frac{eq_s}{R_{si}} + \sum_{s=1}^{S} \sum_{i=1}^{2N} \frac{Z_m q_s}{R_{sm}} + E_{VW}$$

(18)

For a particular nuclear configuration the energy is obtained by solving the corresponding pseudo-Schrödinger equation (eq. 19), where the effective Hamiltonian, Hef, includes the QM Hamiltonian of the solute and the solute-solvent coupling term. It should be stressed that every nuclear movement requires a self-consistent field calculation to be performed. This requires an enormous computational effort, which explains the use of very simple Hamiltonians.

$$\hat{H}^{ef} \Psi = E\Psi$$

(19)

The calculation of free energy of solvation can be done using either FEP or TI theory [24]. As noted before, in practical simulations the annihilation of the interactions between the solute and solvent molecule is controlled by a parameter (λ) that modulates the contribution of the QM/MM term to the total Hamiltonian (eq. 20). Indeed, electrostatic decoupling [25] is usually performed in the annihilation process.

$$\hat{H}^{ef}(\lambda) = \hat{H}^{QM} + \hat{H}^{MM} + \lambda H^{QM/MM}$$

(20)

Quantum discrete models are very attractive from a conceptual point of view and have a wide range of potential applications. It should be emphasized that the QM treatment of the solute avoids the errors in the intramolecular energy terms inherent to force fields. Indeed, solute-solvent polarization effects can be considered. Moreover, the solvent is represented at a discrete level, allowing for analysis of specific solute-solvent interactions. Finally, the sampling of solute (and solvent) configurational space is enabled by MD and MC methods. The usefulness of these methods is, nevertheless, limited by their computational cost, which makes it necessary to use non-polarizable force fields for the solvent and simple Hamiltonians for the solute. It also limits the extension of the configurational sampling. However, these simplifications may affect the accuracy of the results. An additional source of uncertainty arises from the different nature of the van der Waals interactions in classical and quantum discrete models [43], and ideally the transfer of van der Waals parameters from classical force-fields should be avoided. The newest methods, which use a specific parametrization of van der Waals interactions for solutes, should improve the quality of results [42b,43].

Quantum continuum methods

These methods combine the QM treatment of the solute with a continuum description of the solvent. As in classical continuum methods, the free energy of solvation is determined from the addition of three contributions: cavitation, dispersion+repulsion, and electrostatic. The steric term, which is considered independent of the solute electron distribution, is computed for a given nuclear configuration by using classical equations, often parametrized from discrete simulations or from fitting to experimental data [44,45]. A large number of models are available, and several of them have been implemented in *ab initio* and semiempirical QM packages. The main differences between these methods lie in: i) the formalism used to account for the electrostatic free energy of solvation; ii) the shape of the solute/solvent interface; and iii) the procedure adopted for the steric contribution to the free energy of solvation.

The electrostatic term is essentially determined according to the same principles defined for classical continuum models: the dielectric continuum reacts against the solute charge distribution generating a reaction field, which in turn interacts with the solute. Nevertheless, because of the QM treatment of the solute, i) a rigorous representation of the charge distribution is achieved, ii) the mutual solute<->solvent polarization can be accurately incorporated, and iii) the solvent-induced changes in the molecular properties of the solute can be evaluated.

The solvent reaction field is generally described in terms of a multipole expansion or by an apparent surface charge, even though other formalisms are available [30]. The reaction field is introduced into the solute Hamiltonian (eq. 21) in a way analogous to that used for discrete QM methods (see above). It is worth noting that in continuum methods the solute wavefunction and the solvent reaction field are coupled *via* the perturbation operator, R, which usually requires the use of self-consistent strategies. Only few of the large number of quantum continuum methods will be outlined here. The reader is addressed to the review by Tomasi and Persico [30] for a thorough explanation of the different models.

$$\left(\hat{H}^0 + \hat{R}\right)\Psi^{sol} = E_{aq}\Psi^{sol} \tag{21}$$

Methods based on multipole expansions can be easily implemented within the QM framework following the formalisms developed by Rivail [7f,31a,46], Tapia [31d,47], Katritzky and Zerner (48), among others [30,49]. The simplest method corresponds to the Bell-Onsager model, where the solute charge distribution is represented by a simple dipole and the solute cavity is spherical. Important improvements arise from the inclusion of higher order terms in the multipolar expansion. In this context, it is important to note the high-level treatment developed by Rivail and coworkers. The method, implemented in classical, semiempirical and *ab initio* QM frameworks, uses a multipole expansion up to the 7th order to represent the solute charge distribution. Moreover, extension to multicenter expansions is also considered. Further refinements also stem from the extension of this method to improved cavity models [46e]. In addition, the code developed in Nancy on the basis of the formalism of reaction field factors has been extended to the Density Functional theory [46f], and also allows the possibility of performing post-Hartree-Fock calculations [7f,46g].

Another method based on multipole expansions is AMSOL [45a-c], developed by Cramer and Truhlar. AMSOL uses semiempirical Hamiltonians for the solute, and the molecular-shaped cavity is built up from the Born radii of the atoms. Steric factors are considered from an empirical linear relationship with the solvent accessible area (eq. 22), where the tension parameters (σ_i) are obtained from empirical fitting. The electrostatic component is based on the GBM method. Empirically fitted parameters are also used in the calculation of the coulombic integrals (γ_{kl}) to guarantee the quality of the results.

$$\Delta G_{ster} = -\sum_{i=1}^{N} \sigma_i A_i \tag{22}$$

AMSOL was parametrized using a large series of molecules. The final RMS error in the fitting was less than 1 kcal/mol [45a-c]. Early versions of the method may underestimate the electrostatic free energy of solvation due to the use of Mulliken charges, whose shortcomings are well known [50]. However, this problem has been largely corrected in the newest versions of the method [50c-d]. The recent extension of AMSOL to non-aqueous solvents increases the potential applications of this program [51], which is distributed by QCPE.

Among the algorithms based on the apparent surface charge, we limit our attention to i) the Polarizable Continuum Model (PCM) developed by Pisa's group [35a,b], and ii) COSMOS (conductor-like screening model), developed by Klamt and Schüürman [52]. The PCM method, also denoted as MST (Miertus-Scrocco-Tomasi) in other versions developed by our group, is available in both semiempirical and *ab initio* [35a,b,45d-f,53] formalisms, while COSMOS is available only at the semiempirical level.

The newest versions of the PCM method uses the scaled particle theory [54], modified in a suitable way to deal with cavities adapted to the molecular shape, to compute the cavitation contribution. The van der Waals component can be determined using different classical formalisms, which range from polynomial expansions [44a] to linear relationships with molecular surface [45d-f]. In the latest case the tension parameters (eq. 23) are parametrized from experimental data [45d-f]. Molecular-shaped algorithms are used to determine the solute/solvent interface [55]. The electrostatic free energy is computed according to eq. 24, where the reaction field operator ($\hat{V}_R$) is evaluated from eqs. 25-26. It is worth noting that the apparent surface charge is determined *via* the electrostatic potential (both solute and solvent contributions are included), which avoids the use of truncated expansions. Proper attention is also paid to the tails in the electronic distribution outside the cavity.

$$\Delta G_{VW} = -\sum_{i=1}^{N} \zeta_i A_i \tag{23}$$

$$\Delta G_{ele} = \left\langle \Phi^{sol} \left| \hat{H}^0 + \hat{V}^0 \right| \Phi^{sol} \right\rangle - \left\langle \Phi^0 \left| \hat{H}^0 \right| \Phi^0 \right\rangle - $$
$$ -\frac{1}{2}\left[\left\langle \Phi^{sol} \left| \hat{V}_R \right| \Phi^{sol} \right\rangle + \int \rho_{nuc} V_\sigma(s)ds \right] \tag{24}$$

$$\hat{V}_R = \sum_{i=1}^{M} \frac{\sigma_i S_i}{|r_i - r|} \tag{25}$$

$$\sigma_i = -\frac{\varepsilon - 1}{4\pi\varepsilon}\left(\frac{\delta(V_\sigma + V_\rho)}{\delta n}\right)_i \tag{26}$$

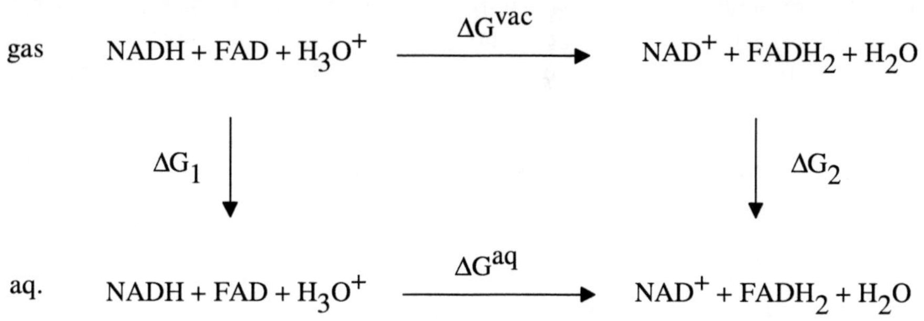

gas $NADH + FAD + H_3O^+$ $\xrightarrow{\quad \Delta G^{vac} \quad}$ $NAD^+ + FADH_2 + H_2O$

$\Delta G_1 \downarrow$ $\downarrow \Delta G_2$

aq. $NADH + FAD + H_3O^+$ $\xrightarrow{\quad \Delta G^{aq} \quad}$ $NAD^+ + FADH_2 + H_2O$

$$\Delta G^{aq} = \Delta G^{vac} + \Delta G_2 - \Delta G_1 = \Delta G^{vac} + \Delta\Delta G_{2-1}$$

$$\Delta G^{aq} = -71 \text{ kcal/mol} + 55 \text{ kcal/mol} = -16 \text{ kcal/mol}$$

$$\Delta G^{aq} \text{ (exp from redox potentials)} = \text{from } -15 \text{ to } -5 \text{ kcal/mol [a]}$$

[a] Depending on protein enviroment

Figure 5. *Thermodynamic cycle for the electron transfer between NADH and FAD. Values in gas phase and in aqueous solution were determined from AM1 and AM1/MST calculations. The cavity for charged species was reduced following the standard procedure (see text and ref. 45g).*

The PCM method has been successfully applied to a large variety of phenomena in solution, like the description of free energies of hydration, solvent effects in chemical reactions, and solvent-induced changes in molecular properties [30 and references therein]. Besides several *ab initio* implementations at the HF and multiconfigurational levels [35a,b,45d,53a-c] and a recent implementation in the Density Functional framework [53d], different semiempirical versions are available [45e-f,56]. The MST version developed by our group, which is available within the MOPAC-93 package [57], yields to free energies of hydration with an expected RMS around 1 kcal/mol for neutral solutes. Recently, the method has been extended to non-aqueous solvents [6b,58]. For carbon tetrachloride and chloroform the average RMS error in the free energy of solvation is found to be less than 0.5 kcal/mol.

The combination of fast semiempirical calculations with accurate solvation calculations makes the semiempirical MST very powerful in scientific areas like biochemistry, where the size of the molecules prevents the use of *ab initio* methods. An example is shown in Figure 5, where AM1/MST calculations are used to study the NADH->FAD electron transfer process in aqueous solution. The agreement between calculations and experiment is reasonable considering the noise in the calculation, the intrinsic shortcomings of semiempirical calculations, and the uncertainties in the experimental measurements. It is clear that semiempirical SCRF calculations cannot provide quantitatively accurate estimates for these processes. However, the results suggest that these calcula-

tions can be used as a powerful tool to gain qualitative insight into complex biochemical processes.

COSMO introduces a novel approach to the study of a system in solution. In this method, Klamt and Schüürman developed a formalism based on the replacement of the dielectric continuum medium by a conductor, which facilitates the treatment of screening effects. The exact solution of the screening problem for the conductor is then corrected by a factor (eq. 24) for the application to a dielectric medium. Eq. 25 gives the total energy of the screened system. In this equation, Q is the generalized charge vector containing the point charges of the nuclei and of the electron densities elements, whereas the matrix A collects the interactions between the charges of the solute with the apparent charges spread over the cavity, and B the interactions between the different apparent charges. A molecular shape surface is defined for the solute cavity. COSMO has been recently implemented in the MOPAC-93 [57] package.

$$f(\varepsilon) = \frac{\varepsilon - 1}{\varepsilon + \frac{1}{2}} \qquad (27)$$

$$\Delta E = -\frac{1}{2} Q \ B \ A^{-1} \ B \ Q \qquad (28)$$

Recently, a new approach based on COSMO for incorporating solvent effect for solutes of arbitrary shape has been reported. This approach is called GCOSMO [59] and has been implemented in both *ab initio* and nonlocal Density Functional levels of theory by modifying the GAUSSIAN 92/DFT [60] program. In GCOSMO the solvent reaction field is included directly in the SCF procedure, and hence both the solute electron distribution and the solvent reaction field are converged simulatneously.

All these methods work in the framework of quantum mechanical codes ranging from semiempirical to post-Hartree-Fock levels. Because of the lack of a microscopic treatment of the solvent, which is represented by a continuum medium, the computational requirements for the study of solvation phenomena do not differ substantially from those required for an isolated molecule. Indeed, most of these codes are easily available, specially when thay are distributed in standard computational programs of widespread use. For instance, Gaussian-94 [61] includes different SCRF methods, like a simple one based on the Onsager model, and different versions of the high level MST algorithm, which differ in the definition of the solute cavity. Undoubtedly, this facilitates the study of a broad range of processes in solution by non-expert users, and it can lead to a significant improvement in our level of knowledge of chemical procecess in solution.

Conclusion

The theoretical representation of condensed phases can now be achieved by a large variety of techniques, which are at the disposal of computational chemists. Continuum or discrete representations of the solvent can be combined with quantum or classical treatments of the solute. Each technique has its strengths and shortcomings, and the proper selection of the appropriate method to be used in the study of a particular system is probably the most important decision for the study of a chemical process in solution.

Acknowledgments This work was supported by the Dirección general de Investigación Científica y Técnica (DGICYT; PB93-0779 and PB94-0940) and by the Centre de Supercomputació de Catalunya (CESCA, Mol. Recog. Project-95). We are indebted to Prof. G. Shields for assistance in the preparation of the manuscript. We thank Prof. J. Tomasi and Prof. J.-L. Rivail for sending us preprints of their research work. We thank also the help of many colleagues and coworkers whose names are in the reference section.

References

1. (a) Beak, P. *Acc.Chem.Res.*, **1977**, *10*, 186 (b) Katritzky, A.R. Handbook of Heterocylic Chemistry. Pergamon, Oxford, 1985 pp. 121-123 (c) Cieplak, P.; Bash, P.; Singh, U.C.; Kollman, P.A. *J. Am. Chem. Soc.*, **1987**, *109*, 6283 (d) Worth, G.A.; King, P.M.; Richards, W.G.

Biochim.Biophys.Acta, **1989**, *993*, 134 (e) Cramer, C.J.; Truhlar, D.G. *J. Am. Chem. Soc.*, **1993**, *115*, 8810 (f) Orozco, M.; Luque, F.J. *J. Am. Chem. Soc.*, **1995**, *117*, 1378.

2. (a) Beak, P.; Fry, F.S.; Lee.J.; Steele, F. *J. Am. Chem. Soc.*, **1976**, *98*, 171 (b) Mason, S.F. *J.Chem.Soc.*, **1957**, 4874.

3. (a) Luque, F.J.; Negre, M.J.; Orozco, M. *J. Phys. Chem.*, **1993**, *97*, 4386 (b) Gao, J.; Luque, F.J.; Orozco, M. *J. Chem. Phys.*, **1993**, *98*, 2975 (c) Cramer, C.J.; Truhlar, D.G. *Chem. Phys. Lett.*, **1992**, *74*, 198.

4. (a) Alagona, G.; Ghio, C.; Igual, J.; Tomasi, J. *J.Am.Chem.Soc.*, **1989**, *111*, 3417 (b) Luque, F.J.; Orozco, M.; Bhadane, P.K.; Gadre, S.R. *J. Phys. Chem.*, **1993**, *97*, 9380 (c) Luque, F.J.; Orozco, M. *J. Chem. Soc. Perkin Trans II*, **1993**, 683 (d) Orozco, M.; Luque, F.J. *Biopolymers*, **1993**, *33*, 1851 (e) Luque, F.J.; Orozco, M.; Bhadane, P.K.; Gadre, S.R. *J. Chem. Phys.*, **1994**, *100*, 6718 (f) Luque, F.J.; Gadre, S.R.; Bhadane, P.K.; Orozco, M. *Chem. Phys. Lett.*, **1995**, *232*, 509.

5. (a) Bayliss, N.S.; McRae, E.G. *J. Phys. Chem.*, **1954**, *58*, 1006 (b) Ruiz-López, M.F.; Rinaldi, D.;Rivail, J.L. *J. Mol.Struc. (Theochem)*, **1986**, *148*, 61 (c) Ruiz-López, M.F.; Rinaldi, D.; Rivail, J.L. *Chem. Phys.*, **1986**, *110*, 403 (d) Karelson, M.; Zerner, M.C. *J.Am.Chem.Soc.*, **1990**, *112*, 9405 (e) Fox, T.; Rösch, N. *Chem. Phys. Lett.*, **1992**, *191*, 33 (f)Aguilar, M.A.; Olivares delValle, F.J.; Tomasi, J. *J. Chem. Phys.*, **1993**, *98*, 7375.

6. (a) Giesen, D.G.; Storer, J.W.; Cramer, C.J.; Truhlar, D.G. *J.Am.Chem.Soc.*, **1995**, *117*, 1057 (b) Luque, F.J.; Alemán, C.; Bachs, M., Orozco, M. *J. Comp. Chem.*, in press.

7. (a) Jorgensen,W.L. in: *Computer Simulation of Biomolecular Systems. Theoretical and Experimental Applications.*, W.F. van Gunsteren and P.K.Weiner (eds), ESCOM, Leiden, 1989, pp. 60-72 (b) Reichartd, C. *Solvents and Solvent Effects in Organic Chemistry*. VCH, New York, 1990, pp. 98-104 (c) Jorgensen, W.L. *Chemtracts Org. Chem.*, **1991**, *4*, 91 (d)*Structure and Reactivity in Aqueous Solution*. C.J. Cramer and D.G. Truhlar (eds), ACS Symp. Ser. 568, Washington, 1994 (e) Whitnell, R.M.; Wilson, K.R. in: *Reviews in Computational Chemistry*. K.Lipkowitz and D.Boyd (eds), Vol 4, Chapter 3, VCH, New York, 1993 (f) Rivail, J.L.; Rinaldi, D. in: *Computational Chemistry, Review of Current Trends*. J. Leszczynski (ed), World Scientif Publishing, in press (g) Cramer, C.J.; Truhlar, D.G. in: *Solvent Effects and Chemical Reactivity*. O.Tapia and J. Bertrán (eds), Kluwer, Dordrecht, in press

8. (a) Jorgensen,W.L.; Buckner, J.K.; Huston, S.E.; Rosky, P.J. *J. Am. Chem. Soc.*, **1987**, *109*, 1891 (b) Buckner, J.K.; Jorgensen, W.L. *J. Am. Chem. Soc.*, **1989**, *111*, 2507.

9. (a) *Gas Phase Ion Chemistry*. M. T. Bowers (ed), Vol. 2, Academic Press, New York, 1979 (b) Olivella, S.;

Urpí, F.; Vilarrasa, J. *J. Comp. Chem.*, **1984**, *5*, 230 and references therein.

10. Pearson, R.G. *J. Am. Chem. Soc.*, **1986**, *108*, 6109.

11. (a) Ingold, C.K. *Structure and Mechanism in Organic Chemistry*. Cornell University Press, Ithaca, New York, 1969 (b) Parker, A.J. *Chem. Rev.*, **1969**, *69*, 1 (c) Nibbering, N.M. in: *Kinetics of Ion-Molecule Reactions*. P. Ausloows (ed), Plenum, New York, 1979 (d) Chandrasekhar, J.; Smith, S.F.; Jorgensen, W.L. *J. Am. Chem. Soc.*, **1985**, *107*, 164 (e) Madura, J.D.; Jorgensen, W.L. *J. Am. Chem. Soc.*, **1986**, *108*, 2517 (f) Evanseck, J.D.; Blake, J.F.; Jorgensen, W.L. *J. Am. Chem. Soc.*, **1987**, *109*, 2349 (g) Howard, A.; Kollman, P. *J. Am. Chem. Soc.*, **1988**, *110*, 7195 (h) Arad, D.; Langridge, R.; Kollman, P. *J. Am. Chem. Soc.*, **1990**, *112*, 491 (i) Orozco, M.; Canela, E.I.; Franco, R. *J. Org. Chem.*, **1991**, *55*, 2630 (j) Gao, J. *J. Am. Chem. Soc.*, **1991**, *113*, 7796 (k) Sola, M.; Lledos, A.; Duran, M.; Bertran, J.; Abboud, J.M. *J. Am. Chem. Soc.*, **1991**, *113*, 2873 (l) Abboud, J.M.; Notario, R.; Bertran, J.; Sola, M. *Prog.Phys.Org.Chem.*, **1993**, *19*, 1.

12. (a) White, W.N.; Wolfarth, E.F. *J. Org. Chem.*, **1970**, *35*, 2196 (b) Cramer, C.J.; Truhlar, D.G. *J. Am. Chem. Soc.*, **1992**, *114*, 8794 (c) Severance, D.L.; Jorgensen, W.L. *J. Am. Chem. Soc.*, **1992**, *114*, 10966 (d) Gao,J. *J. Am. Chem. Soc.*, **1994**, *116*, 1563 (e) Gajewski, J.J.; Brichford, N.L. in: Structure and Reactivity in Aqueous Solution. C.J .Cramer and D.G. Truhlar (eds), Chapter 16, ACS Symp. Series 568, Washington, 1994.

13. (a) Breslow, R. *Acc. Chem. Res.*, **1991**, *24*, 159 (b) Blokzijl, W.; Blandamer, M.J.; Engberts, J. *J. Am. Chem. Soc.*, **1991**, *113*, 4241 (c) Blokzijl, W.; Engberts, J. in: *Structure and Reactivity in Aqueous Solution*. C.J. Cramer and D.G. Truhlar (eds), Chapter 21, ACS Symp. Series 568, Washington, 1994.

14. Kool, E.T.; Breslow, R. *J. Am. Chem. Soc.*, **1988**, *110*, 1596.

15. (a) Bonaccorsi, R.; Cimiraglia, R.; Tomasi, J.; Miertus, S. *J. Mol. Struct. (Theochem)*, **1982**, *87*, 181 (b) Reynolds, C.A. *J. Am. Chem. Soc.*, **1992**, *112*, 7545.

16. (a) Sclove, D.B.; Pazos, J.F.; Camp, R.L.; Greene, F.D. *J. Am. Chem. Soc.*, **1970**, *92*, 7488 (b) Cordes, M.H.J.; Berson, J.A. *J. Am. Chem. Soc.*, **1992**, *114*, 11010 (c) Lim, D.; Hrovat, D.A.; Borden, W.T.; Jorgensen, W.L. *J. Am. Chem. Soc.*, **1994**, *116*, 3494.

17. (a) Baigrie, L.M.; Cox, R.A.; Slebocka-Tilk, H.; Tencer, M.; Tidwell, T.T. *J. Am. Chem. Soc.*, **1985**, *107*, 3640 (b) Coitiño, E.L.; Tomasi, J.; Ventura, O. *J. Chem. Soc. Perkin Trans. II.*, **1994**, *90*, 1.

18. (a) Kygoku, Y.; Lord, R.C.; Rich, A. *Science*, **1966**, *154*, 518 (b) Sikue, T.N.; Schellman, J.A. *J. Mol. Biol.*, **1968**, *33*, 61 (c) Nakano, N.I.; Igarashi,S. *Biochemistry*, **1970**, *9*, 577 (d) Yanson, I.K.; Teplisky, A.B.; Sukhodub, L.F. *Biopolymers*, **1979**, *18*, 1149 (e) Cieplak, P.; Kollman, P.A. *J. Am. Chem. Soc.*, **1988**,

110, 3734 (f) Dang, L.X.; Kollman, P.A. *J. Am. Chem. Soc.*, **1990**, *112*, 503.

19. (a) McCammon, J.A.; Karplus, M. *Acc. Chem. Res.*, **1983**, *16*, 187 (b) Jorgensen, E.L. *Acc. Chem. Res.*, **1989**, *22*, 184 (c) Beveridge, D.L.; DiCapua, F.M. *Ann. Rev. Biophys. Chem.*, **1989**, *18*, 431 (d) Kollman, P.A.; Merz, K.M. *Acc. Chem. Res.*, **1990**, *23*, 246 (e) Kollman, P.A. *Chem. Rev.*, **1993**, *93*, 2395 (f) van Gunsteren, W.F.; Beutler, T.C.; Fraternali, F.; King, P.M.; Mark, A.E.; Smith, P.E. in: *Computer Simulation of Biomolecular Systems.Theoretical and Experimental Applications*. W.F. van Gunsteren, P.K.Weiner and A.J.Wilkinson (eds), Vol. 2, Chapter 13 (Part III), ESCOM, Leiden, 1993.

20. (a) Allinger, N.L. *J. Am. Chem. Soc.*, **1977**, *99*, 8127 (b) Weiner, S.J.; Kollman, P.A.; Case, P.A.; Singh, U.C.; Ghio, C.; Alagona, G. *J. Am. Chem. Soc.*, **1984**, *106*, 765 (c) Nilson, L.; Karplus, M. *J. Comp. Chem.*, **1986**, *7*, 591 (d) Tirado-Rives, J.; Jorgensen,W.L. *J.Am. Chem. Soc.*, **1988**, *110*, 1657 (e) Dauber-Ogusthorpe, P.; Roberts, U.A.; Osguthorpe, D.J.; Wolf, J.; Genest, M.; Hagler, A.T. *Proteins*, **1988**, *4*, 31 (f) Allinger, N.L.; Yuh, Y.H.; Lii, J. *J. Am. Chem. Soc.*, **1989**, *111*, 8551 (g) van Gunsteren,W.F.; Berendsen, H.J.C. GROningen Molecular Simulation (GROMOS), ETH, Zurich, 1992 (h) Orozco, M.; Aleman, C.; Luque, F.J. *Models in Chemistry*, **1993**, *130*, 695 (i) Gelin, B.R. in: *Computer Simulation of Biomolecular Systems.Theoretical and Experimental Applications*. W.F. van Gunsteren, P.K.Weiner and A.J.Wilkinson (eds), Vol. 2, Chapter 5 (Part II), ESCOM, Leiden, 1993.

21. (a) Allen, M.P.; Tildesley, D.J. *Computer Simulations of Liquids*. Clarendon Press, Oxford, 1987 (b) McCammon, J.A.; Harvey, S.C. *Dynamics of Proteins and Nucleic Acids*. Cambridge University Press, Cambridge, 1987 (c) Brooks, C. L.; Karplus, M.; Pettitt, B.M. *Proteins: a Theoretical Perspective of Dynamics, Structure and Thermodynamics*. Wiley, NewYork, 1988 (d) van Gunsteren,W.F. in: *Computer Simulation of Biomolecular Systems.Theoretical and Experimental Applications*. W.F. van Gunsteren, P.K.Weiner and A.J.Wilkinson (eds), Vol. 2, Chapter 1 (Part I), ESCOM, Leiden, 1993 (e) Frenkel, D. in: *Computer Simulation of Biomolecular Systems.Theoretical and Experimental Applications*. W.F. van Gunsteren, P.K.Weiner and A.J.Wilkinson (eds), Vol. 2, Chapter 2 (Part I), ESCOM, Leiden, 1993 (f) Dodd, L.R.; Boone, T.D.; Theodorou, D.N. *Mol. Phys.*, **1993**, *78*, 961. (g) van Gunsteren,W.F.; Luque, F.J.; Timms, D.; Torda, A.E. *Annu. Rev. Biophys. Biomol. Struct.*, **1994**, *23*, 847

22. (a) Böttcher, C. J. *Theory of Electric Polarization*. Elsevier, Amsterdam, 1952 (b) Åqvist, J.; Medina, C.; Sammuelsson, J.-E. *Protein Eng.*, **1994**, *7*, 385.

23. Carlson,H.; Jorgensen, W.L. *J. Phys. Chem.*, in press.

24. (a) Kirkwood, J. *J. Chem. Phys.*, **1935**, *3*, 300 (b) Zwanzing, R. *J. Chem. Phys.*, **1954**, *22*, 1420 (c)

Valleau, J.P.; Torrie, G.M. in: *Modern Theoretical Chemistry* B.Berne (ed), Vol. 5, Plenum, New York, 1977, pp. 169-194 (d) Postma, J.P.M.; Berendsen, H.J.C.; Haak, J.R. *Faraday Symp. Chem. Soc.*, **1982**, *17*, 55 (e) Warshel, A. *J. Phys. Chem.*, **1982**, *86*, 2218 (f) Tembe, B.L.; McCammon, J.A. *J. Comp. Chem.*, **1984**, *8*, 281 (g) Jorgensen, W.L.; Ravimohan, C. *J. Chem. Phys.*, **1985**, *83*, 3050.

25. Bash, P.A.; Singh, U.C.; Langridge, R.; Kollman, P. *Science*, **1987**, *236*, 564.

26. Straatsma, T.P.; Zacharias, M.; McCammon, J.A. in: *Computer Simulation of Biomolecular Systems. Theoretical and Experimental Applications.* W.F. van Gunsteren, P.K.Weiner and A.J.Wilkinson (eds), Vol.2, Chapter 14 (Part V). ESCOM, Leiden, 1993.

27. Mezei, M. *J. Chem. Phys.*, **1985**, *86*, 7084.

28. (a) Jorgensen, W.L.; Briggs, J.M.; Contreras, M.L. *J. Phys. Chem.*, **1990**, *94*, 1683 (b) Rao, B.G.; Singh, U.C. *J. Am. Chem. Soc.*, **1991**, *113*, 4381 (c) Essex, J.W.; Reynolds,C.A.; Richards, W.G. *J. Am. Chem. Soc.*, **1992**, *114*, 3634.

29. (a) Carlson, H.; Nguyen, T.B.; Orozco, M.; Jorgensen, W.L. *J. Comp. Chem.*, **1993**, *14*, 1240 (b) Orozco, M.; Jorgensen, W.L.; Luque, F.J. *J. Comp. Chem.*, 1993, 14, 1498.

30. Tomasi, J.; Persico, M. *Chem. Rev.*, **1994**, *94*, 2027.

31. (a) Rivail, J.L.; Rinaldi, D. *Chem. Phys.*, **1976**, *18*, 223 (b) Tomasi, J.; Bonaccorsi, R.; Cammi, R.; Olivares del Valle, F.J. *J. Mol. Struct. (Theochem)*, **1991**, *234*, 401 (c) Tomasi, J.; Alagona, G.; Bonaccorsi, R.; Ghio, C.; Cammi, R. in: *Theoretical Models of Chemical Bonding.* Z. Maksic (ed), Vol.4, Springer, Berlin, 1991 (d) Tapia, O. *J. Math. Chem.*, **1992**, *10*, 139 (e) Miertus, S.; Frecer, V. *J. Math. Chem.*, **1992**, *10*, 183 (f) Angyan, J.G. *J. Math. Phys.*, **1992**, *10*, 94 (g) Cramer C.J.; Truhlar,D.G. in: *Reviews in Computational Chemistry*, K.B. Lipkowitz and D.B.Boyd (eds), Vol. 6, VCH, New York, 1994 (h) Tomasi, J. in: *Structure and Reactivity in Aqueous Solution.* C.J. Cramer and D.G. Truhlar (eds), Chapter 2, ACS Symp. Series 568, Washington, 1994.

32. Born, M. Z. *Phys.*, **1920**, *1*, 45.

33. (a) Bell, R.P. *Trans. Faraday Soc.*, **1931**, *27*, 797 (b) Onsager, L. *J. Am. Chem. Soc.*, **1936**, *58*, 1486.

34.- Still, W.C.; Tempczyk, A.; Hawley, R.C.; Hendrickson, T. *J. Am. Chem. Soc.*, **1990**, *112*, 6127.

35. (a) Miertus, S.; Scrocco, E.; Tomasi, *J. Chem. Phys.*, **1981**, *55*, 117 (b) Miertus, S.; Tomasi, *J. Chem. Phys.*, **1982**, *65*, 239 (c) Montagnani, R.; Tomasi, J. *J.Mol.Struct. (Theochem)*, **1993**, *279*, 131.

36. (a) Varnek, A.A.; Wipff, G.; Glebov, A.S.; Feil,D. *J. Comp. Chem.*, **1995**, *16*, 1 (b) Luque, F.J.; Bofill, J.M.; Orozco, M. *J. Chem. Phys.*, in press.

37. (a) Warwicker, J.; Watson, H.C. *J. Mol. Biol.*, **1982**, *157*, 671 (b) Gilson, M.; Honig, B. *Nature*, **1987**, *330*, 84 (c) Gilson, M.; Sharp, K.; Honig, B.*J. Comp. Chem.*, **1988**, *9*, 326 (d) Gilson, M.K.; Honig, B.H. *Proteins*, **1988**, *4*, 7 (e) Davis, M.E.; McCammon, J.A. *J. Comp. Chem.*, **1989**, *10*, 387 (f) Tannor, D.J.; Marten, B.; Murphy, R.; Friesner, R.A.; Sitkoff, D.; Nicholls, A.; Ringnalda, M.; Goddard, W.A., III; Honig, B. *J. Am. Chem. Soc.*, **1994**, *116*, 11875.

38. (a) Jean-Charles, A.; Nicholls, A.; Sharp, K.; Honig, B.; Tempczyk, A.; Hendrickson, T.; Still, W.C. *J. Am. Chem. Soc.*, **1991**, *113*, 1454 (b) Alkorta, I.; Villar, H.; P•rez, J.J. *J. Comp. Chem.*, **1993**, *14*, 620.

39. (a) Klapper, I.; Hagstrom, R.; Fine, R.; Sharp, K.; Honig, B. *Proteins*, **1986**, *1*, 47 (b) Nicholls, A.; Honig, B. *J. Comp. Chem.*, **1991**, *12*, 435.

40. (a) Warshel, A.; Levitt, M. *J. Mol. Biol.*, **1976**, *103*, 227 (b) Field, M.J.; Bash, P.A.; Karplus, M. *J. Comp. Chem.*, **1990**, *11*, 700 (c) Warshel, A. *Computer Modeling of Chemical Reactions in Enzymes and Solutions.* Wiley, New York, 1991 (d) Luzhkov, V.; Warshel, A. *J. Comp. Chem.*, **1992**, *13*, 199 (e) Gao, J. *J. Phys. Chem.*, **1992**, *96*, 537 (f) Gao ,J.; Xia, X. *Science*, **1992**, *258*, 631 (g) Field, M. in: *Computer Simulation of Biomolecular Systems. Theoretical and Experimental Applications.* W.F. van Gunsteren, P.K.Weiner and A.J.Wilkinson (eds), Vol. 2, Chapter 4 (Part I), ESCOM, Leiden 1993 (h) Warshel,A. *Chem. Rev.,* **1993**, *93*, 2523 (i) Gao, J. in: *Reviews in Computational Chemistry*, K.B.Lipkowitz and D.B.Boyd (eds),Vol 6, VCH, New York, 1994.

41. (a) Dewar, M.J.S.; Zoebisch, E.G.; Horsley, E.F.; Stewart, J.J.P. *J. Am. Chem. Soc.*, **1985**, *107*, 3902 (b) Stewart, J.J.P., *J. Comp. Chem.*, **1989**, *10*, 209.

42. (a) Stanton, R.V.; Hartsough, D.S.; Merz, K. *J. Phys. Chem.*, **1993**, *97*, 11868 (b) Freindorf, M.; Gao, J. *J. Comp. Chem.*, in press. (c) Tuñón, I.; Martins-Costa, M.T.C.; Millot, C.; Ruiz-López, M.F. *Chem. Phys.Lett.*, **1995**, *241*, 450 (d) Tuñón, I.; Martins-Costa, M.T.C.; Millot, C.; Ruiz-López, M.F.; Rivail, J.L. *J. Comp. Chem.* **1996**, *17*, 19.

43. Alhambra, C.; Luque, F.J.; Orozco, M. *J. Phys. Chem.*, **1995**, *99*, 3084.

44. (a) Floris, F.M.; Tomasi, J.; Pascual-Ahuir, J.L. *J. Comp. Chem.*, **1991**, *12*, 784 (b) Silla, E.; Tuñón, I.; Villar, F.; Pascual-Ahuir, J.L. *J. Mol. Struct.(Theochem)*, **1992**, *254*, 369.

45. (a) Cramer, C.J.; Truhlar, D.G.*J. Am. Chem. Soc.*, **1991**, *113*, 8305 (b) Cramer, C.J.; Truhlar, D.G. *Science*, **1992**, *256*, 213 (c) Cramer, C. J.; Truhlar, D.G. *J. Comp. Chem.*, **1992**, *13*, 1089 (d) Bachs, M.; Luque, F.J.; Orozco, M. *J. Comp. Chem.*, **1994**, *15*, 446 (e) Luque, F.J.; Bachs, M.; Orozco, M. *J. Comp. Chem.*, **1994**, *15*, 847 (f) Orozco, M.; Bachs, M.; Luque, F.J. *J. Comp. Chem.*, **1995**, *16*, 563 (g) Orozco, M.; Luque, F.J. *Chem. Phys.*, **1994**, *182*, 237.

46. (a) Rinaldi, D.; Rivail, J.-L. *Theor. Chim. Acta*, **1973**, *32*, 57 (b) Rinaldi, D.; Ruiz-López, M.F.; Rivail, J.-L. *J. Chem. Phys.*, **1983**, *78*, 834 (c) Rinaldi, D.; Rivail,

J.-L.; Rguini, N.J. *J. Comp. Chem.*, **1992**, *13*, 675 (d) Dillet, V.; Rinaldi, D.; Angyan, J.G.; Rivail, J.-L. *Chem. Phys. Lett.*, **1993**, *202*, 18 (e) Dillet, V.; Rinaldi, D.; Rivail, J.L. *J. Phys. Chem.*, **1994**, *98*, 5034 (f) Ruiz-López, M.F.; Bohr, F.; Martins Costa, M.T.C.; Rinaldi, D. *Chem. Phys. Lett.*, **1994**, *221*, 109 (g) Rinaldi, D.; Rivail, J.L., to be published.

47.   (a) Tapia, O.; Goschinski, O. *Mol. Phys.*, **1975**, *29*, 1654 (b) Tapia, O.; Sussman, F.; Poulain, E. J. *Theor. Biol.*, **1978**, *71*, 49.

48.-   (a) Katritzky, A.R.; Zerner, M.C.; Karelson, M.M. *J. Am. Chem. Soc.*, **1986**, *108*, 7213 (b) Karelson, M.M.; Tamm, T.; Katritzky, A.R.; Cato, S.J.; Zerner, M.C. *Tetrahedron Comput. Methodol.*, **1989**, *2*, 295 (c) Karelson, M.M.; Zerner, M.C. *J. Am. Chem. Soc.*, **1990**, *112*, 7828 (d) Thompson, M.A.; Zerner, M.C. *J. Am. Chem. Soc.*, **1990**, *112*, 9405 (e) Karelson, M.M.; Zerner, M.C. *J. Phys. Chem.*, **1992**, *96*, 6949.

49.   (a) Chudinov, G.E.; Napolov, D.V.; Basilevsky, M.V. *Chem. Phys.*, **1992**, *160*, 41 (b) Rauhut, G.; Clark, T.; Steinke, T. *J. Am. Chem. Soc.*, **1993**, *115*, 9174 (c) Hoshi, H.; Chûjo, R.; Inoue, Y.; Sakurai, M. *J. Mol. Struct. (Theochem)*, **1988**, *49*, 267 (d) Chen, J.L.; Noodleman, L.; Case, D.A., Bashford, D. *J. Phys. Chem.*, **1994**, *98*, 11059 (e) Rashin, A.A.; Young, L.; Topol, J.A. *Biophys. Chem.*, **1994**, *51*, 359 (f) Rashin, A.A.; Bukatin, M. A.; Andzelm, J.; Hagler, A.T. *Biophys. Chem.*, **1994**, *51*, 375.

50.   (a) Mulliken, R.S. *J. Chem. Phys.*, **1955**, *23*, 1833 (b) Singh, U.C.; Kollman, P. *J. Comp. Chem.*, **1984**, *5*, 129 (c) Orozco, M.; Luque, F.J. *J. Comp. Chem.*, **1990**, *11*, 909 (c) Storer, J.W.; Giesen, D.J.; Hawkins, G.D.; Lynch, G.C.; Cramer, C.J.; Truhlar, D.G.; Liotard, D.A. in: *Structure and Reactivity in Aqueous Solution.* C.J. Cramer and D.G. Truhlar (eds), Chapter 3, ACS Symp. Series 568, Washington, 1994 (d) Storer, J.W.; Giesen, D.J.; Cramer, C.J.; Truhlar, D.G. *J. Comput-Aided Mol. Design*, in press.

51.   Giesen, D.J.; Storer, J.W.; Cramer, C.J.; Truhlar, D.G. *J. Am. Chem. Soc.*, **1995**, *117*, 1057.

52.   Klamt, A.; Schüürmann, G. *J. Chem. Soc. Perkin Trans. II*, **1993**, 799.

53.   (a) Olivares del Valle, F.J.; Tomasi, *J. Chem. Phys.*, **1991**, *150*, 139 (b) Olivares del Valle, F.J.; Aguilar, M.A. *J. Comp. Chem.*, **1992**, *13*, 115 (c) Olivares del Valle, F.J.; Aguilar, M.A. *J. Mol. Struct. (Theochem)*, **1993**, *280*, 25 (d) Fortunelli, A.; Tomasi, *J. Chem. Phys. Lett.*, **1994**, *231*, 34.

54.   Pierotti, R.A. *Chem.Rev.*, **1976**, *76*, 717.

55.   Pascual-Ahuir, J.L.; Silla, E.; Tomasi, J.; Bonaccorsi, R. *J. Comp. Chem.*, **1987**, *8*, 778.

56.   (a) Chudinov, G.E.; Napolov, D.V.; Basilevsky, M.V. *Chem. Phys.*, **1992**, *160*, 160, 41 (b) Wang, B.; Ford, G.P. *J. Chem. Phys.*, **1992**, *97*, 4162.

57.   Stewart, J.J.P. MOPAC-93 Computer Program. Stewart Comp. Ltd. 1994.

58.   Luque, F.J.; Zhang, Y.; Alemán, C.; Bachs, M.; Gao, J.; Orozco, M. *J. Phys. Chem.*, in press.

59.   Truong, T.N; Stefanovich, E.V. *Chem. Phys. Lett*, **1995**, *240*, 253

60.   Frisch, M.J.; Trucks, G.W.; Schlegel, H.B.; Gill, P.M.W.; Johnson, B.G.; Wong, M.W.; Foresman, J.B.; Robb, M.A.; Head-Gordon, M.; Replogle, E.S.; Gomperts, R.; Andres, J.L.; Raghavachari, K.; Binkley, J.S.; Gonzalez, C.; Martin, R.L.; Fox, D.J.; DeFress, D.J.; Baker, J.; Stewart, J.J.P.; Pople, J.A. GAUSSIAN 92/DFT, Revision G.3, Gaussian, Pittsburgh, 1993.

61.   Frisch, M.J.; Trucks, G.W.; Schlegel, H.B.; Gill, P.M.W.; Johnson, B.G.; Robb, M.A.; Cheeseman, J.R.; Keith, T.A.; Petersson, G.A.; Montgomery, K.; Raghavachari, K.; Al-Laham, M.A.; Zakrzewski, V.G.; Ortiz, J.V., Foresman, J.B.; Ciolowski, J.; Stefanov, B.B.; Nanayakkara, A.; Challacombe, M.; Peng, C.Y.; Ayala, P.Y.; Chen, W.; Wong, M.W.; Andres, J.L.; Reploge,E.S.; Gomperts, R.; Martin, R.L.; Fox,D.J.; Binkley, J.S.; Defrees, D.J.; Baker, J.; Stewart, J.J.P.; Head-Gordon, M.; Gonzalez, C.; Pople, J.A. GAUSSIAN 94 Revision A.1, Gaussian Inc., Pittsburgh, 1994.

J. Mol. Model. **1996**, 2, 16 - 25

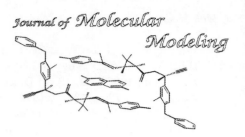

Similarity Study on Peptide γ-turn Conformation Mimetics

Ibon Alkorta*, Maria Luisa Suarez, Rosario Herranz, Rosario González-Muñiz and M. Teresa García-López

Instituto de Química Médica (C.S.I.C.), Juan de la Cierva, 3, E-28006 Madrid, Spain (ibon@pinar1.csic.es)

Received: 25 July 1995 / Accepted: 17 October 1995 / Published: 5 March 1996

Abstract

The ability of a series of structures to mimic the geometric and electronic properties of an ideal γ-turn has been studied. Initially, an exhaustive conformational analysis was carried out using the molecular dynamics technique at high temperature followed by minimization. Additionally, each minimum was optimized with the semi-ab initio molecular orbital method SAM1. Then, the unique minima found have been superimposed with ideal γ-turns, classic and inverse, using the SEAL program which takes into account steric and electronic parameters for the superpositions and finally, three molecular similarity indices were determined for each superposition. These indices consider the general steric and electronic characteristics of the structures, as well as, the position of the carbon atoms that correspond to the $C\alpha^i$ and $C\alpha^{i+2}$ in the peptide chain.

Keywords: γ-turn, peptidomimetics, similarity indices

Introduction

The incorporation of peptide secondary structure mimetics into small bioactive peptides, which leads to restricted analogues, is a well established approach to provide information on the biologically active conformations, and to develop stable, effective and selective receptor ligands [1]. Of special interest are those mimetics which force linear peptide sequences into various defined reverse turn conformations [2].

In recent years, β-turns, as the reverse turns most frequently found in peptides, have been the main focus of attention in the search of conformation mimetics [3]. Little attention has been given, however, to the study of γ-turns. These turns are characterized by a 3->1 hydrogen bond between the CO group of amino acid residue i and the NH group of amino acid residue i+2, as shown in Fig. 1. Two types of γ-turns exist, the classic γ-turn with (φ,ψ) values generally in the range (70 to 95, -75 to -45), and the inverse γ-turn (-95 to

-70, 45 to 75) [4] . Although these turns are less frequent in proteins than β-turns, a recent analysis of 54 proteins with high resolution X-ray crystal structures has shown the existence of ten classic γ-turns [4], and approximately ten-fold more of inverse γ-turns [5]. On the other hand, it has also been suggested that γ-turns are present in the solution conformation of several peptides, including bradykinin [6], substance P [7], cyclosporin [8, 9], vasopresin [10], and cyclic somatostatin analogues [11]. Additionally, it has been proposed that enkephalins assume a γ-turn conformation when binding to membranes and to the δ-opioid receptor [12-14].

Recently, a few heterocyclic systems have been used to lock three amino acid residues into a γ-turn conformation. Thus, 2-oxo-2,3,6,7-tetrahydro-1H-azepides and 2,4-dioxo-hexahydro-1,5-diazepines have been successfully incorporated into fibrinogen receptor antagonists, inhibitors of platenet aggregation [15, 16], and into a HIV-1 protease substrate [17], respectively. On the other hand, the incorporation of 2-oxo-piperidines, as γ-turn mimetics, into the bradykinin sequence support the presence of a reverse turn into the bioactive conformaction of this peptide [18].

Extensive quantum mechanic studies on tripeptide amino acid models (**1**, **2**) in vacuum and aqueous solution [19-21], especially for Ac-Ala-NHMe, have shown that the γ-turn conformation is an energy minimum, being in some cases the absolute minimum [22, 23]. However, when larger peptides have been studied this conformation lost its preponderant role [24]. Molecular modelling studies carried out on γ-turn mimetics have been focused on its structural fitting with specific disposition of a peptide. Thus, the lowest energy conformation of a simplified model of the aforementioned 2-oxo-2,3,6,7-tetrahydro-1H-azepines (**3**) provides a good fit of a cyclic pentapetide fibrinogen antagonist [16]. Likewise, the low energy conformation of simplified 2-oxopi-peridines (**4**) gives a good overlay with the crystal structure of a γ-turn [18].

In this article, the electronic and geometric characteristics of a series of these hetetocyclic systems (**3**, **4**), as well as those corresponding to the model for 2-oxo-piperazines (**5**), previously described as conformationally constrained tripeptide analogues [25], and other seven (**6-11**) and six membered systems (**12-14**) [26-28] have been compared with those of the ideal inverse and classic γ-turns. For this purpose, three similarity indices have been calculated, the first two compare the generic geometric and electronic behaviour of these structures and the third one, described here for the first time, considers the similarity in the position of the at-

oms that correspond to Cα^i and Cα^{i+2} in the peptide chain. The aim of this study is to provide a rational basis to analyze the similarity of γ-turn mimetics already described in the literature and to design new structures with the peptide γ-turn conformations.

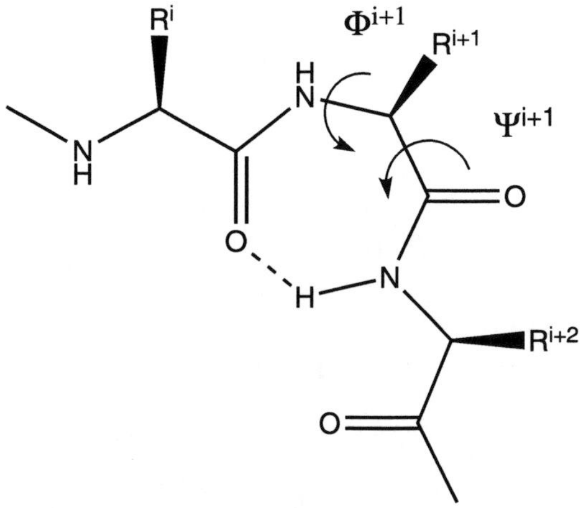

Methods

The inverse and classic γ-turn conformations for the Ac-Ala-NHMe compound (**1**) have been obtained by fully optimization starting from the characteristic average value angles of these structures (Fig. 1) with the semi-ab initio method SAM1 [29], included in the Ampac 5.0 program [30]. The PRECISE keyword has been used to increase 100 times the geometric and electronic convergency criteria and the atomic charges have been derived to reproduce the molecular electrostatic potential (MEP) generated in four van der Waals layers of the molecule [31].

In all the structures (**1-14**), a thorough conformational search has been carried out using the molecular dynamics (MD) technique at high temperature and minimization in vacuo (e=1) with the Insight II program [32]. The MD procedures have been carried out heating the molecules at 1500 K increasing the temperature 10 K each 0.15 ps. and equilibrating at this temperature during 20 ps. Finally, 75 ps. of simulation have been carried out, storing 300 structures at equal intervals. Each structure has been minimized with the cff91 [33] force field using initially the steepest descents minimization methods followed by the conjugate gradient until the gradient was bellow 0.0001 kcal/Å. The minima obtained have been compared and the repeated ones eliminated.

The unique minima have been fully optimized with the SAM1 method and their atomic charges have been generated as described above for compound **1**. Again, the new minima obtained have been compared in order to eliminate the repeated ones. In the case of molecules **4** and **13** which have only one chiral center, the conformations of the enantiomeric compound have been automatically generated with an in house program that creates a mirror image of each conformation.

The superposition of the different conformations obtained for each molecule with the classic and inverse γ-turn models have been carried out using the SEAL program [34]. One hundred different starting positions generated using a Monte Carlo algorithm have been used to find the best simultaneous steric and electronic molecular superpositions of the molecules. The best five superpositions have been stored.

The determination of the similarity of the molecules in the superposition disposition has been done using an in house program which calculates three different similarity indices.

	Classic	Inverse
	Av. Value	Av. value
Φ^{i+1}	75	-79
Ψ^{i+1}	-64	69

Figure 1. *Drawing of γ-turn with their corresponding average angles.*

Figure 2. *(next page) Compounds used in this study. The asterisks indicate that the two possible enantiomers of this center have been studied.*

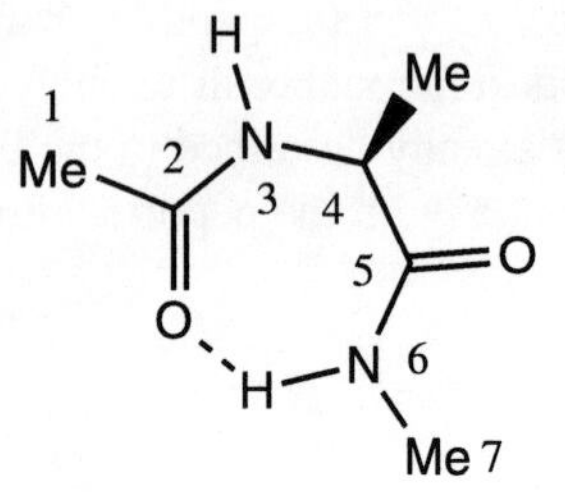

1

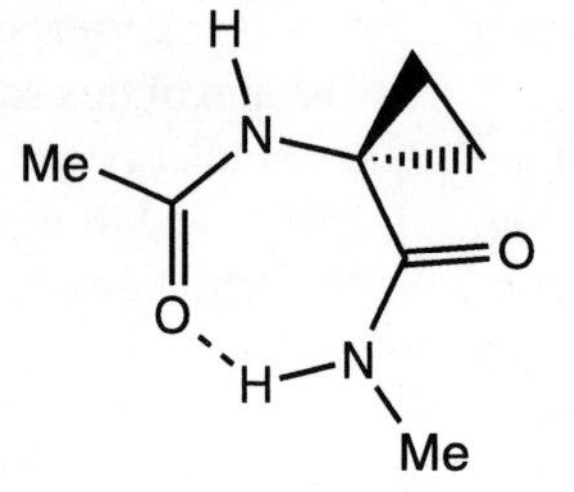

2

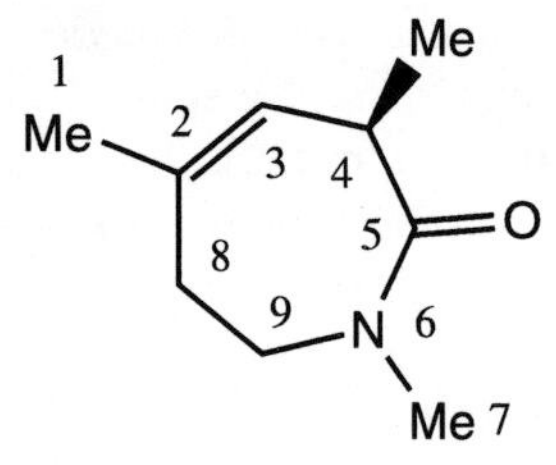

3

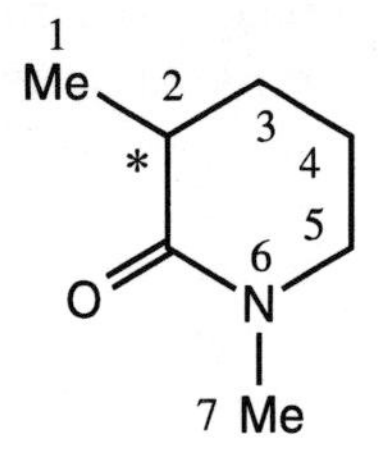

4

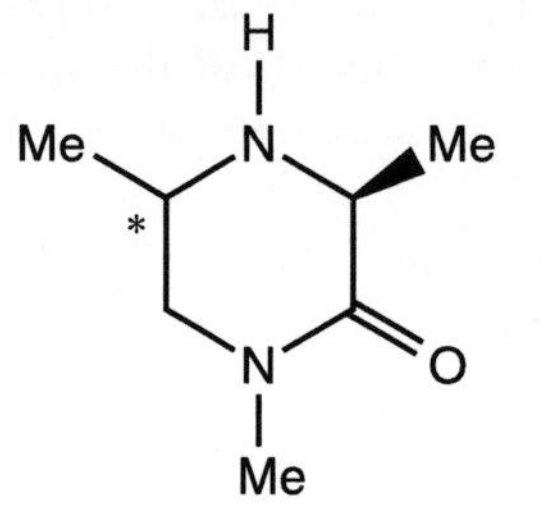

5

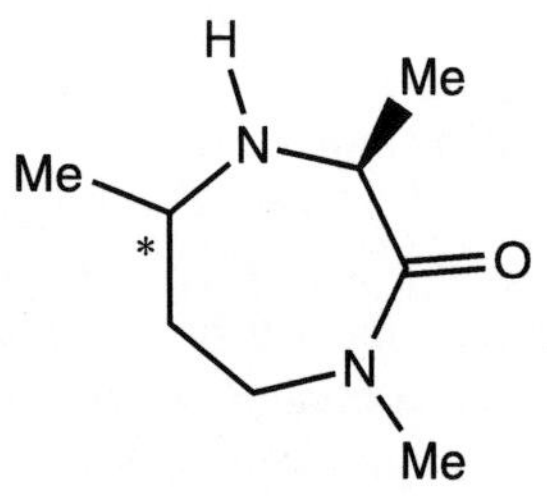

6

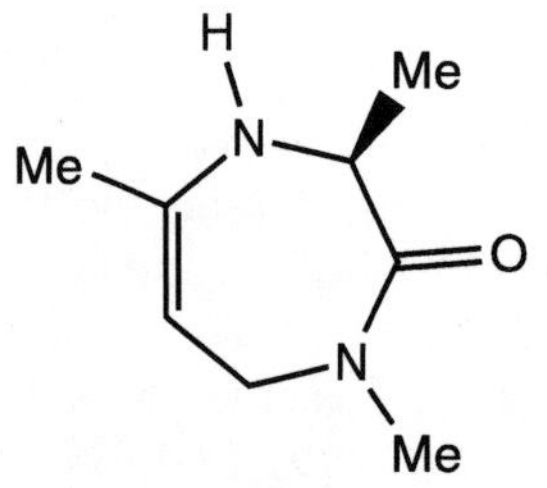

7

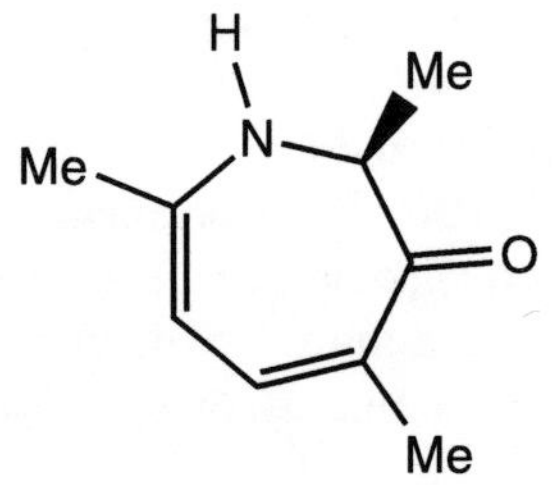

8

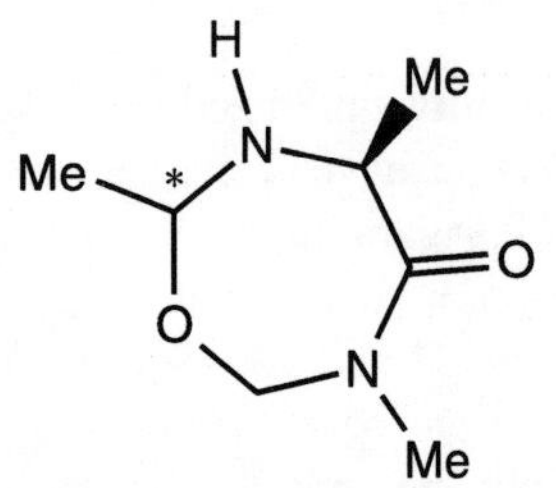

9

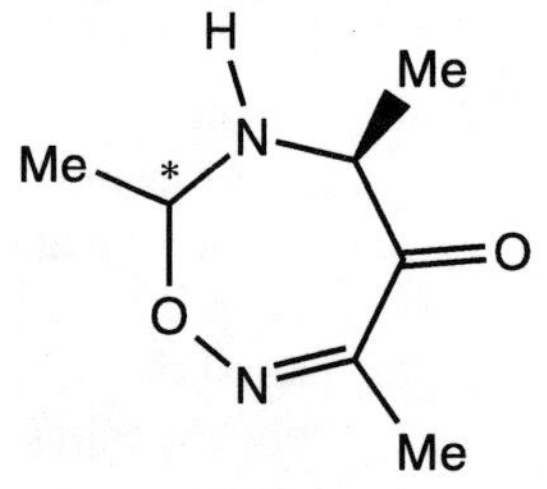

10

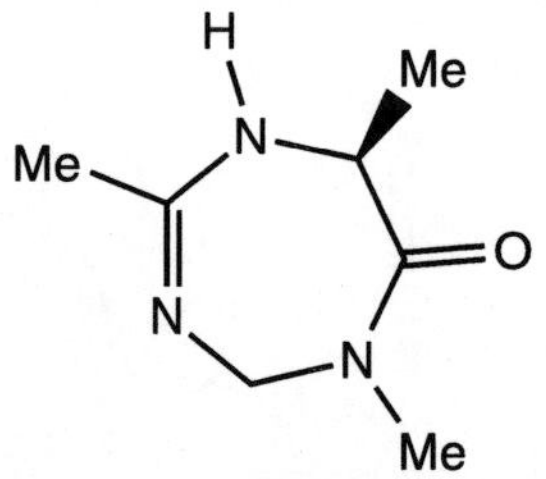

11

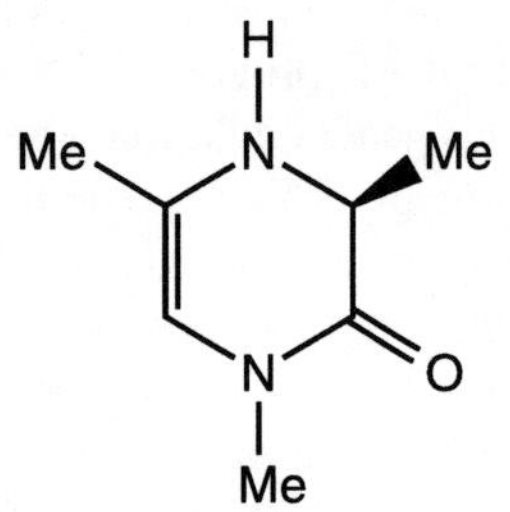

12

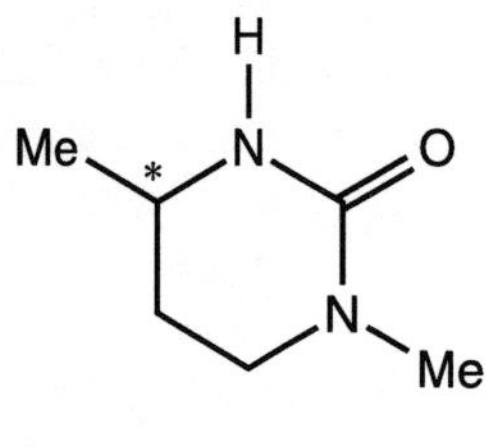

13

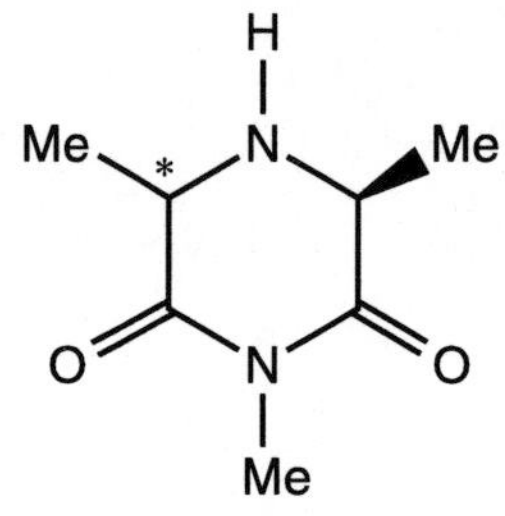

14

The first of them indicates the electrostatic similarity on a three dimensional grid that extends 5 Å from the largest molecule in each direction, the density of points considered being 8 per Å^3. The MEP for each molecule was derived from the atomic charges on all the points of the grid except those that are inside the van der Waals volume of any of the two structures assumed in each superposition. The van der Waals volume of the molecules has been defined using the radii values reported by Gavezzotti [35]. Finally, the similarity has been calculated applying a numerical solution of the Carbo index [36]:

$$R_{ab} = \frac{\sum MEP_a^i * MEP_b^i}{\sqrt{\sum MEP_a^{i2}} * \sqrt{\sum MEP_b^{i2}}} \quad (1)$$

where MEP_a^i and MEP_b^i indicate the value of the MEP on the same grid point, i, generated for molecules A and B. The maximum value of this index is 1 when the MEPs of two molecules are the same. The minimum value is -1 and corresponds to the hypothetical case when the MEP of one molecule is the negative of the MEP of the other molecule for all the points considered.

A second index, evaluates the shape similarity counting the number of grid points inside the individual and common van der Waals volumes and using the following formula:

$$S_{ab} = \frac{V_{ab}}{Min(V_a, V_b)} \quad (2)$$

where V_a and V_b correspond to the number of points that are inside molecules A and B, respectively, and V_{ab} those which are common to both molecules. The denominator indicates that only the smallest volume of the two molecules is used and thus hypothetical superposition of a subset of a molecule with the whole molecule would provide a maximum value of 1.

The third index evaluates the similar disposition of $C\alpha^i$ and $C\alpha^{i+2}$ atoms of **1** (1 and 7 in Fig. 2), and those corresponding to the peptidomimetic model compounds:

$$D_{ab} = e^{-1/n * \sum_{i=1}^{n} d(C_a^i, C_b^i)} \quad (3)$$

where n corresponds to the number of atom compared, in this case 2, and $d(C_a^i, C_b^i)$ is the distance between each pair of atoms compared. This index may be considered as a measure of the ability of the corresponding structure to keep the correct disposition of the peptide backbone attached to it when it is used as a building block. The value of this parameter is 1 when the atoms compared have the same coordinates and rapidly diminishes as the sum of the distance increases.

The total similarity index of the superposition of molecules A and B can be calculated as the sum of eq. (1)-(3):

$$T_{ab} = R_{ab} + S_{ab} + D_{ab} \quad (4)$$

These indices take into account the similarity of each conformation with a γ-turn. In order to have a similarity value that could be associated to all the conformations of a given molecule, an overall molecular similarity index was developed using the following equation:

$$P_{ab} = \frac{\sum R_{ab}^i * p^i}{\sum p^i} + \frac{\sum S_{ab}^i * p^i}{\sum p^i} + \frac{\sum D_{ab}^i * p^i}{\sum p^i} \quad (5)$$

or

$$P_{ab} = \frac{\sum T_{ab}^i * p^i}{\sum p^i} \quad (6)$$

where p^i is the relative population of conformer i calculated using the Boltzmann distribution equation.

Table 1. *Number of minima found for each compound.*

Comp.	Config. [a]	MD & Minim.	SAM1
1		22	13
2		13	10
3		4	4
4	S	5	4
5	R	4	3
5	S	7	6
6	R	12	11
6	S	15	10
7		6	4
8		5	4
9	R	16	12
9	S	16	12
10	R	14	5
10	S	12	7
11		8	2
12		1	1
13	S	2	2
14	R	4	4
14	S	4	4

[a] Configuration of the center indicated with an asterisk in Figure 2

Results and Discussion

The number of minima found, first in the molecular dynamics/minimization procedure, and then with the SAM1 methods, are included in Table 1. As can be seen, some of the minima found with the molecular mechanics method converge to the same minima with SAM1, reducing the total number of minima found with the last procedure. This tendency has already been described for other semiempirical methods [37]. With respect to the energetic values of the minima, in most of the cases the absolute minimum found with the molecular mechanics method is the same as that found with the SAM1 methods, or corresponds to a minimum of small relative energy. Regarding the number of minima found, as expected, cyclization to a seven membered ring reduces slightly the total number of minima, this reduction being larger in the case of compounds with endocyclic double bonds. The six membered compounds show less degrees of freedom and consequently a smaller number of minima.

The similarity indices of the structures studied have been divided in three tables. The first one (Table 2) includes the

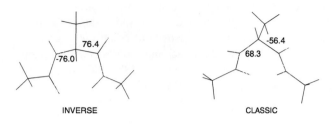

Figure 3. *Structure of 1 used as model of classic and inverse γ-turns.*

best similarity values for each compound compared with the model γ-turns, classic and inverse. Table 3 shows the similarity indices of the absolute minimum of each compound, and in Table 4 are gathered the overall molecular indices, P_{ab}.

In general, the compounds studied can be classified in three different groups: open chain (**1** and **2**), seven membered rings (**3, 6-11**) and six membered rings (**4, 5, 12-14**). The two compounds included in the first group present different capacities to adopt direct and inverse γ-turns, while the in-

Table 2a. *Best similarity indices found for each compound studied when compared with a model of inverse γ-turn.*

Comp.	Conf.[a]	R	S	D	T	E rel.[b]
1		1.00	1.00	1.00	3.00	0.72
2		0.83	0.86	0.78	2.48	0.00
3		0.63	0.92	0.82	2.37	3.38
4	R	0.56	0.76	0.13	1.47	0.82
4	S	0.61	0.77	0.15	1.54	0.00
5	R	0.59	0.82	0.55	1.98	1.06
5	S	0.65	0.74	0.29	1.69	0.77
6	R	0.58	0.90	0.82	2.31	2.35
6	S	0.66	0.88	0.75	2.29	2.12
7		0.76	0.91	0.87	2.55	0.00
8		0.63	0.83	0.64	2.11	0.00
9	R	0.73	0.85	0.72	2.30	5.79
9	S	0.59	0.89	0.68	2.16	0.00
10	R	0.42	0.83	0.71	1.97	0.86
10	S	0.53	0.72	0.11	1.37	2.67
11		0.84	0.89	0.81	2.56	0.00
12		0.28	0.50	0.10	0.89	0.00
13	R	0.44	0.73	0.67	1.84	0.52
13	S	0.45	0.66	0.66	1.78	0.00
14	R	0.45	0.63	0.07	1.16	3.67
14	S	0.46	0.71	0.24	1.42	3.05

[a] *Configuration of the center indicated with an asterisk in Figure 2*
[b] *Relative energy (kcal/mol) of the corresponding conformer.*

Table 2b. *Best similarity indices found for each compound studied when compared with a model of classic γ-turn.*

Comp.	Conf.[a]	R	S	D	T	E rel. [b]
1		1.00	1.00	1.00	3.00	1.55
2		0.85	0.87	0.85	2.57	0.01
3		0.70	0.89	0.72	2.32	3.75
4	R	0.71	0.81	0.14	1.67	1.28
4	S	0.69	0.68	0.18	1.55	1.28
5	R	0.66	0.72	0.57	1.96	2.03
5	S	0.65	0.72	0.29	1.67	0.00
6	R	0.72	0.86	0.76	2.35	3.69
6	S	0.60	0.86	0.68	2.16	2.80
7		0.83	0.93	0.83	2.59	1.50
8		0.65	0.84	0.57	2.07	1.35
9	R	0.72	0.80	0.63	2.15	4.97
9	S	0.69	0.84	0.73	2.27	4.84
10	R	0.60	0.73	0.10	1.45	2.35
10	S	0.58	0.82	0.79	2.20	2.67
11		0.90	0.90	0.80	2.61	0.47
12		0.48	0.67	0.31	1.47	0.00
13	R	0.57	0.65	0.58	1.81	0.00
13	S	0.63	0.74	0.63	2.01	0.52
14	R	0.62	0.68	0.50	1.82	3.21
14	S	0.66	0.76	0.45	1.88	3.05

[a] *Configuration of the center indicated with an asterisk in Figure 2*
[b] *Relative energy (kcal/mol) of the corresponding conformer.*

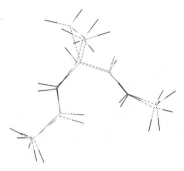

2 + inverse γ -turn

2 + classic γ -turn

Figure 4. *Superposition of the conformations of 2 with the models of inverse (left) and classic (right) γ-turns that provide the best similarity indices .*

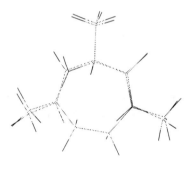

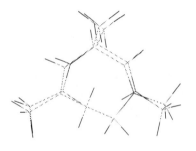

3 + inverse γ -turn

3 + classic γ -turn

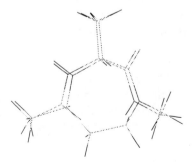

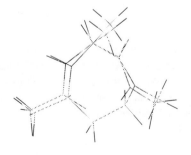

6(R) + inverse γ-turn

6(R) + classic γ -turn

Figure 5. *Superposition of the conformations of 3 and 6(R) with the models of inverse (left) and classic (right) γ-turns that provide the best similarity indices .*

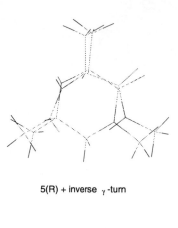

5(R) + inverse γ-turn

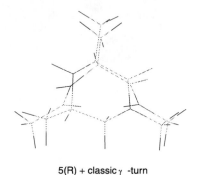

5(R) + classic γ -turn

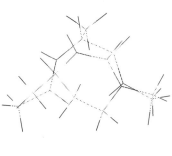

13(S) + inverse γ-turn

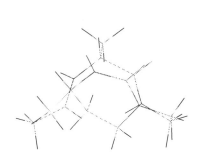

13(S) + classic γ -turn

Figure 6. *Superposition of the conformations of 5(R) and 13(S) with the models of inverse (left) and classic (right) γ-turns that provide the best similarity indices .*

verse γ-turn conformation of **1** corresponds to a minima with a small relative energy, the classic γ-turn being energetically less favourable. This difference is based on the disposition of the methyl group in position 4, while in the inverse γ-turn conformation it is in equatorial disposition, in the classic one it is axial (Fig. 3). The introduction of larger groups in this position, as is the case of the other amino acids except glycine, should increase its relative energy and consequently diminish its tendency to adopt classic γ-turns. This conclusion is in good agreement with the observed experimental tendency of peptides to adopt classic and inverse γ-turn conformations.

On the other hand, compound **2** shows similar ability to adopt both types of γ-turn, since the presence of the cyclopropane forces the molecule to have one methylene group in equatorial disposition while the other one is in axial orientation. Thus, the conformational study of this compound indicates the presence of two degenerated absolute minima corresponding to both types of γ-turns (Fig. 4).

Compounds which include a seven membered ring (**3, 6-11**) present, in general, conformations with good similarity to both classic and inverse γ-turn (Fig. 5). However, the conformations that better mimic classic γ-turn have larger relative energies due to the axial disposition of the methyl group

in position 4, except in the case of **9**(R). This fact, as in the case of **1**, reduces their overall molecular indices.

Regarding stereoisomer in position 2 of the compounds with a seven membered ring (**6, 9** and **10**), those that allow a simultaneous equatorial disposition of the methyl groups in positions 2 and 4 are the most similar to the inverse γ-turn model. Thus, **6**(R) has a better molecular similarity index than its S isomer, while in the case of **9**, the best index corresponds to the isomer S in agreement with their ability to locate the two methyl groups in equatorial disposition. Structures with double bonds in the 2-8 or 2-3 positions (**7, 8** and **11**) show good similarity indices with inverse γ-turns since the methyl group in position 2 is in a pseudo-equatorial disposition.

The similarity indices shown for these compounds indicate that the atoms involved in the typical hydrogen bond of γ-turns could be substituted by other groups as CH$_2$, CH and N without an important loss in their ability to mimic these turns. This fact provides a basis to design compounds with limited flexibility maintaining their similarity with γ-turns.

The good results obtained for the similarity of **3** with the inverse γ-turn model are in good agreement with the experimental HIV-1 protease inhibitory activity of compounds that use this structure as building block to induce γ-turn conformation [16].

The compounds containing a six membered ring (**4, 5, 12-14**) show similar values for both classic and inverse g-turns. In general, the similarity indices are, smaller than those

Table 3a. *Similarity indices for the absolute minimum of each compound studied when compared with a model of inverse γ-turn.*

Comp.	Conf. [a]	R	S	D	T
1		0.69	0.80	0.56	2.05
2		0.83	0.86	0.78	2.48
3		0.57	0.90	0.78	2.26
4	R	0.54	0.68	0.13	1.37
4	S	0.61	0.77	0.15	1.54
5	R	0.63	0.76	0.47	1.86
5	S	0.44	0.69	0.38	1.52
6	R	0.52	0.89	0.72	2.15
6	S	0.51	0.65	0.26	1.43
7		0.76	0.91	0.87	2.55
8		0.63	0.83	0.64	2.11
9	R	0.44	0.66	0.28	1.38
9	S	0.59	0.89	0.68	2.16
10	R	0.41	0.62	0.15	1.19
10	S	0.35	0.67	0.09	1.12
11		0.84	0.89	0.81	2.56
12		0.28	0.50	0.10	0.89
13	R	0.43	0.65	0.15	1.25
13	S	0.45	0.66	0.66	1.78
14	R	0.35	0.66	0.12	1.14
14	S	0.24	0.65	0.06	0.95

[a] Configuration of the center indicated with an asterisk in Figure 2

Table 3b. *Similarity indices for the absolute minimum of each compound studied when compared with a model of classic γ-turn.*

Comp.	Conf. [a]	R	S	D	T
1		0.49	0.59	0.18	1.27
2		0.85	0.87	0.85	2.57
3		0.51	0.77	0.13	1.42
4	R	0.66	0.64	0.07	1.38
4	S	0.65	0.64	0.10	1.40
5	R	0.56	0.69	0.31	1.57
5	S	0.65	0.72	0.29	1.67
6	R	0.63	0.65	0.22	1.51
6	S	0.55	0.61	0.05	1.21
7		0.32	0.75	0.34	1.42
8		0.32	0.71	0.33	1.37
9	R	0.62	0.81	0.46	1.89
9	S	0.63	0.62	0.21	1.48
10	R	0.56	0.75	0.07	1.39
10	S	0.47	0.63	0.09	1.20
11		0.43	0.73	0.14	1.31
12		0.48	0.67	0.31	1.47
13	R	0.57	0.65	0.58	1.81
13	S	0.56	0.58	0.13	1.29
14	R	0.71	0.62	0.43	1.77
14	S	0.56	0.67	0.34	1.58

[a] Configuration of the center indicated with an asterisk in Figure 2

obtained in the comparison of the seven membered systems with the inverse γ-turn model but similar to that obtained from its comparison with the classic γ-turn model. An in depth analysis of the results show that the six membered compounds provide, in general, good steric and electronic indices especially in the comparison with the classic γ-turn. However, the results obtained for the similarity index that measure the disposition of atoms 1 and 7 (D) are lower, except for **13** with this similarity index over 0.6 (Fig. 6).

The experimental data showed that only one of the enantiomers of **4** was useful as building block in the synthesis of compounds with affinity for the bradykinin receptor [18]. Even though the authors were not able to identify which enatiomer was the active one, our calculation indicates that if bradykinin adopts an inverse γ-turn the S conformer should be the active one.

Finally, the 2-oxopiperazines (**5**), which have recently been synthesized in our laboratories as conformationally constrained tripeptide analogues, provide the best similarity indices for all studied six membered systems, which indicates that they could be successfully used as building blocks which mimic γ-turn conformations.

Conclusion

The results here reported indicate that in the case of structures with peptidic skeleton, the stability of γ-turn conformations can be modulated with the substituent attached to $C\alpha^{i+1}$.

Compounds with a seven membered ring show good overall molecular similarity indices when compared to an ideal inverse γ-turn; however, their similarity with the classic γ-turn is much smaller. This tendency is due to the different stability of the conformer that better mimics each kind of γ-turn conformation.

The compounds that contain a six membered ring provide good overall steric and electronic similarity with both, classic and inverse γ-turns. However, these compounds are not able to position the atoms which correspond to the $C\alpha^i$ and $C\alpha^{i+2}$ in the peptide chain in the same disposition as found in the model γ-turns.

Acknowledgements This work was supported by a grant of the Comisión de Ciencia y Tecnología (SAF-94-0705). I.A. is indebted for a Ministerio de Educación y Ciencia contract.

Table 4a. *Overall molecular similarity index, P, and its components (eq. (5)) of the studied molecules when compared with a model of inverse γ-turn.*

Comp.	Conf. [a]	PR	PS	PD	P
1		0.82	0.75	0.59	2.16
2		0.82	0.66	0.44	1.92
3		0.85	0.51	0.67	2.03
4	R	0.69	0.53	0.13	1.35
4	S	0.76	0.60	0.14	1.51
5	R	0.77	0.62	0.48	1.87
5	S	0.70	0.50	0.35	1.56
6	R	0.88	0.53	0.70	2.12
6	S	0.68	0.53	0.28	1.48
7		0.90	0.73	0.83	2.47
8		0.82	0.60	0.59	2.02
9	R	0.68	0.49	0.26	1.44
9	S	0.87	0.58	0.65	2.09
10	R	0.67	0.41	0.26	1.34
10	S	0.67	0.36	0.10	1.13
11		0.84	0.61	0.65	2.11
12		0.51	0.28	0.10	0.90
13	R	0.68	0.44	0.31	1.43
13	S	0.65	0.48	0.54	1.68
14	R	0.66	0.37	0.12	1.15
14	S	0.59	0.36	0.13	1.08

[a] Configuration of the center indicated with an asterisk in Figure 2

Table 4b. *Overall molecular similarity index, P, and its components (eq. (5)) of the studied molecules when compared with a model of classic γ-turn.*

Comp.	Conf. [a]	PR	PS	PD	P
1		0.62	0.56	0.24	1.41
2		0.79	0.64	0.46	1.90
3		0.75	0.52	0.17	1.44
4	R	0.70	0.68	0.10	1.48
4	S	0.65	0.67	0.12	1.44
5	R	0.69	0.58	0.31	1.58
5	S	0.70	0.63	0.29	1.61
6	R	0.67	0.63	0.27	1.57
6	S	0.63	0.56	0.09	1.28
7		0.77	0.36	0.38	1.52
8		0.73	0.35	0.36	1.44
9	R	0.76	0.62	0.41	1.79
9	S	0.65	0.63	0.26	1.53
10	R	0.73	0.56	0.08	1.37
10	S	0.64	0.48	0.10	1.22
11		0.79	0.58	0.35	1.71
12		0.68	0.48	0.32	1.47
13	R	0.68	0.56	0.45	1.69
13	S	0.63	0.59	0.28	1.51
14	R	0.62	0.69	0.41	1.72
14	S	0.68	0.57	0.20	1.45

[a] Configuration of the center indicated with an asterisk in Figure 2

References

1. Gante, J. *Angew. Chem., Int. Ed., Engl.* **1994**, *33*, 1699-1720.
2. Huffman, W.F. *Medicinal Chemistry of the 21st Century*; Wermuth, C.G. ed., Blackwell Scientif Publications, Oxford, 1992; pp 247-257
3. Giannis, A.; Kolter, T. *Angew. Chem. Int. Ed. Engl.* **1993**, *32*, 1244-1267.
4. Milner-White, E.J.; Ross, B.M.; Ismail, R.; Belhadj-Mostefa, K.; Poet, R. *J. Mol. Biol.* **1988**, *204*, 777-782.
5. Milner-White, E.J. *J. Mol. Biol.* **1990**, *216*, 385-397.
6. Cann, J.R.; Stewart, J.M.; Matsueda, G.R. *Biochem.* **1973**, *12*, 3780-3788.
7. Levian-Teitelbaum, D.; Kolodny, N.; Chorev, M.; Selinger, Z.; Gilon, C. *Biopolymers* **1989**, *28*, 51-64.
8. Loosli, H.R.; Kessler, H; Oschkinat, H.; Weber, H.P.; Petcher, T.J.; Widmer, A. *Hel. Chim. Acta* **1985**, *68*, 682-703.
9. Lautz, J.; Kessler, H.; Kaptein, R.; van Gusteren, W.F. *J. Comp-Aid. Mol Design* **1987**, *1*, 219-241.
10. Hagler, A.T.; Osguthorpe, D.J.; Dauber-Osguthorpe, P.; Hempel, J.C. *Science* **1985**, *227*, 1309-1315.
11. Vander Elst, P.; Gondol, D.; Wynants, C.; Tourwé, D.; Van Binst, G. *Int. J. Peptide Protein Res.* **1987**, *29*, 331-346.
12. Nikiforovich, G.V.; Balodis, J. *FEBS Lett.* **1988**, *227*, 127-130.
13. Mammi, N.J.; Hassan, N.; Gordon, M. *J. Am. Chem. Soc.* **1985**, *107*, 4008-4013.
14. Milon, A.; Mayazawa, T.; Higashijima, T. *Biochem.* **1990**, *29*, 65-75.
15. Callahan, J.F.; Newlander, K.A.; Burgess, J.L.; Eggleston, D.S.; Nichols, A.; Wong, A.; Huffman, W.F. *Tetrahedon* **1993**, *17*, 3479.
16. Callahan, J.F.; Bean, J.W.; Burgess, J.L.; Eggleston, D.S.; Hwang, S.M.; Kopple, K.D.; Koster, P.F.; Nichols, A.; Peishoff, C.E.; Samanen, J.M.; Vasko, J.A.; Wong, A.; Huffman, W.F. *J. Med. Chem.* **1992**, *35*, 3970-3972.
17. Newlander, K.A.; Callahan, J.F.; Moore, M.L.; Moore, T.L.; Tomaszek, T.A.; Huffman, W.F. *J. Med. Chem.* **1993**, *36*, 2321-2331.
18. Sato, M.; Lee, J.Y.H.; Nakanishi, H.; Johnson, M.E.; Chrusciel, R.A.; Kahn, M. *Biochem. Biophys. Res. Comm.* **1992**, *187*, 999-1006.

19. Rommel-Möhle, K.; Hofmann, H.J. *J. Mol. Struct. (Theochem)* **1993**, *285*, 211-219.

20. McAllinster, M.A.; Perczel, A.; Császár, P.; Viviani, W.; Rivail, J.L.; Csizmadia, I.G. *J. Mol. Struct. (Theochem)* **1993**, *288*, 161-179.

21. Shang, H.S.; Head-Gordon, T. *J. Am. Chem. Soc.* **1994**, *116*, 1528-1532.

22. Balaji, V.N.; Ramnarayan, K.; Fai Chan, M.; Rao, S.N. *Pep. Res.* **1994**, *7*, 60-71.

23. Shang, H.S.; Head-Gordon, T. *J. Am. Chem. Soc.* **1994**, *116*, 1528-1532.

24. Cheung, M.; McGovern, M.E.; Jin, T.; Zhao, D.C.; McAllister, M.A.; Farkas, O.; Perczel, A.; Császár, P.; Czismadia, I.G. *Theochem*, **1994**, *38*, 151.

25. Herranz, R.; Suárez-Gea, M.L.; García-López, M.T. *Peptides: Chemistry, Structure and Biology: Proceedings of the Thirteenth American Peptide Symposium*; Hodges, R. S.; Smith, J.A. ed., ESCOM, Leiden, 1994; pp 301-303.

26. *Comprehensive Heterocyclic Chemisty*, Katritzky, A.R.; Rees, C.W. ed., Pergamon Press, Oxford, 1984 ; Vol. 7, pp 491, 608, 642 and 640.

27. Barlin, G.B. *The Pyrazines*, Vol.41 of *The Chemistry of Heterocyclic Compounds*, Weissberger, A.; Taylor, E.C. ed., John Wiley & Sons, Inc, New York , 1982.

28. Brown, D.J. *The Pyrimidines* Vol. 52 of *The Chemistry of Heterocyclic Compounds*, Weissberger, A.; Taylor, E.C. ed., John Wiley & Sons, Inc, New York , 1994.

29. Dewar, M.J.S.; Jie, C.; Yu, J. *Tetrahedron* **1993**, *49*, 5003-5038.

30. Ampac 5.0, 1994 Semichem, 7128 Summit, Shawnee, KS 66216.

31. Besler, B.H.; Merz, K.M.; Kollman. P.A. *J. Comput. Chem.* **1990**, *11*, 431.

32. Biosym Technologies, Insight II (version 2.2.0), San Diego, CA, USA.

33. Maple, J.R.; Dinur, U.; Hagler, A.T. *Proc. Nat. Acad. Sci. USA,* **1988**, 85, 5350.

34. Kearsley, S.E.; Smith, G.M. SEAL, QCPE program 634.

35. Gavezzotti. A. *J. Am. Chem. Soc.* **1983**, *105*, 5220.

36. Carbo, R.; Leyda, L.; Arnau, M. *Int. J. Quantum Chem.* **1980**, *17*, 1185-1189.

37. Alkorta, I.; Villar, H.O.; Cachau, R.E. *J. Comput. Chem.* **1993**, *14*, 571-578.

J. Mol. Model. **1996**, 2, 26

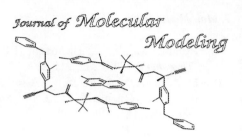

Book review

Computer Aided Molecular Design — Applications in Agrochemicals, Materials, and Pharmaceuticals

Edited by Charles H. Reynolds, M. Katherine Holloway, and Harold K. Cox. Washington, D. C., The American Chemical Society, 1995. x + 428 pp. 15.5 x 23 cm. ISBN 0-8412-3160-5. US $ 99.95.

This book was developed from a symposium of the American Chemical Society, sponsored by the Division of Computers In Chemistry and The Division of Agrochemicals, at the 207th National meeting of the ACS, held in San Diego, California, March 13 - 17, 1994. The book addresses the editors' stated goal of providing answers to two questions, which might be asked by practicing scientists — viz. (1) What new product have arisen from Computer Aided Molecular Design (CAMD) ? and (2) How can the many non-specialist scientists apply modeling and computational techniques to their work ?

The first question is the harder to answer. Molecular modeling / computational chemistry / CAMD techniques have become a part of the team effort of most industrial research projects. The general field has matured so much in the past ten years, that its application is as commonplace as, say, spectroscopic methods. Thus the attribution of a breakthrough discovery exclusively to CAMD is difficult to do in practice, although in most cases, the rest of the team would ratify the importance of the computational component of the work.

To answer the second question, the book is organized into 28 chapters, which are deemed to be case studies in the application of computational chemistry techniques to real-world research problems, in three sections: (1) Pharmaceuticals, (2) Agrochemicals, and (3) Materials.

Volume 589 of the ACS Symposium Series continues the excellent tradition of its predecessors, by offering the reader comprehensive, and current information on subjects of importance to specialists incomputational chemistry as well as their experimental colleagues. Each of the 28 chapters is written by well-known leaders in their respective fields who provide fascinating demonstrations of their extensive expertise. The breadth of computational topics covered makes the volume important to those wishing to gain insight outside of their core area of expertise. The utility of the work is all the more apparent when one considers that most of the techniques described and applied in each of the three topical sections could just as well be applied in either of the other two.

The volume contains three indices: An author index, an affiliation index, and a subject index. Each chapter contains an extensive bibliography. The price is reasonable. It is recommended for individuals, researchers, and institutional libraries.

Emile M. Bellott
Director of Computational Chemistry
Pharm-Eco Laboratories, Inc
128 Spring Street Lexington
Massachusetts, 02173-7800
USA
Tel: (617) 861-9303 Fax: (617) 861-9386
(ebellott@pharmeco.com)

J. Mol. Model. **1996**, 2, 27 - 45

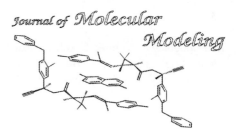

The Erythrocyte/Brain Glucose Transporter (GLUT1) May Adopt a Two-Channel Transmembrane α/β Structure.

Philippe Ducarme*, Mehdi Rahman, Laurence Lins and Robert Brasseur.

Centre de Biophysique Moléculaire Numérique, Faculté des Sciences Agronomiques de Gembloux, chemin des Déportés, 2, B-5030 Gembloux (pducarme@fsagx.ac.be)

Received: 29 September 1995 / Accepted: 17 January 1996 / Published: 29 March 1996

Abstract

There are two models of topology for the membrane domains of the erythrocyte/brain facilitative glucose transporter, GLUT1. The first is composed of 12 membrane-spanning α-helices, the second of 16 membrane-spanning β-strands. We have used Jähnig's and Eisenberg's methods to identify possible transmembrane segments (10 spanning α-helices and 4 β-strands). The topology proposed is more consistent with available experimental data from FTIR, CD and mapping experiment than the previous models . We suggest that GLUT1 might form two channels, one of which is responsible for glucose transport. This agrees with the theoretical and experimental arguments. Finally, an analysis of the mutation periodicity and of the mean hydrophobicity for the GLUT family is provided in order to evaluate the packing of the protein in the membrane.

Keywords: GLUT1, glucose transport, structure, transmembrane protein, α/β structure, modeling
Running Title: GLUT1 transmembrane α/β model
Abbreviations: B16: Fischbarg *et al.* model [4]; CD: Circular Dichroïsm; FTIR: Fourier Transform Infrared; h1-h12: the transmembrane segments of the model of Mueckler *et al.*[3]; HB: model described in this paper; hb1-hb14: the transmembrane segments (α and β) of our model; HB12: Mueckler *et al.* model [3]; RX: X-Rays; Sw: Swiss-Prot references

Introduction

GLUT1 is a membrane protein, present in both red blood cells and the brain that transports glucose through the lipid bilayer. It is one of a family of mammalian facilitative hexose transporters, GLUT1 to GLUT7, that are unequally distributed among cell types [1,31]. The primary sequences of these proteins are very similar (50-76% similarity between GLUT1 to GLUT5 human isoform) [1]. GLUT-proteins transport substrates other than glucose (e.g. GLUT1, GLUT2 and GLUT3 transport galactose, GLUT2 transports fructose) with specific kinetics [2]. GLUT5 is really a fructose transporter and the corresponding DNA of GLUT6 is a pseudogene. Very little is known about GLUT7.

GLUT1 gene (492 amino-acids) was sequenced in 1985 [3]. Two models for its topology have been proposed. The first is widely accepted and is based on a hydropathy analysis of the sequence [3]. The method used assigns, as membrane spanning domains, non-overlapping segments of 21 residues with an average hydropathy of more than 0.42 (consensus normalized scale). The model (named H12 in this paper) consists of twelve membrane-spanning α-helices (h1 to h12). More recently, Fischbarg et al. [4] suggested that this model is not consistent with the recognition of the region Ile386-Ala405 by an antibody on the extracellular side of the membrane. They used an algorithm detecting β-strands to generate a model composed of 16 β-strands (named B16), forming a porin-like structure (β-barrel). The algorithm uses a function equal to a level-headed sum of the average

* *To whom correspondence should be addressed*

hydrophobicity (Kyte & Doollitle scale), the hydrophobic moment (id.) and the turn propensity (Chou & Fasman scale) and predicts β-strands where the function is greater than a threshold. However, this model appears to be incompatible with various experimental data (see below).

No three-dimensional structure has been published for the H12 model. Mueckler et al. suggested a channel formed by the five most amphipathic helices, with no defined role for the rest of the protein [3]. According to the B16 model, Fischbarg et al. proposed that GLUT1 works like a porin, i.e. forms an open channel whose entry is controlled by extramembrane loops [4].

The object of this paper is to identify transmembrane segments (topology) in agreement with the experimental data and to analyze the general frame of the transmembrane parts of GLUT1 (topography). We have used Jähnig's [5] and Eisenberg's [6] algorithms together with Chou & Fasman turn propensity [27] to identify membrane spanning domains. In order to analyze the topography of the protein, we developed a visual method based on the representation of the mutation periodicity and the mean hydrophobicity of protein alignment. Indeed, because GLUT1 is thought to form a channel, the residues pointing into the lumen of the channel should generally be more hydrophilic and better conserved than the residues facing the lipids. Moreover, we propose, on the basis of experimental data, that GLUT1 forms two channels.

Materials and methods

The software used is *PC-PROT+*: Protein Analysis (R. Brasseur), *WinMGM*: Molecular Graphic Manipulation (M.Rahman,[24]), WinDNA (M.Rahman) for hydro-phobicity analysis and WHEEL (Ph.Ducarme) for topography studies. CLUSTAL [25] was used for the alignments (Gap fixed=10; Gap vary.=10) and PhDhtm [33] for neural network based prediction The sequences of the GLUT family were obtained from the Swiss-Prot database (release 26, July 1993).

1. Sequence analysis methods

We analyzed the sequence of GLUT1 with the methods of Eisenberg [6, 28] and Jähnig [5]. These methods both seek stretches of amino acids sufficiently hydrophobic to span the membrane (i.e. it is known that protein residues within the membrane are statistically more hydrophobic than the extramembrane ones). In Jähnig's method, the hydrophobicity is averaged for a stretch of 19 (H_{19}) or 7 (H_7) residues corresponding to an α span and α β span, respectively. H_α 11 is a level-headed average function designed to seek amphipathic helices that can occur in membrane channel structures. In the Eisenberg method, in addition to mean hydrophobicity one calculates the hydrophobic moment, which is a measure of the homogeneity of the hydrophobicity in a segment (i.e. a high moment means that all hydrophobic residues are on the same side of the helix). In the plot of the hydrophobic moment versus mean hydrophobicity, several zones that correspond to a particular behavior of the segment (globular, transmembrane, surface, etc.) have been described.

Turn propensity was calculated as described in [27]. The method uses statistically derived tables to estimate the probability of a segment of 4 residues to be structured as a turn. The results obtained for GLUT1 alone were then confirmed by alignment of a consensus turn propensity function :

hydrophobic moment	mutation moment	prediction method
high(>1)	medium or low(<0.4)	use of hydrophobic moment only.
medium (<1 and >0.5)	low(<0.1)	use of hydrophobic moment only
medium or low(<0.5)	high(>0.4)	use of mutation moment only
low (<0.5)	medium (<0.4 and >0.1)	use of mutation moment only
medium (<1 and >0.5)	medium (<0.4 and >0.1)	*vectorial average* of normalized moments
high(>1)	high(>0.4)	*vectorial average* of normalized moments
low(<0.5)	low(<0.1)	results are not significant

Table 1. *Methodology used for prediction of the orientation of transmembrane segments. Empirical thresholds are based on the study of bacteriorhodopsin.*

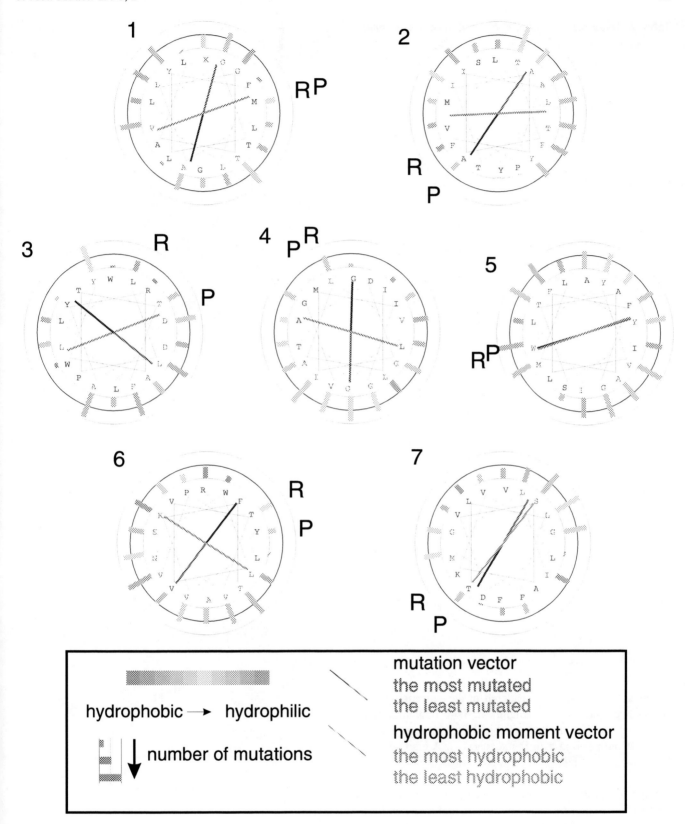

Figure 1. *Representations of the seven helices of the bacteriorhodopsin (see text for description). P is the prediction of the side which point to the lumen of the channel. R shows the lumen of the channel in the crystallographic structure*

Table 2. *Values of the hydrophobic and mutation moments for bacteriorhodopsin homologous sequences.*

Bacteriorhodopsin alignment		
helix	mutation moment	hydrophobic moment
1	.0281	.6899
2	.4100	.2921
3	.2550	1.5111
4	.4933	1.039
5	.8705	.03897
6	.4300	1.1933
7	.3670	1.7982

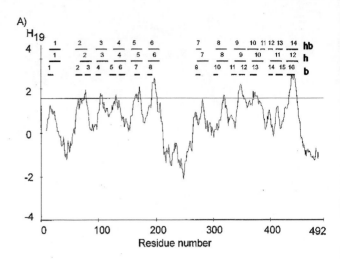

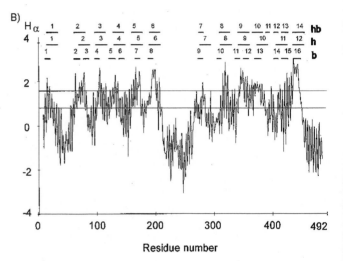

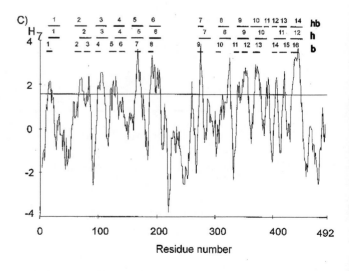

$f_n = \Pi_a (tp_{a,n})^{1/A}$ where $tp_{a,n}$ is the turn propensity of the position n in the sequence a and A the total number of sequences. The propensity was assigned to zero where the position n corresponds to a gap for one or more sequences in the alignment.

2. Alignment analysis method

In order to evaluate the packing of the transmembrane helices, we used an Edmunson-Wheel representation (i.e. a schematic view of the backbone with the helix normal to the drawing plane). On this representation, we superimposed in front of each residue heavy lines, the colors of which are representative of the mean hydrophobicity defined as :

$$|H_x| = \frac{1}{S}\sum_{i=1}^{S} H_{x,i}$$

where x is the position considered in the sequence, s, the sequence considered in the alignment, S, the total number of alignments and $H_{x,s}$, the hydrophobicity (normalized Fauchère scale) at the position x of the sequence s. A gradual scale from orange-red (hydrophobic) to green (hydrophilic) is used.

The lengths of the heavy lines are proportional to the number of mutations, M, corrected following the method of Donnelly *et al.*(see *treatment of outliers and ramps* in [29]) considering a window size of 18 for the corrections. These corrections enhance the legibility of the graphics. The green/orange-red vector shows the direction of the hydrophobic moment of Eisenberg (pointing as the orange-red side) and the blue/purple vector corresponds to what we call the mutation vector (pointing to the purple side). This vector is de-

Figure 2. *Plots of Jähnig functions A)H_{19}, B)H and C)H_7 applied to theGLUT1 sequence. Transmembrane segments predicted by model H12 (h1 to h12,[3]), B16 (b1 to b16, [4]) and HB (hb1 to hb14, model proposed in this paper) have been added for comparison. Threshold values are drawn on each plot.*

fined exactly as the hydrophobic moment, except here the hydrophobicity is replaced by the corrected number of mutations. Finally, residues poorly conserved (M>mean M for the helix) are shown in purple, the others (M<mean M for the helix) in blue.

We tested this representation on bacteriorhodopsin. In figure 1, the predicted and experimentally determined buried faces of the 7 helices are denoted by P and R respectively. We aligned the sequence of bacteriorhodopsin (Sw :

BACR_HALHA) with three homologous sequences (Sw: BACS_HALHA, BACH_NATPH, BACH_HALSP). Predictions have been made following the methodology described in table 1. The exact values of the moments are shown in table 2.

Predictions are in very good agreement with the RX data from crystallization showing that the method, although very simple, seems reliable enough to be applied to proteins with unknown structures.

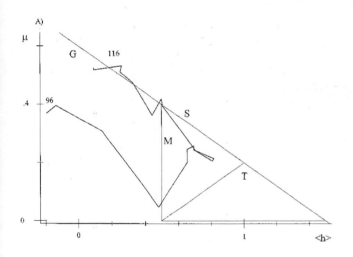

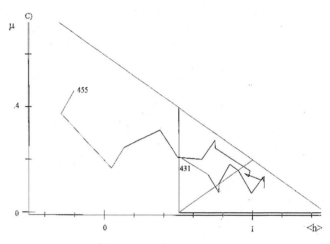

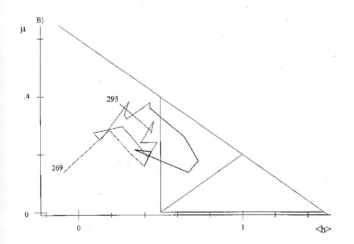

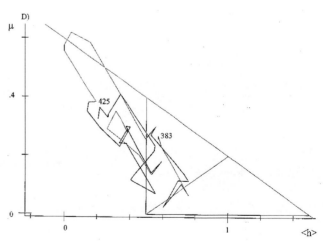

Figure 3. *Eisenberg's plots -hydrophobic moment (μ) versus hydrophobicity(<H>)- of predicted transmembrane segments A) hb3, B) hb 7, C) hb12, D) hb11, hb12 and hb13. The line connects the values of each residue along the primary sequence. Numbers of the two terminal residues of the segment considered are indicated. According to Eisenberg, segments buried in the membrane are expected to be located in the membrane or transmembrane zones of the plot (M or T, cf. fig 2A). G, S and R stand for globular, surface and receptor.*

```
  1    MEPSSKLTG   RLMLAVGGAV   LGSLQTGYNT   GVINAPQKVI   EEFYNQTWVH
                   MTMMM        MMM                       SSS

 51    RYGESILPTT   LTTLWSLSVA   IFSVGGMIGS   FSVGLFVNRF   GRRNSMLMMN
                    MMMMM        MMMMMSMMMM   MTM   SS               M

101    LLAFVSAVLM   GFSKLGKSFE   MLILGRFIIG   VYCGLTTGFV   PMYVGEVSPT
       MMMMMMMMMS   SSS  SS      MMSMSSS      SMMMMMMMMM   MM

151    AFRGALGTLH   QLGIVVGILI   AQVFGLDSIM   GNKDLWPLLL   SIIFIPALLQ
                M   MMMMMTTTTM   MTMMM                SSM   TTTTMMSMSM

201    CIVLPFCPES   PRFLLINRNE   ENRAKSVLKK   LRGTADVTHD   LQEMKEESRQ
       MMM                   R   RRRRR R                          RRRRRRR

251    MMREKKVTIL   ELFRSPAYRQ   PILIAVVLQL   SQQLSGINAV   FYYSTSIFEK
       RRRR                      MMM

301    AGVQQPVYAT   IGSGIVNTAF   TVVSLFVVER   AGRRTLHLIG   LAGMAGCAIL
                    MM  MM MMM   MMMM                SS        M   MMMMMMMMMM

351    MTIALALLEQ   LPWMSYLSIV   AIFGFVAFFE   VGPGPIPWFI   VAELFSQGPR
       MMTMMMMMM            MMMMT   TTTTMMMMM   MMMMMTMTMM   MMM SS

401    PAAIAVAGFS   NWTSNFIVGM   CFQYVEQLCG   PYVFIIFTVL   LVLFFIFTYF
              MMM            M   M   M        MMTMMTTTTT   TTTTTMMM

451    KVPETKGRTF   DEIASGFRQG   GASQSDKTPE   ELFHPLGADS   QV
                    SS  S
```

Figure 3e: *Summary of the analysis shown in figures 3a-d. Globular residues correspond to a white space.*

Results and discussion

1. Analysis of the GLUT1 primary sequence.

• Hydropathy

Function H_{19} of Jähnig's method is used to identify segments able to form hydrophobic membrane-spanning α-helices (figure 2A, which also shows transmembrane segments predicted previously). Only nine stretches scored higher than the threshold specified by Jähnig (H_{19} = 1.6, Kyte&Doolittle scale). The predicted transmembrane -helices corresponding to these peaks were named hb2 to 6, hb8 to 10 and hb14. Other previ-

ously predicted α-helices gave hydrophobic peaks but scored below the threshold value (especially h1 and h11).

According to Jähnig's method, the sequence was further analyzed with the H_{α} function (figure 2B). This plot shows that the segment corresponding to h1 oscillates between the two critical values (0.8 and 1.6) and could be an amphipathic or a weakly hydrophobic helix. This segment was thus named hb1. Segment h11 could also be considered as a transmembrane amphipathic helix. However, this structure has not been retained because of H_7, the turn propensity function and the Eisenberg's plot described below.

H_7 analysis (figure 2C) divided the segment including h11 into three highly hydrophobic peaks. These regions formed

A)   **Extracellular**

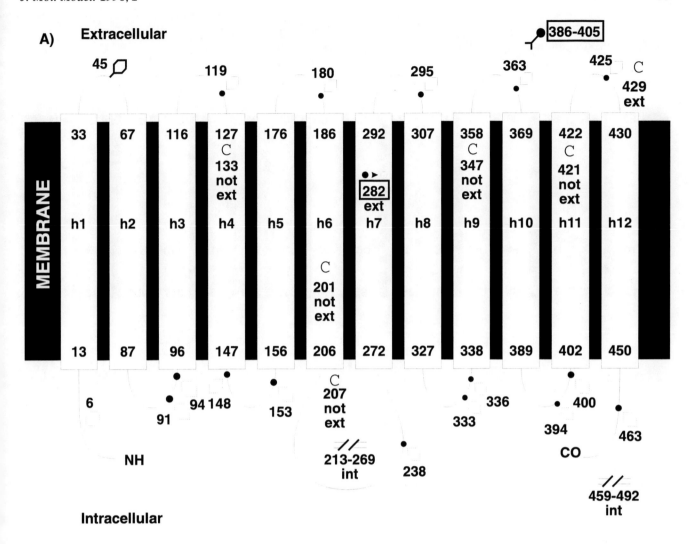

Figure 4a. *Membrane insertion model H12 [3]*

β-strands (named hb11, hb12 and hb13) and not an α-helix. The segment surrounding h7 gave two peaks (figure 2C). The segment Phe[213]-Arg[269] is intracellular [7] and therefore the first N-terminal of these two peaks cannot be a membrane-spanning segment. The second peak was considered as a membrane-spanning -strand and named hb7 . Replacement of residue Gln[282] by Leu reduces the affinity for the outside-specific ligand 2-*N*-4(1-azi-2,2,2-trifluoroethyl)benzoyl-1,3-bis(D-mannos-4-yloxy)-2-propylamine (ATB-BMPA) and has little effect on the transport activity [8]. Gln[282] is likely, therefore, to be extracellular. This is in line with the occurrence of a β-strand.

We used Eisenberg's plots to confirm this model. This method confirms that segments corresponding to h2, h3 (see figure 3A), h4 to h6, h8 to h10 and h12 (3C) can all form transmembrane helices. Only 3 residue of segment h7 (3B) appeared to be in the transmembrane region, the rest being assigned as globular. The region surrounding h11 (3D) formed 3 clusters of (trans)membrane residues separated by two seg-

ments which lie some distance from the (trans)membrane region in the plot. Consequently regions h7 and h11 are unlikely to be structured as helices. The segment hb1 includes a few membrane (M) residues (3E) but the plot is ambiguous and therefore we cannot confirm that this is a transmembrane helix.

These analyses of GLUT1 led us to propose the model HB depicted by figure 4C. The exact limits of the segments hb1 to hb14 (our model) are based on Jähnig, Eisenberg and Edmunson-Wheel plots (data not shown) with the exception of segments hb7, hb12 and hb13, three of the -strands proposed by Fischbarg *et al* [4]. The H12 and B16 models are shown in figures 4A and 4B. Besides hb2 and hb10, the α-helices of our model (4C) are very similar to those of the H12 model (4A). It is worth noting that part of the segment, Pro[383]-Pro[387], which is thought to be important for the protein flexibility, [9, 10] forms an intracellular loop in our model.

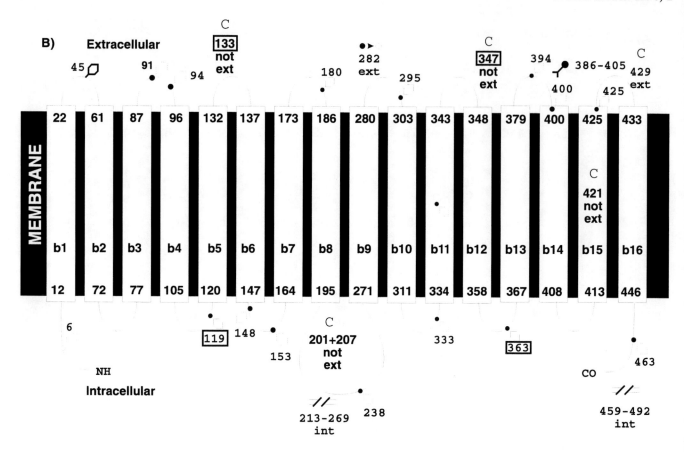

Figure 4b. *Membrane insertion model B16 [4]. See also figure 4c.*

• Turn propensity

In the turn propensity plot of GLUT1 (figure 5A), four very high peaks from residues 381 to 441 can be noted. This profile matches our topology model, each peak corresponding to a short loop separating the transmembrane segments. This is an interesting point because this region of the protein seems to be critical for transport [1]. A consensus turn propensity calculation (figure 5B) was performed on the GLUT-protein alignment as described in material and methods. These high turn propensity regions were found in all the isoforms of the alignment, suggesting that they are a general property of the GLUT family.

2. Confrontation of the topological models with experimental data.

• Secondary structure

FTIR and CD studies disagree about the exact ratio of α to β structures in GLUT1[11, 12, 13, 14] but the most recent studies [12,14] suggest that it is predominantly α-helical with a significant proportion of β-strands. Moreover, papain digestion, associated with FTIR, suggests that the intramembranous part of GLUT 1 is largely α-helical although some β-strands

are present [7]. This is partially confirmed by a new deconvolution algorithm which suggests the presence of transmembrane α-helices (by opposition to extramembrane helices) [14]. The H12 and B16 models are incompatible with these observations because they are *all alpha* or *all beta* models. Conversely, from a qualitative point of view, the HB model matches these data. If we assume that the two large loops (Pro^{36}-Thr^{60} between hb1/hb2 and Pro^{208}-Pro^{271} between hb6/hb7), the carboxy- and the amino-terminal segments are mostly α-helical, the proportion of α-helical structure in our model is about 65%. The transmembrane proportion of β-strands is about 8%.

• Mapping experiments

We superimposed several mapping experiments on the three topological models so as to evaluate their quality (figure 4). HB is the only model fully compatible with the experimental data despite none of the data being used to construct the model. This verification is a strong argument in its favor. The B16 model does not match the experimental results for positions 119 and 363, which are glycosylated after insertion of a glycosylation site [16], suggesting that these sites are extracellular. Moreover Cys^{133} and Cys^{347} are not extracellular [26]. The H12 model cannot explain the recognition of segment 386-405 by an antibody on the extracellular side of the membrane and does not agree with the localization of Gln^{282} in the extracellular space.

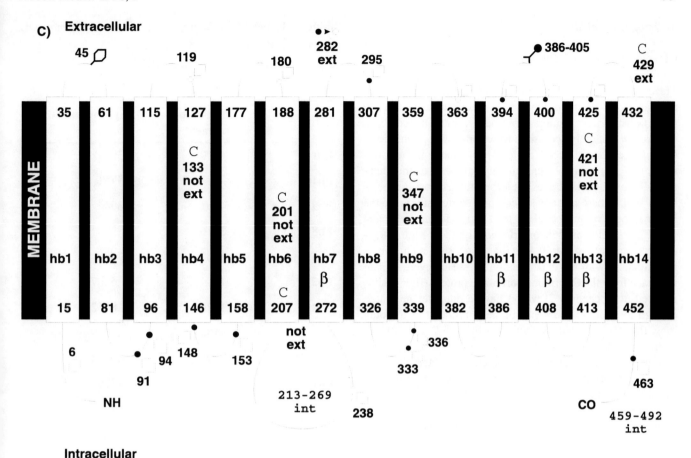

Figure 4c. *Membrane insertion model HB [this paper].The exact segment limits for model H12 are those supplied by SwissProt databank (GTR1_HUMAN, release of sept.93). Mapping experiments are indicated with the residues involved. Boxed segments are those inconsistent with the model.*

• Extension of the model to other proteins

The hydropathy analysis was performed on several isoforms of GLUT1 (GLUT2-4) in different species and on other transporters belonging to the same family as GLUT1 (Sugar Transporter Family) (samples in figure 6). Although GLUT3 and GLUT4 show strong similarity(except for hb4), all the other protein plots are clearly different and lead to distinct predictions. This suggests that the structure of these proteins may also differ. We do not think that this works against the α/β– model because there is no reason why all these transporters should have the same structure. Although it is generally admitted that similar sequences correspond to the same structure, recent studies have shown that one mutation can turn the prions structure from α to β [32]. Thus, it may be possible that the GLUT1,3 and 4 proteins, although sharing the biggest part of their structures with similar proteins, have some widely modified, particular domains in common.

• Neural network prediction

We used a neural network-based prediction program (PhDhtm,[33]) in order to confirm the HB model (figure7). This program considers only two states for the structure : helix (H) and loop (L) and provides a *reliability index* for the prediction (0-9). Helix hb 4 should not be considered because it is clear from the hydropathy plots that this part of the transporter is not similar in GLUT1, 3 and 4. Except for the segments hb1, hb11, hb12 and hb13, the prediction is in good agreement with our model and Mueckler's model. The helical segment predicted near hb7 is too small (12 residues) to span the membrane. This may suggest that the loop following this segment lies in the lumen of the channel (which is thought to be hydrophilic) or that this part of the molecule is not structured as an helix. The predicted helix corresponding to the three β-strands of hb11, hb12 and hb13 is shorter (15 residues) and has lower reliability than the other predicted transmembrane helices. This is not consistent with our model, but as the prediction program knows only two structural states we must keep in mind that, if this part of the protein is actually structured in β-strands, the prediction may be false. Finally, hb1 is not predicted at all. This may be due to the hydrophilic property of this segment already shown above.

3. Topography analysis

• Hypothesis of the double channel.

The construction of the topographical model required an overall framework of the protein to be determined, in order to identify which transmembrane segments are associated. Since the protein is largely accessible to water -80% of deuterium/ hydrogen exchange occurs rapidly [12] - most of the transmembrane segments are presumably involved in the channel. Moreover, GLUT1 does not let small ions cross the membrane so that there must be a steric or electric gate. Since the external loops of our model are too short to form an efficient gate for a single channel constructed with the 14 transmembrane segments (about 40 Å in diameter), such a channel is incompatible with the experimental data. However, a channel made from the C-terminal part of the protein (Ile^{272}-Val^{492}, referred to as channel 2) would have a diameter just large enough for glucose and hence be more efficient in stopping ions. Besides, the glycerol facilitator of *E.Coli* (GlpF) has a similar length (281 residues). This suggests that channel 2 contains enough residue to form an effective facilitator.

Furthermore, most of GLUT1 mutants affecting glucose transport map in the C-terminal part of the protein [1]. The N-terminal part of GLUT1 should also form an amphipathic structure as indicated by the high deuterium/hydrogen exchange mentioned above [12]. As GLUT1 is thought to come

Table 3. *Values of the hydrophobic and mutation moments for the alignment of figure 7.*

	GLUT1, 3 and 4 alignment	
helix	mutation moment	hydrophobic moment
hb14	.4290	.3931
hb10	.2287	.3751
hb9	.4361	.3944
hb8	.2716	1.5021
hb6	.3766	.7690
hb5	.1694	1.6267
hb4	.1626	1.8152
hb3	.4186	2.2679
hb2	.2396	1.3135
hb1	.1397	.4279
h7	.0374	1.7248
h11	.1696	1.5135

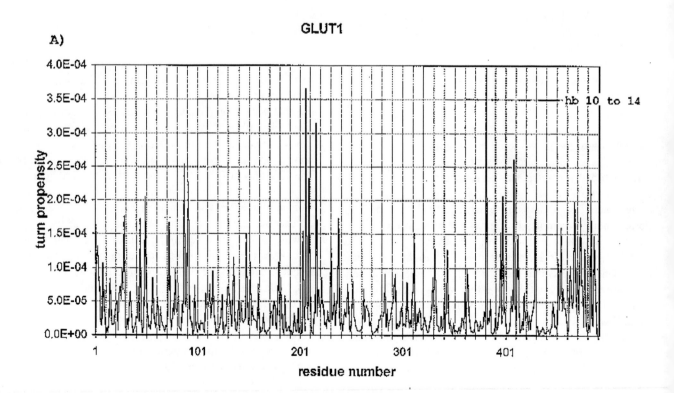

Figure 5 a. *Turn propensity function of GLUT1(27). The threshold value established by Chou and Fasman for predicting a turn is 0.75 E-04. Transmembrane segments in our model have been added around the high propensity regions (see text).*

B)

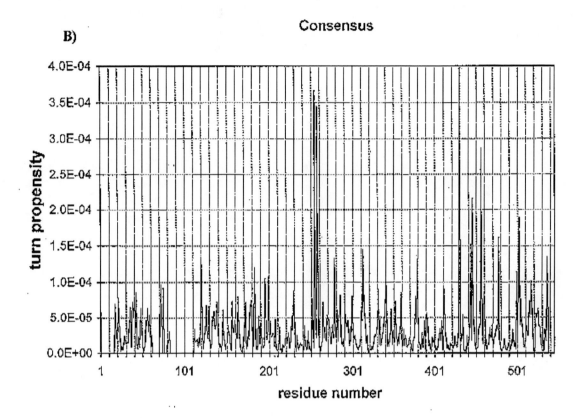

Figure 5b. *Consensus turn propensity function for the alignment of figure 7. Compared to fig. 4A, several peaks are reduced (e.g. near residues 220) whereas four high peaks remain between residues 380 to 440.*

from a genetic duplication [17] and as hydropathy plots show a symmetry between both halves of the protein, the N-terminal (Met^1-Cys^{207}) part of the molecule could also form a channel (named channel 1). This channel, in the case of the proline/betaine cotransporter of E.*coli* (swprot : PROP_ECOLI), which also belongs to the sugar transporter family and probably has a similar global structure, may also transport a substrate. Channel 1 of GLUT1 may thus be a degenerate form of a cotransport channel still interacting with the C-terminal part of the molecule but which does not transport cosubstrate. Another possibility (not exclusive) is that channel 1 serves as a water channel. Indeed, rat glucose transporters GLUT1, GLUT2 and GLUT3 allow the passage of water through the membrane [30]. The large charged loop between helices hb1 and hb2 (Pro^{36}-Thr^{60}) may thus act as an electric filter preventing small ions passing through the channel. Such a structure (a water channel associated with a glucose channel) may act as a osmotic regulator.

Cope *et al.* [18] showed that when the N- or C-terminal halves of GLUT1 are expressed in a Sf9 cell membrane, they do not recognize ATB-BMPA or cytochalasine B. In cells producing both amino- and carboxy-terminal halves, the ligand labeling is restored. This strongly suggests that the

two parts of the molecule can fold independently. This seems incompatible with the model B16 which is a porin-like structure and with the model H12 where the channel is composed of helices of both the first and the second parts of the molecule (h3, h5, h7, h8 and h11[3]). This experiment also suggests that the two channels strongly interact because the amino-terminal part of the molecule, which does not bind ligands, restores ligand binding to of the carboxy-terminal part [18]. Thus, the functions of the two halves of the protein are likely to be coupled. This could explain why the mutation of Gln^{161} in the N-terminal part of the molecule strongly reduces the rate of conformational change [22] even if this residue is not included in channel 2, which transports glucose.

• Topographic analysis of the GLUT family alignment

We aligned the available sequences of GLUT1, 3 and 4 (figure 8). Representations of the 10 helices of our model and of helices h11 and h7 of Mueckler's model are shown in figure 9. Exact values of the moments are given in table 3. Following the methodology described in table 1, the side of the helices predicted to point to the lumen of the channel are denoted by P.

helix hb1: Mutation vector (MV) is medium and the hydrophobic vector(HV) is low. This may reflect that helix hb1 is surrounded by other helices so that there are no really privileged direction. This is in line with the hydrophobicity plots.

Figure 6. *Plots of Jähnig functions of A) GLUT2 from human, B)GLUT3 from mouse, C) GLUT4 from mouse and D) HXT1 from yeast.*

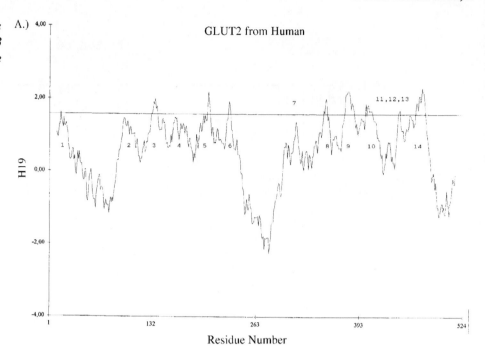

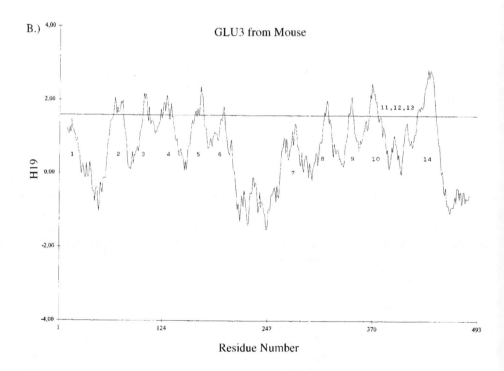

helix hb2: MV is low (the helix is very well conserved) and does not agree with the high HV. This suggests that hb2 plays an important functional role and may be part of a channel.

helix hb3: MV and HV are high and consistent. These are the typical properties corresponding to pore-forming helix.

helix hb4 and hb5 and hb8: MV is medium and HV is high. These helices seem to make part of the pore but without any important functional role.

helix hb6: MV and HV are medium.

helix hb9 and hb 14: HV low but MV high. This suggests that hb9 and hb14 play a particular structural role, because although they are almost entirely hydrophobic, one face of these helices is especially conserved.

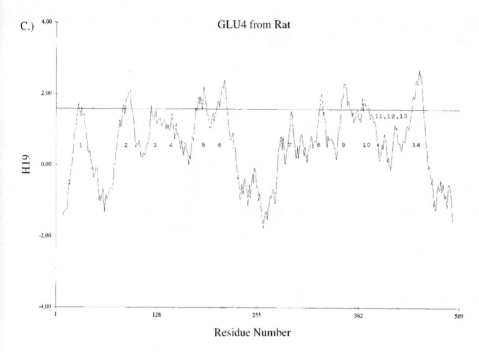

C.)

GLU4 from Rat

Residue Number

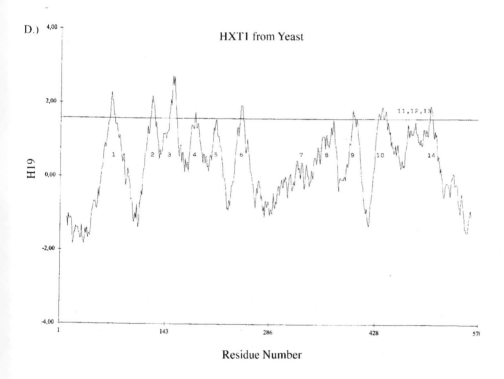

D.)

HXT1 from Yeast

Residue Number

helix hb10 : HV low and MV medium. Same as hb9.

At the moment we cannot provide a topographic model for the C-terminal channel because we have no idea of what a α/β-transmembrane structure could look like. No transmembrane α/β–structure has been identified by X-ray crystallography and this proposal may therefore seem speculative. However, transmembrane structures are so poorly docu-mented that α/β-structures cannot be excluded. Moreover, both experimental and theoretical analyses suggest that such a structure is found in the acetylcholine receptor [15]. If such a structure does exist, it is more likely to be folded, at least in part, in the cytoplasm before being inserted as the inser-tion of a non H-bonded -strand in the membrane is energeti-cally very unfavorable. We do not know how such a process would take place so we can hardly provide a model for a α/β-structure.

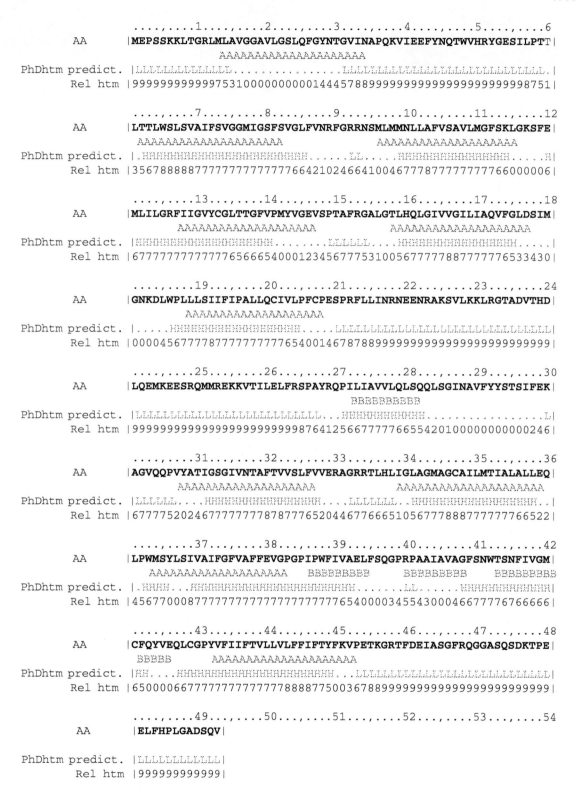

```
          ....,....1....,....2....,....3....,....4....,....5....,....6
       AA |MEPSSKKLTGRLMLAVGGAVLGSLQFGYNTGVINAPQKVIEEFYNQTWVHRYGESILPTT|
             AAAAAAAAAAAAAAAAAAAAAA
PhDhtm predict. |LLLLLLLLLLLLL................LLLLLLLLLLLLLLLLLLLLLLLLLLLLL.|
      Rel htm |99999999999975310000000000014445788999999999999999999998751|

          ....,....7....,....8....,....9....,...10....,...11....,...12
       AA |LTTLWSLSVAIFSVGGMIGSFSVGLFVNRFGRRNSMLMMNLLAFVSAVLMGFSKLGKSFE|
             AAAAAAAAAAAAAAAAAAAAAAA      AAAAAAAAAAAAAAAAAAAAA
PhDhtm predict. |.HHHHHHHHHHHHHHHHHHHHHHH.....LL....HHHHHHHHHHHHHHH....H|
      Rel htm |35678888877777777777776642102466410046777877777777766000006|

          ....,...13....,...14....,...15....,...16....,...17....,...18
       AA |MLILGRFIIGVYCGLTTGFVPMYVGEVSPTAFRGALGTLHQLGIVVGILIAQVFGLDSIM|
             AAAAAAAAAAAAAAAAAAAAAA      AAAAAAAAAAAAAAAAAAAAAA
PhDhtm predict. |HHHHHHHHHHHHHHHHHHHH........LLLLLL....HHHHHHHHHHHHHHHHH.....|
      Rel htm |67777777777777656666540001234567775310056777788777777533430|

          ....,...19....,...20....,...21....,...22....,...23....,...24
       AA |GNKDLWPLLLSIIFIPALLQCIVLPFCPESPRFLLINRNEENRAKSVLKKLRGTADVTHD|
             AAAAAAAAAAAAAAAAAAAAAA
PhDhtm predict. |.....HHHHHHHHHHHHHHHHHHH....LLLLLLLLLLLLLLLLLLLLLLLLLLLLLLLL|
      Rel htm |00004567778777777777765400146787889999999999999999999999999|

          ....,...25....,...26....,...27....,...28....,...29....,...30
       AA |LQEMKEESRQMMREKKVTILELFRSPAYRQPILIAVVLQLSQQLSGINAVFYYSTSIFEK|
                                              BBBBBBBBBB
PhDhtm predict. |LLLLLLLLLLLLLLLLLLLLLLLLLLLL...HHHHHHHHHH..............L|
      Rel htm |99999999999999999999999998764125667777766554201000000000246|

          ....,...31....,...32....,...33....,...34....,...35....,...36
       AA |AGVQQPVYATIGSGIVNTAFTVVSLFVVERAGRRTLHLIGLAGMAGCAILMTIALALLEQ|
             AAAAAAAAAAAAAAAAAAAAAA        AAAAAAAAAAAAAAAAAAAAAA
PhDhtm predict. |LLLLLL...HHHHHHHHHHHHHHHH....LLLLLLL.HHHHHHHHHHHHHHHHHH..|
      Rel htm |67777520246777777787877765204467766651056777888777777766522|

          ....,...37....,...38....,...39....,...40....,...41....,...42
       AA |LPWMSYLSIVAIFGFVAFFEVGPGPIPWFIVAELFSQGPRPAAIAVAGFSNWTSNFIVGM|
             AAAAAAAAAAAAAAAAAAAAAAA  BBBBBBBBBB     BBBBBBBBB    BBBBBBBBB
PhDhtm predict. |.HHHH...HHHHHHHHHHHHHHHHHHHHHHH......LL...HHHHHHHHHH|
      Rel htm |45677000877777777777777777777765400003455430004667776766666|

          ....,...43....,...44....,...45....,...46....,...47....,...48
       AA |CFQYVEQLCGPYVFIIFTVLLVLFFIFTYFKVPETKGRTFDEIASGFRQGGASQSDKTPE|
             BBBBB    AAAAAAAAAAAAAAAAAAAAAA
PhDhtm predict. |HH....HHHHHHHHHHHHHHHHHHH...LLLLLLLLLLLLLLLLLLLLLLLLLLLLLLL|
      Rel htm |65000066777777777777788887750036788999999999999999999999999|

          ....,...49....,...50....,...51....,...52....,...53....,...54
       AA |ELFHPLGADSQV|

PhDhtm predict. |LLLLLLLLLLLL|
      Rel htm |999999999999|
```

Figure 7. *Transmembrane helices prediction from PhDhtm (33) are in purple. H:helix, L:loop, (.):no prediction, Rel htm lines indicate the reliability of the prediction from 0 to 9. Transmembrane segments of our model are in red. A: structure, B: structure. The sequences submitted to PhDhtm are those from fig.8. GLUT1 sequence is shown (AA).*

```
* :=>   match across all seqs.
. :=>   conservative substitutions
```

```
GTR1_HUM  M————EPSSKKLTGRLMLAVGGAVLGSLQFGYNTGVINAPQKVIEEFYNQ
GTR1_BOV  M————EPTSKKLTGRLMLAVGGAVLGSLQFGYNTGVINAPQKVIEEFYNQ
GTR1_MOU  M————DPSSKKVTGRLMLAVGGAVLGSLQFGYNTGVINAPQKVIEEFYNQ
GTR1_RAB  M————EPSSKKVTGRLMLAVGGAVLGSLQFGYNTGVINAPQKVIEEFYNQ
GTR1_RAT  M————EPSSKKVTGRLMLAVGGAVLGSLQFGYNTGVINAPQKVIEEFYNQ
GTR3_HUM  M————GTQKVTPALIFAITVATIGSFQFGYNTGVINAPEKIIKEFINK
GTR3_MOU  M————GTTKVTPSLVFAVTVATIGSFQFGYNTGVINAPETILKDFLNY
GTR4_HUM  MPSGFQQIGSE-DGEPPQQRVTGTLVLAVFSAVLGSLQFGYNIGVINAPQKVIEQSYNE
GTR4_MOU  MPSGFQQIGSDVKDGEPPRQRVTGTLVLAVFSAVLGSLQFGYNIGVINAPQKVIEQSYNA
GTR4_RAT  MPSGFQQIGSE-DGEPPQQRVTGTLVLAVFSAVLGSLQFGYNIGVINAPQKVIEQSYNA
          *                ..*   *..*.   *..**.*****.******......   *
```

```
GTR1_HUM  TWVHRYG——ESILPTTLTTLWSLSVAIFSVGGMIGSFSVGLFVNRFGRRNSMLMMNLL
GTR1_BOV  TWVQRYG——EPIPPATLTTLWSLSVAIFSVGGMIGSFSVGLFVNRFGRRNSMLMMNLL
GTR1_MOU  TWNHRIG——EPIPSTTLTTLWSLSVAIFSVGGMIGSFSVGLFVNRFGRRNSMLMMNLL
GTR1_RAB  TWIHRYG——ERILPTTLTTLWSLSVAIFSVGGMIGSFSVGLFVNRFGRRNSMLMMNLL
GTR1_RAT  TWNHRYG——ESIPSTTLTTLWSLSVAIFSVGGMIGSFSVGLFVNRFGRRNSMLMMNLL
GTR3_HUM  TLTDKGNAPPSEVL——LTSLWSLSVAIFSVGGMIGSFSVGLFVNRFGRRNSMLIVNLL
GTR3_MOU  TLEERLEDLPSEGL——LTALWSLCVAIFSVGGMIGSFSVGLFVNRFGRRNSMLLVNLL
GTR4_HUM  TWLGRQGPEGPSSIPPGTLTTLWALSVAIFSVGGMISSFLIGIISQWLGRKRAMLVNNVL
GTR4_MOU  TWLGRQGPGGPDSIPQGTLTTLWALSVAIFSVGGMISSFLIGIISQWLGRKRAMLANNVL
GTR4_RAT  TWLGRQGPGGPDSIPQGTLTTLWALSVAIFSVGGMISSFLIGIISQWLGRKRAMLANNVL
          *     . .     . .    **.**.*.**********.** .*.. ...**...**   *.*
```

```
GTR1_HUM  AFVSAVLMGFSKLGKSFEMLILGRFIIGVYCGLTTGFVPMYVGEVSPTAFRGALGTLHQL
GTR1_BOV  AFVSAVLMGFSKLGKSFEMLILGRFIIGVYCGLTTGFVPMYVGEVSPTELRGALGTLHQL
GTR1_MOU  AFVAAVLMGFSKLGKSFEMLILGRFIIGVYCGLTTGFVPMYVGEVSPTALRGALGTLHQL
GTR1_RAB  AFVSAVLMGFSKLAKSFEMLILGRFIIGVYCGLTTGFVPMYVGEVSPTALRGALGTLHQL
GTR1_RAT  AFVSAVLMGFSKLGKSFEMLILGRFIIGVYCGLTTGFVPMYVGEVSPTALRGALGTLHQL
GTR3_HUM  AVTGGCFMGLCKVAKSVEMLILGRLVIGLFCGLCTGFVPMYIGEISPTALRGAFGTLNQL
GTR3_MOU  AIIAGCLMGFAKIAESVEMLILGRLLIGIFCGLCTGFVPMYIGEISPTALRGAFGTLNQL
GTR4_HUM  AVLGGSLMGLANAAASYEMLILGRFLIGAYSGLTSGLVPMYVGEIAPTHLRGALGTLNQL
GTR4_MOU  AVLGGALMGLANAVASYEILILGRFLIGAYSGLTSGLVPMYVGEIAPTHLRGALGTLNRL
GTR4_RAT  AVLGGALMGLANAAASYEILILGRFLIGAYSGLTSGLVPMYVGEIAPTHLRGALGTLNQL
          *  .. .**. .   * *.*****..** ..** .*.****.**..** .***.***..*
```

```
GTR1_HUM  GIVVGILIAQVFGLDSIMGNKDLWPLLLSIIFIPALLQCIVLPFCPESPRFLLINRNEEN
GTR1_BOV  GIVVGILIAQVFGLDSIMGNQELWPLLLSVIFIPALLQCILLPFCPESPRFLLINRNEEN
GTR1_MOU  GIVVGILIAQVFGLDSIMGNADLWPLLLSVVFVPALLQCILLPFCPESPRFLLINRNEEN
GTR1_RAB  GIVVGILIAQVFGLDSIMGNEDLWPLLLSVIFVPALLQCIVLPLCPESPRFLLINRNEEN
GTR1_RAT  GIVVGILIAQVFGLDSIMGNADLWPLLLSVIFIPALLQCILLPFCPESPRFLLINRNEEN
GTR3_HUM  GIVVGILVAQIFGLEFILGSEELWPLLLGFTILPAILQSAALPFCPESPRFLLINRKEEE
GTR3_MOU  GIVVGILVAQIFGLDFILGSEELWPGLLGLTIIPAILQSAALPFCPESPRFLLINKKEED
GTR4_HUM  AIVIGILIAQVLGLESLLGTASLWPLLLGLTVLPALLQLVLLPFCPESPRYLYIIQNLEG
GTR4_MOU  AIVIGILVAQVLGLESMLGTATLWPLLLALTVLPALLQLILLPFCPESPRYLYIIRNLEG
GTR4_RAT  AIVIGILVAQVLGLESMLGTATLWPLLLAITVLPALLQLLLLPFCPESPRYLYIIRNLEG
          .**.***.**..**. ..*..***.**. . .**.**   **.******.* *  .. *.
```

```
GTR1_HUM  RAKSVLKKLRGTADVTHDLQEMKEESRQMMREKKVTILELFRSPAYRQPILIAVVLQLSQ
GTR1_BOV  RAKSVLKKLRGTADVTRDLQEMKEESRQMMREKKVTILELFRSAAYRQPILIAVVLQLSQ
GTR1_MOU  RAKSVLKKLRGTADVTRDLQEMKEEGRQMMREKKVTILELFRSPAYRQPILIAVVLQLSQ
GTR1_RAB  RAKSVLKKLRGNADVTRDLQEMKEESRQMMREKKVTILELFRSPAYRQPILSAVVLQLSQ
GTR1_RAT  RAKSVLKKLRGTADVTRDLQEMKEEGRQMMREKKVTILELFRSPAYRQPILIAVVLQLSQ
GTR3_HUM  NAKQILQRLWGTQDVSQDIQEMKDESARMSQEKQVTVLELFRVSSYRQPIIISIVLQLSQ
GTR3_MOU  QATEILQRLWGTSDVVQEIQEMKDESVRMSQEKQVTVLELFRSPNYVQPLLISIVLQLSQ
GTR4_HUM  PARKSLKRLTGWADVSGVLAELKDEKRKLERERPLSLLQLLGSRTHRQPLIIAVVLQLSQ
GTR4_MOU  PARKSLKPLTGWADVSDALAELKDEKRKLERERPMSLLQLLGSRTHRQPLIIAVVLQLSQ
GTR4_RAT  PARKSLKRLTGWADVSDALAELKDEKRKLERERPLSLLQLLGSRTHRQPLIIAVVLQLSQ
          *   *. * *   **   ..*.*.* .. .*. ...*.*.   .. **.. ..******
```

```
GTR1_HUM  QLSGINAVFYYSTSIFEKAGVQQPVYATIGSGIVNTAFTVVSLFVVERAGRRTLHLIGLA
GTR1_BOV  QLSGINAVFYYSTSIFEKAGVQQPVYATIGSGIVNTAFTVVSLFVVERAGRRTLHLIGLA
GTR1_MOU  QLSGINAVFYYSTSIFEKAGVQQPVYATIGSGIVNTAFTVVSLFVVERAGRRTLHLIGLA
GTR1_RAB  QLSGINAVFYYSTSIFEKAGVQQPVYATIGSGIVNTAFTVVSLFVVERAGRRTLHLIGLA
GTR1_RAT  QLSGINAVFYYSTSIFEKAGVQQPVYATIGSGIVNTAFTVVSLFVVERAGRRTLHLIGLA
GTR3_HUM  QLSGINAVFYYSTGIFKDAGVQEPIYATIGAGVVNTIFTVVSLFLVERAGRRTLHMIGLG
GTR3_MOU  QLSGINAVFYYSTGIFKDAGVQEPIYATIGAGVVNTIFTVVSLFLVERAGRRTLHMIGLG
GTR4_HUM  QLSGINAVFYYSTSIFETAGVGQPAYATIGAGVVNTVFTLVSVLLVERAGRRTLHLLGLA
GTR4_MOU  QLSGINAVFYYSTSIFESAGVGQPAYATIGAGVVNTVFTLVSVLLVERAGRRTLHLLGLA
GTR4_RAT  QLSGINAVFYYSTSIFELAGVEQPAYATIGAGVVNTVFTLVSVLLVERAGRRTLHLLGLA
          ***************.**. ***  .* *****.*.*** **.**...**********..**.


GTR1_HUM  GMAGCAILMTIALALLEQLPWMSYLSIVAIFGFVAFFEVGPGPIPWFIVAELFSQGPRPA
GTR1_BOV  GMAGCAVLMTIALALLERLPWMSYLSIVAIFGFVAFFEVGPGPIPWFIVAELFSQGPRPA
GTR1_MOU  GMAGCAVLMTIALALLERLPWMSYLSIVAIFGFVAFFEVGPGPIPWFIVAELFSQGPRPA
GTR1_RAB  GMAACAVLMTIALALLEQLPWMSYLSIVAIFGFVAFFEVGPGPIPWFIVAELFSQGPRPA
GTR1_RAT  GMAGCAVLMTIALALLEQLPWMSYLSIVAIFGFVAFFEVGPGPIPWFIVAELFSQGPRPA
GTR3_HUM  GMAFCSTLMTVSLLLKDNYNGMSFVCIGAILVFVAFFEIGPGPIPWFIVAELFSQGPRPA
GTR3_MOU  GMAVCSVFMTISLLLKDDYEAMSFVCIVAILIYVAFFEIGPGPIPWFIVAELFSQGPRPA
GTR4_HUM  GMCGCAILMTVALLLLERVPAMSYVSIVAIFGFVAFFEIGPGPIPWFIVAELFSQGPRPA
GTR4_MOU  GMCGCAILMTVALLLLERVPAMSYVSIVAIFGFVAFFEIGPGPIPWFV-AELFSQGPRPA
GTR4_RAT  GMCGCAILMTVALLLLERVPSMSYVSIVAIFGFVAFFEIGPGPIPWFIVAELFSQGPRPA
          **  *...**..* *  .     **...* **. ***** ********. ***********


GTR1_HUM  AIAVAGFSNWTSNFIVGMCFQYVEQLCGPYVFIIFTVLLVLFFIFTYFKVPETKGRTFDE
GTR1_BOV  AIAVAGFSNWTSNFIVGMCFQYVEQLCGPYVFIIFTVLLVLFFIFTYFKVPETKGRTFDE
GTR1_MOU  RIAVAGFSNWTSNFIVGMCFQYVEQLCGPYVFIIFTVLLVLFFIFTYFKVPETKGRTFDE
GTR1_RAB  AVAVAGFSNWTSNFIVGMCFQYVEQLCGPYVFIIFTVLLVLFFIFTYFKVPETKGRTFDE
GTR1_RAT  AVAVAGFSNWTSNFIVGMCFQYVEQLCGPYVFIIFTVLLVLFFIFTYFKVPETKGRTFDE
GTR3_HUM  AMAVAGCSNWTSNFLVGLLFPSAAHYLGAYVFIIFTGFLITFLAFTFFKVPETRGRTFED
GTR3_MOU  AIAVAGCCNWTSNFLVGMLFPSAAAYLGAYVFIIFAAFLIFFLIFTFFKVPETKGRTFED
GTR4_HUM  AMAVAGFSNWTSNFIIGMGFQYVAEAMGPYVFLLFAVLLLGFFIFTFLRVPETRGRTFDQ
GTR4_MOU  AMAVAGFSNWTCNFIVGMGFQYVADRMGPYVFLLFAVLLLGFFIFTFLKVPETRGRTFDQ
GTR4_RAT  AMAVAGFSNWTCNFIVGMGFQYVADAMGPYVFLLFAVLLLGFFIFTFLRVPETRGRTFDQ
          .****  .***.**..*. *  ..   *.***..*. .*. *. **...****.****..


GTR1_HUM  IASGFRQGGA—SQSDKTPEELFHPLGA——DSQV
GTR1_BOV  IASGFRQGGA—SQSDKTPEELFHPLGA——DSQV
GTR1_MOU  IASGFRQGGA—SQSDKTPEELFHPLGA——DSQV
GTR1_RAB  IASGFRQGGA—SQSDKTPEELFHPLGA——DSQV
GTR1_RAT  IASGFRQGGA—SQSDKTPEELFHPLGA——DSQV
GTR3_HUM  ITRAFEGQAHGADRSGKDGVMEMNSIEPAKETTTNV
GTR3_MOU  IARAFEGQAHSG—KGPAGV-ELNSMQPVKETPGNA
GTR4_HUM  ISAAFHRTPSLLEQEVKPSTEL-EYLGP——DEND
GTR4_MOU  ISAAFRRTPSLLEQEVKPSTEL-EYLGP——DEND
GTR4_RAT  ISATFRRTPSLLEQEVKPSTEL-EYLGP——DEND
          *.  .*                .        . .  .         ..
```

Figure 8. *Alignment of glucose transporters belonging to the sugar transporter family by CLUSTAL (Gap fixed=10; Gap vary.=10). The different species considered are HUMan, MOUse, RAT, RABbit and BOVine. The asterisk (*) marks perfectly conserved residues, the point (.) marks conservative substitutions.*

• Mechanism of transport

Pawagi & Deber suggested that glucose produces a change in the conformation of the region surrounding Trp[388] [9] which leads to the transfer of this residue from an aqueous domain to or near the membrane. Although this region is rich in proline, directed mutagenesis suggests that the cis/trans isomerization of proline residues is not critical for transport [19]. In our model, Trp[388] lies in segment hb11 and poly-Pro segment separates helix hb10 from β-strand hb11. We suggest that segment hb11 could initiate the molecular movement induced by glucose and transmit it, via the poly-Pro segment to helix hb10 which is considered as forming together with segment hb9 an hydrophobic region closing the outward portion of the channel during glucose transport. On the one hand, Holman and Rees [20] have reported that Trp[388] is one of two which may be located near the ligand binding site. On the other hand a multi-β-structure, contrary to α-helices associations, could be very sensitive to exchange of hydrogen bonds which is thought to occur at the beginning of glucose translocation [21, 23]. Hydrogen bonds could be

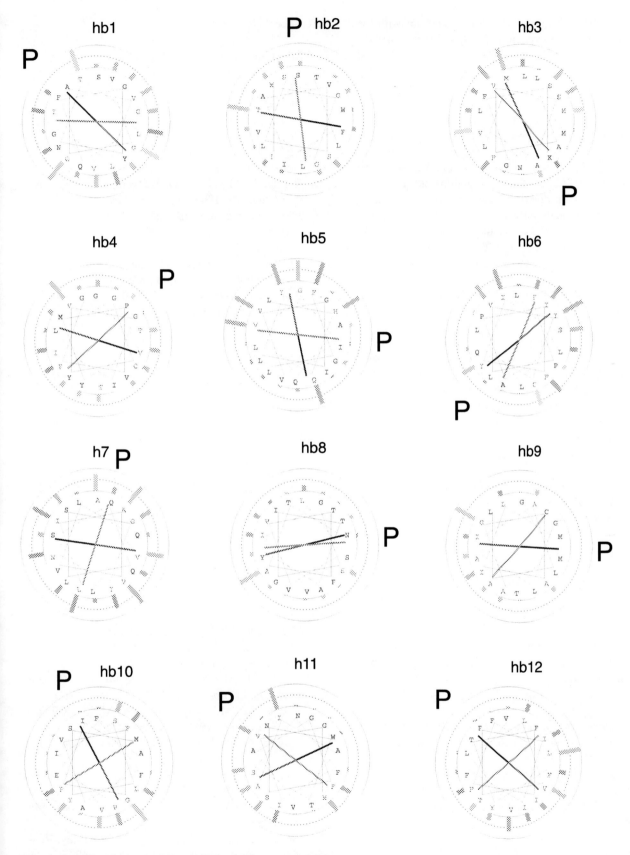

Figure 9. *Representation of the 10 helices of our model (hb) and helices 7 and 11 of Mueckler's model (see text for description). P is the prediction of the side which point to the lumen of the channel.*

J. Mol. Model. **1996,** 2

formed with the backbone of the protein and not with the side chains. This could explain why only a few residues capable of hydrogen bonding have been reported to be critical for glucose transport [22]. It may seem improbable that glucose could initiate such a movement in the membrane by interacting with a β-strand. However, hb11 has very special properties : it is highly hydrophobic, surrounded by very short loops, one of which has high conformational flexibility and could interact through its charge, with the next transmembrane segment. Segment hb14 is particularly rich in Ala (Arg[400]Pro Ala Ala Ieu Ala Val Ala Gly[407]). These amino-acids have particular hydrophobic properties: as its accessible surface is small it does not strongly influence the hydropathic environment. Thus segment hb14 could move without causing an important modification of energy. This could be critical for the working of a channel such as GLUT1.

Conclusion

In this paper we have presented a new model for GLUT1 that seems to agree better than previous models with experimental evidence. The use of this model together with the earlier models may help in the conception of new experiments or theoretical analyses. In fact, as the three models are based on different methods which each have their own limitations, they should be regarded as tools and not as exact predictions. In this perspective the new hypothesis suggested by our model are that:

a) GLUT1 is composed of 14 transmembrane segments.
b) it is an α/β-protein.
c) it forms two channels instead of one.
d) molecular movement could be initiated by a β-strand and transmitted via a poly-P segment to the hydrophobic cleft formed by helices hb10 and hb9.

Availability

A public version of WHEEL is available for academic purposes at: http://www.fsagx.ac.be/info_faculte/info_bp/presentation_bp.html.

Acknowledgments R.B. is Directeur de la Recherche of the Belgian Fonds National de la Recherche Scientifique. We are grateful to the Association Française de Lutte contre la Mucovisidose for financial support. This work was also supported by the Fonds de la Recherche Scientifique Médicale (grant n°2.4534.95).We thank Philippa Talmud for reading the manuscript.

References

1. Silverman, M. *Annu.Rev.Biochem.* **1991,** *60,* 757-794.
2. Bell, G. I., Burant, C. F., Takeda, J., Gould, G. W. *J. Biol. Chem.* **1993,** *268,* 19161-19164.
3. Mueckler, M., Caruso, C., Baldwin, S. A., Panico, M., Blench, I., Morris, H. R., Allard, W. J., Lienhard, G. E., Lodish, H. F. *Science* **1985,** *229,* 941-45.
4. Fischbarg, J., Cheung, M., Czegledy, F., Li, J., Iserovitch, P., Kuang, K., Hubbard, J., Garner, M., Rosens, O. M., Golde, D. W., Vera, J. C. *Proc. Natl. Acad. Sci.USA* **1993,** *90,* 11658-11662.
5. Jähnig, F. Structure *TIBS* **1990,** *15,* 93-95.
6. Eisenberg, D., Schwarz, E., Komaromy, M., Wall, R. *J. Mol. Biol.* **1984,** *179,* 125-142.
7. Cairns, M. T., Alvarez, J., Panico, M., Giggs, A. F., Morris, H. R., Chapman, D., Baldwin, S. A. *Biochem. Biophys. Acta* **1987,** *905,* 295-310.
8. Hashiramoto, M., Kadowaki, T., Clark, A. E., Muraoka, A., Momomura, K., Sakura, H., Tobe, K., Akanuma, Y.,Yazaki,Y., Holman, G. D., Kasuga, M.*J. Biol. Chem.* **1992,** *267,* 17502-17507.
9. Pawagi, A. B., Deber, C. M. *Biochemistry* **1990,** *29,* 950-955.
10. Tamori, Y., Hashiramoto, M., Clark, A. E., Mori, H., Muraoka, A., Kadowaki, T., Holman, G. D., Kusaga, M. *J. Biol. Chem.* **1994,** *269,* 2982-2986.
11. Chin, J. J., Jung, E. K. Y., Jung, C. Y. *J. Biol. Chem.* **1986,** *261,* 7101-7104.
12. Alvarez, J., Lee, D. C., Baldwin, S. A., Chapman, D. *J. Biol. Chem.* **1987,** *262,* 3502-3509.
13. Chin, J. J., Jung, E. K. Y., Jung, C. Y. *Proc. Natl. Acad. Sci. USA* **1987,** *84,* 4113-4116.
14. Park, K., Perczel, A., Fasman, G. D. *Protein Sci.* **1992,** *1,* 1032-1049.
15. Hucho, F., Görne-tschelnokow, U., Strecker, A. *TIBS* **1994,** *19,* 383-387.
16. Hresko, R. C., Kruse, M., Strube, M., Mueckler, M. *J. Biol. Chem.* **1994,** *269,* 20482-20488.
17. Szkutnicka, K., Tschopp, J. F., Andrews, L., Cirillo, V. P. *J. Bacteriology* **1989,** *171,* 4486-4493.
18. Cope, D. L., Holman G. D., Baldwin, S. A., Wolstenholme, A. J. *Biochem. J.* **1994,** *300,* 291-294.
19. Wellner, M., Monden, I., Muckler, M. M., Keller, K. *Eur. J. Biochem.* **1995,** *227,* 454-458.
20. Holman, G. D., Rees, W. D. *Biochim. Biophys. Acta* **1987,** *897,* 395-405.
21. Walmsley, A. R., Lowe, A. G. *Biophys. Biochim. Acta* **1987,** *901,* 229-238.
22. Mueckler, M.,Weng,W., Kruse, M. *J. Biol. Chem.* **1994,** *269,* 20533-20538.
23. Barnett, J. E. G., Homan, G. D., Chalkley, R. A., Munday, K. A. *Bioch. J.* **1975,** *145,* 417-429.
24. Rahman, M., Brasseur, R. *J. Mol.Graphics* **1994,** *12,* 212-218.
25. Higgins, D. G., Sharp, P. M. *Gene* **1988,** *73,* 237-244.

26. Wellner, M., Monden, I., Keller, K. *FEBS* **1992**, *309*, 293-296.
27. Prevelige, P., Fasman, G. D. in *Prediction of protein structure and the principles of protein conformation* G. D. Fasman. (eds.) Plenum Press, New York and London. 391-416.
28. De Loof, H., Rosseneu, M., Brasseur, R., Ruysschaert, J. M. *Proc. Natl. Acad. Sci. USA* **1986**, *83*, 2295-2299.
29. Donnely, D., Overington, J. P., Ruffle, S.V., Nugent, J. H. A., Blundell, T. L. *Protein Science* **1993**, *2*, 55-70.
30. Fischbarg, J. F., Kuang, K., Vera, J. C., Arant, S., Silverstein, S. C., Loike, J., Rosen, O. M. *Proc. Natl. Acad. Sci. USA* **1990**, *87*, 3244-3247.
31. Mueckler, M., Homan, G. *Nature* **1995**, *337*, 100-101.
32. Pusiner, S.B. *Science* **1991**, *252*, 1515-1521.
33. Rost, B., Casadio, R., Fariselli, P., Sander, C. *Protein Science* **1995**, *4*, 521-533.

J. Mol. Model. **1996**, 2, 46 - 50

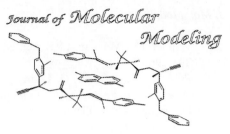

Structural Evaluation of Distant Homology. A 3-D Model of the Ligand Binding Domain of the Nicotinic Acetylcholine Receptor Based on Acetylcholinesterase: Consistency with Experimental Data

Ajita A. Bhat and E. Will Taylor*

Computational Center for Molecular Structure and Design and Department of Medicinal Chemistry,
The University of Georgia, Athens, GA 30602-2352, Tel: (706)-542-5391, Fax: (706)-542-5358
(WTAYLOR@RX.UGA.EDU)

Received:18 November 1995 / Accepted: 5 February 1996 / Published: 12 March 1996

Abstract

Acetylcholine is a ligand for both acetylcholinesterases and nicotinic acetylcholine receptors. Hence, at least some local sequence and structural similarities between the acetylcholinesterases and the receptors which recognize acetylcholine (ACh) might be expected. Peterson [2] produced an alignment of the ACh binding region between these two types of ACh-binding molecules, featuring a number of well conserved residues. The extent of this region of sequence similarity suggests the possible existence of a common ancenstral ACh binding module. To attempt to further validate Peterson's sequence alignment we have built a homology model of the ACh binding domain of the human neuromuscular nicotinic acetylcholine receptor based on the structure of acetylcholinesterase from *Torpedo californica*. Using this 3-D model we have examined the residues which were previously shown to interact with the endogenous ligand by various methods (mapping, site-directed mutagenesis). The consistency of such data with the model provides further support for a structural similarity and possibly a divergent evolutionary relationship between the ACh-binding domains of these two classes of proteins. Results suggest that this model may be able to contribute to an understanding of the structure and function of the ACh receptor. Using this case as an example, we propose that 3-dimensional computer modeling can be used as a tool to evaluate distant homologies when adequate experimental data (e.g., site-directed mutagenesis) is available.

Keywords: Acetylcholine Receptor, Acetylcholinesterase, Modeling, Receptor, Homology, 3-D Structure
Short Title: Structural Evaluation of Distant Homology

Introduction

Along with exon shuffling, gene duplication and subsequent divergence is a fundamental mechanism of protein evolution. Proteins or protein modules derived from a common ancestor can accumulate numerous point mutations leading to homologous proteins with varying degrees of sequence similarity, in some cases so low as to be insignificant despite conserved structural features. A number of techniques have been developed in order to evaluate such potential distant homologies, including the comparison of patterns of conserved sequence motifs in multiple sequence alignments, and the method of "inverse folding", which can be applied when at least one of the sequences has a known crystal structure [1]. When there is sufficient sequence similarity to at least permit a motif-based alignment of the protein sequences, there is another way to evaluate such distant relationships that has not been widely used. One can evaluate the potential homology by examining the consistency between the 3-D pro-

* *To whom correspondence should be addressed*

tein model based on the proposed alignment and the existing biochemical data, if that data is sufficiently specific in relation to the position of amino acid side chains and ligand or substrate binding sites. In this paper, we will use the method which we call structural evaluation of distant homology (SEDH), to assess a proposed relationship between the nicotinic acetylcholine receptor (nAChR) and acetylcholinesterase (AChase). We will demonstrate the approach by mutating a part of the substrate binding domain of acetylcholinesterase (a known 3-D structure) into the corresponding region of the nicotinic acetylcholine receptor ligand binding domain, based on an alignment previously proposed by Peterson [2]. We will show that the consistency of the model with known ligand binding data supports the proposed structural relationship.

Acetylcholine (ACh) is a peripheral and central nervous system neurotransmitter that acts at several types of ACh receptors, viz. muscarinic and nicotinic. After its action acetylcholine is acted upon by the extracellular enzyme acetylcholinesterase (AChase) whose biological role is termination of the impulse at the nerve endings by rapid hydrolysis of ACh. The nicotinic ACh receptor (nAChR) belongs to the ligand gated ion channel (LGIC) superfamily of proteins. nAChR is the most well studied receptor to date within this large family and is thus a prototype for understanding the binding of neurotransmitter to the receptor and the subsequent structural changes of the receptor. The musarinic ACh receptors, nAChRs and the AChases share the primary biological function of interacting with the acetylcholine molecule.[2] Therefore there is reason to expect that they may have some structural features in common, such as certain amino acid residues to which ACh binds.

The alignment presented by Peterson shows consensus residues among these two types of protein sequences.[2] A slightly truncated version of Peterson's alignment of the region of sequence similarity is shown in Figure 1. This region contains most of the residues involved in substrate binding in AChase and AChR. Based on this information we ask the question: can the known 3-D structure of the *Torpedo californica* AChase from the Brookhaven protein data bank (code 1ace.pdb) be mutated to give a resonable nAChR model that is consistent with the experimental data (i.e. amino acid residues involved in ACh binding)?

Structure of AChases

Within the acetylcholinesterase family, the most well studied members have been the proteins from *Torpedo californica* and *Torpedo marmorata*, where AChase exists in the form of a homodimer consisting of about 540 amino acids. The AChase structure has been resolved by X-ray crystallography at 2.8 Å.[3] AChase has a deep, narrow cavity 20 Å long in the center of the homodimer, also known as the "active site gorge".[4] An extensive lining made of aromatic residues (including Y70, W84, W114, Y121, Y130) forms the binding pocket for the quaternary amine which permits different possible orientations for ACh and various sites for agonist and antagonist interactions. The model proposed by J. Sussmann et al.[3] features an oxyanion hole formed by the main chain nitrogens of G118, G119 and A201 (not shown in the figure) which interact with the carbonyl oxygen of ACh. The conserved G117 and G118 probably also play a role in making the chain very flexible, facilitating the binding for its interaction.[3] (Figure 2).

```
Neuromuscular nAChrs:
    135         145         155         165         175         185         195         205         215     223
1  THFPFDQQNCIMKLGIWTYDGTKVSISPESDRPDLSTFMESGEWVMKDYRGWKHWVYYTCCPDTPYLDITYHFIMQRIPLYFVVNVIIPCL
2  THFPFDQQNCIMKLGIWTYDGTKVSISPESDRPDLSTFMESGEWVMKDYRGWKHWVYYTCCPDTPYLDITYHFIMQRIPLYFVVNVIIPCL
3  THFPFDEQNCSMKLGTWTYDGSVVAINPESDQPDLSNFMESGEWVIKEARGWKHWVFYSCCPTTPYLDITYHFVMQRLPLYFIVNVIIPCL
4  THFPFDEQNCSMKLGTWTYDGSVVAINPESDQPDLSNFMESGEWVIKESRGWKHSVTYSCCPDTPYLDITYHFVMQRLPLYFIVNVIIPCL

Acetycholinesterases:
    62          72          82          92          102         112         122         134       142       151
1  STYPNNCQQYVDEQFPGFSGSEMWNPNREMSEDCLYLNIWVPSPRPKSTTVMVWIYGGGFYSGSSTLDVYNGKYLAYTEEVVLVSLSYRVG
2  STYPNNCQQYVDEQFPGFPGSEMWNPNREMSEDCLYLNIWVPSPRPKSATVMVWIYGGGFYSGSSTLDVYNGKYLAYTEEVVLVSLSYRVG
```

Figure 1: *Comparison of AChase and nAChR sequences (after Peterson, 1989 [2]). Conserved residues are shown in red (in several cases they are offset by one position as indicated). In nAChR, residues involved in agonist binding are shown in green and those involved in antagonist binding are shown in blue. Amino acid Y190, involved in both agonist and ant-* *agonist binding, is shown in cyan. Amino acids involved in ACh binding in AChase are colored in magenta. Amino acid Y130 in AChase is conserved and takes part in binding. **1:** Torpedo californica **2:** Torpedo marmorata **3:** Mouse **4:** Human*

Figure 2. *Interaction between ACh and AChase in the experimental (X-ray) AChase crystal structure (1ace.pdb). The peptide backbone is shown as a Cα trace. ACh is shown in red without hydrogens. The side chains of some important residues that influence binding or interact withACh are shown (1-7). 1:Y70 2: W84 3: W114 4: G118 5: G119 6: Y121 7: Y130*

Structure of nAChR

The nAChR is a transmembrane protein consisting of four homologous subunits, $\alpha_2\beta\gamma\delta$. These subunits are arranged in the shape of a torus with a centrally located ion channel. The subunits are structurally related, implying that they may have evolved from a common ancestor.[5] Ligands of various structural classes are proposed to bind to the large region on the extracellular receptor surface towards the N-terminal end, i.e. the α subunit.

Residues Involved In Ligand Binding to nAChR: Experimental Data

Among these nAChR subunits mainly the α subunits have cholinergic binding sites. Some of these binding sites are different for agonists and competitive antagonists, as shown by site-directed mutagenesis and mapping.[6,8] The two main ligand binding segments are 134-153 and 181-200.[9]

Site-directed mutagenesis and photo-affinity labelling studies show that the antagonist binding site (e.g. for α-Bungarotoxin) is not a single residue or a continuous segment but rather discontinuous short segments (C142, H186, V188, Y189, Y190, P194, D195, Y198),[6-8] probably folded together within the AChR structure.[9] Residues involved in agonist binding are Y190, C192, C193, Y198.[6,8]

Thus residues involved in binding lie primarily within the α subunit including and flanking the region of the two cysteines 192 and 193 which form a disulfide bond, creating an unusual cis-peptide linkage.[10, 11] These two cysteines are conserved in all α subunits of the nicotinic acetylcholine receptors.[12] In the past, theoretical calculations have been done which have predicted that the L-Cys-L-Cys peptide bond can occur only in a cis-peptide form, [13, 14] which is supported by X-ray data on model peptides.[10] It has been suggested this Cys-Cys bond may act as a molecular switch to control receptor activation. Furthermore since proline residues have been observed to have a higher probability of having the cis isomer of the preceeding peptide bond compared to other residues,[15] the peptide bond between Cys-193 and Pro-194 has been suggested to be in the cis form.[8] Thus there is the potential for at least one or possibly two consecutive cis-peptide bonds in this region of AChR. These potential cis peptide bonds must be given due consideration when modeling this region of the AChR.

Materials and Methods

Sybyl 6.04 (Tripos Associates, St. Louis, MI 63144-2913) was used to model the receptor protein. The AChase structure from *Torpedo californica* containing the ACh molecule was obtained from the Brookhaven Protein data bank (PDB file 1ace), figure 2. The MUTATE command was used to mutate the residues between amino acids 61 and 151 according to the human neuromuscular nAChR. The rest of the molecule was truncated using the SPLIT command. Residues were renumbered from 133 to 223, to correspond with the nAChR sequence. Possibly due to an unusual bend in the protein backbone around position 198 involving proline 197 (not present in AChase), the phenyl ring of tyrosine 198 was not oriented to fit in the protein globular structure and faced the side opposite to that of ACh. The protein backbone structure between the residues Pro 197 and Asp 200 was thus remodeled using the BIO LOOP feature. From the protein model obtained, two versions were made, one containing a cis peptide bond between Cys 192 and Cys 193 (mono-cis) and the other containing two cis peptide bonds, between Cys192-Cys193 and between Cys 193-Pro 194 (di-cis). Both the protein strucutres were separately minimized using the following procedure: the amino acids Cys 192 to Pro 194 were defined as the 'hot region' using the function ANNEAL in Sybyl. A 'hot radius' of 4.5 Å and an 'interesting region' of 10 Å was defined. Minimization was carried out using steepest descent using KOLLMAN_ALL atom force field. The ACh molecule and the backbone was then defined as AGGREGATE and only the side chains were minimized using the steepest descent method with the KOLLMAN_ALL atom force field. Slowly the constraints were relieved and the entire molecule was minimized using the KOLLMAN_ALL atom force field, using the conjugate-gradient method with KOLLMAN charges, a non-bonded cutoff of 8 Å, 1,4-scaling equal to 0.5 and ΔE of 0.1kcal/mol. The residues which assist in agonist binding and those involved in the interaction with the competitive antagonists are shown in figure 3.

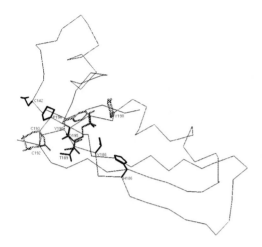

Figure 3. *Locations of known ACh-interacting residues in the nAChR model. The peptide backbone is shown as a Cα trace. ACh is shown in red without hydrogens. Residues which have been shown to interact with agonist ligands are numbered below as 5-7, 10 whereas residues numbered below as 1-5, 8, 9 have been shown to interact with antagonist ligands. The color scheme for the ligands is the same as that used in figure 1. In this model, all these residues clearly cluster around the ACh molecule, consistent with experimental results (see text). Residues C192, C193 and P194 have cis peptide bonds between them (see figure 4). 1: C142 2: H186 3: V188 4: T189 5: Y190 6: C192 7: C193 8: P194 9: D195 10:Y198*

Results and Discussion

We confirmed the alignment with the BESTFIT program [16] (gap weight 1.5/2.5, gap length 0.3/.25) using the sequences of Torpedo californica AChase versus those of Torpedo and human AChR. This gave pairwise alignments that were very close to Peterson's alignment in the region of interest (between amino acids T133 and L223). After minimization of the mono-cis and di-cis protein fragments we found that the protein with two cis peptide bonds (di-cis) was more stable by >5 kcal/mol, than the protein with only one cis peptide bond (mono-cis) between the two cysteines forming a disulfide bond. We also made another model (not shown) di-trans with trans peptide bonds between Cys192-Cys193-Pro194 which was also >5 kcal/mol higher in energy relative to the di-cis model. Considering the energy differences we concluded that the di-cis model is a more valid representation of the active domain, a result consistent with earlier observations.[9] A backbone fit of the AChR model to the AChase enzyme gave a weighted root mean squared difference of 1.6945 Å. Root mean square difference of the back-

bone of the two versions of the AChRs was 0.4188 Å which differed only at the region between Cys 192 and Pro 194 (Figure 4). Among all the interactive residues within the AChR, only one ligand-binding residue has been mutated from the *Torpedo californica* to the human spieces i.e., Y189T. However from the model even the side chain of threonine 189 is seen in close contact with the ACh molecule.

The idea was then to test this model by comparing with the experimental data available in the literature, such as site-directed mutagenesis. Many experiments have been performed in the past [7-13] where ligand binding sites of the AChR were identified. An attempt was made to rationalize the location and function of the binding sites to those given by the model, in order to validate any possible distant homology. It is immediately obvious from the model that, as shown in figure 3, all the residues previously shown to interact with cholinergic ligands are clustered in the immediate vicinity of the predicted ACh binding site. This suggests that the model is at least qualitatively correct.

Furthermore, the model can account for specific changes in activity and building associated with specific amino acid mutations. By mutagenesis Chaturvedi et. al.[8] and O'Leary et. al.[17] have indicated that the aromaticity of the Tyr 190 and Tyr 198 of nAChR were important since they form an electronegative subsite through the formation of a tyrosinate anion which could attract the positive nitrogen of acetylcholine. This is clearly seen in the model where the aromatic rings and their hydroxyl groups face the nitrogen and interact with it. In the AChase structure (1ace.pdb) the glycines 117, 118 and 119 are close to the ACh molecule, which correspond to T189 and Y190 of AChR in the alignment. The model also correctly predicts that Cys 192, Cys 193 and Pro 194 in AChR form a turn at the tip of the beta sheet as pointed out earlier by Chaturvedi et. al.[8] The rigid conformation induced by the three amino acids comprising the disulfide bridge and the cis form of proline (as shown in Figure 4) may uniquely contribute to the ability of this region of the nAChR to bind ligands and contribute in the process of channel gating.[17]

The sidechains of His 186, Val 188 and Asp 195 in nAChR point toward the ligand in the binding domain in this model. Since they are quite a distance (>6 Å) from ACh, they are not involved in agonist binding but could be readily involved in binding larger molecules like α-Bungarotoxin, which again, is precisely consistent with the experimental data [7, 9]. Mutations of H186A, V188T probably decrease the sidechain length and branching required for binding of the antagonist α-Bungarotoxin to its binding site.

The consistency of the model with the experimental results suggests that these two classes of ACh-binding proteins do have some common structural features, supporting the possibility that their ligand binding domains may have evolved from a common ancestor. The results suggest that the Torpedo AChase protein structure in the Brookhaven protein data bank (figure 2) is a reasonable starting point for

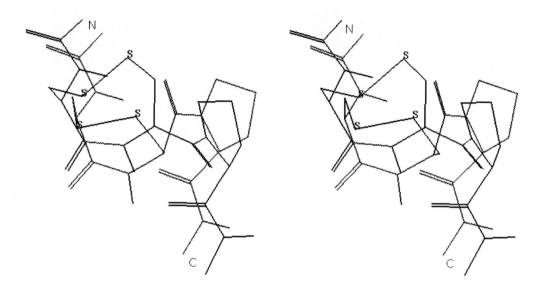

Figure 4. *Stereoview comparing the two versions of the protein model involving the Cys-Cys-Pro sequence, and the potential cis-peptide bonds (highlighted in yellow). The mono-cis model is shown in red and the di-cis model is shown in green. The amino and the carboxy ends of the fragments are denoted with the letters N and C respectively. The models differ only in the peptide bond between Cys 193 and Pro 194. Both the mono-cis model and the di-trans model (not shown) are >5kcal/mol higher in energy relative to the di-cis model.*

homology modeling of the ligand binding domain of the human ACh receptor (figure 3).

With this example, we have illustrated the approach of this method, which we have tentatively called structural evaluation of distant homology (SEDH). When adequate biochemical data (mutagenesis etc.) are available, this method may complement existing approaches like inverse folding, as a tool for validating speculative homology models.

References

1. Godzik, A.; Kolinski, A.; Skolnick, J. *J. Comput. Aided. Mol. Des.* **1993**, *7*, 397.
2. Peterson, G.L. *J. Neurochem. Res.* 1989, 22, 488.
3. Sussman, J.L.; Harel, M.; Silman, I. Chem. Biol. Inter. 1993, 87, 187.
4. Sussman, J.L.; Harel, M.; Frolow, F.; Oefner, C.; Goldman, A.; Toker, L.; Silman, I. Science, 1991, 25, 872.
5. Raftery, M.A.; Hunkapiller, M.W.; Strader, C.D.; Hood, L.E. Science, 1980, 208, 1454.
6. Sine, S.M.; Quiram, P.; Papanikolaou, F.; Kreienkamp, H.; Taylor, P. J. Biol. Chem., 1994, 269, 8808.
7. Conti-Tronconi, B.M.; Diethelm, B.M.; Wu, X.; Tang, F.; Bertazzon, T. Biochem, 1991, 30, 2575.
8. Chaturvedi, V.; Donnelly-Roberts, D.L.; Lentz, T.L. Biochem, 1993, 32, 9570.
9. Conti-Tronconi, B.M.; Tang, F.; Diethelm, B.M.; Spencer, S.R.; Reinhardt-Maelicke, S.; Maelicke, A. Biochem, 1990, 29, 6221.
10. Capasso, S.; Mattia, C.; Mazzarella, L. Acta Cryst B, 1977, 33, 2080.
11. Kao, P.N.; Karlin, A. *J. Biol. Chem.*, **1986**, *261*, 8085.
12. Blount, P.; Merlie, J.P. *J. Cell. Biol.* **1990**, *111*, 2613.
13. Ramachandran, G.N.; Sasisekharan, V. *Adv. Protein. Chem.*, **1968**, *23*, 284.
14. Chandrasekaran, R.; Balasuramanian, R. *Biochim. Biophys. Acta*, **1969**, *188*, 1.
15. McArthur, M.W.; Thornton, J.M. *J. Mol. Biol.*, **1991**, *218*, 397.
16. Program Manual for the Wisconsin Package, Ver. 8, September 1994, Genetics Computer Group, 575 Science Drive, Madison, WI 53711.
17. O'Leary, M.E.; White, M.M. *J. Biol. Chem.*, **1992**, *267*, 8360.

J. Mol. Model. **1996**, 2, 51 - 61

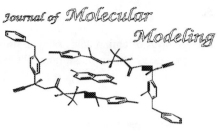

Conformational Behavior and Flexibility of Terminally Blocked Cysteine and Cystine

Zdenek Kríz, Jaroslav Koca*

Department of Organic Chemistry; Faculty of Science; Masaryk University; CR–611 37 Brno, Czech Republic
(zdenek@chemi.muni.cz; jkoca@chemi.muni.cz)

Per H. J. Carlsen

Institute of Organic Chemistry; University of Trondheim, Norwegian Institute of Technology, N–7034 Trondheim – NTH,
Norway (per.carlsen@kjemi.unit.no)

Received: 18 October 1995 / Accepted:11 March 1996 / Published:22 April 1996

Abstract

Conformational potential energy hypersurfaces, *PES*, for the terminally blocked L-Cysteine, L,L-Cystine and
D,L-Cystine have been analyzed by means of molecular mechanics in combination with the programs ROSE,
CICADA, PANIC and COMBINE. Low energy conformations and conformational transitions, conformational
channels, have been located. Global and fragmental flexibility and conformational softness have been calculated for each conformer as well as for the entire molecule. The *PES* analyses were used for simulation of
conformational movement based on Boltzmann probability of the points obtained on the *PES*. Boltzmann travelling revealed interesting correlated conformational movement where three or even more dihedral angles changed
simultaneously. It could be shown that conformational behavior and flexibility were strongly influenced by the
absolute configurations of the amino acids in the peptides.

Keywords: Conformational search, Molecular mechanics, Potential Energy (hyper)surface, D,L-form of amino acid, Conformational flexibility.

Introduction

Amino acids are interesting and important targets for the
computational chemist because they contain a variety of intramolecular interactions, are conformationally labile, and
are of a tractable size for calculations. Detailed knowledge
of the conformational properties of disulphide bridges is essential for studies of all molecules in this group, e.g. for the
understanding of numerous extracellular properties of biologically active oligopeptides [1]. Recently, we have reported
a study [2] of the conformational behavior of the 20 common amino acids. We have shown that the ability of single
residue to form a specified secondary peptide structure is not
only dependent on low energy conformers, but also on the
flexibilities of the conformers. In that study, it was shown
that the conformational behavior of L-Cysteine exhibited the
largest deviations between calculated and X-ray data of real
proteins. This may be ascribed to the disulfide bridges present
in real proteins. The above problem prompted us to elaborate on the study of a small molecules containing disulfide
bridges, i. e., terminally blocked Cystine. We have chosen
two forms of Cystine for our study, L,L– and D,L-Cystine.
The reason is that peptides with a combination of L-forms

* *To whom correspondence should be addressed*

and D-forms of amino acids play an important role in transport processes within membranes. D-forms of amino acids play a key role in the transport activity (e.g., Gramicidin A). Another motivation for studying the D-form of Cystine is that D-Cysteine is included in cyclic analogues of enkephalins, which have specific activity to receptors in the brain. The comparison of the conformational behavior of a single residue and Cystine as a part of cyclic peptide will follow this work.

The present study has three objectives.

1. To continue the studies on relationships between conformational behavior and flexibility of amino-acids as separate entities and as part of peptides and proteins.
2. To test the ability of the programs ROSE, CICADA, PANIC and COMBINE to perform conformational potential energy hypersurface PES analyses.
3. To test performance of the software PMMX [3] in describing the conformational space of single amino acid residues and short peptides. Conformational properties of amino acid residues L-Cysteine and Cystine obtained by the molecular mechanics method have been reported in the literature[4, 5, 6]. These published data there were used as a starting point for our calculations.

Method

All computations were performed with the ROSE [7] (version 2.9) and CICADA [8] (version 3.0) software interfaced with molecular mechanics program PMMX which is the MMX program [9] specially parametrized for peptides using parameters obtained from the literature [10]. Final analysis of data and exibility computations were performed with PANIC [11] (version 3.0). Conformational movements have been simulated by the program COMBINE [12]. All the molecular mechanics calculations were performed for a dielectric constant $\varepsilon = 78.5$ in order to mimic an aqueous environment. The approach is expressed by the following steps.

1. We have extracted from the literature [4, 5, 6] structural data for the rotatable backbone bonds of the low energy conformers for each terminally blocked amino acid. For the side chain we used the ±gauche and trans geometry for the torsion χ_1, χ_2 and ±90 for the disulphide bridge of Cystine (for more details see Figure 5). These values have been used as input data for the program ROSE which generates low energy conformers by a combinatorial algorithm. The Cartesian coordinates of conformers have been generated by ROSE for L-Cysteine, L,L- and D,L- Cystine, respectively.
2. The structures generated by ROSE were next fully relaxed and minimized. The conformations with relative energies less than 1.5 kcal·mol⁻¹ were then selected and used as starting set for the CICADA calculation. CICADA performs a systematic search for low energy conformational interconversions, i.e., conformational channels on the PES. The following parameters were used during the

CICADA calculations for all molecules. The limiting difference for dihedral angle (Dmin) above which two conformers are considered as different was set to 30 degrees. This means that two conformers are considered identical if all their dihedral angles differ by less than Dmin. The smallest energy differences (Emin) taken into account was fixed at 0.1 kcal·mol⁻¹. An energy cutoff of 30 kcal·mol⁻¹ has been applied. This means that no interconversion with a transition energy higher than 30 kcal·mol⁻¹ has been taken into account. For more details about CICADA parameters see [8].

3. CICADA results were then analyzed by the program PANIC. The theory used by the program is based on a graph theoretical model of the *PES* [13, 14]. The program produces a list of minima and transition states along the *PES*. Interconversion pathways and their transition energies are generated by the program. Moreover, PANIC calculates absolute flexibility ϕ, and relative flexibility f, and con-formational softness f^0.
4. All the results were used for the program COMBINE [12] which simulates travelling on the *PES* based on the Boltzmann distribution of conformers. The geometrical parameters for structures along the Boltzmann trajectories were extracted by the program ANALYSE [15]. The same software was used to elucidate correlated conformational movement. The drawings were made by the "in house" software TULIP and FROG ([16]).

The approach is schematically demonstrated in Figure 1. The calculations have been performed on i386 based PC running

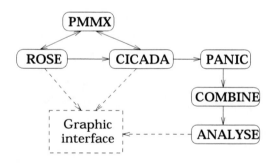

Figure 1. *Scheme of the communication between single programs. Energies of the structures are computed by PMMX which is an MMX program specially parametrized for peptides. The program ROSE is used for generation of starting conformations by a combinatorial algorithm. CICADA performs travelling along the PES. Results from the program CICADA are used to PES and flexibility analysis which is performed by the program PANIC. Results from the programs CICADA and PANIC serve for the simulation of the conformational analysis which is performed by COMBINE. ANALYSE is used to extract geometry parameters of structures obtained by a COMBINE run.*

under the operating system UNIX (ROSE), on SGI Indy running operating system IRIX 5.1 and SGI POWER Challenge L running operating system IRIX 6.0 (all other calculations).

Flexibility

Detailed knowledge of the low energy regions of the potential energy surface is essential in order to quantify conformational flexibility. For that purpose, the following values have previously been defined: f and ϕ for relative and absolute flexibility [17], respectively, and f^0 for the conformational softness [18]. Relative and absolute flexi- bilities can be evaluated for single conformers, for molec- ular fragments as well as for an entire molecule. As a consequence, the relative flexibility f of conformer i is evaluated as a product of three terms, thermodynamic p_i, kinetic k_i, and geometrical g_i:

$$_i f = p_i \cdot k_i \cdot g_i \tag{1}$$

Each term can be defined as follows:
• the thermodynamic contribution p_i is the probability of each conformation at absolute temperature (T), calculated using the Boltzmann distribution.

$$p_i = exp\left(-E_i \big/ KT\right) \Big/ \sum_{j=1}^{n} exp\left(-E_j \big/ KT\right) \tag{2}$$

where K is the Boltzman constant and E_i the relative energy of conformer i.
• the kinetic term k_i is expressed by the Arrhenius rate constant term as follows:

$$k_i = \sqrt[n_i]{\prod_{j=1}^{n_i} exp\left(-E_j \big/ RT\right)} \tag{3}$$

where R is the gas constant.
• the geometrical term g_i is computed by:

$$g_i = \left[\sum_{j=1}^{n_i} \sum_{k=1}^{m} \left|t_k^j - t_k^i\right|\right] \Big/ 2\pi m n_i \tag{4}$$

where n_i is the number of transition states (torsion angle t^j) surrounding the conformation being considered (torsion angle t^i) and m is the number of dihedral angles followed. This term is therefore the sum of geometrical movements from the conformer i via the various transition states measured in the dihedral angle space, and normalised by the number of transition states and the number of rotatable bonds. By analogy, the absolute flexibility; $_i\phi$, of the conformer i is expressed by:

$$_i\phi = p_i \sum_{j=1}^{n_i} g_{ij} \cdot exp\left(-E_j \big/ KT\right) \tag{5}$$

where g_{ij} is analogous to g_i but normalised only by the number of rotatable bonds.

From the conformer flexibility values, it is then possible to calculate the relative f and absolute ϕ flexibility of the molecule as a summation over all conformers. While the absolute flexibility value reflects the best (lowest energy) conformational interconversions, the relative flexibility expresses "an average conformational interconversion". Both flexibility values increase when the energy barriers between the conformers decrease and both are temperature dependent. For

Table 1. *Global descriptors of the CICADA analysis of residues [a]*

| Residue | Number of torsions | | PES description | | |
	Global	Active	points	conform. [b]	pathways
L-Cysteine	6	4	260	49(40)	448
L,L-Cystine	13	9	4999	1550(37)	6776
D,L-Cystine	13	9	5000	1473(293)	7188

[a] Active torsions: torsions driven during the calculation; point on PES is any nuclear configuration of importance and thus saved by the program (i.e. energy minimum,

transition state). For details of CICADA and its parameters see Ref. [7].
[b] Numbers in parentheses: number of conformers (energy minima) with relative energies less than 1.5 kcal/mol.

detailed descriptions see references [17, 19]. Another value of interest is the conformational softness f^0 which reflects the total number of interconversions and also their energy barriers. It can be evaluated for the entire molecule using:

$$f^0 = \left[n/ln(n - 0.67523)\right] \cdot$$
$$\left[\sum_{i=1}^{n}\sum_{j=1}^{k} p_i \, exp\left(-E_j/RT\right)\Big/ln(k + 1224.84)\right] \qquad (6)$$

where n represents the total number of conformers, i. e., (energy minima) and k the total number of interconversions [17].

Results and discussion

Global descriptors of the CICADA analyses are summarized in Table 1. It is revealed that the number of low energy conformers (local energy minima on the *PES*), as well as the number of interconversion pathways (conformational channels), are very strongly dependent on the number of rotat-

able bonds. It is seen by comparison of both forms of Cystine that the total number of the points on the *PES* as well as the number of the conformations that have been found by program CICADA are comparable. This means that both CICADA runs (for D,L and L,L forms) are comparable. In contrast the total number of conformations with relative energy less than 1.5 kcal·mol⁻¹ is much higher for the D,L-form. This fact is also reflected in computed flexibilities that are summarized in Table 2 for successive torsions as well as for the entire molecules.

It is seen from this table that the flexibilities of the investigated molecules are dependent on their sizes. The computed relative flexibility of L-Cystine is much more higher than that of L,L-Cystine and also higher than that of D,L-Cystine. The differences between the computed absolute flexibilities are smaller because the absolute flexibility reflects only the lowest energy interconversion pathways. It is also seen that a significant part of the flexibility has been localized to the torsions φ and ψ around the backbone. On the other hand, the lowest computed flexibility has been concentrated on the disulphide bridges of the Cystine. The largest differences between flexibility of torsions are seen for L,L-Cystine.

Torsion	L–Cysteine Relative (×10⁵) Absolute (×10³)	L,L–Cystine Relative (×10⁵) Absolute (×10³)	D,L–Cystine Relative (×10⁵) Absolute (×10³)
φ	12.0100	0.1114	0.4406
	3.6640	0.2601	0.8030
ψ	13.5100	0.1933	0.9440
	9.0410	1.2510	2.3950
χ_1	9.6100	0.0676	0.2200
	0.6630	0.0446	0.2841
χ_2	6.2400	0.0859	0.2975
	3.1900	0.0841	0.3284
χ_3	–	0.0279	0.1160
	–	0.0339	0.1858
χ_{21}	–	0.0811	0.3147
	–	0.0812	0.3494
χ_{22}	–	0.0648	0.2138
	–	0.0493	0.2670
ϕ_2	–	0.1283	0.5632
	–	0.2427	0.8149
ψ_2	–	0.2204	0.8232
	–	1.2550	2.1900
Total flexibility	41.5800	0.9995	4.0230
	16.6400	3.3410	7.7280

Table 2. *Relative and absolute fragmental flexibilities of single residues. (For denotation see Figure 2 and 5).*

The terminally blocked form of the L-Cysteine, N-acetyl-N'-methylamide of L-Cysteine, has been investigated. The usual notation ϕ, ψ, χ_1, χ_2 is used to describe the rotatable bonds. The description is illustrated in Figure 2.

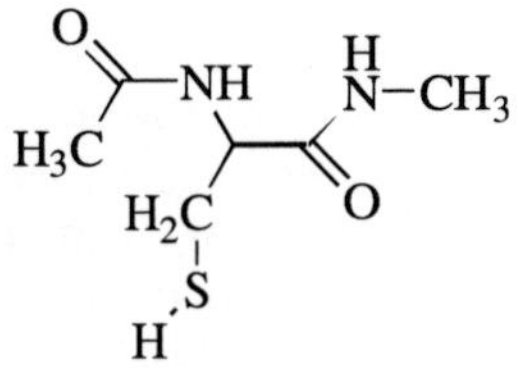

Figure 2. *Schematic representation of the Cysteine studied here along with the labelling of torsional angles of interest.* (ϕ) *-C-N-C^α–C'- ,* (ψ) *-N-C^α-C'-N- ,* (χ_1) *-N-C^α-C^β-S^γ- ,* (χ_2) *-C^α-C^β-S^γ-H-*

The results from the conformational behavior analyses of the molecule are summarized in Tables 3 and 4. Table 3 shows geometries and energies of the best 5 conformations, from the relative energy point of view, which were found by the program CICADA. For the synopsis, the geometry parameters are combined with the letter code according to Zimmerman et al. [5].

It is seen from this table that the global conformational minimum is in the α-helix region with dihedral angle values $\phi = -69°$ and $\psi = -45°$. Calculated flexibility data are summarized in Table 4.

The table shows that the global minimum (conformation number 29) exhibits the highest relative flexibility. The highest absolute flexibility has been calculated for two confor-

Table 3. *Selected conformations found for L-Cysteine (for denotation see Figure 2).*

[a]	[b]	Torsion				Energy
		ϕ	ψ	χ_1	χ_2	(kcal/mol)
1	AGt	-69	-45	-64	179	0.0671
2	CGT	-70	127	-64	-180	0.1546
29	AGG	-69	-45	-62	-60	0.0000
34	AGg	-69	-46	-65	63	0.0002
61	FGg	-71	132	-66	63	0.0889

[a] *Numbering of structures (conformations) as generated by the software.*
[b] *Letter code is composed of Zimmerman notation [5] (first character) and letter abbreviation for the side chain. For denotation see Scheme 1. Starred letters in the original Zimmerman code are substituted by small letters.*

Table 4. *Relative and absolute flexibilities of selected conformations of L-Cysteine [a]*

Conf.	p_i	k_i	g_i	f_T	ϕ_T
1	0.5519	0.1091	0.5152	0.3103	0.2335
2	0.4762	0.1260	0.5088	0.3052	0.2180
29	0.6181	0.1054	0.5023	0.3273	0.1635
34	0.6179	0.0965	0.5062	0.3020	0.1599
61	0.5320	0.0844	0.5194	0.2331	0.1494

$(p_i \times 10, k_i \times 10, g_i \times 10, f_T \times 10^4, \phi_T \times 10^2)$.
[a] *For geometry and energy see Table 3*

mations (1 and 2) with relative energies 0.0671 and 0.1546 kcal·mol^{-1}, respectively. The first one adopts a geometry similar to the global minimum, only with torsion χ_2 in the trans geometry. The second one belongs to the β-pleated sheet region. The ϕ–ψ plot for the conformations and transition states obtained is shown in Figure 3.

The areas designated by α and β show the combinations of ϕ, ψ torsions to form secondary structure α–helix and β–pleated sheet. This figure illustrates the location of conformations and transition states on the *PES*. The inter-conversion pathways between the conformations obtained could be estimated from the positions of the transition states. Figure 4 shows the relative energy and the values of torsions χ_1 and χ_2 for obtained conformations and transition states.

Energy barriers for rotation of the χ_1, χ_2 torsions are clearly seen from Figure 4. The transition states with torsion $\pm$gauche and trans geometry correspond to transition states on the interconversion pathways within other parts of the *PES*. It is

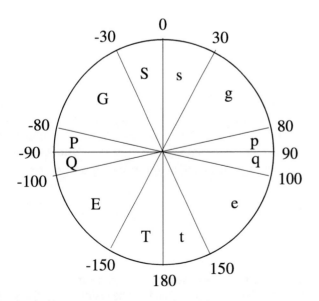

Scheme 1. *Assignment of letter codes for the side chain dihedral angles.*

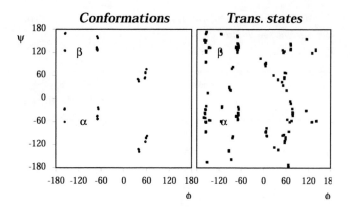

Figure 3. *Plots of the conformers (left part) and transition states (right part) found by CICADA as a function of the ϕ and ψ torsion angles of L-Cysteine. Areas α and β indicate the possible combinations of torsions ϕ and ψ in peptides to form the secondary structure of α-helix and β-pleated sheet, respectively.*

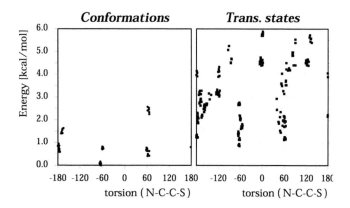

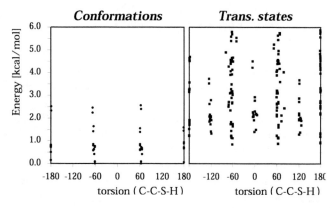

Figure 4. *Energy vs geometry plot for dihedral angles χ₁ (upper figure) and χ₂ (bottom figure) of L-Cysteine. Data are shown for all the conformers (left part) and the transition states (right part) found by CICADA in an energy window of 6 kcal/mol.*

seen from Figure 4 that the rotation of torsion χ_1 is energetically unfavorable compared with a rotation of the terminal -SH group. These observations are reflected in the computed absolute flexibilities of the torsions investigated. Our results are compared with a conformational analysis of L–Cysteine, that was performed by the program Sybyl using the AMBER molecular mechanics force field and starting with X-ray data obtained from Cambridge Structural Database (CSD) [20]. We have found a preference for the –gauche geometry for the χ_1 torsion, whereas the +gauche geometry is preferred by X–ray analysis. Conformations with +gauche geometry of the χ_1 torsion are found at 0.4 kcal·mol⁻¹ higher energy in our results relative to the conformations in –gauche χ_1 geometry. The population of torsion χ_2 is also in a good agreement with the X–ray analysis.

Cystine

Two structures of the Cystine have been selected: D – Cys – S – S – L – Cys and L – Cys – S – S – L – Cys in the terminally blocked form (for denotation see Figure 5).

Geometry parameters of the best 5 conformations for L,L-Cystine are summarized in Table 5. It is seen that the global minimum (conformation number 590) adopts an α–helix geometry in both parts of the molecule. The table shows that the preferred value for the disulphide bridge is about 90°. The distribution of torsions χ_1, χ_2, χ_{21} and χ_{22} is similar to that of corresponding torsions in the L–Cysteine. Our results are in a good agreement with results obtained by X–ray structural analysis of peptides containing Cystine and molecular mechanics and statistic analysis studies[20]. The results of relative and absolute flexibility are summarized in Table 6.

Figure 5. *Schematic representation of the Cystine studied here along with the labelling of torsional angles of interest.* (ϕ) -C-N-Cᵅ-C'- , (ψ) -N-Cᵅ-C'-N- , (χ₁) -N-Cᵅ-Cᵝ-Sᵞ- , (χ₂) -Cᵅ-Cᵝ-Sᵞ-Cᵞ²- , (χ₃) -Cᵝ-Sᵞ-Sᵞ²-Cᵝ²- , (ϕ₂) -C²-N²-Cᵅ²-C'²- , (ψ₂) -N²-Cᵅ²-C'²-N²- ,(χ₂₁) -N²-Cᵅ²-Cᵝ²-Sᵞ²- , (χ₂₂) -Cᵅ²-Cᵝ²-Sᵞ²-Sᵞ- .

The same parameters as for L,L–Cystine are summarized for the D,L–form in Tables 7 and 8. The results for D,L-Cystine are similar to those obtained for L,L–Cystine, except for

Table 5. *Selected conformations found for L,L-Cystine (for denotation see Figure 5).*

Conf. no. [a]	Zimmerman notation [b]	Torsion angles									Energy (kcal=mol)
		ϕ	ψ	χ_1	χ_2	χ_3	χ_{22}	χ_{21}	ϕ_2	ψ_2	
553	AGGpgGA	-76	-44	-55	-67	88	65	-73	-77	-54	0.0015
566	AGGpgGC	-74	-43	-55	-67	89	65	-74	-76	124	0.5939
569	FGGpgGA	-76	133	-56	-67	88	65	-73	-76	-54	0.0900
590	AGgpGGA	-77	-55	-72	65	88	-67	-55	-77	-44	0.0000
596	AGgpGGF	-77	-55	-72	65	88	-67	-55	-77	133	0.0888

[a] *Numbering of structures as generated by the software.*
[b] *Letter code is composed of Zimmerman notation [5] (first and last character) and letter abbreviation for the side chain. For denotation see Scheme 1. Stared letters in the original Zimmerman code are substituted by small letters.*

Table 6. *Relative and absolute exibilities of selected conformations of L,L-Cystine [a]*

Conf.	p_i	k_i	g_i	f_T	ϕ_T
553	0.4423	0.0216	0.3122	0.2982	0.5883
566	0.1628	0.0132	0.3273	0.0702	0.1114
569	0.3810	0.0156	0.3217	0.1910	0.4747
590	0.4435	0.0238	0.3188	0.3368	0.5846
596	0.3817	0.0273	0.3082	0.3215	0.4859

$(p_i \times 10, k_i \times 10, g_i \times 10, f_T \times 10^4, \phi_T \times 10^2)$.
[a] *For geometry and energy see Table 5.*

dihedral angle values for torsions ϕ and ψ, which have opposite signs. It is shown that the global minimum (conformation 1473) exhibits lower relative and absolute flexibility than conformations with relative energy 0.4 kcal·mol^{-1} (e.g. conformations 120 and 176). This means that the global minimum is located in a relatively narrow valley on the *PES* and conformations 120 and 176 are located on energetically higher but shallower parts of the *PES*.

The dependency of the relative energies of conformations and transition states on the values for torsion χ_3 is shown for L,L– and D,L–Cystine in Figure 6.

This figure can be used to rationalize the different behaviors of the two forms of Cystine. It is seen that for the L,L-form values of χ_3 about +90° are preferred, whereas for the D,L-form, both values (+90° and –90°) are populated. The barriers of rotation are lower for the D,L-form than for the L,L-form. This is also reflected by the flexibilities obtained, which are about 6 times higher for the D,L-form than for the L,L-form.

Table 7. *Selected conformations found for D,L-Cystine. (For denotation see Figure 5).*

Conf. no. [a]	Zimmerman notation [b]	Torsion angles									Energy (kcal=mol)
		ϕ	ψ	χ_1	χ_2	χ_3	χ_{22}	χ_{21}	ϕ_2	ψ_2	
120	cgTggGA	69	-129	64	-178	77	66	-74	-78	-51	0.4395
152	agTggGA	68	46	64	-178	78	66	-74	-78	-50	0.3182
176	agTggGC	68	46	65	-179	78	65	-75	-79	117	0.4224
236	ggTqGGF	136	59	69	-152	94	-57	-58	-72	134	0.2642
514	cggqGGF	73	-124	61	63	90	-67	-56	-76	133	0.2463
568	aggQGGC	75	44	56	66	-91	-65	-62	-72	127	0.1895
708	agGGtGA	77	50	74	-66	-78	178	-65	-68	-46	0.3175
1473	agGQgtC	76	54	72	-67	-90	69	177	-74	102	0.0000
1489	agGQgtD	76	53	72	-67	-90	69	175	-155	99	0.1687
1847	cggqGGA	72	-127	61	64	90	-66	-56	-76	-44	0.1878

[a] and [b] see Table 5.

Table 8. *Relative and absolute exibilities of selected conformations of D,L-Cystine [a]*

Conf.	p_i	k_i	g_i	f_T	ϕ_T
120	0.0540	0.0685	0.3206	0.1185	0.1566
152	0.0662	0.0885	0.3075	0.1803	0.2394
176	0.0556	0.0609	0.3412	0.1154	0.1750
236	0.0726	0.1214	0.2975	0.2622	0.1329
514	0.0748	0.0178	0.4557	0.0606	0.4224
568	0.0823	0.0534	0.3133	0.1376	0.6084
708	0.0663	0.0676	0.3149	0.1413	0.1699
1473	0.1133	0.0195	0.2970	0.0656	0.1072
1489	0.0852	0.0141	0.2925	0.0352	0.0609
1847	0.0826	0.0360	0.3055	0.0907	0.4372

$(p_i \times 10, k_i \times 10, g_i \times 10, f_T \times 10^4, \phi_T \times 10^2)$.
[a] For geometry and energy see Table 7.

Simulation of the conformational movement

The results of the conformational analysis obtained were next used for simulation of travelling along the *PES*. Both simulations were performed using the program COMBINE [12] with the same parameters. The global minimum was used as a starting point. The number of steps was set to 1 000 000 and the temperature to 300 K. After each 1000 steps a new starting conformer has been selected. We have focused our attention on the side chain torsions χ_1, χ_2, χ_3, χ_{22} and χ_{21}. The changes of torsions χ_1, χ_2 and χ_3 within a selected 300 steps window are shown in Figures 7 and 8.

The differences in flexibilities of the L,L– and D,L–forms are clearly seen from these pictures. It is also seen that the more flexible torsions result in a larger number of changes relative to the more rigid torsions. It is evident (at least for

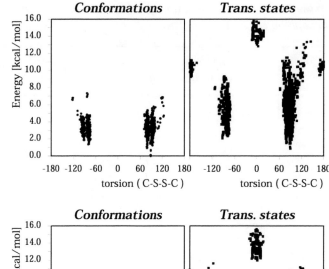

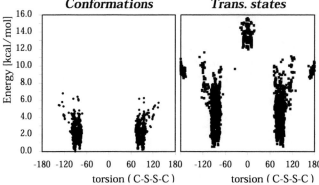

Figure 6. Energy vs geometry plot for dihedral angle χ_3 of L,L-Cystine (upper picture) and D,L-Cystine (bottom picture). Data are shown for all the conformers (left part) and the transition states (righ part) found by CICADA in an energy window of 16 kcal/mol.

this particular case) that the more rigid torsion can evoke changes of flexible torsions. This is seen from the correlation between torsion changes. The data have been calculated for the representative torsions and are summarized in Table 9 for L,L-Cystine and in Table 10 for D,L-Cystine, respectively. It can be seen that a smaller number of torsional changes occure in the L,L–form than in the D,L–form. No significant correlation between the torsional changes is observed for

Table 9. *Matrix of correlated movement for L,L-Cystine [a] for 1000000 steps simulation at 300 K.*

Tors.	No. of changes [b]	ϕ_1	ψ_1	χ_1	χ_2	χ_3	χ_{22}	χ_{21}	ϕ_2	ψ_2
ϕ_1	48967	100.0	5.7	0.3	1.4	0.0	1.2	0.0	0.0	0.0
ψ_1	134122	2.1	100.0	0.2	0.7	0.0	0.4	0.0	0.0	0.2
χ_1	1522	8.9	15.8	100.0	2.3	0.0	8.4	0.0	0.1	2.0
χ_2	6899	9.9	13.6	0.5	100.0	0.0	0.9	1.4	4.8	9.3
χ_3	3	0.0	0.0	0.0	0.0	100.0	100.0	100.0	66.7	100.0
χ_{22}	4876	12.0	11.0	2.6	1.3	0.1	100.0	1.6	3.4	12.3
χ_{21}	1556	0.1	0.7	0.0	6.4	0.2	4.9	100.0	8.3	23.3
ϕ_2	42860	0.0	0.0	0.0	0.8	0.0	0.4	0.3	100.0	6.6
ψ_2	141604	0.0	0.2	0.0	0.5	0.0	0.4	0.3	2.0	100.0

[a] and [b] see Table 10.

Table 10. *Matrix of correlated movement for D,L-Cystine [a] for 1,000,000 steps simulation at 300 K.*

Tors.	No. of changes [b]	Torsions								
		ϕ_1	ψ_1	χ_1	χ_2	χ_3	χ_{22}	χ_{21}	ϕ_2	ψ_2
ϕ_1	67784	100.0	32.3	3.6	3.3	2.6	2.3	3.2	0.7	8.2
ψ_1	166303	13.2	100.0	3.5	3.5	1.6	2.9	3.2	2.0	4.0
χ_1	8711	28.1	67.7	100.0	78.7	34.1	59.4	74.4	24.2	68.8
χ_2	11094	20.2	52.1	61.8	100.0	25.7	47.4	57.3	18.0	55.6
χ_3	3020	58.6	86.8	98.4	94.4	100.0	55.0	99.9	3.6	90.0
χ_{22}	13878	11.3	34.8	37.3	37.9	12.0	100.0	37.3	12.9	39.8
χ_{21}	7973	27.3	66.9	81.3	79.7	37.8	64.8	100.0	27.6	79.7
ϕ_2	75842	0.6	4.5	2.8	2.6	0.1	2.4	2.9	100.0	21.0
ψ_2	152067	3.7	4.4	3.9	4.1	1.8	3.6	4.2	10.5	100.0

[a] Correlation is in percentage. The matrix is not symmetrical as sometimes rotation around one bond induces rotation around another one but not "vice versa".

[b] Energy change larger than 30 degrees in one step is taken into account.

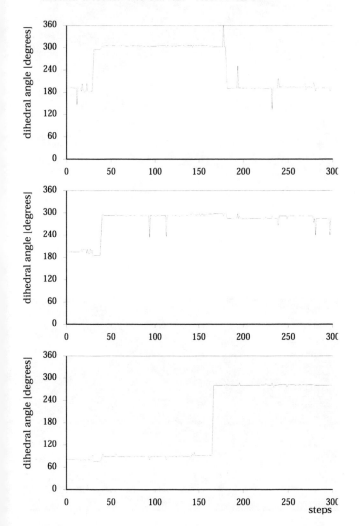

Figure 7. *History of selected dihedral angles during the simulation of conformational movement on L,L-Cystine within the step window 0–300 at the thermodynamic temperature 300 K (χ_1 (upper part), χ_2 (middle part) and χ_3 (bottom part)).*

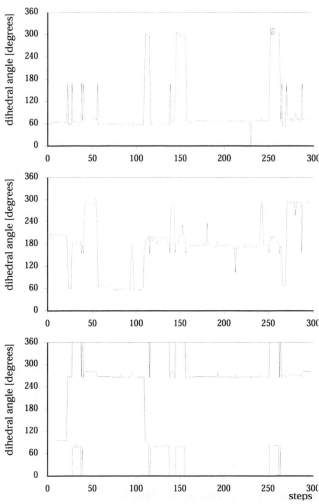

Figure 8. *History of selected dihedral angles during the simulation of conformational motion on D,L-Cystine within the step window 0–300 at the thermodynamic temperature 300 K (χ_1 (upper part), χ_2 (middle part) and χ_3 (bottom part)).*

J. Mol. Model. **1996,** 2

L,L-Cystine (cf. Table 9). A correlation not only between neighboring torsions, but also between relative distant torsions of molecule, e.g., between χ_3 and ψ, has been observed for D,L-Cystine.

This together with the observation of relatively stable conformations with higher relative energy within the simulations, leads to the presumption of the existence of hydrogen bonds in these conformations. We have followed the atom distances during the entire simulation period and found conformations stabilized by intramolecular hydrogen bonds. They are given in Figure 9.

No low energy conformations of L,L-Cystine have been found with intramolecular hydrogen bond.

Conclusions

Conformational potential energy hypersurface, PES, for terminally blocked L–Cysteine, L,L–Cystine and D,L-Cystine have been analyzed by means of molecular mechanics in combination with the programs ROSE, CICADA, PANIC and COMBINE.

- It has been observed that flexibility decreases as the size of the molecule increases. The reason is that the molecule becomes more organized. This feature is not generally observed for all molecules. For example, hydrocarbons exhibit a quite opposite behavior [21]. For molecules of the same size: D,L-Cystine is much more flexible than the structurally better organized L,L–form.
- The results obtained are in good agreement with experimental (X-ray) data and show that PMMX program is well parametrized for peptides.
- The interesting feature of strongly correlated conformational movement has been observed for D,L-Cystine. This phenomenon could be general for the D,L combination and may play a role, i.e. in processes within membranes.
- The above results confirm the ability of the software ROSE, CICADA, PANIC and COMBINE to describe the conformational behavior of flexible molecules.

Acknowledgment: The research has partially been supported by the Grant Agency of the Czech Republic, grant No: 203/94/0522 and by Chem–Consult, Hundhamaren, Norway. This financial support is gratefully acknowledged. The authors thank the Czech Academic Supercomputer Centre in Brno for providing them access to computer facilities. We would also like to thank BIOSYM Technologies, Inc., for providing the Academic licence for the INSIGHT II software.

Supplemetary material MMX input files of all the structures introduced in Tables as well as complete tables of low energy conformers are available in electronic form from the author.

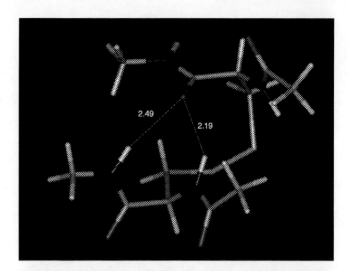

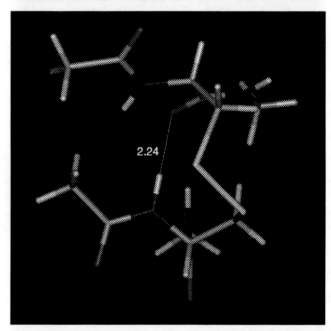

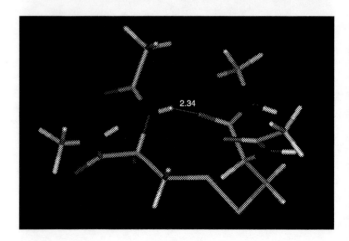

Figure 9. *Conformers with hydrogen bonding found for D,L-Cystine. The numbers indicate lengths of the hydrogen bond (Å).*

References

1. Ramakrishnan, C.; Srinivasan, N.; Sowdhamini, R.; Balaram, P. *Int. J. Pept. Protein Res.* **1990**, *36*, 147-155.
2. Koca, J.; Kríz, Z.; Carlsen, P. H.J. *J. Mol. Struct. (Theochem)* **1994**, *306*, 157.
3. Koca, J.; Carlsen, P. H. J. *J. Mol. Struct. (Theochem)* **1993**, *291*, 271.
4. Görbitz, C. H. J. *Phys. Org. Chem.* **1994**, *7*, 259.
5. Zimmerman, S. S.; Pottle, M. S.; Némethy, G.; Scheraga, H. A. *Macromolecules* **1977**, *10*, 1.
6. Gronet, S.; O'Hair, A. J. *J. Am. Chem. Soc.* **1995**, *117*, 2071.
7. Koca, J.; Carlsen, P. H.J. *J. Mol. Struct. (Theochem)* **1992**, *257*, 131.
8. Koca, J.; Carlsen, P. H.J. *J. Mol. Struct. (Theochem)* **1994**, *308*, 13.
9. Gajewski, J. J.; Gilbert, K. E. MMX is an extended and improved version of Allinger's MM2. The program is available from Serena software, Box 3076, Bloomington, IN 47402-3076.
10. Wolfe, S.; Weaver, D. F.; Yang, K. *Can. J. Chem.* **1988**, *66*, 2687.
11. Koca, J.; Carlsen, P. H. J. The package including ROSE, DAISY and PANIC is available from CHEM- CONSULT, Bautaveien 4, N-7562 Hundhamaren, Norway.
12. Koca, J. *J. Mol. Struct. (Theochem)* **1995**. in press.
13. Koca, J. *Theor. Chim. Acta* **1991**, *80*, 29.
14. Koca J. *Theor. Chim. Acta* **1991**, *80*, 51.
15. Koca, J. unpublished software.
16. Kríz, Z. unpublished software.
17. Koca, J. *J. Mol. Struct.* **1993**, *291*, 255.
18. Koca, J.; Carlsen, P. H.J. *J. Mol. Struct. (Theochem)* **1992**, *257*, 105.
19. Koca, J.; Pérez, S.; Imberty, A. *J. Comp. Chem.* **1995**, *16*, 296.
20. Görbitz, C. H. *Acta Chem. Scand.* **1990**, *44*, 584.
21. Czernek, J.; Koca, J. unpublished results **1995**.

J. Mol. Model. **1996**, 2, 62 - 69

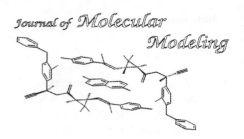

A Semiempirical Transition State Structure for the First Step in the Alkaline Hydrolysis of Cocaine.
Comparison between the Transition State Structure, the Phosphonate Monoester Transition State Analog, and a Newly Designed Thiophosphonate Transition State Analog

Edward C. Sherer, Gordon M. Turner, Tricia N. Lively, Donald W. Landry†,* and George C. Shields*

Department of Chemistry, Lake Forest College, Lake Forest, Illinois 60045 USA (george@sgbq.bq.ub.es)

† Department of Medicine, Columbia University College of Physicians and Surgeons, New York, NY 10032 USA

Received: 30 January 1996 / Accepted: 29 March 1996 / Published: 19 April 1996

Abstract

Semiempirical molecular orbital calculations have been performed for the first step in the alkaline hydrolysis of the neutral benzoylester of cocaine. Successes, failures, and limitations of these calculations are reviewed. A PM3 calculated transition state structure is compared with the PM3 calculated structure for the hapten used to induce catalytic antibodies for the hydrolysis of cocaine. Implications of these calculations for the computer-aided design of transition state analogs for the induction of catalytic antibodies are discussed.

Keywords: cocaine, electrostatic potential, ester hydrolysis, molecular orbital calculations, transition state, catalytic antibodies, transition state analogs

Introduction

The rational production of antibodies able to catalyze a particular reaction was first described by Lerner and Benkovic, [1, 2] and by Schultz, [3] in 1986. Their insight stemmed from a recognition of a structural similarity between an enzyme transforming a substrate and an antibody binding to a molecule that resembles the transition state structure of the substrate. [4] Such an antibody was found to stabilize the transition state over the ground state structure and lower the energy of activation for the indicated reaction. To elicit a catalytic antibody, the molecule that mimics the transition state, the so-called transition state analog (TSA), must be rendered immunogenic by being tethered as a hapten to a larger "carrier" molecule such as bovine serum albumin. Only a subpopulation of anti-analog antibodies will be catalytic and, in order to analyze the individual antibodies of a polyclonal immune response, monoclonal antibodies must be prepared and purified. The art of generating catalytic antibodies lies in the design of the hapten which must be chemically stable yet mimic the structural and electronic properties of the transition state of the indicated reaction.

One of us (DWL) has generated a catalytic antibody for the hydrolysis of the benzoylester of cocaine [5] and recently reported an improved catalyst that enhances the rate by 23,000-fold. [6] The hapten used for the generation of this catalytic antibody is based on the phosphonate monester structure, a structure used extensively for eliciting artificial esterases. [1, 3, 7, 8] In general, researchers designing catalytic antibodies for ester hydrolysis have focused on the general structure of the tetrahedral anionic intermediate as a model for the phosphonate hapten. [1, 3, 7, 8] The formation of the anionic intermediate is the first step in the alkaline hydrolysis of esters [9, 10 see Fig 1], and this is the rate-limiting step in solution. [11-13]

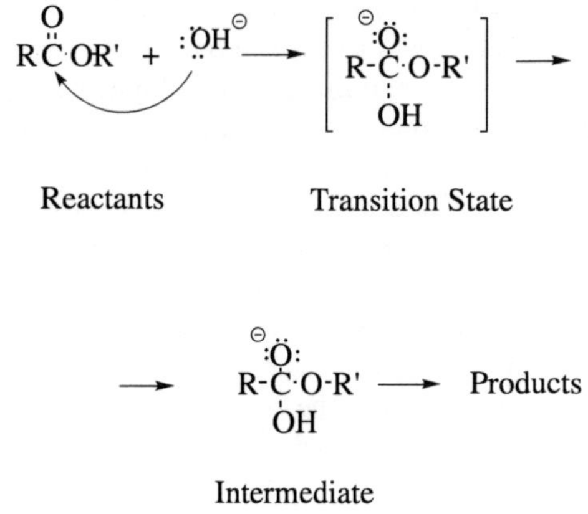

Reactants   Transition State

Intermediate

Figure 1. *The first step in the alkaline hydrolysis of esters.*

In our previous work extensive semiempirical and *ab initio* computations were performed to model the first step in the alkaline hydrolysis of the model ester systems, methyl acetate and methyl benzoate. [14, 15] By comparison with high level *ab initio* calculations at the HF and MP2 levels, it was found that the PM3 method showed promise for modeling transition state structures for ester hydrolysis. Other group's efforts to use quantum mechanics to calculate transition states in order to aid in understanding catalytic antibody work include studies on the chorismate-prephenate rearrangement, [16] the Diels-Alder reaction, [17] the aldol reaction, [18] and the ene reaction between malieimide and 1-butene. [19]

In this article the results of semiempirical calculations are presented for the alkaline hydrolysis of the neutral cocaine benzoylester, for the phosphonate hapten that has been used to generate a catalytic antibody that catalyzes cocaine hydrolysis, [5, 6] and for a newly designed thiophosphonate analog.

Method

The AM1, [20] MNDO, [21] and PM3 [22] semiempirical molecular orbital methods, implemented within the SPARTAN software package, [23] have been used to search for the transition state for the alkaline hydrolysis of cocaine. Starting structures were based on previous work, [14, 15] and careful searches for the transition state structure were made with each method. Because of the difficulties in obtaining a transition state structure, the methodology used in our calculations will be described in the Discussion section of this paper. In addition the energetics of the critical first step were evaluated using the SM3 method [24] and compared with the methyl acetate and methyl benzoate systems.

All geometry-optimized structures were characterized as stationary points, and as minima or maxima, at all levels of theory in this study. The display of normal modes of vibration within SPARTAN allowed verification that the calculated transition structure connected reactants with the anionic tetrahedral intermediate. Molecular electrostatic potentials (MEPs) were calculated using the NDDO approach within SPARTAN. All calculations were carried out on Silicon Graphics Indy and Indigo² workstations at Lake Forest College.

Results

The critical first step in the alkaline hydrolysis of esters is presented in Figure 1. Figure 2 displays the PM3 calculated transition state structure that connects the reactants and tetrahedral intermediate (see Figure 1). Figure 3 displays a PM3 calculated phosphonate analog that was the basis for the hapten used by Landry and co-workers to induce catalytic antibodies. [5, 6] The MEPs, superimposed on the electron density, for the transition state structure and the phosphonate analog are shown in Figure 4. The molecular framework of each structure is visible beneath the MEPs. Red indicates regions with the most negative electrostatic potential while blue indicates regions with the least negative electrostatic potential. Figure 5 shows the neutral cocaine ester and its highest occupied molecular orbital (HOMO). Cocaine and its lowest occupied molecular orbital (LUMO) are displayed in Figure 6. The thiophosphonate transition state analog, and the MEP superimposed on the electron density, are displayed in Figure 7. Table 1 contains the energetics for the reaction

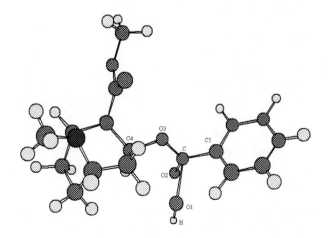

Figure 2. *The PM3 calculated transition state structure for the first step in the alkaline hydrolysis of the neutral cocaine benzoylester.*

displayed in Figure 1, for cocaine and the previously determined ester systems. [14, 15] Geometric parameters for the transition state structure and the phosphonate hapten are listed in Table 2. Close contacts are shown for the carbon that is being attacked by the hydroxyl anion in the transition state structure and for the phosphate in the hapten molecule.

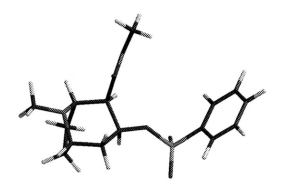

Figure 3. *The PM3 calculated phosphonate analog, PA2, the hapten used to induce catalytic antibodies (ref 5,6).*

Discussion

Transition State Structure (TS)

Previous computational work on the hydrolysis of the esters methyl acetate and methyl benzoate guided our search for the PM3 transition state structure (TS) for the hydrolysis of cocaine. [14, 15] For methyl acetate, the PM3 method located two transition states, one with a linear geometry and one with a bent geometry. The linear and bent geometries were characterized by the C (acetate methyl carbon)-C (acetate ester carbon)-O -C (methyl carbon) torsion angle of the transition state structure. Hartree-Fock (HF) *ab initio* calculations at the 3-21G/HF and 3-21+G/HF levels, starting from the two PM3 transition state structures, converged on a linear TS. Compared with the 3-21+G/HF structure, the linear PM3 TS was superior to the 3-21G/HF TS. [14] For methyl benzoate, the PM3 method found one linear TS. [15] AM1 calculations were not successful for obtaining a reasonable TS for either model system.

PM3 calculations were performed on the neutral cocaine ester, the hydroxide anion, and the intermediate formed after the first step in the alkaline hydrolysis of the cocaine ester. The anionic tetrahedral intermediate had a heat of formation of -216 kcal/mol, a C (acetate ester carbon)—O (hydroxyl oxygen) reaction bond distance of 1.414 Å, and a C-C-O-C torsion angle of 133.9°. The procedure that was successful for finding a PM3 TS was obtained through five steps. In step one the intermediate was used for a constrained transition state optimization, where the C—O reaction bond distance was subject to the constraint of being fixed to 2.30 Å. In addition the converge option was selected to aid SCF convergence. The output file for this calculation revealed that after 18 cycles of constrained transition state optimization, the gradient dropped to 0.007 a.u. and the heat of formation climbed to -165.4 kcal/mol. Further cycles increased the gradient. In the second step the same procedure was repeated, but reduced to 18 optimization cycles, to obtain the lowest gradient/highest heat of formation structure. In the third step a frequency calculation was performed on the 18 cycle output structure, by numerical differentiation of analytical gradients using central differences. One imaginary frequency was found (-360 cm^{-1}) and examination of this normal mode revealed that this vibration connected the cocaine ester and hydroxyl anion reactants with the tetrahedral anionic intermediate. In step four the Hessian matrix, wavefunction, and structure from the previous step were used for a transition state optimization, again with the converge option but with the C—O constraint removed. The output structure from this final optimization converged, had a heat of formation of -183.2 kcal/mol, and was used for a frequency calculation in step five. One imaginary frequency was obtained (-273 cm^{-1}) and examination of this mode revealed that this was the correct TS for this reaction. Furthermore, geometry optimizations performed from the TS, purturbed slightly along the transition state mode (first contracted, and second expanded), led first to the stable cocaine intermediate and second to the initial ester and hydroxyl anion reactants. The final TS is displayed in Figure 2.

This structure was used as the input structure for AM1 and MNDO transition state calculations. In addition the same strategy outlined above was used starting from the AM1 and MNDO minimized tetrahedral intermediates. Also, the linear synchronous transition (LST) method was used within SPARTAN to generate many different starting models for AM1 and MNDO transition state optimization, with the same results as described previously. [14] Efforts with these other semiempirical methods were not successful.

AM1 vs PM3

Failure of the MNDO method to find a TS is not surprising, given the known overestimation of the repulsive core-core potentials. [21,25] In order to reduce the repulsive forces between atoms separated by van der Waals distances, Dewar and co-workers modified the core repulsion function (CRF) by adding two to four parameterized Gaussian terms for each atom. This did indeed reduce the repulsive forces and has been found to be more useful for finding transition state structures than MNDO. In the PM3 Hamiltonian, two Gaussians were added per element in the CRF, and all of the Gaussian parameters were optimized within the PM3 framework. [22] The central difference between AM1 and PM3 is the parameterization. Why then is it possible for PM3 calculations to find transition states for the first step in the alkaline hydrolysis of cocaine ester, methyl acetate, [14] and methyl benzoate, [15] while AM1 calculations fail for these same systems? The answer does seem to reside in the difference between the parameterizations. It has been demonstrated previously that the PM3 parameterization is the reason why PM3 is better than AM1 for modeling intermolecular hydrogen bonding between neutral molecules, a consequence of the reduction of the overall repulsive forces between atoms by the parameterization of the attractive and repulsive Gaussian terms added to the CRF. [25] While the MNDO Hamiltonian drastically overestimates repulsions between atoms separated by van der Waals distances, and the AM1 Hamiltonian slightly overestimates repulsions between non-bonded atoms, the PM3 Hamiltonian underestimates repulsions between non-bonded atoms. This can be seen by the slight underestimation of PM3 determined hydrogen bond lengths [25] and by the formation of close atom-atom contacts such as in the methane dimer, where the H,H separation distance is 1.72 Å. [26]

There are other examples where the PM3 method allows non-bonded atom-atom distances to be too small. [13, 14, 26] The methyl benzoate TS has a hydrogen that is covalently attached to a carbon on the benzoyl ring that is 1.65 Å from the incoming hydroxyl oxygen. Likewise, for the cocaine transition state structure shown in Figure 2, the same nearest hydrogen from the benzoyl ring is separated by 1.709 Å from the incoming hydroxyl oxygen. In addition the closest hydrogen on the tropane ring is 1.736 Å away from the incoming hydroxyl oxygen. These close contacts stabilize the PM3 TS, but are repulsive interactions within the MNDO and AM1 Hamiltonians. Figure 2 shows a view of the cocaine ester transition state where the closeness of the two hydrogen atoms to the hydroxyl anion is apparent. Thus the PM3 method is successful for obtaining transition state structures for alkaline ester hydrolysis because of the stabilizing influence of non-bonded atoms, a consequence of the parameterization.

It is of interest to compare small basis set ab initio HF calculations on transition state structures of the methyl acetate system. We were not able to obtain a TS for methyl acetate using STO-3G/HF theory. Therefore the use of STO-

Table 1. *Energetics for the first step in the alkaline hydrolysis of neutral cocaine, methyl acetate (ref. 14),and methyl benzoate (ref. 15) esters (see Figure 1). Gas phase results were obtained from the PM3 geometry-optimized structures. Solution results were obtained by SM3 single point calculations on the gas phase geometries. All energies in kcal/mol.*

System	E_A (gas phase)	E_A (solution)	ΔH_{rxn} (gas phase)	ΔH_{rxn} (solution)
methyl acetate	-15.6	23.4	-50.6	2.6
methyl benzoate	-22.3	21.8	-55.0	0.6
cocaine ester	-28.5	24.6	-61.4	0.4

3G/HF methods for larger ester TS structures will probably not be useful. The 3-21G/HF transition state structure has two hydrogen—oxygen close contacts. The methyl hydrogen closest to the hydroxyl oxygen is separated by a distance of 1.827 Å while the closest acetate methyl hydrogen is separated by a distance of 2.009 Å. By comparison these same separation distances are 2.301 and 2.459 Å from 3-21+G/HF results and 2.44 and 1.83 Å from PM3 calculated structures. Thus improving the PM3 cocaine transition state structure will require ab initio Hartree-Fock geometry optimizations that include diffuse functions. For the cocaine system, using Gaussian 94 [28] on an Indigo2 R4400 processor, one SCF calculation on the cocaine anion intermediate at the 3-21G/HF level requires 41.5 CPU minutes, and one SCF calculation at the 3-21+ G/HF level requires 11.3 CPU hours. The fact that the PM3 structures are as good or better than the 3-21G/HF structures gives confidence in the use of the PM3

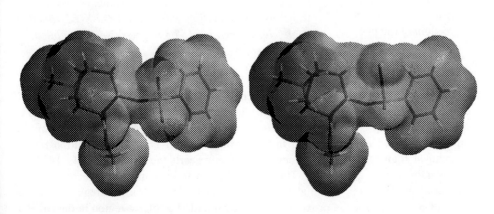

Figure 4. *The PM3 calculated molecular electrostatic potential (MEP), mapped onto the transparent electron density surface, and displayed for the transition state structure and the phosphonate analog. The molecular framework of each molecule is visible through the MEPs. The phosphonate analog, PA2, is displayed on the right side of the figure. Red indicates areas with the most negative electrostatic potential.*

TS structure for modeling alkaline hydrolysis of the neutral cocaine molecule.

Energetic comparison of cocaine and smaller ester systems

Table 1 contains the PM3 estimates for the activation energies and enthalpies of reaction for the first step in the alkaline hydrolysis of the three ester systems. Single point calculations were made (SM3//PM3) as our previous work revealed that SM3 geometry optimization did not result in substantial differences in these estimates. [14, 15] The observed activation energies for alkaline hydrolysis of esters in solution range from 12 to 15 kcal/mol. [29, 30] The large cocaine benzoylester has a negative activation energy for gas phase hydrolysis, which changes to a positive barrier in solution, a direct result of the requirement to desolvate the hydroxide anion in solution. The dramatic lowering of the energy of the hydroxyl anion in solution also changes ΔH_{rxn} for the first step in alkaline hydrolysis, from a large exothermic reaction in the gas phase to approximately thermal-neutral in solution. The PM3 and SM3 values in Table 1 for cocaine parallel those for methyl acetate and methyl benzoate. These estimates reveal that the benzoyl ester of cocaine is not a difficult ester to catalyze, and it should be possible to achieve large rates of catalysis if a catalytic antibody can be made from an appropriate TSA.

Comparison between the PM3 transition state structure and the phosphonate analog hapten

Table 2 shows the structural parameters for the transition state structure (TS) and two phosphonate analogs. The first analog, PA1, is the global minimum on the PM3 potential energy surface. The second analog, PA2 (Figure 3), was obtained by constraining the C4-O3-P-C5 torsion angle to -155.3°, the same value as the transition structure. This structure is just 0.4 kcal/mol higher in energy than PA1. Geometry optimization of PA2, after removing the torsion angle constraint, yields the PA1 minimum. The two phosphonate analogs have similar bond distances and bond angles. The RMS deviation between PA1 and PA2 is 0.530 Å. The RMS deviation between TS and PA1 is 1.127 Å and between TS and PA2 is 1.097 Å. The conformational freedom of the phenyl ring results in the production of many antibodies. Antibodies generated by the phosphonate TSA were originally screened to find two that had significant effects on hydrolysis. [5] As PA2 is the PM3 structure that is most faithful to the PM3 transition state structure, the MEPs of these two structures have been compared in Figure 4.

The TS MEP has its maximum negative potential around the hydroxyl oxygen (-189.8 kcal/mol, reddest region), a negative potential of approximately -151 kcal/mol around O2, and its least negative region of potential about the central carbon region on the hydroxyl hydrogen (-89 kcal/mol).

Table 2. *Comparison of the PM3 geometries of a transition state structure for the first step in the alkaline hydrolysis of the neutral cocaine benzoyl ester and the phosphate analog used to induce catalytic antibodies for cocaine hydrolysis.*

PARAMETER [a]	STRUCTURE [b]		
	TS	PA1	PA2
Bond Distance (Å)			
O3-C4	1.426	1.379	1.378
X-O3	1.387	1.792	1.792
X-C5	1.502	1.908	1.908
X-O2	1.222	1.501	1.497
X-O1	2.313	1.498	1.501
O1-H	0.939	NA	NA
Bond Angle (degrees)			
C4-O3-X	117.3	118.0	119.6
O3-X-O2	118.7	106.4	106.3
O3-X-O1	95.5	106.7	107.2
X-O1-H	102.0	NA	NA
O2-X-O1	96.6	123.6	123.7
O2-X-C5	125.9	110.4	110.6
O1-X-C5	96.3	110.4	110.2
O3-X-C5	111.9	95.5	94.6
Torsion Angle (degrees)			
C4-O3-X-C5	-155.3	171.7	-155.3 [c]
H-O1-X-C5	-115.6	NA	NA
H-O1-X-O2	11.7	NA	NA
C4-O3-X-O2	44.5	58.5	91.7
C4-O3-X-O1	-56.2	-75.1	-42.4

[a] X = C for TS, P for PA1, PA2

[b] TS is the transition state structure, PA1 is the global minimum transition state analog structure, PA2 is the transition state analog with the C4-O3-X-C5 dihedral angle constrained to -155.3 degrees, NA not applicable.

[c] constrained to -155.3 degrees.

The ester oxygen has a potential of approximately -140 kcal/mol while the bluest region in the TS MEP has a potential of -30.28 kcal/mol. PA2 has its most negative potential on the oxygen farthest from the ester O3 (-183.5 kcal/mol), with the other oxygen having nearly the same potential (-182 kcal/mol). The ester O3 atom is surrounded by electrostatic potential, with the most negative value in this region being approximately -152 kcal/mol. The bluest region in the MEP of

Figure 5. *The PM3 calculated neutral cocaine ester and its highest occupied molecular orbital.*

PA2 has a value of -31.4 kcal/mol. It is apparent that the phosphonate analog is more symmetric than the TS, and a clear strategy to improve the fidelity of the TSA to the TS is to design a TSA with asymmetric charge about the central atom. An example of this type of approach will be given in the next section.

Implications for the design of better transition state analogs

The TS for a gas phase reaction is usually different from the TS for the same reaction in solution and different from the same reaction catalyzed by an enzyme. Catalytic antibodies are induced by a TSA that resembles the TS, yet since a single TSA can elicit a multi-step catalytic activity it can be argued that the absolute fidelity of the TSA to the TS of an uncatalyzed reaction is not essential. For instance, Benkovic et al reported an interesting series of experiments where a phosphonamide TSA was used to induce a catalytic antibody (NPN43C9) that catalyzed both p-nitrophenylester hydroly-

sis and p-nitroanilide amide cleavage. [31] Two points emerge from this work that are relevant to this discussion. First, the ester was catalyzed in a multi-step process, with a striking resemblance to the serine protease pathway. Both NPN43C9 and the serine proteases use a series of steps around an isoenergetic intermediate to achieve ester or amide cleavage. Second, the ultimate limit of ester hydrolysis was the rate of product dissociation from the catalytic antibody. This second point reveals that an important factor in hapten design should be to ensure that the TSA does not retain all the prominent structural features of the products, avoiding products that bind tightly to the catalytic antibody. Indeed, it has been shown that an analog that merely incorporates a relatively higher energy conformer is sufficient to produce efficient catalytic antibodies. [32]

Clearly there are many factors involved in the formation and the inner workings of catalytic antibodies that need to be understood. The TSA is in an aqueous environment, and water must be displaced from the TSA for antibody induction. Thus the gas phase TSA, and by extension the gas phase TS, appear to be reasonable starting points for the computational problem of TSA design. The use of MEPs as a design aspect for modeling the TSA from the TS has precedent. A recent study of a TSA that is an inhibitor of an enzyme reveals that MEPs provide a clear explanation for the inhibition in this system. [33] In this investigation, the inhibitor formycin 5'-phosphate binds to the AMP nucleosidase enzyme more tightly than the AMP substrate. Comparison of the MEPs of the inhibitor and substrate show a clear change in the electrostatic potential, a change that is matched by the MEP for the AM1 calculated transition state. The transition state for AMP hydrolysis is characterized by new positive electrostatic

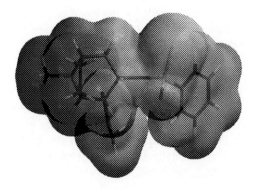

Figure 7. *The PM3 calculated molecular electrostatic potential (MEP), mapped onto the transparent electron density surface, and displayed for the computationally designed thiophosphonate analog. The molecular framework is visible through the MEP. In this analog one of the oxygens on the traditional phosphonate TSA has been replaced through sulfur.*

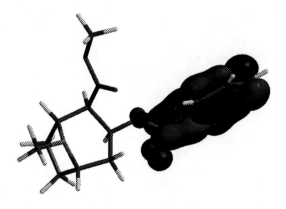

Figure 6. *The PM3 calculated neutral cocaine ester and its lowest occupied molecular orbital.*

potential in the adenine ring as a result of protonation by the enzyme. This is closely matched by the protonated pyrazole ring of formycin 5'-phosphate at position N7. The MEPs provide a clear explanation for the inhibition of AMP nucleosidase by the formycin 5'-phosphate TSA. This work illustrates that the use of MEPs of transition state structures to aid in the design of haptens is an idea that should be pursued.

The fact that a given TSA, such as the one used in the study by Benkovic et al, catalyzes a different reaction in a multi-step fashion highlights the complexity of antibody catalysis. One way that computational chemists can help is to use quantum mechanics to design rigorous TSAs, based on a property such as the MEP that is obtainable from the wave function, and then work with experimentalists to try and correlate the results with the quantum mechanical property. It remains an open question as to whether a rigorous TSA will generate a superior catalytic antibody, or whether methods such as bait and switch, [34, 35] or heterologous immunization, [36] are the optimal ways of generating antibody catalysts. An example of a designed TSA is the thiophosphonate analog displayed in Figure 7. In this analog one of the oxygens on the traditional phosphonate TSA has been replaced with sulfur. The effect of this substitution on the MEP is pronounced; the thiophosphonate MEP is asymmetric and appears to be a better match with the MEP of the TS structure. Differences remain, however, and work in our laboratories directed towards a quantitative comparison of the MEPs is underway. The thiophosphonate analog has been synthesized for use as a hapten for the induction of catalytic antibodies. Experimental work to characterize the molecular structure of this analog and its ability to induce catalytic antibodies is in progress. In addition we will pursue ab initio calculations to test the ability of PM3 to model this system and to compare the use of ab initio and semiempirical methods for modeling the crystal structures of the analogs.

The second point made by Benkovic et al was that the ultimate limit in ester hydrolysis in their multi-pathway system was the rate of product dissociation from the catalytic antibody. By comparing the MEPs of TSAs that do not structurally resemble the products with the MEPs of the TS, it may be possible to design haptens that are accurate TSAs yet do not contain the structural elements of the products. This in turn may lead to a catalytic antibody that has less product inhibition. This type of approach has been used by Miyashita et al to generate catalytic antibodies for the hydrolysis of a nonbioactive chloramphenicol monoester. [32] These ideas need to be tested through the interplay of computational predictions and experimental results using the computationally designed haptens. As an example of a way to use calculations to try and avoid slow product dissociation, consider the possibility of intramolecular catalysis. Figures 5 and 6 show that the PM3 HOMO is delocalized on the tropane nitrogen and most of the tropane ring, while the PM3 LUMO is delocalized over the benzoyl ester and the benzoyl ring for the neutral cocaine free base. Thus the electrons in the HOMO

on the tropane N are poised for an attack on the LUMO, provided that the cocaine molecule can adopt a higher energy conformation that brings the tropane nitrogen and the benzoyl carbon into close proximity. Calculations can explore the conformational space of cocaine, find a TS for intramolecular catalysis, and model potential TSAs. As the TSAs for this system correspond to a high energy conformation of cocaine, it would be expected that there would be much less product inhibition of the cocaine molecule (Figs 5 & 6) with the induced catalytic antibodies for intramolecular catalysis, similar to previous work. [32]

Conclusions

The neutral cocaine ester system is large and an anionic semiempirical transition state structure was only found for the PM3 method, a consequence of the parameterization of the method that underestimates non-bonded atom repulsion. Nevertheless, improvement on this anionic structure through ab initio calculations is going to require transition state optimization at the 3-21+G/HF level, as diffuse functions are essential for the proper description of ester hydrolysis [14, 15]. The PM3 calculated phosphonate transition state analog has an electrostatic potential comparable to the values for the PM3 calculated transition state structure, but the analog potential is more symmetric about the central P, while the transition state potential is asymmetric about the central carbon atom. A newly designed thiophosphonate analog highlights the potential of the interplay between experiment and theory to enhance understanding of induction of catalytic antibodies by transition state analogs. The PM3 method has potential for the initial design of different analogs, and the experimental results obtained from using the computationally designed analogs should help unravel the most important quantum mechanical features in hapten design.

Acknowledgement. This work was supported by Lake Forest College, an NIH AREA grant , and an NSF ILI grant (GCS). This work was also supported by the Executive Office of the President of the United States, Office of National Drug Control Policy (DWL). We thank Gary Nalley for technical assistance. GCS acknowledges a sabbatical award from the Spanish Ministry of Education and Science and gratefully appreciates the hospitality of Professors Modesto Orozco and Javier Luque at the University of Barcelona while writing this manuscript.

Supplementary Materail Available: PDB files for all species are attached to this publication.

cocaine.pdb: PM3 geometry-optimized structure of neutral cocaine.
cocaine_intermediate.pdb: PM3 geometry-optimized structure for the anionic intermediate formed in the critical first step in the alkaline hydrolysis of cocaine.

pa1.pdb: PM3 geometry-optimized structure of the phosphonate TSA used to induce antibodies for the hydrolysis of cocaine.

pa2.pdb: PM3 geometry-optimized structure for the phosphonate analog used to induce catalytic antibodies for the hydrolysis of cocaine, with the C-O-P-C torsion angle constrained to the same value as the PM3 calculated transition state, −155.3°.

thio_phos_analog.pdb: PM3 geometry-optimized structure for the designed thiophosphonate analog. This analog has a MEP closer to the MEP for the PM3 TS structure than either of the phosphonate analogs, PA1 and PA2.

ts.pdb: PM3 geometry-optimized transition state structure for the critical first step in the alkaline hydrolysis of cocaine.

References

1. Tramontano, A.; Janda, K.D.; Lerner, R.A. *Science* **1986** *234*, 1566-1570.
2. Benkovic, S.J. *Annu. Rev. Biochem.* **1992** *61*, 29-54.
3. Pollack, S.J.; Jacobs, J.W.; Schultz, P.G. *Science* **1986** *234*, 1570-1573.
4. Lerner, R.A.; Benkovic, S.J.; Schultz, P.G. *Science* **1991** *252*, 659-667.
5. Landry, D.W.; Zhao, K.; Yang, G.X.-Q.; Glickman, M.; Georgiadis, T.M. *Science* **1993** *259*, 1899-1901.
6. Yang, G.; Chun, J.; Arakawa-Uramoto, H.; Gawinowicz, M.A.; Zhao, K.; Landry, D.W., *J. Am. Chem. Soc.* in press.
7. Tramontano, A.; Ammann, A.A.; Lerner, R.A. *J. Am. Chem. Soc.* **1988** *110*, 2282-2286.
8. Janda, K.D.; Benkovic, S.J.; Lerner, R.A. *Science* **1989** *244*, 437-440.
9. Bender, M.L.; Heck, H.d'A. *J. Am. Chem. Soc.* **1967** *89*, 1211-1220.
10. Shain, S.A.; Kirsch, J.F. *J. Am. Chem. Soc.* **1968** *90*, 5848-5854.
11. O'Leary, M.H.; Marlier, J.F. *J. Am. Chem. Soc.* **1979** *101*, 3300-3306.
12. Marlier, J.F. *J. Am. Chem. Soc.* **1993** *115*, 5953-5956.
13. Dewar, M.J.S.; Storch, D.M. *J. Chem. Soc. Chem. Commun.* **1995** 94-96.
14. Sherer, E.C.; Turner, G.M.; Shields, G.C. *Int. J. Quantum Chem., Quantum Biol. Symp.* **1995** *22*, 83-93.
15. Turner, G.M.; Sherer, E.C.; Shields, G.C. *Int. J. Quantum Chem., Quantum Biol. Symp.* **1995** *22*, 103-112.
16. Wiest, O.; Houk, K.N. *J. Org. Chem.* **1994** *59*, 7582-7584.
17. Gouverneur, V.E.; Houk, K.N.; de Pascual-Teresa, B.; Beno, B.; Janda, K.D.; Lerner, R.A. *Science* **1993** *262*, 204-208.
18. Teraishi, K.; Saito, M.; Fujii, I.; Nakamura, H. *Tetrahedron Lett.* **1992** *33*, 7153-7156.
19. Yliniemela, A.; Konschin, H.; Neagu, C.; Pajunen, A.; Hase, T.; Brunow, G.; Teleman, O. *J. Am. Chem. Soc.* **1995** *117*, 5120-5126.
20. Dewar, M.J.S.; Thiel, W. *J. Am. Chem. Soc.* **1977** *99*, 4899-4907; 4907-4917.)
21. Dewar, M.J.S.; Zoebisch, E.G.; Healy, E.F.; Stewart, J.J.P. *J. Am. Chem. Soc.* **1985** *107*, 3902-3989.
22. Stewart, J.J.P. *J. Comp. Chem.* **1989** *10*, 209-221; 221-264.
23. SPARTAN version 3.1, Wavefunction, Inc., 18401 Von Karman Ave., #370, Irvine, CA 92715 USA.
24. Cramer, C.J.; Truhlar, D.G. *J. Comp. Chem.* **1992** *13*, 1089-1097.
25. Jurema, M.W.; Shields, G.C. *J. Comp. Chem.* **1993** *14*, 89-104.
26. Buß, V.; Messinger, J.; Heuser, N. *QCPE Bull.* **1991** *11(1)*, 5.
27. Kirschner, K.N.; Shields, G.C. *J. Mol. Struc. (THEOCHEM)* **1996**, *362*, 297 - 304.
28. Gaussian 94 (revision B.1), M.J. Frisch, G.W. Trucks, H.B. Schlegel, P.M.W. Gill, B.J. Johson, M.A. Robb, J.R. Cheeseman, T.A. Keith, G.A. Petersson, J.A. Montgomery, K. Raghavachari, M.A. Al-Laham, V.G. Zakrzewski, J.V. Ortiz, J.B. Foresman, J. Cioslowski, B.B. Stefanov, A. Nanayakkara, M. Challacombe, C.Y. Peng, P.Y. Ayala, W. Chem, M.W. Wong, J.L. Andres, E.S. Replogle, R. Gomperts, R.L. Martin, D.J. Fox, J.S. Brinkley, D.J. Defrees, J. Baker, J.J.P. Stewart, M. Head-Gordon, C. Gonzales, and J.A. Pople, Gaussian, Inc., Pittsburgh, PA 1995.
29. Amis, E.S.; Siegel, S. *J. Am. Chem. Soc.* **1950** *72*, 674-677.
30. Rylander, P.N.; Tarbell, D.S. *J. Am. Chem. Soc.* **1950** *72*, 3021-3025.
31. Benkovic, S.J.; Adams, J.A.; Borders, Jr., C.L.; Janda, K.D.; Lerner, R.A. *Science* **1990** *250*, 1135-1139.
32. Miyashita, H.; Karaki, Y.; Kikuchi, M.; Fujii, I. *Proc. Natl. Acad. Sci. USA* **1993** *90*, 5337-5340.
33. Ehrlich, J.I.; Schramm, V.L. *Biochem.* **1994** *33*, 8890-8896.
34. Janda, K.D.; Weinhouse, M.I.; Schloeder, D.M.; Lerner, R.A.; Benkovic, S.J. *J. Am. Chem. Soc.* **1990** *112*, 1274-1275.
35. Janda, K.D.; Weinhouse, M.I.; Danon, T.; Pacelli, K.A.; Schloeder, D.M. *J. Am. Chem. Soc.* **1991** *113*, 5427-5434.
36. Suga, H.; Ersoy, O.; Tsumuraya, T.; Lee, J.; Sinskey, A.J.; Masamune, S. *J. Am. Chem. Soc.* **1994** *116*, 487-494.

J. Mol. Model. **1996**, 2, 70 - 102

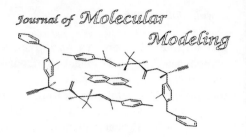

Davydov Soliton Dynamics in Proteins: I. Initial States and Exactly Solvable Special Cases

Wolfgang Förner

Chair for Theoretical Chemistry and Laboratory of the National Foundation of Cancer Research at the Friedrich-Alexander University Erlangen-Nürnberg, Egerlandstr. 3, D-91058 Erlangen, FRG. (foerner@pctc.chemie.uni-erlangen.de)

Received: 10 October 1995 / Accepted: 25 March 1996 / Published: 10 May 1996

Abstract

For the Davydov Hamiltonian several special cases are known which can be solved analytically. Starting from these cases we show that the initial state for a simulation using Davydov's $|D_1>$ approximation has to be constructed from a given set of initial lattice displacements and momenta in form of a coherent state with its amplitudes independent of the lattices site, corresponding to Davydov's $|D_2>$ approximation. In the $|D_1>$ ansatz the coherent state amplitudes are site dependent. The site dependences evolve from this initial state exclusively via the equations of motion. Starting the $|D_1>$ simulation from an ansatz with site dependent coherent state amplitudes leads to an evolution which is different from the analytical solutions for the special cases. Further we show that simple construction of such initial states from the expressions for displacements and momenta as functions of the amplitudes leads to results which are inconsistent with the expressions for the lattice energy. The site-dependence of coherent state amplitudes can only evolve through the exciton-phonon interactions and cannot be introduced already in the initial state. Thus also in applications of the $|D_1>$ ansatz to polyacetylene always $|D_2>$ type initial states have to be used in contrast to our previous suggestion [W. Förner, *J. Phys.: Condens. Matter* **1994**, *6*, 9089-9151, on p. 9105]. Further we expand the known exact solutions in Taylor serieses in time and compare expectation values in different orders with the exact results. We find that for an approximation up to third order in time (for the wave function) norm and total energy, as well as displacements and momenta are reasonably correct for a time up to 0.12-0.14 ps, depending somewhat on the coupling strengh for the transportless case. For the oscillator system in the decoupled case the norm is correct up to 0.6-0.8 ps, while the expectation values of the number operators for different sites are reasonably correct up to roughly 0.6 ps, when calculated from the third order wave function. The most important result for the purpose to use such expansions for controlling the validity of ansatz states is, however, that the accuracy of S(t) and H(t) (constant in time, exact values known in all cases) is obviously a general indicator for the time region in which a given expansion yields reliable values also for the other, physically more interesting expectation values.

Keywords: Proteins, Davydov Model, Special Cases, Expansion of Exact Solutions

Introduction

The most recent and best review of the state of art in Davydov soliton theory was given by Scott [1], the leading expert in the field. The problem which Davydov [2-5] attempted to solve with his mechanism was the storage and transport of energy through protein chains. The energy which is to be transported or stored in biological systems is released by the hydrolysis of adenosinetriphosphate (ATP) molecules which amounts to about 0.4 eV (see [2-5] and [1] for further details and references). In Davydov's opinion the best candidate for storing this energy in proteins is the amide-I vibration, which is essentially of C=O stretch type, because one quantum of this vibration has an energy of 0.205 eV, roughly half of the energy released by ATP hydrolysis. From this starting point Davydov developed his physical model for the energy transport. In α–helical proteins the C=O groups of a turn in the helix form hydrogen bonds to the N-H groups in the following turn. As indicated in the following sketch (see section II) these hydrogen bonds form chains parallel to the helix axis and perpendicular to the covalent backbone. There are always three parallel chains of this kind in an α–helix. Within such a chain the C=O oscillators are coupled via their transition dipole moment with each other, where next neighbor coupling is by far the most important term. This type of coupling is a linear one and makes the system dispersive, i.e. an amide-I vibrational quantum at a site would not remain localized, but would be distributed over the complete chain within a few picoseconds (ps).

As next step Davydov considered the fact that the chain of coupled hydrogen bonds forms a phonon system with the peptide units vibrating against each other in the potential due to the hydrogen bonds. These hydrogen bonds are approximated by a harmonic potential. Since the excitation energy of the amide-I oscillators is naturally dependent on the length of the hydrogen bond in which the C=O group takes part, the system of amide-I oscillators is coupled to the acoustic phonon system of the hydrogen bonded chain (the so-called lattice). Considering a linear dependence of the amide-I excitation energy on the length of the hydrogen bond, the coupling constant can be estimated experimentally. Attempts for the theoretical determination of this constant failed (leading mostly even to values with the wrong sign) due to the use of too small atomic basis sets and the lack of electronic correlation in the ab inito Hartree-Fock calculations performed so far (see [1] for a discussion and references). However, the experimental estimates place its value between 35 and 62 pN.

From these considerations Davydov constructed his model Hamiltonian which contains just that details of the protein α–helix which are the most important ones (constructive and destructive) involved in the transport and storage of energy via amide-I vibrations. The Hamiltonian is given in more details in section II. Due to the coupling of the dispersive amide-I system to the lattice, the nonlinear forces occurring can prevent the distribution of an initially localized amide-I

excitation over the chain. If the dispersive and the nonlinear forces are balancing each other, the excitation will remain localized on a small number of sites at each time due to the nonlinearity, and the whole system of amide-I excitation together with its stabilizing lattice distortion can move through the system due to dispersion. In other words a solitary wave or a soliton could be formed. However, up to now such solitons have not been observed directly in proteins. Only in acetanilide (ACN) which forms single crystals and contains hydrogen bonded chains of C=O groups as in proteins, pinned solitons (which do not move) of the Davydov-type could be observed spectroscopically by Careri's group (see again Scott's review [1] for a detailed discussion). Since proteins are aperiodic and do not form single crystals an observation of Davydov solitons, if present there, is more or less impossible up to now. Even accurate measurements of the constants appearing in the model is not possible. Therefore it is very important to study the dynamics in the Davydov model theoretically as a function of the parameter values, the degree of disorder and temperature to be able to obtain information whether the formation of solitons is possible at all for reasonable windows in the parameter space or not. Especially it is of utmost importance to obtain approximate solutions of the Schrödinger equation for the Davydov Hamiltonian as close as possible to the unknown exact solutions. This work deals with the latter problem and especially the ansatz states proposed by Davydov for this purpose are investigated. Further we propose a propagation scheme in the conlusion, because in that way the inclusion of temperature effects into the theory is more straightforward than in the case of an ansatz treatment.

These basic concepts of the Davydov soliton mechanism for energy transport in proteins [2-5], as well as the different attempts to include the effects of finite temperature into the model [4-13] and the controversy about thermal stability of protein solitons is discussed in the introduction of Ref. [6]. Therefore we do not want to elaborate on these points here. The extensive discussion on the validity of the different ansatz states used in the literature [14-23] is also reviewed there [6]. The ideas on which the Davydov mechanism is based are nowadays extended also to other systems in more or less similar ways. Davydov himself e.g. used a bisoliton concept to explain high-T_c superconductivity in materials containing copperoxide, and a Hamiltonian similar to that for the description of energy transport in proteins for the explanation of electron transport (electrosoliton) which is important in biological redox processes where proteins serve as catalysators. A wide variety of applications of these ideas is collected and dicussed again by Scott in his review [1].

In a series of papers we dealt mainly with ansatz states which include quantum effects in the lattice into the description and with the inclusion of effects of finite temperature into these theories [6,20, 24-27]. Since already at 0K the |D$_1$> ansatz is still an approximation, one would like to have a numerical estimate of the errors introduced by this approximate ansatz. Therefore, we presented in Ref. [27] expectation values of several operators in the state |δ> which represents the error of the |D₁> state if it is substituted into the time dependent Schrödinger equation:

$$\left[i\hbar\left(\partial/\partial t - \hat{H}_D\right)\right]\left|D_1\right\rangle = J\left|\delta\right\rangle \ .$$

J is one of the parameters in the Hamiltonian (see below). For an exact solution J|δ>=0 would be required. We compared these expectation values with the corresponding ones in the state $\left(\hat{H}_D/J\right)\left|D_1\right\rangle$ to get a numerical estimate of the errors occurring. For the sake of comparison the same was done also for the semiclassical so-called |D₂> ansatz [2]. In this study we found that the errors introduced in these expectation values compared to those in the corresponding $\left(\hat{H}_D/J\right)\left|D_1\right\rangle$ state are negligible. Since the set of basis states is incomplete when using the |D₁> ansatz, this does not ensure a good quality of the |D₁> approximation, however, it could be expected, that the lack of basis states, if important, should lead to larger errors also within the basis space actually employed, than those we found numerically.

Since we are extending at present the application of |D₁> type ansatz states also to the polyacetylene case [28] it seems to be desirable to obtain some more detailed informations on the limitations of this ansatz. For this purpose we want to expand the exact solution $\left|\Phi\right\rangle = \exp\left[-i\hat{H}_D t/\hbar\right]\Phi_0\rangle$ for the Davydov Hamiltonian $\left(\hat{H}_D\right)$, where |Φ₀> is the initial state, in a Taylor series in time and compare the results with those from a |D₁> simulation. Attempts into this direction have been reported previously by Cruzeiro-Hansson, Christiansen and Scott [29]. However, they restricted their considerations to a dimer and found that second order terms can be neglected only for times much smaller than 0.1 ps. Further they give no comparisons to approximate simulations and for the case of N sites they give a system of equations, but they draw no numerical conclusions from it.

In order to be able to work numerically with such an expansion, we need informations on the time scales in which the different orders are correct. For this purpose we study in the present paper the performance of such expansions for analytically known solutions for some special cases of the Davydov Hamiltonian. We give the analytical solutions and their expansions in Taylor serieses in time. Then we compare norms, total energies, displacements, momenta and expectation values of the number operators for the wave functions in different orders with those obtained from the exact solutions. Further we draw some conclusions on the form of initial states, necessary for reliable |D₁> simulations for these special cases. In the following paper (*J. Mol. Model.*, accepted) we will discuss the results for the complete Davydov Hamiltonian, based on the results of this work.

Finally in the third paper of this series we will present applications of dynamics, obtained with the methods discussed. Specifically we will present vibrational spectra which can be computed directly from simulations obtained with states of |D₁> type. Since the Davydov mechanism was introduced to explain energy storage and transport in proteins, first of all the question of the existence of such solitons in proteins is of utmost important. Our third paper will also deal with this problem. We want to present simulations including temperature effects, and a detailed study on the initial states, from which solitons are formed. Then we need to explain, why in infrared and Raman spectra of polypeptides no signs of solitons in the amide-I region are found, although theoretically they exist. Further, our model after some extensions can be used to study also coupling of the amide-II vibration (where the N-H bonds are stretched) to optical and acoustical lattice phonons. The reason, why such features should be included also is, that experimentally unusual features in the spectra of polypeptides in the amide-II region were found, and still lack an explanation (see [34] for a short review and further references). It is also of importance to apply the model to acetanilide (in modified form, since there the C=O stretching vibration is coupled to optical phonons), because in this case at low temperature the normal amide-I band vanishes and a new solitonic band appears in the Raman spectra (see [1] for discussion and references). Thus the acetanilide case could give additional insight, up to what extent the Davydov model is able to explain measured spectra, especially as function of temperature.

Davydov's Hamiltonian and the |D₁> Approximation

The Hamiltonian, as well as the form of the |D₁> approximation have been discussed extensively in the literature. However, for the purpose of clearcut definitions in the following, we repeat the basic formulas here. The Davydov Hamiltonian for our problem [2] reads as

$$\hat{H}_D = \sum_n \left[E_0 \hat{a}_n^+ \hat{a}_n - J\left(\hat{a}_n^+ \hat{a}_{n+1} + \hat{a}_{n+1}^+ \hat{a}_n\right) \right.$$
$$\left. + \frac{\hat{p}_n^2}{2M} + \frac{W}{2}\left(\hat{q}_{n+1} - \hat{q}_n\right)^2 + \chi \hat{a}_n^+ \hat{a}_n \left(\hat{q}_{n+1} - \hat{q}_n\right) \right] \quad (1)$$

In equ. (1) $\hat{a}_n^+$ ($\hat{a}_n$) are the usual boson creation (annihilation) operators [4] for the amide-I oscillators at sites n (see sketch at the top of the following page).

From infrared spectra the ground state energy of an isolated amide-I oscillator can be deduced (E_0=0.205 eV). Usually for all parameters in equ. (1) site-independent mean values are used. The average value for the coupling of the transition dipole moments of neighboring amide-I oscillators is J=0.967 meV. The average spring constant of the hydrogen bonds is taken usually to be W=13 N/m, as measured in crystalline formamide. $\hat{p}_n$ is the momentum and $\hat{q}_n$ the position operator of unit n. The mass M of a peptide unit is taken as the mean value of the masses of the units in myosine

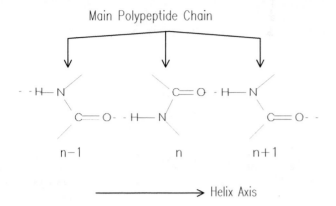

Main Polypeptide Chain

n−1   n   n+1

Helix Axis

(M=114m_p; m_p is the proton mass). The energy of the CO stretching vibration in hydrogen bonds is a function of the length r of the hydrogen bond (E=E_0+χr). For χ the experimental estimates are between 35 pN and 62 pN. Ab initio calculations on formamide dimers usually lead to χ=30-50 pN, however, with small basis set ab initio calculations even negative values for χ were obtained (see e.g. [1] for a review and references).

The one-particle Hamiltonian [2,3], where one-particle refers to the quanta of the amide-I vibration, in second quantized form is given by

$$\hat{H}_D = \sum_n \left[E_0 \hat{a}_n^+ \hat{a}_n - J\left(\hat{a}_{n+1}^+ \hat{a}_n + \hat{a}_n^+ \hat{a}_{n+1}\right)\right]$$

$$+ \sum_k \hbar\omega_k \left[\hat{b}_k^+ \hat{b}_k + \frac{1}{2} + \sum_n B_{nk}\left(\hat{b}_k + \hat{b}_k^+\right)\hat{a}_n^+ \hat{a}_n \right]$$

(2)

$$B_{nk} = \frac{\chi}{\omega_k}\frac{1}{\sqrt{2M\hbar\omega_k}}\left[U_{n+1,k} - U_{nk}\right]$$

$b^{\wedge+}_k$ ($b^{\wedge}_k$) are creation (annihilation) operators for acoustic phonons of wave number k. The translational mode has to be excluded from all summations. Note that in the simulations presented we use again the asymmetric interaction model where only the coupling of the oscillator n to the hydrogen bond between n and n+1 in which the oscillator takes part is considered. ω_k denotes the eigenfrequency of the normal mode k and $\underline{\underline{U}}$ contains the normal mode coefficients. $\underline{\underline{\omega}}$ and $\underline{\underline{U}}$ are obtained by diagonalization of the matrix $\underline{\underline{V}}$ with elements

$$V_{nm} = \frac{W}{M}\left\{2\delta_{nm} - \left(1-\delta_{nN}\right)\delta_{m,n+1} - \left(1-\delta_{n1}\right)\delta_{m,n-1} -\right.$$

$$\left. -\delta_{n1}\delta_{mN} - \delta_{nN}\delta_{m1}\right\}$$

(3)

$$\left(\underline{\underline{U}}^+\underline{\underline{V}}\,\underline{\underline{U}}\right)_{kk'} = \omega_k^2\delta_{kk'} \quad ; \quad \underline{\underline{U}}^+\underline{\underline{U}} = \underline{\underline{U}}\,\underline{\underline{U}}^+ = \underline{\underline{1}}$$

The form of $\underline{\underline{V}}$ implies that we use cyclic boundary conditions and N units.

First of all we rewrite our Hamiltonian into the form

$$\hat{H}_D = \hat{H} + E_0\sum_n \hat{a}_n^+ \hat{a}_n + \frac{1}{2}\sum_k \hbar\omega_k \equiv \hat{H} + \hat{D} \quad ;$$

$$\left[\hat{H},\hat{D}\right] = 0$$

(4)

For the exact solution the time dependent Schrödinger equation holds:

$$i\hbar\frac{\partial}{\partial t}|\Phi> = \hat{H}_D|\Phi>$$

(5)

Now we factorize our exact wave function as

$$|\Phi> = \exp\left[-\frac{it}{\hbar}D\right]|\psi> \quad ; \quad D \equiv E_0 + \frac{1}{2}\sum_k \hbar\omega_k$$

(6)

Then we obtain (D is a time independent real scalar)

$$\hat{H}_D|\Phi> = \exp\left[-\frac{it}{\hbar}D\right]\cdot\left(\hat{D}|\psi> + \hat{H}|\psi>\right)$$

$$i\hbar\frac{\partial}{\partial t}|\Phi> = D\cdot\exp\left[-\frac{it}{\hbar}D\right]|\psi> + \exp\left[-\frac{it}{\hbar}D\right]\cdot i\hbar\frac{\partial}{\partial t}|\psi>$$

(7)

We know that the exact wave function can be written in the form

$$|\psi> = \sum_n \exp\left[\hat{S}_n(t)\right]a_n(t)\hat{a}_n^+|0>$$

(8)

where $a_n(t)$ is a complex scalar and |0> the vacuum state. It is known that the generator $\hat{S}_n(t)$ contains only phonon operators and complex scalars (see [16] for details). Therefore we can write

$$\hat{D}|\Phi> = \exp\left[-\frac{it}{\hbar}D\right]\cdot\left(E_0\sum_{nm}\exp\left[\hat{S}_n(t)\right]a_n(t)\hat{a}_m^+\hat{a}_m\hat{a}_n^+|0> +\right.$$

$$\left. +\frac{1}{2}\sum_k \hbar\omega_k|\psi>\right)$$

(9)

$$\hat{a}_m^+\hat{a}_m\hat{a}_n^+|0> = \hat{a}_m^+\left(\delta_{nm} + \hat{a}_n^+\hat{a}_m\right)|0>$$

$$\Rightarrow \hat{D}|\Phi> = D\cdot\exp\left[-\frac{it}{\hbar}D\right]|\psi>$$

Together with equ. (7) this leads to

$$i\hbar\frac{\partial}{\partial t}|\Phi> = \exp\left[-\frac{it}{\hbar}D\right]\cdot\left(D + i\hbar\frac{\partial}{\partial t}\right)\cdot|\psi>$$

$$\hat{H}_D|\Phi> = \exp\left[-\frac{it}{\hbar}D\right]\cdot\left(D + \hat{H}\right)\cdot|\psi>$$

(10)

and therefore from

$$ih\frac{\partial}{\partial t}|\Phi\rangle = \hat{H}_D|\Phi\rangle = \left(E_0\sum_n \hat{a}_n^+\hat{a}_n + \frac{1}{2}\sum_k \hbar\omega_k + \hat{H}\right)|\Phi\rangle$$

(11a)

follows

$$|\Phi\rangle = \exp\left[-\frac{it}{\hbar}\left(E_0 + \frac{1}{2}\sum_k \hbar\omega_k\right)\right]\cdot|\psi\rangle$$

(11b)

where |ψ> has to obey the Schrödinger equation

$$i\hbar\frac{\partial}{\partial t}|\psi\rangle = \hat{H}|\psi\rangle$$

(12)

with the simplified Hamiltonian

$$\hat{H} = -J\sum_n \left(\hat{a}_{n+1}^+\hat{a}_n + \hat{a}_n^+\hat{a}_{n+1}\right) +$$

$$+\sum_k \hbar\omega_k\left[\hat{b}_k^+\hat{b}_k + \sum_n B_{nk}\left(\hat{b}_k + \hat{b}_k^+\right)\hat{a}_n^+\hat{a}_n\right]$$

(13)

A more simple form of this proof is given in Appendix E. Note, that the zero-point energies in the exponential prefactors are present, whether the coupling between amide-I oscillators and the phonons exists or not, since also if χ=0 holds, the lattice is still present in the Hamiltonian. Thus a remark of Kapor [30] on this topic does not apply. The |D₁> ansatz for |ψ> has the form

$$|D_1\rangle = \sum_n a_n(t)\hat{U}_n\hat{a}_n^+|0\rangle$$

(14)

where the coherent state operators are given by

$$\hat{U}_n|0\rangle_p = \exp\left[-\frac{1}{2}\sum_k |b_{nk}(t)|^2\right]\cdot\exp\left[\sum_k b_{nk}(t)\hat{b}_k^+\right]|0\rangle_p$$

$$= \exp\left\{\sum_k\left[b_{nk}(t)\hat{b}_k^+ - b_{nk}^*(t)\hat{b}_k\right]\right\}|0\rangle_p$$

(15)

Note, that the second equality holds only if the operator acts on the phonon vacuum |0>_p, and that in our notation |0>=|0>_e|0>_p, where |0>_e is the vacuum state for the amide-I oscillators (exciton vacuum). A simpler form of this ansatz is the |D₂> state which is a product state:

$$|D_2\rangle = \sum_n a_n(t)\hat{a}_n^+|0\rangle_e\hat{U}|0\rangle_p$$

$$\hat{U}|0\rangle_p = \exp\left[-\frac{1}{2}\left|\sum_k b_k(t)\right|^2\right]\cdot\exp\left[\sum_k b_k(t)\hat{b}_k^+\right]|0\rangle_p$$

$$= \exp\left\{\sum_k b_k(t)\hat{b}_k^+ - b_k^*(t)\hat{b}_k\right\}|0\rangle_p$$

(16)

The $b_{nk}(t)$ and the $b_k(t)$, respectively, are the coherent state amplitudes and $|a_n(t)|^2$ is the probability to find an amide-I quantum at site n. These are the quantities which have to be determined.

The equations of motion for these quantities can be obtained with the Euler-Lagrange equations of the second kind (see [6,17,20,24-27]). Note, that with the Hamiltonian method as used previously by Davydov and others incorrect equations are obtained in case of the |D₁> state [17]. The final equations of motion for the |D₁> ansatz are

$$i\hbar\dot{a}_n = -\frac{i\hbar}{2}\sum_k\left(\dot{b}_{nk}b_{nk}^* - \dot{b}_{nk}^*b_{nk}\right)a_n +$$

$$+\sum_k \hbar\omega_k\left[B_{nk}\left(b_{nk} + b_{nk}^*\right) + |b_{nk}|^2\right]a_n +$$

$$-J\left(D_{n,n+1}a_{n+1} + D_{n,n-1}a_{n-1}\right)$$

(17a)

$$i\hbar\dot{b}_{nk} = \hbar\omega_k\left(b_{nk} + B_{nk}\right) - J\left[D_{n,n+1}\left(b_{n+1,k} - b_{nk}\right)\frac{a_{n+1}}{a_n} + \right.$$

$$\left. + D_{n,n-1}\left(b_{n-1,k} - b_{nk}\right)\frac{a_{n-1}}{a_n}\right]$$

where the coherent state overlaps are given by

$$D_{nm} = \exp\left[-\frac{1}{2}\sum_k\left(|b_{nk} - b_{mk}|^2 + b_{nk}b_{mk}^* - b_{nk}^*b_{mk}\right)\right]$$

(17b)

Mechtly and Shaw [16] have shown, that for initial conditions $a_n(0)=\delta_{n1}$ and $b_{nk}(0)=0$ the small time behaviour of the system is given by

$$a_n(t) \rightarrow \left(2i\frac{J^2M}{\hbar^2W}\right)^{n-1}\frac{t^{n-1}}{(n-1)!}$$

$$b_{nk}(t) \rightarrow -i\left[\sum_{m=1}^n W_{mk}Q_k\right]\frac{t}{n}$$

(18a)

$$W_{mk} = \sqrt{2} \cdot cos\left[k\left(m+\tfrac{1}{2}\right)a\right] \; ; \; Q_k = \sqrt{\frac{4\pi\alpha\omega_k}{(N+1)a}}$$

$$4\pi\alpha \equiv \frac{\chi^2}{2\hbar W}\sqrt{\frac{M}{W}} \; ; \; a \equiv \frac{\hbar}{2J}\sqrt{\frac{W}{M}}$$

where in their case the eigenfrequencies are

$$\omega_k = \frac{2}{a} sin\left[\frac{ka}{2}\right] \; ; \; k = \frac{\pi j}{(N+1)a} \; ; \; j = 1,2,...,N \qquad (18b)$$

Thus two factors in one of the terms in equ. (17a) where $a_n(t)$ occurs in the denominator have for small t the behaviour

$$\lim_{t\to 0}\frac{a_{n-1}(t)}{a_n(t)} \sim \frac{t^{n-2}}{t^{n-1}} = \frac{1}{t}$$

$$\lim_{t\to 0}\left[b_{n-1,k}(t)-b_{nk}(t)\right]\sim t \qquad (18c)$$

and therefore the product of both approaches a constant as t approaches zero. The other term has the behaviour

$$\lim_{t\to 0}\frac{a_{n+1}(t)}{a_n(t)} \sim \frac{t^n}{t^{n-1}} = t \qquad (18d)$$

and vanishes when t approaches zero. Thus in principle the denominators $a_n(t)$ in equ. (17a) pose no difficulties, although if they vanish for t approaching zero. However, in [16] it is reported that instabilities are encountered when the short time solutions are incorporated into a program. To avoid such problems we follow the suggestion given in [16] and all a_n which vanish in the initial state are put to $a_n(0)=x$, where x is a small, physically insignificant number, e.g. x=0.005 [16].

The equations for the $|D_2\rangle$ state, transformed from the normal mode to the coordinate representation, are

$$i\hbar\dot{a}_n = -J\left(a_{n+1}+a_{n-1}\right)+\chi\left(q_{n+1}-q_n\right)a_n$$

$$\dot{p}_n = W\left(q_{n+1}-2q_n+q_{n-1}\right)+\chi\left(\left|a_n\right|^2 - \left|a_{n-1}\right|^2\right)$$

$$\dot{q}_n = \frac{p_n}{M} \qquad (19)$$

The numerical solution of all these equations can be accomplished with the help of a fourth order Runge-Kutta method. Note, that the lattice parts of equ. (19) are not entirely classical as their form might suggest, but the q_n's and p_n's have to be viewed as expectation values of the corresponding quantum mechanical operators rather than as classical variables. However, the $|D_2\rangle$ state is the exact solution for $\hat{H}_p$ if the operators of the displacements and momenta are replaced by real numbers $q_n(t)$ and $p_n(t)$, respectively [31,32].

Initial States

In this subchapter we want to discuss the question of the correctness of the $|D_1\rangle$ state in the decoupled (χ=0) and in the transportless (or small polaron) case (J=0). Especially it is interesting to investigate whether or not this poses restrictions on the form of the initial state. This is an important problem, since for the initial state (time t=0) we have the physical situation that a set of coefficients for the amide-I oscillators in the wave function, $\{a_n(0)\}$, and a set of displacements and momenta $\{q_n(0),p_n(0)\}$ is given. The question is now, how to compute coherent state amplitudes from these sets of initial values.

The Small Polaron Limit

First we want to discuss the transportless case, also called the small polaron limit (J=0). Since here we have an excitation which is not transported along the chain, but deforms the lattice, it can be called a polaron and further if the initial excitation is localized, it is called a small polaron. Brown et al. stated in 1986 (Ref. [14], second paper) that with $|D_1\rangle$ dynamics incorrect values for the displacements and consequently also for the phonon energy are obtained. This result is due to the use of the Hamiltonian method introduced by Davydov [2,3] which yields incorrect equations of motion in the $|D_1\rangle$ case [17]. In his paper from 1988 (Ref. [14], last paper), Brown concluded that $|D_1\rangle$ satisfies the Schrödinger equation in the small polaron limit, but derives no equations of motion. Again in 1988 (Ref. [14], third paper) Brown et al. stated that an ansatz treatment yields correct displacements but incorrect phonon energies. This statement was based on the $|D_2\rangle$ ansatz, where it is certainly correct, however, since no direct reference to $|D_2\rangle$ was made there, it could lead to the impression, that any ansatz treatment would be plagued by the same problem, what is not the case. To avoid any misunderstandings, we give in Appendix A a complete derivation of the exact solution for the small polaron limit which is a $|D_1\rangle$ state, together with the expressions for the relevant expectation values.

Assuming now a set of initial conditions for the lattice $\{q_n(0),p_n(0)\}$, then the lattice energy is clearly given by

$$E_{lat}(0) = \sum_n\left\{\frac{W}{2}\left[q_{n+1}(0)-q_n(0)\right]^2 + \frac{p_n^2(0)}{2M}\right\} \qquad (20)$$

Let us construct now from these initial conditions a set $\{b_{nk}(0)\}$ of coherent state amplitudes following the suggestion given by us in [28]:

$$P_{nn} = \left|a_n(0)\right|^2 \qquad (21)$$

$$b_{nk}(0) = \frac{1}{P_{nn}}\sqrt{\frac{M\omega_k}{2\hbar}}U_{nk}q_n(0)+\frac{i}{P_{nn}}\sqrt{\frac{1}{2\hbar M\omega_k}}U_{nk}p_n(0)$$

which clearly gives the correct displacements and momenta back, and thus leads also to a correct exciton-lattice interaction term, which is linear in the phonon operators. However, substitution into that part of equ. (A23) which represents the energy of the decoupled lattice as derived in Appendix A leads to

$$E_{lat} = \sum_n \frac{1}{P_{nn}} \left[\frac{M}{2} \sum_k U_{nk} \omega_k^2 U_{nk} q_n^2(0) + \frac{p_n^2(0)}{2M} \sum_k U_{nk}^2 \right]$$

$$= \sum_n \frac{1}{P_{nn}} \left[W q_n^2(0) + \frac{p_n^2(0)}{2M} \right]$$

(22)

which obviously differs from equ. (20) and is therefore incorrect. Analytical expressions for $\underline{\omega}$ and $\underline{U}$ (in real representation) are given in Appendix B. From that we conclude that the choice for $b_{nk}(0)$ is consistent with the exciton-lattice interaction but not with the lattice energy. The reason is, that the averaged equations which the $b_{nk}(0)$ have to obey

$$\sum_n \left\{ P_{nn} Re[b_{nk}(0)] - \sqrt{\frac{M\omega_k}{2\hbar}} U_{nk} q_n(0) \right\} = 0$$

$$\sum_n \left\{ P_{nn} Im[b_{nk}(0)] - \sqrt{\frac{1}{2M\hbar\omega_k}} U_{nk} p_n(0) \right\} = 0 \qquad (23)$$

$$\sum_n \left\{ \hbar\omega_k |b_{nk}(0)|^2 P_{nn} - \frac{W}{2}[q_{n+1}(0) - q_n(0)]^2 - \frac{p_n(0)^2}{2M} \right\} = 0$$

simply do not contain enough information to determine the $b_{nk}(0)$ uniquely. Further the most obvious choice for a solution, namely to set each term in the first two sums individually to zero (from this equ. (21) follows) does not yield a solution which could fulfill all three equations.

However, we can derive a consistent choice for the initial conditions, if we assume the $b_{nk}(0)$ as site independent $(b_{nk}(0)=b_k(0))$, i.e. if we choose the initial state to be of $|D_2>$ form. From this initial state the site dependence of the b's present in the evolution governed by the complete Hamiltonian evolves naturally from the $|D_1>$ equations of motion. If we set $b_{nk}(0)=b_k(0)$ in the above equations, we get instead of a weighted average simply factors $\sum_n P_{nn}=1$, and thus the initial values

$$Re[b_{nk}(0)] = \sum_{n'} \sqrt{\frac{M\omega_k}{2\hbar}} U_{n'k} q_{n'}(0)$$

(24)

$$Im[b_{nk}(0)] = \sum_{n'} \sqrt{\frac{1}{2M\hbar\omega_k}} U_{n'k} p_{n'}(0)$$

With this choice we obtain again the correct values for $q_n(0)$ and $p_n(0)$ back and thus also the correct exciton lattice interaction. Further, this ansatz yields also the correct lattice energy:

$$E_{lat} = \sum_{nk} \hbar\omega_k |b_{nk}(0)|^2 P_{nn} =$$

$$= \frac{M}{2} \sum_{nn'} \sum_k U_{n'k} \omega_k^2 U_{nk} q_{n'}(0) q_n(0) +$$

$$+ \frac{1}{2M} \sum_{nn'} \sum_k U_{n'k} U_{nk} p_{n'}(0) p_n(0) = \qquad (25)$$

$$= \frac{W}{2} \sum_{nn'} V_{n'n} q_{n'}(0) q_n(0) + \sum_n \frac{p_n^2(0)}{2M}$$

where $\underline{\underline{V}}$ is defined in Appendix B.

In the case of v bosons occupying the same state the initial variables are given by

$$Re[b_{nk}(0)] = \frac{1}{v} \sum_{n'} \sqrt{\frac{M\omega_k}{2\hbar}} U_{n'k} q_{n'}(0)$$

$$Im[b_{nk}(0)] = \frac{1}{v} \sum_{n'} \sqrt{\frac{1}{2M\hbar\omega_k}} U_{n'k} p_{n'}(0) \qquad (26)$$

However, here we have [28] alternatively

$$q_n = v \sum_{n'k} \sqrt{\frac{2\hbar}{M\omega_k}} U_{nk} |a_{n'}|^2 Re[b_{n'k}]$$

$$p_n = v \sum_{n'k} \sqrt{2M\hbar\omega_k} U_{nk} |a_{n'}|^2 Im[b_{n'k}]$$

$$E_{lat} = v^2 \sum_{nk} \hbar\omega_k |b_{nk}|^2 |a_n|^2 ; \quad \sum_n |a_n|^2 = 1 \qquad (27)$$

Therefore it is obvious, that in order to obtain a correct initial state we have to start from a $|D_2>$ like ansatz at t=0, calculated from the initial set of displacements and momenta. Otherwise we would obtain an incorrect lattice energy and consequently an incorrect time evolution.

The Decoupled Case

In the decoupled case ($\chi=0$) we have the Hamiltonian

$$\hat{H}_{DC} = \hat{J} + \hat{H}_{lat}$$
$$\hat{J} \equiv -J\sum_n \left(\hat{a}_n^+ \hat{a}_{n+1} + \hat{a}_{n+1}^+ \hat{a}_n \right)$$
$$\hat{H}_{lat} \equiv \sum_k \hbar\omega_k \hat{b}_k^+ \hat{b}_k \qquad (28a)$$

Since the two parts of the Hamiltonian are independent, the exact solution is a product of a state for the excitons with a state for the phonons. The separation ansatz leads to

$$|\psi_l\rangle\left[i\hbar\frac{\partial}{\partial t}|\psi_e\rangle - \hat{J}|\psi_e\rangle\right] =$$
$$-|\psi_e\rangle\left[i\hbar\frac{\partial}{\partial t}|\psi_l\rangle - \hat{H}_{lat}|\psi_l\rangle\right] \qquad (28b)$$

which can only be fulfilled if both sides of the equation vanish independently. Here $|\psi_e\rangle$ is the exciton and $|\psi_l\rangle$ the lattice (phonon) state. The phonon operator is a sum of operators for each normal mode and its solution is a product of coherent states for each mode:

$$|\psi_l\rangle = \prod_k |\beta_k\rangle = e^{\sum_k \left[b_k(t)\hat{b}_k^+ - b_k^*(t)\hat{b}_k \right]}|0\rangle =$$
$$= e^{-\frac{1}{2}\sum_k |b_k(t)|^2} e^{\sum_k b_k(t)\hat{b}_k^+}|0\rangle \qquad (29)$$

Thus the exact solution of

$$i\hbar\frac{\partial}{\partial t}|\psi_l\rangle = \hat{H}_{lat}|\psi_l\rangle \qquad (30)$$

is given by our ansatz if

$$b_k(t) = b_k(0)e^{-i\omega_k t} \qquad (31)$$

where $b_k(0)$ has to be constructed from the initial displacements and momenta as described in the previous chapter. The exact solution of the Schrödinger equation for the oscillator system is given in Appendix C. It is obvious that the exact solution is a $|D_2\rangle$ state which is identical to a $|D_1\rangle$ state with site independent $b_{nk}(t)$.

Thus we have to show, that from the equations of motion for the $|D_1\rangle$ ansatz the $b_{nk}(t)$ remain site independent if we start from a site independent initial state in the case $\chi=0$:

$$i\hbar\dot{b}_{nk} = \hbar\omega_k b_{nk} - J\left[\left(b_{n-1,k} - b_{nk} \right)D_{n,n-1}\frac{a_{n-1}}{a_n} + \right.$$
$$\left. + \left(b_{n+1,k} - b_{nk} \right)D_{n,n+1}\frac{a_{n+1}}{a_n} \right] \qquad (32)$$

Thus we obtain

$$-\frac{i\hbar}{2}\sum_k \left(\dot{b}_{nk}b_{nk}^* - \dot{b}_{nk}^* b_{nk} \right)a_n = -\sum_k \hbar\omega_k |b_{nk}|^2 a_n +$$
$$+\frac{J}{2}\sum_k \left[\left(b_{n-1,k}b_{nk}^* - |b_{nk}|^2 \right)D_{n,n-1}\frac{a_{n-1}}{a_n} + \right.$$
$$+ \left(b_{n+1,k}b_{nk}^* - |b_{nk}|^2 \right)D_{n,n+1}\frac{a_{n+1}}{a_n} +$$
$$+ \left(b_{n-1,k}^* b_{nk} - |b_{nk}|^2 \right)D_{n,n-1}^*\frac{a_{n-1}^*}{a_n^*} +$$
$$\left. + \left(b_{n+1,k}^* b_{nk} - |b_{nk}|^2 \right)D_{n,n+1}^*\frac{a_{n+1}^*}{a_n^*} \right]a_n$$

$$(33)$$

and together with equ. (17) we have

$$i\hbar\dot{a}_n = -J\left(D_{n,n-1}a_{n-1} + D_{n,n+1}a_{n+1} \right) +$$
$$+\frac{J}{2}\sum_k \left[\left(b_{n-1,k}b_{nk}^* - |b_{nk}|^2 \right)D_{n,n-1}\frac{a_{n-1}}{a_n} + \right.$$
$$+ \left(b_{n+1,k}b_{nk}^* - |b_{nk}|^2 \right)D_{n,n+1}\frac{a_{n+1}}{a_n} +$$
$$+ \left(b_{n-1,k}^* b_{nk} - |b_{nk}|^2 \right)D_{n,n-1}^*\frac{a_{n-1}^*}{a_n^*} + \qquad (34)$$
$$\left. + \left(b_{n+1,k}^* b_{nk} - |b_{nk}|^2 \right)D_{n,n+1}^*\frac{a_{n+1}^*}{a_n^*} \right]a_n$$

At the beginning of the first time step, i.e. at t=0 we can replace $b_{nk}(0)$ by $b_k(0)$ from our initial state and the equations of motion yield

$$D_{nm}(0)=1 \quad ; \quad i\hbar\dot{a}_n(0) = -J\left[a_{n+1}(0) + a_{n-1}(0) \right]$$
$$i\hbar\dot{b}_{nk}(0) = \hbar\omega_k b_k(0) \qquad (35)$$

Thus after the first time step (t=τ) we obtain

$$D_{nm}(\tau)=1 \quad ; \quad a_n(\tau)=a_n(0)+\frac{iJ}{\hbar}\left[a_{n-1}(0)+a_{n+1}(0)\right]\tau$$

$$b_{nk}(\tau)=b_k(0)(1-i\omega_k\tau)$$

$$(36)$$

Obviously the b's remain site independent after the first time step, thus consequently for all times, and therefore the time simulation with the equations of motion for the $|D_1\rangle$ state yields the correct solution of $|D_2\rangle$ form, provided the initial state is of this form. Note, that for site-independent b's also the, for the decoupled case artificial coupling terms in equ.'s (32) and (34) vanish and that further we have $D_{nm}(t)=1$.

The Complete Hamiltonian

After we have shown, that in both special cases, the initial state has to be of $|D_2\rangle$ form, i.e. with site independent coherent state amplitudes $b_{nk}(0)$, if a simulation using a $|D_1\rangle$ ansatz should lead to the correct analytical solution, we have to ask how such an initial state evolves in time when we apply the complete Hamiltonian. Since in this case, the basis space of $|D_1\rangle$ is incomplete we have to study the errors developing through the use of this ansatz. For this purpose we can use the fact that [17]

$$\left(i\hbar\frac{\partial}{\partial t}-\hat{H}_D\right)|D_1\rangle=J|\delta\rangle \qquad (37)$$

where the form of the error state $|\delta\rangle$ as function of $\{a_n(t),b_{nk}(t)\}$ as computed in a $|D_1\rangle$ simulation is known [17]. In our previous work [26] we have derived expressions for the expectation values of different operators for the two states $\left(\hat{H}/J\right)|D_1\rangle$ and $|\delta\rangle$ and compared them in numerical calculations. Such a procedure can serve as the appropriate tool to answer our above mentioned question. Therefore we introduced cyclic boundary conditions into our program [26] and performed simulations for χ=0,10,20,30,40 pN. For χ=0 we found that all the expectation values computed for the state $|\delta\rangle$ from a numerical $|D_1\rangle$ simulation vanish, as it is to be expected because in this case $|D_1\rangle$ yields the exact solution for site independent amplitudes $b_{nk}(0)$.

To be more precise we applied as initial conditions for the oscillator system

$$a_n(0)=R\cdot sech\left[\frac{(n-o)X^2}{4WJ}\right] \qquad (38)$$

where R is a normalization constant, J=0.967 meV, W=13 N/m, X=62 pN (note, that X is used only in the initial state for

all cases, in contrast to χ), and the maximum of the function was put at site o=11 in a chain of N=21 units. The displacements and momenta of the lattice were chosen such that kinetic and potential energy each equal $0.5(N-1)k_BT$ for T=300K. This energy was distributed according to Bose-Einstein statistics on the normal modes, excluding the translation:

$$q_n(0)=\sum_k U_{nk}\sqrt{\frac{e_k}{W\sum_{n'}\left(U_{n'k}-U_{n'+1,k}\right)^2}}$$

$$p_n(0)=\sum_k U_{nk}\sqrt{\frac{e_k}{M\sum_{n'}U_{n'k}^2}} \qquad (39)$$

$$e_k=(N-1)k_BT\hbar\omega_k f_k/S$$

$$f_k=\left[e^{\frac{\hbar\omega_k}{k_BT}}-1\right]^{-1} \quad ; \quad S=\sum_k \hbar\omega_k f_k$$

where M=114 m_p, and k_B is Boltzmann's constant. For the fourth order Runge-Kutta simulations we used a time step of 1 fs and a total simulation time of 100 ps, corresponding to 100,000 time steps. In this period the error in total energy was typically less than 5-10 peV (much less than 0.1 % of the exciton phonon interaction energies) and the error in norm around 1 ppb (parts per billion).

In Fig. 1 we show the norms

$S_H(t)=\left\langle\left(\hat{H}/J\right)D_1\left|\left(\hat{H}/J\right)D_1\right\rangle\right.$ and $S_E(t)=\langle\delta|\delta\rangle$ for three values of χ, namely 0, 10, and 20 pN. Note, that for these coupling strengths no solitons are present in the system. Here we see, that for the decoupled case (Fig.1a) the norm of the error state is below machine accuracy, since here we obtain the analytical solution from our simulations. The norm of $\left(\hat{H}/J\right)|D_1\rangle$ decreases slightly in the course of the simulation. When we switch on the coupling to a small value of χ=10 pN (Fig.1b), the increase of $S_H(t)$ with time becomes rather large, up to more than 35. However, the increase of the error $S_E(t)$ is much smaller, between 0 and 0.27, indicating the accuracy of the $|D_1\rangle$ simulation. For the larger coupling χ=20 pN (Fig. 1c) $S_H(t)$ increases by a factor of 10 up to values around 350. The increase of the error is much smaller, namely up to a value of 1.5. If we increase χ further up to 30 and 40 pN (not shown here), the increase of $S_H(t)$ per 10 pN increase in coupling remains the same, while the maximum values of $S_E(t)$ start to converge to maximum values around 2.2-2.3. When increasing the coupling further we reach the values already discussed in [26]. Thus it is obvious, that the error in the norm starts to increase smoothly with χ increasing from its value of 0 where $|D_1\rangle$ is the exact

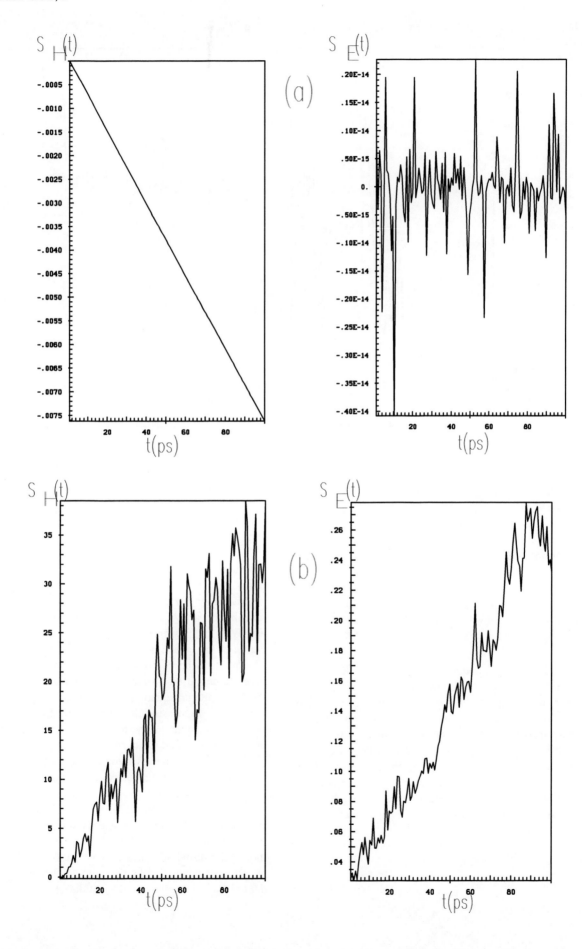

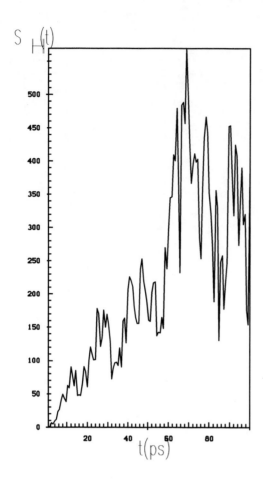

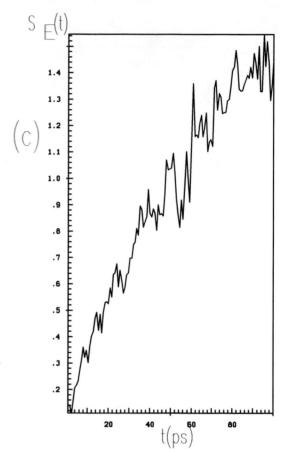

Figure 1: *The norms* $S_H(t) = \left\langle \left(\hat{H}/J\right)D_1 \middle| \left(\hat{H}/J\right)D_1 \right\rangle$ *and*

$S_E(t) = <\delta|\delta>$ *for the two states under consideration as function of time for three values of the coupling constant:*

(a) $\chi=0$ pN (S_H relative to $S_H(0)=675,827.93106$)
(b) $\chi=10$ pN (S_H relative to $S_H(0)=676,851.60694$)
(c) $\chi=20$ pN (S_H relative to $S_H(0)=677,877.52615$)

solution, when the simulation is started from an initial $|D_2>$ like state. Fig. 2 shows the probabilities $N_n(t)=<D_1|\hat{a}_n^+\hat{a}_n|D_1>$ to find an amide-I quantum at site n as function of time for the three values of the coupling constant. The results show that in all three cases no soliton is formed from the initial sech-distribution. The dispersion of this distribution becomes less regular with increasing coupling strength, because the phonons, coupled with increasing strength to the oscillators, are initially excited with an energy corresponding to a temperature of 300K, which leads to a large aperiodicity in the lattice coordinates.

In Fig. 3 finally we show the expectation values

$N_n^H = \left\langle \left(\hat{H}/J\right)D_1 \middle| \hat{a}_n^+\hat{a}_n \middle| \left(\hat{H}/J\right)D_1 \right\rangle$ and $N_n^E(t)=<\delta|\hat{a}_n^+\hat{a}_n|\delta>$ of

the number operators again for the three values of the coupling constant.

Naturally, for the decoupled case the error is less than machine accuracy, because in this case our evolution is exact. With increasing coupling strength, we find a slight increase of the errors, however, about six orders of magnitude smaller than the corresponding values of $N_n^H(t)$. Thus also here the deviation of our results from the exact solution of the Schrödinger equation is very small, even more or less negligible, when using the $|D_1>$ ansatz starting from an initial $|D_2>$ like state. We do not discuss the expectation values of displacement and momentum operators here, because they behave rather similar as those of the number operators we have just studied. Thus our conclusion drawn from the analytically solvable special cases as discussed in the previous subchapters remain valid also when we apply the complete Hamilton operator in numerical simulations.

Therefore, we can savely draw as final conclusion, that indeed the initial state has to have the form of a $|D_2>$ state, i.e. site independent coherent state amplitudes $b_{nk}(0)$, in order to be able to obtain reliable results from $|D_1>$ simulations.

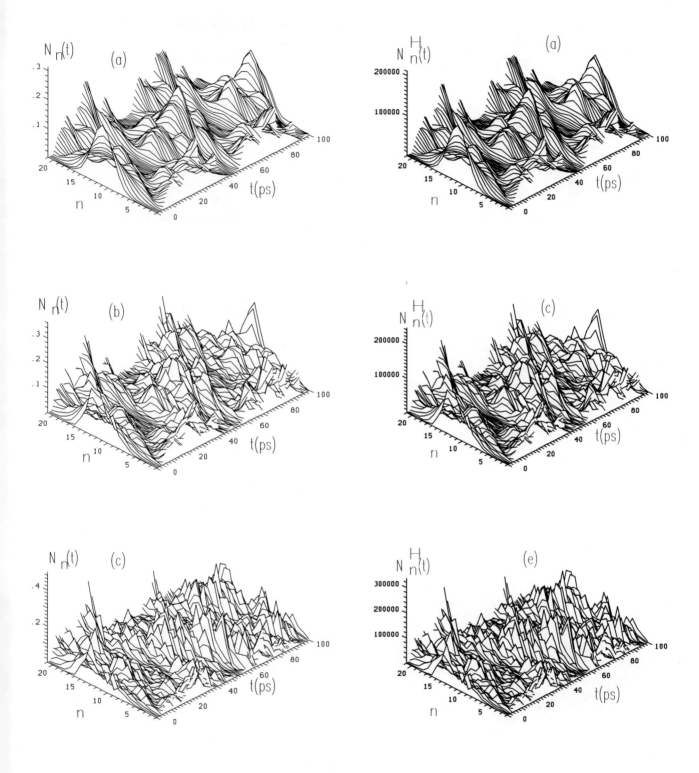

Figure 2: *The probability $N_n(t)=<D_1|\hat{a}_n^+\hat{a}_n|D_1>$ to find an amide-I vibrational quantum at site n as function of time for three values of the coupling constant:*
(a) $\chi=0$ pN   (b) $\chi=10$ pN (c) $\chi=20$ pN

Figure 3: *The expectation values*

$$N_n^H = \left\langle \left(\hat{H}/J\right)D_1 \left| \hat{a}_n^+ \hat{a}_n \right| \left(\hat{H}/J\right)D_1 \right\rangle$$ (a,c,e) as functions of

site and time for three values of the coupling constant:

(a,b) $\chi=0$ pN   (c,d) $\chi=10$ pN   (e,f) $\chi=20$ pN

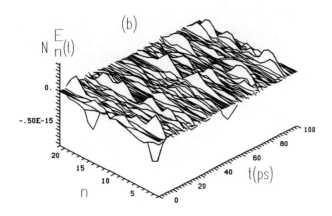

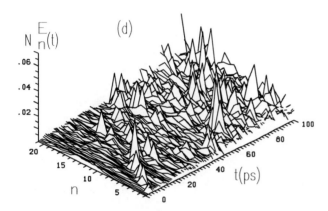

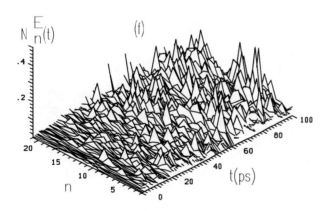

Figure 3 (continued): *The expectation values* $N^E_n(t) = <\delta|\hat{a}_n^+\hat{a}_n|\delta>$ *(b,d,f) as functions of site and time for three values of the coupling constant:*

(a,b) $\chi=0$ pN   (c,d) $\chi=10$ pN   (e,f) $\chi=20$ pN

Expansions of Exact Solutions

Since in the following paper we want to use for the general case expansions of the exact wave functions in polynomials in time, it is very important to get a reliable tool to judge up to which time a given order of the expansion can be considered as reliable. The best way to do so is the expansion of exactly known solutions in such polynomials and to compare the results of different orders with the exact states. There are in principle two ways to compute such expansions. The simplest one is to expand directly the exact solution. The other one, which has to be used in cases where the exact solution is not known, is to expand the formally exact wave function

$$\left|\psi(t)\right\rangle = e^{T\hat{H}}\left|\psi(0)\right\rangle \quad ; \quad T \equiv -\frac{it}{\hbar}$$

$$\Rightarrow \left|\psi(t)\right\rangle = \sum_{\nu=0}^{\infty} \frac{T^\nu}{\nu!}\hat{H}^\nu\left|\psi(0)\right\rangle \tag{40}$$

Thus a given order μ can be written as

$$\left|\psi(\mu)\right\rangle \equiv \sum_{\nu=0}^{\mu} \frac{T^\nu}{\nu!}\hat{H}^\nu\left|\psi(0)\right\rangle = \sum_{\nu=0}^{\mu} \frac{T^\nu}{\nu!}\left|\psi_\nu\right\rangle$$

$$\left|\psi_\nu\right\rangle = \hat{H}\left|\psi_{\nu-1}\right\rangle = \hat{H}^\nu\left|\psi_0\right\rangle \quad ; \quad \left|\psi_0\right\rangle = \left|\psi(0)\right\rangle \tag{41}$$

In the present work we used both methods, just to avoid errors in the calculation. Then expectation values for any operator $\hat{O}$ can be computed from expectation values in lower order by

$$\left|\psi(\mu)\right\rangle = \left|\psi(\mu-1)\right\rangle + \frac{T^\mu}{\mu!}\left|\psi_\mu\right\rangle \tag{42}$$

$$O(\mu) \equiv \left\langle\psi(\mu)\left|\hat{O}\right|\psi(\mu)\right\rangle = O(\mu-1)+$$

$$+\frac{1}{\mu!}\left[(T^*)^\mu\left\langle\psi_\mu\left|\hat{O}\right|\psi(\mu-1)\right\rangle + T^\mu\left\langle\psi(\mu-1)\left|\hat{O}\right|\psi_\mu\right\rangle\right]+$$

$$+\frac{|T|^{2\mu}}{(\mu!)^2}\left\langle\psi_\mu\left|\hat{O}\right|\psi_\mu\right\rangle$$

In case of $\hat{O}$ being hermitian, the second term reduces to

$$\frac{2}{\mu!}Re\left[T^\mu\left\langle\psi(\mu-1)\left|\hat{O}\right|\psi_\mu\right\rangle\right] \tag{43}$$

Further the expectation value can be expanded in any case to yield

$$\left\langle \psi_\mu \left| \hat{O} \right| \psi(\mu-1) \right\rangle = \sum_{v=0}^{\mu-1} \frac{T^v}{v!} \left\langle \psi_\mu \left| \hat{O} \right| \psi_v \right\rangle \tag{44}$$

The properties of the different orders can further be used to transform expectations values of the Hamiltonian to overlap integrals:

$$\left\langle \psi_\mu \left| \hat{H} \right| \psi_v \right\rangle = \left\langle \psi_{\mu+1} \middle| \psi_v \right\rangle = \left\langle \psi_\mu \middle| \psi_{v+1} \right\rangle \tag{45}$$

We do not want to elaborate here on more details of the sometimes tedious calculations and proceed to the two cases under consideration.

The Oscillator System in the Decoupled Case

Here the relevant expectation values to be computed are

$$S(\mu,t) = \left\langle \psi(\mu) \middle| \psi(\mu) \right\rangle \;, H(\mu,t) = \left\langle \psi(\mu) \left| \hat{H} \right| \psi(\mu) \right\rangle \text{ and}$$

$$N_n(\mu,t) = \left\langle \psi(\mu) \left| \hat{a}_n^+ \hat{a}_n \right| \psi(\mu) \right\rangle \text{. These functions have to be}$$

compared with the corresponding results obtained from the exact solution, namely S(t)=1, H(t)=0 and N_n(t) as given in Appendix C for an initial state, where the excitation is localized at one site o, i.e. $\left| \psi(0) \right\rangle = \left| \psi_0 \right\rangle = \hat{a}_o^+ \left| 0 \right\rangle$. The Hamiltonian in spatial representation is given by

$$\hat{J} = -J \sum_n \left(\hat{a}_n^+ \hat{a}_{n+1} + \hat{a}_{n+1}^+ \hat{a}_n \right)$$

$$\hat{J} \hat{a}_n^+ \left| 0 \right\rangle = -J \left(\hat{a}_{n-1}^+ + \hat{a}_{n+1}^+ \right) \left| 0 \right\rangle \tag{46}$$

Thus we can directly write down the first three orders of the exact solution

$$\left| \psi_0 \right\rangle = \hat{a}_o^+ \left| 0 \right\rangle$$

$$\left| \psi_1 \right\rangle = -J \left(\hat{a}_{o-1}^+ + \hat{a}_{o+1}^+ \right) \left| 0 \right\rangle \tag{47}$$

$$\left| \psi_2 \right\rangle = J^2 \left(\hat{a}_{o-2}^+ + 2\hat{a}_o^+ + \hat{a}_{o+2}^+ \right) \left| 0 \right\rangle$$

$$\left| \psi_3 \right\rangle = -J^3 \left(\hat{a}_{o-3}^+ + 3\hat{a}_{o-1}^+ + 3\hat{a}_{o+1}^+ + \hat{a}_{o+3}^+ \right) \left| 0 \right\rangle$$

Obviously in any order μ the excitation is at maximum transported only μ sites away from the initial excitation. Since we can also perform a direct expansion of the state to any

arbitrary order, it seems to be more advantegeous to use for this purpose the normal mode representation (Appendix C)

$$\hat{J} = -J\underline{\hat{a}}^+ \underline{\underline{X}} \underline{\hat{a}} = -J\underline{\hat{c}}^+ \underline{\underline{\lambda}} \underline{\hat{c}} = -J \sum_k \lambda_k \hat{c}_k^+ \hat{c}_k \quad ; \quad \underline{\hat{a}} = \underline{\underline{R}} \underline{\hat{c}} \tag{48}$$

$$R_{nk} = \frac{1}{\sqrt{N}} e^{\frac{2\pi i}{N} nk} \quad ; \quad \lambda_k = 2 \cdot \cos\left(\frac{2\pi}{N} k\right)$$

$$\left| \psi_0 \right\rangle = \sum_k c_k(0) c_k^+ \left| 0 \right\rangle$$

With the transformations (here $a_n(0)=\delta_{no}$)

$$c_k(0) = \sum_n R_{nk}^* a_n(0) \quad ; \quad a_n(t) = \sum_k R_{nk} c_k(t) \tag{49}$$

we can write down the solution in any order in time by expansion of the exact wave function as

$$\left| \psi(\mu) \right\rangle = \sum_n a_n(\mu,t) \hat{a}_n^+ \left| 0 \right\rangle =$$

$$= \frac{1}{N} \sum_{nk} e^{\frac{2\pi i}{N} k(n-o)} \sum_{v=0}^\mu \frac{1}{v!} \left(i \frac{J\lambda_k}{\hbar} t \right)^v \hat{a}_n^+ \left| 0 \right\rangle \tag{50}$$

The expectation values of the number operators are consequently

$$N_n(\mu,t) = \left| a_n(\mu,t) \right|^2 =$$

$$= \frac{1}{N^2} \sum_{kk'} e^{\frac{2\pi i}{N}(k-k')(n-o)} \sum_{v,v'=0}^\mu \frac{\lambda_k^v (-\lambda_{k'})^{v'}}{v!v'!} \left(i\frac{J}{\hbar} t \right)^{v+v'} \tag{51}$$

However, in the form $N_n(\mu,t)=|a_n(\mu,t)|^2$, where $a_n(\mu,t)$ is actually calculated from equ. (50), the expectation values are most easily programmed for arbitrary order. From (51) we obtain

$$S(\mu,t) = \sum_n N_n(\mu,t) = \frac{1}{N} \sum_k \sum_{v,v'=0}^\mu \frac{(-1)^{v'}}{v!v'!} \left(i\frac{J\lambda_k}{\hbar} t \right)^{v+v'} \tag{52}$$

The expectation value of the Hamiltonian is obtained as

$$H(\mu,t) = \sum_{nn'} a_n(\mu,t) a_{n'}^*(\mu,t) \left\langle 0 \left| \hat{a}_{n'} \hat{J} \hat{a}_n^+ \right| 0 \right\rangle =$$

$$= -\frac{J}{N} \sum_{nn'} a_n(\mu,t) a_{n'}^*(\mu,t) \sum_k \lambda_k e^{\frac{2\pi i}{N} k(n-n')} \tag{53}$$

Substitution of the explicit values for the coefficients yields the final expression

$$H(\mu,t) = -\frac{J}{N}\sum_{k}\lambda_k\sum_{\nu,\nu'=0}^{\mu}\frac{(-1)^{\nu'}}{\nu!\nu'!}\left(i\frac{J\lambda_k}{\hbar}t\right)^{\nu+\nu'} \tag{54}$$

Note that all the double sums over the different orders in (52) and (54) are real, because each term $T_{\nu\nu'}$ occurs twice. The $T_{\nu\nu}$ are anyway real. Therefore we always obtain combinations $(T_{\nu\nu'}+T_{\nu'\nu})$ which can only be imaginary, if one of the indices is odd and the other one even. However, just in this case due to the minus signs we have $T_{\nu\nu'}=-T_{\nu'\nu}$, and thus all imaginary contributions vanish, and the sum can be restricted to terms where $(\nu+\nu')$ is even. Therefore we can write after some reordering

$$H(\mu,t) = -\frac{J}{N}\sum_{\substack{\nu,\nu'=0\\(\nu+\nu')even}}^{\mu}\frac{(-1)^{\nu'}}{\nu!\nu'!}\left(i\frac{Jt}{\hbar}\right)^{\nu+\nu'}\sum_{k}\lambda_k^{\nu+\nu'+1} \tag{55}$$

Since we have

$$\sum_{k=1}^{N}\lambda_k^n = \sum_{k}\left(e^{\frac{2\pi i}{N}k}+e^{-\frac{2\pi i}{N}k}\right)^n = \sum_{m=0}^{n}\binom{n}{m}\sum_{k=1}^{N}e^{\frac{2\pi i}{N}k(n-2m)} =$$

$$= N\sum_{m=0}^{n}\binom{n}{m}\delta_{n,2m} = \begin{cases} 0 & ; \text{ if n is odd} \\ \\ \dfrac{N\cdot n!}{\left[(n/2)!\right]^2} & ; \text{ if n is even or 0} \end{cases} \tag{56}$$

and our summation is restricted to even values of $(\nu+\nu')$, we have in the k-summation only odd exponents $(\nu+\nu'+1)$ at λ_k, and thus $H(\nu,t)=0$. This result can be used to check together with equation (53) the correctness of the program. Since in $S(\mu,t)$ we have in the k-summations only even exponents $(\nu+\nu')$ at the λ_k, this quantity does not vanish:

$$S(\mu,t) = \sum_{\substack{\nu,\nu'=0\\(\nu+\nu')even}}^{\mu}\frac{(-1)^{(3\nu'+\nu)/2}(\nu+\nu')!}{(\nu!)+(\nu'!)\left\{\left[(\nu+\nu')/2\right]!\right\}^2}\left(\frac{Jt}{\hbar}\right)^{\nu+\nu'} \tag{57}$$

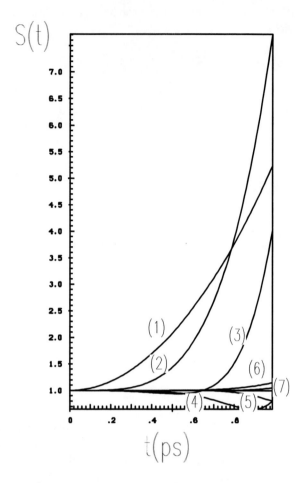

Figure 4: *The norm $S(\mu,t)$ of the state vector in the decoupled case for different orders μ ($\mu=1$-7) of the wave function (indicated by numbers at the different curves) as function of time (the exact value is $S(t)=1$).*

Note, that $(3\nu'+\nu)/2$ is integer, because $(\nu'+\nu)$ is even. Finally, from equ. (52) we obtain

$$S(t) = \lim_{\mu\to\infty}S(\mu,t) = \frac{1}{N}\sum_{k=1}^{N}e^{i\frac{J\lambda_k}{\hbar}t}e^{-i\frac{J\lambda_k}{\hbar}t} = 1 \tag{58}$$

We have calculated these expectation values for the case $J=0.967$ meV up to the 7th order of the state vector, as described above. In all these calculations, the total energy was in absolute value smaller than 10^{-17} eV, i.e. it is vanishing within machine accuracy. In Fig. 4 we show the norms $S(\mu,t)$. We draw all computed orders in one plot. The time covered was 1 ps. It is obvious, that starting from third order, the norm is reasonably correct up to a time of roughly 0.6-0.8 ps, while in larger orders it is correct in the full interval of 1 ps. In Fig. 5 we show the time evolution of the expectation values of the number operators $N_n(t)$ for the exact state vector (Fig. 5a).

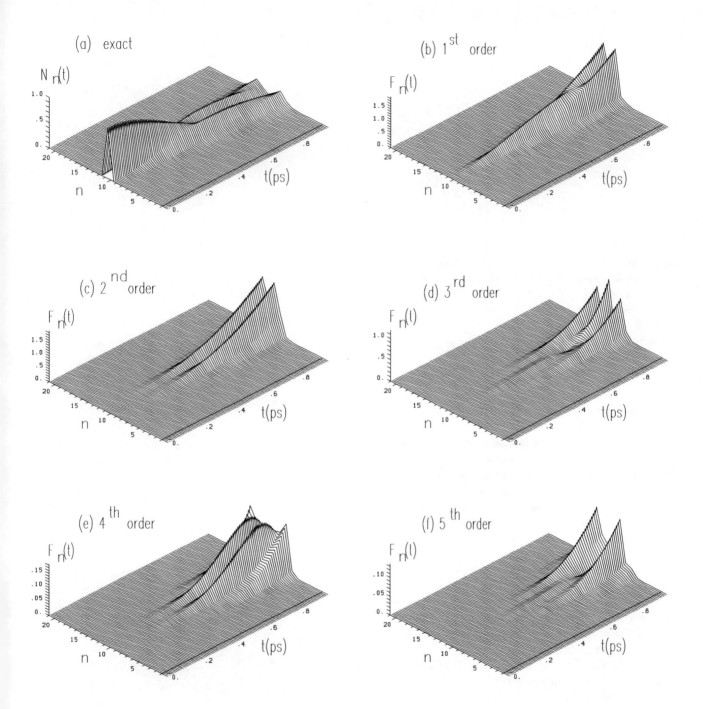

Figure 5. (continued on next page)

Obviously, within 1 ps the initially localized excitation spreads only over a few sites left and right of the initial excitation site. In the further parts of Fig. 5 we plot the errors of the corresponding expectation values $N_n(\mu,t)$, $F_n(\mu,t)=|N_n(t)-N_n(\mu,t)|$, for the state vectors in the different orders μ in time. It is clear that in first or second order, only in a small time interval of 0.1 ps (1st order) or 0.3 ps (2nd order) the errors are reasonably small, while in third order already a time interval of 0.6 ps is covered. In the higher orders, μ=4-7, the fine structure of the transport evolves, and the errors are reasonably small over the whole interval of 1 ps. In the highest (7th) order the maximum error within this time is around 0.02. Therefore, if we want to compare results as obtained from the $|D_1\rangle$ state with those from an expansion as described (see following paper) we are only able to do that up to roughly 0.6 ps time if we want to restrict ourselves to a third order expansion.

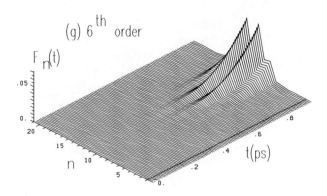

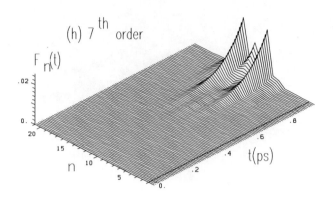

Figure 5: *The expectation values of the number operators $N_n(t)$ calculated from the exact state vector and the errors $F_n(\mu,t)=|N_n(t)-N_n(\mu,t)|$ for different orders μ of the state vector as function of site and time in the decoupled case:*

(a) $N_n(t)$ (b) $\mu=1$ (c) $\mu=2$ (d) $\mu=3$
(e) $\mu=4$ (f) $\mu=5$ (g) $\mu=6$ (h) $\mu=7$

The Phonon System in the Small Polaron Limit

The conclusion drawn in the last subchapter holds only for a freely dispersing excitation in the oscillator subsystem. In the complete system this time evolution is perturbed by the interaction with the phonons. Thus we have to study also, how well the lattice is described by such an expansion. To this end we turn now to the small polaron limit with a localized excitation at a site o (o=11 in our case of a chain with 21 units). Via the interaction of strength χ this localized excitation interacts with the initially unexcited lattice ($b_{nk}(0)=0$). It excites a shock wave in the lattice which travels roughly with the speed of sound through the chain. The excitation itself remains at its initial site, because J=0. The exact solution for this case $|\omega\rangle$ is given in equ. (A14) in Appendix A. The time evolution of the displacements and momenta, as computed from the exact state vector, for our case (W=13 N/m, χ=62 pN and M=114 m_p) is shown in Fig. 6.

Our Hamiltonian in this case is given in equ. (A1) in Appendix A. The expansion of the wave function in a power series in time is obtained both by direct Taylor expansion of all the time dependent terms in the exact solution, equ. (A14), and ordering according to the powers of t, as well as by succesive action of the Hamiltonian on the initial state, where one has to commute the annihilation operators for phonons occurring in the operator through the expression for the preceding order until they act on the vacuum and vanish. This has to be done, until the final form of the state contains only phonon creation operators. The calculation is rather simple, but lengthy.

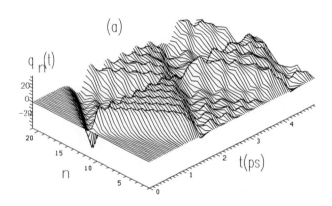

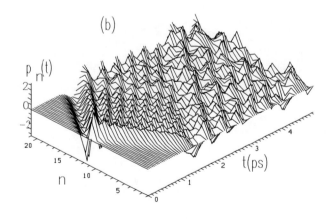

Figure 6: *The time evolution of displacements $q_n(t)$ (in mÅ, part a) and momenta $p_n(t)$ (in meVps/Å, part b) in the small polaron limit as computed from the exact state vector for χ=62 pN (the plots of the time evolution for χ=35 pN are very similar, only the absolute values of $q_n(t)$ and $p_n(t)$ are of roughly half the magnitude).*

Therefore we show here only the final results, which are identical in both procedures described:

$$\left|\psi(\mu)\right\rangle = \sum_{v=1}^{\mu} \frac{T^v}{v!}\left|\psi_v\right\rangle \quad ; \quad T = -\frac{it}{\hbar}$$

(59)

$$\left|\psi_v\right\rangle = \hat{\omega}\left|\psi_{v-1}\right\rangle \quad ; \quad \left|\psi_0\right\rangle = \hat{a}_o^+\left|0\right\rangle$$

From this the components of the state vector in the first three orders μ are obtained as

$$\left|\psi_0\right\rangle = \hat{a}_o^+\left|0\right\rangle$$

$$\left|\psi_1\right\rangle = \hat{\omega}\left|\psi_0\right\rangle = \hat{y}_o^+\hat{a}_o^+\left|0\right\rangle \quad ; \quad \hat{y}_o^+ \equiv \sum_k \hbar\omega_k B_{ok}\hat{b}_k^+$$

$$\left|\psi_2\right\rangle = \hat{\omega}\left|\psi_1\right\rangle =$$

$$= \left[\left(\hat{y}_o^+\right)^2 + \sum_k \left(\hbar\omega_k\right)^2 B_{ok}\left(B_{ok} + \hat{b}_k^+\right)\right]\hat{a}_o^+\left|0\right\rangle$$

(60)

$$\left|\psi_3\right\rangle = \hat{\omega}\left|\psi_2\right\rangle = \left[\left(\hat{y}_o^+\right)^3 + \right.$$

$$\left. + \sum_k \left(\hbar\omega_k\right)^2 B_{ok}\left(B_{ok} + \hat{b}_k^+\right)\left(3\hat{y}_o^+ + \hbar\omega_k\right)\right]\hat{a}_o^+\left|0\right\rangle$$

The relevant expectation values in this case are the norm of the state $S(\mu,t)=\langle\psi(\mu,t)|\psi(\mu,t)\rangle$, the expectation value of the Hamiltonian and those of the phonon operators

$$S(\mu,t) = \left\langle\psi(\mu,t)\middle|\psi(\mu,t)\right\rangle$$

$$H(\mu,t) = \left\langle\psi(\mu,t)\middle|\hat{\omega}\middle|\psi(\mu,t)\right\rangle$$

$$B_k(\mu,t) = \left\langle\psi(\mu,t)\middle|\hat{b}_k\middle|\psi(\mu,t)\right\rangle$$

(61)

$$q_n(\mu,t) = \sum_k \sqrt{\frac{2\hbar}{M\omega_k}}U_{nk}\,\text{Re}\left[B_k(\mu,t)\right]$$

$$p_n(\mu,t) = \sum_k \sqrt{2M\hbar\omega_k}U_{nk}\,\text{Im}\left[B_k(\mu,t)\right]$$

The calculation of these expectation values is again rather tedious, so we refer the reader for some details and the ex-

plicit expressions to Appendix D and turn now to the numerical results.

As mentioned above we show in Figure 6 displacements and momenta as computed from the exact solution for $\chi=62$ pN. The corresponding plots for $\chi=35$ pN (not shown here), are looking very similar, however, in that case the absolute values are roughly half of that for the larger coupling. The results show the usual shock wave in the lattice, caused by the localized excitation at site o (o=11 in chains of 21 units). The wave clearly disperses and becomes enhanced when its front passes the initial excitation for the second time after roughly 2-3 ps. Figure 7, which shows the time evolution of $H(\mu,t)$ and of $S(\mu,t)$ indicates clearly, that the third order wave function is reasonably accurate on a time scale which is much smaller than that for the corresponding third order function in the decoupled oscillator system discussed before. The reason for this is that the characteristic times of the oscillations in the lattice are much smaller than the characteristic time of the oscillator system, as can be seen from Table 1.

Table 1: *The characteristic times of the different lattice oscillations, $T_k = 1/\omega_k$, and for the amide-I oscillator subsystem, $T_0 = \hbar/J$, for the chain discussed in the main text (the different coupling constants do not influence these times).*

k	T_k(ps)	T_o(ps)
1,20	0.406	
2,19	0.205	
3,18	0.139	
4,17	0.107	
5,16	0.089	0.681
6,15	0.077	
7,14	0.070	
8,13	0.065	
9,12	0.062	
10,11	0.061	
21	∞	

For the case of the smaller coupling ($\chi=35$ pN, m=114 m$_p$) the third order curve is close to the corresponding exact one for a time of roughly 0.12-0.14 ps and up to 0.12 ps for the larger coupling ($\chi=62$ pN). This is due to the fact that for larger interaction the shock wave has a larger amplitude, which is excited on the same time scale. Thus with increasing coupling it becomes more difficult to describe the exact curves with a low order wave function. In Figure 8 we show the evolution of the displacements and momenta for sites o and o+1. In our very small region of time only these two sites are excited to a non-negligible extent. Already at sites

J. Mol. Model. **1996**, 2

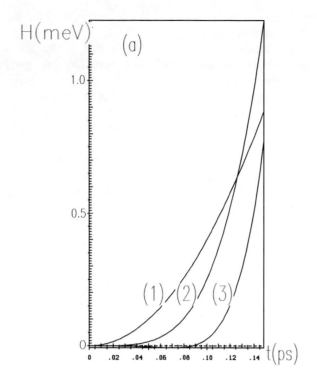

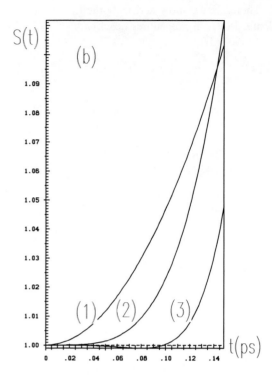

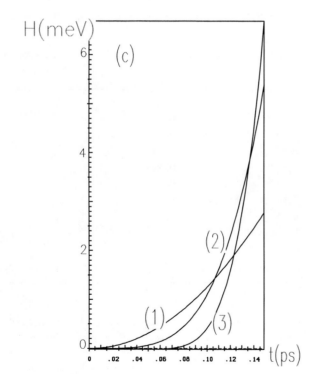

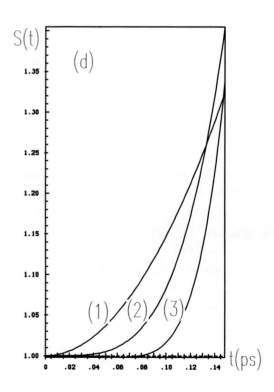

Figure 7: *The functions H(μ,t) (in meV) and S(μ,t) in the small polaron limit (the graphs corresponding to the different orders are marked by μ)*
(a) H(μ,t), χ=35 pN   (b) S(μ,t), χ=35 pN
(c) H(μ,t), χ=62 pN   (d) S(μ,t), χ=62 pN

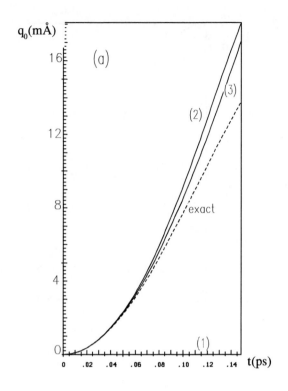

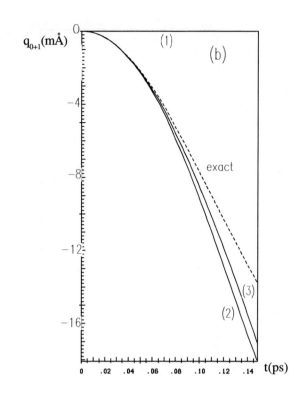

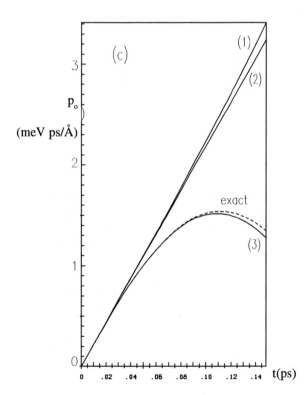

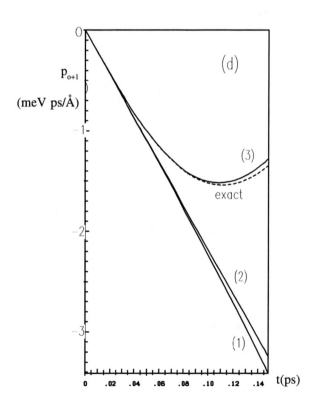

Figure 8 (continued on next page): *The displacements $q_o(\mu,t)$ and $q_{o+1}(\mu,t)$ together with the corresponding exact curves (in mÅ) and the momenta $p_o(\mu,t)$ and $p_{o+1}(\mu,t)$ together with the corresponding exact curves (in meVps/Å) in the small polaron limit (o=11, N=21):*

(a) $q_o(\mu,t)$; $\chi=35\ pN$
(b) $q_{o+1}(\mu,t)$; $\chi=35\ pN$
(c) $p_o(\mu,t)$; $\chi=35\ pN$
(d) $p_{o+1}(\mu,t)$; $\chi=35\ pN$
(e) $q_o(\mu,t)$; $\chi=62\ pN$
(f) $q_{o+1}(\mu,t)$; $\chi=62\ pN$
(g) $p_o(\mu,t)$; $\chi=62\ pN$
(h) $p_{o+1}(\mu,t)$; $\chi=62\ pN$

J. Mol. Model. **1996,** *2*

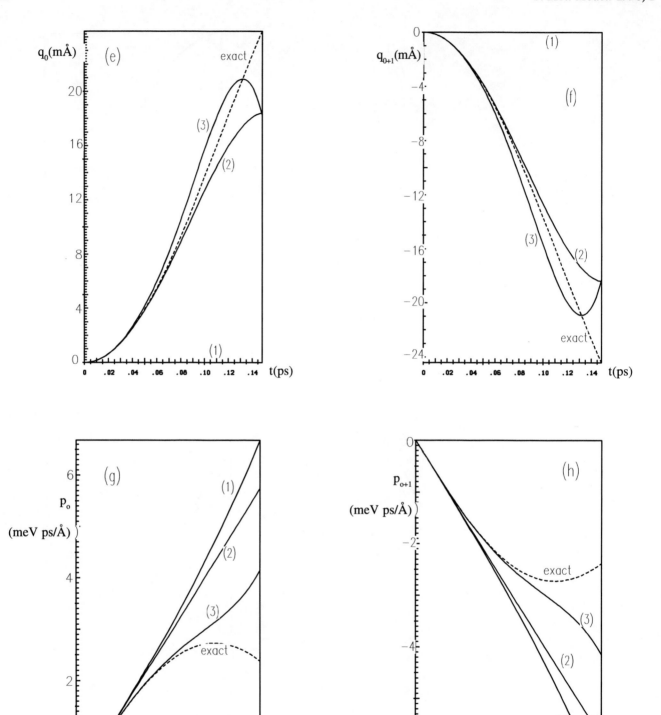

Figure 8 (continued)

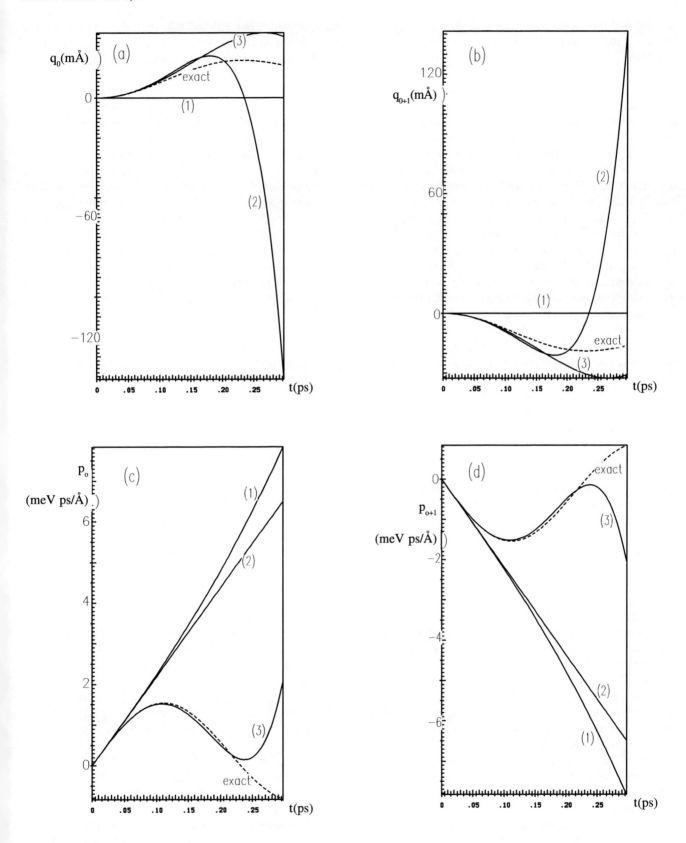

Figure 9: *As Figure 8 but for a longer time ($\chi=35$ pN):*
(a) $q_o(\mu,t)$ (b) $q_{o+1}(\mu,t)$
(c) $p_o(\mu,t)$ (d) $p_{o+1}(\mu,t)$

o-1 and o+2 the excitations within 0.15 ps are negligible (a maximum of 0.1 mÅ for q_{o-2}, of 1.8 mÅ for q_{o-1}, of 0.05 meV ps/Å for p_{o-2}, and of 0.7 meV ps/Å for p_{o-1} in case of the smaller coupling). Thus on our small time scale we deal mainly with a dimer. In Figure 8a to d the displacements and momenta for the small coupling case are shown. The reliability of the third order curves in this case obviously corresponds to that of the total energy and the norm as discussed above. Thus also for displacements and momenta the third order wave function is a reasonable approximation to the exact one up to 0.12-0.14 ps in the small coupling case and up to around 0.12 ps for the larger coupling (Figure 8e to h). The situation for the momenta is somewhat strange, because at least for the small coupling the momenta in third order are reasonably correct on a larger time scale than the displacements, as can be seen in Fig. 9, and even the displacements are qualitatively tolerable up to more than 0.2 ps, although the norm and total energy of the third order state differ to a quite large extent from their correct values. However, this does not hold for larger coupling constants, like χ=62 pN, because in this case the unphysical increase of the factors containing explicit powers of t starts to dominate earlier. This is due to the fact, that for increasing coupling shock waves with increasing amplitudes are excited on the same time scale as for smaller couplings.

Conclusion

We have expanded for the exactly solvable transportless and the decoupled case of the Davydov Hamiltonian the exact wave functions in power serieses in time. From this one can conclude up to what times such expansions of the exact wave functions in the general case are reliable. This information is important for studies of the short time behaviour of the $|D_1\rangle$, or also generally of other ansatz states. We found that in the case of coupling between a localized excitation of an amide-I oscillator in a cyclic chain (small polaron or transportless limit) to the lattice phonons a third order expansion of the exact wave function gives reliable results for the relevant variables of the lattice up to 0.12 ps for the larger coupling of χ=62 pN, and up to about 0.12-0.14 ps for a smaller coupling constant of χ=35 pN. The calculation of higher orders becomes very tedious and is for this reason not feasible. In the case of a decoupled oscillator system a third order wave function yields reasonable expectation values for about 0.6-0.8 ps. However, for our expansion of the exact wave function in the general case, which is the subject of the following paper, we have to restrict the study of its short time behaviour to the smaller one of the two time scales. Only in this case we can be confident that expectation values computed with a third order expansion are reasonable. More important, we have seen that the accuracy of S(t) and H(t) (constant in time for exact solutions and initial values known) parallels completely that of the other, physically more interesting expectation values for a given expansion in both systems studied. Thus S(t) and H(t) are indicators for the time up to which an expansion yields reliable results. In any case, one has to conclude that expansions of the kind used here are not useful for the investigation of long time dynamics, since to this end very large orders would be required, which are prohibitively tedious to compute. Therefore one cannot get rid in this way of the requirement to use ansatz states for such purposes.

Further we found that, although the $|D_1\rangle$ ansatz state contains site dependent coherent state amplitudes, the initial state has to be constructed in form of a $|D_2\rangle$ wave function, i.e. with site independent amplitudes, from the initial set of lattice displacements and momenta. Otherwise in case of the decoupled limit $|D_1\rangle$ dynamics would not lead to the exact solution. Further, the known requirements for computation of correct amplitudes b_{nk} from a given set of displacements and momenta do not lead to a unique set of b_{nk}'s and only the use of site independent b_k's yields consistent values of the lattice energy.

In conlusion, the present work lead to the foundations for a thorough investigation of the very small time behaviour of $|D_1\rangle$ dynamics (or generally for all ansatz states) in comparison to the exact wave function, which is the subject of the following paper.

Acknowledgement The financial support of the „Deutsche Forschungsgemeinschaft" (project no. Fo 175/3-1) and of the „Fond der Chemischen Industrie" is gratefully acknowledged.

References

1. Scott, A.C. *Phys. Rep.* **1992**, *217*, 1.
2. Davydov, A.S.; Kislukha, N.I. *Phys. Stat. Sol.* **1973**, *B59*, 465; Davydov, A.S. *Phys. Scripta* **1979**, *20*, 387.
3. Davydov, A.S. *Zh. Eksp. Teor. Fiz.* **1980**, *78*, 789; *Sov. Phys. JETP* **1980**, *51*, 397.
4. Scott, A.C. *Phys. Rev.* **1982**, *A26*, 57; *Physica Scripta* **1984**, *29*, 279; Mac Neil, L.; Scott, A.C. *Physica Scripta* **1984**, *29*, 284; *Phil. Trans. Roy Soc. London* **1985**, *A315*, 423.
5. Förner, W.; Ladik, J. In *Davydov's Soliton Revisited*; Christiansen, P.L.; Scott, A.C., Eds.; NATO ASI, Series B -Physics, Vol. 243; Plenum, New York (1991), 267.
6. Förner, W. *J. Phys.: Condensed Matter* **1993**, *5*, 803.
7. Halding, J.; Lomdahl, P.S. *Phys. Lett.* **1984**, *A124*, 37.
8. (a) Lomdahl, P.S.; Kerr, W.C. *Phys. Rev. Lett.* **1985**, *55*, 1235; and in *Davydov's Soliton Revisited*, Christiansen, P.L.; Sott, A.C., Eds. NATO ASI, Series B - Physics, Vol. 243, Plenum, New York (1991), 259.

   (b) Kerr, W.C.; Lomdahl, P.S. *Phys. Rev.* **1987**, *B35*, 3629.

   (c) Kerr, W.C.; Lomdahl, P.S. In *Davydov's Soliton Revisited*, Christiansen, P.L.; Scott, A.C., Eds.; NATO ASI,

Series B - Physics, Vol. 243, Plenum, New York (1991), 23.

9. Lawrence, A.F.; McDaniel, J.C.; Chang, D.B.; Pierce, B.M.; Birge, R.R. *Phys. Rev.* **1986**, *A33*, 1188.

10. Bolterauer, H. In *Structure Coherence and Chaos in Dynamical Systems*, Christiansen, P.L.; Parmentier, R.D., Eds.; Proc. MIDIT 1986 Workshop, Manchester University Press, Englsnd (1986); Bolterauer, H. In *Davydov's Soliton Revisited*, Christiansen, P.L.; Scott, A.C., Eds.; NATO ASI, Series B - Physics, Vol. 243, Plenum, New York (1991), 99, and 309.

11. Cottingham, J.P.; Schweitzer, J.W. *Phys. Rev. Lett.* **1989**, *62*, 1792; Schweitzer, J.W.; Cottingham, J.P. In *Davydov's Soliton Revisited*, Christiansen, P.L.; Scott, A.C., Eds.; NATO ASI, Series B - Physics, Vol. 243, Plenum, New York (1991), 285.

12. Motschmann, H.; Förner, W.; Ladik, J. *J. Phys.: Cond. Matter* **1989**, *1*, 5083.

13. (a) Förner, W. *J. Phys.: Cond. Matter* **1991**, *3*, 4333. (b) Förner, W. *J. Comput. Chem.* **1992**, *13*, 275.

14. Brown, D.W.; Linderberg, K.; West, B.J. *Phys. Rev.* **1986**, *A33*, 4104, 4110; *Phys. Rev.* **1987**, *B35*, 6169; *Phys. Rev.* **1988**, *B37*, 2946; Brown, D.W. *Phys. Rev.* **1988**, *A37*, 5010.

15. Cruzeiro, L.; Halding, J.; Christiansen, P.L.; Skovgaard, O.; Scott, A.C. *Phys. Rev.* **1988**, *A37*, 880.

16. Mechtly, B.; Shaw, P.B. *Phys.Rev.* **1988**, *B38*, 3075.

17. Skrinjar, M.J.; Kapor, D.V; Stojanovic, S.D. *Phys. Rev.* **1988**, *A38*, 6402.

18. Förner, W. *Phys. Rev.* **1991**, *A44*, 2694; *J. Mol. Struct. (Theochem)* **1993**, *282*, 223.

19. Förner, W. *Nanobiology* **1992**, *1*, 413.

20. Förner, W. *J. Phys.: Cond. Matter* **1992**, *4*, 1915.

21. Wang, X.; Brown, D.W.; Linderberg, K. *Phys. Rev. Lett.* **1989**, *62*, 1796.

22. Scott, A.C. Presented at the *Conference on Nonlinear Sciences: The Next Decade*, Los Alamos, National Laboratory, on May 21, 1990.

23. Brown, D.W.; Ivic, Z. *Phys. Rev.* **1989**, *B40*, 9876.

24. Förner, W. *J. Phys.: Cond. Matter* **1993**, *5*, 823.

25. Förner, W. *J. Phys.: Cond. Matter* **1993**, *5*, 3883.

26. Förner, W. *J. Phys.: Cond. Matter* **1993**, *5*, 3897.

27. Förner, W. *Physica* **1993**, *D68*, 68.

28. Förner, W. *J. Phys.: Cond. Matter* **1994**, *6*, 9089.

29. Cruzeiro-Hansson, L.; Christiansen, P.L.; Scott, A.C. In *Davydov's Soliton Revisited*, Christiansen, P.L.; Scott, A.C., Eds.; NATO-ASI Series B - Physics, No. 243; Plenum Press New York (1991), 325.

30. Kapor, D. Remark in *Davydov's Soliton Revisited*, Christiansen, P.L.; Scott, A.C., Eds.; Nato ASI Series No. 243, Plenum Press, New York (1991), 29.

31. Cruzeiro-Hansson, L.; Okhonin, V.A.; Khlebopros, R.G.; Yassievich, I.N. *Nanobiology* **1992**, *1*, 395.

32. Cruzeiro-Hansson, L. *Phys. Rev. Lett.* **1994**, *73*, 2927.

33. Förner, W. *J. Mol. Struct. (Theochem)* **1996**, *362*, 101.

34. Barthes, M. In *Nonlinear Excitations in Biomolecules*, Peyrard, M., Ed.; Springer, Berlin, Heidelberg, New York, (1995), 209.

Appendix A: Treatment of the Small Polaron Limit Using the |D₁> Ansatz

The Hamiltonian $\hat{H}_{sp}$ in this case and the |D₁> ansatz are given by

$$\hat{H}_{SP} \equiv \hat{\omega} = \sum_{nk} \hbar\omega_k B_{nk}\left(\hat{b}_k^+ + \hat{b}_k\right)\hat{a}_n^+\hat{a}_n + \sum_k \hbar\omega_k \hat{b}_k^+ \hat{b}_k$$

$$|D_1\rangle = \sum_n a_n(t)\hat{a}_n^+|0\rangle_e|\beta_n\rangle \tag{A1}$$

$$|\beta_n\rangle = \exp\left[-\tfrac{1}{2}\sum_k |b_{nk}(t)|^2\right]\cdot\exp\left[\sum_k b_{nk}(t)\hat{b}_k^+\right]|0\rangle_p$$

First of all we have to show that our ansatz satisfies the Schrödinger equation for the Hamiltonian given in (A1). The left hand side of the equation is readily calculated and yields

$$i\hbar\frac{\partial}{\partial t}|D_1\rangle = i\hbar\sum_n\left\{-\sum_k\left[\dot{b}_{nk}\left(\tfrac{1}{2}b_{nk}^* - \hat{b}_k^+\right) + \tfrac{1}{2}\dot{b}_{nk}^* b_{nk}\right]a_n + \dot{a}_n\right\}|\beta_n\rangle\hat{a}_n^+|0\rangle_e \tag{A2}$$

To eliminate the time derivatives in (A2) we need the equations of motion for the |D₁> ansatz under the condition J=0. From equ. (17) in the main text follows

$$i\hbar\dot{a}_n = -\frac{i\hbar}{2}\sum_k\left(\dot{b}_{nk}b_{nk}^* - \dot{b}_{nk}^* b_{nk}\right)a_n + \sum_k \hbar\omega_k\left[B_{nk}\left(b_{nk} + b_{nk}^*\right) + |b_{nk}|^2\right]a_n \tag{A3}$$

$$i\hbar\dot{b}_{nk} = \hbar\omega_k\left(b_{nk} + B_{nk}\right) \tag{A4}$$

Substitution of (A4) into (A3) yields

$$i\hbar\dot{a}_n = \tfrac{1}{2}\sum_k \hbar\omega_k B_{nk}\left(b_{nk} + b_{nk}^*\right)a_n \tag{A5}$$

The use of (A5) together with (A4) leads to the final form of (A2)

$$i\hbar\frac{\partial}{\partial t}|D_1\rangle = \sum_{nk} \hbar\omega_k\left[B_{nk}b_{nk} + \left(b_{nk} + B_{nk}\right)\hat{b}_k^+\right]|\beta_n\rangle a_n\hat{a}_n^+|0\rangle_e \tag{A6}$$

With the help of the eigenvalue equation for coherent states

$$\hat{b}_k|\beta_n\rangle = b_{nk}|\beta_n\rangle \tag{A7}$$

we obtain (A8) for the different terms on the right hand side of the Schrödinger equation and comparison with (A6) shows that the equation is fulfilled, and thus |D₁> together with the equations of motion (A4) and (A5) is the exact solution for the small polaron limit:

$$\hat{\omega}|D_1\rangle = \sum_n\left[\sum_{mk} \hbar\omega_k B_{mk}\left(\hat{b}_k^+ + \hat{b}_k\right)\hat{a}_m^+\hat{a}_m + \sum_k \hbar\omega_k\hat{b}_k^+\hat{b}_k\right]|\beta_n\rangle a_n\hat{a}_n^+|0\rangle_e =$$

$$= \sum_{nk} \hbar\omega_k\left[B_{nk}\hat{b}_k^+ + B_{nk}b_{nk} + b_{nk}\hat{b}_k^+\right]|\beta_n\rangle a_n\hat{a}_n^+|0\rangle_e = i\hbar\frac{\partial}{\partial t}|D_1\rangle \tag{A8}$$

The explicit form of this solution is obtained by direct integration of (A4):

$$\int_{b_{nk}(0)}^{b_{nk}(t)}\frac{db'_{nk}}{b'_{nk} + B_{nk}} = -i\omega_k\int_0^t dt' \tag{A9}$$

and thus

$$b_{nk}(t) = b_{nk}(0)e^{-i\omega_k t} + B_{nk}\left(e^{-i\omega_k t} - 1\right) \tag{A10}$$

Further, together with (A10), we can integrate (A5)

$$i\hbar \int_{a_n(0)}^{a_n(t)} \frac{da'_n}{a'_n} = \frac{1}{2}\sum_k \hbar\omega_k B_{nk} \int_0^t \left[b_{nk}(t') + b^*_{nk}(t')\right] dt' \tag{A11}$$

which yields the general solution for the transportless case

$$|D_1\rangle = \sum_n \exp\left\{\sum_k \left[b_{nk}(t)\hat{b}^+_k - b^*_{nk}(t)\hat{b}_k\right]\right\} a_n(t)\hat{a}^+_n |0\rangle \tag{A12}$$

$$a_n(t) = a_n(0) \cdot \exp\left\{-i\sum_k B^2_{nk}\left[\sin(\omega_k t) - \omega_k t\right]\right\} \cdot \exp\left\{-i\sum_k B_{nk}\left\{\text{Re}\left[b_{nk}(0)\right]\sin(\omega_k t) + \text{Im}\left[b_{nk}(0)\right]\left[1 - \cos(\omega_k t)\right]\right\}\right\}$$

$$b_{nk}(t) = b_{nk}(0)e^{-i\omega_k t} + B_{nk}\left(e^{-i\omega_k t} - 1\right)$$

where Re[x] (Im[x]) denotes the real (imaginary) part of a complex number x. For later considerations we need also the special case, that we start from an undistorted lattice and an excitation localized at one site o, i. e. $b_{nk}(0)=0$ and $a_n(0)=\delta_{no}$:

$$a_n(t) = e^{-i\sum_k B^2_{ok}\left[\sin(\omega_k t) - \omega_k t\right]}\delta_{no} \quad ; \quad b_{nk}(t) = B_{nk}\left(e^{-i\omega_k t} - 1\right) \tag{A13}$$

with the total state vector

$$|\omega\rangle = e^{-i\sum_k B^2_{ok}\left[\sin(\omega_k t) - \omega_k t\right]} \cdot e^{\sum_k \left[b_{ok}(t)\hat{b}^+_k - b^*_{ok}(t)\hat{b}_k\right]} \hat{a}^+_o |0\rangle$$

$$b_{ok}(t) = B_{ok}\left(e^{-i\omega_k t} - 1\right) \tag{A14}$$

For the computation of expectation values we concentrate first on the same case as Brown et al. [14], namely an undistorted lattice $b_{nk}(0)=0$, but an arbitrary excitation $a_n(0)$. Then the lattice energy is given by

$$E_{lat} = \sum_k \hbar\omega_k \left\langle D_1 \left| \hat{b}^+_k \hat{b}_k \right| D_1 \right\rangle = \sum_{nk} \hbar\omega_k |a_n|^2 |b_{nk}|^2 \tag{A15}$$

With the help of

$$|a_n(t)|^2 = |a_n(0)|^2 \equiv P_{nn} \quad ; \quad |b_{nk}(t)|^2 = 2B^2_{nk}\left[1 - \cos(\omega_k t)\right] \tag{A16}$$

we obtain

$$E_{lat} = 2\sum_{nk} \hbar\omega_k B^2_{nk} P_{nn}\left[1 - \cos(\omega_k t)\right] \tag{A17}$$

which is identical to the exact result given by Brown et al. (Ref. [14], second paper). The exciton-lattice interaction energy is

$$E_{int} = \sum_{nk} \hbar\omega_k B_{nk} \left\langle D_1 \left| \left(\hat{b}^+_k + \hat{b}_k\right)\hat{a}^+_n \hat{a}_n \right| D_1 \right\rangle = 2\sum_{nk} \hbar\omega_k B_{nk} \text{Re}\left[b_{nk}\right]|a_n|^2 = 2\sum_{nk} \hbar\omega_k B^2_{nk} P_{nn}\left[\cos(\omega_k t) - 1\right] = -E_{lat} \tag{A18}$$

Therefore, as to be expected, the total energy is conserved and vanishes in this case. The expectation values of the phonon operators are

$$B_k \left\langle D_1 \middle| \hat{b}_k \middle| D_1 \right\rangle = \sum_n |a_n|^2 b_{nk} = \sum_n P_{nn} B_{nk} \left(e^{-i\omega_k t} - 1 \right) \quad ; \quad B_k^* = \left\langle D_1 \middle| \hat{b}_k^+ \middle| D_1 \right\rangle$$

(A19)

From this we obtain the displacements as

$$q_n(t) = \sum_k \sqrt{\frac{2\hbar}{M\omega_k}} U_{nk} \, \mathrm{Re}[B_k] = \sum_{mk} \sqrt{\frac{2\hbar}{M\omega_k}} P_{mm} U_{nk} B_{mk} \left[\cos(\omega_k t) - 1 \right]$$

(A20)

where matrix $\underline{U}$ contains the normal mode coefficients (see Appendix B for details). This is again identical to the exact quantity given by Brown et al. Finally the momenta are given by

$$p_n(t) = \sum_k \sqrt{2M\hbar\omega_k} U_{nk} \, \mathrm{Im}[B_k] = -\sum_{mk} \sqrt{2M\hbar\omega_k} P_{mm} U_{nk} B_{mk} \sin(\omega_k t)$$

(A21)

For the general case we have

$$E_{lat} = \sum_{nk} \hbar\omega_k \left\{ |b_{nk}(0)|^2 + 2B_{nk}^2 \left[1 - \cos(\omega_k t) \right] + 2B_{nk} \left[x_{nk} \left[1 - \cos(\omega_k t) \right] - y_{nk} \sin(\omega_k t) \right] \right\} \cdot P_{nn}$$

$$E_{int} = \sum_{nk} \hbar\omega_k \left\{ 2B_{nk}^2 \left[\cos(\omega_k t) - 1 \right] + 2B_{nk} \left[x_{nk} \cos(\omega_k t) + y_{nk} \sin(\omega_k t) \right] \right\} \cdot P_{nn}$$

(A22)

where the abbreviations $x_{nk} = \mathrm{Re}[b_{nk}(0)]$ and $y_{nk} = \mathrm{Im}[b_{nk}(0)]$ were used. Finally the conserved total energy $E_{tot} = E_{lat} + E_{int}$ is given by

$$E_{tot} = \left\langle D_1 \middle| \hat{\omega} \middle| D_1 \right\rangle = \sum_{nk} \hbar\omega_k \left\{ |b_{nk}(0)|^2 + B_{nk} \left[b_{nk}(0) + b_{nk}^*(0) \right] \right\} \cdot P_{nn}$$

(A23)

Appendix B: Normal Mode Coefficients for the Lattice

The solution of the classical normal mode problem for a cyclic chain of oscillators leads to the problem of diagonalization of a matrix $\underline{V}$:

$$\ddot{\underline{q}} = -\underline{V} \, \underline{q} \Rightarrow \underline{U}^+ \ddot{\underline{q}} = -\underline{U}^+ \underline{V} \, \underline{U} \, \underline{U}^+ \underline{q} \Rightarrow \underline{U}^+ \ddot{\underline{q}} = -\underline{\omega}^2 \underline{U}^+ \underline{q} \quad ; \quad \omega_{kk'}^2 \equiv \omega_k^2 \delta_{kk'}$$

$$V_{nm} = \frac{W}{M} \left\{ 2\delta_{nm} - (1 - \delta_{nN}) \delta_{m,n+1} - (1 - \delta_{n1}) \delta_{m,n-1} - \delta_{n1} \delta_{mN} - \delta_{nN} \delta_{m1} \right\}$$

(B1)

For this purpose we split the matrix in the form

$$\underline{V} = \frac{W}{M} \left(2 \cdot \underline{1} - \underline{X} \right)$$

$$X_{nm} = (1 - \delta_{nN}) \delta_{m,n+1} + (1 - \delta_{n1}) \delta_{m,n-1} + \delta_{n1} \delta_{mN} + \delta_{nN} \delta_{m1}$$

(B2)

Instead of the eigenvalue problem $\underline{\underline{V}}\underline{u}_k = \omega_k^2 \underline{u}_k$ we solve the related problem $\underline{\underline{X}}\underline{u}_k = \lambda_k \underline{u}_k$. The relation is, that both matrices have the same eigenvectors and the relation of the eigenvalues is

$$\underline{\underline{V}}\underline{u}_k = \frac{W}{M}\left(2\cdot\underline{\underline{1}}-\underline{\underline{X}}\right)\underline{u}_k = 2\frac{W}{M}\underline{u}_k - \frac{W}{M}\lambda_k\underline{u}_k = \frac{W}{M}\left(2-\lambda_k\right)\underline{u}_k \Rightarrow \omega_k^2 = \frac{W}{M}\left(2-\lambda_k\right) \tag{B3}$$

Since $\underline{\underline{X}}$ is a reducible representation of the rotation group C_N we can write down its eigenvector matrix $\underline{\underline{U}}$ without further calculation:

$$\underline{\underline{U}} \equiv \left(\underline{u}_1 \quad \underline{u}_2 \quad \cdots \quad \underline{u}_N\right) \quad ; \quad U_{nk} = \frac{1}{\sqrt{N}}e^{\frac{2\pi i}{N}nk} \quad ; \quad \Rightarrow U_{nk} = U_{N+1,k} \quad ; \quad U_{nk} = U_{n,N+k} \quad ; \quad U_{n,N-k} = U_{nk}^*$$

$$\sum_k U_{nk}U_{n'k}^* = \delta_{nn'} \quad ; \quad \sum_n U_{nk}^* U_{nk'} = \delta_{kk'} \tag{B4}$$

Since in our cyclic system, the choice of the numbering is arbitrary, we assume n and k to run from 1 to N. The eigenvalues are obtained by explicitly performing the matrix product

$$\left(\underline{\underline{U}}^+ \underline{\underline{X}}\underline{\underline{U}}\right)_{kk'} = \lambda_k \delta_{kk'} \tag{B5}$$

using the relation

$$\delta_{kk'} = \frac{1}{N}\sum_{n=1}^{N} e^{\frac{2\pi i}{N}n(k-k')} \tag{B6}$$

From this calculation the eigenvalues of $\underline{\underline{X}}$ and $\underline{\underline{V}}$ are found to be

$$\lambda_k = 2\cdot\cos\left(\frac{2\pi}{N}k\right) \quad ; \quad \omega_k^2 = 2\frac{W}{M}\left[1-\cos\left(\frac{2\pi}{N}k\right)\right] \tag{B7}$$

Using $1-\cos(2\alpha)=2\sin^2(\alpha)$ we obtain

$$\omega_k = 2\sqrt{\frac{W}{M}}\cdot\sin\left(\frac{\pi}{N}k\right) \tag{B8}$$

Obviously with exception of k=N each eigenvalue is doubly degenerate (we concentrate here and in the rest of the paper on odd numbers N), namely $\omega_k=\omega_{N-k}$, which is easily shown, using the trigonometric relation $\sin(\alpha-\beta)=\sin(\alpha)\cos(\beta)-\cos(\alpha)\sin(\beta)$. Therefore, any linear combination of the two eigenvectors belonging to the eigenvalues ω_k and ω_{N-k} yields also eigenvectors of $\underline{\underline{V}}$. Thus from the set of degenerate eigenvectors, we can form a new set of real and orthonormal eigenvectors by

$$\varphi_{nk}^{(1)} = \frac{1}{\sqrt{2}}\left(U_{nk} + U_{n,N-k}\right) = \sqrt{\frac{2}{N}}\cdot\cos\left(\frac{2\pi}{N}nk\right) \tag{B9}$$

$$\varphi_{nk}^{(2)} = -\frac{i}{\sqrt{2}}\left(U_{nk} - U_{n,N-k}\right) = \sqrt{\frac{2}{N}}\cdot\sin\left(\frac{2\pi}{N}nk\right)$$

$$k = 1,\ldots,\frac{N-1}{2}$$

Since the new eigenvectors belong still to degenerate eigenvalues, we can form again another set of orthonormal eigenvectors by the linear combination:

$$\psi_{nk}^{(1,2)} = \frac{1}{\sqrt{2}}\left(\varphi_{nk}^{(1)} \pm \varphi_{nk}^{(2)}\right)$$

(B10)

leading to

$$\cos(\alpha) + \sin(\alpha) = \sqrt{2}\cos\left(\alpha - \frac{\pi}{4}\right) \Rightarrow \psi_{nk}^{(1)} = \sqrt{\frac{2}{N}}\cos\left(\frac{2\pi}{N}nk - \frac{\pi}{4}\right)$$

(B11a)

and

$$\cos(\alpha) - \sin(\alpha) = \sqrt{2}\cos\left(\alpha + \frac{\pi}{4}\right) \Rightarrow \psi_{nk}^{(2)} = \sqrt{\frac{2}{N}}\cos\left(\frac{2\pi}{N}nk + \frac{\pi}{4}\right)$$

(B11b)

where k runs again from 1 to (N-1)/2. To get rid of the unpleasent fact of having two functions for one k, we use in the second function (B11b) the index k'=N-k instead of k, but with k' running from [(N-1)/2+1] to (N-1). After performing this substitution and using the relation $\cos(\alpha+\beta)=\cos\alpha\cos\beta - \sin\alpha\sin\beta$ with $\alpha=2\pi n$ we arrive at

$$\psi_{nk}^{(1)} = \sqrt{\frac{2}{N}}\cos\left(\frac{2\pi}{N}nk - \frac{\pi}{4}\right) \quad ; \quad k = 1,\ldots,\frac{N-1}{2}$$

$$\psi_{nk'}^{(2)} = \sqrt{\frac{2}{N}}\cos\left(\frac{2\pi}{N}nk' - \frac{\pi}{4}\right) \quad ; \quad k' = \left(\frac{N-1}{2}+1\right),\ldots,(N-1)$$

(B12)

Since $\cos\left[2\pi n - (\pi/4)\right] = 1/\sqrt{2}$ we have now our final set of real eigenvalues and we substitute it for the eigenvector matrix $\underline{\underline{U}}$:

$$\omega_k = 2\sqrt{\frac{W}{M}}\cdot\sin\left(\frac{\pi}{N}k\right)$$

$$U_{nk} = \sqrt{\frac{2}{N}}\cos\left(\frac{2\pi}{N}nk - \frac{\pi}{4}\right) \quad ; \quad k = 1,\ldots,N \quad ; \quad N \text{ odd}$$

$$\underline{\underline{U}}^+\underline{\underline{U}} = \underline{\underline{U}}\ \underline{\underline{U}}^+ = \underline{\underline{1}}$$

(B13)

This form of the normal mode coefficients can also be used to develop an ab inito Hartree-Fock Crystal Orbital formalism based on real numbers only. However, in this case it does not lead to a complete block-diagonalization, because it leaves the pairwise degenerate sets unresolved [33].

Appendix C: Analytical Solution of the Oscillator System in the Decoupled Case

The Hamiltonian for the oscillator system in the decoupled ($\chi=0$) case reads as

$$\hat{J} = -J\sum_n\left(\hat{a}_n^+\hat{a}_{n+1} + \hat{a}_{n+1}^+\hat{a}_n\right)$$

(C1)

With the ansatz

$$|\psi_e\rangle = \sum_n a_n(t)\hat{a}_n^+|0\rangle_e$$

(C2)

the Schrödinger equation can be exactly solved and is transformed to

$$-\frac{i\hbar}{J}\dot{\underline{a}} = \underline{\underline{X}}\,\underline{a} \tag{C3}$$

where $\underline{\underline{X}}$ is the same matrix as defined in Appendix B, equ. (B2). Therefore we have again the same eigenvector matrix $\underline{\underline{R}}$ and eigenvalues λ_k:

$$\lambda_k = 2\cos\left(\frac{2\pi}{N}k\right) \quad ; \quad R_{nk} = \frac{1}{\sqrt{N}}e^{\frac{2\pi i}{N}nk} \quad ; \quad k = 1,\ldots,N \tag{C4}$$

However, here we keep its complex form and name it $\underline{\underline{R}}$. Then with the transformation $\underline{\underline{R}}^+\underline{a} = \underline{c}$ our equation becomes

$$-\frac{i\hbar}{J}\dot{\underline{c}} = \underline{\underline{\lambda}}\,\underline{c} \quad ; \quad \lambda_{kk'} = \lambda_k\delta_{kk'} \tag{C5}$$

which is simple to integrate:

$$\int_{c_k(0)}^{c_k(t)}\frac{dc_k'}{c_k'} = i\frac{J}{\hbar}\lambda_k\int_0^t dt' \Rightarrow c_k(t) = c_k(0)e^{i\frac{J}{\hbar}\lambda_k t} \tag{C6}$$

The backtransformation is simply done by multiplying the result for $\underline{c}(t)$ from the left with $\underline{\underline{R}}$ and replacing $\underline{c}(0)$ by $\underline{\underline{R}}^+\underline{a}(0)$:

$$\underline{a}(t) = \underline{\underline{R}}\,\underline{c}(t) = \underline{\underline{R}}\cdot e^{i\frac{J}{\hbar}\underline{\underline{\lambda}}t}\underline{\underline{R}}^+\underline{a}(0) \quad ; \quad \left(e^{i\frac{J}{\hbar}\underline{\underline{\lambda}}t}\right)_{kk'} = e^{i\frac{J}{\hbar}\lambda_k t}\delta_{kk'} \tag{C7}$$

or without matrix notation:

$$a_n(t) = \frac{1}{N}\sum_{n'=1}^{N}\sum_{k=1}^{N} e^{\frac{2\pi i}{N}k(n-n')}\cdot e^{2i\frac{Jt}{\hbar}\cos\left(\frac{2\pi}{N}k\right)}\cdot a_{n'}(0) \tag{C8}$$

Thus the total wave function is given by

$$|\psi_e\rangle = \frac{1}{N}\sum_{n,n'=1}^{N}\sum_{k=1}^{N} e^{\frac{2\pi i}{N}k(n-n')}\cdot e^{2i\frac{Jt}{\hbar}\cos\left(\frac{2\pi}{N}k\right)}\cdot a_{n'}(0)\hat{a}_n^+|0\rangle_e \tag{C9}$$

The expectation values N_n of the number operators $\hat{a}_n^+\hat{a}_n$ are given by

$$N_n(t) = |a_n(t)|^2 = \frac{1}{N^2}\sum_{kk'}\sum_{n'n''} e^{\frac{2\pi i}{N}n(k-k')}\cdot e^{\frac{2\pi i}{N}(k'n''-kn')}\cdot e^{2i\frac{Jt}{\hbar}\left[\cos\left(\frac{2\pi}{N}k\right)-\cos\left(\frac{2\pi}{N}k'\right)\right]}\cdot a_{n'}(0)a_{n''}^*(0) \tag{C10}$$

For calculation of the norm of the state we have to sum N_n over n. Then the first exponential can be summed over n and yields together with the factor 1/N the Kronecker symbol $\delta_{kk'}$. When the sum over k' is performed the time dependent exponent vanishes. Then the summation over k can be performed, leading together with the second factor 1/N to $\delta_{n'n''}$. Thus the norm is conserved and equals the norm of the initial state.

The total energy is given by

$$E_{tot} = -J\sum_n \left(a_n^*(t)a_{n+1}(t) + a_n(t)a_{n+1}^*(t)\right) = -2J\sum_n \text{Re}\left[a_n^*(t)a_{n+1}(t)\right] = -2J\sum_n \text{Re}\left[a_n^*(0)a_{n+1}(0)\right] \tag{C11}$$

where the last equality is obtained with the help of the same summation procedure as above, with the only difference that instead of $\delta_{n'n''}$ in this case a factor $\delta_{n'',n'+1}$ is obtained.

For later use we want to give finally the wave function coefficients for the special case of an initial excitation localized at just one site o, i.e. $a_n(0)=\delta_{on}$:

$$|J\rangle = \frac{1}{N}\sum_{nk} e^{\frac{2\pi i}{N}k(n-o)} \cdot e^{2i\frac{Jt}{\hbar}\cos\left(\frac{2\pi}{N}k\right)} \cdot \hat{a}_n^+ |0\rangle_e \tag{C12}$$

In this case the total energy vanishes and N_n is given by

$$N_n(t) = \frac{1}{N^2}\sum_{kk'} e^{\frac{2\pi i}{N}(n-o)(k-k')} \cdot e^{2i\frac{Jt}{\hbar}\left[\cos\left(\frac{2\pi}{N}k\right)-\cos\left(\frac{2\pi}{N}k'\right)\right]} \tag{C13}$$

Appendix D: The Relevant Expectation Values for the Phonon System in the Small Polaron Limit

First of all we compute the norm of the states in different order. In 0th order we have simply S(0,t)=1, and in 1st order

$$S(1,t) = 1 + \sum_k \left(B_{ok}\omega_k t\right)^2 \tag{D1}$$

Then in 2nd order we obtain

$$S(2,t) = \left\langle \psi(1,t) + \frac{1}{2}T^2\psi_2 \middle| \psi(1,t) + \frac{1}{2}T^2\psi_2 \right\rangle =$$

$$= S(1,t) - \frac{t^2}{\hbar^2}\text{Re}\left[\langle\psi_0|\psi_2\rangle\right] + \frac{t^3}{\hbar^3}\text{Im}\left[\langle\psi_1|\psi_2\rangle\right] + \frac{1}{4}\frac{t^4}{\hbar^4}\langle\psi_2|\psi_2\rangle = \frac{3}{4}\left[\sum_k (B_{ok}\omega_k t)^2\right]^2 + \frac{1}{4}\sum_k B_{ok}^2(\omega_k t)^4 \tag{D2}$$

In this way we simplify all expectation values to the corresponding one of the preceding order and a series of expectation values between the $|\psi_\mu\rangle$. Further we calculate S(3,t) to

$$S(3,t) = -\frac{1}{3}S(2,t) + \frac{5}{12}\left[\sum_k (B_{ok}\omega_k t)^2\right]^3 + \frac{5}{18}\left[\sum_k B_{ok}^2(\omega_k t)^3\right]^2 + \frac{5}{12}\sum_k B_{ok}^2(\omega_k t)^4 \sum_{k'}(B_{ok'}\omega_{k'}t)^2 + \frac{1}{36}\sum_k B_{ok}^2(\omega_k t)^6 \tag{D3}$$

For calculation of the expectation values of the Hamiltonian we split the expression into two terms leading to

$$H(\mu,t) = \sum_k \hbar\omega_k \left\langle \psi(\mu,t)\middle|\hat{b}_k^+\hat{b}_k\middle|\psi(\mu,t)\right\rangle + 2\sum_k \hbar\omega_k B_{ok}\,\text{Re}\left[B_k(\mu,t)\right] \tag{D4}$$

This leads to H(0,t)=0 and further to

$$H(1,t) = \sum_k \hbar\omega_k (B_{ok}\omega_k t)^2$$

$$H(2,t) = \frac{5}{4}\sum_k \hbar\omega_k (B_{ok}\omega_k t)^2 \sum_{k'}(B_{ok'}\omega_{k'}t)^2 + \frac{1}{4}\sum_k \hbar\omega_k B_{ok}^2(\omega_k t)^4 \tag{D5}$$

$$H(3,t) = -\frac{1}{3}H(2,t) + \frac{7}{12}\sum_k \hbar\omega_k B_{ok}^2(\omega_k t)^4 \sum_{k'}(B_{ok'}\omega_{k'}t)^2 +$$

$$+\sum_k \hbar\omega_k (B_{ok}\omega_k t)^2 \left\{ \frac{35}{12}\left[\sum_{k'}(B_{ok'}\omega_{k'}t)^2\right]^2 + \frac{35}{36}\sum_{k'}B_{ok'}^2(\omega_{k'}t)^4 \right\} + \frac{1}{36}\sum_k \hbar\omega_k B_{ok}^2(\omega_k t)^6$$

Further we give the expectation values of the phonon annihilation operators, where $B_k(0,t)=0$:

$$B_k(1,t) = -iB_{ok}\omega_k t$$

$$B_k(2,t) = -iB_{ok}\omega_k t\left[1 + \frac{1}{2}\sum_{k'}(B_{ok'}\omega_{k'}t)^2\right] + \frac{1}{2}B_{ok}\omega_k t\sum_{k'}B_{ok'}^2(\omega_{k'}t)^3 - \frac{1}{2}B_{ok}(\omega_k t)^2\left[1 - \frac{1}{2}\sum_{k'}(B_{ok'}\omega_{k'}t)^2\right] \tag{D6}$$

$$B_k(3,t) = B_k(2,t) + \frac{i}{2}B_{ok}\omega_k t\left\{\sum_{k'}(B_{ok'}\omega_{k'}t)^2 - \frac{1}{2}\left[\sum_{k'}(B_{ok'}\omega_{k'}t)^2\right]^2 - \frac{1}{6}\sum_{k'}B_{ok'}^2(\omega_{k'}t)^4\right\} -$$

$$-\frac{i}{6}B_{ok}(\omega_k t)^2\sum_{k'}B_{ok'}^2(\omega_{k'}t)^3 + \frac{i}{6}B_{ok}(\omega_k t)^3\left[1 - \frac{1}{2}\sum_{k'}(B_{ok'}\omega_{k'}t)^2\right] +$$

$$+B_{ok}\omega_k t\left\{\sum_{k'}B_{ok'}^2(\omega_{k'}t)^3\left[-\frac{2}{3} + \frac{5}{6}\sum_{k''}(B_{ok''}\omega_{k''}t)^2\right] + \frac{1}{12}\sum_{k'}B_{ok'}^2(\omega_{k'}t)^5\right\} + \frac{1}{36}B_{ok}(\omega_k t)^3\sum_{k'}B_{ok'}^2(\omega_{k'}t)^3 +$$

$$+\frac{1}{2}B_{ok}(\omega_k t)^2\left[-\sum_{k'}(B_{ok'}\omega_{k'}t)^2 + \frac{1}{2}\left[\sum_{k'}B_{ok'}(\omega_{k'}t)^2\right]^2 + \frac{1}{6}\sum_{k'}B_{ok'}^2(\omega_{k'}t)^4\right]$$

Finally note, that

$$B_k^*(\mu,t) = \left\langle\psi(\mu,t)\left|\hat{b}_k^+\right|\psi(\mu,t)\right\rangle \tag{D7}$$

holds.

Appendix E: Separation of the Phase Factor of the Exact Solution

We have found that the operators $\hat{D}$ and $\hat{H}$ commute [see equ.(4)]. Further we know that the initial single exciton state must be of the form

$$\left|\Phi(0)\right\rangle = \sum_n \hat{B}_n(0) a_n(0) \hat{a}_n^+ \left|0\right\rangle \tag{E1}$$

where the operator $\hat{B}_n(0)$ creates the initial set of displacements and momenta from the phonon vacuum. Thus $\hat{B}_n(0)$ can contain only complex scalars and phonon operators. Then

$$e^{T\hat{D}}\left|\Phi(0)\right\rangle = \sum_{\nu=0}^{\infty} \frac{T^\nu}{\nu!} \hat{D}^\nu \hat{B}_n(0) a_n(0) \hat{a}_n^+ \left|0\right\rangle \quad ; \quad T \equiv \left(-\frac{it}{\hbar}\right) \tag{E2}$$

holds. Since

$$\hat{D}\hat{a}_n^+\left|0\right\rangle = \left[\sum_m E_0 \hat{a}_m^+ \hat{a}_m + \frac{1}{2}\sum_k \hbar\omega_k\right]\hat{a}_n^+\left|0\right\rangle = \left[E_0 + \frac{1}{2}\sum_k \hbar\omega_k\right]\hat{a}_n^+\left|0\right\rangle \tag{E3}$$

we have

$$\hat{D}^\nu \hat{a}_n^+\left|0\right\rangle = \left[E_0 + \frac{1}{2}\sum_k \hbar\omega_k\right]^\nu \hat{a}_n^+\left|0\right\rangle \tag{E4}$$

and therefore

$$e^{T\hat{D}}\left|\Phi(0)\right\rangle = e^{T\left(E_0 + \frac{1}{2}\sum_k \hbar\omega_k\right)}\left|\Phi(0)\right\rangle \tag{E5}$$

Finally we obtain

$$\left|\Phi(t)\right\rangle = e^{T\left(\hat{H}+\hat{D}\right)}\left|\Phi(t)\right\rangle = e^{T\hat{H}} e^{T\hat{D}}\left|\Phi(0)\right\rangle = e^{-\frac{it}{\hbar}\left(E_0 + \frac{1}{2}\sum_k \hbar\omega_k\right)} e^{T\hat{H}}\left|\Phi(0)\right\rangle \tag{E6}$$

which is the same separation as discussed in the main text.

J. Mol. Model. **1996**, 2, 103 - 135

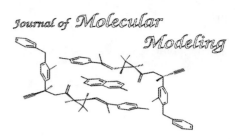

Davydov Soliton Dynamics in Proteins: II. The General Case

Wolfgang Förner

Chair for Theoretical Chemistry and Laboratory of the National Foundation of Cancer Research at the Friedrich-Alexander University Erlangen-Nürnberg, Egerlandstr. 3, D-91058 Erlangen, FRG. (foerner@pctc.chemie.uni-erlangen.de)

Received: 10 October 1995 / Accepted: 25 March 1996 / Published: 10 May 1996

Abstract

We performed long time simulations using the $|D_1\rangle$ approximation for the solution of the Davydov Hamiltonian. In addition we computed expectation values of the relevant operators with the state $\left(\hat{H}_D/J\right)|D_1\rangle$ and the deviation $|\delta\rangle$ from the exact solution over long times, namely 10 ns. We found that in the very long time scale the $|D_1\rangle$ ansatz is very close to an exact solution, showing expectation values of the relevant physical observables in the state $\left(\hat{H}_D/J\right)|D_1\rangle$ being about 5-6 orders of magnitudes larger than in the deviation state $|\delta\rangle$. In the intermediate time scale of the ps range such errors, as known from our previous work, are somewhat larger, but still more or less negligibly. Thus we also report results from an investigation of the very short time (in the range 0-0.4 ps) behaviour of the $|D_1\rangle$ state compared with that of an expansion of the exact solution in powers of time t. This expansion is reliable for about 0.12 ps for special cases as shown in the previous paper. However, the accuracy of the exactly known value of the norm and the expectation value of the Hamiltonian finally indicates up to what time a given expansion is valid, as also shown in the preceding paper. The comparison of the expectation values of the operators representing the relevant physical observables, formed with the third order wave function and with the corresponding results of $|D_1\rangle$ simulations has shown, that our expansion is valid up to a time of roughly 0.10-0.15 ps. Within this time the second and third order corrections turned out to be not very important. This is due to the fact that our first order state contains already some terms of the expansion, summed up to inifinite order. Further we found good agreement of the results obtained with our expansion and those from the corresponding $|D_1\rangle$ simulations within the time of about 0.10 ps. At later times, the factors with explicit powers of t in second and third order become dominant, making the expansion meaningless. Possibilities for the use of such expansions for larger times are described. Alltogether we have shown (together with previous work on medium times), that the $|D_1\rangle$ state, although of approximative nature, is very close to an exact solution of the Davydov model on time scales from some femtoseconds up to nanoseconds. Especially the very small time region is of importance, because in this time a possible soliton formation from the initial excitation would start.

Keywords: Proteins, Davydov Model, Nonlinear Dynamics, Expansion of Exact Solutions, Ansätze

Introduction

In the Introduction to the preceding paper [1], we discussed already the basic concepts of the Davydov soliton mechanism for energy transport in proteins [2-5]. Therefeore we refer the reader to that paper and references therein, and especially to the most recent and best review of the state of art in Davydov soliton theory given by Scott [6]. In our preceding paper [1] we studied special cases of the Davydov Hamiltonian. In these cases it turned out that the |D₁> ansatz state of Davydov [3] represents the exact solution, if the initial state is restricted to |D₂> [2] like states with site independent coherent states amplitudes. Further we found that expansions of the exact wave functions in polynomial serieses in time and truncation of these polynomials after the third order yield reliable results for times up to 0.10-0.12 ps for the lattice in the small polaron limit, and 0.6-0.8 ps for the amide-I oscillators in the decoupled case. More important, we found that the accuracy of the norm and of the expectation value of the Hamiltonian (total energy) corresponds exactly to that of the expectation values of the exciton number operators in the decoupled case and that of the displacement and momentum operators in the small polaron limit. Thus norm and total energy are a measure of the time scale within which a given expansion yields reliable results. In the present work which is based on [1] and makes use also of the explicit forms of the wave functions discussed there we determine similar expansions of the exact solution of the full Davydov Hamiltonian up to the third order and compare them on the very short time scale with the results of |D₁> simulations in order to get quantitative informations on the reliability of that ansatz which is not an exact solution of the time dependent Schrödinger equation. Further, using concepts developed in [7] and applied to a medium time scale in the range of a few picoseconds in [8], we study the behaviour of the errors introduced by the use of the |D₁> ansatz on large time scales of the order of nanoseconds.

This study completes our previous one [8], where we concluded that in the subspace of basis functions spanned by the |D₁> ansatz, the errors are negligible within times of a few picoseconds. From this result we concluded, that on this time scale the |D₁> ansatz should be rather close to an exact solution, because we expect that if the exact solution would contain important contributions from basis states not included in |D₁>, this should cause large errors also in the space spanned by |D₁>. However, the very small time scale is an important one, because in that time a possible soliton formation from a localized initial state starts, especially since the lattice is initially in equilibrium and only driven by the interaction with the localized initial excitation in the chain of amide-I oscillators. In case of a possible soliton formation exactly these initial displacements of the lattice formed in the first few hundredths of a picosecond stabilizes the amide-I excitation against dispersion. To investigate this range of time we expand the exact solution $|\Phi\rangle = \exp\left[-i\hat{H}_D t/\hbar\right] \Phi_0\rangle$ for the

Davydov Hamiltonian ($\hat{H}_D$), where |Φ₀> is the initial state, in a Taylor series in time and compare the results with those from a |D₁> simulation. Attempts into this direction have been reported previously by Cruzeiro-Hansson, Christiansen and Scott [9]. However, they restricted their considerations to a dimer and found that second order terms can be neglected only for times much smaller than 0.1 ps. Further they give no comparisons to approximate simulations and for the case of N sites they give a system of equations, but draw no numerical conclusions from it.

Davydov's Hamiltonian and the |D₁> Approximation

The Hamiltonian, as well as the form of the |D₁> approximation have been discussed extensively in the literature and in the preceding paper [1]. However, for the purpose of clearcut definitions in the following, we repeat here the form of the Hamiltonian as it is used extensively in this work. The full Hamiltonian can be written as

$$\hat{H}_D = \hat{H} + \hat{D} \quad ; \quad \hat{D} \equiv E_0 \sum_{n=1}^{N} \hat{a}_n^+ \hat{a}_n + \frac{1}{2} \sum_{k=1}^{N-1} \hbar\omega_k \tag{1}$$

where E_0 is the amide-I excitation energy, $\hat{a}_n$ ($\hat{a}_n^+$) are annihilation (creation) operators of amide-I vibrational quanta at site n in the chain and ω_k are the eigenfrequencies of the decoupled lattice. All definitions and details can be found in [1]. As also shown in [1] the exact state vecor is then

$$\left|\Phi\right\rangle = e^{-\frac{it}{\hbar}\left(E_0 + \frac{1}{2}\sum_{k=1}^{N-1} \hbar\omega_k\right)} \left|\psi\right\rangle \tag{2}$$

where |ψ> obeys the time dependent Schrödinger equation

$$i\hbar \frac{\partial}{\partial t} \left|\psi\right\rangle = \hat{H} \left|\psi\right\rangle \tag{3}$$

with the simplified Hamiltonian

$$\hat{H} = \hat{J} + \hat{\omega} \tag{4}$$

where

$$\hat{J} = -J \sum_{n=1}^{N} \left(\hat{a}_{n+1}^+ \hat{a}_n + \hat{a}_n^+ \hat{a}_{n+1}\right) \tag{5}$$

Here J is the coupling constant between two neighboring amide-I oscillators. Further

$$\hat{\omega} = \sum_{k=1}^{N-1} \hbar\omega_k \left[\hat{b}_k^+ \hat{b}_k + \sum_{n=1}^{N} B_{nk} \left(\hat{b}_k + \hat{b}_k^+ \right) \hat{a}_n^+ \hat{a}_n \right]$$

(6)

$$B_{nk} \equiv \frac{\chi}{\omega_k} \frac{1}{\sqrt{2M\hbar\omega_k}} \left[U_{n+1,k} - U_{nk} \right] \quad ; \quad k \neq N$$

where $\hat{b}_k$ ($\hat{b}_k^+$) is the annihilation (creation) operator for an acoustical lattice phonon k, M is the mass of a site and χ the coupling constant between the amide-I oscillators (excitons) and the lattice. The matrix $\underline{\underline{U}}$ contains the normal mode coefficients in the real representation for the decoupled lattice. Again the details are derived in paper [1]. We use for the present study cyclic boundary conditions and chains with an odd number of sites (N), thus n=N+1 equals n=1 and n=0 equals n=N. The $|D_1\rangle$ ansatz for $|\psi\rangle$ has the form (the restriction at the sums over k just excludes the translational mode)

$$|D_1\rangle = \sum_{n=1}^{N} a_n(t) \hat{U}_n \hat{a}_n^+ |0\rangle$$

(7)

$$\hat{U}_n |0\rangle_p = \exp\left[-\frac{1}{2} \sum_{k=1}^{N-1} |b_{nk}(t)|^2 \right] \cdot \exp\left[\sum_{k=1}^{N-1} b_{nk}(t) \hat{b}_k^+ \right] |0\rangle_p$$

$$= \exp\left\{ \sum_{k=1}^{N-1} \left[b_{nk}(t) \hat{b}_k^+ - b_{nk}^*(t) \hat{b}_k \right] \right\} |0\rangle_p$$

Note, that the last equality holds only if the operator acts on the phonon vacuum $|0\rangle_p$, and that in our notation $|0\rangle = |0\rangle_e |0\rangle_p$, where $|0\rangle_e$ is the vacuum state for the amide-I oscillators (exciton vacuum). The $b_{nk}(t)$ are the coherent state amplitudes and $|a_n(t)|^2$ is the probability to find an amide-I quantum at site n. The equations of motion for these quantities can be obtained with the Euler-Lagrange equations of the second kind (see again [1] and references therein for all details):

$$i\hbar\dot{a}_n = -\frac{i\hbar}{2} \sum_{k=1}^{N-1} \left(\dot{b}_{nk} b_{nk}^* - \dot{b}_{nk}^* b_{nk} \right) a_n +$$

$$+ \sum_{k=1}^{N-1} \hbar\omega_k \left[B_{nk} \left(b_{nk} + b_{nk}^* \right) + |b_{nk}|^2 \right] a_n -$$

$$- J \left(D_{n,n+1} a_{n+1} + D_{n,n-1} a_{n-1} \right)$$

(8)

$$i\hbar\dot{b}_{nk} = \hbar\omega_k \left(b_{nk} + B_{nk} \right) - J \left[D_{n,n+1} \left(b_{n+1,k} - b_{nk} \right) \frac{a_{n+1}}{a_n} + \right.$$

$$\left. + D_{n,n-1} \left(b_{n-1,k} - b_{nk} \right) \frac{a_{n-1}}{a_n} \right]$$

where the coherent state overlaps are given by

$$D_{nm} = \exp\left[-\frac{1}{2} \sum_{k=1}^{N-1} \left(|b_{nk} - b_{mk}|^2 + b_{nk} b_{mk}^* - b_{nk}^* b_{mk} \right) \right]$$

(9)

Long Time Simulations

In this sub-chapter we want to study the important question of the long-time behaviour of the errors introduced by the $|D_1\rangle$ ansatz. One cannot exclude that these errors, which are very small for intermediate times of some ps (see [1] and references therein for details), might increase with time. In our case, the basis space of $|D_1\rangle$ is incomplete, since it is not an exact solution. However, from the magnitude of the errors introduced within the basis space of $|D_1\rangle$, we can estimate the importance of the basis states missing in the ansatz. If these missing states would be important in the exact solution, one would expect, that the errors made by the ansatz should be rather large already within the subset of the basis space spanned by $|D_1\rangle$. For the numerical investigation of these errors we can use the fact that [7]

$$\left(i\hbar \frac{\partial}{\partial t} - \hat{H}_D \right) |D_1\rangle = J |\delta\rangle$$

(10)

where the form of the error state $|\delta\rangle$ as function of $\{a_n(t), b_{nk}(t)\}$ as computed in a $|D_1\rangle$ simulation is known [7]. In our previous work [8] we have derived expressions for the expectation values of different operators for the two states $\left(\hat{H}_D / J \right) |D_1\rangle$ and $|\delta\rangle$ and compared them in numerical calculations. Such a procedure can serve as the appropriate tool to answer our above mentioned question.

As initial state state we use the same one as described in detail in paper [1], section III.3. It consists of a sech-function for the coefficients in the oscillator part and a lattice, populated with phonons according to a temperature of 300K. We performed calculations for different exciton-phonon coupling constants, namely χ = 60 pN, 120 pN, 180 pN, and 240 pN, and chains of length N=21 units. For the fourth order Runge-Kutta simulations we used in case of χ = 60 pN a time step of 0.1 fs and a total simulation time of 10 ns, corresponding to 10^8 time steps. In this period the error in total energy (0.796 eV) was typically between 0 and -7 peV (the exciton-phonon interaction energy was between 0.5 and -4.5 meV) and the error in norm between 0 and -0.006 ppb (parts per billion). The computation time was 2.48 minutes of CPU (Central Processing Unit) time for the simulation of 1 ps and thus 413 hours of CPU time for the complete simulation on an IBM RISC/6000-320H workstation. In the other three calculations we used a time step of 1 fs for the same total simulation time and in these cases the absolute values of the errors in total energy were less than 3 μeV (χ=120 pN, exciton-phonon interaction energy E_{ep} between 0.5 and -12 meV), less than 44 μeV (χ=180 pN, exciton-phonon interaction energy E_{ep} between -5 and -25 meV) and less than 400 μeV (χ=240 pN, exciton-phonon interaction energy E_{ep} between -2 and -45 meV), respectively. The absolute values of the errors in the norms were less than 4 ppm (parts per million, χ=120 pN), 80 ppm (χ=180 pN) and 750 ppm (χ=240 pN). From this it is obvious that a time step of 1 fs is sufficient to obtain correct results for simulation times of 10 ns. Also for the case of χ=60 ps the larger time step caused no significant changes in the results.

Figure1 shows the time evolution of the probability to find an amide-I quantum at a site n for the three different cases. While we observe a complete dispersion of the initial sech-distribution in case of the two smaller couplings, the initial excitation remains localized in the range of the initial distribution up to 10 ns in case of the larger couplings. For χ=60 pN a considerable fraction of the excitation localizes itself at more or less a single site close to the center of the initial excitation after 2-3 ns and remains there until the end of the simulation. Such spontaneous localizations cannot occur for the larger coupling, because in these cases the thermal disorder in the lattice is more strongly coupled to the oscillator system. In Figure 2 we show the norms

$$S_H(t) = \left\langle \left(\hat{H}/J\right)D_1 \middle| \left(\hat{H}/J\right)D_1 \right\rangle$$ for the four values of the cou-

pling constant. Obviously the norms show a decreasing tendency for increasing coupling, namely from maximum values of roughly 7000 (χ=60 pN) to 1200 (χ=240 pN). For the smallest coupling the function shows a fast oscillation around 5000 when time increases, while for the largest coupling a

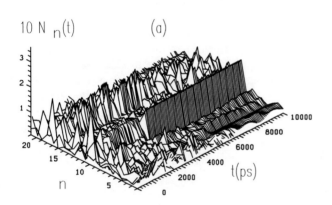
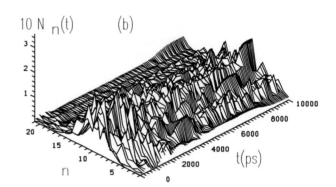
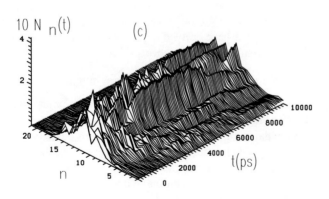
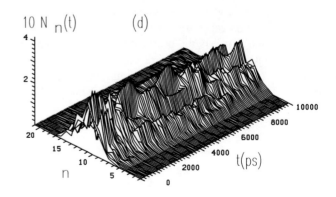

Figure 1: *Long-time evolution of* $N_n(t)=<D_1|\hat{a}_n^+\hat{a}_n|D_1>$ *for four values of the exciton-lattice coupling constant.*
(a) χ=60 pN   *(b) χ=120 pN*
(c) χ=180 pN   *(d) χ=240 pN*

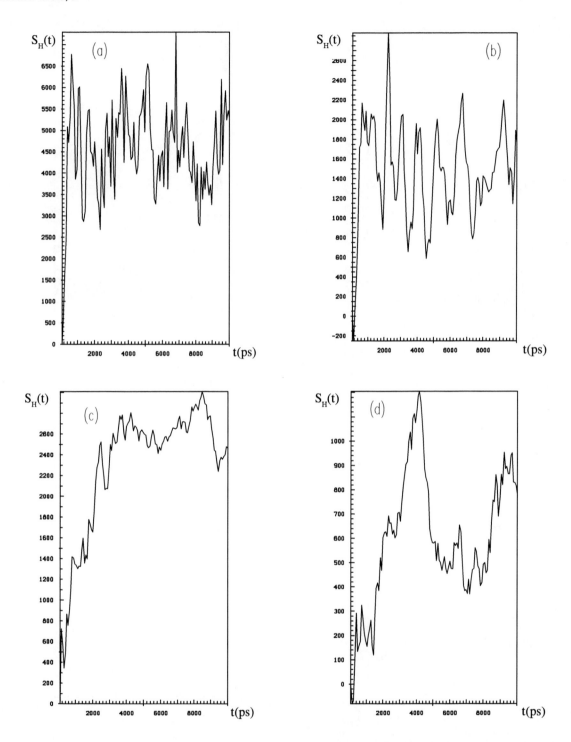

Figure 2: *The norm* $S_H(t) = \left\langle \left(\hat{H}/J\right)D_1 \middle| \left(\hat{H}/J\right)D_1 \right\rangle$ *as function*

of time for four values of the coupling constant (note that t=80 ps is the first time in the simulation where our program prints out intermediate results, however, in the plots for our long time simulations t=0 and t=80 ps are indistinguishable):

(a) χ= 60 pN (S_H relative to S_H(t=80 ps)=684,827.548170)
(b) χ=120 pN (S_H relative to S_H(t=80 ps)=688,883.505744)
(c) χ=180 pN (S_H relative to S_H(t=80 ps)=694,750.770186)
(d) χ=240 pN (S_H relative to S_H(t=80 ps)=710,240.923822)

very slow oscillation around 800 starts after roughly 3 ns. Fig. 3 shows $S_E(t)=\langle\delta|\delta\rangle$ for the different couplings. In all cases the error remains about 8 orders of magnitude smaller than $S_H(t)$, indicating that within the $|D_1\rangle$ basis space no significant errors of the norm of the state occur. Note, that the $S_H(t)$-plots are drawn relative to $S_H(t=80$ ps). Moreover, in all cases $S_E(t)$ increases within the first 500 to 1000 ps to values around 2.3 and afterwards decreases to a small amplitude oscillation around 2.03 and 2.00, independent of the value of the coupling constant. Even after 10 ns the mean value of $S_E(t)$ still seems to decrease further very slowly. Therefore

J. Mol. Model. **1996**, 2

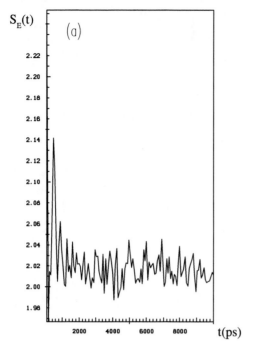
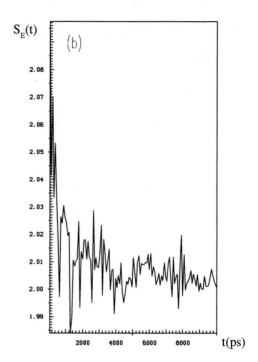
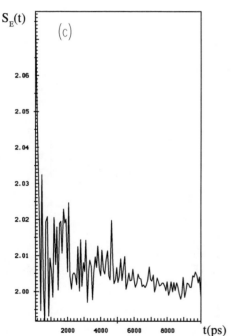
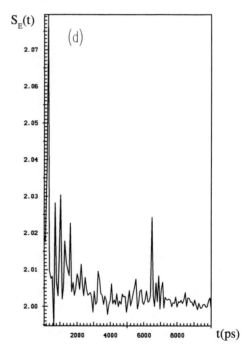

Figure 3: *The norm $S_E(t)=<\delta|\delta>$ as function of time for four values of the coupling constant:*

(a) $\chi= 60\ pN$ *(b) $\chi=120\ pN$*
(c) $\chi=180\ pN$ *(d) $\chi=240\ pN$*

we can conclude, at least on the basis of the norms, that the error introduced by the incomplete basis space of the ansatz decreases in time, and that the $|D_1>$ state becomes more accurate in the long time limit.

Figure 4 shows the time evolution of the number operators for the oscillators, $N_n^H(t)= \left\langle \left(\hat{H}/J\right)D_1 \left| \hat{a}_n^+ \hat{a}_n \right| \left(\hat{H}/J\right)D_1 \right\rangle$ and Figure 5 those for the error state $N_n^E(t)=<\delta|\hat{a}_n^+\hat{a}_n|\delta>$. Also here the errors show no tendency to increase with increasing time, in contrast, they have a constant order of magnitude through all 10 ns. They follow closely the time evolution of $N_n^H(t)$, however, being 6 to 7 orders of magnitude smaller, thus as in case of the norms, the deviations are completely negligibly, also over times as large as 10 ns. The same holds

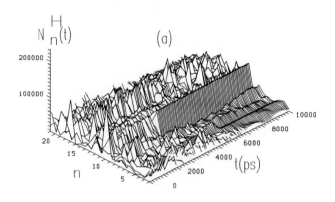

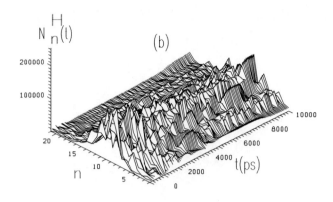

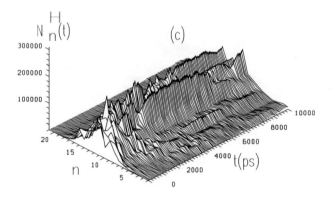

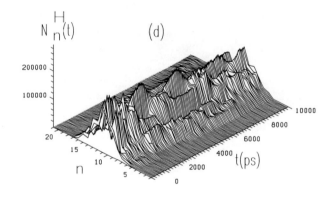

Figure 4:

The expectation values $N_n^H(t) = \left\langle (\hat{H}/J)D_1 \middle| \hat{a}_n^+ \hat{a}_n \middle| (\hat{H}/J)D_1 \right\rangle$

as functions of site and time for the different values of the coupling constant (also here and in Fig. 5-9 the first line drawn in the plot is that at t=80 ps, also the time distance between two lines is 80 ps):

(a) χ=60 pN   *(b)* χ=120 pN
(c) χ=180 pN   *(d)* χ=240 pN

for the expectation values of the displacement and momentum operators. $q_n^H(t) = \left\langle (\hat{H}/J)D_1 \middle| \hat{q}_n \middle| (\hat{H}/J)D_1 \right\rangle$ and $q_n^E(t) = \left\langle \delta \middle| \hat{q}_n \middle| \delta \right\rangle$, respectively, as well as $p_n^H(t) = \left\langle (\hat{H}/J)D_1 \middle| \hat{p}_n \middle| (\hat{H}/J)D_1 \right\rangle$ and $p_n^E(t) = \left\langle \delta \middle| \hat{p}_n \middle| \delta \right\rangle$, respectively, not shown here, exhibit a quasi-random behaviour as it is to be expected because of the „thermal" excitation in the initial state. However, also here the errors closely

follow the corresponding pictures computed from the state $(\hat{H}_D/J)D_1\rangle$, and are about 5 to 6 orders of magnitude smaller, which does not change in the large time scales of our simulations.

Therefore we can conclude from our results, that for large times $|D_1>$ is close to the exact solution, with nearly negligible deviations. Further we know from our previous simulations on intermediate time scales in the range of ps, that in this region the errors are somewhat larger, although still more or less negligible [8]. In the next section we study the time around 0.1 ps, where a possible soliton formation would start from localized initial states.

Expansion of the Exact Wavefunction

We start, as in our preceding paper [1], from the well-known ansatz for the exact solution of the Schrödinger equation

$$|\psi\rangle = e^{-\frac{it}{\hbar}(\hat{J}+\hat{\omega})}|\psi_0\rangle \qquad (11)$$

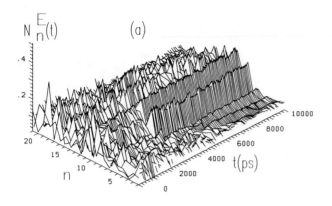

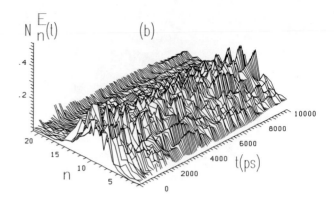

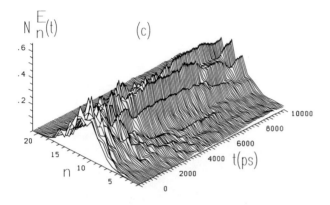

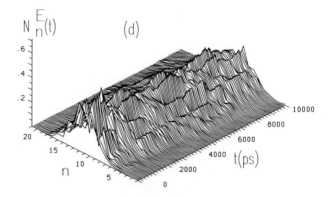

Figure 5: *The expectation values $N_n^E(t)=<\delta|\hat{a}_n^+\hat{a}_n|\delta>$ as functions of site and time for the different values of the coupling constant:*

(a) $\chi=60\ pN$ *(b) $\chi=120\ pN$*
(c) $\chi=180\ pN$ *(d) $\chi=240\ pN$*

Note here that the two parts of the Hamiltonian do not commute

$$\left[\hat{\omega},\hat{J}\right]\equiv\hat{\omega}\hat{J}-\hat{J}\hat{\omega}\equiv\hat{C}=\sum_{n=1}^{N}\sum_{k=1}^{N-1}\hbar\omega_k B_{nk}\left(\hat{b}_k^+ +\hat{b}_k\right)\hat{\alpha}_n$$

(12)

$$\hat{\alpha}_n\equiv J\left[\left(\hat{a}_{n+1}^+ +\hat{a}_{n-1}^+\right)\hat{a}_n -\hat{a}_n^+\left(\hat{a}_{n+1}+\hat{a}_{n-1}\right)\right]$$

As model system we use a cyclic chain of N (N odd) units and an initial excitation of an amide-I oscillator at site o. The lattice is initially in its groundstate. Thus we have

$$\left|\psi_0\right\rangle=\hat{a}_o^+\left|0\right\rangle$$

(13)

Then the exponential in the exact wave function is expanded in a Taylor series yielding

$$\left|\psi(t)\right\rangle=\sum_{\nu=0}^{\infty}\frac{T^\nu}{\nu!}\left(\hat{\omega}+\hat{J}\right)^\nu\left|\psi_0\right\rangle \quad ; \quad T\equiv\left(-\frac{it}{\hbar}\right)$$

$$\left|\psi(t)\right\rangle=\left[1+T\left(\hat{J}+\hat{\omega}\right)+\frac{T^2}{2}\left(\hat{J}^2+\hat{J}\hat{\omega}+\hat{\omega}\hat{J}+\hat{\omega}^2\right)+ \right.$$

$$+\frac{T^3}{6}\left(\hat{J}^3+\hat{J}^2\hat{\omega}+\hat{\omega}\hat{J}^2+\hat{\omega}^2\hat{J}+ \right.$$

$$\left.\left.+\hat{J}\hat{\omega}^2+\hat{\omega}\hat{J}\hat{\omega}+\hat{J}\hat{\omega}\hat{J}+\hat{\omega}^3\right)+...\right]\psi_0\rangle$$

(14)

From this expansion we can immediately extract two series

$$\left|T_1\right\rangle=\sum_{\nu=0}^{\infty}\frac{T^\nu}{\nu!}\hat{J}^\nu\left|\psi_0\right\rangle=e^{T\hat{J}}\left|\psi_0\right\rangle=\left|J\right\rangle$$

$$\left|T_2\right\rangle=\sum_{\nu=0}^{\infty}\frac{T^\nu}{\nu!}\hat{\omega}^\nu\left|\psi_0\right\rangle=e^{T\hat{\omega}}\left|\psi_0\right\rangle=\left|\omega\right\rangle$$

(15)

which are the exact solutions of the separate Schrödinger equations for the two operators. They are derived in detail in paper [1] and can be cast into the form

$$|J\rangle = \sum_{n=1}^{N} a_n^J(t)\hat{a}_n^+ |0\rangle \quad ; \quad |\omega\rangle = \hat{U}_o a_o^\omega(t)\hat{a}_o^+ |0\rangle$$

$$\hat{U}_o = U_o(t)e^{\sum_{k=1}^{N-1} b_{ok}(t)\hat{b}_k^+} \quad ; \quad U_o(t) = e^{-\frac{1}{2}\sum_{k=1}^{N-1}|b_{ok}(t)|^2} \tag{16}$$

where the translational mode of the lattice ($\omega_N=0$) has to be excluded. Note here that

$$\hat{U}_o|0\rangle \equiv |\beta_o\rangle \quad ; \quad \hat{b}_k|\beta_o\rangle = b_{ok}(t)|\beta_o\rangle$$

$$\langle\beta_o|\hat{b}_k^+ = \langle\beta_o|b_{ok}^*(t)$$

$$\langle\beta_o|0\rangle = \langle0|\beta_o\rangle = U_o(t) \quad ; \quad \langle\beta_o|\beta_o\rangle = 1 \tag{17}$$

holds and the time dependent coefficients are given by [1]

$$a_n^J(t) = \frac{1}{N}\sum_{k=1}^{N} e^{\frac{2\pi i}{N}k(n-o)} \cdot e^{2i\frac{Jt}{\hbar}\cos\left(\frac{2\pi}{N}k\right)}$$

$$\sum_{n=1}^{N}|a_n^J(t)|^2 = 1 \tag{18}$$

$$a_o^\omega(t) = e^{-i\sum_{k=1}^{N-1} B_{ok}^2[\sin(\omega_k t)-\omega_k t]} \quad ; \quad |a_o^\omega(t)|^2 = 1$$

$$b_{ok}(t) = B_{ok}\left(e^{-i\omega_k t}-1\right) \quad ; \quad U_o(t) = e^{\sum_{k=1}^{N-1} B_{ok}^2[\cos(\omega_k t)-1]}$$

Note, that in case of the amide-I oscillators also the mode with k=N has to be included. Then with

$$|\psi(0)\rangle = |\psi(1)\rangle \equiv |J\rangle + |\omega\rangle - |\psi_0\rangle \tag{19}$$

our exact wave function can be written as

$$|\psi(t)\rangle = |\psi(0)\rangle + \sum_{\nu=2}^{\infty}\frac{T^\nu}{\nu!}\hat{\Omega}_\nu|\psi_0\rangle$$

$$\hat{\Omega}_\nu \equiv \left(\hat{\omega}+\hat{J}\right)^\nu - \hat{\omega}^\nu - \hat{J}^\nu \tag{20}$$

Further the different orders µ are defined as

$$|\psi(\mu)\rangle = |\psi(0)\rangle + \sum_{\nu=2}^{\mu}\frac{T^\nu}{\nu!}|\psi_\nu\rangle \quad ; \quad |\psi_\nu\rangle \equiv \hat{\Omega}_\nu|\psi_0\rangle$$

$$\tag{21}$$

$$\hat{\Omega}_1 = 0$$

$$\hat{\Omega}_2 = \hat{\omega}\hat{J} + \hat{J}\hat{\omega}$$

$$\hat{\Omega}_3 = \hat{\omega}\hat{J}^2 + \hat{J}^2\hat{\omega} + \hat{J}\hat{\omega}^2 + \hat{\omega}^2\hat{J} + \hat{\omega}\hat{J}\hat{\omega} + \hat{J}\hat{\omega}\hat{J}$$

$$\ldots$$

In this paper we want to restrict the expansion to µ=3, since the calculation of higher order terms becomes too tedious. This is also the reason why such expansions are only useful for the study of the very short time behaviour of exact solutions, but not for general simulations, where a large amount of high order terms would be necessary to obtain reliable results for times say in the ps-scale. In our previous paper [1] we had studied in detail the time scale on which such a third order expansion is valid. For comparisons with the corresponding results of |D_1> simulations we are interested in the norm of the states of different order in time, the expectation values H(µ,t) of the Hamiltonian, $N_n(\mu,t)$ of the exciton number operators and $B_k(\mu,t)$ of the phonon annihilation operators. From the latter ones we can compute easily expectation values $q_n(\mu,t)$ of the displacement and $p_n(\mu,t)$ of the momentum operators:

$$H(\mu,t) = \left\langle\psi(\mu,t)\left|\hat{\omega}+\hat{J}\right|\psi(\mu,t)\right\rangle$$

$$N_n(\mu,t) = \left\langle\psi(\mu,t)\left|\hat{a}_n^+\hat{a}_n\right|\psi(\mu,t)\right\rangle$$

$$B_k(\mu,t) = \left\langle\psi(\mu,t)\left|\hat{b}_k\right|\psi(\mu,t)\right\rangle$$

$$B_k^*(\mu,t) = \left\langle\psi(\mu,t)\left|\hat{b}_k^+\right|\psi(\mu,t)\right\rangle \tag{22}$$

$$S(\mu,t) = \left\langle\psi(\mu,t)|\psi(\mu,t)\right\rangle = \sum_{n=1}^{N} N_n(\mu,t)$$

$$q_n(\mu,t) = \sum_{k=1}^{N-1}\sqrt{\frac{2\hbar}{M\omega_k}}U_{nk}\,\mathrm{Re}\left[B_k(\mu,t)\right]$$

$$p_n(\mu,t) = \sum_{k=1}^{N-1}\sqrt{2M\hbar\omega_k}\,U_{nk}\,\mathrm{Im}\left[B_k(\mu,t)\right]$$

where $\underline{U}$ is the eigenvector matrix of the decoupled lattice in real representation and is discussed in detail in Appendix B of the preceding paper [1]. The explicit expressions for these expectation values are given in the Appendices, since they are rather massy.

The first order wave function vanishes with our choice of $|\psi(0)\rangle$, and thus we proceed directly to the second order correction which is given by

$$|\psi_2\rangle = \hat{\Omega}_2|\psi_0\rangle = \left(\hat{J}\hat{\omega} + \hat{\omega}\hat{J}\right)|\psi_0\rangle$$

$$\hat{\omega}\hat{J}|\psi_0\rangle = -J\sum_{k=1}^{N-1}\hbar\omega_k\hat{b}_k^+\left(B_{o-1,k}\hat{a}_{o-1}^+ + B_{o+1,k}\hat{a}_{o+1}^+\right)|0\rangle$$

$$\hat{J}\hat{\omega}|\psi_0\rangle = -J\sum_{k=1}^{N-1}\hbar\omega_k B_{ok}\hat{b}_k^+\left(\hat{a}_{o-1}^+ + \hat{a}_{o+1}^+\right)|0\rangle \tag{23}$$

$$\left(\sum_{k=1}^{N-1}\hbar\omega_k\hat{b}_k^+\hat{b}_k|\psi_0\rangle = \sum_{k=1}^{N-1}\hbar\omega_k\hat{b}_k^+\hat{b}_k\hat{J}|\psi_0\rangle = 0\right)$$

Leading finally to

$$|\psi_2\rangle = -J\sum_{k=1}^{N-1}\hbar\omega_k\hat{b}_k^+\left(A_{o-1,k}\hat{a}_{o-1}^+ + A_{o+1,k}\hat{a}_{o+1}^+\right)|0\rangle =$$

$$= \left(\hat{\Theta}_{-1}\hat{a}_{o-1}^+ + \hat{\Theta}_{+1}\hat{a}_{o+1}^+\right)|0\rangle$$

$$\tag{24}$$

$$\hat{\Theta}_{\pm 1} \equiv -J\sum_{k=1}^{N-1}\hbar\omega_k A_{o\pm 1,k}\hat{b}_k^+ \quad ; \quad A_{o\pm 1,k} \equiv B_{ok} + B_{o\pm 1,k}$$

The full second order wave function is then

$$|\psi(2)\rangle = |\psi(0)\rangle - \frac{t^2}{2\hbar^2}|\psi_2\rangle \tag{25}$$

The third order correction is given by

$$|\psi_3\rangle = \hat{\Omega}_3|\psi_0\rangle =$$

$$= \left(\hat{\omega} + \hat{J}\right)\left(\hat{\omega}\hat{J} + \hat{J}\hat{\omega}\right)|\psi_0\rangle + \hat{\omega}\hat{J}^2|\psi_0\rangle + \hat{J}\hat{\omega}^2|\psi_0\rangle =$$

$$= \left(\hat{\omega} + \hat{J}\right)|\psi_2\rangle + \hat{\omega}\hat{J}^2|\psi_0\rangle + \hat{J}\hat{\omega}^2|\psi_0\rangle \tag{26}$$

With the definition

$$\hat{y}_n = \sum_{k=1}^{N-1}\hbar\omega_k B_{nk}\hat{b}_k \tag{27}$$

we obtain for the different terms, where partially results of the preceding paper are used:

$$\hat{\omega}\hat{J}^2|\psi_0\rangle =$$

$$= J^2\sum_{k=1}^{N-1}\hbar\omega_k\hat{b}_k^+\left(B_{o-2,k}\hat{a}_{o-2}^+ + 2B_{ok}\hat{a}_o^+ + B_{o+2,k}\hat{a}_{o+2}^+\right)|0\rangle \tag{28}$$

$$\hat{J}\hat{\omega}^2|\psi_0\rangle =$$

$$= -J\left[\left(\hat{y}_o^+\right)^2 + \sum_{k=1}^{N-1}\left(\hbar\omega_k\right)^2 B_{ok}\left(B_{ok} + \hat{b}_k^+\right)\right]\left(\hat{a}_{o-1}^+ + \hat{a}_{o+1}^+\right)|0\rangle$$

Further the action of our operators on $|\psi_2\rangle$ yields

$$\hat{J}|\psi_2\rangle = J^2\sum_{k=1}^{N-1}\hbar\omega_k\hat{b}_k^+\left[A_{o-1,k}\hat{a}_{o-2}^+ + \right.$$

$$\left. + \left(A_{o-1,k} + A_{o+1,k}\right)\hat{a}_o^+ + A_{o+1,k}\hat{a}_{o+2}^+\right]|0\rangle \tag{29}$$

and

$$\hat{\omega}|\psi_2\rangle = -J\left\{\sum_{k,k'=1}^{N-1}\hbar\omega_k\hbar\omega_{k'}\hat{b}_k^+\hat{b}_{k'}^+ \cdot\right.$$

$$\cdot\left[B_{o-1,k}A_{o-1,k'}\hat{a}_{o-1}^+ + B_{o+1,k}A_{o+1,k'}\hat{a}_{o+1}^+\right]|0\rangle +$$

$$+ \sum_{k=1}^{N-1}\left(\hbar\omega_k\right)^2\left[\left(B_{o-1,k} + \hat{b}_k^+\right)A_{o-1,k}\hat{a}_{o-1}^+ + \right.$$

$$\left.\left. + \left(B_{o+1,k} + \hat{b}_k^+\right)A_{o+1,k}\hat{a}_{o+1}^+\right]|0\rangle\right\} \tag{30}$$

Collecting all the terms, we can write $|\psi_3\rangle$ in the form

$$|\psi_3\rangle = \sum_{v=-2}^{2}(-1)^v\hat{\Gamma}_v\hat{a}_{o+v}^+|0\rangle \tag{31}$$

where the operators $\hat{\Gamma}_v$ contain only phonon creation operators:

$$\hat{\Gamma}_0 \equiv J^2\sum_{k=1}^{N-1}\hbar\omega_k D_k\hat{b}_k^+ \quad ; \quad \hat{\Gamma}_{\pm 2} \equiv J^2\sum_{k=1}^{N-1}\hbar\omega_k D_k^{(\pm)}\hat{b}_k^+$$

$$\hat{\Gamma}_{\pm 1} \equiv J\sum_{k=1}^{N-1}\hbar\omega_k\left[\hbar\omega_k\left(E_k^{(\pm)} + F_k^{(\pm)}\hat{b}_k^+\right) + \right.$$

$$\left. + \sum_{k'=1}^{N-1}\hbar\omega_{k'}G_{kk'}^{(\pm)}\hat{b}_k^+\hat{b}_{k'}^+\right] \tag{32}$$

and the real scalar quantities:

$$D_k \equiv B_{o-1,k} + 4B_{ok} + B_{o+1,k}$$

$$D_k^{(\pm)} \equiv B_{ok} + B_{o\pm1,k} + B_{o\pm2,k}$$

$$E_k^{(\pm)} \equiv B_{ok}^2 + \left(B_{ok} + B_{o\pm1,k}\right)B_{o\pm1,k} \qquad (33)$$

$$F_k^{(\pm)} \equiv 2B_{ok} + B_{o\pm1,k}$$

$$G_{kk'}^{(\pm)} \equiv B_{ok}B_{ok'} + B_{o\pm1,k}\left(B_{ok'} + B_{o\pm1,k'}\right)$$

With the help of these abbreviations the calculation of expectation values, as given in the Appendices, as well as their programming can be considerably simplified. The total third order wave function is then given by

$$\left|\psi(3)\right\rangle = \left|\psi(2)\right\rangle + i\frac{t^3}{6\hbar^3}\left|\psi_3\right\rangle =$$

$$= \left|\omega\right\rangle + \left|J\right\rangle - \left|\psi_0\right\rangle - \frac{t^2}{2\hbar^2}\left|\psi_2\right\rangle + i\frac{t^3}{6\hbar^3}\left|\psi_3\right\rangle \qquad (34)$$

where the sum of the first three terms is also denoted as |ψ(0)>.

With the help of the expectation values as given in the Appendices we performed calculations using cyclic chains of N=21 units for the so-called standard parameters (W=13 N/m, M=114 m_p, J=0.967 meV) and two values of the exciton-phonon coupling constant χ=35 pN and 62 pN as in the preceding paper. In the initial state the lattice is in its equilibrium, i.e. $b_{nk}(0)=0$, and the amide-I excitation is localized at site o=11, i.e. $a_n(0)=\delta_{no}$. As mentioned above, the accuracies of S(t) and H(t) are direct measures of the maximal time a given order of the expansion of the wave function is valid for, we concentrate first on these two functions. The expectation value of the Hamiltonian H(t) (Figure 6 a,c) remains very close to its exact value up to roughly 0.10-0.15 ps in case of the third order expansion. After that the terms which include explicitly powers of t obviously dominate and lead to a fast, unphysical increase. In case of the second order this increase starts somewhat later in time and is less steep. The deviations from the exact value in the first order are rather small and increase very slowly, due to the fact that in first order no explicit powers of t occur. The overall picture for the norm S(t) is qualitatively the same. Also in this case the deviations are tolerable up to a time of about 0.10-0.15 ps. From this, as we have seen in the first paper, we can conlude that also the other expectation values should be reliable at least up to roughly 0.1 ps.

In Figure 7 we show the physically more interesting expectation values of the number operators, displacement and momentum operators for the units o (o=11), where in the initial state the excitation is localized and o+1 for a time of 0.4 ps and the two coupling constants under consideration.

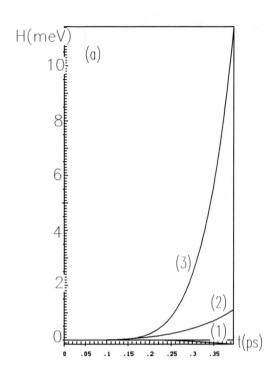

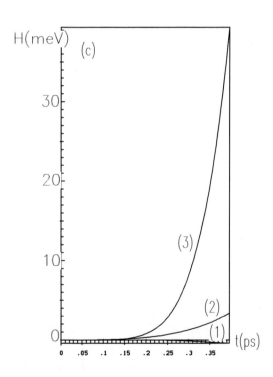

Figure 6 (continues next page): *The functions H(μ,t) (in meV, relative to H(t)=H(0)=0) and S(μ,t) (relative to S(t)=S(0)=1; the graphs corresponding to the different orders are marked by μ):*

(a) H(μ,t), χ=35 pN   *(b) S(μ,t), χ=35 pN*
(c) H(μ,t), χ=62 pN   *(d) S(μ,t), χ=62 pN*

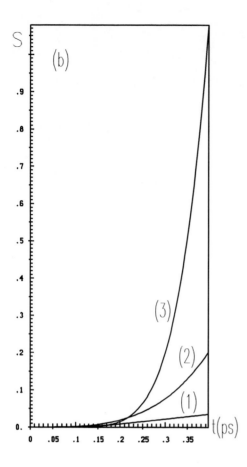

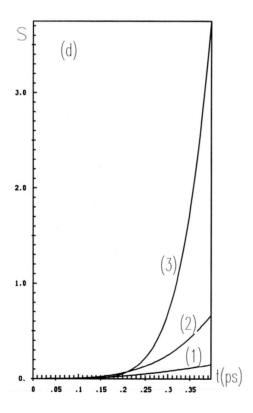

In the rather short simulation time at sites n<o-1 or n>o+2 no important dynamics evolve. In all the figures the results of the corresponding |D₁> simulations are plotted as dashed lines and those of the expansion as solid lines with the order indicated at them. The time step in the simulations was 4 fs. Thus we have 100 time steps exactly at the times where we computed the expectation values for the expansion. In these simulations the absolute value of the errors in total energy were less than 6 peV (exciton-phonon interaction energy between 0 and -2.4 meV) and the absolute values of the errors in the norm less than 1 ppb (parts per billion). As mentioned previously in paper I, we need as initial state for |D₁> simulations the form

$$a_n(0) = \frac{\delta_{no} + x(1-\delta_{no})}{\sqrt{1+(N-1)|x|^2}} \tag{35}$$

where N is the number of sites in the chain, o the initial excitation site and x a small, real scalar. This is necessary to avoid numerical problems due to a_n occuring in the denominators in the equations of motion. However, if we use $x=5\cdot10^{-3}$, which is physically irrelevant in long time simulations, we obtain for very short times between 0 and 0.1 ps spurious minima e.g. in the expectation values of the number operators $N_n(t)$ for n<o-1 and n>o+1 of a depth of about 5 ppm. These spurious minima, not found in the expansions, can be avoided if x is reduced to $x=5\cdot10^{-5}$.

The first six parts of Figure 7 show the relevant expectation values for the smaller coupling constants. It is obvious, that up to a time of roughly 0.15 ps the |D₁> results agree perfectly well with those from the three orders of the expansion, which in this region of time do not differ very much from each other. In most cases of differences (Fig. 7e,f) obviously the second order starts to deviate from the first one and then the third order correction brings the curve again closer to the first order. After about 0.2 ps the explicit factors

Figure 7 (following pages): *The expectation values of the number operators $N_o(\mu,t)$ and $N_{o+1}(\mu,t)$ together with the corresponding |D₁> results, the displacements $q_o(\mu,t)$ and $q_{o+1}(\mu,t)$ together with the corresponding |D₁> results (in mÅ) and the momenta $p_o(\mu,t)$ and $p_{o+1}(\mu,t)$ together with the corresponding |D₁> results (in meVps/Å; o=11, N=21). The |D₁> curves are given as dashed lines, the solid lines are marked with numbers to indicate the different orders μ.*

(a) $N_o(\mu,t)$; χ=35 pN   *(b) $N_{o+1}(\mu,t)$; χ=35 pN*
(c) $q_o(\mu,t)$; χ=35 pN   *(d) $q_{o+1}(\mu,t)$; χ=35 pN*
(e) $p_o(\mu,t)$; χ=35 pN   *(f) $p_{o+1}(\mu,t)$; χ=35 pN*
(g) $N_o(\mu,t)$; χ=62 pN   *(h) $N_{o+1}(\mu,t)$; χ=62 pN*
(i) $q_o(\mu,t)$; χ=62 pN   *(j) $q_{o+1}(\mu,t)$; χ=62 pN*
(k) $p_o(\mu,t)$; χ=62 pN   *(l) $p_{o+1}(\mu,t)$; χ=62 pN*
(m) $q_{o+2}(\mu,t)$; χ=35 pN   *(n) $p_{o+2}(\mu,t)$; χ=35 pN*

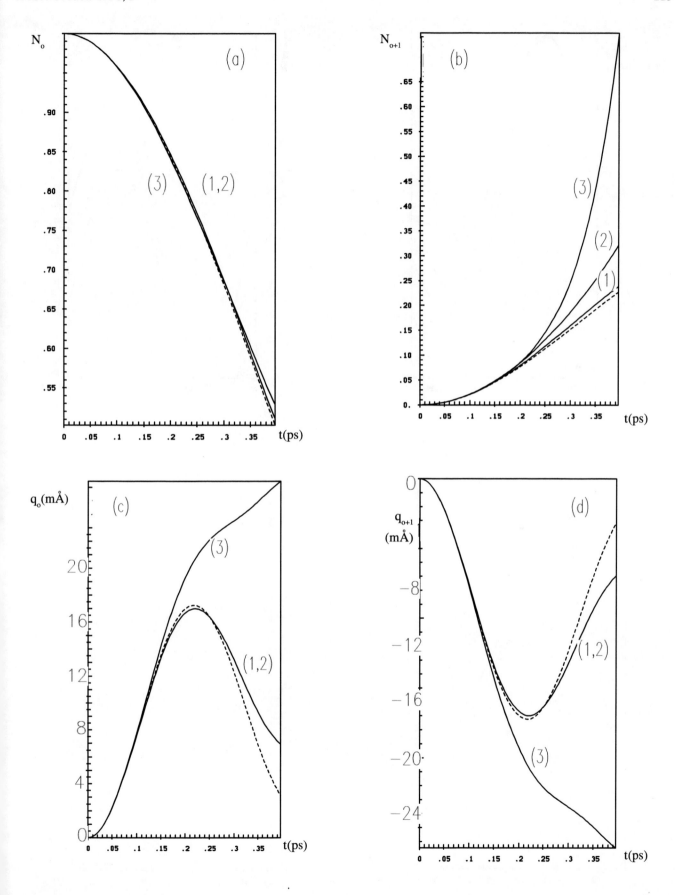

Figure 7a-d

J. Mol. Model. **1996**, 2

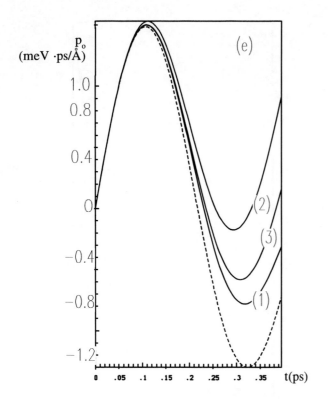

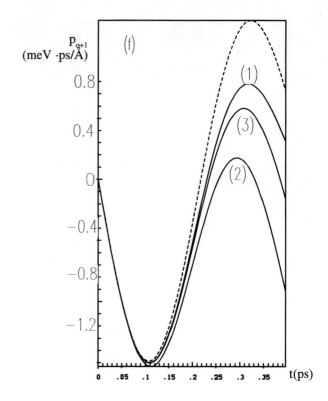

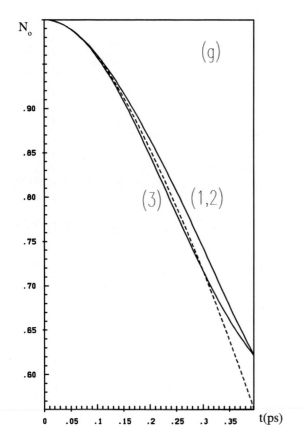

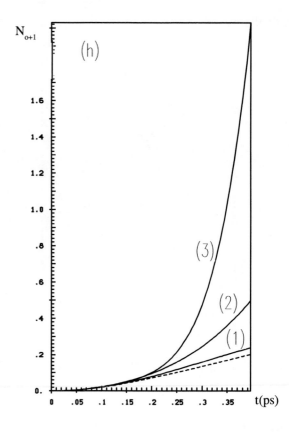

Figure 7e-h

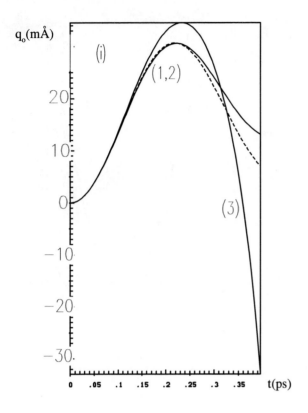

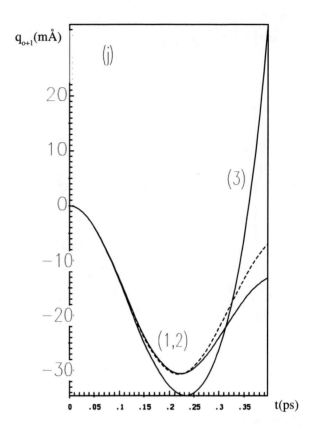

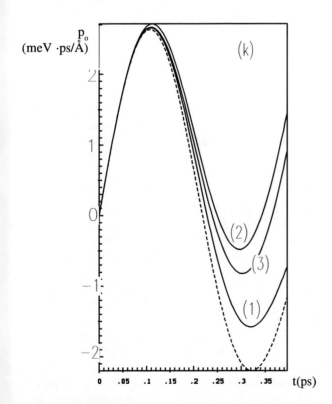

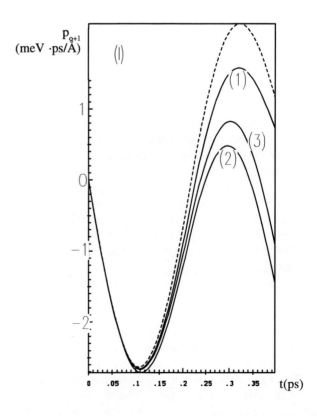

Figure 7i-l

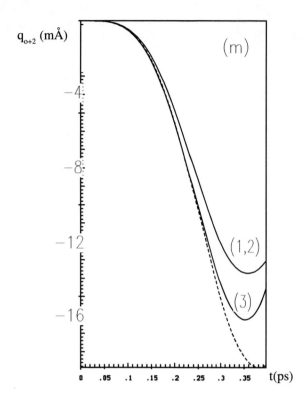

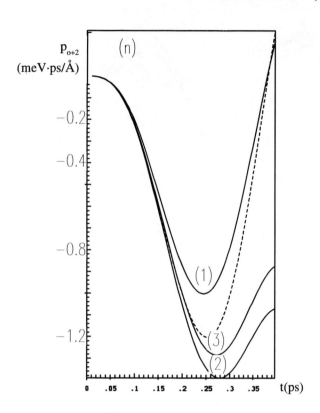

Figure 7m-n

with powers of t especially in the third order curves start to dominate and make the expansion unreliable. To obtain an „exact" wave function for larger times higher orders of the expansion would be necessary. One might wonder, why for N_o, q_o and q_{o+1} the second order curves coincide completely with the first order ones. For N_o the reason is simple, as equation (A9) shows: second order corrections simply show up only for sites o-1 and o+1, but not for site o. The fact that for the q's also the second order corrections vanish, while this is not the case for the p's implies that the second order corrections in the expectation values $B_k(2)$ must be purely imaginary as equation (22) indicates. Equation (B6) shows that the only complex factors in the expression for the corrections are $a_{o-1}{}^J$ and $a_{o+1}{}^J$. We looked at the numerical values for these coefficients, and indeed, within the first 0.4 ps their real part is less than 10^{-16} and their imaginary part varies between 0 and 0.5. Therefore it is clear that the second order corrections influence only the momenta but not the displacements. The situation for the larger coupling constant (Fig. 7 g-l) is similar to that for the smaller one, therefore we don't want to discuss it in detail. The most important result of both calculations is, that the $|D_1\rangle$ results agree very well with those from the expansion within 0.10-0.15 ps, the time in which the expansion can be considered as „exact" solution. This time is also the most important one for a possible soliton formation, because the lattice is driven only by exciton-phonon coupling in these first 100-150 fs where the excitation starts to move from the initial excitation site to its

nearest neighbors. During this process the lattice deforms in a way which stabilizes the excitation to an extent, that a soliton can be formed or not. In Fig. 7g we show for completeness q_{o+2} and p_{o+2}. In the short time intervall their values are smaller than the ones discussed above, however, here the first order becomes worse and the $|D_1\rangle$ curves are nearly identical to the third order results up to roughly 0.25 ps.

Conclusion

In an attempt to study the properties of the $|D_1\rangle$ approximation numerically, we have performed long time simulations over a period of 10 ns and computed the relevant expectation values of the deviation state and compared them to those computed from the state $\left(\hat{H}_D/J\right)|D_1\rangle$. This study complements our previous investigations of the medium time scale in the order of 100 ps [8]. The expectation values of the deviation state, which were already neglegible in the medium time scale turned out to decrease even in the course of time.

Further we expanded the formally exact solution of the Davydov Hamiltonian in a Taylor series in the time t, to assess the very short time behaviour also. We found that such an expansion around t=0 up to third order is valid within a time of 0.10-0.15 ps. Further the second and third order corrections turned out to be more or less negligible in this range of time. This is probably due to the fact, that as first order we chose already a state in which some of the terms in the expansion are summed to infinite order, resulting in the solutions of the decoupled oscillator system and the small polaron limit, respectively. Therefore we conclude that on this time-scale the two limiting cases govern the dynamics of the system. However, for larger times the first order becomes incorrect, because in the small polaron limit, starting from a localized initial excitation, only the initial excitation site is affected by the exciton-phonon interaction, while due to the dipole interaction the amide-I excitation spreads over the neighboring sites also within the $|\psi(1)\rangle$ state. For the time in which the expansion is valid, however, the results obtained from the $|D_1\rangle$ simulation agree very well with it. Thus, together with the long-time results and our previous work on medium time scales, we conclude that the $|D_1\rangle$ approximation must be very close to the exact solution for times from 0 up to 10 ns.

As it was to be expected from the beginning, an expansion of the exact solution around a single point in time, i.e. t=0, cannot replace methods using ansatz states for simulation on longer time scales. One could think to compute simply higher orders of the expansion. The draw back of this approach is, that for longer time rather high orders would be necessary, leading to prohibitively complicated expressions. Also it is known, that attempts to expand wave functions only around a single point in time usually lead into problems, when they are applied for longer times. However, there is another possibility to use the expansion method also up to larger times, namely to use a given expansion only for a small time interval, say τ, and use the state obtained as initial state for a further expansion around t=τ and so forth. If the time interval is small enough, e.g. τ=0.05-0.10 ps, the expansion could even be restricted to the first order. However, in this case the first order becomes more complicated.

Assume time steps $\ell\tau$ with ℓ=0,1,... and 0<t<τ, then for ℓ=0 we have the same expansion as discussed above. However, at t=τ we have the new initial state

$$\left|\psi_0(1,\tau)\right\rangle = \sum_n \left(a_n^J(\tau) - \delta_{no}\right)\hat{a}_n^+|0\rangle + a_o^\omega(\tau)\hat{U}_o(\tau)\hat{a}_o^+|0\rangle$$

$$(36)$$

Then our first order state in the second time interval is

$$\left|\psi(1,\tau+t)\right\rangle = \left(e^{-\frac{it}{\hbar}\hat{\omega}} + e^{-\frac{it}{\hbar}\hat{j}} - 1\right)\left|\psi_0(1,\tau)\right\rangle \qquad (37)$$

where the different terms can be derived from the exact special case solutions given in detail in paper I. For the small polaron contribution we obtain

$$(38)$$

$$\left|\omega(\tau+t)\right\rangle = e^{-\frac{it}{\hbar}\hat{\omega}}\left|\psi_0(1,\tau)\right\rangle =$$

$$= e^{-\frac{it}{\hbar}\hat{\omega}}\sum_n \left(a_n^J(\tau) - \delta_{no}\right)\hat{a}_n^+|0\rangle + e^{-\frac{it}{\hbar}\hat{\omega}}a_o^\omega(\tau)\hat{U}_o(\tau)\hat{a}_o^+|0\rangle$$

where the first term can be obtained from equation (A12) of paper I by insertion of the new initial conditions

$$a_n(0) = a_n^J(\tau) - \delta_{no} \quad ; \quad b_{nk}(0) = 0 \qquad (39)$$

and the second term also from equation (A12) by insertion of

$$a_n(0) = e^{-i\sum_k B_{ok}^2\left[\sin(\omega_k\tau) - \omega_k\tau\right]}\cdot\delta_{no} \qquad (40)$$

$$b_{nk}(0) = B_{ok}\left(e^{-i\omega_k\tau} - 1\right)\delta_{no}$$

A similar decomposition can be performed for the oscillator part:

$$\left|J(\tau+t)\right\rangle = e^{-\frac{it}{\hbar}\hat{j}}\left|\psi_0(1,\tau)\right\rangle =$$

$$= e^{-\frac{it}{\hbar}\hat{j}}\sum_n \left(a_n^J(\tau) - \delta_{no}\right)\hat{a}_n^+|0\rangle + a_o^\omega(\tau)\hat{U}_o(\tau)e^{-\frac{it}{\hbar}\hat{j}}\hat{a}_o^+|0\rangle$$

$$(41)$$

where both terms can be computed from equation (C9) of paper I. For coherent state operators, as occurring in the second part one only needs to note that the exponential operator for the oscillator system commutes with them. Thus for time steps $\ell\tau$ with ℓ>1 with arbitrary coefficients $d_n(\ell\tau)$ one has to treat cases like this as a superposition of exact solutions for the decoupled oscillator system:

$$e^{-\frac{it}{\hbar}\hat{j}}\sum_n d_n(\ell\tau)\hat{U}_n(\ell\tau)\hat{a}_n^+|0\rangle =$$

$$= \sum_n \hat{U}_n(\ell\tau)d_n(\ell\tau)\left[e^{-\frac{it}{\hbar}\hat{j}}\hat{a}_n^+|0\rangle\right] \qquad (42)$$

where all the terms can be calculated individually for each n as an exact solution of the decoupled oscillator system with an initial excitation localized at site n, leading to

$$e^{-\frac{it}{\hbar}\hat{j}}\sum_n d_n(\ell\tau)\hat{U}_n(\ell\tau)\hat{a}_n^+|0\rangle =$$

$$= \frac{1}{N}\sum_{nmk}\hat{U}_n(\ell\tau)d_n(\ell\tau)e^{\frac{2\pi i}{N}k(m-n)}e^{2i\frac{Jt}{\hbar}\cos\left(\frac{2\pi}{N}k\right)}\hat{a}_m^+|0\rangle \qquad (43)$$

In this way for any time step the contributions to $|\psi(1,\ell\tau+t)\rangle$ can be calculated, using always $|\psi(1,\ell\tau)\rangle$ as initial state, and thus the first order approximation can be propagated through larger times. However, the time step τ has to be chosen small enough, that the first order is a reliable representation of the exact solution. According to the present work τ should be around 0.05-0.10 ps. Investigations along this line are in progress, which, as we hope, will lead at least for a few picoseconds to a state which is nearly identical to the exact solution. However, the expressions obtained become more complicated at each time step (see Appendix D), especially for the computation of expectation values. A simpler possibility would be, to calculate at each τ from $|\psi(1,\ell\tau)\rangle$ the set $\{a_n(\ell\tau),q_n(\ell\tau),p_n(\ell\tau)\}$ and to construct a $|D_2\rangle$ like state from them, which in turn could be used as initial state for the next period. However, in this approximation one could miss the quantum mechanical phase mixing between phonons and excitons, which is described by $|D_1\rangle$ like states. This possibility has to be checked by numerical calculations.

The final step of these investigations will be the introduction of temperature into such expansion methods and to compare the results with the usually used methods for the treatment of temperature effects, e.g. Davydov's method, which uses an averaged Hamiltonian or our lattice population ansatz, where the lattice is populated with thermal phonons prior to the start of the simulation. For this purpose we would have to start with an initial state of the form [8]

$$\left|\psi_v(0)\right\rangle = \sum_n a_n(0)\hat{a}_n^+ \left|0\right\rangle_e \left|v\right\rangle$$

$$\left|v\right\rangle = \prod_k \frac{\left(\hat{b}_k^+\right)^{v_k}}{\sqrt{v_k!}}\left|0\right\rangle_p \tag{44}$$

where v denotes any one of the possible phonon distributions in the lattice. Then in the usual way we can write down the exact time evolution as

$$\left|\psi_v(t)\right\rangle = e^{-\frac{it}{\hbar}\hat{H}}\left|\psi_v(0)\right\rangle \tag{45}$$

with expectation values for an arbitrary operator

$$A_v(t) = \left\langle \psi_v(t)\left|\hat{A}\right|\psi_v(t)\right\rangle = \left\langle \psi_v(0)\left|e^{\frac{it}{\hbar}\hat{H}}\hat{A}e^{-\frac{it}{\hbar}\hat{H}}\right|\psi_v(0)\right\rangle \tag{46}$$

Finally a thermal average results in the desired expectation value at a temperature T

$$A(t,T) = \sum_v \rho_v(T)A_v(t)$$

$$\rho_v(T) = \frac{\left\langle v\left|e^{-\frac{\hat{H}_p}{k_BT}}\right|v\right\rangle}{\sum_\mu \left\langle \mu\left|e^{-\frac{\hat{H}_p}{k_BT}}\right|\mu\right\rangle} \tag{47}$$

where k_B is Boltzmann's constant and $\hat{H}_p$ is the phonon part of the Hamiltonian. Then the final expansion is given by

$$A(t,T) = \sum_v \rho_v(T) \sum_{k,l=0}^{\infty} \frac{(-1)^l\left(\frac{it}{\hbar}\right)^{k+l}}{k!l!} \cdot$$
$$\left\langle \psi_v(0)\left|\hat{H}^k\hat{A}\hat{H}^l\right|\psi_v(0)\right\rangle \tag{48}$$

We want to use such an expansion again up to the third order in both k and l for the case of the small polaron limit and to compare it then to the results of the different models for inclusion of temperature effects into the theory.

In the third and final paper of this series we will present, on the basis of the discussions in this work and paper I, applications of the $|D_1\rangle$ model to proteins with emphasis on the question, whether or not Davydov solitons are stable in such systems at 0K and at physiological temperatures. Further we will present vibrational spectra of proteins, calculated from the dynamics as obtained with our model.

Acknowledgement The financial support of the „Deutsche Forschungsgemeinschaft" (project no. Fo 175/3-1) and of the „Fond der Chemischen Industrie" is gratefully acknowledged.

References

1. Preceding paper: Förner, W. *J. Mol. Model.* **1996,** *2*, 70.

2. Davydov, A.S.; Kislukha, N.I. *Phys. Stat. Sol.* **1973,** *B59*, 465; Davydov, A.S. *Phys. Scripta* **1979,** *20*, 387.

3. Davydov, A.S. *Zh. Eksp. Teor. Fiz.* **1980,** *78*, 789; *Sov. Phys. JETP* **1980,** *51*, 397.

4. Scott, A.C. *Phys. Rev.* **1982,** *A26*, 57; *Physica Scripta* **1984,** *29*, 279; Mac Neil, L.; Scott,A.C. *Physica Scripta* **1984,** *29*, 284; *Phil. Trans. Roy Soc. London* **1985,** *A315*, 423.

5. Förner, W.; Ladik, J. In *Davydov's Soliton Revisited*; Christiansen, P.L.; Scott, A.C., Eds.; NATO ASI, Series B -Physics, Vol. 243; Plenum, New York (1991), 267.

6. Scott, A.C. *Phys. Rep.* **1992**, *217*, 1.

7. Skrinjar, M.J.; Kapor, D.V; Stojanovic, S.D. *Phys. Rev.* **1988**, *A38*, 6402.

8. Förner, W. *J. Phys.: Cond. Matter* **1993**, *5*, 3897.

9. Cruzeiro-Hansson, L.; Christiansen, P.L.; Scott, A.C. In *Davydov's Soliton Revisited*, Christiansen, P.L.; Scott, A.C., Eds.; NATO-ASI Series B - Physics, No. 243; Plenum Press New York (1991), 325.

Appendix A: Expectation Values of the Exciton Number Operators and the Norm

The expectation values of the number operators for the excitons in the different orders m of the wave function are given by

$$N_n(\mu,t) = \left\langle \psi(\mu,t) \left| \hat{a}_n^+ \hat{a}_n \right| \psi(\mu,t) \right\rangle \tag{A1}$$

and the norm of the states can be obtained by direct calculation of $S(\mu,t) = \langle\psi(\mu,t)|\psi(\mu,t)\rangle$ or by summation of the $N_n(\mu,t)$ over all sites n, since the total number of excitons equals 1. The zeroth order results in

$$N_n(0) = \left\langle \psi(0) \left| \hat{N}_n \right| \psi(0) \right\rangle = \left\langle \omega + J - \psi_0 \left| \hat{N}_n \right| \omega + J - \psi_0 \right\rangle \tag{A2}$$

which is easily evaluated and yields (since there is no first order correction $N_n(1)=N_n(0)$ and $|\psi(0)\rangle=|\psi(1)\rangle$ holds)

$$N_n(1) = 2\left\{1 + \operatorname{Re}\left[\left(a_o^\omega\right)^*\left(a_o^J - 1\right)\right]U_o - \operatorname{Re}\left[a_o^J\right]\right\}\delta_{no} + \left|a_n^J\right|^2$$

$$U_o = e^{-\frac{1}{2}\sum\limits_{k=1}^{N-1}|b_{ok}|^2} = e^{\sum\limits_{k=1}^{N-1}B_{ok}^2[\cos(\omega_k t)-1]} \tag{A3}$$

Summation over n yields the norm:

$$S(1) = \sum_{n=1}^{N} N_n(1) = 3 - 2\operatorname{Re}\left[\left(a_o^\omega\right)^*\left(1 - a_o^J\right)\right]U_o - 2\operatorname{Re}\left[a_o^J\right] \tag{A4}$$

In second order we have to evaluate

$$N_n(2) = \left\langle \psi(2) \left| \hat{N}_n \right| \psi(2) \right\rangle = \left\langle \psi(1) - \frac{t^2}{2\hbar^2}\psi_2 \left| \hat{N}_n \right| \psi(1) - \frac{t^2}{2\hbar^2}\psi_2 \right\rangle = N_n(1) - \frac{t^2}{\hbar^2}\operatorname{Re}\left[\left\langle \psi(1) \left| \hat{N}_n \right| \psi_2 \right\rangle\right] + \frac{t^4}{4\hbar^4}\left\langle \psi_2 \left| \hat{N}_n \right| \psi_2 \right\rangle \tag{A5}$$

Due to the phonon operators in $|\psi_2\rangle$ we have

$$\left\langle J \left| \hat{N}_n \right| \psi_2 \right\rangle = \left\langle \psi_0 \left| \hat{N}_n \right| \psi_2 \right\rangle = 0 \tag{A6}$$

Further, since in $\langle\omega|$ only the exciton operator for site o occurs, while in $|\psi_2\rangle$ only those for sites o+1 and o-1 are present

$$\left\langle \omega \left| \hat{N}_n \right| \psi_2 \right\rangle = \left\langle \psi(1) \left| \hat{N}_n \right| \psi_2 \right\rangle = 0 \tag{A7}$$

holds. Together with

$$\left\langle \psi_2 \left| \hat{N}_n \right| \psi_2 \right\rangle = J^2 \sum_{k=1}^{N-1} (\hbar\omega_k)^2 \left(A_{o-1,k}^2 \delta_{n,o-1} + A_{o+1,k}^2 \delta_{n,o+1}\right) \tag{A8}$$

we obtain

$$N_n(2) = N_n(1) + \left(\frac{Jt}{2\hbar}\right)^2 \sum_{k=1}^{2N-1} (\omega_k t)^2 \left(A_{o-1,k}^2 \delta_{n,o-1} + A_{o+1,k}^2 \delta_{n,o+1}\right) \tag{A9}$$

which summed over n yields the norm

$$S(2) = \sum_{n=1}^{N} N_n(2) = S(1) + \left(\frac{Jt}{2\hbar}\right)^2 \sum_{k=1}^{2N-1} (\omega_k t)^2 \left(A_{o-1,k}^2 + A_{o+1,k}^2\right) \tag{A10}$$

The third order is more complicated and needs evaluation of

$$N_n(3) = \left\langle \psi(2) + \frac{i}{6}\left(\frac{t}{\hbar}\right)^3 \psi_3 \left| \hat{N}_n \right| \psi(2) + \frac{i}{6}\left(\frac{t}{\hbar}\right)^3 \psi_3 \right\rangle = N_n(2) - \frac{t^3}{3\hbar^3} \operatorname{Im}\left[\left\langle \psi(2)\hat{N}_n \middle| \psi_3 \right\rangle\right] + \frac{t^6}{36\hbar^6} \left\langle \psi_3 \middle| \hat{N}_n \middle| \psi_3 \right\rangle$$

$$\left\langle \psi(2)\hat{N}_n \middle| \psi_3 \right\rangle = \left\langle \omega + J - \psi_0 \middle| \hat{N}_n \middle| \psi_3 \right\rangle - \frac{t^2}{2\hbar^2} \left\langle \psi_2 \middle| \hat{N}_n \middle| \psi_3 \right\rangle \tag{A11}$$

Using the well known commutation relations of our operators we obtain

$$\left\langle \omega \middle| \hat{N}_n \middle| \psi_3 \right\rangle = J^2 \left(a_o^\omega\right)^* \sum_{k=1}^{N-1} \hbar\omega_k D_k b_{ok}^* U_o \delta_{no} \quad ; \quad \left\langle \psi_0 \middle| \hat{N}_n \middle| \psi_3 \right\rangle = 0$$

$$\left\langle J \middle| \hat{N}_n \middle| \psi_3 \right\rangle = -J \sum_{n=1}^{N-1} (\hbar\omega_k)^2 \left(E_k^{(+)} \left(a_{o+1}^J\right)^* \delta_{n,o+1} + E_k^{(-)} \left(a_{o-1}^J\right)^* \delta_{n,o-1} \right) \tag{A12}$$

Collecting the terms yields

$$\left\langle \psi(1) \middle| \hat{N}_n \middle| \psi_3 \right\rangle = J \sum_{k=1}^{N-1} \hbar\omega_k \left[J \cdot D_k \left(a_o^\omega\right)^* b_{ok}^* U_o \delta_{no} - \hbar\omega_k \left(E_k^{(-)} \left(a_{o-1}^J\right)^* \delta_{n,o-1} + E_k^{(+)} \left(a_{o+1}^J\right)^* \delta_{n,o+1} \right) \right] \tag{A13}$$

Together with the fact that

$$\left\langle \psi_2 \middle| \hat{N}_n \middle| \psi_3 \right\rangle = J^2 \sum_{k=1}^{N-1} (\hbar\omega_k)^3 \left(A_{o-1,k} F_k^{(-)} \delta_{n,o-1} + A_{o+1,k} F_k^{(+)} \delta_{n,o+1} \right) \tag{A14}$$

is real and

$$\left\langle \psi_3 \middle| \hat{N}_n \middle| \psi_3 \right\rangle = \sum_{v=-2}^{2} \left\langle 0 \middle| \hat{\Gamma}_v^+ \hat{\Gamma}_v \middle| 0 \right\rangle \delta_{n,o+v} \tag{A15a}$$

$$\left\langle \psi_3 \middle| \hat{N}_{o\pm2} \middle| \psi_3 \right\rangle = J^4 \sum_{k=1}^{N-1} \left(\hbar\omega_k D_k^{(\pm)} \right)^2 \tag{A15b}$$

$$\left\langle \psi_3 \left| \hat{N}_{o\pm1} \right| \psi_3 \right\rangle = J^2 \left\{ \left[\sum_{k=1}^{N-1} (\hbar\omega_k)^2 E_k^{(\pm)} \right]^2 + \sum_{k=1}^{N-1} \left[(\hbar\omega_k)^2 F_k^{(\pm)} \right]^2 + \sum_{k,k'=1}^{N-1} (\hbar\omega_k \hbar\omega_{k'})^2 \left[\left(G_{kk'}^{(\pm)} \right)^2 + G_{kk'}^{(\pm)} G_{k'k}^{(\pm)} \right] \right\}$$

$$\left\langle \psi_3 \left| \hat{N}_o \right| \psi_3 \right\rangle = J^4 \sum_{k=1}^{N-1} (\hbar\omega_k D_k)^2 \tag{A15c}$$

we finally obtain

$$N_n(3) = N_n(2) + \delta_{n,o-2} \left\{ \frac{1}{36} \left(\frac{Jt}{\hbar} \right)^4 \sum_{k=1}^{N-1} \left(\omega_k t D_k^{(-)} \right)^2 \right\} + \delta_{n,o-1} \left\{ \frac{1}{3} \frac{Jt}{\hbar} \sum_{k=1}^{N-1} (\omega_k t)^2 E_k^{(-)} \operatorname{Im} \left[\left(a_{o-1}^J \right)^* \right] + $$

$$+ \frac{1}{36} \left(\frac{Jt}{\hbar} \right)^2 \left[\left(\sum_{k=1}^{N-1} (\omega_k t)^2 E_k^{(-)} \right)^2 + \sum_{k=1}^{N-1} \left((\omega_k t)^2 F_k^{(-)} \right)^2 + \sum_{k,k'=1}^{N-1} (\omega_k t \cdot \omega_{k'} t)^2 \left(\left(G_{kk'}^{(-)} \right)^2 + G_{kk'}^{(-)} G_{k'k}^{(-)} \right) \right] \right\} + $$

$$+ \delta_{no} \left(\frac{Jt}{\hbar} \right)^2 \sum_{k=1}^{N-1} (\omega_k t) \left\{ \frac{1}{36} \left(\frac{Jt}{\hbar} \right)^2 \omega_k t D_k^2 + \frac{1}{3} \operatorname{Im} \left[a_o^\omega b_{ok} \right] D_k U_o \right\} + \delta_{n,o+1} \left\{ \frac{1}{3} \frac{Jt}{\hbar} \sum_{k=1}^{N-1} (\omega_k t)^2 E_k^{(+)} \operatorname{Im} \left[\left(a_{o+1}^J \right)^* \right] + $$

$$+ \frac{1}{36} \left(\frac{Jt}{\hbar} \right)^2 \left[\left(\sum_{k=1}^{N-1} (\omega_k t)^2 E_k^{(+)} \right)^2 + \sum_{k=1}^{N-1} \left((\omega_k t)^2 F_k^{(+)} \right)^2 + \sum_{k,k'=1}^{N-1} (\omega_k t \cdot \omega_{k'} t)^2 \left(\left(G_{kk'}^{(+)} \right)^2 + G_{kk'}^{(+)} G_{k'k}^{(+)} \right) \right] \right\} + \tag{A16}$$

$$+ \delta_{n,o+2} \left\{ \frac{1}{36} \left(\frac{Jt}{\hbar} \right)^4 \sum_{k=1}^{N-1} \left(\omega_k t D_k^{(+)} \right)^2 \right\} $$

The norm S(3) is then simply given by summation of $N_n(3)$ over the sites n:

$$S(3) = \sum_{n=1}^{N} N_n(3) \tag{A17}$$

The explicit expression for S(3) is obtained from (A16) by replacing on the right hand side the term $N_n(2)$ with S(2) and by leaving out the Kronecker δ's.

Appendix B: Expectation Values of the Phonon Operators

In this Appendix we want to calculate the expectation values of the phonon annihilation operators. Note that their complex conjugates are the expectation values of the phonon creation operator. These expectation values are

$$B_k(\mu) = \left\langle \psi(\mu) \left| \hat{b}_k \right| \psi(\mu) \right\rangle \tag{B1}$$

The first expectation value in this series, $B_k(0) = B_k(1)$ is given by

$$B_k(1) = \left\langle \psi(1)\left|\hat{b}_k\right|\psi(1)\right\rangle = \left\langle \omega + J - \psi_0\left|\hat{b}_k\right|\omega + J - \psi_0\right\rangle = \left\langle \omega + J - \psi_0\left|\hat{b}_k\right|\omega\right\rangle$$

$$\left\langle \omega\left|\hat{b}_k\right|\omega\right\rangle = \left|a_o^\omega\right|^2 \left\langle \beta_o\left|\hat{b}_k\right|\beta_o\right\rangle = b_{ok}$$

$$\left\langle J\left|\hat{b}_k\right|\omega\right\rangle = a_o^\omega\left(a_o^J\right)^* \left\langle 0\left|\hat{b}_k\right|\beta_o\right\rangle = a_o^\omega\left(a_o^J\right)^* b_{ok}U_o \tag{B2}$$

$$\left\langle \psi_0\left|\hat{b}_k\right|\omega\right\rangle = a_o^\omega b_{ok}U_o$$

$$B_k(1) = b_{ok}\left\{1 + a_o^\omega\left[\left(a_o^J\right)^* - 1\right]U_o\right\}$$

For $B_k(2)$ we have to evaluate

$$B_k(2) = \left\langle \psi(2)\left|\hat{b}_k\right|\psi(2)\right\rangle = \left\langle \psi(1) - \frac{t^2}{2\hbar^2}\psi_2\left|\hat{b}_k\right|\psi(1) - \frac{t^2}{2\hbar^2}\psi_2\right\rangle$$

$$= B_k(1) - \frac{t^2}{2\hbar^2}\left[\left\langle \psi(1)\left|\hat{b}_k\right|\psi_2\right\rangle + \left\langle \psi_2\left|\hat{b}_k\right|\psi(1)\right\rangle\right] + \frac{t^4}{4\hbar^4}\left\langle \psi_2\left|\hat{b}_k\right|\psi_2\right\rangle \tag{B3}$$

where the terms are

$$\left\langle \psi(1)\left|\hat{b}_k\right|\psi_2\right\rangle = \left\langle \omega + J - \psi_0\left|\hat{b}_k\right|\psi_2\right\rangle = \left\langle J\left|\hat{b}_k\right|\psi_2\right\rangle = -J\hbar\omega_k\left[\left(a_{o-1}^J\right)^* A_{o-1,k} + \left(a_{o+1}^J\right)^* A_{o+1,k}\right]$$

$$\left\langle \psi_2\left|\hat{b}_k\right|\psi(1)\right\rangle = \left\langle \psi_2\left|\hat{b}_k\right|\omega + J - \psi_0\right\rangle = 0 \tag{B4}$$

and

$$\left\langle \psi_2\left|\hat{b}_k\right|\psi_2\right\rangle = J^2\sum_{k'k''}\hbar\omega_{k'}\hbar\omega_{k''}\left(A_{o-1,k'}A_{o-1,k''} + A_{o+1,k'}A_{o+1,k''}\right)\cdot\left\langle 0\left|\hat{b}_{k'}\hat{b}_k\hat{b}_{k''}^+\right|0\right\rangle = 0 \tag{B5}$$

and thus the final result reads as

$$B_k(2) = B_k(1) + \frac{1}{2}\frac{Jt}{\hbar}\omega_k t\left[\left(a_{o-1}^J\right)^* A_{o-1,k} + \left(a_{o+1}^J\right)^* A_{o+1,k}\right] \tag{B6}$$

For the third order wave function we have to compute

$$B_k(3) = \left\langle \psi(3)\left|\hat{b}_k\right|\psi(3)\right\rangle = \left\langle \psi(2) + \frac{i}{6}\frac{t^3}{\hbar^3}\psi_3\left|\hat{b}_k\right|\psi(2) + \frac{i}{6}\frac{t^3}{\hbar^3}\psi_3\right\rangle$$

$$= B_k(2) + \frac{i}{6}\frac{t^3}{\hbar^3}\left[\left\langle \psi(2)\left|\hat{b}_k\right|\psi_3\right\rangle - \left\langle \psi_3\left|\hat{b}_k\right|\psi(2)\right\rangle\right] + \frac{1}{36}\frac{t^6}{\hbar^6}\left\langle \psi_3\left|\hat{b}_k\right|\psi_3\right\rangle \tag{B7}$$

The two mixed terms can be reduced to

$$\hat{b}_k \left| J - \psi_0 \right\rangle = 0 \Rightarrow \left\langle \psi_3 \left| \hat{b}_k \right| \psi(2) \right\rangle = \left\langle \psi_3 \left| \hat{b}_k \right| \omega \right\rangle - \frac{1}{2} \frac{t^2}{\hbar^2} \left\langle \psi_3 \left| \hat{b}_k \right| \psi_2 \right\rangle$$

$$\left\langle \psi(2) \left| \hat{b}_k \right| \psi_3 \right\rangle = \left\langle J + \omega - \psi_0 \left| \hat{b}_k \right| \psi_3 \right\rangle - \frac{1}{2} \frac{t^2}{\hbar^2} \left\langle \psi_2 \left| \hat{b}_k \right| \psi_3 \right\rangle \qquad (B8)$$

Now we have to evaluate the individual expectation values:

$$\left\langle \psi_3 \left| \hat{b}_k \right| \omega \right\rangle = \sum_{\nu=-2}^{2} (-1)^\nu a_o^\omega \left\langle 0 \left| \hat{\Gamma}_\nu^+ \hat{a}_{o+\nu} \hat{b}_k \hat{a}_o^+ \hat{U}_o \right| 0 \right\rangle = a_o^\omega b_{ok} \left\langle 0 \left| \Gamma_0^+ \right| \beta_o \right\rangle = J^2 a_o^\omega b_{ok} U_o \sum_{k'=1}^{N-1} \hbar \omega_{k'} D_{k'} b_{ok'} \qquad (B9)$$

$$\left\langle \psi_3 \left| \hat{b}_k \right| \psi_2 \right\rangle = \sum_{\nu=-2}^{2} (-1)^\nu \left\langle 0 \left| \hat{\Gamma}_\nu^+ \hat{a}_{o+\nu} \hat{b}_k \left(\hat{\Theta}_{-1} \hat{a}_{o-1}^+ + \hat{\Theta}_{+1} \hat{a}_{o+1}^+ \right) \right| 0 \right\rangle = - \left[\left\langle 0 \left| \hat{\Gamma}_{-1}^+ \hat{b}_k \hat{\Theta}_{-1} \right| 0 \right\rangle + \left\langle 0 \left| \hat{\Gamma}_{+1}^+ \hat{b}_k \hat{\Theta}_{+1} \right| 0 \right\rangle \right]$$

$$= J^2 \hbar \omega_k \left[A_{o-1,k} \sum_{k'=1}^{N-1} \left(\hbar \omega_{k'} \right)^2 E_{k'}^{(-)} + A_{o+1,k} \sum_{k'=1}^{N-1} \left(\hbar \omega_{k'} \right)^2 E_{k'}^{(+)} \right]$$

Collecting the terms, this yields

$$-\frac{i}{6} \frac{t^3}{\hbar^3} \left\langle \psi_3 \left| \hat{b}_k \right| \psi(2) \right\rangle = \frac{i}{6} \left(\frac{Jt}{\hbar} \right)^2 \left\{ -a_o^\omega b_{ok} U_o \sum_{k'=1}^{N-1} (\omega_{k'} t) D_{k'} b_{ok'} + \right.$$

$$\left. + \frac{1}{2} (\omega_k t) \left[A_{o-1,k} \sum_{k'=1}^{N-1} (\omega_{k'} t)^2 E_{k'}^{(-)} + A_{o+1,k} \sum_{k'=1}^{N-1} (\omega_{k'} t)^2 E_{k'}^{(+)} \right] \right\} \qquad (B10)$$

The contributions to the next term are

$$\left\langle J \left| \hat{b}_k \right| \psi_3 \right\rangle = \sum_{\nu=-2}^{2} (-1)^\nu \left(a_{o+\nu}^J \right)^* \left\langle 0 \left| \hat{b}_k \hat{\Gamma}_\nu \right| 0 \right\rangle$$

$$= J^2 \hbar \omega_k \left[\left(a_{o-2}^J \right)^* D_k^{(-)} + \left(a_o^J \right)^* D_k + \left(a_{o+2}^J \right)^* D_k^{(+)} \right] - J (\hbar \omega_k)^2 \left[\left(a_{o-1}^J \right)^* F_k^{(-)} + \left(a_{o+1}^J \right)^* F_k^{(+)} \right]$$

$$\left\langle \psi_0 \left| \hat{b}_k \right| \psi_3 \right\rangle = \left\langle 0 \left| \hat{b}_k \hat{\Gamma}_0 \right| 0 \right\rangle = J^2 \hbar \omega_k D_k \qquad (B11)$$

$$\left\langle \omega \left| \hat{b}_k \right| \psi_3 \right\rangle = \left(a_o^\omega \right)^* \left\langle \beta_o \left| \hat{b}_k \hat{\Gamma}_0 \right| 0 \right\rangle = J^2 \hbar \omega_k \left(a_o^\omega \right)^* D_k U_o$$

and therefore

$$\left\langle \omega + J - \psi_0 \left| \hat{b}_k \right| \psi_3 \right\rangle = -J (\hbar \omega_k)^2 \left\{ \left(a_{o-1}^J \right)^* F_k^{(-)} + \left(a_{o+1}^J \right)^* F_k^{(+)} \right\} +$$

$$+ J^2 \hbar \omega_k \left\{ \left(a_{o-2}^J \right)^* D_k^{(-)} + \left[\left(a_o^\omega \right)^* U_o + \left(a_o^J \right)^* - 1 \right] D_k + \left(a_{o+2}^J \right)^* D_k^{(+)} \right\} \qquad (B12)$$

Further

$$\left\langle \psi_2 \left| \hat{b}_k \right| \psi_3 \right\rangle = \sum_{\nu=-2}^{2} (-1)^{\nu} \left\langle 0 \left| \left[\hat{\Theta}_{-1}^+ \hat{a}_{o-1} + \hat{\Theta}_{+1}^+ \hat{a}_{o+1} \right] \hat{b}_k \hat{a}_{o+\nu}^+ \hat{\Gamma}_\nu \right| 0 \right\rangle = - \left[\left\langle 0 \left| \hat{\Theta}_{-1}^+ \hat{b}_k \hat{\Gamma}_{-1} \right| 0 \right\rangle + \left\langle 0 \left| \hat{\Theta}_{+1}^+ \hat{b}_k \hat{\Gamma}_{+1} \right| 0 \right\rangle \right]$$

$$= J^2 \hbar \omega_k \sum_{k'=1}^{N-1} (\hbar \omega_{k'})^2 \left[A_{o-1,k'} \left(G_{kk'}^{(-)} + G_{k'k}^{(-)} \right) + A_{o+1,k'} \left(G_{kk'}^{(+)} + G_{k'k}^{(+)} \right) \right] \qquad (B13)$$

and thus

$$\frac{i}{6} \frac{t^3}{\hbar^3} \left\langle \psi(2) \left| \hat{b}_k \right| \psi_3 \right\rangle = \frac{i}{6} \left\{ -\frac{Jt}{\hbar} (\omega_k t)^2 \left[\left(a_{o-1}^J \right)^* F_k^{(-)} + \left(a_{o+1}^J \right)^* F_k^{(+)} \right] + \right.$$

$$+ \left(\frac{Jt}{\hbar} \right)^2 \omega_k t \left[\left(a_{o-2}^J \right)^* D_k^{(-)} + \left(\left(a_o^\omega \right)^* U_o + \left(a_o^J \right)^* - 1 \right) D_k + \left(a_{o+2}^J \right)^* D_k^{(+)} \right] -$$

$$\left. - \frac{1}{2} \left(\frac{Jt}{\hbar} \right)^2 \omega_k t \sum_{k'=1}^{N-1} (\omega_{k'} t)^2 \left[A_{o-1,k'} \left(G_{kk'}^{(-)} + G_{k'k}^{(-)} \right) + A_{o+1,k'} \left(G_{kk'}^{(+)} + G_{k'k}^{(+)} \right) \right] \right\} \qquad (B14)$$

The final expectation value is

$$\left\langle \psi_3 \left| \hat{b}_k \right| \psi_3 \right\rangle = \sum_{\nu,\mu=-2}^{2} (-1)^{\nu+\mu} \left\langle 0 \left| \hat{a}_{o+\nu} \hat{\Gamma}_\nu^+ \hat{b}_k \hat{\Gamma}_\mu \hat{a}_{o+\mu}^+ \right| 0 \right\rangle = \sum_{\nu=-2}^{2} \left\langle 0 \left| \hat{\Gamma}_\nu^+ \hat{b}_k \hat{\Gamma}_\nu \right| 0 \right\rangle \qquad (B15)$$

where

$$\left\langle 0 \left| \hat{\Gamma}_{\pm 2}^+ \hat{b}_k \hat{\Gamma}_{\pm 2} \right| 0 \right\rangle = J^4 \sum_{k'k''} \hbar \omega_{k'} \cdot \hbar \omega_{k''} D_{k'}^{(\pm)} D_{k''}^{(\pm)} \left\langle 0 \left| \hat{b}_{k'} \hat{b}_k \hat{b}_{k''}^+ \right| 0 \right\rangle = 0$$

$$\left\langle 0 \left| \hat{\Gamma}_{\pm 1}^+ \hat{b}_k \hat{\Gamma}_{\pm 1} \right| 0 \right\rangle = J^2 (\hbar \omega_k)^2 F_k^{(\pm)} \sum_{k'=1}^{N-1} (\hbar \omega_{k'})^2 E_{k'}^{(\pm)} + J^2 \hbar \omega_k \sum_{k'=1}^{N-1} (\hbar \omega_{k'})^3 F_{k'}^{(\pm)} \left(G_{kk'}^{(\pm)} + G_{k'k}^{(\pm)} \right) \qquad (B16)$$

$$\left\langle 0 \left| \hat{\Gamma}_0^+ \hat{b}_k \hat{\Gamma}_0 \right| 0 \right\rangle = J^4 \sum_{k'k''} \hbar \omega_{k'} \hbar \omega_{k''} D_{k'} D_{k''} \left\langle 0 \left| \hat{b}_{k'} \hat{b}_k \hat{b}_{k''}^+ \right| 0 \right\rangle = 0$$

and thus

$$\frac{1}{36} \frac{t^6}{\hbar^6} \left\langle \psi_3 \left| \hat{b}_k \right| \psi_3 \right\rangle = \frac{1}{36} \left(\frac{Jt}{\hbar} \right)^2 \omega_k t \sum_{k'=1}^{N-1} (\omega_{k'} t)^2 \left\{ \omega_k t \left[F_k^{(-)} E_{k'}^{(-)} + F_k^{(+)} E_{k'}^{(+)} \right] + \right.$$

$$\left. + \omega_{k'} t \left[F_{k'}^{(-)} \left(G_{kk'}^{(-)} + G_{k'k}^{(-)} \right) + F_{k'}^{(+)} \left(G_{kk'}^{(+)} + G_{k'k}^{(+)} \right) \right] \right\} \qquad (B17)$$

Then $B_k(3)$ is given by

$$
\begin{aligned}
B_k(3) = B_k(2) + \frac{i}{6}\Bigg\{ &-\frac{Jt}{\hbar}(\omega_k t)^2\Big[\big(a_{o-1}^J\big)^* F_k^{(-)} + \big(a_{o+1}^J\big)^* F_k^{(+)}\Big] + \\
&+\left(\frac{Jt}{\hbar}\right)^2 \omega_k t\Big[\big(a_{o-2}^J\big)^* D_k^{(-)} + \big(\big(a_o^\omega\big)^* U_o + \big(a_o^J\big)^* - 1\big)D_k + \big(a_{o+2}^J\big)^* D_k^{(+)}\Big] - \\
&-\frac{1}{2}\left(\frac{Jt}{\hbar}\right)^2 \omega_k t \sum_{k'=1}^{N-1}(\omega_{k'}t)^2\Big[A_{o-1,k'}\big(G_{kk'}^{(-)} + G_{k'k}^{(-)}\big) + A_{o+1,k'}\big(G_{kk'}^{(+)} + G_{k'k}^{(+)}\big)\Big]\Bigg\} + \\
&+\frac{i}{6}\left(\frac{Jt}{\hbar}\right)^2\Bigg\{-a_o^\omega b_{ok} U_o \sum_{k'=1}^{N-1}(\omega_{k'}t)D_{k'}b_{ok'} + \frac{1}{2}(\omega_k t)\Big[A_{o-1,k}\sum_{k'=1}^{N-1}(\omega_{k'}t)^2 E_{k'}^{(-)} + A_{o+1,k}\sum_{k'=1}^{N-1}(\omega_{k'}t)^2 E_{k'}^{(+)}\Big]\Bigg\} + \\
&+\frac{1}{36}\left(\frac{Jt}{\hbar}\right)^2 \omega_k t \sum_{k'=1}^{N-1}(\omega_{k'}t)^2\Big\{\omega_k t\big[F_k^{(-)}E_{k'}^{(-)} + F_k^{(+)}E_{k'}^{(+)}\big] + \omega_{k'}t\big[F_{k'}^{(-)}\big(G_{kk'}^{(-)} + G_{k'k}^{(-)}\big) + F_{k'}^{(+)}\big(G_{kk'}^{(+)} + G_{k'k}^{(+)}\big)\big]\Big\}
\end{aligned}
\tag{B18}
$$

The expectation values of the displacement and momentum operators can be computed simply from $B_k(\mu)$ and $B_k^*(\mu)$ as decribed in the main text.

Appendix C: Expectation Values of the Hamiltonian

Since the Hamilton operator is hermitian, we can write for the expectation value for the first order wave function, omitting the vanishing contributions of the total of 18:

$$
H(0) = H(1) = \big\langle \omega + J - \psi_0\big|\hat\omega + \hat J\big|\omega + J - \psi_0\big\rangle = 2\,\mathrm{Re}\Big[\big\langle\omega|\hat\omega|J\big\rangle - \big\langle\omega|\hat\omega|\psi_0\big\rangle + \big\langle\omega|\hat J|J\big\rangle - \big\langle J|\hat J|\psi_0\big\rangle\Big]
\tag{C1}
$$

where

$$
\begin{aligned}
\big\langle J|\hat J|\psi_0\big\rangle &= -J\Big[\big(a_{o-1}^J\big)^* + \big(a_{o+1}^J\big)^*\Big] \\
\big\langle\omega|\hat\omega|J\big\rangle &= a_o^J\big(a_o^\omega\big)^* U_o \sum_{k=1}^{N-1}\hbar\omega_k B_{ok}b_{ok}^* \\
\big\langle\omega|\hat\omega|\psi_0\big\rangle &= \big(a_o^\omega\big)^* U_o \sum_{k=1}^{N-1}\hbar\omega_k B_{ok}b_{ok}^* \\
\big\langle\omega|\hat J|J\big\rangle &= -J\big(a_o^\omega\big)^*\big(a_{o-1}^J + a_{o+1}^J\big)U_o
\end{aligned}
\tag{C2}
$$

and thus we obtain finally for H(1)

$$
H(1) = 2U_o \sum_k \hbar\omega_k B_{ok}\,\mathrm{Re}\Big[a_o^\omega b_{ok}\big(\big(a_o^J\big)^* - 1\big)\Big] + 2J\Big\{\mathrm{Re}\big[a_{o-1}^J + a_{o+1}^J\big] - U_o\,\mathrm{Re}\Big[\big(a_o^\omega\big)^*\big(a_{o-1}^J + a_{o+1}^J\big)\Big]\Big\}
\tag{C3}
$$

For the second order we need the following expectation values for $|\psi_2\rangle$ [with equation (29)]

$$
\big\langle\psi_2|\hat\omega|\psi_2\big\rangle = J^2 \sum_{k=1}^{N-1}(\hbar\omega_k)^3\big(A_{o-1,k}^2 + A_{o+1,k}^2\big)
\tag{C4}
$$

$$\langle\psi_2|\hat{J}|\psi_2\rangle = 0 \Rightarrow \langle\psi_2|\hat{H}|\psi_2\rangle = \langle\psi_2|\hat{\omega}|\psi_2\rangle$$

and the total function H(2) is given by

$$H(2) = H(1) - \frac{t^2}{\hbar^2}\mathrm{Re}\left[\langle\psi(1)|\hat{H}|\psi_2\rangle\right] + \frac{t^4}{4\hbar^4}\langle\psi_2|\hat{H}|\psi_2\rangle \tag{C5}$$

Since of the six individual expectation values contained in $\langle\psi(1)|\hat{H}|\psi_2\rangle$ four are vanishing we arrive at

$$\langle\psi(1)|\hat{H}|\psi_2\rangle = \langle\omega|\hat{J}|\psi_2\rangle + \langle J|\hat{\omega}|\psi_2\rangle$$

$$\langle\omega|\hat{\omega}|\psi_2\rangle = \langle J|\hat{J}|\psi_2\rangle = \langle\psi_0|\hat{\omega}|\psi_2\rangle = \langle\psi_0|\hat{J}|\psi_2\rangle = 0 \tag{C6}$$

where

$$\langle\omega|\hat{J}|\psi_2\rangle = J^2\left(a_o^\omega\right)^* U_o \sum_{k=1}^{N-1}\hbar\omega_k b_{ok}^*\left(A_{o-1,k} + A_{o+1,k}\right) \tag{C7}$$

$$\langle J|\hat{\omega}|\psi_2\rangle = -J\sum_{k=1}^{N-1}\left(\hbar\omega_k\right)^2\left[A_{o-1,k}B_{o-1,k}\left(a_{o-1}^J\right)^* + A_{o+1,k}B_{o+1,k}\left(a_{o+1}^J\right)^*\right] \tag{C8a}$$

Thus finally we obtain

$$H(2) = H(1) + \frac{1}{4}\left(\frac{Jt}{\hbar}\right)^2\sum_{k=1}^{N-1}\hbar\omega_k\left(\omega_k t\right)^2\left(A_{o-1,k}^2 + A_{o+1,k}^2\right) -$$

$$-\left(\frac{Jt}{\hbar}\right)^2 U_o\sum_{k=1}^{N-1}\hbar\omega_k\left(A_{o-1,k} + A_{o+1,k}\right)\mathrm{Re}\left[b_{ok}a_o^\omega\right] + J\sum_{k=1}^{N-1}\left(\omega_k t\right)^2\left\{A_{o-1,k}B_{o-1,k}\,\mathrm{Re}\left[a_{o-1}^J\right] + A_{o+1,k}B_{o+1,k}\,\mathrm{Re}\left[a_{o+1}^J\right]\right\} \tag{C8}$$

For the third order correction we have to evaluate

$$H(3) = \left\langle\psi(2) + \frac{i}{6}\frac{t^3}{\hbar^3}\psi_3\middle|\hat{H}\middle|\psi(2) + \frac{i}{6}\frac{t^3}{\hbar^3}\psi_3\right\rangle = H(2) - \frac{1}{3}\frac{t^3}{\hbar^3}\mathrm{Im}\left[\langle\psi(2)|\hat{H}|\psi_3\rangle\right] + \frac{1}{36}\frac{t^6}{\hbar^6}\langle\psi_3|\hat{H}|\psi_3\rangle \tag{C9}$$

$$\langle\psi(2)|\hat{H}|\psi_3\rangle = \langle\psi(1)|\hat{H}|\psi_3\rangle - \frac{1}{2}\frac{t^2}{\hbar^2}\langle\psi_2|\hat{H}|\psi_3\rangle$$

Since

$$\langle\psi_2|\hat{H}|\psi_3\rangle = \langle 0|\left[\hat{\Theta}_{-1}^+\hat{a}_{o-1} + \hat{\Theta}_{+1}^+\hat{a}_{o+1}\right]\left(\hat{\omega} + \hat{J}\right)\sum_{\nu=-2}^{2}(-1)^\nu\hat{\Gamma}_\nu\hat{a}_{o+\nu}^+|0\rangle \tag{C10}$$

is obviously real, and thus $\mathrm{Im}\left[\langle\psi_2|\hat{H}|\psi_3\rangle\right] = 0$, it remains to calculate

$$H(3) = H(2) - \frac{1}{3}\frac{t^3}{\hbar^3}\mathrm{Im}\left[\langle\omega + J - \psi_0|\hat{H}|\psi_3\rangle\right] + \frac{1}{36}\frac{t^6}{\hbar^6}\langle\psi_3|\hat{H}|\psi_3\rangle \tag{C11}$$

where we obtain for $|\psi_0\rangle$

$$\hat{H}|\psi_0\rangle = -J\left(\hat{a}_{o-1}^+ + \hat{a}_{o+1}^+\right)|0\rangle + \sum_{k=1}^{N-1} \hbar\omega_k B_{ok} \hat{b}_k^+ \hat{a}_o^+ |0\rangle \tag{C12}$$

$$\langle\psi_o|\hat{H}|\psi_3\rangle = J\left(\langle 0|\hat{\Gamma}_{-1}|0\rangle + \langle 0|\hat{\Gamma}_{+1}|0\rangle\right) + \sum_{k=1}^{N-1} \hbar\omega_k B_{ok} \langle 0|\hat{b}_k \hat{\Gamma}_o|0\rangle = J^2 \sum_{k=1}^{N-1} (\hbar\omega_k)^2 \left(E_k^{(-)} + B_{ok} D_k + E_k^{(+)}\right)$$

Note, that $\langle\psi_0|\hat{H}|\psi_3\rangle$ is real. The action of the Hamiltonian on $|\omega\rangle$ yields

$$\hat{H}|\omega\rangle = -J a_o^\omega \left(\hat{a}_{o+1}^+ + \hat{a}_{o-1}^+\right)|\beta_o\rangle + a_o^\omega \sum_{k=1}^{N-1} \hbar\omega_k \left[B_{ok} b_{ok} + (B_{ok} + b_{ok})\hat{b}_k^+\right]\hat{a}_o^+|\beta_o\rangle \tag{C13}$$

and thus

$$\langle\omega|\hat{H}|\psi_3\rangle = \left(a_o^\omega\right)^* J\left[\langle\beta_o|\hat{\Gamma}_{-1}|0\rangle + \langle\beta_o|\hat{\Gamma}_{+1}|0\rangle\right] + \left(a_o^\omega\right)^* \sum_{k=1}^{N-1} \hbar\omega_k \left[B_{ok} b_{ok}^* \langle\beta_o|\hat{\Gamma}_0|0\rangle + \left(B_{ok} + b_{ok}^*\right)\langle\beta_o|\hat{b}_k\hat{\Gamma}_0|0\rangle\right] \tag{C14}$$

The explicit evaluation of the expectation values results in

$$\langle\omega|\hat{H}|\psi_3\rangle = J^2 \left(a_o^\omega\right)^* U_o \sum_{k=1}^{N-1} \hbar\omega_k \left\{ \sum_{k'=1}^{N-1} \hbar\omega_{k'} b_{ok}^* b_{ok'}^* \left[G_{kk'}^{(-)} + G_{kk'}^{(+)} + B_{ok} D_{k'}\right] + \right.$$
$$\left. + \hbar\omega_k \left[\left(E_k^{(-)} + E_k^{(+)}\right) + b_{ok}^*\left(F_k^{(-)} + F_k^{(+)}\right) + \left(B_{ok} + b_{ok}^*\right)D_k\right]\right\} \tag{C15}$$

Finally, the action of the Hamiltonian on $|J\rangle$ leads to

$$\hat{H}|J\rangle = \sum_{n=1}^{N}\left[-J\left(a_{n+1}^J + a_{n-1}^J\right) + a_n^J \sum_{k=1}^{N-1} \hbar\omega_k B_{nk} \hat{b}_k^+\right]\hat{a}_n^+|0\rangle \tag{C16}$$

and thus the expectation value of the Hamiltonian is given by

$$\langle J|\hat{H}|\psi_3\rangle = \sum_{\nu=-2}^{2} (-1)^\nu \left\{-J\left[\left(a_{o+\nu-1}^J\right)^* + \left(a_{o+\nu+1}^J\right)^*\right]\langle 0|\hat{\Gamma}_\nu|0\rangle + \sum_{k=1}^{N-1} \hbar\omega_k B_{o+\nu,k}\left(a_{o+\nu}^J\right)^*\langle 0|\hat{b}_k\hat{\Gamma}_\nu|0\rangle\right\} \tag{C17}$$

which yields

$$\langle J|\hat{H}|\psi_3\rangle = -J\sum_{k=1}^{N-1} (\hbar\omega_k)^3 \left[\left(a_{o-1}^J\right)^* B_{o-1,k} F_k^{(-)} + \left(a_{o+1}^J\right)^* B_{o+1,k} F_k^{(+)}\right] +$$
$$+ J^2 \sum_{k=1}^{N-1} (\hbar\omega_k)^2 \left[\left(a_{o-2}^J\right)^* B_{o-2,k} D_k^{(-)} + \left(a_{o-2}^J + a_o^J\right)^* E_k^{(-)} + \left(a_o^J\right)^* B_{ok} D_k + \left(a_o^J + a_{o+2}^J\right)^* E_k^{(+)} + \left(a_{o+2}^J\right)^* B_{o+2,k} D_k^{(+)}\right]$$

$$\tag{C18}$$

and thus

$$H'(3) = -\frac{1}{3}\frac{t^3}{\hbar^3} \text{Im}\left[\langle\psi(1)|\hat{H}|\psi_3\rangle\right] = \frac{1}{3}\left(\frac{Jt}{\hbar}\right)^2 U_o \sum_{k=1}^{N-1} \omega_k t \left\{\sum_{k'=1}^{N-1} \hbar\omega_{k'} \text{Im}\left[a_o^\omega b_{ok} b_{ok'}\right]\left[G_{kk'}^{(-)} + G_{kk'}^{(+)} + B_{ok} D_{k'}\right] + \right.$$

$$+ \hbar\omega_k\left[\text{Im}\left[a_o^\omega\right]\left(E_k^{(-)} + E_k^{(+)}\right) + \text{Im}\left[a_o^\omega b_{ok}\right]\left(F_k^{(-)} + F_k^{(+)}\right) + \left(\text{Im}\left[a_o^\omega\right]B_{ok} + \text{Im}\left[a_o^\omega b_{ok}\right]\right)D_k\right]\right\} -$$

$$- \frac{J}{3}\sum_{k=1}^{N-1}(\omega_k t)^3\left[\text{Im}\left[a_{o-1}^J\right]B_{o-1,k}F_k^{(-)} + \text{Im}\left[a_{o+1}^J\right]B_{o+1,k}F_k^{(+)}\right] +$$

$$+ \frac{J}{3}\frac{Jt}{\hbar}\sum_{k=1}^{N-1}(\omega_k t)^2\left[\text{Im}\left[a_{o-2}^J\right]B_{o-2,k}D_k^{(-)} + \text{Im}\left[a_{o-2}^J + a_o^J\right]E_k^{(-)} + \right.$$

$$\left. + \text{Im}\left[a_o^J\right]B_{ok}D_k + \text{Im}\left[a_o^J + a_{o+2}^J\right]E_k^{(+)} + \text{Im}\left[a_{o+2}^J\right]B_{o+2,k}D_k^{(+)}\right]$$

(C19)

Finally we have to calculate the three parts of the expectation value of the Hamiltonian with the state $|\psi_3\rangle$. As first step we evaluate the exciton-phonon interaction part:

$$\sum_{n=1}^{N} B_{nk}\hat{a}_n^+\hat{a}_n|\psi_3\rangle = \sum_{n=1}^{N} B_{nk}\sum_{\nu=-2}^{2}(-1)^\nu \hat{\Gamma}_\nu \hat{a}_n^+\hat{a}_n\hat{a}_{o+\nu}^+|0\rangle = \sum_{\nu=-2}^{2}(-1)^\nu B_{o+\nu,k}\hat{\Gamma}_\nu\hat{a}_{o+\nu}^+|0\rangle$$

(C20)

$$\sum_{n=1}^{N} B_{nk}\left\langle\psi_3\left|\left(\hat{b}_k^+ + \hat{b}_k\right)\hat{a}_n^+\hat{a}_n\right|\psi_3\right\rangle = \sum_{\nu=-2}^{2} B_{o+\nu,k}\left\langle 0\left|\hat{\Gamma}_\nu^+\left(\hat{b}_k^+ + \hat{b}_k\right)\hat{\Gamma}_k\right|0\right\rangle = 2\sum_{\nu=-2}^{2} B_{o+\nu,k}\,\text{Re}\left[\left\langle 0\left|\hat{\Gamma}_\nu^+\hat{b}_k\hat{\Gamma}_\nu\right|\right\rangle\right]$$

The expectation values occurring in (C20) had been calculated already in Appendix B and thus we can write directly

$$\left\langle\psi_3\left|\sum_{n=1}^{N}\sum_{k=1}^{N-1}\hbar\omega_k B_{nk}\left(\hat{b}_k^+ + \hat{b}_k\right)\hat{a}_n^+\hat{a}_n\right|\psi_3\right\rangle =$$

$$= 2J^2\sum_{k,k'=1}^{N-1}(\hbar\omega_k)^2(\hbar\omega_{k'})^2\left\{\hbar\omega_k\left[B_{o-1,k}F_k^{(-)}E_{k'}^{(-)} + B_{o+1,k}F_k^{(+)}E_{k'}^{(+)}\right] + \right.$$

$$\left. + \hbar\omega_{k'}\left[B_{o-1,k}F_{k'}^{(-)}\left(G_{kk'}^{(-)} + G_{k'k}^{(-)}\right) + B_{o+1,k}F_{k'}^{(+)}\left(G_{kk'}^{(+)} + G_{k'k}^{(+)}\right)\right]\right\}$$

(C21)

The phonon part yields

$$\left\langle\psi_3\left|\sum_{k=1}^{N-1}\hbar\omega_k\hat{b}_k^+\hat{b}_k\right|\psi_3\right\rangle = \sum_{k=1}^{N-1}\hbar\omega_k\sum_{\nu=-2}^{2}\left\langle 0\left|\hat{\Gamma}_\nu^+\hat{b}_k^+\hat{b}_k\hat{\Gamma}_\nu\right|0\right\rangle$$

$$\left\langle 0\left|\hat{\Gamma}_{\pm2}^+\hat{b}_k^+\hat{b}_k\hat{\Gamma}_{\pm2}\right|0\right\rangle = J^4(\hbar\omega_k)^2\left(D_k^{(\pm)}\right)^2$$

(C22)

$$\left\langle 0\left|\hat{\Gamma}_{\pm1}^+\hat{b}_k^+\hat{b}_k\hat{\Gamma}_{\pm1}\right|0\right\rangle = J^2(\hbar\omega_k)^2\left\{(\hbar\omega_k)^2\left(F_k^{(\pm)}\right)^2 + \sum_{k'=1}^{N-1}\left[\hbar\omega_{k'}\left(G_{kk'}^{(\pm)} + G_{k'k}^{(\pm)}\right)\right]^2\right\}$$

$$\left\langle 0\left|\hat{\Gamma}_o^+ \hat{b}_k^+ \hat{b}_k \hat{\Gamma}_0\right|0\right\rangle = J^4\left(\hbar\omega_k\right)^2 D_k^2$$

Finally we have to evaluate the expectation value of the operator $\hat{j}$:

$$\hat{j}\left|\psi_3\right\rangle = -J\sum_{\mu=-2}^{2}(-1)^\mu \hat{\Gamma}_\mu\left(\hat{a}_{o+\mu-1}^+ + \hat{a}_{o+\mu+1}^+\right)\left|0\right\rangle \tag{C23}$$

From this we obtain

$$\begin{aligned}
\left\langle \psi_3\left|\hat{j}\right|\psi_3\right\rangle &= -J\sum_{\nu,\mu=-2}^{2}(-1)^{\nu+\mu}\left\langle 0\left|\hat{\Gamma}_\nu^+ \hat{\Gamma}_\mu \cdot \left(\hat{a}_{o+\nu}\hat{a}_{o+\mu-1}^+ + \hat{a}_{o+\nu}\hat{a}_{o+\mu+1}^+\right)\right|0\right\rangle = \\
&= J\sum_{\nu=-1}^{2}\left[\left\langle 0\left|\hat{\Gamma}_{\nu-1}^+ \hat{\Gamma}_\nu\right|0\right\rangle + \left\langle 0\left|\hat{\Gamma}_\nu^+ \hat{\Gamma}_{\nu-1}\right|0\right\rangle\right] = 2J\sum_{\nu=-1}^{2}\mathrm{Re}\left[\left\langle 0\left|\hat{\Gamma}_{\nu-1}^+ \Gamma_\nu\right|0\right\rangle\right] = \\
&= 2J^4\sum_{k=1}^{N-1}\left(\hbar\omega_k\right)^3\left[D_k^{(-)}F_k^{(-)} + D_k\left(F_k^{(-)} + F_k^{(+)}\right) + D_k^{(+)}F_k^{(+)}\right]
\end{aligned} \tag{C24}$$

Then the complete expectation value, multiplied with the appropriate factor is given by

$$\begin{aligned}
H''(3) &= \frac{1}{36}\frac{t^6}{\hbar^6}\left\langle \psi_3\left|\hat{H}\right|\psi_3\right\rangle = \frac{1}{36}\frac{t^6}{\hbar^6}\left\langle \psi_3\left|\hat{j}+\hat{\omega}\right|\psi_3\right\rangle = \\
&= \frac{J}{18}\left(\frac{Jt}{\hbar}\right)^3\sum_{k=1}^{N-1}(\omega_k t)^3\left[D_k^{(-)}F_k^{(-)} + D_k\left(F_k^{(-)} + F_k^{(+)}\right) + D_k^{(+)}F_k^{(+)}\right] + \\
&\quad + \frac{1}{36}\left(\frac{Jt}{\hbar}\right)^2\sum_{k=1}^{N-1}\hbar\omega_k(\omega_k t)^2\left\{\left(\frac{Jt}{\hbar}\right)^2\left[\left(D_k^{(-)}\right)^2 + D_k^2 + \left(D_k^{(+)}\right)^2\right] + \right. \\
&\quad \left. + (\omega_k t)^2\left[\left(F_k^{(-)}\right)^2 + \left(F_k^{(+)}\right)^2\right] + \sum_{k'=1}^{N-1}(\omega_{k'}t)^2\left[\left(G_{kk'}^{(-)} + G_{k'k}^{(-)}\right)^2 + \left(G_{kk'}^{(+)} + G_{k'k}^{(+)}\right)^2\right]\right\} + \\
&\quad + \frac{1}{18}\left(\frac{Jt}{\hbar}\right)^2\sum_{k,k'=1}^{N-1}(\omega_k t)^2(\omega_{k'}t)^2\left\{\hbar\omega_k\left[B_{o-1,k}F_k^{(-)}E_{k'}^{(-)} + B_{o+1,k}F_k^{(+)}E_{k'}^{(+)}\right] + \right. \\
&\quad \left. + \hbar\omega_{k'}\left[B_{o-1,k}F_{k'}^{(-)}\left(G_{kk'}^{(-)} + G_{k'k}^{(-)}\right) + B_{o+1,k}F_{k'}^{(+)}\left(G_{kk'}^{(+)} + G_{k'k}^{(+)}\right)\right]\right\}
\end{aligned} \tag{C25}$$

Then our final result is given by

$$H(3) = H(2) + H'(3) + H''(3)$$

$$H(2) = \left\langle \psi(2)\left|\hat{H}\right|\psi(2)\right\rangle \qquad\qquad \text{[equ. (C8)]} \tag{C26}$$

$$H'(3) = -\frac{1}{3}\frac{t^3}{\hbar^3}\mathrm{Im}\left[\left\langle \psi(1)\left|\hat{H}\right|\psi_3\right\rangle\right] \qquad\qquad \text{[equ. (C19)]}$$

$$H''(3) = \frac{1}{36}\frac{t^6}{\hbar^6}\left\langle \psi_3\left|\hat{H}\right|\psi_3\right\rangle \qquad\qquad \text{[equ. (C25)]}$$

Appendix D: Propagation of the First Order Wave Function to Larger Times

In this Appendix we want to show for some time steps τ the explicit formulas for the dynamics as they result from the calculation of the individual terms. The index $\mu=1$ we drop from the states in the following. We consider steps $\ell\tau$, $\ell=0,1,...$ and times t with $0 \le t \le \tau$. As already mentioned in the main text we have for the first period, $\ell=0$:

$$\left|\psi(t)\right\rangle = \sum_n \left(a_n^J(t)-\delta_{no}\right)\hat{a}_n^+\left|0\right\rangle + a_o^\omega(t)\hat{U}_o(t)\hat{a}_o^+\left|0\right\rangle \tag{D1}$$

From this we obtain for $t=\tau$:

$$\left|\psi(\tau)\right\rangle = \sum_n \left(a_n^J(\tau)-\delta_{no}\right)\hat{a}_n^+\left|0\right\rangle + a_o^\omega(\tau)\hat{U}_o(\tau)\hat{a}_o^+\left|0\right\rangle \tag{D2}$$

which is the initial state for the first order term in the second period, $\ell=1$:

$$\left|\psi(\tau+t)\right\rangle = \left(e^{-\frac{it}{\hbar}\hat{\omega}} + e^{-\frac{it}{\hbar}\hat{J}} - 1\right)\left|\psi(\tau)\right\rangle \tag{D3}$$

Before explicitly writing down the states resulting from equation (D3), we want to define some quantities to keep the final formulas shorter:

$$\varphi_k(t) \equiv e^{2i\frac{Jt}{\hbar}\cos\left(\frac{2\pi}{N}k\right)} \quad ; \quad k=1,...,N$$

$$\gamma_n(t) \equiv e^{-i\sum_{k=1}^{N-1}B_{nk}^2\left[\sin(\omega_k t)-\omega_k t\right]}$$

$$\zeta_{nk}(t) \equiv B_{nk}\left(e^{-i\omega_k t}-1\right) \quad ; \quad k=1,...,N-1 \tag{D4a}$$

$$b_{nk}^{(0)}(t) \equiv b_{nk}(t)$$

$$\hat{U}_n^{(j)}(t) \equiv e^{-\frac{1}{2}\sum_{k=1}^{N-1}\left|b_{nk}^{(j)}(t)\right|^2} e^{\sum_{k=1}^{N-1}b_{nk}^{(j)}(t)\hat{b}_k^+} \quad ; \quad \hat{U}_n^{(0)}(t) \equiv \hat{U}_n(t)$$

$$\vartheta_n^{(j)}(t) \equiv e^{-i\sum_{k=1}^{N-1}\Omega_{nk}^{(j)}(t)} \tag{D4b}$$

$$\Omega_{nk}^{(j)}(t) \equiv \text{Re}\left[b_{nk}^{(j)}(\tau)\right]\sin(\omega_k t) + \text{Im}\left[b_{nk}^{(j)}(\tau)\right]\left[1-\cos(\omega_k t)\right]$$

This yields for the first term, together with the expressions for the small polaron limit from paper I:

$$e^{-\frac{it}{\hbar}\hat{\omega}}\left|\psi(\tau)\right\rangle = \sum_n a_n^{(1)}(t)\hat{U}_n^{(1)}(t)\hat{a}_n^+\left|0\right\rangle + a_o^{(2)}(t)\hat{U}_o^{(2)}(t)\hat{a}_o^+\left|0\right\rangle$$

$$a_n^{(1)}(t) = \left[a_n^J(\tau)-\delta_{no}\right]\gamma_n(t) \quad ; \quad b_{nk}^{(1)}(t) = \zeta_{nk}(t) \tag{D4c}$$

$$a_o^{(2)}(t) = a_o^\omega(\tau)\gamma_o(t)\vartheta_o^{(0)}(t) \quad ; \quad b_{ok}^{(2)}(t) = b_{ok}(\tau)e^{-i\omega_k t} + \zeta_{nk}(t)$$

Further we act with the second operator on the initial state, observing that the exponential operator and coherent state operators commute with each other. This yields:

$$e^{-\frac{it}{\hbar}\hat{j}}|\psi(\tau)\rangle = \sum_n \left[a_n^{(3)}(t) + \hat{U}_o(\tau)a_n^{(4)}(t) \right]\hat{a}_n^+|0\rangle$$

$$a_n^{(3)}(t) = \frac{1}{N}\sum_{n'}\sum_{k=1}^{N} e^{\frac{2\pi i}{N}k(n-n')}\varphi_k(t)\left[a_{n'}^J(\tau) - \delta_{n'o} \right] \tag{D5}$$

$$a_n^{(4)}(t) = \frac{a_o^\omega(\tau)}{N}\sum_{k=1}^{N} e^{\frac{2\pi i}{N}k(n-o)}\varphi_k(t)$$

Collecting the terms and substracting the initial state leads to

$$|\psi(\tau+t)\rangle = \sum_n \left[d_n^{(1)}(t) + a_n^{(1)}(t)\hat{U}_n^{(1)}(t) + a_n^{(4)}(t)\hat{U}_o(\tau) \right]\hat{a}_n^+|0\rangle + \left[a_o^{(2)}(t)\hat{U}_o^{(2)}(t) - a_o^\omega(\tau)\hat{U}_o(\tau) \right]\hat{a}_o^+|0\rangle$$

$$d_n^{(1)}(t) = a_n^{(3)}(t) - a_n^J(\tau) + \delta_{no} \tag{D6}$$

This yields directly the initial state for the third period:

$$|\psi(2\tau)\rangle = \sum_n \left[d_n^{(1)}(\tau) + a_n^{(1)}(\tau)\hat{U}_n^{(1)}(\tau) + a_n^{(4)}(\tau)\hat{U}_o(\tau) \right]\hat{a}_n^+|0\rangle + \left[a_o^{(2)}(\tau)\hat{U}_o^{(2)}(\tau) - a_o^\omega(\tau)\hat{U}_o(\tau) \right]\hat{a}_o^+|0\rangle \tag{D7}$$

From this state we obtain

$$e^{-\frac{it}{\hbar}\hat{j}}|\psi(2\tau)\rangle = \sum_n \left[d_n^{(2)}(t) + a_n^{(5)}(t)\hat{U}_o(\tau) + a_n^{(6)}(t)\hat{U}_o^{(2)}(\tau) + \sum_{n'} f_{nn'}^{(1)}(t)\hat{U}_{n'}^{(1)}(\tau) \right]\hat{a}_n^+|0\rangle \tag{D8}$$

where the coefficients are given by

$$d_n^{(2)}(t) = \frac{1}{N}\sum_{n'}\sum_{k=1}^{N} d_{n'}^{(1)}(\tau)e^{\frac{2\pi i}{N}k(n-n')}\varphi_k(t)$$

$$f_{nn'}^{(1)}(t) = \frac{1}{N}\sum_{k=1}^{N} a_{n'}^{(1)}(\tau)e^{\frac{2\pi i}{N}k(n-n')}\varphi_k(t) \tag{D9}$$

$$a_n^{(5)}(t) = \frac{1}{N}\sum_{k=1}^{N}\left[\sum_{n'} a_{n'}^{(4)}(\tau)e^{\frac{2\pi i}{N}k(n-n')} - a_o^\omega(\tau)e^{\frac{2\pi i}{N}k(n-o)} \right]\varphi_k(t)$$

$$a_n^{(6)}(t) = \frac{a_o^\omega(\tau)}{N}\sum_{k=1}^{N} e^{\frac{2\pi i}{N}k(n-o)}\varphi_k(t)$$

Further we can write

$$e^{-\frac{it}{\hbar}\hat{\omega}}|\psi(2\tau)\rangle = \sum_{j=3}^{7}\sum_n e_n^{(j)}(t)\hat{U}_n^{(j)}(t)\hat{a}_n^+|0\rangle \tag{D10}$$

with the different coherent state amplitudes

$$b_{nk}^{(3)}(t)=\zeta_{nk}(t) \quad ; \quad b_{nk}^{(4)}(t)=b_{nk}^{(1)}(\tau)e^{-i\omega_k t}+\zeta_{nk}(t) \quad ; \quad b_{nk}^{(5)}(t)=b_{ok}(\tau)e^{-i\omega_k t}\delta_{no}+\zeta_{nk}(t)$$

$$b_{nk}^{(6)}(t)=\left[b_{ok}^{(2)}(\tau)e^{-i\omega_k t}+\zeta_{ok}(t)\right]\delta_{no} \quad ; \quad b_{nk}^{(7)}(t)=\left[b_{ok}(\tau)e^{-i\omega_k t}+\zeta_{ok}(t)\right]\delta_{no} \tag{D11}$$

and coefficients

$$e_n^{(3)}(t)=d_n^{(1)}(\tau)\gamma_n(t) \quad ; \quad e_n^{(4)}(t)=a_n^{(1)}(\tau)\vartheta_n^{(1)}(t)\gamma_n(t) \quad ; \quad e_n^{(5)}(t)=a_n^{(4)}(\tau)\vartheta_o^{(0)}(t)\gamma_o(t)$$

$$e_n^{(6)}(t)=a_o^{(2)}(\tau)\vartheta_o^{(2)}(t)\gamma_o(t)\delta_{no} \quad ; \quad e_n^{(7)}=-a_o^{\omega}(\tau)\vartheta_o^{(0)}(t)\gamma_o(t)\delta_{no} \tag{D12}$$

Now we can write down the state vector for the third interval:

$$\left|\psi(2\tau+t)\right\rangle=\sum_n\left[\sum_{j=3}^7 e_n^{(j)}(t)\hat{U}_n^{(j)}(t)+d_n^{(2)}(t)+a_n^{(5)}(t)\hat{U}_o(\tau)+a_n^{(6)}(t)\hat{U}_o^{(2)}(\tau)-d_n^{(1)}(\tau)-a_n^{(1)}(\tau)\hat{U}_n^{(1)}(\tau)-\right.$$

$$\left.-a_n^{(4)}(\tau)\hat{U}_o(\tau)+\sum_{n'}f_{nn'}^{(1)}(t)U_{n'}^{(1)}(\tau)\right]\hat{a}_n^+\left|0\right\rangle-\left[a_o^{(2)}(\tau)\hat{U}_o^{(2)}(\tau)-a_o^{\omega}(\tau)\hat{U}_o(\tau)\right]\hat{a}_o^+\left|0\right\rangle \tag{D13}$$

From this expression we can compute now the initial state for the fourth period and so on. It is obvious that with each period the expressions for the state vectors become more complicated. However, the calculation of expectation values from these states is rather simple, because they are all just superpositions of free exciton, $|D_1\rangle$- and $|D_2\rangle$-type states. The problem is that their derivation becomes lengthy and tedious. Currently we try to find out whether or not it is possible to establish a kind of recursive algorithm for this task.

J. Mol. Model. **1996**, 2, 136 – 148

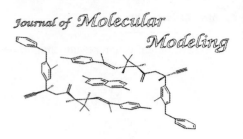

Molecular Parameterisation and the Theoretical Calculation of Electrode Potentials

David R. Lowis[1] and W. Graham Richards*

Oxford Centre for Molecular Sciences and Physical Chemistry Laboratory, South Parks Road, Oxford OX1 3QZ, U.K. Tel: (01865) 275406. Fax: (01865) 275410. (gr@vax.ox.ac.uk)

1. Present address: Centre for Molecular Design, Washington University School of Medicine, 700 S. Euclid Ave. St Louis, MO 63110-1012, U.S.A. Tel: (314) 362-2272. Fax: (314) 362-0234. (dave@ibc.wustl.edu)

Received: 2 January 1996 / Accepted: 2 May 1996 / Published: 28 May 1996

Abstract

The difference in reduction potentials between ortho and para-benzoquinones has been calculated. The employs gas phase *ab initio* and semi-empirical computations in combination with free energy perturbation theory applied to gas and solution phase Monte Carlo simulations. The effects on calculated results of altering solute electrostatic parameterisation in solution phase simulations is examined. Atom centred charges derived from the molecular electrostatic potentials, MEPs, from optimised *ab initio* wavefunctions and charges generated by consideration of hydrogen bonded complexes are considered. Parameterisation of hydroxyl torsions in hydroquinone molecules is treated in a physically realistic manner. The coupled torsional system of the ortho-hydrobenzoquinone molecule is described by a potential energy surface calculated using gas phase AM1 semi-empirical computations rather than the simple torsional energy functions frequently employed in such calculations. Calculated differences in electrode potentials show that the electrostatic interactions of quinone and hydroquinone molecules in aqueous solution are not well described by atom centred charges derived from *ab initio* calculated MEPs. Moreover, results in good agreement with the experimental reduction potential difference can be obtained by employing high level *ab initio* calculations and solution phase electrostatic parameters developed by consideration of hydrogen bonded complexes.

Keywords: Electrode Potential, Electrostatic parameters, Quinone, Free energy perturbation, *ab initio*, Semi Empirical

Introduction

The use of Monte Carlo [1] (MC) and molecular dynamics [2] (MD) statistical mechanical simulations has become increasingly widespread and routine in the study of biomolecular and organic systems. Such methods suffer from two fundamental problems: the extent of configurational space searched and the representation of system energetics through classical potential energy functions [2] and their associated parameters. Many methods, such as umbrella sampling [3], preferential sampling [4] and double wide sampling [5] have been employed to alleviate the problem of non-ergodic phase space searching. Due to limitations in computational power, molecular mechanics force fields used to represent both inter and intramolecular interactions are still simplistic in their functional form. Much effort has been expended in the development of so called transferable parameters [6 - 10] used to characterise non-bonded van der Waals and electrostatic interactions. Such parameterisations have met with some

---

* *To whom correspondence should be addressed*

success in the calculation of solution phase physical properties. Nevertheless, the reliability and transferability of parameter sets between different systems must still be considered questionable, especially with respect to the atom centred charges used to represent electrostatic interactions.

Intermolecular interactions in most commonly used force fields comprise electrostatic and Lennard-Jones van der Waals terms [8 - 10] Non-bonded electrostatics are, in general, represented as simple coulombic interactions between point charges placed on atomic centres. Such atom centred charges may be derived in a number of different ways. This paper investigates the effects on calculated quinone redox potentials of changing molecular electrostatic parameterisation used in MC simulations in order to assess the quality of the parameters used. Redox potential results calculated using atom centred charges obtained from 1) fitting to *ab initio* computed molecular electrostatic potentials (MEPs) [11] and 2) those obtained by considering hydrogen bonded complexes of the quinones are their respective hydroquinones, an OPLS methodolgy [7], are compared.

The difference in reduction potentials between ortho and para-benzoquinones is calculated using a combination of gas phase *ab initio*, semi-empirical and solution and gas phase MC simulations. Free energy perturbation [12] (FEP) theory is used in the calculation of solution phase free energy differences. The quinone system was selected because of the availability of experimental data [13], the relative simplicity of the systems involved and the fact that anthraquinone compounds are likely candidates for bioreductive anti-cancer agents [14]. Additionally, an earlier study[15], which employed a similar methodology but utilised MD simulations, calculated the difference in redox potentials between these compounds to an apparent accuracy of 20mV. The problem of torsional parameterisation in flexible molecules is also addressed. The coupled torsional system of ortho-hydro-

benzoquinone is parameterised to a semi-empirically calculated potential energy surface rather than the more usual simple torsional energy functions.

Methods

A simple thermodynamic cycle (figure 1) approach was used to calculate the difference in electrode potentials between ortho and para-benzoquinones. The difference in standard reduction potentials between ortho and para-benzoquinones, ΔE, is related to this solution phase free energy difference $\Delta\Delta G_{aq}$ by equation 1.

$$\Delta\Delta G_{aq} = -nF(\Delta E^\circ) \tag{1}$$

where n is the number of electrons involved in the redox process, F is Faraday's constant and ΔE° the difference in reduction potentials between ortho and para-benzoquinones.

The aqueous phase free energy term $\Delta\Delta G_{aq}$ may be calculated from the thermodynamic cycle employing Hess' law (equation 2).

$$\Delta\Delta G_{aq} = \Delta\Delta G_{gas} + (\Delta G_3 - \Delta G_1) - (\Delta G_2 - \Delta G_4) \tag{2}$$

see figure 1 for explanation of terms.

$\Delta\Delta G_{gas}$ represents the gas phase free energy difference between the ortho-hydrobenzoquinone / para-benzoquinone system (top left of cycle) and the ortho-benzoquinone / para-hydrobenzoquinone system (top right of cycle). Terms $\Delta G_3 - \Delta G_1$ and $\Delta G_2 - \Delta G_4$ represent the differences in the solvation free energies between the reduced and oxidised forms of ortho and para-benzoquinones respectively

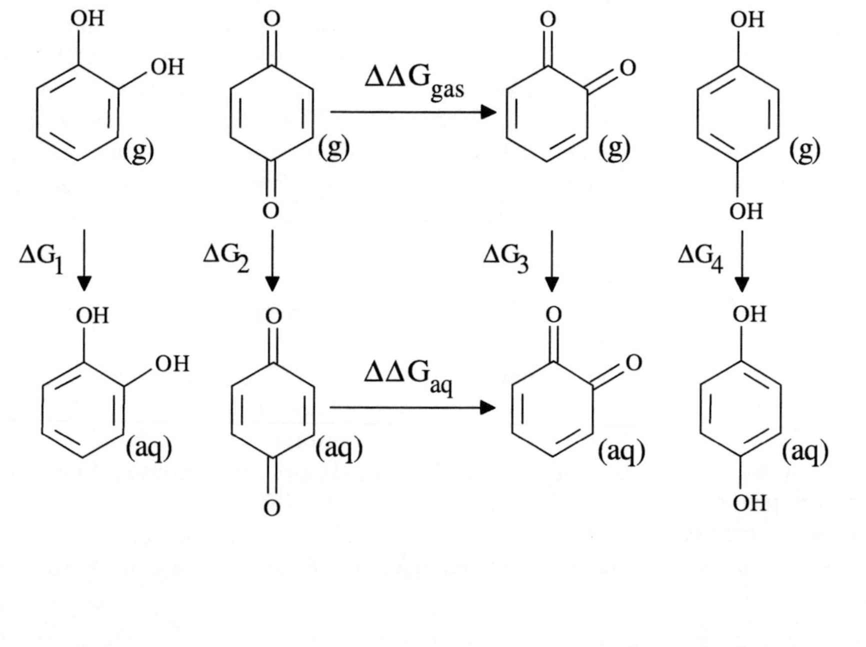

Figure 1. *Thermodynamic cycle used to calculate the difference in reduction potentials between ortho and para-benzoquinones.*

Calculation of the gas phase free energy difference, $\Delta\Delta G_{gas}$. The quantity $\Delta\Delta G_{gas}$, can be constructed from a number of terms as depicted in equation 3.

$$\Delta\Delta G_{gas} = \Delta\Delta H_{0K} + \Delta\Delta H_{ZPE} + \Delta\Delta H_{thermal} + T\Delta\Delta S \qquad (3)$$

explanation of terms in text.

$\Delta\Delta H_{0K}$ is the difference in gas phase internal energies at 0K. This quantity was computed using *ab initio* calculations employing the Gaussian 92 program [16]. Full restricted Hartree-Fock (RHF) geometry optimisations were carried out on ortho- and para-benzoquinones and their respective hydroquinones at the STO-3G [17], 3-21G [18] and 6-31G* [19] basis set levels. All computations involved the calculation of atom centred charges through the CHELPG [11] algorithm in order to provide molecular electrostatic parameters for the solution phase simulations. Finally MP2/6-31G*]20] single point calculations were performed on the RHF/ 6-31G* optimised geometries to obtain accurate gas phase energies and therefore $\Delta\Delta H_{0K}$. The quantities $\Delta\Delta H_{ZPE}$, the difference in molecular zero point energies, $\Delta\Delta H_{thermal}$, the difference in molecular thermal enthalpies, and $T\Delta\Delta S$, the difference in molecular entropies at 298K, were obtained from gas phase semi-empirical calculations. Single point AM1 [21] calculations were run on the RHF/6-31G* optimised molecular geometries of all quinone and hydroquinone species in order to obtain normal vibrational frequency modes and thermodynamic quantities ($\Delta\Delta H_{thermal}$, and $T\Delta\Delta S$). Analysis of the computed vibrational frequency modes through equation 4 produced molecular zero point energies which, along with the calculated thermodynamic quantities were used to correct the *ab initio* calculated 0K internal energies to free energies at 298K (equation 3). All semi-empirical calculations were run using the MOPAC [22] program, version 6.

$$Z.P.E. = 0.5\, h\, \Sigma_i\, \omega_i / 2\pi \qquad (4)$$

where Z.P.E. is the molecular zero point energy, h is Planck's constant and ω_i the normal vibrational modes of the molecule.

The calculation of $\Delta\Delta G_{gas}$ through equation 3 is valid only if the molecules are assumed to maintain a rigid geometry. The energy barriers to internal rotation for the hydroxyl groups of the hydroquinones are, however, low in comparison with available thermal energy at 298K [13]. Thus, representation of the hydroquinones as rigid molecules in both the gas and solution phase is physically unrealistic. The contribution made to $\Delta\Delta G_{gas}$ by variation in hydroquinone hydroxyl torsions was found from MC simulations and FEP theory.

Gas phase mutations of ortho-hydrobenzoquinone to ortho-benzoquinone and para-hydrobenzoquinone to para-benzoquinone were performed (see figure 2) and the free energy changes calculated using FEP theory. Each perturbation was divided into 21 windows [23] to allow the precise evaluation and rapid convergence of calculated free energy

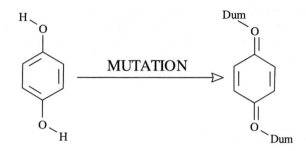

Figure 2. *Mutation of para-hydrobenzoquinone to the quinone. During the mutation parameters which describe the hydroquinone, bond lengths, angles, dihedrals and van der Waals and electrostatic parameters are slowly changed to those which describe the quinone. Hydroquinone hydroxyl hydrogen atoms are shrunk into massless, chargeless dummy atoms, denoted Dum in the figure.*

changes. The progress of the perturbations was described by a coupling parameter, λ, the hydroquinone being described by $\lambda = 0$ and the quinone by $\lambda = 1$ and the perturbation between the two by l being increased from 0 to 1 in increments of 0.05. Each of the mutations was performed in a series of 22 MC simulations designed to give two estimates (forward, hydroquinone -> quinone, and backward, quinone -> hydroquinone) of the gas phase free energy change. Both quinones and hydroquinones were represented by all atom models in their RHF/3-21G optimised. Quinone molecules were treated as rigid moieties. Only hydroquinone torsional degrees of freedom were sampled and only torsional contributions to the free energy changes were evaluated. Details of the hydroquinone torsional parameterisation can be found in the later section dealing with the subject. Metropolis sampling [1] was employed throughout all simulations. Dihedral angles were allowed to vary by up to $10°$ per MC step in order to allow an acceptance ratio for torsional moves of around 40%. All simulations were run using the BOSS program, version 3.1 [24], on a Hewlett Packard workstation. Each simulation consisted of 10×10^3 configurations of equilibration the first 3×10^3 of which were at constant volume, followed by 16×10^3 configurations of data collection. All simulations were run in the isothermal-isobaric ensemble at 298 K and 1 atmosphere pressure.

Calculation of solution phase free energy differences

The differences in the free energies of solvation between the hydroquinones and the quinones, $\Delta G_3 - \Delta G_1$ and $\Delta G_2 - \Delta G_4$ for ortho and para systems respectively, were calculated employing MC statistical mechanics simulations and free energy perturbation theory. The difference in solvation free energies between ortho-and ortho-benzoquinone was computed by mutating the hydroquinone to the quinone (figure

2) in dilute aqueous solution. Application of the FEP equation [25] (equation 5) to collated simulation data was used to calculate the free energy change. A similar computation was performed for the para-quinone system

$$\Delta G = -kT \ln \langle \exp(-\Delta H_{AB}/kT)\rangle_A \qquad (5)$$

the FEP equation. k is Boltzmanns constant, T the temperature, ΔH_{AB} the difference in internal energies between systems A and B both being in the same configurational state. The brackets $\langle\rangle_A$ represent an average taken over configurations of state A.

Differences in solvation free energies comprise contributions from two sources: changes in intermolecular interactions upon solvation and changes in intramolecular interactions upon solvation of the species involved. This second quantity, which involves the polarisation of solute molecules by solvent, is notoriously difficult to calculate and has been ignored in this study. Hence, all solution phase simulations involved the evaluation of intermolecular energies only. Intramolecular interactions are described in the BOSS program in the usual Lennard-Jones plus Coulombic format (equation 6).

$$\Delta E = \Sigma_i^a \Sigma_j^b (A_{ij}/r^{12}_{ij} - C_{ij}/r^6_{ij} + q_i q_j \varepsilon^2/r_{ij}) \qquad (6)$$

explanation of terms in text.

The interaction energy between two molecules is thus characterised by the sum over all pairwise interactions between atoms i and j on the two molecules a and b. The terms A and C in equation 6 are related to the Lennard-Jones e and s parameters such that $A_{ii} = 4\varepsilon_i\sigma^{12}_i$ and $C_{ii} = 4\varepsilon_i\sigma^6_i$. s and e Lennard Jones parameters for all atoms were taken from the OPLS force field [6].

Intermolecular electrostatic interactions are represented in the BOSS program as coulombic interactions between point charges placed on atom centres. Atom centred charges computed in the RHF *ab initio* calculations described in the previous section were used to provide three sets (STO-3G, 3-21G and 6-31G*) of electrostatic parameters for each molecule. In addition to these a further set of electrostatic parameters was specially developed for the quinone system through the consideration of hydrogen bonded complexes of the quinone and hydroquinone molecules. This method is comparable to that used to develop OPLS parameters for the nucleotide bases [7].

Electrostatic parameter development

OPLS parameters for both united and all atom representations have been derived through pure liquid simulations [6] and reproduction of *ab initio* calculated geometries and binding energies of hydrogen bonded complexes of the solutes in question [7]. It was decided to use this second, simpler approach to derive electrostatic parameters for the quinone /

hydroquinone molecules under study. 11 hydrogen bonded complexes (figure 3) of the four molecules involved were considered. Binding energies for these complexes were obtained from RHF/6-31G* *ab initio* calculations. A full restricted Hartree-Fock geometry optimisation of a single water molecule was performed at the 6-31G* basis set level. Hartree-Fock geometry optimisations of the hydrogen bond lengths and angles of complexes A through K were performed at the 6-31G* basis set level. All other internal coordinates were held fixed at the values obtained from RHF/6-31G* optimisations of the isolated molecules. It should be noted that RHF/6-31G* and RHF/3-21G optimised molecular geometries are practically identical.

In the case of the quinone molecules themselves only one hydrogen bonded complex exists for each (A and B, figure 3). It is therefore possible to reproduce hydrogen bond geometries and energies of these complexes with many different sets of atom centred charges. In order to combat this effect quinone-water complexes were set up with an OH water bond in direct line with the O=C carbonyl bond of the quinone. RHF/6-31G* optimisations of the bond length only were performed on these complexes, denoted A2 and B2, to provide second quinone complexes which could be used in the parameterisation process. Binding energies for the hydrogen bonded complexes were obtained by subtracting the RHF/6-31G* energies of the isolated constituent molecules from the RHF/6-31G* energy of the optimised complex.

In order to reproduce *ab initio* calculated geometry and binding energy data using atom centred charges, the hydrogen bond lengths and angles of complexes A through K were optimised employing the BOSS intermolecular potential function (equation 6). This process was performed using the SOPLS program [26] which employs a simplex minimisation method. Output from the SOPLS code includes the complex energy, optimised geometry as well as coulombic and van der Waals interaction energies between individual atom pairs. A directed trial and error procedure was utilised to obtain electrostatic parameters which yielded hydrogen bond geometries and binding energies for complexes A through K in good agreement with the *ab initio* calculated values.

The hydrogen bond lengths and angles of complexes A through K and the hydrogen bond lengths of complexes A2 and B2 were minimised using the SOPLS program. Quinone and hydroquinone molecules were represented in their RHF/6-31G* optimised geometry with initial electrostatic parameters being taken as Mulliken charges obtained from the RHF/6-31G* *ab initio* calculations on the isolated molecules. OPLS parameters were employed to represent van der Waals interactions throughout. Water was represented as the TIP4P model [27] which was to be used in the solution phase simulations. Only intermolecular interactions were evaluated. Quinone and hydroquinone electrostatic parameters were altered based on the results of these minimisations using individual atom pair interaction energies as a guide. This process was repeated until agreement in hydrogen bond geometries and interaction energies between the *ab initio* and SOPLS calculations

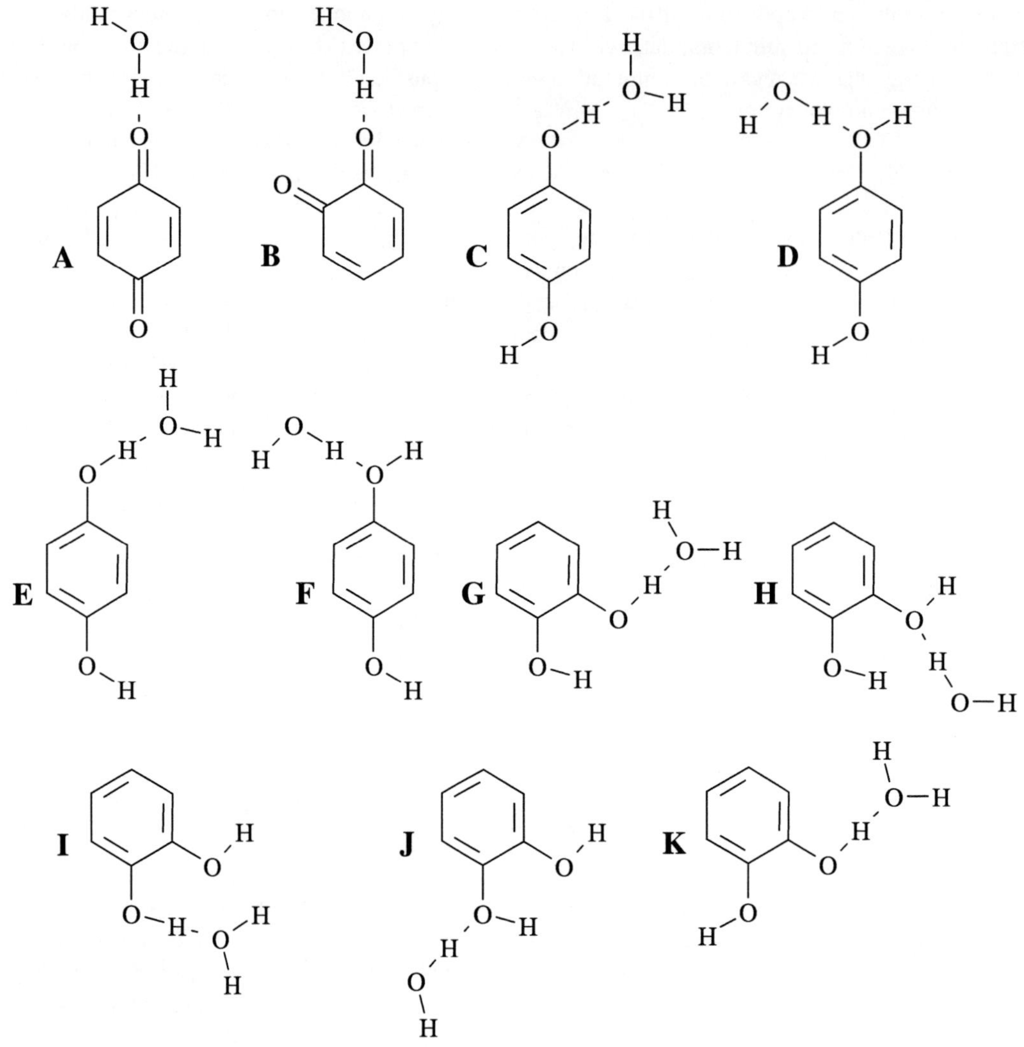

Figure 3. *Complexes used in the derivation of optimised electrostatic parameters for ortho and para benzoquinones and hydrobenzoquinones. Hydrogen bond angles were defined as the angle between the atoms water oxygen, water hydrogen and quinone/hydroquinone oxygen where water acts as a hydrogen bond donor and between atoms water hydrogen, water oxygen and hydroquinone hydrogen where water acts as a hydrogen bond acceptor.*

over all complexes could no longer be improved. Values of OPLS electrostatic parameters for atoms in chemically similar environments to those in the quinone and hydroquinone molecules were used as a check on the parameters obtained to ensure there were no drastic discrepancies.

Torsional parameterisation

As has been mentioned, the hydroxyl torsions of ortho and para-hydrobenzoquinones have low energy barriers to rotation. It would therefore be inappropriate to employ rigid molecular geometries for their representation. The BOSS program allows for sampling over torsional degrees of freedom employing a Fourier series [28] (equation 7) to describe the energetics of the torsional system.

$$V(\phi) = V(0) + V(1) [1 - \cos \phi] / 2 + V(2) [1 - \cos 2\phi] / 2 + V(3) [1 + \cos 3\phi] / 3 \qquad (7)$$

where $V(\phi)$ is the torsional energy when the dihedral angle has value ϕ, $V(0)$, $V(1)$, $V(2)$ and $V(3)$ are the fourier coefficients.

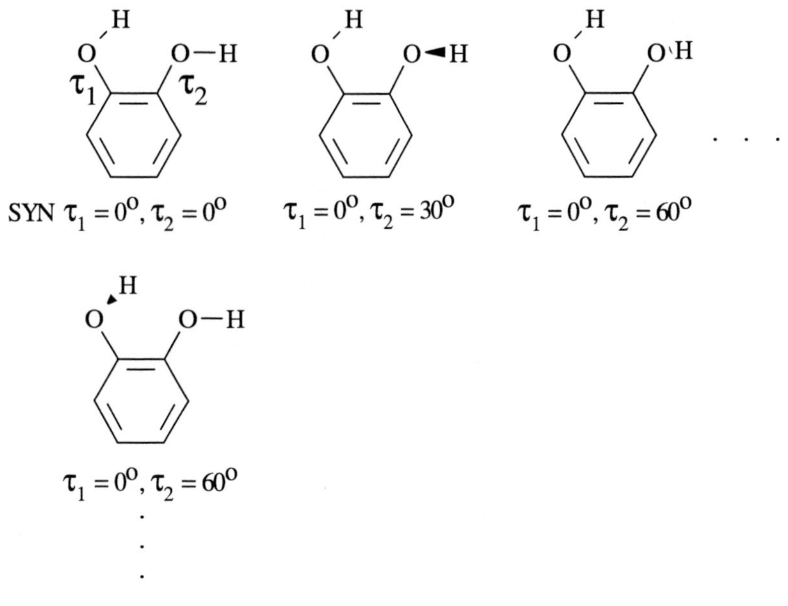

Figure 4. *Conformational forms of ortho-hydrobenzoquinone upon which AM1 semi-empirical calculations were performed in order to parameterise the coupled hydroxyl torsional system of the molecule.*

In the case of para-hydrobenzoquinone the hydroxyl groups are remote from each other and can be approximated to two isolated torsional systems. In this case the hydroxyl torsions can be described adequately by the Fourier series. In the case of ortho-hydrobenzoquinone, however, the hydroxyl torsions are situated on adjacent carbon atoms of the aromatic ring and can therefore interact producing a coupled torsion system with complex energetic behaviour. Such behaviour cannot be described adequately by the use of Fourier series which, through symmetry arguments, should be identical for each hydroxyl torsion. Furthermore, attempts to reproduce the AM1 coupled dihedral energy surface by allowing 1-4 and 1-5 intramolecular interactions between the hydroxyl groups and the hydroquinone ring atoms met with no success.

The energetic properties of the coupled hydroxyl torsion system of ortho-hydrobenzoquinone were investigated using AM1 semi-empirical calculations. Using the RHF/6-31G* geometry as a starting point, single point AM1 calculations were performed at $30°$ intervals of both hydroxyl torsions (τ_1 and τ_2) such that calculations were performed at $\tau_1 = 0°$ $\tau_2 = 0°$, $\tau_1 = 30°$ $\tau_2 = 0°$, $\tau_1 = 60°$ $\tau_2 = 0°$,..... $\tau_1 = 0°$ $\tau_2 = 30°$, $\tau_1 = 30°$ $\tau_2 = 30°$ ... $\tau_1 = 180°$ $\tau_2 = 180°$ (see figure 4). The resultant energy surface was used explicitly to parameterise the coupled torsion system. The BOSS source code was modified to allow it to read the torsional energy surface and interpolate between calculated points thus allowing torsional sampling directly according to the calculated energy surface.

Simulations

Ortho and para-hydrobenzoquinones were mutated to their respective quinone forms in aqueous solution. Each perturbation was split into 21 windows to allow the precise evaluation and rapid convergence of calculated free energy changes.

The progress of the perturbations was described by a coupling parameter, λ, such that, the reduced quinone form being described by $\lambda = 0$, the oxidised form by $\lambda = 1$ and the perturbation from hydrobenzoquinone to benzoquinone by λ being increased from 0 to 1 in increments of 0.05.

Each mutation was performed in a series of 22 MC statistical mechanical simulations designed to give two estimates of the (forward, hydroquinone -> quinone, and backward, quinone -> hydroquinone) of the difference in solvation free energies between hydroquinone and quinone. All mutations were performed in a periodic box containing 504 water molecules. The TIP4P model was employed for interactions involving solvent whilst all quinone and hydroquinone molecules were represented by all atom models in their RHF/3-21G optimised geometries. Parameters describing van der Waals interactions were taken from the OPLS force field whilst CHELPG calculated atom centred charges calculated from the MEPs of the RHF/3-21G optimised wavefunctions were employed for electrostatic interactions. Electrostatic parameters on symmetry related atoms were taken as averaged values over those atoms. Ortho-hydrobenzoquinone torsional parameterisation was employed as described in the above section. Para-hydrobenzoquinone torsions were parameterised by setting V(2) in the BOSS Fourier series to 13.99 kJmol^{-1}, the experimental torsional energy barrier height for phenol [13]. Quinone molecules were treated as rigid. Each simulation consisted of 3.3×10^6 configurations of equilibration, the first 3×10^5 of which were at constant volume, followed by 6.5×10^6 configurations of data collection. These two mutations were repeated employing the optimised electrostatic parameters discussed above.

The effects of altering the electrostatic parameterisation were examined by mutating the electrostatic parameters of the reduced and oxidised forms of both ortho and para-benzoquinones from the RHF/3-21G MEP fitted charges, used

J. Mol. Model. **1996,** *2*

Table 1 *Results from ab initio and semi empirical gas phase calculations.*

Calculation method	Quantity	Value [kJmol^{-1}]
AM1 Semi-empirical	$\Delta\Delta H_{ZPE}$	0.028
	$\Delta\Delta H_{thermat}$	-0.144
	$T\,\Delta\Delta S$	-0.293
ab initio, RHF/STO-3G	$\Delta\Delta H_{0K}$	13.271
ab initio, RHF/3-21G	$\Delta\Delta H_{0K}$	45.414
ab initio, RHF/6-31G*	$\Delta\Delta H_{0K}$	46.407
ab initio, MP2/6-31G*	$\Delta\Delta H_{0K}$	40.168

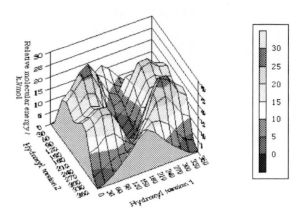

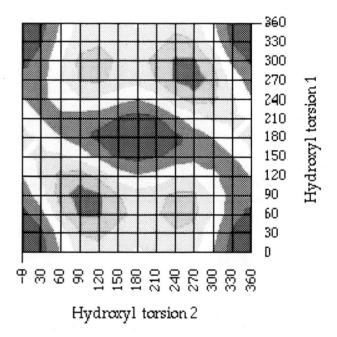

Figure 5. *3D and contour plots of the AM1 potential energy surface used to parameterise the coupled torsional system of ortho-hydrobenzoquinone.*

in the previous structural mutations, to each of the other two *ab initio* derived electrostatic parameter sets. Each of the 8 necessary perturbations was performed in a series of 11 simulation windows designed to give two estimates (forward and backward) of the free energy change. All simulations were performed in a periodic box of 504 water molecules. The TIP4P model was employed for all solvent interactions whilst hydroquinone and quinone molecules were represented in their RHF/3-21G geometries. Van der Waals parameters for the quinone and hydroquinone molecules were taken from the OPLS force field. Free energy changes from these simulations can be combined with the free energy changes calculated from the hydroquinone to quinone mutations to evaluate the values of $\Delta G_3 - \Delta G_1$ and $\Delta G_2 - \Delta G_4$ for different electrostatic parameter sets. Each charge mutation simulation consisted of 2.3×10^6 configurations of equilibration, the first 3×10^5 of which were at constant volume, followed by 5×10^6 configurations of data collection.

Metropolis and preferential sampling were employed throughout all simulations. Spherical cutoffs for intermolecular interactions were applied at 8.5 Å for both solvent-solvent and solute-solvent interactions with the cutoff criterion such that if any atom of the solute lay within the cutoff then interactions with all solute atoms were included in the calculation. Hydroxyl torsions were allowed to vary up to 10° per MC step. Parameters which determine the range of solute molecule movement were set in order to achieve an overall configuration acceptance rate of approximately 40%. Simulations involving the para system were run using BOSS, version 3.1 whilst those involving the ortho system were run using a version of BOSS modified to allow dihedral sampling according to the calculated AM1 energy surface.

Intramolecular contributions to all solution phase free energy changes caused by the inclusion of energy terms to represent torsional energy barriers were estimated by application of the FEP equation to dihedral angle distributions from each simulation window. These intramolecular contributions were subsequently removed from calculated solution phase free energy changes.

Results and Discussion.

Semi-empirical and ab initio gas phase

Results from the *ab initio* and semi-empirical gas phase calculations are displayed in table 1.

It can be seen from table 1 that thermodynamic quantities calculated semi-empirically, employing the AM1 hamiltonian make only a minor correction to *ab initio* gas phase energy differences. The maximum correction to $\Delta\Delta G_{gas}$ being only 0.293 kJmol^{-1} for the entropic contribution. Results from the RHF/STO-3G *ab initio* calculations give markedly lower values for the difference in 0K gas phase energies than do the other basis sets. 3-21G and 6-31G* RHF computations show close agreement differing by less than 1 kJmol^{-1}. The effects of electron correlation, as found from the MP2/6-31G* calculations reduce the difference in gas phase energies between the species by 6.239 kJmol^{-1} from the RHF/6-31G* value.

Torsional parameterisation and gas phase simulations

The results from the AM1 calculations on the conformational forms of ortho-hydrobenzoquinone are shown in figure 5 in both 3D and contour form.

The coupled dihedral energy surface and contour plots displayed show global minima at $\tau_1 = 0°$, $\tau_2 = 0°$ and $\tau_1 = 180°$, $\tau_2 = 180°$ (minima at $\tau_1 = 0°$, $\tau_2 = 360°$, $\tau_1 = 360°$, $\tau_2 = 0°$ and $\tau_1 = 360°$, $\tau_2 = 360°$ are identical to the $\tau_1 = 0°$, $\tau_2 = 0°$ global minimum). These two global minima represent the two degenerate syn forms of ortho-hydrobenzoquinone in which both hydroxyl groups point in the same direction (either to the left or the right - see figure 4). A steep sided valley running approximately (τ_1, τ_2) 0°, 0° -> 120°, 30° -> 180°, 120° -> 180°, 180° -> 180°, 240° ->240°, 330° -> 360°, 360° is also evident on the energy surface. This valley represents the minimum energy route for interconversion between the two degenerate syn forms of ortho-hydrobenzoquinone. The highest energy point in this valley (at approximately $\tau_1 = 150°$, $\tau_2 = 330°$ or $\tau_1 = 210°$, $\tau_2 = 30°$) corresponds to the transition state for the interconversion between the two syn forms and is some 17.2 kJmol^{-1} above the syn energy minima. The valley in the energy surface demonstrates that the lowest energy path for syn-syn interconversion is available through concerted motion of the two adjacent hydroxyl groups. The conformation of the ortho-hydrobenzoquinone molecule at this saddle point correlates well with the rotational saddle point structure for the interconversion of syn forms calculated from an RHF/3-21G *ab initio* transition state search [15]. A subsidiary minimum is also observable on the energy surface at

$\tau_1 = 0°$, $\tau_2 = 180°$ corresponding to the anti conformation (in which the hydroxyl groups 'point' away from each other).

It was this energy surface which was used to parameterise the hydroxyl torsions of ortho-hydrobenzoquinones in MC simulations. Free energy results from both gas phase mutations are given in table 2.

It can be seen from this table that free energy changes calculated from forward and backward perturbations are in excellent agreement with hystereses of less than 0.1 kJmol^{-1} for both systems. Standard deviations on incremental free energy changes from each individual simulation window were small indicating good convergence the calculated quantity. The free energy change for the mutation of para-hydrobenzoquinone to para-benzoquinone in the gas phase, evaluating only torsional energy terms, is -6.87 kJmol^{-1}. The free energy change for the ortho mutation is -7.48 kJmol^{-1}. Thus the torsional contribution to $\Delta\Delta G_{gas}$ is -7.48 - (-6.87) = -0.61 kJmol^{-1}. Calculated values for $\Delta\Delta G_{gas}$ including the torsional contribution are displayed in table 3.

Dihedral angle distributions from para-hydrobenzoquinone perturbation showed that both hydroxyl rotational minima (where the hydroxyl group is coplanar with the aromatic ring) were sampled for both hydroxyl groups. Examination of the coupled dihedral angle distribution for ortho-hydrobenzoquinone perturbation showed that although the most sampled conformations were in and around the syn

Table 2 *Free energy results from the gas phase mutations hydroquinone -> quinone for both ortho and para systems.*

Mutation	$\Delta\Delta G_{torsion}$ [kJmol^{-1}]		
	forward	backward	average
2hpbq -> pbq	-6.86	-6.88	-6.87 ± 0.02
2hobq -> obq	-7.47	-7.49	-7.48 ± 0.02

Quantity	*ab initio* calculation	Value [kJmol^{-1}]
$\Delta\Delta G_{gas}$ (from equation 3)	RHF/STO-3G	13.448
	RHF/3-21G	45.591
	RHF/6-31G*	46.584
	MP2/6-31G*	40.345
$\Delta\Delta G_{torsion}$		-0.61
$\Delta\Delta G_{gas}$ (including torsional contribution)	RHF/STO-3G	12.838
	RHF/3-21G	44.981
	RHF/6-31G*	45.956
	MP2/6-31G*	39.735

Table 3 *Calculated values of $\Delta\Delta G_{gas}$ from gas phase ab initio, semi-empirical and gas phase perturbation calculations.*

Table 4 *Optimised hydrogen bond binding energies and geometries calculated by both ab initio and SOPLS methods.*

Complex	ab initio			BOSS / SOPLS		
	Binding energy [kJmol^{-1}]	Hydrogen bond length [Å]	Hydrogen bong angle [°]	Binding energy [kJmol^{-1}]	Hydrogen bond length [Å]	Hydrogen bong angle [°]
A	-25.16	2.09	151.4	-25.20	1.76	167.3
A2	-14.87	2.13	(180.0)*	-20.72	1.75	(180.0)*
B	-25.83	2.09	152.5	-25.83	2.01	152.5
B2	-13.60	2.17	(180.0)*	-19.62	1.99	(180.0)*
C	-27.60	1.96	137.2	-27.38	2.06	144.8
D	-18.78	2.09	160.6	-19.26	1.74	175.0
E	-27.07	1.97	138.3	-27.13	2.06	147..9
F	-18.09	2.09	160.0	-18.13	1.74	174.5
G	-30.65	1.93	134.2	-30.81	2.04	130.6
H/I	-15.94	2.01	123.0	-18.50	2.12	93.6
J	-19.54	2.08	160.0	-20.22	1.78	172.4
K	-27.13	1.96	139.7	-27.00	2.04	137.0

minimum, the anti conformation and the degenerate syn conformation were also sampled.

Electrostatic parameterisation

Optimised hydrogen bond lengths, angles and binding energies found from *ab initio* and molecular mechanics (the BOSS force field, SOPLS program) are displayed in table 4.

Examination of table 4 indicates that binding energies of the complexes calculated using SOPLS are in good agreement with those calculated by *ab initio* methodology. The SOPLS calculated binding energy of the para-benzoquinone-water complex, A, differs from the *ab initio* value by only 0.04 kJmol^{-1}. SOPLS and *ab initio* calculated hydrogen bond lengths show a disparity of 0.33 Å and hydrogen bond angles by 15.9°. Inspection of the SOPLS calculated data for the A2 complex shows poorer agreement with *ab initio* calculated values than does the complex A. The binding energy of the A2 complex is, however, smaller than that for the A complex and so the qualitative trend in binding energies between the two has been reproduced.

SOPLS calculated binding energy and hydrogen bond geometry data for the ortho-benzoquinone-water complex, B, show excellent agreement with the *ab initio* calculated values. Calculated binding energies and hydrogen bond angles are identical to two and one decimal places respectively whilst the SOPLS calculated hydrogen bond length is only 0.08 Å shorter than the *ab initio* calculated value. As in the case of the para-benzoquinone-water complexes, A and A2, agreement between SOPLS and *ab initio* calculated binding

energies and hydrogen bond geometry for the B2 complex is poorer than for the B complex. Again, however, the trend in binding energies between B and B2 complexes is qualitatively reproduced.

Binding energies of complexes D, F and J (where water acts as a hydrogen bond donor) calculated through SOPLS all lie in the range -21.0 kJmol^{-1} to -18.0 kJmol^{-1} compared with the range -20.0 kJmol^{-1} to -18.0 kJmol^{-1} for *ab initio* optimised complexes. The average discrepancy between SOPLS and *ab initio* calculated binding energies for complexes D, J and F is only 0.37 kJmol^{-1}. Although SOPLS calculated binding energies show excellent agreement with *ab initio* calculated values, SOPLS hydrogen bond lengths are all shorter than their *ab initio* counterparts by 0.33 Å on average. SOPLS calculated hydrogen bond lengths for complexes D, F and J show excellent agreement with each other matching the observed *ab initio* trend. Hydrogen bond angles calculated using SOPLS for complexes D, F and J are all within 3° of each other, a trend seen in the *ab initio* results. SOPLS hydrogen bond angles for these complexes, however, tend to be larger than the *ab initio* calculated values by approximately 14°.

It should be noted that in both the *ab initio* optimisations and the SOPLS minmisations complexes H and I converge on the same hydrogen bonded structure with water acing as a hydrogen bond acceptor. The H / I complex shows the poorest agreement between SOPLS and *ab initio* calculated binding energies and hydrogen bond geometries within the hydroquinone complexes. The ortho-hydrobenzoquinone-water binding energy in this complex differs between the two

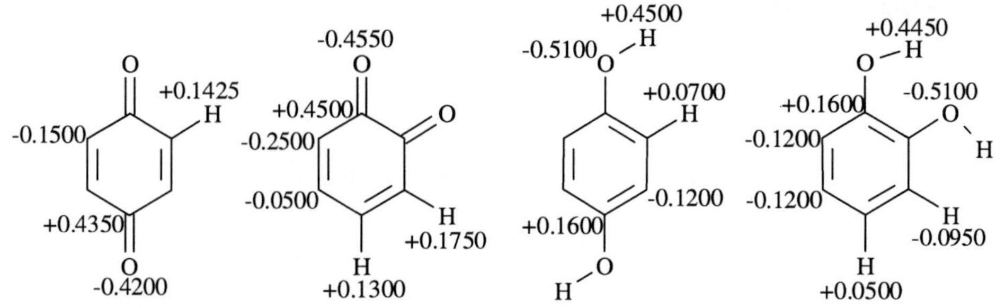

Figure 6. *Optimised electrostatic parameters for ortho and para benzoquinones and hydrobenzoquinones.*

calculation methods by 2.56 kJmol^{-1} whilst hydrogen bond length and angle differ by 0.11 Å and 29.4°. Although this agreement is relatively poor the SOPLS calculated binding energy is lower than all but the F complex and the hydrogen bond angle is the smallest out of all complexes. This is in reasonable agreement with the trends seen in the *ab initio* calculated values.

The SOPLS and *ab initio* calculated binding energies for complexes C, E, G, and K (where water acts as a hydrogen bond acceptor) all show close agreement with the average difference being 0.143 kJmol^{-1} . The largest discrepancy in these figures is 0.22 kJmol^{-1} for the C complex. The trend in *ab initio* calculated binding energies for these complexes is also reproduced in the SOPLS calculations; the binding energies of complexes C, E and K all lie in the range -28.0 kJmol^{-1} to -27.0 kJmol^{-1} with the G complex having the largest binding energy of -30.65 kJmol^{-1}. Hydrogen bond lengths for complexes C, E, G and K calculated through SOPLS are larger in magnitude than the *ab initio* calculated values by approximately 0.1 Å. SOPLS calculated hydrogen bond angles for complexes C, E, G and F all show reasonable agreement with *ab initio* calculated values.

With the exception of A2, B2 and H/I complexes, SOPLS binding energies show remarkably good agreement with those calculated using *ab initio* methodology. Hydrogen bonded geometries calculated using SOPLS show deviation from *ab initio* calculated values, however, discrepancy is not large and many of the trends seen in the *ab initio* results are reproduced. The average differences between *ab initio* and SOPLS calculated binding energies are 2.95 kJmol^{-1} for para-benzoquinone complexes, 3.01 kJmol^{-1} for ortho-benzoquinone complexes, 0.20 kJmol^{-1} for para-hydrobenzoquinone complexes and 0.88 kJmol^{-1} for ortho-hydrobenzoquinone complexes. Thus, although the results of the SOPLS calculations on complexes A through K do not agree perfectly with the *ab initio* results it can be asserted that the use of the optimised electrostatic parameters in con-

junction with OPLS van der Waals parameters do adequately represent intermolecular quinone and hydroquinone interactions with the respect to hydrogen bonding. Optimised electrostatic parameters are displayed in figure 6.

Free energy and electrode potential results

Free energy results from the solution phase mutations, both molecular and electrostatic, are given in table 5. Overall solution phase free energy changes were calculated by averaging the results from forward and backward perturbation runs.

In all cases it can be seen that hystereses between forward and backward perturbations are small indicating the efficiency of phase space sampling in all simulations. Standard deviations on incremental free energy changes for individual simulation windows in all perturbations were small (< 10% of the free energy change per window, average) indicating good convergence of calculated free energy changes. Dihedral angle distributions for para-hydrobenzoquinone and coupled dihedral angle distributions for ortho-hydrobenzoquinone show that although sampling over torsional phase space was not exhaustive, torsional space was sampled correctly according to the energy expression employed for the particular system in question. Analysis of dihedral and coupled dihedral angle distribution in all mutations involving hydrobenzoquinones, through the FEP equation, was used to assess the intramolecular torsional contribution to calculated free energy changes. These values are also displayed in table 5.

Intramolecular contributions to solution phase free energy changes for both ortho and para molecular perturbations show good agreement between simulations involving different electrostatic parameter sets (3-21G and SOPLS derived). This agreement indicates torsional sampling in these simulations has been thorough within the energetically allowed regions of torsional space. Furthermore, in the case of the para system this intramolecular contribution (-6.58 kJmol^{-1} for 3-21G electrostatic parameters and -6.96 kJmol^{-1} for SOPLS optimised electrostatic parameters) is very similar to the free energy change for the gas phase perturbation, 6.86 kJmol^{-1} . This similarity demonstrates similar torsional sampling in both gas and solution phase simulations. The

Table 5 *Solution phase free energy results and intramolecular torsional contributions to them for molecular perturbations (hydroquinone -> quinone) and electrostatic perturbations.*

Mutation (electrostatic parameters)		$\Delta\Delta G$ [kJmol^{-1}] (inter + intra)	Intramolecular contribution [kJmol^{-1}]	$\Delta\Delta G$ [kJmol^{-1}] (inter only)
2hpbq -> pbq, (3-21G)		19.73 ± 0.5	-6.58	26.31
2hobq -> obq, (3-21G)		12.86 ± 041	-11.52	24.38
2hpbq -> pbq, (SOPLS)		4.55 ± 1.22	-6.96	11.51
2hobq -> obq, (SOPLS)		-17.89 ± 1.80	-11.12	-6.77
3-21G -> STO3G	2hpbq	31.58 ± 0.35	negligible	31.58
	pbq	19.07 ± 0.34	negligible	19.07
	2hobq	33.14 ± 0.39	negligible	33.14
	obq	20.79 ± 0.25	negligible	20.79
2hpbq -> pbq, (STO-3G)		7.22	-6.58	13.80
2hobq -> obq, (STO-3G)		0.51	-11.52	12.03
3-21G -> 6-31G*	2hpbq	-3.54 ± 0.01	negligible	-3.54
	pbq	-13.98 ± 0.18	negligible	-13.98
	2hobq	0.12 ± 0.10	negligible	0.12
	obq	-11.97 ± 0.16	negligible	-11.97
2hpbq -> pbq, (6-31G*)		9.29	-6.58	15.87
2hobq -> obq, (6-31G*)		0.77	-11.52	12.29

intramolecular torsional contributions to the difference in solvation free energies between ortho-hydrobenzoquinone and ortho-benzoquinone are some 4.04 kJmol^{-1} and 3.63 kJmol^{-1} (for perturbations involving 3-21G and SOPLS derived electrostatic parameters respectively) greater in magnitude than the free energy result from the ortho system gas phase perturbation. The discrepancy between the gas and solution phase values is indicative of differing torsional sampling in the gas and solution phases. Indeed, examination of the coupled dihedral angle distributions from the gas and solution phase simulations revealed more extensive torsional sampling in the gas phase simulation.

In general, the intramolecular torsional contributions to the differences in solvation free energies derived from the dihedral angle distributions favours the oxidised, quinone, form.

Analysis of dihedral angle distributions from electrostatic perturbations involving hydrobenzoquinones revealed negligible intramolecular contributions to charge mutations free energy changes. This result was expected as the torsional parameterisation of the hydroquinones remains constant throughout the charge perturbations. Furthermore, alteration in sampling of torsional phase space due to changes in electrostatic parameterisation appears to be insignificant. Thus, calculated charge mutation free energies could be combined

directly with differences in solvation free energies calculated through the earlier molecular mutations to ascertain the effect on solvation free energies differences of altering electrostatic parameterisation.

Differences in solvation free energies (ΔG_3 - ΔG_1 and ΔG_2 - ΔG_4 from the thermodynamic cycle, figure 1) between the quinones and there respective reduced forms, as described by different electrostatic parameters, were found by subtracting the intramolecular torsional contribution from calculated free energy changes.

Free energy results show, in general, that solvation appears to stabilise the hydroquinone over the quinone form. Only for the ortho quinone system where electrostatics are described by optimised electrostatic parameters is this found not to be the case. It can be seen from table 5 that the calculated difference in solvation free energies between the quinones and their reduced forms varies quite considerably depending on the elctrostatic parameters employed. This results demonstrates the sensitive dependence of FEP calculated free energy changes on electrostatic parameterisation. It is also apparent that when electrostatic parameters derived from *ab initio* derived MEPs are employed in solution phase simulations the difference between the solvation free energy differences calculated for ortho and para-benzoquinones remains almost constant (1.77 kJmol^{-1}, 1.93 kJmol^{-1} and 3.58

kJmol^{-1} for STO-3G, 3-21G and 6-31G* derived charges respectively). This same quantity found from perturbations involving the SOPLS optimised electrostatic parameters is calculated as 18.28 kJmol^{-1}. Thus it appears that calculated differences in solvation free energies are merely scaled by the alteration of electrostatic parameters derived from one *ab initio* basis set to another.

Differences in reduction potentials, ΔE, between ortho and para-benzoquinones, as calculated through equations 1 and 2 are displayed along with the experimental value in table 6.

It is apparent from table 6 that electrode potential results generally in best agreement with experiment are those found when $\Delta\Delta G_{gas}$ is calculated employing RHF/STO-3G *ab initio* calculations and atom centred charges derived from *ab initio* calculated MEPs are used in solution phase simulations. In these cases the average error in ΔE is 38.9 mV corresponding to an error in $\Delta\Delta G_{aq}$ (figure 1) of 7.5 kJmol^{-1}. Although these results seem reasonable the STO-3G basis set used to calculate the major contributant to $\Delta\Delta G_{gas}$ represents the lowest level of *ab initio* theory used in this study and therefore the least accurate. The agreement with experiment of results obtained from the combination of STO-3G $\Delta\Delta G_{gas}$ and *ab initio* derived electrostatic parameters has been tentatively ascribed to a cancellation of errors.

Electrode potential results derived from values of $\Delta\Delta G_{gas}$ found from RHF/3-21G, RHF/6-31G* and MP2/6-31G* calculations and simulations where *ab initio* derived electrostatic parameters have been employed show poor agreement with the experimental answer of -92.2 mV. The average error over these results is 120.9 mV corresponding to an error in $\Delta\Delta G_{aq}$ of 23.3 kJmol^{-1}. Of these results, those derived from use of the MP2/6-31G* value of $\Delta\Delta G_{gas}$ are in best agreement with experiment having an average error of 101.1 mV (corresponding to a 19.5 kJmol^{-1} error in $\Delta\Delta G_{aq}$). The MP2

level of *ab initio* theory in conjunction with the 6-31G* basis set should be capable of performing energy caclulations in good agreement with experiment on molecules as small as the quinones and hydroquinones and it was expected that use of MP2/6-31G* values for $\Delta\Delta G_{gas}$ would give the most accurate results. It is noticeable that, given a particular value of $\Delta\Delta G_{gas}$, the value of the calculated reduction potential shows little dependence on the basis set employed generate the electrostatic parameters through *ab initio* calculations. This demonstates that atom centred charges derived from different *ab initio* basis sets provide similar relative descriptions of the aqueous ortho-hydrobenzoquinone / para-benzoquinone and para-hydrobenzoquinone / ortho-benzoquinone systems. Overall, when *ab initio* CHELPG derived charges are employed to represent the electrostatic interactions of the quinone and hydroquinone molecules in aqueous solution, results obtained are in poor agreement with the experimental value.

Due to the high level of *ab initio* theory used in the MP2/6-31G* calculations it is assumed that errrors in the calculated value of $\Delta\Delta G_{aq}$ and hence ΔE are due to errors in the calculation of solvation free energy differences. Errors in the calculation of solvation free energy differences could be caused by either a poor sampling of phase space, poor convergence of free energy changes or the convergence of free energy changes to the wrong values. Given the small hystereses in calculated free energy changes and the small standard deviations on incremental free energy results it appears that it is the latter reason which accounts for the observed errors in calculated quantites. If this is the case then it appears that atom centred charges derived from *ab initio* wavefunctions using the CHELPG algorithm provide an inaccurate description of molecular electrostatics in aqueous solution.

Examination of the reduction potential results calculated using data from simulations where the SOPLS / optimised electrostatic parameters were employed shows a much better general agreement with experiment than do the other results. With the exception of the results derived using the RHF/STO-3G $\Delta\Delta G_{gas}$ value, all calculated values of ΔE lie within 52 mV of the experimental result. The best result is obtained when the value of $\Delta\Delta G_{gas}$ obtained from the MP2/6-31G* calculations is employed and is only 18.9 mV from the ex-

Table 6 *Calculated differences in reduction potentials between ortho and para-benzoquinones. Reduction potential results are shown for all different sets of electrostatic parameters employed in solution phase simulations and all different ab initio basis sets used in the calculation of $\Delta\Delta G_{gas}$. Also shown is the experimental value.*

Electrostatic parameters	$\Delta\Delta G_{gas}$ (RHF/STO-3G)	$\Delta\Delta G_{gas}$ (RHF/3-21G)	$\Delta\Delta G_{gas}$ (RHF/6-31G*)	$\Delta\Delta G_{gas}$ (MP2/6-31G*)
STO-3G MEP	-57.3 mV	-223.8 mV	-229.0 mV	-196.7 mV
3-21G MEP	-56.5 mV	-223.0 mV	-228.2 mV	-195.9 mV
6-31G* MEP	-47.9 mV	-214.4 mV	-219.6 mV	-187.3 mV
Optimised / SOPLS	+28.3 mV	-138.2 mV	-143.4 mV	-111.1 mV
Experimental [14]		-92.2 mV		

perimental value corresponding to an error in $\Delta\Delta G_{aq}$ of only 3.6 kJmol^{-1}.

Conclusions.

For the most part, the difference in solution phase reduction potentials between ortho and para-benzoquinones evaluated employing high level *ab initio* calculations and the free energy perturbation theory within Monte Carlo simulations shows poor agreement with the experimental value. This poor agreement arises due to the poor quality of the intermolecular force field parameters applied to describe the solute molecules in solution phase simulations. The best agreement betwen calculated and experimental reduction potential differences arises when electrostatic parameters derived specifically for the sytem in question are employed to describe solute molecules and *ab initio* calculations are performed at the MP2 level. Development of such parameters and the ensuing MC simulations constitute a time consuming process. It should be noted that the use of free energy perturbation theory within Monte Carlo simulations in conjunction with high level *ab initio* calculations is capable of producing results in good agreement with experiment. However, as a routine method of obtaining such quantities FEP theory is limited and more epeditious methods are required to allow the computation of solution phase reduction potentials to become routine.

Acknowledgements: D. R. L. would like to acknowledge the financial assistance of Lilly Research as part of a C. A. S. E studentship.

References.

1.   N. Metropolis and S. Ulam, *J. A. Stat. Ass.*, **1949**, 44, 335.
2.   M. P. Allen and D. J. Tildesley, *Computer Simulations of Liquids*, **1987**, Oxford University Press, Oxford, UK.
3.   Torrie G. M. and Valleau J. P., *J. Comput. Phys.*, **1977**, 23, 187.
4.   W. L. Jorgensen, *J. Phys. Chem.*, **1983**, 87, 5304.
5.   W. L. Jorgensen, *BOSS (Version 2.8) Users Manual*, **1990**, Dept. of Chemistry, Yale University, New Haven, CT, USA.
6.   W. L. Jorgensen and J. Tirado-Rives, *J. Am. Chem. Soc.*, **1988**, 110, 1657.
7.   J. Paranata, S. J. Wierschke and W. L. Jorgensen, *J. Am. Chem. Soc.*, **1991**, 113, 2810.
8.   W. L. Jorgensen, E. R. Laird, T. B. Nguyen and J. Tirado-Rives, *J. Comp. Chem.*, **1993**, 14, 206.
9.   S. J. Weiner, P. A. Kollman, D. T. Nguyen and D. A. Case, *J. Comp. Chem.*, **1985**, 7, 230.
10. U. Burkert and N. L. Allinger, *ACS Monograph*, **1982**, 177.
11. C. M. Brenenman and K. B. Wiberg, *J. Comp. Chem.*, **1990**, 11, 361.
12. D. L Beveridge and F. M. DiCapua in *Computer Simulations of Biomolecular Systems: Theoretical and Experimental Applications.*, **1989**, Eds. W. F. van Gunsteren and P. K. Weiner, ESCOM, Leiden, The Netherlands.
13. D. G. Lister, J. N. MacDonald and N. L. Owen, *Internal rotation and inversion.*, **1978**, Academic Press Inc., London, UK.
14. H. W. Moore, *Science*, **1977**, 197, 527.
15   C. A. Reynolds, *J. Am. Chem Soc.*, **1990**, 112, 7545.
16. M. J. Frisch, M. Head-Gordon, H. B. Schlegel, K. Raghavachari, J. S. Binkley, C. Gonzales, D. J. Defrees, D. J. Fox, R. A. Whiteside, R. Seeger, C. F. Melius, J. Baker, R. Martin, L. R. Khann, J. J. P. Stewart, E. M. Fluder, S. Topiol and J. A. Pople, *Gaussian 92*, **1992**, Gaussian Inc., Pittsburgh, PA, USA.
17. W. J. Hehre, R. Ditschfield, R. F. Stewart and J. A. Pople, *J. Chem. Phys.*, **1969**, 52, 2769.
18. J. S. Binkley, J. A. Pople and W. J. Hehre, *J. Am. Chem. Soc.*, **1980**, 102, 939.
19. W. J. Hehre, R. Ditschfield and J. A. Pople, *J. Chem. Phys.*, **1972**, 56, 2257 and P. C. Hanrihanran and J. A. Pople, *Theor. Chim. Acta.*, **1973**, 28, 213.
20. J. S. Binkley and J. A. Pople, *Int. J. Quant. Chem.*, **1975**, 9, 229.
21. M. J. S. Dewar, E. G. Zoebisch, E. F. Healy and J. J. P. Stewart, *J. Am. Chem.. Soc.*, **1985**, 107, 3902.
22. J. J. P. Stewart and F. J. Seiler, *MOPAC (Version 6)*, **1988**, United States Air Force Academy, Colorado Springs, CO, USA.
23. C. A. Reynolds in ”*Computer Aided Molecular Design.*”, **1989**, Ed. W. G. Richards, I. B. C., London, UK.
24. W. L. Jorgensen, *BOSS (Version 3.1)*, **1991**, Yale University, New Haven, CT. USA.
25. R. W. Zwanzig, *J. Chem. Phys.*, **1954**, 22, 1420.
26. W. L Jorgensen and D. L. Severance, *SOPLS*, **1994**, Yale University, New Haven, CT., USA.
27. W. L. Jorgensen, J. Chandresekhar, J. D Madura, R. W. Impey and M. L. Klein, *J. Chem. Phys.*, **1983**, 79, 926.
28. W. L. Jorgensen, *BOSS (Version 3.1) Users Manual.*, p 9, **1991**, Yale University, New Haven, CT. USA.

J. Mol. Model. **1996**, 2, 149 – 159

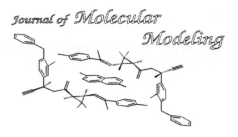

Substrate Specifity of Chymotrypsin.
Study of Induced Strain by Molecular Mechanics

Bernd Kallies*, Rolf Mitzner

Institut für Physikalische und Theoretische Chemie, Universität Potsdam, Am Neuen Palais 10, 14469 Potsdam, Germany
(kallies@serv.chem.uni-potsdam.de)

Received: 5 March 1996 /Accepted: 4 June 1996 /Published: 13 June 1996

Abstract

Acylenzyme intermediates, produced by transfer of the acyl portions of selected natural substrates onto the catalytic serine hydroxyl of the serine protease chymotrypsin, were modeled with the AMBER force field. The obtained structures were used to calculate interaction and deformation energies. A set of 32 geometry variables were extracted out of each structure. They describe deformation effects specific for each substrate. It is shown by statistical analyses, that the interaction and deformation energies correspond to measured substrate reactivities. The extracted geometry variables are able to reproduce this dependency through multivariante statistical methods. These analyses suggest that there exist specific deformations of both the substrate and the enzyme portion, which are related to substrate reactivity. The geometry changes observed for high specific substrates are interpreted in terms of mechanistical requirements of the enzymatic reaction. The obtained model validates the hypothesis of induced strain as possible source of substrate specifity of chymotrypsin.

Keywords: serine protease mechanism, force field calculation, substrate specifity

Introduction

The specifity of biochemical receptors and its sources has long been subject of investigation. It is given that the fit of the three-dimensional structure and the complementarity of the surface properties of a ligand to a receptor site are necessary conditions for the biological activity of the ligand. These conditions may be associated with Fischer's „lock and key" picture [1]. It holds for most receptors, which only immobilize a ligand. Our current understanding of this principle is the source of the great success of drug design by methods using quantitative structure-activity relationships [2].

If a receptor has to perform a chemical reaction with an immobilized ligand like enzymes do, there exist additional aspects of specifity. The specifity of an enzyme can be expressed by reached reaction rates, not by binding coefficients only. At least two hypotheses were developed accessing sources of enzyme specifity. The first uses Pauling's theory of transition state stabilization [3]. This hypothesis is used mainly to explain the catalytic power of enzymes in comparison to the uncatalyzed reaction. It also explains different reaction rates by different interactions between transition states of several substrates and the enzymatic active site. The second „anti-Pauling" hypothesis states induction of conformational strain into the substrate and the enzyme prior to the reaction by use of binding energy, which is produced by substrate immobilization [4]. This strain may be productive in terms of a catalytic effect, if the conformational changes lie on the reaction coordinate. These two additional sources of receptor specifity are not subject of common QSAR studies, since their implementation requires other structure descriptors than those of isolated substrates.

The presented work represents an attempt to develop QSAR for an enzymatic reaction, which base on variables describing the substrates in the active site and structure changes induced by substrate binding. These variables were taken from enzyme-substrate adducts, which were obtained through molecular mechanics. We chose the deacylation reaction of the serine protease chymotrypsin acylated with several natural amino acid substrates as object for our studies.

Background [5]

Serine proteases like trypsin, chymotrypsin or subtilisin, as well as serine esterases, cysteine proteases and some mechanistically related lipases are known to function through a two-step mechanism. After immobilization of an ester or amide substrate the acyl portion of the substrate is transferred onto the hydroxymethyl sidechain of an active site serine residue (acylation reaction). This step forms the first hydrolysis product and an ester intermediate (acylenzyme). The latter species is cleaved in a subsequent step by solvent water, releasing carboxylate and the regenerated enzyme (deacylation reaction). Both steps are assisted by a histidine residue hydrogen-bonded to the serine sidechain and an aspartate residue H-bonded to the histidine sidechain (general base catalysis). Each reaction step should proceed via the usual addition-elimination mode of most acyl-transfer reactions, which involves formation of a short-living tetrahedral intermediate by attack of a nucleophile onto a carbonyl carbon. The formal negative charge resting at the carbonyl oxygen (oxyanion) of these intermediates is stabilized by interactions between the oxyanion and enzyme residues, which form the „oxyanion hole". It is built by NH-portions of enzymatic carboxamides (protein backbone or asparagine sidechain). The stereospecifity of chymotrypsin in discriminating between L- and D-amino acid substrates was explained through various force field calculations [6-8] by interaction of the NH-group of the bound substrate and a C=O group of the protein backbone of the binding pocket.

The rate limiting step of the overall hydrolysis reaction depends on the general carbonyl activity of the substrate. Amides often show rate limiting acylation. In the case of labile ester substrates, deacylation is the rate limiting step. This behaviour enables one to measure individual rate constants for acylation and deacylation. Since large sets of rate constants based on the esterase function of chymotrypsin are published, we chose this system. Among the reactivity data the deacylation and turnover rates of ester substrates are the most reliable ones, so we decided to study this individual reaction.

Premises

Our work is based on several assumptions, which have to be explained first in order to guide through the results. The central idea is that an acylenzyme is formed with every substrate studied, and that this acylenzyme can be handled with com-

mon empirical force fields. We assume that the substrate immobilization and the acylation reaction were successful. This assumption holds not, if the free substrate does not fit the enzymatic active site. We performed no docking studies, but utilized the covalent bond between the substrate and enzyme portion of an acylenzyme and the behaviour of the system induced by it, instead. This bond behaves like a true ester bond through force field calculations. The substrate is already placed in the active site, and it can not leave out. In this case, force field calculations will converge to a structural compromise. It includes deformation of both the substrate and the enzyme portion. The careful study of these conformational changes enables us to detect outliers, for which the assumption of successful binding does not hold. It is the basis for the development of variables for QSAR in our study, too.

Another assumption is related to the role of water molecules. The re-solvation of both the substrate and the binding site (disruption of the hydrate shells and solvation by a new micro-environment) is an important contribution to the free association energy produced during the step of immobilization of a ligand at a receptor site [2]. We did not study the initial association process, but calculated an association energy equivalent for the special case of covalent bound substrates. This quantity we define as the sum of the interaction energy between the enzyme and the substrate part of an acylenzyme and deformation energies of these parts, using their free states as reference (see below). Water molecules were not explicitly incorporated in all these calculations. Since the neglect of hydrate shells might be subject of criticism, we consider the errors produced by this approach. At first we look at geometry changes and deformation energy contributions to calculated association energies. Since the structure of the free enzymatic active site without water molecules should differ from that with water molecules included, the calculated deformation energy of the enzymatic part of an acylenzyme bears an error. But its amount should be equal in all acylenzyme structures, so it does not contribute to the relative relationships we derive in this paper. The assumption of equal geometries of isolated substrates in the gas phase and a solvated state is common, it introduces a neglectable error in calculated substrate deformation energies. A similar effect of the used technique is produced by the neglect of dehydration energies, contributing to the association energy. This assumption bases on always equal dehydration energies of the enzymatic active site and equal dehydration energies of different substrates. Now we consider the interaction energy contribution to the association energy. It should be influenced by water molecules remaining in the active site after immobilization of substrates, which fit not the whole binding pocket. It influences the geometries of acylenzymes, too, which are used for the calculation of deformation energies. We assume, that this contribution to the interaction energy can be modelled by damping of electrostatic forces by a distance dependent dielectric constant. In summary, the neglect of explicit water molecules yields errors, which are either

Name	R₁	R₂	Conformations
Ace-OSer	H	H	1
Ac-Gly-OSer	L-NHCOMe	H	1
Ac-Ala-OSer	L-NHCOMe	Me	1
Ac-Val-OSer	L-NHCOMe	i-C_3H_7	3
Ac-Ile-OSer	L-NHCOMe	$CH(CH_3)$-C_2H_5	9
Ac-Leu-OSer	L-NHCOMe	CH_2-i-C_3H_7	9
Ac-Asn-OSer	L-NHCOMe	CH_2-C(O)NH_2	9
Ac-Phe-OSer	L-NHCOMe	CH_2-Ph	1
Ac-Tyr-OSer	L-NHCOMe	CH_2-Ph-4-OH	1
Ac-Trp-OSer	L-NHCOMe	CH_2-indolyl	2

Table 1.
Description of acylated serine monomers SerO-C(O)-CHR₁R₂

near equal in all structures studied or neglectable in relation to other errors produced by application of molecular mechanics.

Computational Details

Building of acylenzymes and substrates

All studies base on the X-ray structure PDB 1GCT [9] (γ-chymotrypsin acylated with the tetrapeptide Tyr-Ala-Gly-Pro), obtained from the Protein Data Bank [10,11] at Brookhaven National Laboratory. It was modified with SYBYL-6.3 [12] and the AMBER-forcefield [13]. All water molecules included in the original structure were deleted. The rudiment of Ser11 was completed. Other amino acids not visible in the crystal structure were omitted. Hydrogens were added to model protonation states at pH 7. Ser195 acylated with the original tetrapeptide ligand was used as template to define new monomers of a serine residue acylated with various N-acetylated L-amino acids. The monomers used are summarized in table 1. The atom types and charges were chosen consistently with the all-atom model of the AMBER force field. The net atomic charges of the ester group were developed from Mulliken charges taken from AM1 [14] semiempirical quantum mechanical calculations on a methyl acetate patch. They were scaled to give neutral monomers along with charges of other atoms from the AMBER forcefield. This procedure yields the charges -0.484, 0.795 and -0.439 for the carbonyl oxygen (atom type O), carbonyl carbon (atom type C) and ether oxygen (atom type OS), respectively. These charges are consistent with charges of amide groups used in the AMBER force field and with net atomic charge differences between related esters and amides, obtained by ab inito molecular orbital calculations [15]. After deletion of the original ligand and replacement of Ser195 with the new monomers, geometry optimizations followed. We used a distance dependent dielectric constant of 4.0. 1-4-interactions were scaled by a factor of 0.5. A cutoff of nonbonding interactions at 12 Å was applied. Geometry optimizations were done with a conjugate gradient routine

in subsequent steps, terminating after reaching an RMS gradient threshold, like outlined in table 2. Through the first step, a distance constraint of 1.8 Å between Hγ(Ser214) and Oδ1(Asp102) was applied in order to produce a hydrogen bond between these residues. After completion of this step, this constraint was obmitted. The scheme was used to obtain geometries of rotamers of the acyl portions of several flexible substrates in the active site. These conformations are produced by rotation around the Cα-Cβ- and Cβ-Cγ-bonds of the amino acid sidechain of the substrates Val, Leu, Ile and Asn in steps of 120°. The number of conformations studied for each substrate is given in table 1.

The free substrates which would form the acylated enzyme after splitting of a leaving group were modelled in a similar manner. The monomer definitions of acylated Ser195 were used as templates for methyl ester substrates by replacing the serine portion by a methyl group. The free substrates were modelled by conformational analyses using a grid search technique and following full geometry optimizations within the AMBER forcefield.

Calculation of energies

From acylenzyme structures obtained after geometry optimizations we calculated interaction energies, deformation and association energies. All energy calculations were done on reduced models. They contain 66 monomers in each

Table 2. *Optimization scheme for acylenzymes.*

Step	Optimized atoms	$\nabla_{RMS}E$ [a]
1	hydrogen atoms	5
2	amino acid sidechains	1
3	all	0.5
4	119 monomers around Ac-Ser195	0.1
5	66 monomers around Ac-Ser195	0.01

[a] in kcal/(mol·Å).

case (residues 16, 17, 30-33, 40-45, 53-60, 94, 99, 102, 138-143, 146, 151, 160, 172, 182-185, 188-198, 212-222, 224-229 + end groups of 16, 146, 151).

The acylenzymes were splitted into an enzyme and a substrate portion by a symbolic cut of the Ca(Ser195)-Cb(Ser195) bond. The resulting model of the enzymatic active site without substrate contains 66 monomers and 880 atoms. Calculation of the force field energy terms with disregard of one portion yields the force field energy of the remaining part in the acylenzyme.

Interaction energies **EInt** between the enzyme and the acyl portion of the substrate including the full ester group are calculated as difference between the non-bonded energy terms of the reduced acylenzyme model and the sum of the non-bonded energies of the two parts. The difference between total energies would include the bond stretching, angle and torsion bendings located in the cutted Ca(Ser195)-Cb(Ser195)

bond and neighbouring atoms. This „internal contribution" to the interaction energy is always near equal in magnitude in our models.

Deformation energies of the enzyme portion of the acylenzymes **EDefE** were calculated as difference between the total energies of the corresponding 880-atom models in the acylenzyme and the structure minimized without substrate. The deformation energy of a substrate **EDefS** we define as the difference between total energies of the substrate portion in the acylenzyme and in the free substrate, using its weighted averaged energy as reference. The sum of the obtained interaction and deformation energies yields the association energy EAss.

The weighted averaging of energies bases on simple Boltzmann statistics [16]. The energy of a system with N possible non-degenerate states can be calculated from

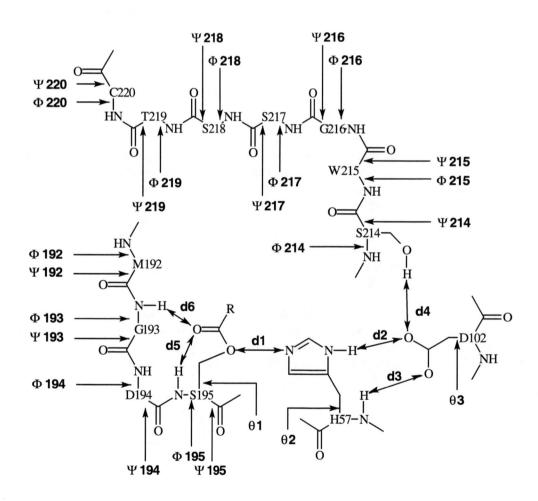

Figure 1. *Definition of geometry variables. Amino acid residues are labeled at Cα with the one-letter code and sequence number. Arrows point to the rotable bond of a dihedral angle. Backbone torsions are defined by* $\Phi_n = C_{n-1} - N_n - C\alpha_n - C_n$, $\Psi_n = N_n - C\alpha_n - C_n - N_{n+1}$.

$$E = \sum_{i=1}^{N} f_i E_i$$

where f_i is a normalized weighting factor for state i, given by

$$f_i = \frac{q_i}{\sum_{i=1}^{N} q_i}$$

The quantity q_i is the individual partition function of state i, which is defined as

$$q_i = \exp(-\Delta E_i / RT)$$

ΔE_i is the energy difference between state i and the ground state measured in J/mol, R is the gas constant and T the temperature. We used T = 298 K. Setting E_i equal to the energy of a particular conformation of a flexible molecule and N equal to the number of possible conformations, E yields the conformational energy of this molecule with respect to statistical occupancies of other than the ground state conformation by neglecting activation barriers. This procedure can also be used to average quantities other than energies such as geometry variables of conformations of a molecule, if the weighting factors were calculated from representative energies.

Statistical analyses

Four different sets of acylenzyme structures were developed. The first contains all 37 structures (set 1). Set 1a is the same like set 1, except outliers. The other sets contain only one representative structure for each substrate. For set 2 structures of the substrates Ac-Ile, Ac-Leu, Ac-Val and Ac-Asn were averaged by their arithmetic mean after exclusion of outliers. Set 3 includes the minimum energy conformations using the total energy of the 66-monomer structure as criterion. Set 4 contains structures which were averaged by Boltzmann statistics as described above. The weighting factors were developed from total energies of the 66-monomer structures. Here only outliers defined as „non-reactive" are omitted. Outliers which are defined as „inactive" are automatically eliminated by very small weighting factors (for the definition of outliers see next section).

All sets contain 39 variables for each structure. In addition to the four energy differences a set of 32 geometry variables was taken out of each acylenzyme structure. These variables are summarized in figures 1 and 2. Three sets of experimental data were added to each set, including measured values of log(1/Km) and log(k3) of p-Nitrophenyl esters of the studied acid portions, and log(kcat) of methyl or ethyl esters. The selected p-Nitrophenylesters of amino acid substrates bear the Benzoyloxycarbonyl (Z) N-protecting group, N-acetylated substrates were chosen in the case of

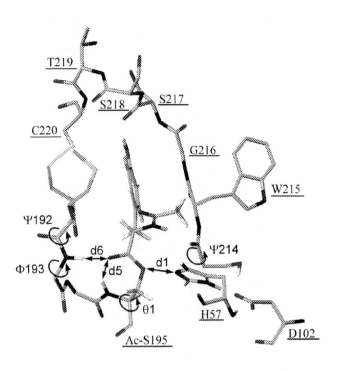

Figure 2. *Active site of chymotrypsin acylated with Ac-Tyr.*

methyl and ethyl esters. The values were taken from [17] and references cited therein.

Statistics performed on these sets included simple linear regression analyses for energies and experimental data and partial least square statistics (PLS) for geometry variables and energies. PLS analyses were done with the QSAR module of SYBYL 6.3. The optimal number of components was determined by 10 subsequent cross-validation runs using the largest possible amount of components as basis and r^2_{PRESS} (q^2) as criterion. Then the analyses were repeated with the number of components which would yield the highest q^2 without further validation of the model. The variables were used unscaled and unweighted. In addition to PLS analyses hierarchical cluster analyses were done, using the complete linkage model. Tables of sets 1 and 4 are given as supplementary material.

Results and Discussion

Rough data

From force field calculations the region which is necessary for enzymatic activity can be determined. Figure 3 shows correlation coefficients for assumed linear dependencies between total energies of several reduced models of acylenzyme structures and the energy of the complete acylenzymes. For our object we find a minimal region of about 30 residues (about 5 Å around acylated Ser195), which contains near the same information like the overall structure. Any further re-

	Set 2	Set 3	Set 4	
log (1/Km) = a · EAss + b, (f=6) [b]				
a	-0.09 ± 0.03	-0.09 ± 0.03	-0.09 ± 0.02	
b	3.3 ± 0.5	3.2 ± 0.6	3.2 ± 0.5	
r² / F	0.673 / 12.34	0.692 / 13.47	0.699 / 13.96	
log (kcat) = a · EAss + b, (f=5)				
a	-0.14 ± 0.04	-0.15 ± 0.04	-0.15 ± 0.04	
b	-2.6 ± 1	-3.0 ± 1	-3.0 ± 1	
r² / F	0.693 / 11.29	0.735 / 13.89	0.739 / 14.16	
log (k3) = a · EAss + b, (f=7)				
a	-0.13 ± 0.04	-0.14 ± 0.04	-0.14 ± 0.04	
b	-2.2 ± 0.8	-2.4 ± 0.8	-2.4 ± 0.8	
r² / F	0.634 / 12.11	0.667 / 14.02	0.660 / 13.56	

Table 3. *Parameters of an assumed linear dependency of reactivities on EAss. [a]*

[a] EAss in kcal/mol.
[b] f: degrees of freedom.

duction yields loss of information. So our base model with 66 monomers represents a good choice for calculating interaction and deformation energies.

We now focus our intention on the detection and explanation of outliers. A close inspection of obtained acylenzyme structures led us to the definition of two different outlier cases. The first case we access through anormal geometries. Among the modelled structures there are 7 cases, where the NH-hydrogen of the substrate acyl portions forms a hydrogen bond to His57. This behaviour is coupled with the complete loss of contact between the carbonyl oxygen of the ester bond to cleave and the oxyanion hole (NH-groups of Gly193/Ser195). The substrate conformations exhibiting these features are characterized by a bad steric contact between their acyl portions and the protein backbone around Met192. Since at least

the accessibility of Nε2(His57) for water molecules is necessary for deacylation, we defined these structures as „nonreactive" outliers and did not include them in our data sets.

The second outlier set was detected after looking on deformation energies of the enzyme portions (see figure 4). There are four structures with unusual high deformations of the enzyme. Since this behaviour should be unfavourable, we define them as „inactive". The high deforming conformations can be used to calculate an allowed and a forbidden substrate volume for chymotrypsin (see figure 5). The obtained forbidden area can be explained by repulsion between the acyl portion and the backbone of Trp215/Gly216. These outliers were produced by simplexing, which we used to reduce highly repulsive contacts prior to geometry optimization.

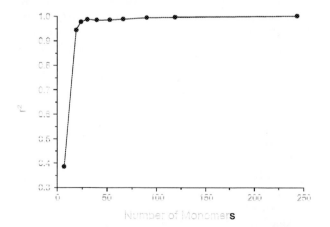

Figure 3. *Correlation coefficients for assumed linear dependencies between total energies of reduced acylenzyme models and the complete structures.*

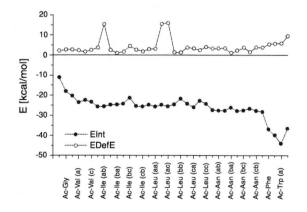

Figure 4. *Interaction energies and deformation energies of the enzyme portions. The order of structures follows that given in table 1.*

Table 4. *Dependency of EAss on geometry variables. [a]*

	Set 2				Set 3				Set 4		
s	2.949				3.845				2.634		
r²	0.887				0.802				0.907		
F	27.51				14.17				34.22		
	x [b]	**coeff.**	**cum.**		**x**	**coeff.**	**cum.**		**x**	**coeff.**	**cum.**
	Ψ192	-0.378	0.163		Ψ214	0.428	0.143		Ψ192	-0.372	0.153
	Ψ214	0.378	0.325		Ψ192	-0.317	0.282		Ψ214	0.418	0.304
	Φ193	0.324	0.468		Φ193	0.268	0.399		Φ193	0.319	0.434
	Ψ216	-0.273	0.567		Ψ216	-0.273	0.500		Ψ216	-0.277	0.532
	θ1	-0.294	0.646		Ψ215	-0.236	0.593		θ1	-0.312	0.616
	Φ215	-0.145	0.715		θ1	-0.266	0.670		Φ215	-0.182	0.691
	Φ214	0.188	0.763		Φ216	0.297	0.736		Φ214	0.198	0.740
	Φ195	0.167	0.801		Φ214	0.236	0.792		Φ216	0.209	0.785
	Φ192	0.191	0.832		Φ217	-0.160	0.837		Φ217	-0.151	0.829
	Φ217	-0.113	0.863		Ψ217	0.188	0.876		Φ195	0.178	0.864

[a] *PLS with 32 independent geometry variables, 2 components.*

[b] *Only the 10 regression coefficients with the highest contribution are listed in the order of their normalized values. Normalization was done with respect to the variance of x_i and y. The last column cumulates normalized coefficients wich where scaled to sum to 1.0.*

After exclusion of these outliers the number of acylenzymes reduces to 26 structures for 10 substrates. They were used to study relationships between energies and geometries.

Energy relationships

From empirical considerations it is believed that a substrate will be much more strained after immobilization than the enzyme itself [18]. On the other hand, deformation of both the substrate and the enzyme becomes possible, when the interaction is strong enough. This possibility was demonstrated in [19] for the association complex of chymotrypsin with N-Ac-Trp-amide by ab initio and semiempirical molecular orbital calculations. Our models can describe such a relationship for the acylenzyme states of various substrates. We obtain a probably linear dependency between the overall deformation energy and the interaction energy (see figure 6).

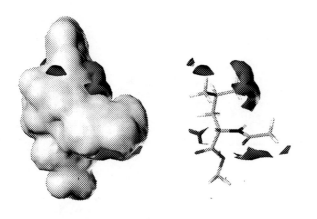

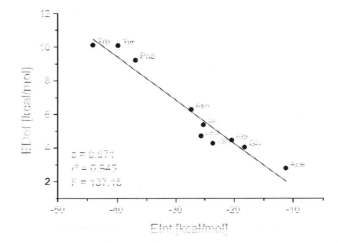

Figure 5. *Left: allowed substrate volume (from 22 „active" conformations, yellow); right: forbidden substrate volume (from 4 „inactive" conformations, red) and included inactive Ac-Leu conformer.*

Figure 6. *Parameters for an assumed linear dependency of overall deformation energies on interaction energies (set 4).*

Table 5. *Selected geometries for high specific substrates.*

Distance	Phe, Tyr, Trp		other substrates	
d1, Å	2.96	± 0.01	3.04	± 0.04
d2, Å	1.901	± 0.005	1.853	± 0.006
d5, Å	1.973	± 0.001	1.94	± 0.02
d6, Å	1.950	± 0.001	2.03	± 0.03

Set 4 is the most predictive in this case. Sets 2 and 3 yield near the same dependency. Sets 1 and 1a bear a high noise level. We do not discuss the latter sets any further in this work.

No linear relationship can be obtained between the energy needed for substrate deformation and that needed for enzyme deformation (see figure 7). We find three groups of different structures here. The first contains substrates with small acyl groups (Ace, Ac-Gly, Ac-Ala) which can not be deformed. The unusual finding that they induce enzyme deformations is an effect of our choice of the reference structure used for calculation of EDefE. The second group contains substrates with aromatic acyl portions (Ac-Phe, Ac-Tyr, Ac-Trp) which exhibit large deformation of both the substrate and the enzyme structure. The third group includes structures with intermediate strength of interaction. They show deformation of the substrate portion rather than of the enzyme part.

These three groups correspond to substrate structures known to have different reactivities in chymotrypsin-catalysed reactions [20]. In order to obtain a quantitative relationship between measured reactivities and calculated energies we assumed a linear dependency. The most predictive model in the majority of cases is the association energy of set 4. The correlation coefficients for all sets and reactivity data are given in table 3. Figure 8 shows the obtained relationship with the deacylation rate constant, for which the most data were available. The correlations are very poor, since the data set is very small and the source and interpretation of experimental data is not out of question [21]. So the derived regression model has to be interpreted in terms of a rough dependency between substrate reactivity and calculated association energies. But it enables one to derive the assumption, that the association energy can be used as a reactivity substitute with some success. We will make use of this working hypothesis, although from outliers of an assumed linear relationship it can be estimated, that a high association energy may not necessarily be coupled with high reaction rates. This behaviour seems to hold for the Ac-Trp substrate, if the measured reactivity is correct. Other substrates may set specific interactions in motion like Ac-Asn, thus reaching a high reaction rate without large deformations of the enzyme.

Geometry analyses

The calculated deformation energies correspond to geometry changes of the enzyme and substrate part of the active site. Performing hierarchical cluster analyses with all 32 geometry variables of set 4, we obtain figure 9. The same clusters of substrate structures is obtained as discussed above. Since geometry variables are dependent from each other, the data set has to be reduced by extraction of principal components. We performed PLS analyses using the association energy as dependent variable. Defining the whole set of 32 geometry variables as X-block, cross-validation runs yielded two principal components. As usual, the scores of cases in

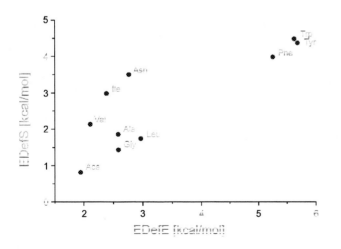
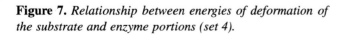

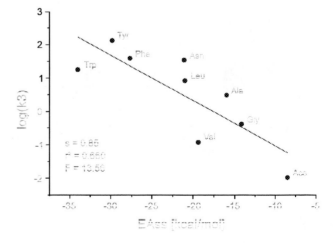

Figure 7. *Relationship between energies of deformation of the substrate and enzyme portions (set 4).*

Figure 8. *Relationship between the deacylation rate constant k3 and calculated association energies (set 4).*

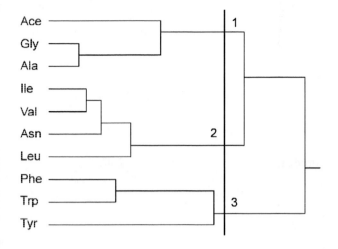

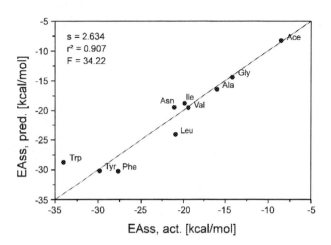

Figure 9. *Result from hierarchical cluster analysis of 32 geometry variables (set 4).*

Figure 10. *Quality of the PLS fit between 32 geometry variables and the association energy (set 4).*

the reduced X-block of the two non-redundant variables have the same structure like obtained from cluster analyses of the whole data set. The obtained fit is presented in figure 10. Again set 4 is the most predictive one, although other averaged sets yield the same relationship. Interestingly, the Ac-Trp substrate is predicted by geometries to yield a smaller association energy than calculated from force field energies. This result corresponds to the measured reactivity of this substrate (see figure 8).

These analyses enable one to state a probable relationship between geometry variables and substrate reactivity. The relationship is determined by dihedral angles mostly, like shown in table 4. The only region which does not contribute to the model is the border of the binding pocket from Ser217 to Cys220. Variables from this region exhibit small variance only.

It was not possible to reduce the set of initial geometry variables to a smaller amount which enables both better interpretation and still high statistical significance. In our models, the distances d1, d2, d5 and d6 bear enough information to discuss sources of high substrate specifity. Hierarchical cluster analyses of these four distances yield clear separation of the three aromatic substrates, while the remaining structures are grouped unsystematically. The clustering is produced by a different behaviour of all four distances in the case of specific aromatic substrates (see table 5).

Relations to the reaction mechanism

We now have to explain the observed specific behaviour of aromatic substrates in terms of the reaction mechanism, which is presented in figure 11. Our current knowledge of the mechanism of deacylation yields at least three conditions necessary for this reaction.

At first, the contact between the carbonyl oxygen of the ester group to cleave and the NH-hydrogens of the protein backbones at Gly193/Ser195 is thought to yield a rate enhancement. This interaction is already present in the reactant state. It becomes stronger in the transition state involving water attack and proton transfer to His57. It reaches its maximum in the tetrahedral intermediate, because the negative charge of the carbonyl oxygen rises during the reaction. Our finding that the hydrogen bond distances d5 and d6 measured in the reactant state are related to reactivity, has to be interpreted by necessary changes of these distances through the reaction event, assuming that the found relationship is true. If it is valid, then elongation of d5 coupled with shortening of d6 occurs through water attack. The induction of these changes in the acylenzyme should yield a reactant state that is closer to the transition state. Experimental results can give some evidence for our hypothesis. In [22] it was shown by spectroscopic studies on adducts of chymotrypsin with several substrates, that one hydrogen bond to the carbonyl oxygen of specific substrates seems to be lost, whereas it remains present for unspecific substrates. Our model predicts d5 to be the broken hydrogen bond. It is the one which has an unfavourable geometry, because it requires formation of a ring with a high steric strain (see figure 2).

The second requirement for a successful deacylation reaction can be developed from the breakdown of the tetrahedral intermediate into carboxylate and reformed active site (see figure 11). During formation of the tetrahedral intermediate Nε2(His57) becomes protonated. This proton is thought to be transferred onto the Ser195 sidechain. Therefore a hydrogen bond must be formed between protonated His57 and acylated Ser195. It can only be formed, if the distance between Oγ(Ser195) and Nε2(His57) is small enough, as indicated by the d1 variable. The dihedral angle θ1 contributing to the PLS model may be responsible for this change.

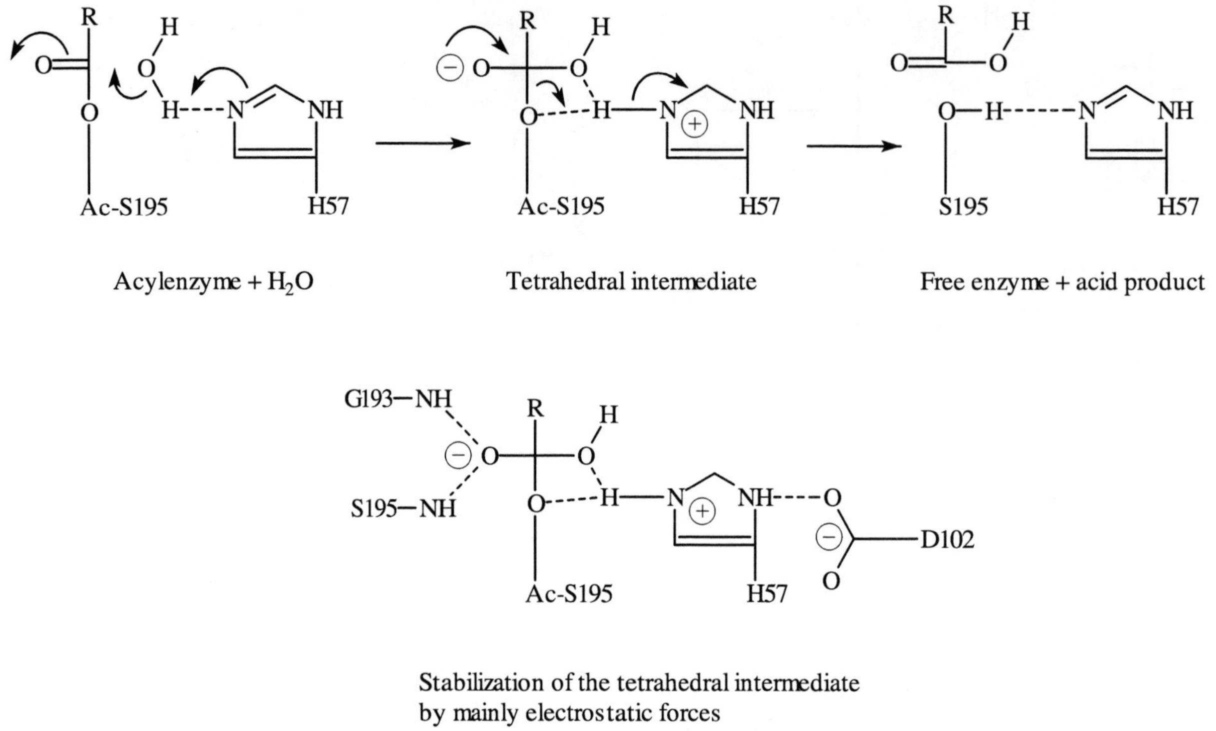

Figure 11. *Mechanism and catalysis of deacylation.*

The third statement has to include the His-Asp interaction, which is described by the d2 variable. There exists experimental evidence for the dependency of reaction rates on changes of His-Asp interactions. Kinetic studies in H_2O/D_2O mixtures enable one to measure the number of protons which can be exchanged between enzymatic residues and the solvent during a reaction. These measurements yielded the conclusion [23], that in the case of specific substrates two protons can be exchanged, whereas the reaction with unspecific substrates involves exchange of one proton only. This result was interpreted by a shortening of the His-Asp distance in the case of specific substrates. A second set of experiments was done to calculate the pKa of His57 by evaluations of pH-profiles of measured reaction rates [24,25]. It was shown that there exists a relationship between the ability of His57 to accept a proton from the solvent and the substrate structure. These results were never explained, but they should represent just another method to access features of the His-Asp dyad which are influenced by substrate binding. From this point of view our finding that the d2 variable is influenced in the case of specific substrates bears some truth. It is possible, that the interaction between Asp102 and the Ser214 sidechain is responsible for this change, because we built a hydrogen bond between these sidechains, and because the dihedral angles $\Psi214/\Phi214$ were shown to contribute to the derived PLS model (see table 4).

Summary

The modelling of various acylenzyme structures and the extraction of representative variables enabled us to develop hypotheses for relationships between events occurring during substrate immobilization and substrate reactivity. The obtained results represent a detailed access to the „induced strain" hypothesis. It was shown that interactions between the substrate and the active site in the binding pocket can influence the position of atoms involved in the following reaction event. The effect on structures of the reaction centre depends on substrate structure. In the case of specific substrates strong interactions lead to unique constellations of catalytic residues, which we related to possible geometry changes during the reaction. Except high reactive substrates, the obtained relationships are too complex to explain differences between low and very low reactivity by geometry variables.

In summary, is was surprising, that modelling experiments using force field simulations of reactant states yielded a hypothesis of the behaviour during the enzymatic reaction. This hypothesis is able to describe specific structure deformations as one possible source of different substrate reactivity by using QSAR techniques, which we applied to uncommon structure descriptors.

Supplementary Material Available

Tables of sets 1 and 4 as comma separated text files, including geometry variables, energies, energy differences and experimental reactivity data.

Bibliographic References

1. Fischer, E. *Ber. Dtsch. Chem. Ges.*, **1894**, *27*, 2985.
2. Kubinyi, H. *QSAR: Hansch Analysis and Related Approaches*. VCH: Weinheim 1993.
3. Pauling, L. *Nature*, **1948**, *161*, 707.
4. Menger, F. M. *Biochemistry*, **1992**, *31*, 5368.
5. Page, M. I.; Williams, A. *Enzyme Mechanisms*. Royal Society of Chemistry: London 1987.
6. DeTar, D. F. *J. Am. Chem. Soc.*, **1983**, *103*, 107.
7. Wipff, G.; Dearing, A.; Weiner, P. K.; Blaney, J. M.; Kollman, P. A. *J. Am. Chem. Soc.*, **1983**, *105*, 997.
8. Bemis, G. W.; Carlson-Golab, G.; Katzenellenbogen, J. A. *J. Am. Chem. Soc.*, **1992**, *114*, 570.
9. Dixon, M. M.; Matthews, B. W. *Biochemistry*, **1989**, *28*, 7033.
10. Bernstein, F. C.; Koetzle, T. F.; Williams, G. J. B.; Meyer, E. F.; Brice, M. D.; Rodgers, J. R.; Kennard, O.; Shimanouchi, T., Tasumi, M. *J. Mol. Biol.*, **1977**, *112*, 535.
11. Abola, E. E.; Bernstein, F. C.; Bryant, S. H.; Koetzle, T. F.; Weng, J. In: *Crystallographic Databases - Information Content, Software Systems, Scientific Applications*; Allen, F. H.; Bergerhoff, G.; Sievers, R. (Eds.); Data Commission of the International Union of Crystallography: Bonn Cambridge Chester 1987; p. 107.
12. TRIPOS Associates, Inc.
13. Weiner, S. J.; Kollman, P. A.; Case, D. A.; Chandra Singh, U.; Ghio, C.; Alagona, G.; Profeta, S., Weiner, P. *J. Am. Chem. Soc*, **1984**, *106*, 765.
14. Dewar, M. J. S.; Zoebisch, E. G.; Healy, E. F., Stewart, J. J. P. *J. Am. Chem. Soc.*, **1985**, *107*, 3902.
15. Kallies, B.; Mitzner, R. *J. Chem. Soc., Perkin Trans. 2*, **1996**, in the press (5/08360E).
16. Atkins, P. W. *Physikalische Chemie*. 1st ed.; VCH: Weinheim 1988.
17. Hansch, C.; Grieco, C.; Silipo, C., Vittoria, A. *J. Med. Chem.*, **1977**, *20*, 1420.
18. Warshel, A. *Computer Modeling of Chemical Reactions in Enzymes and Solutions*. Wiley: New York 1991.
19. Dive, G.; Dehareng, D.; Ghuysen, J. M. *J. Am. Chem. Soc.*, **1994**, *116*, 2548.
20. Schellenberger, V.; Braune, K.; Hoffmann, H. J., Jakubke, H. D. *Eur. J. Biochem.*, **1991**, *199*, 623.
21. Zerner, B., Bender, M. L. *J. Am. Chem. Soc.*, **1964**, *86*, 3669.
22. Whiting, A. K., Peticolas, W. L. *Biochemistry*, **1994**, *33*, 552.
23. Elrod, J. P.; Hogg, J. L.; Quinn, D. M.; Venkatasubban, K. S., Schowen, R. L. *J. Am. Chem. Soc.*, **1980**, *102*, 3917.
24. Hirohara, H.; Philipp, M., Bender, M. L. *Biochemistry*, **1977**, *16*, 1573.
25. Béchet, J. J.; Dupaix, A., Roucous, C. *Biochemistry*, **1973**, *12*, 2566.

J. Mol. Model. **1996**, 2, 160 – 174

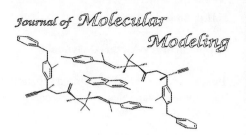

Using Molecular Dynamics to Predict Factors Affecting Binding Strength and Magnetic Relaxivity of MRI Contrast Agents

Yen T. Tan, Richard S. Judson*, and Carl F. Melius

Center for Computational Engineering, MS 9214, Sandia National Laboratories, Livermore, CA 94551-0969, USA
(rsjuds@ca.sandia.gov)

John Toner and Gang Wu

Nycomed R&D, Inc., 466 Devon Park Dr., Wayne, PA 19087-8630, USA

Received: 25 March 96 / Accepted: 29 May 1996 / Published: 13 June 1996

Abstract

We demonstrate the use of molecular dynamics and molecular mechanics methods to calculate properties and behavior of metal-chelate complexes that can be used as MRI contrast agents. Static and dynamic properties of several known agents were calculated and compared with experiment. We calculated the static properties such as the q-values (number of inner shell waters) and binding distances of chelate atoms to the metal ion for a set of chelates with known X-ray structure. The dynamic flexibility of the chelate arms was also calculated. These computations were extended to a series of exploratory chelate structures in order to estimate their potential as MRI contrast agents. We have also calculated for the first time the NMR relaxivity of an MRI contrast agent using a long (5 nsec) molecular dynamics simulation. Our predictions are promising enough that the method should prove useful for evaluating novel candidate compounds before they are synthesized. One novel static property, the projected area of chelate atoms onto a virtual surface centered on the metal ion (gnomonic projection), was found to give an effective measure of how well the chelate atoms use the free space around the metal ion.

Keywords: Diagnostic imaging, contrast reagents, molecular modeling, chelate, molecular flexibility

Introduction

Magnetic resonance imaging (MRI) has become one of the standard diagnostic tools used by physicians. However, the time-consuming nature of this procedure and the related expense make it important to extract the maximum amount of useful information from each scan. Paramagnetic imaging agents[1] are heavily used today to enhance the contrast in MRI scans in specific ways and hence to increase their infor-

mation content. Typical agents are comprised of a paramagnetic ion (typically Mn^{+2} or Gd^{+3}) bound by a chelate, which is itself perhaps bound to a protein or other polymer. The unpaired electron spins on the paramagnetic ion create a local magnetic field that couples to the nuclear spin on nearby water protons, which greatly increases their relaxation rate. MRI scans are typically run near saturation, so quicker relaxation enables more power to be absorbed in the regions of the body containing significant concentrations of the agent, which in turn increases the image contrast.

Successful new MRI contrast agents need to meet at least two design criteria: (1) They have to provide significant contrast enhancement, which translates into a requirement that $1/T_1$ for solutions of the compound be large. (2) They have to be stable with respect to dissociation of the ion-chelate complex. This is due to the high toxicity that is typical of both the free ion and the free chelate. These chelators must bind strongly to the Gd ion even in the presence of other free cations that are normally present in the body. The paramagnetic properties of these agents have been the subject of several recent reviews.[1-5]

In this paper, we demonstrate the use of molecular mechanics and dynamics methods to evaluate chelate structure and function. One factor that is important for the stability of a compound is the dynamic behavior of the chelate arms. What appears from a static structure to be a good binder may in fact be dynamically very mobile. To this end, we give results of several molecular dynamics simulations of chelates in water. These simulations were performed on both known compounds and on several proposed compounds with different binding characteristics. We concentrate on carboxyl groups that can bind in either the bidentate or monodentate conformation to the paramagnetic ion, and examine the structural effect of having two binding oxygen atoms that are equidistant from the ion, vs. a lone binding oxygen atom. An important result is that monodentate carboxyl groups are held more rigidly than are the corresponding bidentate ligands.

Next, we describe a procedure for calculating relaxivities for typical paramagnetic imaging agents and show that reasonably accurate predictions are possible. Relaxivities $(1/T_1)$ are a function of the time-dependent positions of water protons relative to the paramagnetic ion.[6] The procedure involves computing long (5 nsec) trajectories of the chelate-metal compound in a water bath and then calculating certain well known correlation functions of the proton-ion vectors. From approximate analytic models,[6,7] it is known that important factors determining $1/T_1$ include the rotational correlation time of the complex, the rate at which waters exchange between the inner and outer solvation spheres, and the number of waters in the inner shell of the ion. This makes it important to accurately model the dynamical interactions of the metal ion, the chelate, and nearby waters. We include electronic effects with an approximate analytic model.

Finally, we look at the static geometric properties of several chelates based on the recognition that there is a limit to the number of chelate groups that can be employed to bind a metal ion. The number of binding sites plus the number of water molecules in the inner solvation sphere of the ion, denoted by q_{tot}, equals the fixed coordination number of the ion, which is between 6 and 10, depending on the ion. For a given value of q_{tot}, and hence a fixed number of coordinating chelate groups, the binding strength is at least partially determined by how the molecule uses the space around the metal ion. The arms connecting the chelating groups need to be flexible enough to allow all of the groups to be optimally placed around the ion at the proper ion-to-chelate atom dis-

tance. An effective chelate structure should be capable of maximizing the attraction between the positive ion and negative chelating groups while minimizing the repulsive interactions between different chelating groups. For instance, a chelate with a rigid backbone could leave gaps in the coverage of the surface of the metal ion, decreasing the potential binding energy while simultaneously forcing negatively charged groups to be relatively close together. We demonstrate a method to evaluate the effectiveness of a number of common chelating agents whose 3-dimensional structures are known from X-ray data, and show that commonly used MRI compounds make maximal use of the available surface area of the ion.

Several other groups have used molecular mechanics methods to study metal ion chelate complexes. Hancock and co-workers[8-10] have reported extensive molecular mechanics calculations of ion-chelate systems, concentrating on the concept of steric strain. The strain energy is the sum of all bond length, bond angle, torsional distortions, and van der

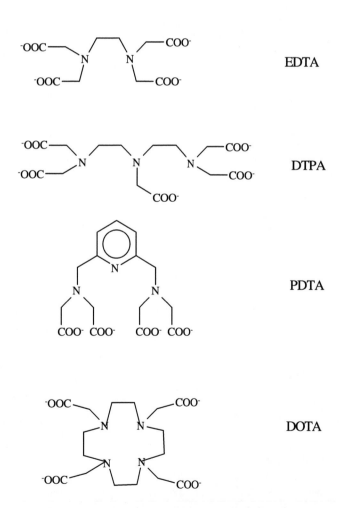

Figure 1. *Chelate structures of MRI contrast agents analyzed in the molecular dynamics simulations. The first panel shows existing compounds. The second panel (following page) shows proposed derivatives of PDTA also examined here.*

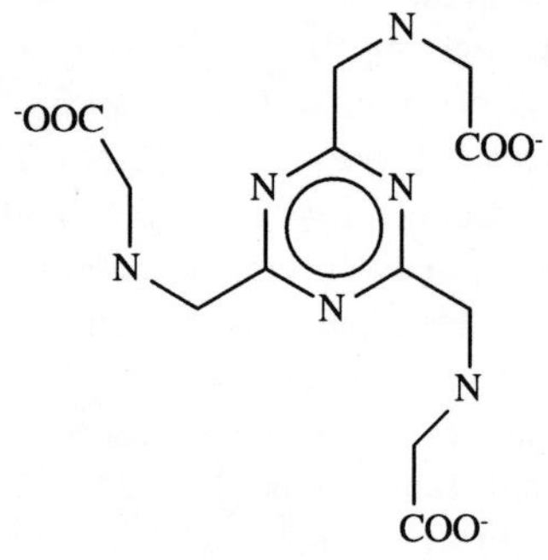

J. Mol. Model. **1996**, *2*

P3A

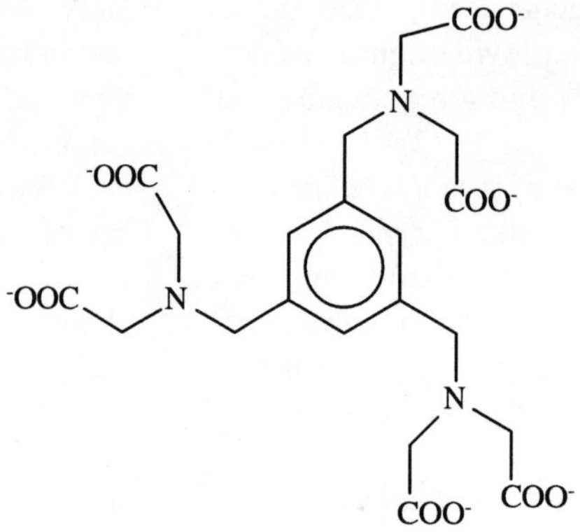

P6A

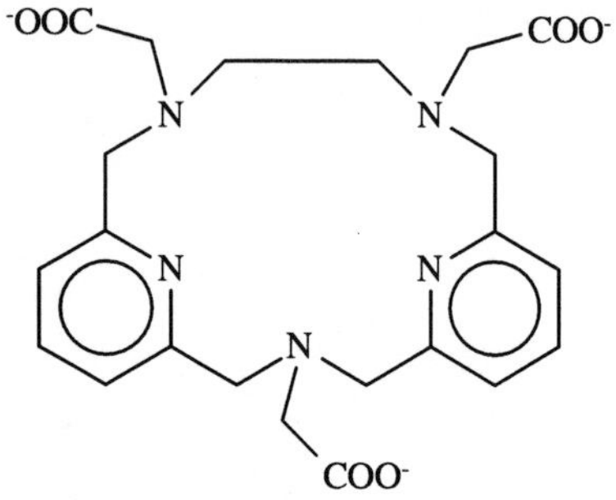

2P3A

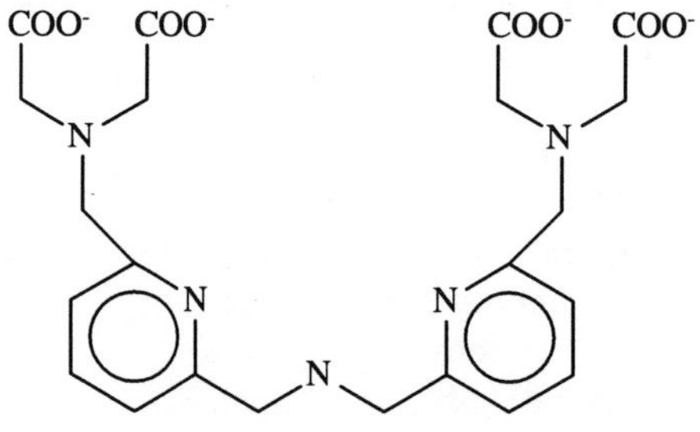

2P4A

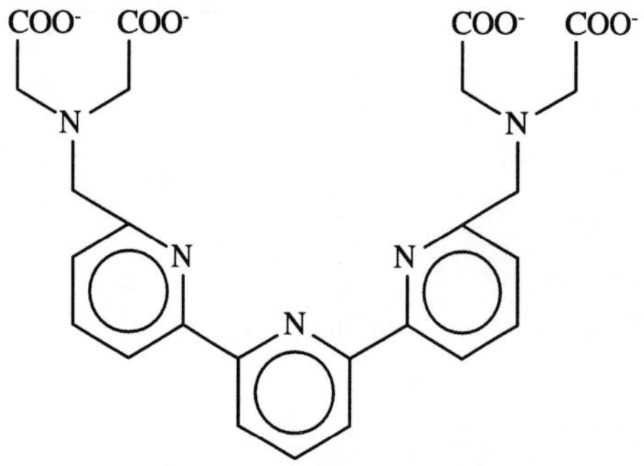

3P4A

Waals interactions in the molecule. Complexes with high strain energy are less likely to form than those with less strain energy. Fossheim, et al.[11,12] performed extensive molecular mechanics and dynamics simulations on lanthanide ion-chelate complexes. Their aim was to study the structure and energetics of free ligands and the corresponding ligand ion complexes, including the important effects of solvation. Among other interesting findings, they report strong correla-tions between molecular mechanics interaction energies and experimental binding constants. They also investigated the dynamics of complexation, i.e. the process whereby the ligand captures an ion in solution. Kumar and Tweedle[13] report similar calculations on a series of cyclic polyamino com-plexes. They find a good correlation between ligand strain energy and the energy of reorganization from the proposed final intermediate (one carboxyl group protonated) to the fully formed, deprotonated ligand-ion complex. Frey and co-workers[14] compare the results of molecular mechanics calculations with luminescence spectroscopic measurements of Eu(III)-ligand complexes. They use molecular mechanics calculations to both determine low energy conformations and to help interpret the spectroscopic results. In particular, they get good agreement for the number of inner shell waters. Another interesting result is that the excitation wavelength for the Eu ion correlates well with the sum of partial atomic charges of the chelating atoms. Hay,[15,16] Rappe, et al.[17] and Badertscher, et al.[18] have developed special purpose molecular mechanics force fields specifically tailored for use with metal-chelate complexes. Hay has published a review of work in this area.[16] With the exception of the work by Fossheim, et al.,[11,12] all of these papers concentrate on static properties of the metal-chelate complexes, while we concentrate on the dynamic properties of these complexes. The calculations we report are also quite different from those in the Fossheim papers.

For the studies in this paper, we have selected a group of chelating agents that contain different numbers of carboxyl and amine binding groups. This group includes two compounds for which much experimental data are available:[5,12,19,20] EDTA **[1]** (ethylene-diamine-tetraacetic acid), and DTPA **[2]** (diethylene-triamine-pentaacetic acid). In addition, calculations were also conducted for DOTA **[3]** (cyclododecane-tetraacetic acid)[21] and a new compound, PDTA **[4]** (pyridine-diamine-tetraacetic acid)[22] and its derivatives. The structures of the chelates used for this study are shown in Figure 1.

X-ray structure data, stability constants and MRI relaxivities are available for DTPA, EDTA, DOTA, and PDTA.[21-27] DTPA has five carboxyl groups and three amine nitrogens while EDTA and PDTA both have four carboxyl groups and two amine nitrogens. DOTA is a cyclic compound with four nitrogens and four carboxyl groups. The thermodynamic stability constants for DTPA and DOTA (Gd(chelate)$\rightarrow$ Gd + chelate) are considerably higher than that of EDTA, as shown in Table 1. [5,26-28] However, EDTA has a higher q-value than DTPA and DOTA and hence a higher MRI relaxivity, $1/T_1$ (see Table 1). Thus EDTA would be more effective as an MRI contrast agent if it were not for the toxic effects arising from its lower stability constant and the resulting release of free Gd ions in the body.

Dynamic Behavior of Chelates

In this section, we describe molecular dynamics calculations performed on a series of chelates to characterize their dynamic behavior of the binding groups. In particular, we examine the difference in dynamic behavior between chelates with mono and bidentate carboxyl groups. We calculate mean distances from the Gd ion to chelating atoms, as well as the standard deviation of this distance. The standard deviation provide a measure of the vibrational flexibility of these in-

Table 1. *Experimental properties of several Gd chelates*

Compound	log K [a]	q [b]	$1/T_1$ (20 Mhz) [c]
Gd^{+3}	NA	9	16.1
Gd(EDTA)$^-$	17.4	3	7.6
Gd(DTPA)$^{2-}$	22.5	1	4.8
Gd(DOTA)$^-$	24.7	1	4.7
Gd(PDTA)$^-$	18.6	2	6.3

[a] *Log K is the thermodynamic stability constant for Gd(Chelate) « Gd + Chelate.[1]*
[b] *q is the number of inner shell waters.*
[c] *Relaxivity at 20 MHz and 25 C.[1]*

teractions. As we will show, several of the exploratory compounds assume a bidentate conformation relative to the metal ion, and their carboxyl groups exhibit great mobility.

A. Molecular Mechanics Details

Molecular dynamics calculations were conducted on EDTA **[1]**, DTPA **[2]**, DOTA **[3]**, and PDTA,**[4]** plus its derivatives. Each molecule was solvated using the SPC/E water model of Berendsen.[29] For most molecules, a 12 Å water sphere was used for solvation and the proper water density was maintained by employing a 13.5 Å repulsive spherical shell having a harmonic force constant of 100. The SETTLE algorithm[30] was used to constrain the waters in their equilibrium conformation. A variety of cutoff schemes were tried, but most failed to adequately treat the high electrostatic forces arising from the Gd^{+3}. The method developed by Levitt was selected for use in these computations.[31-33] An atom-based, non-bonded list is used. The original Coulomb(E_q^0) and van der Waals(E_{VDW}^0) terms are replaced by the following:

$$E_q = E_q^0 \times \left(1 - 2(r/r_c) + (r/r_c)^2\right) \quad (1)$$

and

$$E_{VDW}(r) = E_{VDW}^0(r) - E_{VDW}^0(r_c) - (\partial E_{VDW}^0 / \partial r)(r_c) \times (r - r_c) \quad (2)$$

for $r < r_c$. Both terms are set to zero for $r \geq r_c$, where the value of r_c used is 7.5 Å. Atom pairs are included in the non-bonded list if they are closer than 9.0 Å. The non-bonded list was updated every 10 time steps or 0.01 ps. Berendsen's temperature rescaling method[29] was used with a time constant of 0.1 psec. All runs were performed at 300 K using CCEMD,[34] a general purpose molecular dynamics program based on the MD code of Windemuth and Schulten.[35] We use the CHARMm force field[36] with QUANTA3.3 [37] parameters for the chelates.

Chelate charges were determined from SCF calculations on chelate fragments. Charges for the solvated species were then calculated by using a boundary element method to match the SCF wavefunction to a continuum solvent.[38] The calculations were performed using a modified version of Gaussian 92.[39] Geometry optimizations of chelate fragments binding to the Ca ion were first performed using the 6-31g* basis set. A self-consistent solvation energy calculation was then performed to determine solvation energies and charges. The 6-31g** basis set was used for these calculations. Charges were averaged to equal -0.65 for carboxylate oxygens, to +0.45 for carboxylate carbons, to -0.50 for backbone nitrogens, and to -0.40 for ring nitrogens. The same charge and atom type were used for both of the two carboxylate oxygens, regardless of their conformation relative to the metal ion. Small adjustments in charges, 0.05 or less, were sometimes made to carbons and hydrogens to assure charge neutrality for the metal-chelate complex. (All input files used for these computations, which include charges, atom types, and force field parameters, can be obtained from the authors upon request.)

We attempted to adjust the van der Waals parameters for the ligand and metal atoms in order to have all ligand-metal distances assume their crystal values. However, these distances also depend on the bond angles and distances within the chelate as it wraps around the metal, and without changing these parameters significantly the metal-ligand distance requirements could not be satisfied. In the end we left all force field parameters at their default values, except for the Gd van der Waals parameters, which were adjusted to give the proper Gd-O distances. The final values are 2.0 • for the van der Waals radius and 0.026 kcal/mol for the well depth. Using these parameters, energy minimization of the chelate molecules resulted in Gd-O distances that were very close to those found in the crystal structures of EDTA and DTPA.[21,23-25] The Gd-N distances are too small. After equilibration (discussed in the next paragraph) we recorded the Gd-O and Gd-N distances from these chelates and compared them with the corresponding crystal structure values. The Gd-O distances for the model and the crystal are (2.40,2.42), (2.40,2.38), and (2.30,2.37) for EDTA, DTPA, and DOTA respectively. The corresponding Gd-N distances are (2.45,2.65), (2.61,2.70), and (2.62,2.68). We should note that other authors achieved better Gd-N distances for these and similar chelates using different force field parameterizations.[11,12,14]

The self-diffusion coefficient for water in the simulations had a value of 2.2×10^{-5} cm²/sec which compares well with the quoted experimental value of 2.4×10^{-5} cm²/sec.[29] Prior to the data-gathering runs, the ensembles were equilibrated by annealing to 300 K over an interval of 12 ps with a temperature window of 40 K. Using CCEMD, molecular dynamics runs were conducted for a period of 100 ps to assure adequate statistics for the Gd-ligand non-bonded distances.

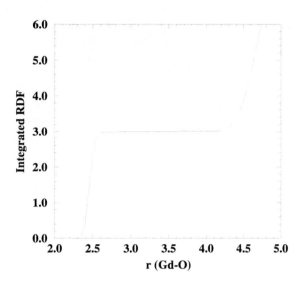

Figure 2. *Integrated radial distribution function for water is plotted as a function of the Gd - water oxygen distance. The inner shell of Gd in Gd-EDTA is shown with three waters (q = 3.0).*

B. Results

From molecular dynamics runs, it is possible to obtain information on the structure of the chelate as well as the distance between specific binding atoms in the chelate and the Gd ion. The MD runs were of 100 psec duration. Inter-atomic distances were collected every 0.1 psec. For the reference compounds, EDTA [1] and DTPA [2], as well as DOTA [3], and PDTA [4], the carboxylate groups were all in the monodentate conformation, with one carboxyl oxygen close to the Gd ion and the other displaced away from it. In EDTA, the near oxygen is at an average distance of 2.41 Å while the far oxygen is at 4.32 Å. From an average of the X-ray data, the corresponding distances are 2.40 Å and 4.48 Å respectively.[21,23,24] For DTPA, the average distance values for the near oxygen is 2.37 Å, which is almost identical to the X-ray distance of 2.38 Å. For DOTA, the near oxygen is 2.37 Å from MD simulations and from X-ray data. The average distance of binding nitrogens from the metal ion in EDTA is 2.41 Å vs. 2.65 in the crystal. For DTPA, the corresponding distances are 2.58 and 2.70. For DOTA, the average MD and crystal distances are 2.59 and 2.68 Å. Although the calculated Gd-N distances are too short, we do not believe that a better parameterized model would significantly alter any of our qualitative conclusions.

During the course of an MD run, the radial distribution function of water molecules around the metal ion was monitored. For bare Gd in water, the calculated average Gd - O$_{water}$ distance equals 2.44 Å. The integrated radial distribution func-

Table 2. *Dynamics data on known and proposed MRI contrast agents.*

Compound	Carboxyl [a]	Amide [b]	q [c]	q_{tot} [d]	$<d_O>$ [e]	$<\sigma_O>$ [f]	$<d_N>$ [g]	$<\sigma_N>$ [h]
EDTA	4-mono	2/2	3.0	9.0	2.41 / 4.32	0.06 / 0.10	2.41	0.05
DTPA	5-mono	3/3	1.0	9.0	2.37 / 4.34	0.06 / 0.10	2.58	0.09
PDTA	4-mono	3/3	2.0	9.0	2.38 / 4.29	0.05 / 0.11	2.56	0.09
DOTA	4-mono	4/4	1.0	9.0	2.37 / 4.32	0.06 / 0.11	2.59	0.09
P3A	1-mono	1/5	2.9	8.9	2.39 / 4.4	0.06 / 0.11	2.62	0.14
	2-bi				2.44 / 2.7	0.08 / 0.59		
P6A	4-mono	0/3	3.0	9.0	2.42 / 4.0	0.08 / 0.22	4.1-5.3	0.16
	1-bi				2.45 / 2.49	0.07 / 0.08		
2P3A	3-mono	5/5	0.9	9.0	2.37 / 4.3	0.04 / 0.22	2.50	0.07
2P4A	4-mono	5/5	0.0	9.0	2.39 / 4.3	0.06 / 0.10	2.55	0.09
3P4A	4-mono	5/5	0.2	9.2	2.39 / 4.3	0.07 / 0.11	2.58	0.10

[a] *Number of mono and bi-dentate carboxyl groups binding to the Gd.*

[b] *Number of nitrogens binding to the Gd / total number of nitrogens.*

[c] *Dynamical value of q. This is the time-averaged number of waters in the inner shell.*

[d] *Total dynamically calculated coordination number of the Gd.*

[e] *Average distance of the carboxyl oxygens from the Gd (near/far).*

[f] *Dynamics variation of d_{Gd-O} (near/far).*

[g] *Average distance of the nitrogens from the Gd.*

[h] *Dynamics variation of d_{Gd-N}*

tion for water molecules surrounding Gd-EDTA is shown in Figure 2. From the plateau of the curve, it is observed that there are three water molecules in the first shell that are directly coordinated to Gd^{+3} and hence $q=3$ (here q denotes the number of first shell waters only) for EDTA. For DTPA, the corresponding q-value obtained from MD runs is $q=1$. The MD results are summarized in Table 2. As expected, EDTA, with four binding carboxyl oxygens and two nitrogens has a q of 3.0, while DTPA, with five binding oxygens and three nitrogens has a q of 1.0.[19] Non-integer values are sometimes obtained when waters exchange in and out of the inner sphere and spend a small portion of the observed time in the bulk solvent. Results for the other molecules are also given in Table 2. PDTA, with four binding carboxyl oxygens and three nitrogens, has a q of 2.0, while DOTA with four oxygens and four nitrogens has a q of 1.0. MD simulations for the reference molecules all show Gd with a coordination of nine and binding distances that are in good agreement with X-ray data. Note that the most effective MRI contrast agents, DTPA and DOTA, both have a q-value of 1 and have all of their carboxylate groups in the monodentate conformation.

We ran simulations of several analogues of PDTA [4], some of which were anticipated to show bidentate behavior.

These molecules have structures that are similar to the PDTA molecule, but have varying numbers of nitrogens and carboxyl groups attached to the pyridine ring(s) (Figure 1). When the number of binding carboxyl and amide groups drops below six (the number for EDTA), sufficient space becomes available around the metal ion for one or more of the carboxyl groups to assume a bidentate configuration, in which the two oxygen atoms are approximately equi-distant from the Gd ion. As shown in Table 2, the bidentate conformation is observed in the calculated results for P3A [5] and P6A [6]. (In our nomenclature, the number preceding the "P" indicates the number of pyridine rings in the molecule, is omitted when that number is 1, and the number preceding the "A" indicates the number of acetic acid groups in the molecule.) In the structures where both monodentate and bidentate conformations occur, the Gd-O distance is calculated to be larger for the bidentate carboxyls, 2.44 Å vs. 2.39 Å in P3A [5], and 2.45 Å vs. 2.42 Å in P6A [6]. This trend is in agreement with X-ray data (discussed in Section IV) that show a distance of 2.46 Å for a structure having bidentate carboxyl oxygens and 2.38 Å for the average monodentate oxygen-metal distance (for EDTA, DTPA, and DOTA). The resulting small increase in the Gd-O distance could decrease the electrostatic component of the interaction energy. Due to ion-induced polarization effects between the carboxyl oxygens,[38] the electrostatic binding strength of the two oxygens in a bidentate conformation to Gd is predicted to be weaker than that from two monodentate oxygens that are parts of two separate carboxyl groups. Thus the monodentate conformation should be stronger in terms of binding per carboxyl oxygen, even in the absence of increased Gd-O distance for the bidentate conformation.

The MD results for the average Gd-O and Gd-N distance and the accompanying dynamic variation values, σ_o, give an indication of the extent of motion of the chelate atom relative to the ion, as shown in Table 2. A large value of s_o implies large flexibility, and conversely, a small σ_o implies little flexibility. When the carboxyl group is in the monodentate conformation, the binding Gd-O distance varies between 2.37 and 2.42 Å with a σ_o of about 0.06 Å. The far oxygen distance is about 4.3 Å and its σ_o is significantly larger (0.11 - 0.22) showing that the outer carboxyl moves more freely than the inner one. In the monodentate conformation, the carboxyl group essentially pivots about the near oxygen. In the bidentate conformation, the size of the σ_o value depends on whether the two oxygen atoms are equidistant from the metal ion or if one is closer than the other (P6A [6] vs. P3A [5]). When the bidentate conformation is imperfect, as in P3A, the oxygen slightly further from the metal ion will have a larger σ_o value and is hence more flexible. This gives an indication of the fluidity of motion found with bidentate oxygens.

Of the exploratory molecules examined, all five structures were able to achieve a total coordination number of 9. For the molecules containing multiple pyridine rings (2P3A [7], 2P4A [8], and 3P4A [9]), all of their carboxyl groups take on a monodentate conformation. These molecules have more binding nitrogens than carboxylate groups, with 5 binding nitrogens and 3 to 4 monodentate oxygens. This differs from the reference chelates that have either more binding carboxyl oxygens than binding nitrogens (EDTA, DTPA, PDTA) or an equal number (DOTA). DTPA, for instance, has three nitrogens and five monodentate oxygens. Of the exploratory structures, 2P3A [7] has a q-value close to one, a characteristic of both DTPA and DOTA, while 2P4A [8] and 3P4A [9] have a q-value close to zero. If $q=1$ is essential for sufficient contrast enhancement of MRI images, 2P3A [7] should be a good candidate. Its usefulness as a potential MRI contrast agent would then depend on whether nitrogens are as effective as carboxyl oxygens in binding to the Gd. Two exploratory molecules, P3A [5] and P6A [6], contained carboxyl groups in the bidentate conformation and a q of 3. When the chelate binding atoms are taken into consideration, a q_{tot} of 9 is attained. These simulations have thus shown that a Gd coordination number of 9 is obtained, both for the reference compounds and the exploratory structures.

Predicting Relaxivity

The relaxivity of a paramagnetic ion complex arises from the interactions between rotational and spin degrees of freedom. In this section, we describe a model of relaxivity that combines a first principles description of the rotational contribution to the relaxation which can be described classically, coupled with a phenomenological description of the purely quantum mechanical electron spin resonance contribution.

A. Theory

The theory of relaxation of protons in the presence of a paramagnetic ion is well understood and is treated in depth in a number of texts.[6,7,40] The basic mechanism for relaxation is that as protons move in the vicinity of a paramagnetic ion,

they feel a time-varying magnetic field that helps couple the proton magnetization to the thermal bath. The most important coupling mechanisms between the proton spin and the paramagnetic electron spin are dipole-dipole and contact coupling. The contact term arises when the proton penetrates the outer electron shell of the ion. In this paper, we only treat S state ions such as Gd^{+3} and Mn^{+2}, for which the contact term can be neglected. The relaxation rate for dipole-dipole coupling, $1/T_1$ equals

$$1/T_1 = \gamma_I^2 \, \gamma_S^2 \, \hbar^2 \, S(S+1) \left(\tfrac{1}{12} J^{(0)}(\omega_I - \omega_S) + \right.$$
$$\left. + \tfrac{3}{2} J^{(1)}(\omega_I) + \tfrac{3}{4} J^{(2)}(\omega_I + \omega_S) \right) \tag{3}$$

where $I=1/2$ is the proton spin, S is the electronic spin of the paramagnetic ion, γ_I and γ_S are the gyromagnetic ratios ω_I and ω_S are the Larmour frequencies for the proton and electron spins, respectively. The spectral densities $J^{(0)}$, $J^{(1)}$ and $J^{(2)}$ are Fourier transforms of certain time correlation functions:

$$J^{(i)}(\omega) = \int_{-\infty}^{\infty} dt \; e^{-i\omega t} \, g^{(i)}(t) \tag{4}$$

The time correlation functions are:

$$g^{(i)}(t) = \overline{\left\langle f^{(i)}(t_0) \; f^{(i)}(t+t_0) \right\rangle} \tag{5}$$

where the brackets indicate an ensemble average over all water protons and the bar an average over time origins t_0. The functions f are given by:

$$f^{(0)}(t) = \left(1 - 3\cos^2\theta\right)/r^3$$

$$f^{(1)}(t) = \sin\theta \; \cos\theta \; e^{-i\varphi} / r^3 \tag{6}$$

$$f^{(2)}(t) = \sin^2\theta \; e^{-2i\varphi} / r^3$$

whose angular parts are simply unnormalized spherical harmonics, $Y_{2m}(\theta,\varphi)$. The paramagnetic ion is placed at the origin of the laboratory coordinate system; the coordinates (r,θ,φ) give the position of a particular proton relative to the ion. The angular and $1/r^3$ factors arise from the radial part of the dipole-dipole Hamiltonian. Already we can see that the rate of relaxation will be a sensitive function of the dynamics of water protons moving in the vicinity of the ion because of terms of the order of $1/r(t)^6$.

Approximate analytic models of the relaxivity can be derived starting from the assumption that each of the correlation functions $g^{(i)}$ will decay exponentially:

$$g^{(i)}(t) = g_0^{(i)} \exp[-t / \tau^{(i)}] \tag{7}$$

One can then perform the Fourier transforms to arrive at the expression

$$1/T_1 = \gamma_I^2 \gamma_S^2 \hbar^2 S(S+1) \left(\frac{1}{12} \frac{g_0^{(0)} \tau^{(0)}}{1+(\omega_I-\omega_S)^2 \tau^{(0)2}} + \right.$$
$$\left. + \frac{3}{2} \frac{g_0^{(1)} \tau^{(1)}}{1+\omega_I^2 \tau^{(1)2}} + \frac{3}{4} \frac{g_0^{(2)} \tau^{(2)}}{1+(\omega_I+\omega_S)^2 \tau^{(2)2}} \right) \tag{8}$$

This can be further simplified by recognizing that the electron Larmour frequency, ω_s, is give by $\omega_s = (\gamma_S/\gamma_I) \times \omega_I$ where (γ_S/γ_I) is 658. Therefore in Eq. 8, $(\omega_s \pm \omega_I)^2 \approx \omega_s^2$. Using this relationship together with the approximations, $\tau^{(0)} = \tau^{(1)} = \tau^{(2)} \equiv \tau_c$, and $\omega_I \tau_c \ll 1$, we arrive at:

$$1/T_1 = \gamma_I^2 \gamma_S^2 \hbar^2 S(S+1) \cdot$$
$$\cdot \left(\left[g_0^{(0)} \Big/ 12 + 3 g_0^{(2)} \Big/ 4 \right] \frac{\tau_c}{1+\omega_s^2 \tau_c^2} + \left[3 g_0^{(1)} \Big/ 2 \right] \tau_c \right) \tag{9}$$

For the special case of a bare ion in water, assuming that on average protons are uniformly distributed about the ion, we can calculate the ratios $g_0^{(0)} : g_0^{(1)} : g_0^{(2)}$ that are 6:1:4. These arise from calculating the integrals

$$r^3 \int_0^\pi d\theta \sin\theta \left| f^{(i)}(r,\theta,\varphi) \right|^2 \tag{10}$$

This leads to the final approximation:

$$1/T_1 = \gamma_I^2 \gamma_S^2 \hbar^2 S(S+1) g_0 \left(\frac{7}{2} \frac{\tau_c}{1+\omega_s^2 \tau_c^2} + \frac{3}{2} \tau_c \right) \tag{11}$$

where g_0 is a constant. So for the bare paramagnetic ion in water, we expect to see a plateau at low frequency which drops off to a second plateau at $\omega \approx 1/\tau_c$. The ratio of the heights of the two plateaus should be 10/3. A further approximation can be included to model the relaxivity due to chelated ions by adding, to Eqs. 9 or 11, a factor which is the fraction of coordination sites open to waters. This accounts for the reduced access of protons to the inner solvation sphere of the ion, where a large portion of the relaxation occurs.

This factor is denoted by q/q_0, where q_0 is 9 in the case of Gd^{+3}. Scaling by this factor is not entirely accurate because outer shell waters also account for a small but significant amount of relaxivity. For a $q=1$ compound, such as DTPA, the outer shell relaxivity accounts for about half of the total.

So far, we have only treated the relaxation of the proton spins and have neglected the simultaneous electron spin relaxation (e.s.r.) of the paramagnetic ion. A first principles treatment of this e.s.r process is difficult, although it has been carried out for several special cases using a variety of approximations for the perturbations due to the motion of nearby nuclear spins.[6,7,40-43] Here we include the effect in an approximate way using the Solomon-Bloembergen-Morgen (SBM) theory.[44,45] Our starting point is Eq. 7 where SBM theory prescribes that the correlation time $\tau^{(i)}$ consists of three components,

$$\frac{1}{\tau^{(i)}} = \frac{1}{\tau_R^{(i)}} + \frac{1}{\tau_M} + \frac{1}{T_e^{(i)}(\omega_S)} \tag{12}$$

where $\tau_R^{(i)}$ is a rotational correlation time, τ_M is the mean lifetime of water protons in the inner shell of the paramagnetic ion, and $T_e^{(i)}$ is the e.s.r correlation time. In SBM theory, the first two correlation times account in a phenomenological manner for the detailed dynamics we directly calculate. The relaxation behavior associated with those terms is built into our numerically calculated correlation functions, $g^{(i)}(t)$. The additional additive term in Eq. (12) can be approximately included into $g^{(i)}(t)$ by multiplying by an exponential term. The correlation function including the e.s.r component is given by[40]

$$\bar{g}^{(i)}(t) = g^{(i)}(t) \exp[-t / T_e^{(i)}(\omega_S)] \tag{13}$$

The functions $\bar{g}^{(i)}(t)$ are then substituted into Eq. 4 to calculate the complete spectral densities. An important property of $T_e^{(i)}$ is its frequency dependence which is given by

$$\frac{1}{T_e^{(i)}} = \frac{1}{5 \tau_{SO}} \left(\frac{1}{1+\omega_s^2 \tau_v^2} + \frac{4}{1+4\omega_s^2 \tau_v^2} \right) \tag{14}$$

for $i=1$ and

$$\frac{1}{T_e^{(i)}} = \frac{1}{5 \tau_{SO}} \left(1.5 + \frac{2.5}{1+\omega_s^2 \tau_v^2} + \frac{1}{1+4\omega_s^2 \tau_v^2} \right) \tag{15}$$

for $i=0$ and 2. The parameters τ_{SO} and τ_v can be determined experimentally by fitting relaxivities to the full SBM equations.[46] Here, we make the approximation that these parameters will change little from one chelate to the next and use values determined for aqueous Gd^{+3} as a universal set. In

J. Mol. Model. **1996,** *2*

point of fact, the addition of this electronic effect makes minor changes in the predicted relaxivities for small, freely rotating chelates. The values used are $\tau_{SO}=132$ psec and $\tau_v=16$ psec.[46] These values were determined for 25 C. Correcting the values of the parameters for the presence of the chelate will change our numerically determined relaxivities slightly, but will not affect any of our qualitative conclusions.

Before giving the numerical results, we will summarize the computational procedure. A long (5 nsec) molecular dynamics run is performed. From the saved coordinates, the values of $f^{(i)}(t)$ (Eq. 6) are calculated and used in Eq. 5 to give the raw correlation functions $g^{(i)}(t)$. The spectral densities, $J^{(i)}(\omega)$, are computed (Eq. 4) using the modified correlation functions, $\bar{g}^{(i)}(t)$, (Eq. 13). Finally, the values of the relaxivities are calculated from Eq. 3. For our calculations, none of the approximations discussed in Eqs. 7-11 are used.

B. Results

In this section, we present relaxivity results for Gd-EDTA and compare these with experimental data. The molecular dynamics parameters are the same as described in the previous section except that a periodic box of 20 Å on a side was used that contained 249 water molecules. The simulation was run for 5 nsec, with the trajectory being saved every 0.1 psec. The calculations reported here took about 600 hrs on an SGI R8000 Power Challenge. During an initial run, one of the three inner shell waters was found to move in and out of the inner shell, with a mean lifetime of about 500 psec. This produced an effective q value of 2.5 and yielded a relaxivity that was too small by a factor of about 2.5/3. Consequently, a

second run was performed in which 3 waters were constrained to remain in the inner shell, by adding a weak restraining bond between the water oxygens and the Gd ion of length 2.4 Å. The subsequent results were obtained from the constrained run.

In Figure 3 we show the correlation functions $g^{(i)}(t)$ for $i=0$, 1 and 2. In each case, there is an initial short time decay followed by a long time tail. The calculated correlation times, $g^{(i)}(t)$, are each approximately 25 psec. This value is found by fitting the initial decay to an exponential. A direct comparison with experiment is difficult because the experimental correlation times, which are derived from a multi-parameter fit to the dispersion data, include the electronic contribution whereas ours do not, i.e. the calculation measures the initial decay of $g^{(i)}(t)$ (Eq. 5) and not of $\bar{g}^{(i)}(t)$ (Eq. 13). Note that inclusion of the electronic component would decrease our computed correlation times. At $t=0$, the ratio $g_0^{(0)}:g_0^{(1)}:g_0^{(2)}$ is 5.8:1:4.1 which is close to the ratio 6:1:4 given by the infinite time average for a freely rotating complex (See Eq. 9).

Figure 4 shows the $1/T_1$ dispersion curve along with the experimental values.[3] Before calculating the Fourier transforms in Eq. 4, the correlation times were multiplied by a function that went smoothly to zero at $t=1.5$ nsec. The experimental relaxivity values are indicated by crosses. The calculated total dispersion is given by the solid line while the calculated inner sphere dispersion is shown by a dot-dashed line. Inner shell contributions are calculated by including only contributions from the three bound waters. At the low frequency end, the ratio of computed values to ex-

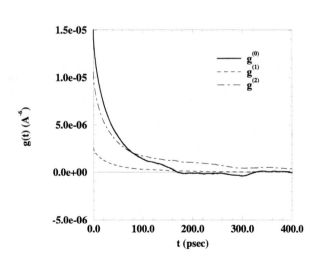

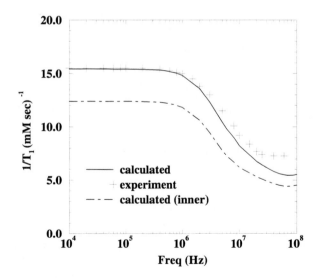

Figure 3. *Time correlation functions $g^{(i)}(t)$ for i=0 (solid curve), i=1 (dash-dot curve) and i=2 (dashed curve). The ratios at t=0 are 5.8:1.0:4.1. The correlation functions were smoothly damped to zero at t=1.5 nsec before calculating the spectra.*

Figure 4. *Comparison between experimental and calculated values of $1/T_1$. Experimental data points are indicated by crosses.[3] The calculated values are given by the solid line. The calculated inner-shell relaxivity is given by the dot-dashed line.*

perimental values is 0.99. At the high frequency end (20 MHz) the ratio (calculated/experiment) is 0.75. For the computed results, the ratio of the height of the low frequency plateau to that at high frequency is 10/3.5, in contrast to the ratio of 10/3 predicted from Eq. 11. The experimental ratio is 10/4.8. Recall that Eq. 11 is derived for the case of a bare ion in water. Nonetheless, our predicted high frequency plateau is low, which indicates that the contribution from $g^{(1)}(t)$ is too small relative to the other components. The 3 inner shell contributions account for about 80% of the relaxivity at both low and high frequencies. Therefore, the entire outer shell contribution is ≈ 0.75 of that from one inner shell water, for EDTA.

Another theoretical prediction is that the correlation functions arising from inner and outer sphere process should behave differently.[7] In particular, the inner sphere correlation functions should decay as a single exponential while the outer sphere contributions should die off as a sum of exponentials with increasing correlation times arising from increasingly distant waters. However, as the distances and correlation times increase, these terms will contribute less due to the $1/r^3$ factor in Eq. 6. Our numerical results for the inner and outer shell contributions to $g^{(0)}(t)$ are shown in Figure 5. The results for the other two components of *g* are identical. The outer shell contribution dies off very quickly with a correlation time of about 22 psec. There is no long time tail as for the inner shell contribution. Physically, this makes sense because the outer shell waters are randomized quickly. In the bulk, for instance, the mean time for a pair of waters to exchange position is only about 10 psec.

In conclusion, the agreement between theory and experiment is quite good given that no adjustable parameters were

used in the calculations. The calculated low frequency plateau almost exactly reproduces the experimental data, which indicates that the long time average structure of the complex is correctly modeled. The high frequency plateau yields information concerning fluctuations about the average structure, which we have modeled somewhat less accurately. The calculated value is somewhat lower than the experimental result at 20 MHz. A variety of approximations have been made in these calculations that could influence the numerical accuracy of the final results. These include using the experimental values for the electronic relaxation parameters of a bare Gd ion. The force field used will obviously affect the rotational correlation times. This will likely be sensitive to the water model used (charges, internal flexibility, etc.). Small differences in the Gd-O distance can have large effects on the relaxivity values, e.g. an 0.05 Å shift will change the values by >10%. Finally, the long time tails of the correlation functions are not fully converged, i.e. we still see some oscillatory behavior even with 5 nsec of statistics. We performed some approximate calculations using a random walker on a sphere that indicate that full convergence may need on the order of $1000 \times \tau_c$, or about 25 nsec. Future work will aim to understand the importance of each of these postulated sources of error and thereby to increase the quantitative accuracy of the calculations.

Static Structures - Gnomonic Projections

In this section, we turn to the description of a static method for determining the efficacy with which different chelate binding groups use the space around a metal ion. Gd^{+3}, used as the model paramagnetic ion in this work, has a nominal coordination number of 9, e.g. 9 water molecules will cluster around the unchelated ion with their oxygens pointing towards it. Ions of the lanthanide series have coordination numbers varying from 8 to 10. Smaller ions of the lanthanide series, such as Er (ionic radius of 0.97 Å) have a typical coordination number of 8, while larger ions, such as Ce and La (ionic radii of 1.14 and 1.16 Å) have a typical coordination number of 10.[47,48] In the presence of a chelate such as EDTA, this difference will manifest itself in terms of differing numbers of first shell water molecules around the lanthanide ion, which may vary from 2 to 3. For instance, a coordination of 8 is found for Er(EDTA) which has two first shell water molecules. A coordination of 10 is achieved in crystalline Ce(EDTA) and La(EDTA) with the replacement of a water molecule by a bidentate carboxylate group which can be shared with an adjoining lanthanide ion. The Gd ion with an ionic radius of 1.00 Å is in the group of lanthanide ions having a coordination number of around 9.

A good chelate will be sufficiently flexible to allow its chelating groups to wrap around the metal ion and fill a number of coordination sites in a low energy conformation. An inflexible chelate on the other hand may be unable to relax into a low energy conformation. Two manifestations of

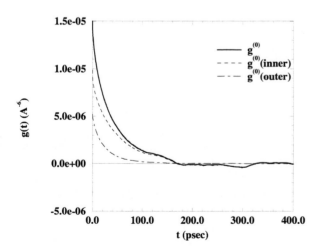

Figure 5. *Total correlation function and inner and outer shell contributions for* $g^{(0)}(t)$*. Note that the outer shell contribution dies off much more quickly than the inner shell component. Comparisons for the other two components of* g *show this same behavior.*

Table 3. *X-ray data on effective areas and binding distances of chelates.*

Chelate	A_{fract} [a]	d_{On}/d_{Of} [b]	d_N [c]	CSD designation
Gd[biacetate]	0.670	2.44 / 2.49	2.62	VEDSEC
Dy[DTPA-Amide](H_2O)$_1$	0.644	2.34/4.40	2.66	VOSBOU
Gd[DTPA-Amide](H_2O)$_1$	0.634	2.37 / 4.45	2.70	VETDON
Gd[DOTA](H_2O)$_1$	0.637	2.36 / 4.43	2.68	KUKGOM
Gd[EDTA](H_2O)$_3$	0.629	2.40 / 4.48	2.65	BIFZEV
Gd(H_2O)$_9$ [d]	0.626	2.42	NA	
Gd[Pyridine-(CO_2^-, CO_2H)](H_2O)$_3$	0.630	2.42 / 4.52	2.55	JOZGUA
Gd[CO_2^--CH_2O-CH_2-CO_2^-]	0.625	2.41 / 4.48	NA	NAOAGD
($H2O$)$_3$(CF_3CO_2)Đ	0.571/0.604	2.39	NA	SERYOD
GdĐ(CF_3CO_2)$_4$ĐGd	0.571/0.604	2.39	NA	SERYOD
Đ(CF_3CO_2)($H2O$)$_3$	0.571/0.604	2.39	NA	SERYOD

[a] A_{fract} *is the fractional surface area taken up by the ligands, defined in Eq. 2.*

[b] *These are Gd distances to the near and far bidentate oxygens in the crystal structure.*

[c] *These are Gd distances to the backbone nitrogens in the crystal structure.*

[d] *This is an average of 2 structures*

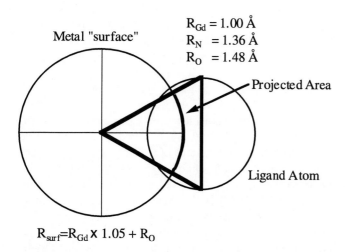

Figure 6. *Diagram showing the construction of a gnomonic projection in cross section. A large sphere is first constructed, centered on the Gd ion, and given a radius* $R_{sph} = 1.05 \times R_{Gd} + R_O$. *Next, a second sphere is centered on each ligand atom. These spheres have the appropriate van der Waals radii for the ligand atoms. For each ligand atom, a cone is drawn with its apex at the center of the metal atom and with its centerline along the line connecting the centers of the Gd and ligand atoms. The cone intersects the ligand atom sphere at a plane containing the center of the ligand atom. The gnomonic projection is then the area of the surface of the metal sphere enclosed by the cone.*

this are that some chelate arms may be forced away from the ion, or chelate arms may be forced to be too close to each other. Each of these will decrease the binding energy, even for a constant value of q_{tot}. Therefore, the effective use of the space surrounding the Gd ion is an important aspect of strong chelate binding. Hancock[8-10] has previously discussed the relationship between chelate flexibility and packing. Another way to analyze this issue is to examine the areas occupied by the chelate oxygen and nitrogen atoms. Here we display and measure these areas using what are termed gnomonic projections. This method involves the projection of the cross-sectional area of a chelate atom onto the surface of a virtual unit sphere whose center is located at the center of the metal ion. We take the radius to be the sum of the ionic radius for Gd and O or N (Figure 6). The ionic radii used here [oxygen (1.36 Å), nitrogen (1.48 Å), Gd (1.00 Å)] were obtained from the comprehensive work of Marcus.[47,48] The radius of the virtual sphere, R_{sph}, is normalized to the nominal Gd - O distance:

$$R_{sph} = 1.05 \times R_{Gd} + R_O \qquad (16)$$

where $R_O = 1.36 Å$ is the radius of an oxygen ligand atom. A normalization factor of 1.05 was used to make the radius of the virtual sphere equal to the average Gd - O distance of 2.40 Å, obtained from the X-ray structure of EDTA.[13-15] Through the use of gnomonic projections, a fractional surface area is obtained. We calculate the area of each first shell atom when projected onto the surface of the virtual sphere, and then divide by the total surface area of the sphere. This factor is further normalized by dividing by the fractional area covered by a hexagonal array of spheres on a plane. Therefore, a value of 1.0 for the fractional projected surface (denoted A_{fract}) is the maximum possible. The final result for the fractional projected surface area is:

$$A_{fract} = \frac{\sqrt{3}}{\pi}\left(1 - \frac{D}{\sqrt{D^2 + R_L^2}}\right) \qquad (17)$$

where D is the distance from the metal ion to the ligand atom and where R_L is the radius of the ligand atom whose overlap area is being calculated. Our numerical results take into account the case where the projections from different ligand atoms overlap, to make sure that those areas are not double counted. Note that A_{fract} decreases with increasing ligand-metal distance, *D*. Figure 6 shows the construction of the gnomonic projection in cross-section.

From the gnomonic projections, the total projected area on the virtual surface is obtained for the chelate oxygens and nitrogens, as well as for the oxygen atoms from first shell water molecules. The Cambridge Structural Database (CSD) was searched for structures containing a paramagnetic ion and molecules containing oxygen and carbon atoms. The search resulted in a database containing 870 crystal structures. From these structures, those containing Mn, Zn, or multiple ions were discarded. Gnomonic projections were obtained for each compound in the remaining group of 98 crystal structures. In Table 3 we give the occupied areas for a series of compounds with either Gd or Dy as the ion.

Gnomonic projections showed DTPA, DOTA and EDTA to have among the highest values of A_{fract} found. These chelates all have a total of nine binding atoms, including first shell waters. The gnomonic projection for these three structures and a bi-acetate compound (CSD designation VEDSEC [10]) are shown in Figure 7. The results are summarized in Table 3, which shows that the fractional projected area, including that from the first shell water oxygens, is between 0.629 and 0.637 for the first three compounds. Their effective projected area was exceeded by only one structure in the database, VEDSEC [10], with a value of 0.670. This compound contains two acetate molecules with their carboxyl groups in a bidentate conformation and a ring structure with 6 nitrogens, for a total of ten binding atoms surrounding the Gd (Figure 7d). As indicated in the diagram [10], the carboxyl groups sit above and below the plane formed by the macrocycle, and the Gd ion sits in the center. Note that the large ring is not planar, but is slightly cupped in an asymmetric fashion. It is difficult to attain higher fractional projected areas because steric effects become significant as more atoms are positioned at proper metal-ligand distances for Gd binding.

To better understand the effective area concept, it is instructive to examine some other compounds with similar or different projected areas (Table 3). Since the fractional projected area for common chelators is close to 0.63, full binding for Gd, with 9 coordination atoms, implies an effective surface area of about 0.07 surface units per atom. Table 3 shows that the fractional projected areas for Gd-chelates can range from 0.604 to 0.670. This range is about equivalent to the area occupied by one water molecule. Compound JOZGUA [11] has a fractional projected area that is comparable to EDTA. For that compound, the Gd ion is bound to 4 monodentate oxygens from two molecules with pyridine rings. These oxygens, combined with 2 ring nitrogens and 3 water molecules, result in a coordination of 9 and a fractional projected area of 0.630. This structure is similar to EDTA, which also contains 4 monodentate carboxylate oxygens, 2 nitrogens, and 3 water molecules. For NAOAGD [12], the Gd ion is surrounded by 6 monodentate oxygens and 3 ether oxygens in 3 separate linear chain molecules. There are no waters or amide nitrogens in this structure. The ether oxygens, however, are located slightly further from the Gd ion than are carboxylate oxygens. The resulting fractional surface area is 0.625, which is somewhat lower than the value for EDTA, reflecting the larger average Gd-O distances.

JOZGUA [11]

NAOAGD [12]

From our molecular dynamics simulations, we saw that the atoms of the first shell around the Gd ion moved in a correlated fashion. It is significant to note that when a water molecule or a chelating atom begins to move away from the Gd ion, other atoms and water molecules will move towards the metal ion so that a constant average distance for all first shell atoms, relative to the Gd ion, is maintained. Water molecules venture furthest away from the Gd ion – to a distance of 3.2 Å, relative to their normal X-ray distance of 2.4 Å. In general, any atomic movement that attempts to change the coordination number of Gd will be offset by the correlated

VEDSEC [10]

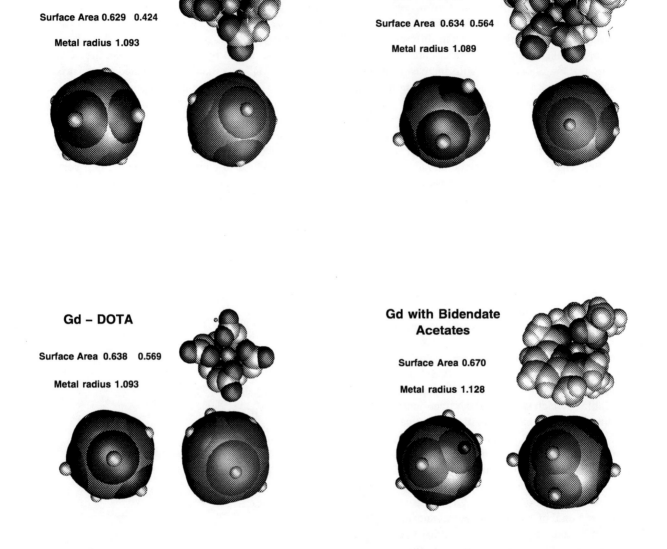

Figure 7. *Examples of gnomonic projections for 4 structures:*
(a) Gd - EDTA
(b) Gd - DTPA
(c) Gd - DOTA
(d) Gd - Bi-acetate (VEDSEC).
A model of the chelate molecule, with the Gd ion at the center,
is shown in the upper right portion of each figure. The lower
half of each figure shows two views of the gnomonic projection
onto a gray sphere. Intersected areas for the carboxylate
oxygen atoms are represented in red, water oxygens in pink,
and those for the nitrogen atoms appear in blue.

motion of other atoms near the Gd ion. We can predict that chelating atoms or water molecules in other compounds will also adjust their position to maintain a coordination number of 9 for Gd.

The one X-ray structure, VEDSEC [10], that shows a coordination of 10 about a Gd has no water molecules and two bidentate carboxylate groups. VEDSEC [10] is able to fit 10 ligand atoms around the metal because a bidentate carboxylate group takes up less space than the corresponding pair of water molecules. (See Figure 7d.) Note that the 10th ligand atom only adds a factor of 0.035 to A_{fract}, which is about half that contributed by typical ligand atoms. The static reasons for this are that the carboxyl oxygens are allowed to draw close to one another because of their chemical binding, and that both oxygens lie slightly further from the metal than do other ligand atoms. It is interesting to compare the footprints of waters in Figs. 7 a-c (pink circles), which tend to stay away from other circles, with those for the bidentate carboxyl ligand (paired red circles) in Figure 7d. Waters have a large footprint because they lie close to the metal, and because of their high mobility relative to the other chelating groups. Figure 7d clearly shows that the carboxyl circles are smaller than those for the water oxygens, because the carboxyls are further from the metal, and that the 2 bidentate carboxyl oxygens lie close together. This results agree with the predictions from our dynamics simulations.

A case at the opposite extreme is that of structure SERYOD [13]. The molecular structure for SERYOD is $(H2O)_3(CF_3CO_2)GdĐ(CF_3CO_2)_4ĐGd(CF_3CO_2)(H2O)_3$ in which four carboxyl groups bind symmetrically to each of the two Gd atoms, forming a barrel. Capping each end are three waters, with their oxygens pointing at the Gd, and one more carboxyl group. One of the carboxylate oxygens in the terminal CF_3CO_2 group is shared with a Gd in the next unit cell and cannot be positioned at or near the normal binding distance of 2.4 Å. Instead it is located at 3.4 Å from the metal ion, so that SERYOD's coordination may be considered as either 8 or 9. Counting the out-of-position oxygen as a binding atom, the resulting projected area is 0.604, with a coordination of 9. If the out-of-position oxygen is excluded, the coordination number becomes 8 and the effective area is 0.571. As in VEDSEC [10], the asymmetrically placed outer carboxyl group contributes only about half of the expected fractional projected area of 0.07 for a tightly binding ligand atom, principally because it cannot get in close to the Gd. These bidentate carboxylate groups increase the projected area but bind in an asymmetric, possibly metastable conformation. This is similar to the case of P3A [5] considered in the previous section. There, the outer carboxyl atom was by far the most mobile ligand atom seen in the dynamics simulations.

From these results, it appears that for structures with a composition similar to those of existing MRI compounds, consisting of binding carboxylate oxygens, nitrogens and water molecules, a chelate with significantly larger effective projected area will be difficult to design. There were no structures in the CSD containing at least one inner shell water that made more effective use of the area around the Gd than the known MRI compounds.

Summary

We have shown that it is possible to use molecular dynamics and mechanics simulations to evaluate the static and dynamic properties of novel chelate structures. By use of these tools, it is now possible to predict q-values and binding conformations for new chelate structures and evaluate their potential usefulness as MRI contrast agents. Additionally, we can calculate the magnetic resonance relaxivity for typical metal-chelate complexes.

A number of MRI chelating agents have been examined using molecular dynamics techniques. It is shown that, as expected, carboxyl groups typically assume the monodentate position in which one oxygen binds to the Gd ion (R(O-Gd) ~ 2.4 Å) and exhibits small rms motion. The other oxygen is further away and exhibits greater motion. The near oxygen is practically stationary while the far oxygen rotates on a lever arms whose pivot is the inner oxygen. We have also seen that motions of the chelating atoms are correlated, meaning that as one atom moves away from the metal ion, others will move closer, so that the average metal-chelate distance remains constant. These motions are a low frequency vibrational mode of the complex. The accuracy of our MD simulations in predicting q-values and carboxylate binding distances for reference chelates has been shown. The use of gnomonic projections provides a semi-quantitative method of evaluating how effectively binding atoms use the space surrounding the paramagnetic ion. As chelate structures, DTPA and DOTA are shown to use the space around the Gd ion most efficiently, with a fractional occupied area of about 0.64, relative to a maximum value found in the CSD of about 0.67. For other proposed structures to bind as efficiently as DTPA to the Gd ion, they will need to have comparable fractional areas, and should at the same time allow sufficient space for at least one water molecule to occupy the inner shell of the Gd ion.

Thus we have demonstrated several methods that can be used to evaluate the effectiveness of potential MRI contrast agents. Flexibility in the molecular backbone and carboxylate arms is important to make effective use of the free space around the Gd ion. This has also been discussed by Hancock.[9-11] For efficient binding to the Gd ion, with its coordination number of 9, the sum of binding carboxyl and amine groups should be 8 or more. An outstanding question is the binding strength of carboxyl groups relative to amine groups. For instance, if a structure had more amine groups than carboxyl groups, would the binding be sufficiently strong? This question needs to be addressed before we can make definitive choices between a set of candidate structures such as those analyzed here.

Several common chelating agents, like DTPA and DOTA, which are currently in clinical use, have q-values of 1. In this paper, molecular simulation methods have been used to predict q-values for proposed structures that vary from 0 to 3. It would be interesting to synthesize and test some of these molecules and examine the relationship between the predicted binding group mobility and q-value, the stability constant, and relaxivity.

Acknowledgments: Many helpful discussions with Dr. Kenneth Kellar, including his assistance with the magnetic relaxivity data, and the assistance of Dr. L. Castonguay in obtaining the CSD structures are gratefully acknowledged. This work was supported by the Department of Energy under contract DE-ACO4-94AL85000 and by a grant from Sterling-Winthrop, Inc.

References

1. Lauffer, R. B. *Chem.Rev.* **1987**, *87*, 901-927.
2. Koenig, S. H. and Brown, R. D. in *Magnetic Resonance Annual 1987*, H. Y. Kressel, Ed., Raven Press, New York, 1987 .
3. Koenig, S. H., Baglin, C., Brown, R. D., and Brewer, C. F. *Magn.Reson.Med.* **1984**, *1*, 496.
4. Koenig, S. H. *Israel J. Chem.* **1988**, *28*, 345.
5. Anderegg, G. in *Critical Evaluation of Equilibrium Constants in Solution*. Vol. 14 Pergamon Press, New York, 1977, 1.
6. Abragam, A., *The Principles of Nuclear Magnetism*, Clarendon Press, Oxford, 1961.
7. Hertz, H. G. in *Water: A Comprehensive Treatise*. Vol. 3, F. Franks, Ed., Plenum Press, New York, 1973, 301-399.
8. Hancock, R. D. *Pure and Applied Chem.* **1986**, *58*, 1445-1452.
9. Hancock, R. D. and Martell, A. E. *Chem.Rev.* **1989**, *89*, 1875-1914.
10. Hancock, R. D. *Acc.Chem.Res.* **1990**, *23*, 253.
11. Fossheim, R. and Dahl, S. G. *Acta Chemica Scand.* **1990**, *44*, 698-706.
12. Fossheim, R., Dugstad, H., and Dahl, S. G. *J.Med.Chem.* **1991**, *34*, 819-826.
13. Kumar, K. and Tweedle, M. F. *Inorg.Chem.* **1993**, *32*, 4193-4199.
14. Frey, S. T., Chang, C. A., Carvalho, J. F., Varadarajan, A., Schultze, L. M., Pounds, K. L., and Jr., W. d. H. *Inorg.Chem.* **1994**, *33*, 2882-2889.
15. Hay, B. *Inorg.Chem.* **1991**, *30*, 2876-2884.
16. Hay, B. *Coordination Chem. Revs.* **1993**, *126*, 177-236.
17. Rappe, A. K., Colwell, K. S., and Casewit, C. J. *Inorg. Chem.* **1993**, *32*, 3438-3450.
18. Badertscher, M., Musso, S., Welti, M., Pretsch, E., Maruizumi, T., and Ha, T.-K. *J. Comput. Chem.* **1990**, *11*, 819-828.
19. Geraldes, C. G. G. C., Brown, R. D., Brucher, E., Koenig, S. H., Sherry, A. D., and Spiller, M. *Mag. Reson. in Med.* **1992**, *27*, 284.
20. Kim, S. H., Pohost, G. M., and Elgavish, G. A. *Bioconjugate Chem.* **1992**, *3*, 20.
21. Ladd, M. F. C., Povey, D. C., and Stace, B. C. *J.Cryst.Mol.Struct.* **1974**, *4*, 313.
22. Kullnig, R. K., (Private Communication), (1993).
23. Ladd, M. F. C., Povey, D. C., and Stace, B. C. *Acta Cryst.* **1973**, *B29*, 2973.
24. Contrait, P. M. *Acta Cryst.* **1972**, *B28*, 781.
25. Shkol'nikova, L. M., Polyanchuk, G. V., Dyatlova, N. M., and Polyakova, I. A. *Zh. Strukt. Khimii.* **1984**, *25*, 103.
26. Sherry, A. D. *J.Less Common Metals* **1989**, *149*, 133.
27. Sherry, A. D., Brown, R. D., Geraldes, C. F. G. C., Koenig, S. H., Kuan, K.-T., and Spiller, M. *Inorg. Chem.* **1989**, *28*, 620.
28. Cacheris, W. P., Nickle, S. K., and Sherry, A. D. *Inorg. Chem.* **1987**, *26*, 958.
29. Berendsen, H. J. C., Grigera, J. R., and Straatsma, T. P. *J.Phys.Chem.* **1987**, *91*, 6269-6271.
30. Miyamoto, S. and Kollman, P. A. *J.Comp.Chem.* **1992**, *13*, 952-962.
31. Leach, A. R. in *Reviews in Computational Chemistry*. Vol. 2, K. B. Lipkowitz and D. B. Boyd, Ed., VCH Publishers, New York, 1991, 1-55.
32. Levitt, M. *Chemica Scripta* **1989**, *29A*, 197.
33. Levitt, M., Hirshberg, M., Sharon, R., and Daggett, V. *Comp.Phys.Comm.* **1995**, *(submitted)*,
34. Judson, R. S., Barsky, D., Faulkner, T., McGarrah, D. B., Meliu, C. F., Meza, J. C., Mori, E., Plantenga, T., and Windemuth, A., Sandia National Laboratories, 1995-SAND 95-8258. *CCEMD - Center for Computational Engineering Molecular Dynamics, Theory and Users' Guide, Version 2.2.* .
35. Windemuth, A. and Schulten, K. *Mol.Sim.* **1991**, *5*, 353-361.
36. Brooks, B. R., Bruccoleri, R. E., Olafson, B. D., States, D. J., Swaminathan, S., and Karplus, M. *J. Comput. Chem.* **1983**, *4*, 187-217.
37. QUANTA/CHARMM, Molecular Simulations, Inc., Waltham MA, 1993.
38. Colvin, M. C. and Melius, C. F., Sandia National Laboratories, 1993-SAND93-8239. *Continuum Solvent Models for Computational Chemistry*.
39. Gaussian 92, Revision A, Frisch, M. J., Trucks, G. W., Head-Gordon, M., Gill, P. M. W., Wong, M. W., Foresman, J. B., Johnson, B. G., Schlegel, H. B., Robb, M. A., Replogle, E. S., Gomperts, R., Andres, J. L., Raghavachari, K., Binkley, J. S., Gonzalez, C., Martin, R. L., Fox, D. J., Defrees, D. J., Baker, J., Stewert, J. J. P., and Pople, J. A., Gaussian, Inc., Pittsburgh PS, 1992.
40. Bertini, I. and Luchinat, C., *NMR of Paramagnetic Molecules in Biological Systems*, Benjamin/Cummings, Menlo Park, CA, 1986.
41. Hwang, L.-P. and Freed, J. *J.Chem.Phys.* **1975**, *63*, 4017-4026.
42. Kowalewski, J., Nordenskiold, L., Benetis, N., and Westlund, P.-O. *Prog.NMR Spec.* **1985**, *17*, 141-185.
43. Rubinstein, M., Baram, A., and Luz, Z. *Mol.Phys.* **1971**, *20*, 67-80.
44. Bloembergen, N., Purcell, E. M., and Pound, R. V. *Phys.Rev.* **1948**, *73*, 679.
45. Bloembergen, N. and Morgan, L. O. *J.Chem.Phys.* **1961**, *34*, 842.
46. Koenig, S. H. and Epstein, M. *J.Chem.Phys.* **1975**, *63*, 2279-2284.
47. Marcus, Y. *J.Sol.Chem.* **1983**, *12*, 271.
48. Marcus, Y. *Chem.Rev.* **1988**, *88*, 1475.

J. Mol. Model. **1996**, 2, 175 – 182

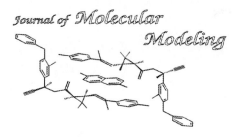

Solvation Effects on the Conformational Behaviour of Gellan and Calcium Ion Binding to Gellan Double Helices

Vivienne L. Larwood, Brendan J. Howlin*, Graham A. Webb

Dept. of Chemistry, University of Surrey, Guildford, Surrey, GU2 5XH, UK (chs1bh@surrey.ac.uk)

Received: 13 March 1996 / Accepted 04 June 1996 / Published: 21 June 1996

Abstract

The contribution of the presence of solvent to the conformations adopted by disaccharide fragments within the repeat unit of gellan have been studied by molecular modelling techniques. Initial conformational energy searches, using a dielectric continuum to represent the solvent, provided starting geometries for a series of molecular dynamics simulations. The solution behaviour from these simulations was subsequently compared to fibre diffraction data of the potassium gellan salt. The present calculations indicate considerable flexibility of the glycosidic linkages, and this is discussed in relation to its effect on gel formation. One of the fragments was solvated with explicit water molecules. These calculations showed the same conformational behaviour as those simulations conducted in implicit solvent.

Finally, a series of molecular dynamics (MD) simulations were performed to study the calcium binding to gellan. The results from this clearly showed a well defined binding site for this ion.

Abbreviations: MM : molecular mechanics; MD : molecular dynamics.
Keywords: polysaccharides, gellan, glycosidic linkage.

Introduction

Gellan is an anionic, microbial polysaccharide with the basic repeat unit of $(1\rightarrow3)$-β-D-Glc$p$-$(1\rightarrow4)$-β-D-Glc$p$A-$(1\rightarrow4)$-β-D-Glc$p$-$(1\rightarrow4)$-α-L-Rha$p$ [1]. A schematic of the gellan repeat unit can be found in Figure 1. Its ability to form firm aqueous gels at low polysaccharide concentration in the presence of specific counter ions has led to its use as a thickening agent in a variety of food applications. Its affects have been studied in food systems as diverse as dairy products [2] to corn syrups [3] and low sugar bakery jams [4].

Several groups have studied the dilute solution behaviour of gellan, using a variety of counter ions, with a view to gaining some insight into the process of gel formation [5-13].

These have mainly concentrated on measuring gel strengths under different conditions of counter ion type and concentration, temperature, polysaccharide concentration and pH. The aim of this work is to study the effect of solvent on the conformational properties of the gylcosidic linkages within the gellan repeat unit and to use a combined molecular mechanics (MM) and molecular dynamics (MD) approach to investigate the co-ordination of divalent ions to the gellan double helix, using Ca^{2+} ions as an example.

Polysaccharides have received relatively little attention from computational studies compared to other macromolecules, such as proteins. Both groups suffer from similar difficulties, such as the current inability to predict conformation based solely on primary structure. However, progress is being made. So far, in the case of polysaccharides, this has

* *To whom correspondence should be addressed*

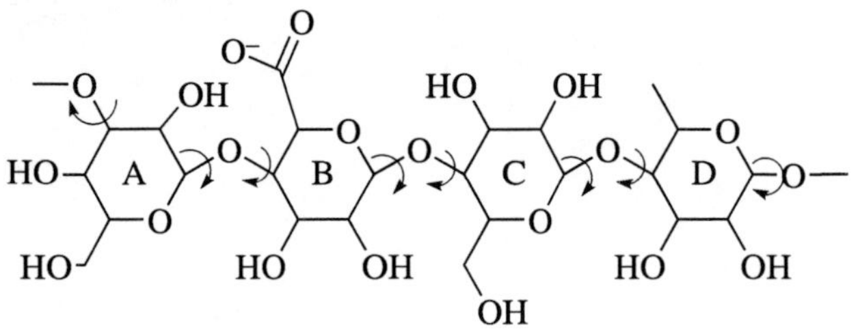

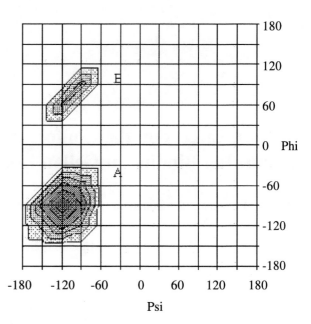

Figure 1. *Schematic representation of gellan with residue labelling.*

mainly concentrated on analysing individual glycosidic linkages. We use this type of approach to validate the model and force field used in subsequent calculations.

We are well aware that modelling the entire time sequence leading to gelation from a disordered aqueous solution is beyond the scope of current simulation methods. It is also impractical at present to simulate models which are of a sufficient size to represent all aspects of gel formation. Therefore we have decided to concentrate on one stage of the pathway, which is where double helices have formed and are in close proximity to each other. In this simulated state the specific interactions with the counter ions can be probed in detail without the need to attempt to model the formation of the gel.

Methodology

General

All calculations were performed on a Silicon Graphics Iris Indigo XZ 4000 workstation. The molecular modelling package INSIGHT II Version 2.3.0 (Biosym Technologies, San Diego, CA, USA) was used to construct all models, and the INSIGHT II interface to Discover Version 2.95 was utilised for the calculations on the structures. The AMBER force field [14] with the modifications for polysaccharides by Homans [15] was used in all simulations.

The explicit image protocol was used in all MD simulations, as was the Verlet algorithm [16]. A step size of 1 femtosecond was used in all of the MD simulations.

All simulations using periodic boundary conditions were conducted under conditions of constant temperature and volume. The temperature is maintained using Berendsen's method [17], which couples the system to a temperature bath.

Conformational Energy Maps

Relaxed conformational energy maps were calculated by restraining the φ and ψ dihedral angles and energy minimising the remaining structure around these dihedral angles. The dihedral angles involved in this study have been shown in Figure 1. For the $(1\rightarrow4)$ linkages φ is defined as H1-C1-O1-

Figure 2. *Conformational energy search for the A-B glycosidic linkage.*

C4', and ψ is defined as C1-O1-C4'-H4', where a prime symbol indicates the following residue. For the $(1\rightarrow3)$ linkages, φ is defined as H1-C1-O1-C3', and ψ is defined as C1-O1-C3'-H3'. Initially, both of the dihedral angles were set to -180°, and the structures were energy minimised. The dihedral angle ψ was then incremented by 30°, while φ was held constant, and the energy minimisation procedure was repeated. This was then repeated until the dihedral angle ψ had swept out 360°; the dihedral angle φ is then incremented by 30° and the whole procedure was again repeated until each of the dihedral angles had been rotated by the full 360° for all values of the other.

Disaccharide Simulations with Dielectric Constant

The minimum energy conformations predicted by these relaxed conformational energy maps were subsequently used as the starting structures for 500 picoseconds MD simulations. This procedure has been described by Stern et al. [18]. Cutoffs for the non-bonding interactions were not used in these calculations with the disaccharide fragments, as the molecules under consideration were relatively small.

A relative dielectric constant of 80 was used to implicitly represent the presence of solvent. Some representation of solvent effects is essential when modelling polysaccharides, due to their strong tendency to form hydrogen bonds. The MD simulations were run at a temperature of 295K. There was an initial 50 picoseconds equilibration; the subsequent 500 picoseconds of the trajectory were used for the data collection.

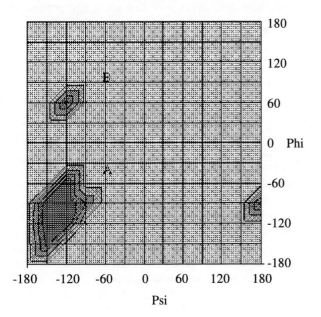

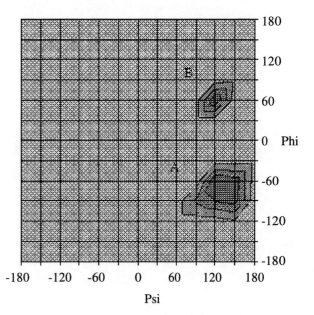

Figure 3. *Conformational energy search for the B-C glycosidic linkage.*

Figure 4. *Conformational energy search for the C-D glycosidic linkage.*

Disaccharide Simulations in Explicit Water

One of the disaccharides also underwent a MD simulation in the presence of explicit water molecules, to determine the effect this may have on the minimised structure from the calculations in vacuum. The disaccharide was solvated using a 15Å³ box of equilibrated water molecules. The TIP3P model of water [19] was used in all simulations. The TIP3P has recently been successfully used in MD simulations with sucrose [20]. The waters were placed randomly around the disaccharide, whilst avoiding any steric over-laps. This resulted in a total of 84 water molecules.

In this particular system, cut-offs were employed for the non-bonding interactions at 15Å, since periodic boundary conditions were required to contain the water molecules. A group based switching function smoothly adjusted the potential to zero over a 1.5Å range.

Double Helix Simulations Without Ions

Gellan gels consist of rigid, crystalline regions of aggregated double helices [21], linked by more flexible segments [9]. The model system has been developed to represent the interactions present within the crystalline regions only. This reduced size model consists of two double helices. The two strands of each double helix are composed of just four pyranose rings. This represents the repeat unit of the polysaccharide chains, and is the minimum required to model the cation binding. These systems are fully solvated with explicit water molecules to model more accurately the small

scale interactions, such as hydrogen bonding, and to include the solvation shell around the ions.

These simulation cells have the dimensions a = 26.0, b = 22.0, c = 35Å with the angles $\alpha = \beta = \gamma = 90°$. This is slightly larger than the unit cell dimensions of a = b = 15.8, c = 28.2Å, to allow for full solvation of the polysaccharide chains. The waters are placed randomly around the chains, in the same way as described for the disaccharide fragments. This resulted in a total of 185 water molecules.

As before, a step size of 1 femtosecond was chosen. The systems are judged to have equilibrated sufficiently when the total fluctuation of the temperature is less than 5° about the target temperature, and the ratio of potential energy to kinetic energy remains constant. These indicate that the system has reached thermal equilibrium.

Double Helix Simulations With Ions

The Ca^{2+} ions are initially placed arbitrarily in positions close to the carboxyl groups between the polysaccharide chains, and it is from these positions that they move into the co-ordination sites proposed by Chandrasekaran [21] during subsequent energy minimisation of the systems. The proposed binding site itself involves two carboxyl groups, one from each of the double helices. Thus the ions are believed to bridge between the double helices, rather than within a double helix.

The systems were first minimised, holding the chains rigidly fixed in space, to enable the water molecules to orient themselves with respect to the other atoms in the system. 10 picoseconds intervals of MD were also used to distribute the

J. Mol. Model. **1996**, *2*

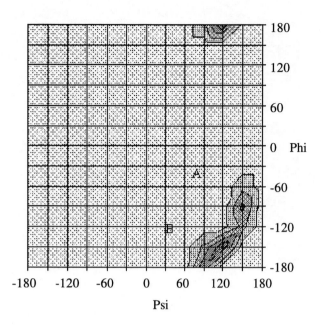

Figure 5. *Conformational energy search for the D-A glycosidic linkage.*

water molecules more evenly around the simulation cell. During this process, the polysaccharides and ions were held in position, and the temperature used was 295K.

The complete systems were then relaxed using steepest descents and conjugate gradient energy minimisation. A total of 200 picoseconds was used for the equilibration process; this included 80 picoseconds each at 100K and 200K and finally 40 picoseconds at 295K. The following 200 picoseconds of the trajectory, at a temperature of 295K, was then collected and analysed. Structures were saved every 500 femtoseconds. The final simulation temperature was 295K. This is the normal temperature of use for gellan products. It should be noted that the actual temperature at which gellan

gels deteriorate depends on the concentration of gellan and the nature and concentration of the counter ions present [7].

Results

Structural data for polysaccharide gels can often be particularly difficult to obtain; a general lack of crystallinity precludes the use of X-ray crystallography, whilst solution NMR often yields only rotationally averaged conformations. In favourable cases, such as gellan, it may be possible to extract fibres in which the molecules are preferentially oriented in the same direction [22]. From such samples it is then possible to obtain fibre diffraction data; although this is not as precise as X-ray crystallography, it can nevertheless provide sufficient details of the structure to determine unit cell dimensions, helix pitch and possibly even the location of specific crystalline water molecules. Such data provide useful starting structures for atomistic simulations. In the case of gellan, this type of data was obtained by Chandrasekaran et al. for both the lithium [23] and the potassium [21] salt of gellan. The fibre diffraction data for the potassium salt were used for the starting structure of the double helices. These data are also used for comparison with calculated results for the disaccharide fragments. These calculations were performed as described in the methodology section.

Conformational Energy Maps

The results from the conformational energy searches and subsequent MD calculations performed here on all four of the glycosidic linkages have been compared to the fibre diffraction values, and listed in Table 1. As can be seen, there is good agreement between experimental and calculated dihedral angles for the A-B, C-D and D-A glycosidic linkages. Some deviations are expected from the fibre diffraction data as the molecules are relaxed. These results are described individually in the following sections. Particular attention has been given to those linkages which showed only moderate agreement with the experimental data.

The relaxed conformational energy map for the A-B glycosidic linkage has been included as Figure 2. Two minimum energy conformations were located; these have been

Table 1. *Comparison between the fibre diffraction data for the glycosidic linkages and the results from both the M.M. and M.D. calculations.*

Dihedral angle	Dihedral angle (°) fibre diffraction data	Dihedral angle (°) conformational energy searches	Dihedral angle (°) molecular dynamics simulations.
O5D-C1D-O3A-C3A	-124	-150	$-180 < \theta < -150: 150 < \theta < 180$
C1D-O3A-C3A-C4A	88	120	$90 < \theta < 120$
O5A-C1A-O4B-C4B	-101	-90	$-100 < \theta < -70$
C1A-O4B-C4B-C5B	-136	-120	$-150 < \theta < -90$
O5B-C1B-O4C-C4C	-154	-90	$-120 < \theta < -30$
C1B-O4C-C4C-C5C	-144	-120	$-180 < \theta < -90$
O5C-C1C-O4D-C4D	-150	-90	$-120 < \theta < -60$
C1C-O4D-C4D-C5D	86	120	$60 < \theta < 120$

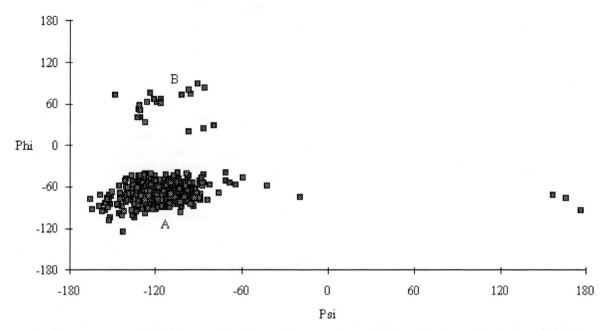

Figure 6. *M.D. simulation of A-B glycosidic linkage, starting from energy well B.*

labelled as energy wells "A" and "B" in Figure 2. The "A" well is located at $\phi = -90°$, $\psi = -120$, which compares favourably with the conformation obtained from the fibre diffraction data of $\phi = -101°$, $\psi = -136$.

The relaxed conformational energy map for the B-C glycosidic linkage revealed two minima. These can be seen in Figure 3. This linkage showed moderate agreement with the fibre diffraction data, almost certainly due to the lack of stabilising interactions from neighbouring residues. This will be discussed in more detail in the section on the MD simulations.

The conformational energy search for the C-D linkage revealed a wide ranging minimum energy well, which is labelled as "A" in Figure 4, and a local minima, "B". The conformation corresponding to "B" is just 1 kcal mol⁻¹ above the global minimum at "A". However, neither of these conformations corresponded particularly well to the fibre diffraction data. This is not entirely unexpected, due to the conformational freedom of the rhamnose residue.

The conformational energy map for the D-A linkage showed close agreement with the fibre diffraction data. The conformational energy map for this particular linkage has been included as Figure 5. Two minimum energy conformations have been labelled as "A" and "B".

Disaccharide Simulations with Dielectric Constant

The conformation corresponding to energy well "A" for the A-B glycosidic linkage, see Figure 2, was then used as the starting point for a MD simulation of the A-B disaccharide, as described in the methodology section. During this MD

simulation the A-B glycosidic linkage remained in this conformation.

The "B" well in Figure 2 represents a conformation in which the HO3B - O5A hydrogen bond is not topologically possible. This hydrogen bond is an important factor in the formation and stabilisation of the gellan helix; its presence and importance were confirmed by the fibre diffraction data of Chandrasekaran et al. [8]. The MD simulation starting from the conformation associated with the "B" energy well underwent an early transition to the "A" well, where it remained for the remainder of the simulation. These results indicate the importance of this particular hydrogen bond, the formation of which appears to be a major driving force in the helix formation. The MD simulation starting at conformation "B" can be seen in Figure 6, and the HO3B - O5A hydrogen bond can be viewed in Figure 7. The lack of this stabilising influence almost certainly also affected the conformational energy map for the B-C glycosidic linkage.

The global, or over-all, energy minimum for the B-C glycosidic linkage, labelled as "A" in Figure 3 was used as the starting conformation for the MD simulation. The linkage remained in this conformation throughout the MD simulation. When the MD simulation was started at conformation "B", a 10 kcal mol⁻¹ energy barrier was crossed early in the simulation, and the linkage remained in the "A" conformation for the remainder of the simulation.

When the crystal structure was used as the starting conformation for a MD simulation for the C-D glycosidic, there was a strong preference for the "B" energy well (see Figure 4). When the "B" energy well conformation was used as the starting conformation, there was a strong preference for the "A" energy well, with only a few transitions to the "B" energy well. This linkage showed only moderate agreement with the fibre diffraction data for the MD simulations; again,

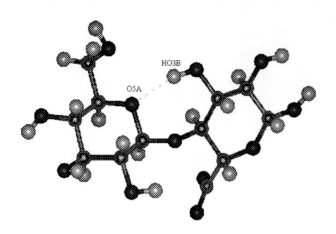

Figure 7. *Representation of HO3B • • • O5A hydrogen bond.*

this is likely to be due to the conformational mobility of the rhamnose residue. The closest low energy conformation to the fibre diffraction data was at $\phi = -150°$, $\psi = 120°$. This was located within 3 kcal mol^{-1} of the minimum energy conformation at "A". However, there were no transitions to this conformation during any of the MD simulations performed.

The fibre diffraction data for the D-A glycosidic linkage is $\phi = -124°$, $\psi = 88°$, which corresponds to the "A" minimum energy well at $\phi = -150°$, $\psi = 120°$ (see Figure 5). Two MD simulations were performed for this linkage. The first used the fibre diffraction data as the starting conformation, due its proximity to the "A" energy well. The linkage remained in this conformation throughout the simulation, with no transitions to any other regions.

The second MD simulation used the "B" energy well conformation as the starting conformation. During this simulation the linkage underwent an early transition, during equilibration, to the "A" energy well, where it remained throughout the simulation. The resulting trajectory for this simulation can be viewed in Figure 8. Thus, the D-A glycosidic linkage adopted a stronger preference for the conformation corresponding to the fibre diffraction data than the other three linkages. This is probably due to the reduced flexibility of $(1 \rightarrow 3)$ compared to $(1 \rightarrow 4)$ linkages. These results emphasise the importance of the occurrence of the $(1 \rightarrow 3)$ linkage in the gellan repeat unit for the over-all stability of the helical conformation.

Disaccharide Simulation in Explicit Water

The B-C glycosidic was also minimised in a 15Å^3 box of explicit water molecules, as described in the methods section. This revealed no differences in the location of the minimum energy conformations, although there were differences in the orientations of some of the hydroxyl groups. These results confirm the validity of using implicit solvent in these calculations on disaccharide fragments.

Double Helix Simulations with Ions.

Gellan gel formation at low polysaccharide concentration is dependent on the presence of counter ions. It is believed that they are responsible for the aggregation of the micro-crystalline regions [9]. A qualitative approach has been adopted here in which the ionic interactions are probed in detail. Experimental data to study these interactions are unavailable at

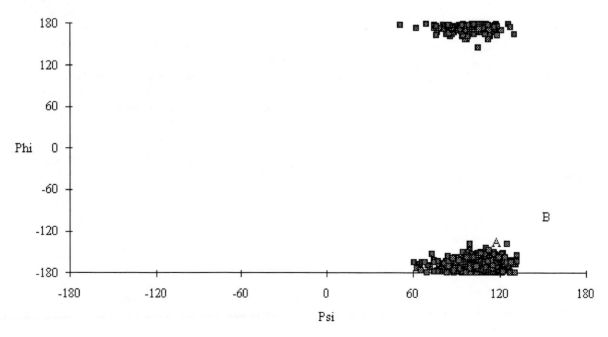

Figure 8. *M.D. simulation of D-A glycosidic linkage, starting from energy well B.*

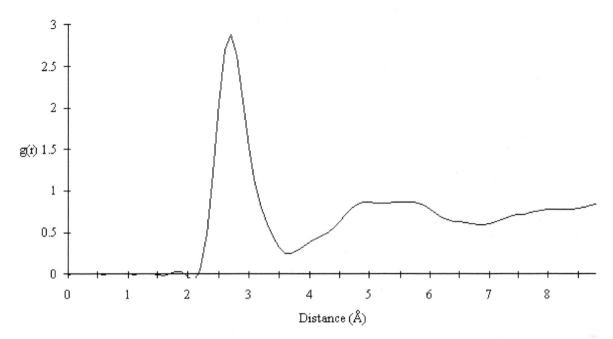

Figure 9. *Pair distribution function for the water molecules around the calcium ion.*

present. The ionic interactions near the carboxyl group are of particular interest, as this is the proposed co-ordination site for the cations. From the MD simulations it should be possible to determine the most favourable positions for the ions to adopt in this region.

Micro-crystalline regions form within aqueous samples of gellan gum. These crystalline regions are thought to be composed of aggregated double helices, based on fibre diffraction data [21]. It is currently believed that the counter ions, which promote gel formation at low polysaccharide concentration, are responsible for the aggregation of these regions [24].

The de-acylated gellan molecule has been reported [21] to form a parallel half-staggered double helix in which each polysaccharide chain is a left-handed 3-fold helix of pitch 56.4 Å. The counter ions known to promote gelation include mono-valent ions such as Li^+, Na^+, K^+ and Cs^+, as well as divalent ions such as Ca^{2+}, Mg^{2+}, Sr^{2+}, Ba^{2+}, Zn^{2+}, Cu^{2+} and Pb^{2+}. Higher concentrations of the monovalent ions are required to produce gels of the strengths produced when divalent counter ions are used. The gel strength for the monovalent ions [7] increases in the order of:

$$Li^+ < Na^+ < K^+ < Cs^+.$$

For divalent ions the corresponding series is:

$$Mg^{2+}, Ca^{2+}, Sr^{2+}, Ba^{2+} < Zn^{2+} < Cu^{2+} < Pb^{2+}.$$

H^+ ions are known to induce gels of higher strengths than even the divalent cations, but the mechanism involved is still unclear. In contrast, large, bulky ions such as tetramethylammonium ions inhibit gel formation; hence they are often included in studies of the dilute solution behaviour of gellan solutions.

The counter ions are believed to be co-ordinated in some way with the carboxyl oxygens of the glucuronic acid residues (labelled as B in Figure 1). In the case of divalent ions, it is believed that the cations bridge directly between two carboxyl oxygens on two double helices [25]. However, the monovalent ions are believed to co-ordinate via water oxygen bridges to the carboxyl oxygens [24].

During the detailed examination of the proposed co-ordination site Ca^{2+} ions were placed near the proposed co-ordination site, which has not been proved experimentally, and a MD simulation was performed, as described earlier. The results from this clearly showed that the ion bridged between the carboxyl oxygens on the two double helices. The average distances during this time were 2.6Å and 2.7Å between the Ca^{2+} ion and the nearest carboxyl oxygen from each double helix. The average distance between these two carboxyl oxygens was 3.1Å. Three water molecules were specifically co-ordinated around the Ca^{2+} ion. The distances between these and the Ca^{2+} were 2.74, 4.64 and 4.92Å respectively.

The van der Waals' and Coulombic interaction energies between the Ca^{2+} ion and the carboxyl oxygens were measured for each binding position of the ion during the MD simulation. This revealed a co-ordination distance of 2.3Å corresponding to the minimum interaction energies observed.

A pair distribution function for the water molecules around the Ca^{2+} ion shows a co-ordination shell of water at about 2.6Å around the ion. This can be seen in Figure 9.

This distance corresponds well to the co-ordination distances reported above between the Ca^{2+} ion and the carboxyl oxygen, and suggests that the water oxygens are replaced by carboxyl oxygens when the ion is co-ordinated with the double helices.

A pair distribution function of the water oxygens around one of the carboxyl oxygens involved in the calcium co-ordination site revealed signs of structuring of the water in this region. Sharp peaks occurred at about 2.8Å, 3.3Å and 3.5Å, and a small peak at 4.4Å, with the continuum beginning at around 5Å. Again, this confirms the important role of the water molecules in these systems. It is also likely that the increase in entropy of the water molecules as they are replaced by the Ca^{2+} ion drives the complex formation, thus leading to gelation. Calculations with monovalent ions are currently in progress, and it will be interesting to compare the structuring of the water in the vicinity of the monovalent bridging ions with that observed here.

Conclusions

In general, the models and force field used here have been shown to reproduce conformations which have been found experimentally. The deviations between experimental crystalline and calculated solvated conformations can be explained in terms of solvation effects and the loss of crystal packing forces.

Several important structural features have been identified. The first of these is the HO3BO5A hydrogen bond, which appears to be one of the main driving forces behind the helix formation. It is interesting to note that this particular feature remains even in the presence of solvent effects.

The second major contribution is the presence of the (13) linkage in the gellan repeat unit. This appears to maintain its stability not only in the crystalline form, but also when solvent is present. Thus it appears to enhance gellan's ability to form, in solution, the helical conformation which is believed to be adopted in the gel state.

We have demonstrated that gellan gel formation is aided by the cation mediated aggregation of double helices. Our results show that divalent cations such as Ca^{2+} can assist in the aggregation of the double helices by directly bridging between carboxyl oxygens. A consequence of this is the formation of ordered micro-crystalline regions, which are necessary for the formation of rigid gels. In addition, some structuring of the water surrounding the binding site was observed.

References

1. Jansson, P.E.; Lindberg B.; Sandford, P.A. *Carbohydrate Research* **1983**, *124*, 135.
2. Graham, H.D. *Journal of Food Science* **1993**, *58* 539.
3. Papageorgiou, M.; Kasapis S.; Richardson, R.K. *Carbohydrate Polymers* **1994**, *25*, 101.
4. Duran, E.; Costell, E.; Izquierdo L.; Duran, L. *Food Hydrocolloids* **1994**, *8*, 373.
5. Crescenzi, V.; Dentini, M.; Coviello T.; Rizzo, R. *Carbohydrate Research* **1986**, *149*, 425.
6. Crescenzi, V.; Dentini, M.; Coviello, T.; Paoletti, S.; Cesàro A.; Delben, F. *Gazetta Chimica Italiana* **1987**, *117*, 611.
7. Grasdalen H.; Smidsrød, O. *Carbohydrate Polymers* **1987**, *7*, 371.
8. Crescenzi, V.; Dentini M.; Dea, I.C.M. *Carbohydrate Research* **1987**, *160*, 283.
9. Gunning A.P.; Morris, V.J. *Int. J. Biol. Macromolecules* **1990**, *12*, 338.
10. Okamoto, T.; Kubota, K.; Kuwahara, N. *Food Hydrocolloids* **1993**, *7*, 363.
11. Tang, J.; Lelievre, J.; Tung, M.A.; Y. Zeng, Y. *Journal of Food Science* **1994**, *59*, 216.
12. Ogawa, E. *Polymer Journal* **1995**, *27*, 567.
13. Moritaka, H.; Nishinari, K.; Taki, M.; Fukuba, H. *Journal of Agriculture and Food Chemistry* **1995**, *43*, 1685.
14. Weiner, S.J.; Kollman, P.A.; Case, D.A.; Chandra, S.U.; Ghio, C.; Alagona, G.; Profeta, S.P.; Weiner, P. *Journal of the American Chemical Society* **1984**, *106*, 765.
15. Homans, S.W. *Biochemistry* **1990**, *29*, 9110.
16. Verlet, L. *Physical Review* **1967**, *159*, 98.
17. Berendsen, H.J.C.; Postma, J.P.M.; van Gunsteren, W.F.; DiNola, A.; Haak, J.R. *Journal of Chemical Physics* **1984**, *81*, 3684.
18. Stern, P.S.; Chorev, M.; Goodman M.; Hagler, A.T. *Biopolymers* **1983**, *22*, 1885.
19. Jorgensen, W.L.; Chandrasekhar, J; Madura, J.D. *Journal of Physical Chemistry* **1983**, *79*, 926.
20. Engelsen, S.B.; Hervé du Penhoat, C.; Pérez, S. *Journal of Physical Chemistry* **1995**, *99*, 13334.
21. Chandrasekaran, R.; Puigjaner, L.C.; Joyce, K.L.; Arnott, S. *Carbohydrate Research* **1988**, *181*, 23.
22. Chandrasekaran, R.; Radha, A. *Trends in Food Science and Technology* **1995**, *6*, 143.
23. Chandrasekaran, R.; Millane, R.P.; Arnott S.; Atkins, E.D.T. *Carbohydrate Research* **1988**, *175*, 1.
24. Chandrasekaran, R.; Radha, A.; Thailambal, V.G. *Carbohydrate Research* **1992**, *224*, 1.
25. Chandrasekaran, R. *Advances in Experimental Medical Biology* **1991**, *302*, 773.

J. Mol. Model. **1996**, 2, 183 – 189

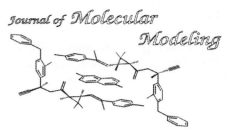

Modeling of Platinum Clusters in H-Mordenite

Maria E. Grillo* **and Maria M. Ramirez de Agudelo**

INTEVEP, S.A., Research and Technological Support Center of Petróleos de Venezuela, Apartado 76343, Caracas 1070A-Venezuela, Tel. (58) 2 9088054, Fax (58) 2 9086527 (meg2@intevep.pdv.com)

Received: 5 December 1995 / Accepted: 9 July 1996 / Published: 16 August 1996

Abstract

The size, location and structure of Pt clusters in H-mordenite have been investigated by molecular mechanics energy minimization and molecular dynamics simulation techniques using the **Catalysis** software of Molecular Simulations (MSI). Lattice energy minimizations are performed to study the effects of the specific framework aluminum positions on the location and stability of monoatomic Pt sites in H-mordenite. The lattice energies relative to the siliceous platinum-aluminosilicate structure reveal that the stability of a single Pt atom in H-mordenite is remarkably influenced by the specific location of the Al atoms in the lattice. At the studied Si/Al ratio of two Al ions per unit cell, a stabilization of the H-mordenite lattice upon Pt deposition is obtained. Moreover, lattice energy calculations on Pt/aluminosilicate mordenites of different metal contents per unit cell have been performed. An optimum size for the aggregate confined to the 12-ring main channel that is almost independent of the Pt content per mordenite unit cell has been found. The structural features of the resulting clusters at the end of molecular dynamics simulations on Pt/alumina-mordenites reflect a strong metal-zeolite interaction. The present results are consistent with a previous molecular dynamics simulation study on the structure of platinum deposited on SiO_2 surfaces.

Keyword: Supported Pt clusters, H-mordenite, molecular mechanics, molecular dynamics
Running Title: Modeling of Pt Clusters in H-Mordenite

Introduction

Over the past few years much effort has been given to the study of the catalytic activity of metal/zeolite systems. The activity of this type of catalyst has been related to the dispersion of the metal cluster as well as the nature of the zeolite used. The understanding of the factors controlling the growth process of the metal clusters in zeolite super cages is important for catalyst design. Recently, quite a number of experimental studies have concentrated on understanding and fine tuning the experimental conditions, leading to the most ac-

tive and selective Pt/H-mordenite catalyst. [1-6] This bifunctional catalyst is widely used for hydroisomerization of light alkanes to obtain isomers that have proved to be octane enhancers of gasoline.

The final location and state of the metal ions after the calcination step will determine the final metal dispersion obtained, as suggested by Homeyer and Sachtler. [1] For instance, clusters of a maximum size of 0.8-1.0 nm are obtained in the H-mordenite main channel when suitable preparation conditions are employed. A cluster in this size range is found by Transmission Electron Microscopy (TEM) measurements on H-mordenite, when the platinum amine complex $[Pt(NH_3)_4]^{2+}$ is calcinated at 300 °C, followed by reduction at 500 °C. [3] Our lattice energy calculations on alumina-mordenites of different Pt contents indicate a maximum cluster size of about 0.8 nm, in agreement with TEM results [3] on Pt/H-mordenite.

In the case of Pt in mordenite, preparation at low reduction temperature, low metal loading and high proton concentration currently result in highly dispersed, even monoatomic, platinum. The monoatomic Pt sites in mordenite were first detected by the H_2 evolution above 300 °C using Temperature-Programmed Desorption (TPD). [4] A later study based on H/D exchange of cyclopentane on Pt/H-mordenite confirms the existence of extremely small Pt particles, possibly monoatomic sites. [5] These isolated Pt atoms are proposed to be located in the side-pockets of the mordenite main channel from the observed stereoselectivity of methylcyclopentane ring-opening catalysis, [4] as well as by Monte Carlo calculations on siliceous mordenite. [7] Furthermore, FTIR studies of CO adsorption suggest that platinum is stabilized by zeolitic protons probably forming $[Pt_1–H_z]^{z+}$ adducts. [6] The present calculations demonstrate that the side pocket is the energetically preferred location of a single Pt atom in alumina-mordenite at the considered Si/Al ratio.

The average number of metal atoms per cluster can be estimated by the xenon adsorption method, as developed by Ryoo et al. [8] This method has been used to estimate the cluster size of several group VIII metals supported in cubic faujasite, e.g. Y zeolite, and of Pt in hexagonal faujasite (EMT). [2,8,9,10] A very narrow distribution in the cluster size has been obtained for 2-10 wt% Pt/NaY samples by extended X-ray absorption fine structure (EXAFS) and xenon adsorption measurements. The Pt cluster filling the super cage is proposed to consist of 50-60 atoms. In hexagonal faujasite (EMT), the EXAFS and Xe adsorption results suggest that a Pt cluster of 20-30 atoms is formed in the smaller three windows super cage. An important result of these studies is that the size of the Pt cluster that just fills a super cage does not change with the metal content, and it is stabilized by the interaction with the cage-wall. This is consistent with the present lattice energy calculations and molecular dynamics

simulations on Pt_x/alumina-mordenites, where x is the number of metal atoms per mordenite unit cell.

In this paper the structure and stability of a series of platinum clusters loaded in alumina-mordenite are modeled by both Molecular Mechanics Energy Minimization (MMEM) and Molecular Dynamics Simulations (MDS) techniques. Furthermore, the influence of the specific Al substitution Tetrahedral sites (T-sites) on the location and stability of monoatomic platinum sites in aluminosilicate mordenite is address by using the MMEM procedure.

Methodology

The **Discover** code of MSI [11] was used to perform **mo**lecular **m**echanics **e**nergy **m**inimizations (MMEM) and **mo**lecular **d**ynamics **s**imulations (MDS) of Pt clusters supported on aluminium substituted at distinct T-sites mordenites. The theoretical aspects and applications of these computational techniques in modeling and predicting crystal structures have been extensively reviewed. [12]

Effects of the Pt content

Molecular Mechanics Energy Minimization. The lattice energy is calculated and minimized using molecular mechanics potential functions (force fields). The functional form of the interatomic potentials implemented in this code include the Coulombic part of the lattice energy calculated exactly by the Ewald summation method and short range forces.

The minimizations are performed in two stages: firstly the aluminosilicate mordenite is relaxed to its minimum energy configuration using a "Consistent Valence Force Field" for the simulation of protonated aluminosilicates included in the **Discover** code (CVFF-CZEO). This force field treats the oxygen atoms as different atom types depending on the particular environment in the aluminosilicate structure and accounts for bridging hydroxyl groups. For a detailed description of the method used to derive the potential parameters as well as the functional form of the potential energy function see references 13 and 14.

In a second stage of the minimization, the energy preferred structure of a Pt cluster inside the aluminosilicate cage is calculated employing a force field implemented in **Discover** which includes metals (CVFF_CALP). This force field accounts for different short-range van der Waals (vdW) forces between the Pt atoms and both Si-O-Si and Si-O-Al sites in the zeolite structure. The potential model used, however, does not contain parameters for the interaction of the zeolitic hydroxyl groups and the Pt atoms. Hence, the present simulations do not account for the polarization effect of the zeolite Brønsted acid sites on the Pt atoms. Nevertheless at the considered Si/Al ratio (two Al ions per unit cell), this omission might not introduce a serious error in determining the effect of the inner zeolite walls interactions with neutral

Table 1. *Interatomic potential parameters of the CVFF_CALP parameter set implemented in Discover. Pt, O, Si and Al refers to the parameters for the platinum atoms and for the oxygen, silicon and aluminum ions in the zeolite structure. The parameters, A_i and B_i, are given in (kcal/mol·$Å^{12}$) and (kcal/mol·$Å^6$), respectively.*

i	A_i	B_i
O	272894.7846	498.8788
Si	3149175.0000	710.0000
Al	3784321.4254	11699.8493
Pt	4576819.9618	16963.3082

Table 2. *Interatomic potential parameter set, PCFF, implemented in Discover. ε_i refers to the potential well depth in kcal/mol, and r_i^* is the interatomic distance in Å.*

parameter		r_i^*	ε_i
o_z		3.3200	0.2400
a_z		0.0001	0.0000
o_b	[a]	5.2191	0.0135
h_b	[a]	1.2149	5.2302
o_{sh}	[b]	3.4618	0.1591
h_{os}	[b]	2.3541	0.0988
o_{as}	[c]	5.2591	0.0129
o_{ah}	[d]	3.7245	0.1026
h_{oa}	[d]	1.2879	3.6860
o_{ss}	[e]	3.4506	0.1622

[a]   *Oxygen and hydrogen atoms in bridging hydroxyl group.*
[b]   *Oxygen and hydrogen atoms in a terminal hydroxyl group connected to silicon.*
[c]   *Oxygen atom bonded to aluminum atom.*
[d]   *Oxygen and hydrogen atoms in a terminal hydroxyl group connected to aluminum.*
[e]   *Oxygen atom between two SiO_4 tetrahedra.*
The pair interaction potential:

$$V\left(r_{ij}\right) = \varepsilon_{ij}\left[2\left(\frac{r_{ij}^*}{r_{ij}}\right)^9 - 3\left(\frac{r_{ij}^*}{r_{ij}}\right)^6\right]$$

The off-diagonal potential parameters take the form:

$$\varepsilon_{ij} = \frac{\sqrt{\varepsilon_{ii}\varepsilon_{jj}}\,2r_{ii}^{*3}r_{jj}^{*3}}{r_{ii}^{*6} + r_{jj}^{*6}}\;;\; r_{ij}^* = \left(\frac{r_{ii}^{*6} + r_{jj}^{*6}}{2}\right)^{\frac{1}{6}}$$

Pt atoms, on the structure an stability of large supported clusters.

The metal-zeolite interaction is calculated using a pair Lennard-Jones (12-6) potential:

$$V\left(r_{ij}\right) = \frac{A_{ij}}{r_{ij}^{12}} - \frac{B_{ij}}{r_{ij}^6} \tag{1}$$

where r_{ij} is the distance between atoms i and j in angstroms. The off-diagonal, heteronuclear interactions are calculated as geometric averages of the form: $A_{ij} = \sqrt{A_i \times A_j}$ and

$B_{ij} = \sqrt{B_i \times B_j}$. The parameters used for the homonuclear

Table 3. *Cell parameters and symmetry information on the mordenite lattice. Percent of Al in the tetrahedral sites, T = Tetrahedral. The distances are given in Å*

	a = 18.094		
	b = 20.516		
	c = 7.524		
	$\alpha = \beta = \gamma = 90°$		
	space group: Cmc21 (No. 36)		

T1	T2	T3	T4
12	5	28	17

interactions are presented in   Table 1. These quantities are related to the potential well depth ε_{ij} and with the interatomic distance r_{ij}^* at which the minimum occurs in a straightforward manner:

$$A_{ij} = \varepsilon_{ij}r_{ij}^{*12} \text{ and } B_{ij} = 2\varepsilon_{ij}r_{ij}^{*6}.$$

The used of a Lennard-Jones potential is a severe approximation in the calculations. However, the success of this approximation in previous studies of the self-diffusion of metallic adsorbates on metallic substrates [15,16,17] and atomic C and O self-diffusion on Pt(111) [18], together with the study of the structural features of platinum deposited on a vitreous silica substrate [19], encourages the used of Lennard-Jones potentials from MSI in this work.

Molecular Dynamics Simulations. The simulation system consists of a single periodically replicated unit cell of H-mordenite containing two aluminium ions, two charge compensating protons and Pt clusters of 20 and 38 atoms per unit cell respectively, $(Pt)_x/H_2Al_2Si_{46}O_{96}$. These structures are previously relaxed to the minimum energy configurations. The equations of motion are integrated with a time step of 0.001 ps in the canonical ensemble (NVT). The system is equilibrated to the desired temperature for 5000 iterations, the dynamics is run for 10000 iterations, equilibrated again for further 5000 iterations and then allowed to converged to a equilibrium structure for 40000 iterations, time over which the measurable properties are averaged. These equilibration and dynamic times proved to be sufficiently long to calculate equilibrium properties of the systems considered.

Effects of the specific Al framework positions

The effect of the position of the Al ions and bridge hydroxyl groups on the location and stability of a single Pt atom in a H-mordenite unit cell is evaluated by MMEM by using a force field potential model, which allows for the interaction

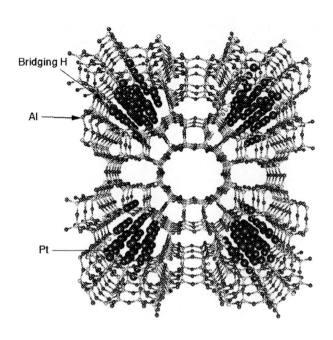

Figure 1. *Schematic diagram of the optimum energy structure of the (Pt)$_{22}$/H-mordenite model system.*

of Brønsted acid sites with Pt atoms. For this purpose, a recently released force field (PCFF) accounting for the interaction of Pt with all different atom types in acidic aluminosilicates defined in the CVFF_CZEO parameter set, has been used (see Section 3.2 and Table 2).

Results and Discussion

Effects of the Pt content

Molecular Mechanics Energy Minimizations. The lattice energies of Pt/alumina-mordenites with a metal content ranging from 20 to 40 atoms per unit cell have been calculated. For the purpose of these calculations, we have considered a mordenite unit cell containing two aluminium ions substituted at T1 and T3 sites and two charge compensating protons. For the mordenite structure, the data of Alberti et al., [20] summarized in Table 3 has been used. The framework has four crystallographically distinct T-sites for Si or Al (T = Tetrahedral).

The Pt was introduced as an fcc type cluster in the centre of the unit cell. The minimization proceeds in two steps: first the mordenite framework atomic positions are adjusted until a minimum energy structure is obtained. In a second step of the optimization, the host structure is held rigid and only the Pt atoms are allowed to relax, until a minimum lattice energy is obtained.

In Table 4, the relative lattice energies corresponding to the optimum structures of (Pt)$_x$/alumina-mordenites with different number of metal atoms, *x*, per unit cell are given. The calculated energies suggest an optimum cluster size of 23

Table 4. *Relative lattice energies, **E**, in kcal/mol of the considered (Pt)$_n$/H$_2$Al$_2$Si$_{46}$O$_{96}$ structural models, (Al at T1 and T3 sites). **n** is the total number of Pt atoms contained per mordenite unit cell. **n$_{mc}$** is the number of Pt atoms confined to the 12-ring main channel. Δ is the energy gap between two succesive model structures.*

model	n	n_{mc}	E^a	Δ
1	20	20	-747.4	
				-107.4
2	22	21	-854.8	
				-514.4
3	30	22	-1369.1	
				-272.0
4	34	22	-1641.2	
				-144.4
5	38	23	-1785.6	
				502.0
6	40	24	-1238.6	

[a] *Relative to the lattice energy of the siliceous mordenite structure.*

atoms located in the 12-ring channel, as a result of the size constraints defined by the cage dimension. This cluster size is nearly independent of the initial number of Pt atoms in the fcc-cluster per unit cell. The remaining Pt atoms are distributed in side pockets. Moreover, the significant energy gap obtained when going from a total Pt content of 38 to 40 atoms per unit cell, is indicative of the maximum metal content per unit cell. Figure 1 shows the resulting structure from the lattice energy minimization of a fcc-type Pt cluster containing 22 atoms per H-mordenite unit cell (model 2).

These results might be related to recent EXAFS and xenon adsorption measurements on 2-10 wt% Pt/NaY and on Pt/EMT zeolites. [10] From these studies it is concluded that the average Pt cluster size in the super cage of NaY zeolite and in the small super cage of EMT is almost independent on the Pt content of the samples.

After the structure optimization, the resulting clusters contained in the mordenite main channel are arranged in a nearly symmetrical manner to just fit the cage dimension. This indicates a significant interaction between Pt and the cage wall, which results in a structure reconstruction of the fcc-metal particles. This metal-zeolite interaction stabilizes an optimum cluster size within the cage and limits the tendency to agglomeration, as suggested by experimental studies on clustering of Pt on FAU types zeolites. [2,10] Moreover, the optimum number of Pt atoms obtained suggests an aggregate of a maximum diameter of 0.8 nm as indicated by TEM results on a Pt/H-mordenite catalysts mentioned above. [3]

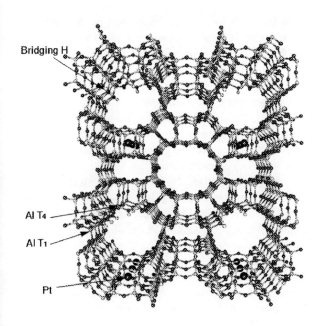

Figure 2. *Schematic diagram of the minimum energy Pt location in H-mordenite substituted at the T1 and T4 framework positions.*

Table 5. *Calculated structural information for the Pt_n/ H-mordenite model structures 1 and 5 (n= 20 and 38 in Table 4) at 723 K. $N_c^{Pt\text{-}Pt}$ and $N_c^{Pt\text{-}O}$ are the coordination numbers of the nearest Pt-Pt and Pt-O pairs, respectively. $d_{Pt\text{-}Pt}$ and $d_{Pt\text{-}O}$ are the bond lengths in Å for the Pt-Pt and Pt-O pairs, respectively. **fcc** refers to the structure prior to the interaction with the support. **final** refers to the structure after the equilibration at 723 K.*

model	5		1	
	fcc	final	fcc	final
$d_{Pt\text{-}Pt}$	2.77	2.66	2.77	2.70
$N_c^{Pt\text{-}Pt}$	12	5	12	5
$d_{Pt\text{-}O}$	1.92	2.78	2.26	2.69
$N_c^{Pt\text{-}O}$	1	3	1	2

The strong reconstruction of Pt fcc-clusters induced by the interaction with the zeolite is fully consistent with calculations reported by Levine et al. [19] on structural changes of Pt clusters upon deposition on an amorphous SiO_2 surface. As in the present study, these authors used a pair-wise additive Lennard-Jones (12-6) potential to describe the Pt-substrate interactions. In all simulated deposition processes, low energy Pt clusters resulted to be highly asymmetric and more disorder than in the starting configurations. With increasing the interaction with the surface prior to growth, a corresponding increase in dispersion (smaller particles) was obtained. Moreover conjugate gradient minimizations by Sachdev et al. of isolated Pt clusters with 5-60 atoms with polyhedral symmetry, demonstrated that the non-magic number clusters relaxed into structures which did not resemble the initial polyhedral symmetry. [21]

Molecular Dynamics Simulation results. In this section, Pt/ alumina-mordenites of different metal content per unit cell are equilibrated and allowed to relax to an equilibrium structure at 723K. Specifically, the minimized structures 1 and 5 of Table 4 containing 20 and 38 metal atoms per unit cell are taken as starting configurations in the molecular dynamics simulations. The coordination numbers of the nearest Pt-Pt and Pt-O pairs are calculated for a central Pt atom from the corresponding radial distribution functions. The calculated structural features for each of the cluster models are displayed in Table 5.

In both models considered, the metal clusters confined to the 12-ring main channel presents the same average coordination number $N_c^{Pt\text{-}Pt}$ for the nearest Pt-Pt pairs and the bond length coincides with a range of 2.68 ± 0.2 Å, see Table 5. This also confirms the above mentioned EXAFS results on Pt supported in NaEMT and NaY zeolites. [2,8,10] The calculated $N_c^{Pt\text{-}Pt}$ value for Pt/H-mordenite compares favourably with those from EXAFS for H-mordenite ($N_c^{Pt\text{-}Pt}$ = 4.2),[22] NaEMT ($N_c^{Pt\text{-}Pt}$ = 6.6) [10] and NaY ($N_c^{Pt\text{-}Pt}$ = 4.9).[2] This value is much smaller than the 12 value in the bulk fcc structure indicating an increase in the metal dispersion (smaller particle size) as a result of the size constraint imposed by the cage dimension.

In the resulting clusters of the models 1 and 5, the Pt-O coordination number ($N_c^{Pt\text{-}O}$) increases from 1 (in the fcc-type clusters) to 2 and 3 respectively, see Table 5. This increase of the Pt-O coordination number value indicates a stronger interaction of platinum with the zeolite oxygens in the low energy structure compare to the initial configuration. The metal redispersion is higher for the Pt cluster model 5 (n=38). A decrease in the coordination number might be seen as a decrease in dimensionality of the Pt crystals, similar to that found by Levine et al. for Pt deposition on SiO_2 surfaces (as shown in Fig. 1). A decrease in $N_c^{Pt\text{-}Pt}$ from 12 in the fcc particles to a cluster of $N_c^{Pt\text{-}Pt}$ = 5 might suggest a change from tridimensional particles to almost bidimensional ones. Thus, this might predict the growth of flakes or even ribbon-type crystals attached and stabilized by the zeolite walls.

Effects of the specific Al framework positions

The framework stability of mordenite containing two aluminium ions, one Pt atom and two charge compensating protons per unit cell has been studied by means of Molecular Mechanics Energy Minimizations (MMEM). In this section,

J. Mol. Model. **1996**, *2*

Table 6. *Relative lattice energies, **E**, in kcal/mol of the six calculated Pt/H-mordenite structural models, (Pt/ $H_2Al_2Si_{46}O_{96}$). Δ is the energy difference between two successive models.*

model	Al T-sites	E [a]	Δ	Pt location
1	T1, T4	-293.3		s.p.
			8.5	
2	T3, T4	-284.79		s.p.
			6.9	
3	T2, T4	-277.88		s.p.
			3.2	
4	T1, T3	-274.73		m.c.
			14.4	
5	T1, T2	-260.31		m.c.
			4.1	
6	T3, T2	-256.25		m.c.

s.p.: side pocket

m.c.: main channel.

[a] Relative to the lattice energy of the siliceous Pt/ mordenite structure.

we concentrate on the effects of the specific location of the two Al atoms within the mordenite framework on the stability of the Pt/aluminosilicate mordenite lattice. It is assumed that the specific locations of the bridging hydroxyl groups within the mordenite lattice have only a secondary effect on the lattice stability. This is based on recent lattice energy calculations on H-mordenite, which report that there is not a single preferred bridging hydroxyl group. Instead, these are found to be distributed over several framework sites. [23] Hence, in this work the two OH groups are distributed over arbitrary chosen lattice sites.

The six possible structural models, combining two Al ions substituted over the four crystallographically different T-sites in the mordenite lattice have been calculated. In all the structures considered, the Pt atom was initially located at the centre of the 12-ring main channel. The effect upon aluminium insertion in the siliceous Pt/mordenite structure is a stabilization of the lattice energy (more negative). In Table 6, the lattice energies of the calculated Pt/H-mordenite structural models relative to siliceous Pt/mordenite are displayed. As reflected by the energy difference between successive models, the position of the Al ions within the framework structure influences significantly the lattice stability of Pt loaded in H-mordenite.

In the three most energetically favoured structural models of Pt/H-mordenite (see Table 6) as well as in the siliceous structure, the minimum energy configuration locates the Pt atom in side pockets. This suggest that at the Si/Al ratio stud-

ied, the preferred Pt location is determined by the more effective dispersive vdW interactions in a smaller pore.

Considering the experimental enrichment of Al in the mordenite T-sites (see Table 3), the obtained trend in relative stabilities of Pt/H-mordenites (see Table 6) indicates that for an alumina-mordenite with aluminium statistically distributed throughout the structure, there is a high probability of finding Pt nucleation sites in side pockets. Obviously, in real cases that would depend on the pretreatment temperatures to which the solid has been subjected.

This result has already been suggested by the stereoselectivity of the methylcyclopentane conversion catalysis on Pt/H-mordenite [4] and predicted by a previous Monte Carlo simulation study of Pt on siliceous mordenite. [6] This might also be related to the formation mechanism suggested by Sachtler et al. for transition metal clusters (Pt, Pd and Ni) in NaY zeolite at high calcination temperatures. In the proposed kinetic path, the bare metal ions migrate after calcination to smaller cages that provide greater charge stabilization. [1]

Upon deposition of a Pt atom on acidic mordenite, a stabilization of all six structural models with respect to the H-mordenite lattice was obtained. The largest stabilization on Pt loading of -17.8 kcal/mol was obtained for the H-mordenite with the Al atoms substituted in the framework positions T1 and T3 (model 4 in table 6). An energy lowering of about 16 kcal/mol was obtained for all structural models involving the T4 Al-substitution site (first three low energy structures in Table 6).

Conclusions

The present lattice energy minimization results suggest that the relative stability of monoatomic platinum sites in aluminosilicate mordenites is related to the specific aluminium insertion T-sites in the framework structure. Nevertheless at the Si/Al ratio studied, the optimum Pt location is determined by the interaction with mordenite walls rather than by the interaction with Brønsted protons, as observed by the calculated effect of aluminium substitution on platinum location in mordenite (see Table 6).

Lattice energy minimizations and molecular dynamics simulations on Pt/alumina-mordenites of different Pt contents seem to indicate that the structural features of the platinum cluster confined to the 12-ring main channel are almost independent of the total Pt content, and strongly dependent upon the surrounding zeolite structural field.

References

1. Homeyer, S. T.; Sachtler, W. M. H. in Stud. Surf. Sci. Catal.; Weitkamp, J.; Karge, H.G.; Pfeifer, H., Hölderich, W. (Eds.) Elsevier: Amsterdam, 1989; Vol. 49, pp 984-975.

2. Ryoo, R.; Cho, S. J.; Pak, C.; Lee, J. Y. *Catal. Lett.* **1993**, *20*, 107.

3. Giannetto, G.; Montes, A.; Alvarez, F.; Guisnet, M. *Revista Soc. Venez. Catal.* **1991**, *5*, 33.

4. Lerner, B. A.; Carvill, B. T.; Sachtler, W. M. H. *J. Mol. Catal.* **1992** , *77*, 99.

5. Lei, G.; Sachtler, W. M. H. *J. Catal.* **1993**, *140*, 601.

6. Zholobenko, V.; Carvill, B.; Lerner, B.; Lei, G.; Sachler, W. M. H. *Proceedings of the 206th National Meeting American Society*, Chicago, IL, **1993**, *735*.

7. Blanco, F.; Urbina-Villalba, G.; Ramirez de Agudelo, M. M. in Stud. Surf. Sci. Catal.; Weitkamp, J.; Karge, H. G.; Pfeifer, H.; Hölderich, W. (Eds.) Elsevier: Amsterdam, 1994; Vol. 84, pp 2162-2155.

8. Ryoo, R.; Cho, S. J.; Pak, C.; Kim, J-G; Ihm, S-K; Lee, J. Y. *J. Am. Chem. Soc.* **1992**, *114*, 76.

9. Kim, J-G; Ihm, S-K; Lee, J. Y.; Ryoo, R. *J. Phys. Chem.* **1991**, *95*, 8546.

10. Ihee, H.; Bécue, T.; Ryoo, R.; Potvin, C.; Manoli, J-M; Djéga-Mariadassou, G. in Stud. Surf. Sci. Catal.; Weitkamp, J.; Karge, H. G.; Pfeifer, H.; Hölderich, W. (Eds.) Elsevier: Amsterdam, 1994; Vol. 84, pp 772-765.

11. Discover Molecular Simulation Program, version 94.1, Molecular Simulations Inc., San Diego, 1995.

12. Catlow, C. R. A. Modelling of Structure and Reactivity in Zeolites; Academic Press: London, 1992.

13. Hill, J-R; Sauer, J. *J. Phys. Chem.* **1994**, *98*, 1238.

14. Hill, J-R; Sauer, J. *J. Phys. Chem.* **1995**, *99*, 9536.

15. Doll, J. D. and McDowell, H. K. *J. Chem. Phys.* **1982**, *77*, 479.

16. Doll, J. D. and McDowell, H. K. *Surf. Sci.* **1983**, *123*, 99.

17. McDowell, H. K. and Doll, J. D. *J. Chem. Phys.* **1983**, *78*, 3219.

18. Doll, J. D.; Freeman, D. L. *Surf. Sci.* **1983**, *134*, 769.

19. Levine, S. M.; Garofalini, S. H. *Surf. Sci.* **1985**, *163*, 59.

20. Alberti, A.; Davoli, P.; Vezzalini, G. Z. *Kryst.* **1986**, *175*, 249.

21. Sachdev, A.; Masel, R. I.; Adams, J. B. *J. of Cat.* **1992**, *136*, 320.

22. Lamb, H. H.; Clayton, M. J.; Otten, M. M.; Reifsnyder, S. N. Presented at the 14th North-American Meeting of the Catalysis Society, Snowbird, Utah. June 11-16, 1995.

23. Schöder, K-P; Sauer, J. Proceedings of the 9th International Zeolite Conference, Montreal 1992; von Ballmoos, R., et al. (Eds.) Butterworth-Heinemann, 1993; pp 694-687.

J. Mol. Model. **1996**, 2, 190 – 204

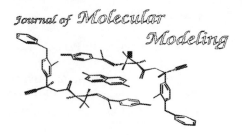

PM3 and AM1 Study on β-N-acetyl-muramic Acid and 3 Murein Related Derivatives

Peter Keller* and Hans Bradaczek

Institut für Kristallographie, Freie Universität Berlin, Takustr. 6, D-14195 Berlin, Germany. Tel.: +49-30-838-3458;
 Fax: +49-30-838-3464 (keller@chemie.fu-berlin.de)

Received: 4 March 1996 / Accepted: 8 August 1996 / Published: 30 August 1996

Abstract

The energetically favoured conformations of β-N-acetyl-Muramic acid, its C6-O-acetylated form, the methylamide and the methyl-glycoside have been investigated using the semiempirical PM3 and AM1 methods. All these compounds are either components or fragmentary structures of the murein network. The atomic coordinates of the starting set of the β-N-acetyl-Muramic acid molecule have been obtained by a PM3 minimization of one saccharide molecule cut out from the murein single strand model proposed by Barnickel at al. [1]. The sidegroups of the derivatives have been introduced by a molecular editor. These conformations served as starting points in conformational space for a grid search by scanning all sidechain torsional angles for non-hydrogen atoms with exception of the N-acetyl group which was held in cisoid position (i.e. N2-H bond is parallel to C1-H and C3-H bond) and only minimized. The PM3 method with an additional amide correction potential and the AM1 method were used. The torsional angle distributions of the lactyl sidechain (free acid and methylamide), the C6-O-acetylated sidechain and the C1-methoxy sidechain have been investigated, showing distinct energetically favoured torsional angle regions. The results are compared to earlier studies on β-N-acetyl-Muramic acid by J.S. Yadav et al. [2,3] who were using the MNDO and PCILO methods and by P.N.S. Yadav et al. [4] who were using the empirical MM2 force-field.

Keywords: Muramic Acid, Derivatives, Quantum Chemistry, Semiempirical Method, Conformational Energy

Introduction

β-N-acetyl-D-Muramic acid (NAM) and its C6-O-acetylated form are present in murein, a component of bacterial cell walls [5,6]. The C6-O-acetylated β-N-acetyl-D-Muramic acid (NAM-O6AC) has been found in the murein of Proteus mirabilis [7], Neisseria gonorrhoeae [8], Staphylococcus aureus [9] and 8 further species, see review [10]. It has been proposed, that O-acetylation is one factor for the reduced lysozyme degradability of murein [8,11]. The methylamide of NAM (NAM-MEAM) simulates the amide linkage between the saccharide and the peptide part in the murein network. Additionally the conformations of β-methylglycoside (NAM-O1ME) have been calculated.

The precise knowledge of the spatial structure of NAM and its derivatives is of interest for the molecular modelling of the murein network and especially for the drug design of muramyl peptide related compounds showing immuno-stimulating, somnogenic and pyrogenic activities [12]. The semiempirical methods AM1 and PM3 seem accurate enough to be applied to larger biological molecules [13,14,15] - larger

---

* *To whom correspondence should be addressed*

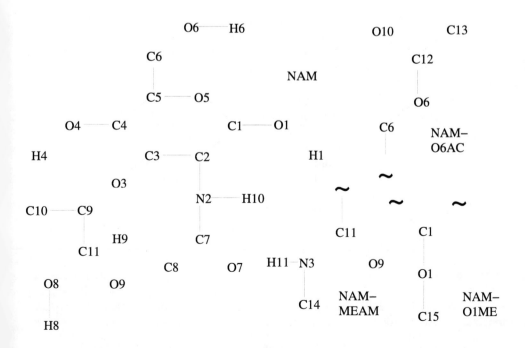

Figure 1. *Chemical formula of NAM and derivatives.*

NAM:	R1 = OH;	R2 = COOH;	
	R3 = OH		
NAM-O6AC:	R1 = O(CO)CH$_3$;	R2 = COOH;	
	R3 = OH		
NAM-MEAM:	R1 = OH;	R2 = COOH;	
	R3 = OH		
NAM-O1ME:	R1 = OH;	R2 = COOH;	
	R3 = OCH$_3$		

in this quantum chemical context means equal or more than 10 non-hydrogen atoms. Their specific ability to find reasonable intramolecular hydrogen bonds makes these methods superior to the original MNDO parametrization.

Methods and Details of Calculations

Molecular modelling methods

In the present calculations the MNDO/PM3 method [16] with empirical correction potential for the torsional angle of the amide bond and the AM1 method [17] are used. The calculations were performed on the IRIS/INDIGO (Silicon Graphics Inc.) cluster of our institute under the operating system IRIX System V.4 using SPARTAN, SGI versions 3.1.1 GL and 4.0.3 GL (Wavefunction, Inc.). Systematic grid search method was used for conformational analysis. From each grid point an energy minimization was started. Conformations with high energies were rejected.

Starting geometries and search parameters. The murein single strand model of Barnickel et al [1] has been used as starting set. The cartesian coordinates were kindly provided by the authors. One single NAM molecule in $^4C_1$ conformation has been cut out from the strand model, followed by a PM3 minimization. The starting conformations of the derivatives have been obtained by using the molecular editor implemented in SPARTAN to create the different side-chains. These conformations served as starting points in conformational space for a grid search by scanning all sidechain torsional angles for non-hydrogen atoms with exception of the N2-acetyl group which was minimized from cisoid position (i.e. N2-H bond is parallel to C1-H and C3-H bond). All amide bonds were set trans. Grid stepsize was set to 120° for all scanned torsional angles. The N2-acetyl group was not included in the angle scan but solely minimized, because X-ray structures of related compounds, like α-N-acetyl-Muramic acid [18], β-N,N'-diacetyl-Chitobiose trihydrate [19] and the co-crystallized complexes of muramyl-dipeptide (MDP) with isolectin I [20] and the β(1→4) connected trisaccharide NAM-NAG-NAM (NAG: N-acetyl-2-Glucos-amine) bound to hen egg white lysozyme [21], show that the favoured position of the N2-acetyl group is cisoid or, in the occurrence of hydrogen bonds, a slightly rotated position of the amide plane from cisoid position.

Figure 2. *Numbering scheme for NAM and derivatives.*

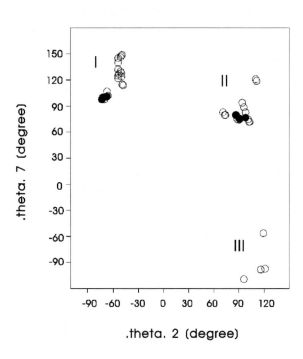

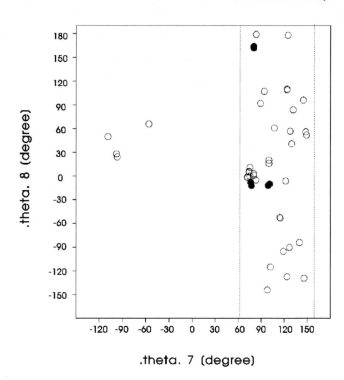

Figure 3. *Scatterplot of the O3 (ether linkage) torsional angles (θ2 vs. θ7) of the 40 NAM-conformers with lowest energy calculated by the PM3 methode with amide correction potential (o) and the 10 NAM-conformers with lowest energy calculated by the AM1 methode (●). The roman numbers denote the energetically favoured torsional angle regions.*

Figure 4. *Scatterplot of the C9 (lactyl-.alpha.-C) torsional angles (θ7 vs. θ8) of the 40 NAM-conformers with lowest energy calculated by the PM3 methode with amide correction potential (o) and the 10 NAM-conformers with lowest energy calculated by the AM1 methode (●). Most of the PM3 conformers are equally distributed in a narrow angle band (dotted lines) θ7 ranging between 70° and 150°. The region III PM3-conformers of the ether linkage (O3) scatterplot are located in a separate area.*

Definitions

Figure 1 shows the chemical formula and figure 2 the numbering scheme for NAM and its derivatives. The definitions of the torsional angles are given in table 1. The torsional angles are defined by a sequence of 4 atoms. The second bond between atom 2 and 3 is the axis of rotation. A positive rotation is then defined as the rotation of the third bond between atom 3 and 4 in a counter-clockwise direction viewed from the moving bond (between atom 3 and 4) towards the fixed bond (between atom 1 and 2). (SPARTAN implemented signum definition)

Results

β-N-acetyl-D-Muramic acid

Listing of low-energy conformers of NAM. Table 2 lists the 10 lowest energy conformers calculated by the PM3 method with amide correction potential and the AM1 method. Additionally, the reference data of J.S. Yadav et al. [2,3] and P.N.S. Yadav et al. [4] are given in table 2. The conformers are numbered with rising energy. To distinguish between PM3 and AM1 conformers, the AM1 conformers are marked with an

apostrophe. The absolute energy difference between the lowest energy conformer of PM3 and AM1 is considerable and amounts to 25 kcal/mol.

Calculated hydrogen bonding patterns. The hydrogen bonds of the first five conformers calculated by both methods are shown in table 3. Normally PM3 gives smaller values for the H-bond distance, ranging between 1.80 Å and 1.83 Å, compared to the AM1 values which range from 2.14 to 2.21 Å. The carboxylic acid group can form stable hydrogen bonds either to the C4-hydroxyl group or to the N2-H of the amide group, leading to a restriction of the conformational flexibility of the lactyl group. In case of hydrogen bonding, the amide group deviates slightly from planarity in some AM1 conformers. PM3 conformers are uneffected because of the usage of the additional empirical amide torsional angle potential.

Torsional angle distribution at the O3 ether linkage. According to our calculations the sterical requirements of the O3 ether linkage between pyranosyl ring and D-lactyl group are quite restrictive. Figure 3 shows the scatterplot (θ2 vs. θ7) of the O3 torsional angles. The 40 lowest PM3 energy

No.	NAM	NAM-O6AC	NAM-MEAM	NAM-O1ME
1	C1-C2-N2-C7			
2	C4-C3-O3-C9			
3	C5-C4-O4-H4			
4	O5-C5-C6-O6			
5	C5-C6-O6-H6	C5-C6-O6-C12	C5-C6-O6-H6	C5-C6-O6-H6
6	O5-C1-O1-H1	O5-C1-O1-H1	O5-C1-O1-H1	O5-C1-O1-C15
7	C3-O3-C9-C11			
8	O3-C9-C11-O9			
9	C9-C11-O8-H8	C9-C11-O8-H8	C9-C11-N3-C14	C9-C11-O8-H8
10	–	C6-O6-C12-O12	–	–

Table 1. *Definition of torsional angles*

NAM: β-N-acetyl-Muramic acid
NAM-O6AC:C6-O-acetylated β-N-acetyl-Muramic acid
NAM-MEAM: Methylamide of β-N-acetyl-Muramic acid
NAM-O1ME: Methylglycoside of β-N-acetyl-Muramic acid

Conf. No.	θ [°]								Energy [kcal/mol]	ΔE [kcal/mol]
	1	**2**	**3**	**4**	**5**	**6**	**7**	**8**		
PM3:										
1	109	-49	-168	72	-62	70	115	-53	-353.60	0.0
2	69	101	-49	79	-63	126	75	0	-353.28	0.32
3	114	-71	-176	72	-63	68	100	20	-352.91	0.69
4	105	-50	-169	72	-62	72	124	109	-352.89	0.71
5	109	116	-173	72	-62	69	-98	28	-352.64	0.96
6	98	-54	-175	73	-62	76	145	96	-352.52	1.08
7	93	74	-48	80	-63	83	80	3	-352.49	1.11
8	72	88	49	75	-64	119	75	11	-352.45	1.15
9	84	103	-50	79	-63	85	72	-2	-352.42	1.18
10	115	-70	178	73	-63	68	98	-144	-352.17	1.43
AM1:										
1'	87	91	-63	171	-69	76	76	-8	-379.49	0.0
2'	87	91	-74	-67	61	77	75	-8	-379.33	0.16
3'	87	91	-63	65	-57	76	75	-8	-378.92	0.56
4'	87	86	-63	171	-69	75	80	164	-378.37	1.12
5'	87	87	-74	-68	62	76	80	163	-378.19	1.30
6'	87	87	-64	65	-57	75	80	162	-377.76	1.73
7'	87	98	-161	157	57	74	77	-12	-377.43	2.06
8'	80	-72	-174	66	-58	67	99	-12	-376.39	3.10
9'	80	-67	-170	-55	61	69	101	-10	-376.32	3.17
10'	80	-73	-174	157	56	65	98	-12	-376.24	3.24

(ref.data) Methode									Energy	Reference
MNDO	113	99	-97	159	-165	-61	98	53	Minimum	[2]
PCILO	173	142	-54	-168	-177	-61	139	-142	Minimum	[3]
MM2	100	-86	56	-179	162	104	-78	-84	Minimum	[4]

Table 2. *The torsional angles and absolute and relative energies of the calculated first 10 low-energy conformers of NAM by PM3 and AM1 methode are listed. θ 9 allways minimized to the value of 180°. Therefore it is not included in the table.*

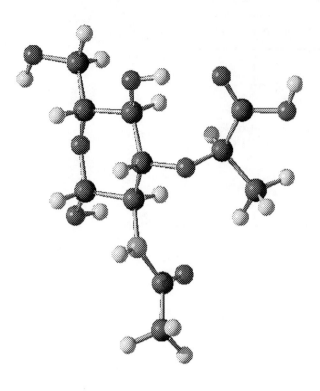

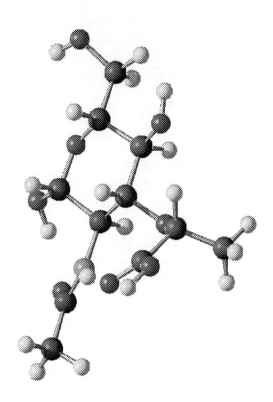

Figure 5. *Molecular graphic of NAM-conformer 1, the energy minimum conformer calculated with the PM3 methode with amide correction potential. This and all following molecular graphics have been created using the program SCHAKAL88 (E. Keller, Universitaet Freiburg, Germany, 1988).*

Figure 6. *Molecular graphic of NAM-conformer 2, the second lowest energy PM3 conformer. The N-acetyl group forms two hydrogen bonds to neighbouring side chains.*

conformers up to a relative energy of 2.4 kcal/mol and the 10 lowest AM1 energy conformers up to a relative energy of 3.2 kcal/mol are shown in the distribution. PM3 and AM1 low energy conformers are found in the torsional angle regions I and II (see figure 3). Considering the corresponding energies, the region I is slightly favoured by PM3, whereas region II is strongly favoured by AM1. A small number of PM3 conformers with relative energies of more than about 1 kcal/mol over minimum are accumulated in a third more diffuse region III (figure 3). No AM1 conformers with reasonable energy can be found there. Ether linkage torsional angle regions (PM3): I : $\theta2$= -60° +/- 15°, $\theta7$= 120° +/- 30° II : $\theta2$= 90° +/- 20°, $\theta7$= 90° +/- 30° III: $\theta2$= 105° +/- 15°, $\theta7$= -80° +/- 30°.

Torsional angle distribution at lactyl α-carbon. Figure 4 shows the torsional angle distribution at C9, the α-C-atom of the lactyl sidechain, as scatterplot. The same conformers as in the previous plot have been taken into account. All PM3 conformers with exception of the earlier mentioned region III conformers are equally distributed in a small angle band, $\theta7$ ranging from 70° to 150°. The carboxylic acid plane may assume any rotational angle ($\theta8$), as long as $\theta7$ is inside the allowed angle band, according to the PM3 results. In the AM1 calculation, we found, that the carboxylic acid plane rota-

tional angle is fixed, allowing only the ~170° and the ~-10° positions. $\theta7$ is restricted to two values, one about 80°, the other around 100°. Thus, the AM1 results indicated a quite rigid geometry of the α C torsional angles.

NAM conformer 1 (figure 5): The calculated PM3 energy of the energy minimum conformer is -353.60 kcal/mol. In the ether linkage torsional angle scattering plot this conformer is inside region I. One hydrogen bond is situated between the hydroxyl group at C4 and the carbonyl oxygen of the acid (O9). The N-acetyl group is positioned strictly cisoid. The C6-O6 bond of the hydroxymethyl group is gauche to C5-O5 bond and trans to C5-C4 bond.

NAM conformer 2 (figure 6): A PM3 energy of -353.28 kcal/mol has been calculated. Conformer 2 is the lowest PM3 energy conformer of region II (ether linkage). The N-acetyl group forms two hydrogen bonds: One between N2-H10 and the carbonyl oxygen of the acid (O9), the other between the carbonyl oxygen of the amide (O7) and the anomeric hydroxyl group. These two hydrogen bonds force the amide plane of the N-acetyl group out of cisoid position, slightly rotated in a clockwise direction (viewed from the moved side-group towards the fixed pyranose ring) by 50°. Thus positioning the N2-C7 bond nearly trans to C2-C3 bond. Like in con-

Table 3. *PM3 calculated H-bonds are generally shorter than AM1 H-bonds and the angles are greater compared to the more acute angles calculated by AM1, PM3 conformer 3 being an exception which may not meet the strict definitions of an hydrogen bond.*

Conf. No.	Atoms	Angle [°]	Distance [Å]
PM3:			
1	O4-H4...O9	168	1.81
2	O1-H1...O7	151	1.80
	N2-H10...O9	164	1.83
3	O4-H4...O9	144	2.48
4	O4-H4...O8	167	1.83
5	O4-H4...O9	161	1.80
AM1:			
1'	O4-H4...O6	128	2.15
	N2-H10...O9	154	2.18
2'	N2-H10...O9	154	2.18
3'	N2-H10...O9	153	2.18
4'	O4-H4...O6	128	2.14
	N2-H10...O8	150	2.21
5'	N2-H10...O8	150	2.21

former 1, the C6-O6 bond is gauche to C5-O5 bond and trans to C5-C4 bond.

NAM conformer 1' (figure 7): The calculated AM1 energy is -379.49 kcal/mol. 25.89 kcal/mol less than the PM3 energy minimum conformer. The lowest AM1 energy conformer is found in region II of the ether linkage plot. The molecule shows two intramolecular hydrogen bonds. One between N2-H10 of the amide and the carbonyl oxygen of the acid (O9). The second between O6 and the hydroxyl group at C4. The amide plane of the N-acetyl group is rotated out of cisoid position in clockwise direction by 48°. The C6-O6 bond is positioned trans to C5-O5 bond and gauche to C5-C4 bond.

C6-O-acetylated β-N-acetyl-D-Muramic acid

Listing of low-energy conformers of NAM-O6AC. The 10 energetically most favoured conformers of NAM-O6AC calculated by PM3 method with an additional empirical amide potential and AM1 method (conformer numbers marked with an apostrophe) are listed in table 4. Torsional angles $\theta 5$ and $\theta 10$ define the newly introduced ester group at C6 (definitions see figure 1 and table 1).

Calculated hydrogen bonds. Table 5 lists the hydrogen bonding parameters of the 5 lowest energy conformers of NAM-O6AC calculated by both methods. According to the PM3 results, the ester group is not involved in hydrogen bond formation. The hydrogen bond between O4-H4 and the

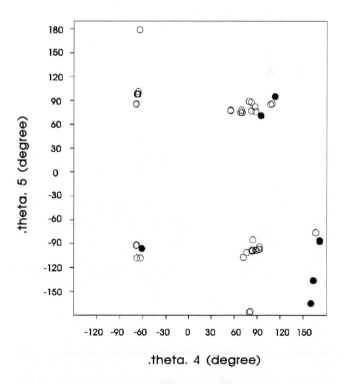

Figure 7. *Molecular graphic of NAM-conformer 1', the energy minimum conformer calculated by AM1 methode.*

Figure 8. *Scatterplot of C6 torsional angles (ester group, $\theta 4$ vs. $\theta 5$) of NAM-O6AC. Definition of symbols see legends of figures 3 and 4.*

Table 4. *The torsional angles and absolute and relative energies of the calculated first 10 low-energy conformers of NAM-O6AC by PM3 and AM1 methode are listed. As in table 2, the values for θ 9 are omitted, because they allways minimized to 180°.*

Conf. No.	θ [°]									Energy [kcal/mol]	ΔE [kcal/mol]
	1	2	3	4	5	6	7	8	10		
PM3:											
1	123	-67	-167	68	75	12	98	-7	177	-393.31	0.0
2	114	-67	-167	83	-99	71	99	-10	178	-393.15	0.16
3	123	-67	-167	79	89	11	98	-7	-179	-393.11	0.20
4	107	-48	-165	-67	98	74	115	-51	-175	-393.09	0.22
5	112	-65	-164	-66	98	72	99	-8	-176	-393.04	0.27
6	123	-67	-167	76	-101	15	98	-7	-5	-392.88	0.43
7	108	-50	-168	83	-98	73	116	-54	178	-392.87	0.44
8	123	-67	-168	55	78	15	98	-7	-1	-392.78	0.53
9	110	-49	-168	70	75	65	115	-53	176	-392.73	0.58
10	122	-67	-166	80	-176	15	98	-7	0	-392.70	0.61
AM1:											
1'	87	91	28	95	71	77	75	-8	-4	-416.91	0.0
2'	87	91	-61	113	95	76	76	-8	-1	-416.88	0.03
3'	87	92	-80	-61	-96	75	76	-8	3	-415.99	0.91
4'	87	86	29	95	71	76	80	164	-4	-415.85	1.05
5'	87	86	-60	113	95	74	80	163	-1	-415.75	1.16
6'	87	92	-61	171	-87	75	75	-8	2	-415.69	1.22
7'	87	91	-51	160	-165	75	75	-8	-2	-415.36	1.55
8'	87	91	-54	163	-136	75	75	-8	-1	-415.31	1.60
9'	87	87	-81	-61	-96	74	81	162	4	-414.90	2.01
10'	87	87	-61	171	-86	74	80	163	2	-414.62	2.29

carbonyl-oxygen of the lactyl group (O9) is similar to the one found in the not O6-acetylated NAM. Whereas the AM1 method favours one hydrogen bonding pattern involving the ester carbonyl-oxygen (O10) of the O-acetyl group. As in the NAM case, the H-bond distances calculated by the PM3 method are considerably shorter (by 0.3 A) than the respective AM1 values.

Differences in lactyl sidechain conformations of NAM-O6AC in comparison to NAM. Ether linkage torsional angles (θ2 and θ7) for NAM-O6AC calculated by the PM3 method are found predominantly in region I. In contrast, the AM1 results are found nearly exclusively in region II. A similar difference, although less pronounced, was observed for the NAM results. Region III conformers have not been found for reasonable energies by neither method. The torsional angle distribution at the lactyl-α-carbon (C9;θ7,θ8) is very similar to the NAM results, but again no region III conformers could be observed.

Torsional angle distribution at C6-O-acetyl sidechain. C6 torsional angle values are widely distributed according to the calculations by both methods , no preference could be detected (figure 8,θ4 vs. θ5). Both methods favour a planar ester group geometry. AM1 results give the cis orientation (Def.: C-O-C=O) for the ester torsional angle θ10, thus allowing an intramolecular hydrogen bond between the ester carbonyl-oxygen (O10) and the neighboured hydroxylgroup at C4. PM3 method slightly favours the trans geometry (θ10) for the ester group by about 0.5 kcal/mol.

AM1 and PM3 energy minimum conformers of NAM-O6AC in detail. The lowest energy conformers calculated by both methods are presented below.

NAM-O6AC conformer 1 (figure 9): The calculated PM3 energy using additional amide potential is -393.31 kcal/mol. The ether linkage belongs to region I. A hydrogen bond between O4-H4 and the carbonyl-oxygen of the lactyl group

Table 5. *PM3 and AM1 calculated hydrogen bonds of NAM-O6AC are listed. In correspondence to the NAM case, the PM3 calculated hydrogen bonds are shorter by 0.3 to 0.4 Å. than the AM1 values.*

Conf. No.	Atoms	Angle [°]	Distance [Å]
PM3:			
1	O4-H4...O9	156	1.81
2	O4-H4...O9	157	1.82
3	O4-H4...O9	156	1.81
4	O4-H4...O9	168	1.81
5	O4-H4...O9	158	1.81
AM1:			
1'	O4-H4...O10	142	2.22
	N2-H10...O9	153	2.20
2'	O4-H4...O10	149	2.13
	N2-H10...O9	154	2.18
3'	O4-H4...O10	140	2.18
	N2-H10...O9	153	2.16
4'	O4-H4...O10	141	2.22
	N2-H10...O8	150	2.23
5'	O4-H4...O10	149	2.13
	N2-H10...O8	150	2.21

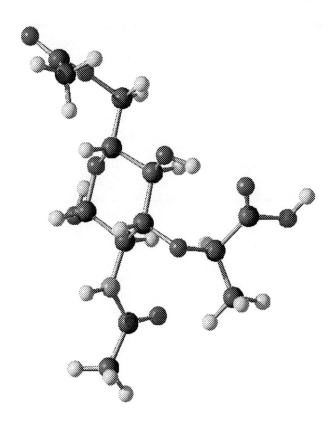

Figure 9. *Molecular graphic of NAM-O6AC conformer 1, the lowest energy conformer by PM3 methode with amide correction potential.*

(O9) is formed. The amide group is strictly cisoid. The ester bond is trans referred to θ10 (C-O-C=O).

NAM-O6AC conformer 1' (figure 10): The calculated AM1 energy is -416.91 kcal/ mol. The ether linkage torsional angles belong to region II. Two hydrogen bonds are formed, one between the N2-H10 of the amide group and the carbonyl-oxygen of the lactyl group (O9), the other between the hydroxyl group at C4 (O4-H4) and the carbonyl-oxygen of the ester group (O10). The amide plane of N-acetyl group is rotated out of cisoid position in a clockwise direction by 50° (viewed from the moved side-group towards the fixed pyranose ring).

Methylamide of β-N-acetyl-D-Muramic acid

Listing of low-energy conformers of NAM-MEAM. Table 6 lists the low-energy conformers of NAM-MEAM calculated by both semiempirical methods. The absolute energy difference between the energy minimum conformers calculated by PM3 and AM1 is 22 kcal/mol.

Calculated hydrogen bonding patterns and amide-amide π-complex structures. The energy minimum conformer calculated by PM3 with additional amide correction potential (conformer 1) shows an interesting chained hydrogen bonding structure (figure 11 and table 7). The C6 hydroxyl group

is directed towards the O4 and the C4 hydroxylgroup is pointing towards the carbonyl oxygen of the amidated lactyl group (O9). This structure is energetically in concurrence to the formation of an amide-amide π-complex found in PM3-conformer 3 with a relative energy rise of only 0.16 kcal/mol. However it must be stated, that the molecules are calculated in vacuo. Thus, such a complex would probably not be stable if intermolecular hydrogen bonds can be formed. The precise geometry of the amide-amide π-complex is given in figure 12. The AM1 energy minimum conformers of NAM-MEAM and NAM form the same stable O4-H4...O6 H-bond. Additionally an intramolecular interaction between both amide groups via a hydrogen bond is calculated for the AM1 minimum conformer of NAM-MEAM.

NAM-MEAM conformations of the amidated lactyl group at C3. The ether linkage torsional angle distribution shows similar features as the free acid form (NAM). However, according to the AM1 results, most energetically reasonable conformers are accumulated in the region II of the O3 torsional angle distribution plot. No AM1 conformers could be found in the region III, as in the NAM case. PM3 results indicate a split of region II into two subregions, subregion IIA (θ2= 100° +/-10°, θ7= 90° +/-10°) and subregion IIB

Table 6. *The torsional angles and absolute and relative energies of the calculated first 10 low-energy conformers of NAM-MEAM by PM3 andAM1 methode are listed. θ9 allways minimized to the value of 180°(PM3) rsp. 180°+/- 4°(AM1) and is not included in the table.*

Conf. No.	θ [°]								Energy [kcal/mol]	ΔE [kcal/mol]
	1	2	3	4	5	6	7	8		
PM3:										
1	119	-67	-170	173	42	65	97	-1	-303.07	0.0
2	109	-50	-171	72	-61	70	117	-48	-302.96	0.11
3	111	113	179	73	-63	71	118	-128	-302.91	0.16
4	117	-67	-169	72	-62	67	97	0	-302.82	0.25
5	106	-50	-161	72	-63	72	123	-114	-302.20	0.87
6	111	-47	-162	-66	-49	64	117	-74	-301.93	1.14
7	124	-68	-169	-65	-49	15	95	1	-301.90	1.17
8	111	-50	-171	-67	-49	65	116	-47	-301.64	1.43
9	116	113	178	-65	-45	65	118	-129	-301.33	1.74
10	109	116	-177	72	-62	69	-102	38	-301.18	1.89
AM1:										
1'	87	89	-62	170	-68	77	76	-8	-325.06	0.0
2'	82	148	179	66	-59	73	107	-126	-325.04	0.02
3'	123	134	177	68	-60	48	153	149	-323.86	1.20
4'	124	135	177	157	60	45	154	148	-323.70	1.36
5'	85	106	-173	66	-58	74	84	-66	-323.60	1.46
6'	87	97	-164	66	-57	76	83	-34	-323.58	1.48
7'	84	141	-71	175	-67	74	111	-120	-323.45	1.61
8'	82	148	-170	-143	-56	71	106	-127	-323.42	1.64
9'	85	93	-65	173	-68	72	86	-139	-322.91	2.15
10'	82	-67	60	71	-62	71	91	-153	-322.84	2.22

(θ2= 140° +/-10°, θ7= 100-150°), plots not shown. Both methods gave very similar results concerning the torsional angle distribution at the lactyl-α-carbon C9 (θ7 vs. θ8). Most of the energetically reasonable conformers are enclosed in a rectangle θ7= 90 - 130°, θ8= -140 - 0° (PM3) rsp. θ7= 75 - 115°, θ8= -150 - 0° (AM1).

NAM-MEAM conformer 1 (figure 11): The calculated PM3 energy of NAM-MEAM conformer 1 amounts to -303.07 kcal/mol. The N2-acetyl group is cisoid and not involved in hydrogen bonding. The amide group of the lactyl sidechain is bonded to a chained hydrogen bond between the hydroxylgroup at C6, the hydroxylgroup at C4 and the carbonyl oxygen of the amide (O6-H6...O4-H4... O9). The ether torsional angles belong to region I. C6-O6 bond is trans to C5-O5 and gauche to C5-C4.

NAM-MEAM conformer 1' (figure 13): NAM-MEAM conformer 1' has an AM1-energy of -325.06 kcal/mol. The

N2-acetyl group is rotated out of cisoid position by 50° (in clockwise direction when viewed from the moving side-group towards the fixed pyranose ring), forming a hydrogen bond to the carbonyl oxygen of the lactyl-amide group (N2-H10...O9). The ether torsional angles belong to region II. C6-O6 bond is trans to C5-O5 and gauche to C5-C4.

Methylglycoside of β-N-acetyl-D-Muramic acid

Listing of low-energy conformers of NAM-O1ME. Table 8 lists the calculated low-energy conformers of NAM-O1ME. θ6 is the glycosidic torsional angle of the methylglycoside. PM3 favours a value of 70° for θ6 . Values of -98° +/- 2° are found in conformer 2,8 and 10. AM1 strongly favours the values of 64° +/- 2° for θ6 . The N2-acetyl group is positioned cisoid by both methods. O3 sidechain conformations are very similar to NAM.

Table 7. *PM3 and AM1 calculated hydrogen bond geometries of NAM-MEAM are listed. Some conformers do not form hydrogen bonded structures, instead conformer 3 (PM3) shows the formation of an amide-amide pi-complex.*

Conf. No.	Atoms	Angle [°]	Distance [Å]
PM3:			
1	O6-H6...O4	143	1.85
	O4-H4...O9	157	1.80
2	O4-H4...O9	167	1.79
3	-		
4	O4-H4...O9	157	1.81
5	-		
AM1:			
1'	O4-H4...O6	128	2.15
	N2-H10...O9	152	2.15
2'	-		
3'	N3-H11...O7	141	2.23
4'	O6-H6...O4	129	2.22
	N3-H11...O7	142	2.22
5'	N3-H11...O7	146	2.17

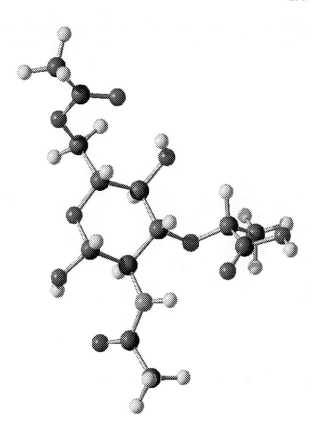

Figure 10. *Molecular graphic of NAM-O6AC conformer 1', the lowest energy conformer by AM1 methode.*

Hydrogen bonding patterns. Hydrogen bonding is less variable. PM3 calculates a hydrogen bond between O4-H4 and the carbonyl oxygen of the lactyl group (O9) in the 4 lowest-energy conformers. AM1 found the hydrogen bond between N2-H10 (amide) and the carbonyl oxygen of the lactyl group in the conformers 1',2',3' and 5'.

NAM-O1ME conformer 1 (figure 14): Calculated PM3 energy = -345.91 kcal/mol. N2-acetyl group is cisoid. Ether linkage belongs to region I. C6-O6 bond is gauche to C5-O5 bond and trans to C5-C4 bond. O1-C15 bond (C15 = aglycan) is situated parallel to C2-H bond. Torsional angle values are very close to NAM PM3-conformer 1.

NAM-O1ME conformer 1' (figure 15): The AM1 energy amounts to -371.34 kcal/mol. Torsional angle values are very close to the corresponding NAM AM1-conformer 1'.

The introduction of the methyl group at O1 had only little effect on the overall structures of the energy minimum conformers calculated by both methods.

Discussion

Semiempirical quantum chemical methods have been successfully applied to monosaccharides and monosaccharide related compounds, e.g. β-D-fructo-pyranose using AM1 and MNDO/M [14], 2-methoxytetrahydropyran using PM3 and AM1 [15]. In the case of β-D-fructopyranose, the AM1 procedure performed better on the intramolecular interactions than the MNDO/M procedure. In respect to the puckering parameters for the ring shape and the average torsional angle around the pyranose ring, the AM1 values were in close agreement with the values of the crystal conformation in contrast to the MNDO/M values. The calculations on β-D-fructopyranose suggest that AM1 handles O-H...O hydrogen bonding better than MNDO/M [14]. Scano and Thomson [15] calculated the heats of formation for 19 large molecules with PM3 and AM1 methods, to test the application of these methods to larger biological molecules. The monosaccharide related compound 2-methoxytetrahydropyran was calculated and compared to experimental results. PM3 gave results of higher accuracy. So it seemed reasonable to apply AM1 and PM3 methods for the calculation of the conformational structure of NAM and its derivatives (NAM-O6AC, NAM-MEAM, NAM-O1ME).

β-N-acetyl-Muramic acid. In this study, distinct hydrogen bonding patterns have been calculated for the NAM molecule in β-form (table 3). Knox and Murthy [18] reported in their crystallographic study on the α-form an intramolecular hydrogen bond between the N-H bond of the amide and

Table 8. *The torsional angles and absolute and relative energies of the calculated first 10 low-energy conformers of NAM-O1ME by PM3 and AM1 methode are listed.*

Conf. No.	θ [°]								Energy [kcal/mol]	ΔE [kcal/mol]
	1	2	3	4	5	6	7	8		
PM3:										
1	113	-49	-168	72	-62	70	115	-53	-345.91	0.0
2	114	-49	-167	70	-65	-98	115	-54	-345.16	0.75
3	118	117	-175	72	-62	69	-97	27	-344.99	0.92
4	112	-50	-171	-75	-70	69	115	-53	-344.33	1.58
5	115	112	-177	72	-63	70	120	-104	-344.30	1.61
6	109	-53	-173	72	-63	70	131	82	-344.27	1.64
7	117	96	-160	72	-62	70	-109	49	-344.26	1.65
8	108	-54	-175	71	-65	-100	144	96	-344.00	1.91
9	112	108	-175	72	-63	70	120	87	-343.89	2.02
10	124	118	-175	70	-65	-96	-97	27	-343.84	2.07
AM1:										
1	116	92	-65	172	-67	64	74	-11	-371.34	0.0
2	116	99	-162	67	-58	65	75	-16	-370.23	1.11
3	120	100	-166	65	-58	66	84	-23	-369.99	1.35
4	116	-78	-173	67	-59	62	92	-5	-369.43	1.91
5	118	93	-77	-69	59	-74	73	-12	-368.97	2.37
6	119	92	-67	173	-65	-73	73	-12	-368.83	2.51
7	125	-75	-174	63	-61	-67	96	-9	-368.25	3.09
8	115	-64	-174	-133	-57	62	106	-15	-368.13	3.21
9	117	-76	-176	-117	-61	61	92	1	-368.10	3.24
10	122	99	-163	64	-60	-72	74	-17	-368.00	3.34

carbonyl oxygen of the lactyl group. The PM3 and AM1 calculations, presented here, also suggest such an hydrogen bond for the β-anomer. The lactyl sidechain backbone geometry is determined by the three torsional angles $\theta 2$, $\theta 7$ and $\theta 8$. These angles are quite restricted to certain values (AM1) or angle ranges (PM3). AM1 strongly favours the angle triplet ($\theta 2$= 90° +/- 4°, $\theta 7$= 80° +/- 5°, $\theta 8$= -10° or +165° +/- 2°) up to a relative energy of 2 kcal/mol, whereas PM3 allows certain angle regions (see figs. 3,4).

The N2-acetyl sidechain torsional angle $\theta 1$ is minimized to the value of 87° according to the AM1 results, for all low energy conformers. PM3 shows a greater variation between strict cisoid position ($\theta 1$ = 109°), if no hydrogen bonds are formed, up to a value of $\theta 1$ = 69°, in case of two hydrogen bonds of the amide group to neighbouring side groups.

Pincus et al. [22] found the anti-cisoid conformation of the N-acetyl group ($\theta 1$= -60°) with the amide carbonyl oxygen coming into close proximity to the hydrogens at C1 and C3 to be energetically favoured for the NAM molecule. For better comparison of our results to the results of empirical conformational energy calculations by Pincus et al. [22], we

used the energy minimum conformer of the authors, for a single point and an energy minimization with PM3 and AM1 methods. We found, that the energy minimum conformer for the NAM molecule proposed by Pincus et al. [22] has a relative energy of more than 10 kcal/mol (PM3) and nearly 20 kcal/mol (AM1). Even after optimization of the Pincus model the energy values were still 5 kcal/mol above our calculated minimum structures. Thus, the anti-cisoid conformation of the N-acetyl group will not be discussed further in this study.

J.S. Yadav et al. [2,3] investigated the conformational structure of NAM using the semiempirical quantum chemical methods MNDO and PCILO. In the lowest MNDO energy conformer of NAM (see table 2, ref. data) the N-acetyl group is in cisoid position ($\theta 1$= 113°) in agreement to our PM3 results. However, MNDO was unable to find hydrogen bonds. Thus, the lactic acid group is not involved in hydrogen-bond formation with its neighbouring side chains, as it is the case with the N-acetyl group [2]. In contrast, the PCILO calculations on NAM showed the formation of hydrogen bonds between C4-hydroxyl group and carboxylic acid (O4-H4...O9 and O4-H4...O6) [3]. In the present study various

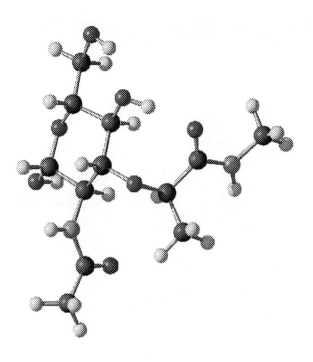

Figure 11. *Molecular graphic of NAM-MEAM conformer 1, the energy minimum conformer calculated by PM3 methode with amide correction potential.*

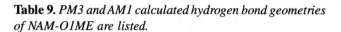

Table 9. *PM3 and AM1 calculated hydrogen bond geometries of NAM-O1ME are listed.*

Conf. No.	Atoms	Angle [°]	Distance [Å]
PM3:			
1	O4-H4...O9	168	1.81
2	O4-H4...O9	168	1.81
3	O4-H4...O9	161	1.79
4	O4-H4...O9	168	1.80
5	–		
AM1:			
1	N2-H10...O9	138	2.24
2	N2-H10...O9	136	2.20
3	N2-H10...O9	133	2.20
4	O4-H4...O9	133	2.13
5	N2-H10...O9	137	2.25

hydrogen bonding patterns have been found (see table 3), including those proposed by J.S. Yadav et al. [3]. The MM2 calculations of P.N.S. Yadav et al. [4] are only of limited value, because, despite the low computation time needed for such calculations, they did no grid search, using only the conformers calculated by Pincus et al. [22] as starting geometries.

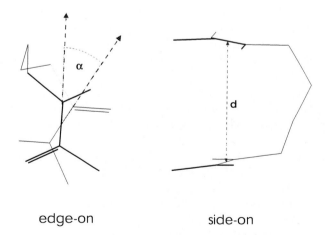

edge-on   side-on

Figure 12. *Wire graphic of the geometry of the amide-amide pi-complex in NAM-MEAM conformer 3 (PM3). Other atoms not belonging to the complex are omitted. The complex is shown edge-on (left) and side-on (right). The bonding parameters are: α = 30°; d= 3.85 A.*

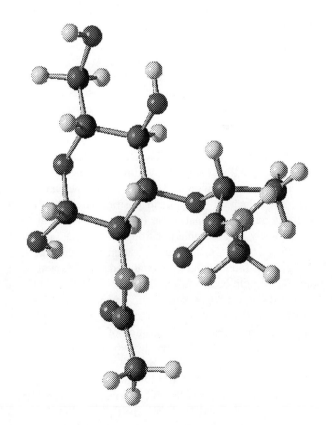

Figure 13. *Molecular graphic of NAM-MEAM conformer 1', the energy minimum conformer by AM1 methode.*

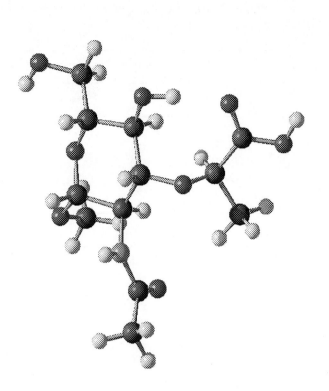

Figure 14. *Molecular graphic of NAM-O1ME conformer 1, the energy minimum conformer calculated by PM3 method with amide correction potential.*

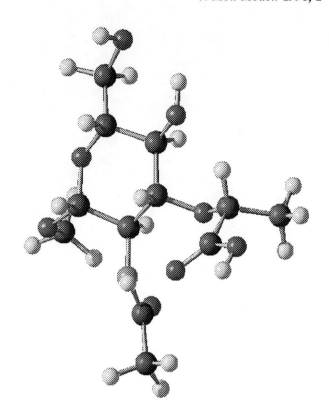

Figure 15. *Molecular graphic of NAM-O1ME conformer 1', the lowest energy conformer by AM1 methode.*

In the following, co-crystallizations of muramyl compounds with proteins, reported in the literature, will be discussed. Y. Bourne et al. [20] achieved the crystallization of muramic acid (not N-acetylated) and muramyl dipeptide in (D,D) configuration, complexed to a legume lectin and determined the X-ray structures. The ether torsional angles ($\theta 2$= 148°, $\theta 7$= 105°) of the muramyl dipeptide residue are situated in the near of region II (figure3). The amide plane of the N-acetyl sidechain is in cisoid position ($\theta 1$= 124°). The hydroxymethyl torsional angle ($\theta 4$ = -41°) is found in an energetically unfavourable position, bond C6-O6 gauche to C5-O5 and to C5-C4. The same position of the hydroxy-methyl group is found in the co-crystallized muramic acid structure. This might be due to a hydrogen bond between O6 and an Asp residue in the peptide chain, reported be Y. Bourne et al. [20] For the isolated NAM molecule, calculated in this study, such a double gauche position of the hydroxymethyl group is energetically strongly disfavoured. Atomic coordinates for the calculation of torsional angles of the compounds described by Y. Bourne et al. [20] have been taken from the Protein Data Bank, Brookhaven National Laboratory, Upton, NY, files 1loc.brk and 1lod.brk. The $\beta(1\rightarrow4)$ linked NAM-NAG-NAM trisaccharide (NAG: N-acetyl-2-Glucos-amine) bound to active-site cleft of hen egg-white lysozyme has been crystallographically studied by N.C.J. Strynadka and M.N.G. James [21]. The B-site NAM was less distorted. The experi-

mental torsional angle value for the N-acetyl group was reported as $\theta 1$= 114°. The same value for $\theta 1$ has been calculated for the NAM conformer 3 using PM3 method (table 2).

C6 O-acetylated β-N-acetyl Muramic acid. The O-acetylation of the NAM residue may have certain biological significance according to the experimental results of S. Gyorffy and A.J. Clarke [23], who investigated the localization of O-acetylation of the P. mirabilis peptidoglycan by immunoelectron microscopy. W.B. Schweitzer and J.D. Dunitz [24] found conformational preferences in the average geometries of carboxylic esters by analysing data from many crystal structures retrieved from a database. For esters of primary alcohols usually the C-C bond is antiperiplanar to the ester C=O bond. In the nomenclature used in this study, $\theta 10$ (C-O-C=O) would be close to 0°. Remarkably, this statistically predicted ester geometry has been calculated for all low-energy NAM-O6AC conformers using the AM1 method, $\theta 10$ ranging between -4° and +5°.

Methylamide of β-N-acetyl Muramic acid. PM3 results showed a concurrence between a chained hydrogen bond geometry, conformer 1, and an amide-amide π-complex, conformer 3. In order to get an estimation of the amide-amide π-complex binding energy PM3 energy has been calculated for an analog complex consisting of two methyl-acetamide molecules in antiparallel orientation. We calculated a π-complex stabilization energy (PM3) of the dimer of 4.74 kcal/

mol. This is quite close to the stabilization energy (PM3) of one to two hydrogen bonds [13]. The methyl-acetamide dimer has a C=N midpoint-to-midpoint distance of d= 3.81 Å, only 0.04 Å shorter than the intramolecular amide-amide π-complex of NAM-MEAM conformer 3 (PM3). The twist angle α (pseudo-torsional angle N-midpoint-midpoint-C) was 0° for the dimer. Such a complex seems to be a real energetic alternative for the diamide compound according to PM3 calculations. AM1 gave no stable amide-amide π complex for the methyl-acetamide dimer, instead minimization led to a hydrogen bonded structure with a stabilization energy of 4.24 kcal/mol. This is a remarkable difference between AM1 and PM3 methods applied on amide-amide structures.

AM1 and PM3 proved to give reasonable molecular conformations for biomolecular structures like the saccharides calculated in this study. However, there are some differences in the hydrogen bond patterns between PM3 and AM1 low energy conformers. Because of the importance of muramyl-peptides as possible supporting treatment of cancer [25] and HIV-infections [26] a further semiempirical study on muramyl-monopeptide and an ab initio study on the peptide part of muramyl-dipeptide are in progress.

Conclusions

β-N-acetyl-Muramic acid - The amide plane of the N-acetyl group is either positioned strictly cisoid, or, in the occurence of one or two possible hydrogen bonds to neighboring sidechains, slightly rotated out of cisoid position by about 50° in clockwise direction, when viewed from the moving sidechain towards the fixed pyranosyl ring. - The torsional flexibility of the ether bridge of the lactyl group is sterically restricted to at least two well defined torsional angle regions I: $\theta2$= -60° +/-15°, $\theta7$= 120° +/- 30° II: $\theta2$= 90° +/-20°, $\theta7$= 90° +/- 30° - PM3 strongly favours the gauche position (C6-O6 in respect to C5-O5) of the hydroxymethyl group, whereas AM1 allows gauche and trans positions.

C6-O-acetylated β-N-acetyl-Muramic acid - The AM1 calculated cis geometry for the ester torsional angle (C-O-C=O) (= $\theta10$) is in good agreement with the statistical average for esters of primary alcohols. - The torsional angle distribution at C6 of the acetylated sidechain shows no sterical preference according to both methods applied.

Methylamide of β-N-acetyl-Muramic acid - PM3 results slightly favour a chained hydrogen bonding structure O6-H6...O4-H4...O9 (O9 = lactyl-carbonyl oxygen) in concurrence to an amide-amide π complex with a distance of 3.85 Å.- AM1 predicts a hydrogen bonded structure between the two amide groups in the molecule.

Methylglycoside of β-N-acetyl-Muramic acid - PM3 favours a value of 70° for O5-C1-O1-C(aglycan) = $\theta6$. AM1 strongly favours the values of 64° +/- 2°. - Overall structure of the energy minimum conformers practically uneffected by the introduction of the methyl group at O1.

Supplementary Material The molecular structures of the calculated energy minimum conformers are available as supplementary material in PDB-file format.

References

1. Barnickel, G.; Naumann, D.; Bradaczek, H.; Labischinski, H.; Giesbrecht , P. International FEMS Symposium on the Murein Sacculus of Bacterial Cell Walls; Hakenbeck, R., Hoeltje, J.-V., Labischinski, H.(Eds.) Walter de Gruyter: Berlin, New York, 1983; pp 61-66.
2. Yadav, J.S.; Labischinski, H.; Barnickel, G.; Bradaczek, H. *J. Theor. Biol.* **1981**, *88*, 441
3. Yadav, J.S.; Barnickel, G.; Bradaczek, H. *J. Theor. Biol.* **1982**, *95*, 151
4. Yadav, P.N.S.; Rai, D.K.; Yadav, J.S. *J. Mol. Struct.* **1989**, *194*, 19
5. Rogers, J.H.; Perkins, H.R.; Ward, J.B. Microbial Cell Walls and Membranes; Chapman & Hall: London, 1980
6. Weidel, W.; Pelzer, H. *Adv. Enzymol.* **1964**, *26*, 193
7. Fleck, J.; Mock, M.; Minck, R.; Ghuysen, J.-M. *Biochim. Biophys. Acta.* **1971**, *233*, 489
8. Perkins, H.R.; Chapman, S.J.; Blundell, J.K. International FEMS Symposium on the Murein Sacculus of Bacterial Cell Walls; Hakenbeck, R., Hoeltje, J.-V., Labischinski, H. (Eds.) Walter de Gruyter: Berlin, New York, 1983; pp 255-260.
9. Burghaus, P.; Johannsen, L.; Naumann, D.; Labischinski, H.; Bradaczek, H.; Giesbracht, P. International FEMS Symposium on the Murein Sacculus of Bacterial Cell Walls; Hakenbeck, R., Hoeltje, J.-V., Labischinski, H. (Eds.) Walter de Gruyter: Berlin, New York, 1983; pp 317-322.
10. Clarke, A.J.; Dupont, C. *Can. J. Microbiol.* **1992**, *38*, 85.
11. Johannsen, L.; Labischinski, H.; Burghaus, P.; Giesbrecht, P. International FEMS Symposium on the Murein Sacculus of Bacterial Cell Walls; Hakenbeck, R., Hoeltje, J.-V., Labischinski, H. (Eds.) Walter de Gruyter: Berlin, New York, 1983; pp 261-266.
12. Johannsen, L. *APMIS* **1993**, *101*, 337.
13. Jurema, M.W.; Shields, G.C. *J. Comput. Chem.* **1993**, *14*, 89.
14. Khalil, M.; Woods, R.J.; Weaver, D.F.; Smith, Jr. V.H. *J. Comput. Chem.* **1991**, *12*, 584.
15. Scano, P.; Thompson, C. *J. Comput. Chem.* **1991**, *12*, 172.
16. Stewart, J.J.P. *J. Comput. Chem.* **1989**, *10*, 209.
17. Dewar, M.J.S.; Zoebisch, E.G.; Healy, E.F.; Stewart, J.J.P. *J. Am. Chem. Soc.* **1985**, *107*, 3902.
18. Knox, J.R.; Murthy, N.S. *Acta Cryst.* **1974**, *B30*, 365.
19. Mo, F. *Acta Chem. Scand. Ser. A* **1979**, *33*, 207.
20. Bourne, Y.; Ayouba, A.; Rouge, P.; Cambillau, C. *J. Biol. Chem.* **1994**, *269*, 9429.

21. Strynadka, N.C.J., James, M.N.G. *J. Mol. Biol.* **1991,** *220*, 401.

22. Pincus, M.R.; Burgess, A.W.; Scheraga, H.A. *Biopolymers* **1976,** *15*, 2485.

23. Gyorffy, S.; Clarke, A.J. *J. Bacteriol.* **1992,** *174*, 5043.

24. Schweizer, W.B.; Dunitz, J. D. *Helv. Chim. Acta.* **1982,** *65*, 1547

25. Zidek, Z.; Frankova, D. *Int. J. Immunopharmacol.* **1995,** *17*, 313

26. Maruyama, Y.; Kurimura, M.; Achiwa, K. *Chem. Pharm. Bull.* **1994,** *42*, 1709.

J.Mol.Model. (electronic publication) – ISSN 0948–5023

J. Mol. Model. **1996**, 2, 205 - 216

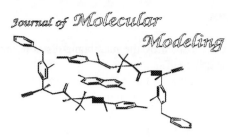

Characterization of a Novel Reverse-orientation Model for a Peptide/MHC Complex Putatively Associated with Type I Diabetes Mellitus

Carol DeWeese, William W. Kwok§, Gerald T. Nepom§, and Terry P. Lybrand*

Center for Bioengineering, University of Washington, Box 351750, Seattle, WA 98195-1750, USA, Tel: 206-685-1515, Fax: 206-616-4387 (lybrand@proteus.bioeng.washington.edu)

§ Virginia Mason Research Center, 1000 Seneca Street, Seattle, WA 98101, USA

Received: 14 June 1996 / Accepted: 18 July 1996 / Published: 2 September 1996

Abstract

Molecular modeling techniques were used to generate structures of several HLA-DQ proteins associated with insulin-dependent diabetes mellitus (IDDM). A peptide fragment from glutamic acid decarboxylase (GAD), a known IDDM autoantigen, binds to certain HLA-DQ molecules positively associated with IDDM. Modeling studies were used to explore possible binding interactions between this GAD peptide and several HLA-DQ molecules. Based on the characterization of anchor pockets in the HLA-DQ binding groove and of peptide side chains, a novel binding mode was proposed. This binding mode predicts the GAD peptide is positioned in the binding groove in the direction opposite the orientation observed for class I proteins and the class II DR1, DR3, and I-E^k proteins. Peptide docking exercises were performed to construct models of the HLA-DQ/peptide complexes, and the resulting models have been used to design peptide binding experiments to test this "reverse-orientation" binding mode. A variety of experimental results are consistent with the proposed model and suggest that some peptide ligands of class II molecules may bind in a reversed orientation within the binding groove.

Keywords: Major Histocompatibility Complex proteins, insulin-dependent diabetes mellitus, peptide docking, molecular modeling.

Introduction

The primary genetic factor associated with insulin-dependent diabetes mellitus (IDDM) susceptibility in humans is the DQ gene region of the human leukocyte antigen (HLA) complex. [1-5] Although genetic susceptibility is widely accepted as the primary factor required for IDDM onset, individuals who are genetically programmed to be susceptible do not develop the disorder until exposed to necessary environmental triggers. [6,7] Studies of human IDDM suggest that viruses, chemicals and toxins, and dietary factors are all potential triggers. [8] However, none have been identified definitively.

The DQ genes, part of the major histocompatibility complex (MHC), encode heterodimeric transmembrane proteins that bind antigenic peptide ligands for presentation to T cell receptors (TCRs) on CD4+ T lymphocytes. A class II MHC protein such as HLA-DQ consists of an α chain and a β chain, encoded by separate genes. The peptide binding region is a

* To whom correspondence should be addressed

J. Mol. Model. **1996**, 2

cleft formed by the membrane-distal portions of both the α and β chains.

Peptide binding in the groove of a class II MHC molecule occurs via noncovalent interactions involving both side chain and backbone atoms of the peptide. The side chains of "anchor residues" in the peptide interact with corresponding "anchor pockets" in the binding groove. The peptide lies flat in the groove and the complex is stabilized by a hydrogen bond network involving primarily MHC side chains and the peptide backbone. Both ends of a class II MHC binding groove are open, allowing the termini of the bound peptide to extend beyond the ends of the groove. [9-14]

Unlike antibodies and T cell receptors, the repertoire of MHC molecules in an individual is limited. Necessity dictates that in order to initiate an immune response to a large number of antigens, each MHC protein must be able to bind multiple peptide ligands. This characteristic has been demonstrated in studies of peptide binding motifs. [15-17] Peptide binding to MHC proteins appears to depend partially on the absence of unfavorable side chains in anchor residue positions, rather than strictly on the presence of specific, favorable anchor residues. The identified peptide binding motifs indicate that anchor residues may be categorized as favorable, impartial, or unfavorable. Peptides containing either favorable or impartial anchor residue side chains bind MHC proteins [15-17], presumably with varying affinities, and the complex is stabilized by a network of hydrogen bonds. Peptide binding is prevented when unfavorable residues are present at anchor positions.

The HLA gene loci are among the most polymorphic within the population. [18-20] Only specific amino acid positions are polymorphic sites (e.g., position 86 of the DQ β chain can be Ala, Gly, or Glu), and high homology exists among the various alleles. In the binding groove region, most DQ α and β chains share at least 90-95% sequence identity. [20] Therefore, the differences observed in binding characteristics among various HLA-DQ haplotypes are due to the effects of amino acid substitutions at the polymorphic sites within the binding groove.

The haplotype HLA-DQ3.2 (allelic designation DQA1*0301-DQB1*0302), is positively associated with IDDM susceptibility in Caucasian populations, while DQ3.1 (DQA1*0301-DQB1*0301) is negatively associated with susceptibility. [3-5,21,22] A third haplotype, DQ3.3 (DQA1*0301-DQB1*0303), is associated with susceptibility in Japanese populations [23], and in the Swedish population, when in heterozygous combination with DQ3.2 or selected other DQ alleles. [24] These three MHC molecules all share a common α chain and have highly homologous β chains (~97-99% sequence identity). In the peptide binding region, DQ3.1 and DQ3.2 differ by four residues at positions 13, 26, 45, and 57. DQ3.3 is identical to DQ3.2 except for a single substitution at position 57. The polymorphic residues and their locations in the antigen binding groove are shown in Figure 1A.

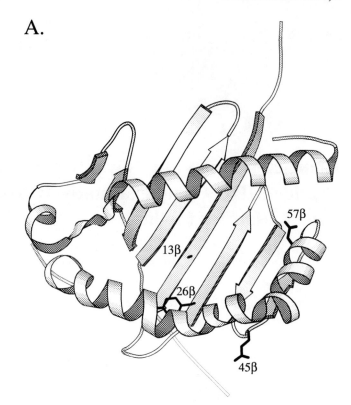

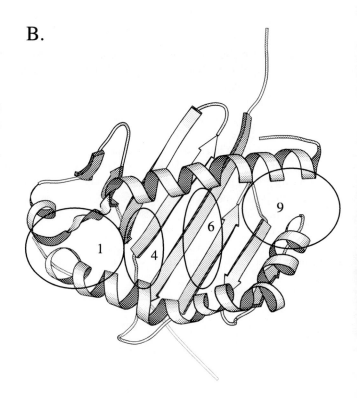

Figure 1. *A) Ribbon structure of DQ3.1 binding groove with the four polymorphic side chains shown. B) Anchor pockets 1, 4, 6, and 9 of the DR1 structure are indicated. Figures 1, 3, and 4 were generated with the program Molscript.[47]*

In spite of the high sequence homology of these three MHC molecules, they exhibit some dramatic differences in peptide binding. In particular, a 13-residue fragment from the 65-kD isomer of glutamic acid decarboxylase (GAD65) binds well to DQ3.2, but binds poorly to DQ3.3 and does not bind to DQ3.1 at all. GAD65 is a known IDDM autoantigen [25], and the 13-residue peptide fragment (designated 34p) that binds well to DQ3.2 exhibits high sequence homology to the most immunogenic peptide derived from the P2-C protein of coxsackievirus B4 [26], a putative environmental trigger of IDDM onset.

We used molecular modeling techniques to construct three-dimensional models for the DQ3.1, DQ3.2, and DQ3.3 MHC molecules, and to generate docked complexes for peptide 34p with each MHC molecule. Based on the peptide docking exercises, we propose a hypothesis to explain the differential binding of peptide 34p to these three MHC molecules, and we predict a "reverse-orientation" binding motif for DQ3.2 with peptide 34p.

Methods

Model construction

A model structure of DQ3.2 was generated using homology modeling techniques based on the structure of DR1 (DRA-DRB1*0101), a similar class II MHC protein encoded by different gene loci. The DR1 structure, which contains a bound peptide from the influenza protein hemagglutinin (HA), was determined by x-ray crystallography. [9] DR1 and DQ3.2 share ~62% sequence identity overall, with ~52% sequence identity in the peptide binding region, and many of the substitutions at polymorphic sites are conservative changes. [20] Chothia and Lesk demonstrated that proteins with greater than 50% sequence identity generally have very similar tertiary structures. [27] Thus, the DR1 structure should be a good template for DQ3.2 model construction. We do expect that there are some structural differences between DR and DQ class II MHC molecules, particularly in two regions. A region in the DQα chain between residues 48-56 (residues 45-53 in DR1 α chain) contains a number of nonconservative substitutions, including a cluster of arginine residues in the DQ molecules. Another interesting region of probable structural variation between DR and DQ proteins is located in the β chain at position 55, where many DQ proteins have a proline substituted in place of the arginine observed in DR1. This substitution yields two adjacent prolines in these DQ molecules, which we predict will likely disrupt the beginning of the helical region in the β1 domain. However, neither of these regions impact the peptide binding groove profoundly in our DQ models. More significantly, the three DQ heterodimers we have modeled are highly homologous, and thus, likely have nearly indentical three-dimensional structures. We have focused our modeling efforts on the differences between these three DQ molecules.

The DQ3.2 model was generated using standard homology modeling techniques. Backbone atoms of the DR1 template were fixed at crystallographic positions and necessary amino acid side chain substitutions were made to generate the DQ3.2 sequence. Since side-chain conformations in proteins with high sequence identity are highly conserved [28,29], DR1 side chain conformations were retained for all homologous sites and conservative substitutions (e.g., Thr for Ser), to the extent possible. For nonconservative substitutions, side-chain atoms were placed initially in the most probable conformation. [30] The models were constructed using MidasPlus [31] and PSSHOW [32] interactive molecular graphics programs.

In addition to the amino acid substitutions, DQ α chains contain three insertions relative to the DR α chain. These insertions are located at the amino terminus of the α chain (positions 1, 2, and 9), far removed from the binding groove region. The amino terminus has an extended chain conformation, and insertions at positions 1 and 2 were made by attaching single residues in an extended conformation to the amino terminus of the protein. The insertion at position 9 was made by breaking the protein backbone at the site of insertion, attaching the appropriate residue to the amino terminus of residue 10, and rotating the fragment comprising residues 1-8 to form a trans peptide bond connecting residues 8 and 9. Side-chain atoms were placed in the most probable conformations.

Once all amino acid substitutions and insertions were made, substituted side chains were adjusted manually to relieve steric clashes. Limited energy minimization was then performed using AMBER 4.0[33] with an all-atom potential function [34] to refine the models. Only the membrane-distal binding groove portion of the molecule was relaxed by minimization; the membrane-proximal region was fixed in the crystallographic conformation during minimization, as this region in DQ3.2 is highly homologous to DR1 (~70%), with few nonconservative substitutions. Conjugate gradient energy minimization was performed in vacuo with a distance-dependent dielectric constant. The model structure was evaluated by calculating side chain packing densities using the program QPACK [35] and by verifying reasonable side-chain conformations. [30] Visual inspection was performed to insure that all polar and charged residues not exposed to solvent had suitable interaction partners to permit formation of hydrogen bonds and salt bridges. The final DQ3.2 model was used as a template for construction of DQ3.1 and DQ3.3 models, using the protocol outlined above.

The DQ3.2 structure generated in this homology modeling exercise has several acidic and basic residues in regions predicted to form key anchor pockets in the peptide binding groove. To assess the probable charge state of these residues, pKa values were calculated for all ionizable residues using Poisson-Boltzmann electrostatics calculations and a pKa calculation protocol developed by Antosiewicz. [36,37] Briefly, this method entails 1) calculation of the self-ionization energy of each titratable group when free in aqueous

J. Mol. Model. **1996,** 2

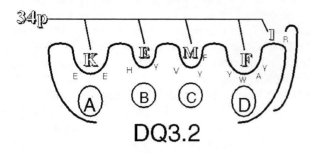

Figure 2. *Diagram of DQ3.2 anchor pockets. The four anchor residues from 34p are shown in the appropriate pockets predicted by the reverse-orientation model.*

solution, 2) calculation of the ionization energy of each titratable group in the neutral protein (i.e., all other titratable groups are held neutral, but partial charges for all atoms in the protein are included in Poisson-Boltzmann calculations), 3) calculation of the interaction energy between all ionizable groups, and 4) a Monte Carlo simulation to determine the lowest energy state(s) from among the 2^M possible ionization states in the protein, where M = number of titratable groups. The electrostatic potentials and electrostatic interaction energies were computed using a finite difference algorithm to solve the linearized Poisson-Boltzmann equation with the UHBD program. [38] A coarse grid lattice (2.5 Å spacing) was used to calculate long-range electrostatic contributions, followed by a focusing technique with successively higher resolution grids (1.20 Å, 0.75 Å, 0.25 Å lattice point spacing) to obtain converged results for short-range electrostatic interactions. All calculations were performed at T = 293 K, pH 7.0, and 150 mM ionic strength with a solvent dielectric of 80.0, a protein dielectric of 20.0, and a 2.0 Å Stern layer. It has been observed in previous studies that a protein dielectric of 20.0 yields good agreement with experimentally measured pK_a values. [36,39] Partial charges and van der Waals radii needed for the calculations were taken from the latest AMBER potential functions. [40]

Peptide docking

Peptide docking involved the manual placement of putative peptide anchor residues in appropriate anchor pockets of the binding groove. To identify possible peptide anchor residues, each anchor pocket was assessed for size, hydrophobicity, and presence of charged and/or polar residues. Anchor residues with complementary properties and appropriate spacing in the peptide were then chosen.

A model of peptide 34p was constructed in extended-chain conformation, with all side chains placed in the most probable conformations. The peptide was docked manually into the DQ3.2 binding groove with the selected anchor residues positioned in the corresponding anchor pockets. Backbone

torsion angles of the peptide were adjusted to accommodate the fit of the peptide in the binding groove, and anchor residue side chains were then rotated to fit well in their respective anchor pockets. Finally, limited energy minimization was performed using the all-atom potential functions to relieve any residual unfavorable contacts. Initially, only the peptide was permitted to relax, while subsequent minimization included the peptide and the binding groove region of the MHC molecule. The membrane-proximal portion of the DQ3.2 molecule was fixed throughout the minimization process.

Results

Construction of a DQ3.2/34p reverse-orientation model.

The nomenclature often used to distinguish each anchor pocket within the binding groove of a class II protein is a 1-4-6-9 scheme, based on the crystal structure of DR1 and its bound ligand, the HA peptide. [9] The amino acid sequence of the HA peptide is

Pro-Lys-**Tyr**-Val-Lys-**Gln**-Asn-**Thr**-Leu-Lys-**Leu**-Ala-Thr

$$\mathbf{1} \quad 2 \quad \mathbf{3} \quad \mathbf{4} \quad 5 \quad \mathbf{6} \quad 7 \quad 8 \quad \mathbf{9}$$

and the anchor residues are shown in bold type. Each anchor pocket is numbered according to the corresponding peptide anchor residue that binds in the pocket, with the first anchor residue designated as position 1. The locations of these pockets within the groove are shown in Figure 1B. For purposes of clarification in this discussion, pockets 1, 4, 6, and 9 will be designated A, B, C, and D, respectively (see Figure 2).

In the DQ3.2 model, pockets A and D are more pronounced than pockets B and C. Pocket A, the largest in the DQ3.2 model, contains primarily polar and charged residues, including two exposed glutamic acids (34α and 86β). The pocket is stabilized by a network of hydrogen bonds and salt bridges formed by the side chains of the residues lining the pocket. Arg55α forms a salt bridge with Glu34α, and Glu86β forms hydrogen bonds with Ser10α and His27α. This is in contrast to the hydrophobic character of the comparable pocket in HLA-DR1. [9] Our electrostatics calculations suggest that both glutamate residues are significantly ionized, even at pH 4.5. We predicted that a positively charged residue from a bound peptide would be a preferred anchor residue for this pocket, forming a charge interaction with one of the glutamates. Figure 3A shows the charged and polar side chains that stabilize pocket A.

Pockets B and C are shallow and less distinct than pockets A and D. Pocket B of DQ3.2 contains primarily hydrophobic residues, plus an exposed histidine side chain and an exposed tyrosine hydroxyl group. We predicted a polar side chain able to form a hydrogen bond with the His or Tyr would be a likely anchor residue for this pocket. Pocket C is predominantly hydrophobic, and we predicted a hydrophobic anchor residue would be needed for this pocket.

Pocket D contains numerous hydrophobic and aromatic residues in DQ3.2. In the DR1 structure, this pocket contains primarily small hydrophobic side chains, plus one salt bridge formed by Arg76α and Asp57β. DQ3.2 contains an homologous Arg at position 79α, but has an alanine at position 57β. This alanine substitution in DQ3.2 disrupts the salt bridge observed in the DR1 crystal structure. Arg79α is located at the end of the binding groove, and its side chain can be easily oriented either into pocket D or out of the binding groove. Since pocket D contains no other polar or charged side chains, we chose to position the side chain in an alternate high probability conformation [30] with the guanidino group pointing away from the pocket, where it has greater solvent accessibility. We predicted a hydrophobic anchor residue for this pocket, with preference perhaps for an aromatic residue in order to form π-stacking interactions in the pocket. Figure 3B shows the aromatic, hydrophobic side chains that stack together to form pocket D of DQ3.2.

After anchor pocket characterization was completed, the peptide was inspected for the presence of complementary anchor residues. The amino acid sequence of peptide 34p is

Ile-Ala-Arg-Phe-Lys-Met-Phe-Pro-Glu-Val-Lys-Glu-Lys

1   2   3   4   5   6   7   8   9   10   11   12   13

The anchor residues of a peptide are usually found in a nine amino acid span approximately in the center of the peptide sequence. [41] In order to achieve an extended backbone conformation, the spacing between anchor residues for

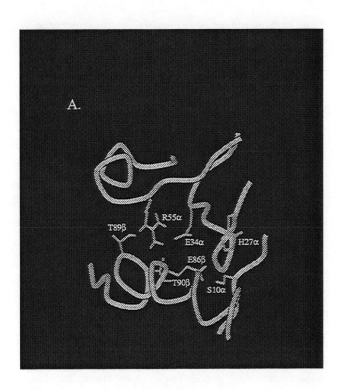

Figure 3. *Top view of A) pocket A in DQ3.2, B) pocket D in DQ3.2, and C) pocket D in DQ3.3 models. The protein backbone is shown in teal. Side chains shown (in orange) in pocket A are: Ser 10α, His 27α, Glu 34α, Arg 55α, Glu 86β, Thr 89β, and Thr 90β. In pocket D, Arg 79α and Ala/Asp 57β are shown in red, and all other side chains (Val 76α, Tyr 9β, Tyr 30β, Tyr 37β, Tyr 60β, and Trp 61β) are shown in orange.*

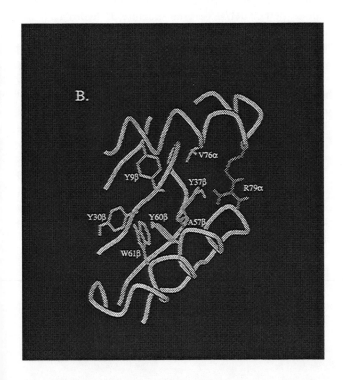

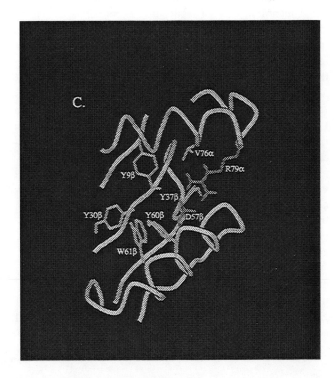

Table 1. *Properties of anchor pockets and the selected anchor residues from peptide 34p in a "reverse-orientation" binding mode.*

Pocket	Pocket Characteristics	Anchor Residue Properties Needed	Anchor Residue from peptide 34p
A	Polar and charged residues, with exposed Glu residues	Positively charged side chain	Lys (res. 11)
B	Primarily hydrophobic, plus exposed His and Tyr residues	Negatively charged or polar side chain	Glu (res. 9)
C	Hydrophobic	Hydrophobic side chain	Met (res. 6)
D	Hydrophobic; numerous aromatic side chains	Hydrophobic, preferably aromatic side chain	Phe (res. 4)

pockets A and D was restricted to seven or eight amino acids (i, i+7 or i, i+8). Anchor residues A and D, which received highest priority, were selected first, followed by anchor residues that complement pockets B and C. A suitable motif in the center of the peptide sequence containing the desired anchor residue properties and appropriate spacing was easily identified. The properties of each pocket, the corresponding properties sought in the anchor residue, and the selected side chains from the peptide are summarized in Table 1.

The anchor residues chosen for pockets A and D (positions 11 and 4, respectively) fit the characteristic profiles for these pockets ideally. The Lys side chain at position 11 is perfectly positioned in pocket A to form hydrogen bonds and charge interactions with the glutamate residues, and the Phe side chain at position 4 packs well with the aromatic residues in pocket D. This motif is comparable to the peptide binding motif identified in pool sequencing experiments for HLA-DQ2 [42], where Lys and Phe are the principle anchor residues in pockets A and D, respectively. Recent studies have also identified three peptides from a dust mite allergen protein that produce an immune response via DQ3.2-restricted TCR activation in transgenic mice. [43] All three peptides identified in this study fit our proposed binding motif well, with a lysine residue and various hydrophobic residues in either an i, i+7 or i, i+8 pattern. All three peptides can be docked easily in our DQ3.2 model, with lysine in pocket A and a hydrophobic anchor in pocket D. One of the peptides also has a glutamate at position i+2 which fits nicely in pocket B, exactly as seen in our DQ3.2-34p complex. Finally, the peptide binding motif for DQ3.2 in pool sequencing studies indicates that a Lys or Arg residue is the preferred primary anchor. [44]

With lysine and phenylalanine residues anchored in pockets A and D, the Glu side chain at position 9 was easily docked in pocket B to form a hydrogen-bonding partner for both the His and Tyr residues. The Met side chain at position 6 is well accommodated by the hydrophobic anchor pocket C. All an-

chor residues were positioned in the pockets while maintaining reasonable backbone torsion angles for the peptide.

In the DR1 crystal structure, the HA peptide backbone is in a polyproline II peptide conformation [9] as is the CLIP peptide in the recent DR3 complex structure [13] and two peptides of single-chain constructs with the mouse class II MHC molecule I-Ek. [14] The peptide backbone conformation in our DQ3.2/34p model complex is similar to these crystal structure complexes, although with a somewhat less pronounced twist that is not a classic polyproline II structure. Figure 4 shows the anchor residues from peptide 34p and key side chains in anchor pockets A, B, C, and D.

Peptide binding to class II proteins depends on interactions between peptide anchor residues and MHC anchor pockets, and also on the formation of a large number of hydrogen bonds between the MHC protein and backbone atoms of the peptide. In the DR1/HA crystal structure, 15 hydrogen bonds between DR1 side chains and the HA peptide backbone are reported [9], and the pattern is similar for the DR3/CLIP and I-E^k complexes. [13,14] In our model, DQ3.2 side chains were oriented to form 12 hydrogen bonds with peptide backbone atoms, most of which are analogous to those observed in the DR1 and DR3 structures. Six nonpolymorphic residues (N62α, N69α, R76α, W61β, N82β, H81β) that participate in nine hydrogen bonds in DR1/HA and DR3/CLIP are conserved in DQ3.2. Seven of these hydrogen bonds are maintained in our model, one additional hydrogen bond is formed with R79α, and one is lost due to the presence of a proline in peptide 34p. In addition to hydrogen bonds involving the peptide backbone, our model exhibits hydrogen bonding and charge interactions between peptide anchor residues and anchor pocket side chains, as discussed above. Figure 5 shows the hydrogen bond interactions between DQ3.2 and the peptide backbone.

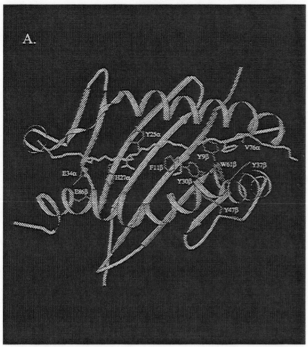

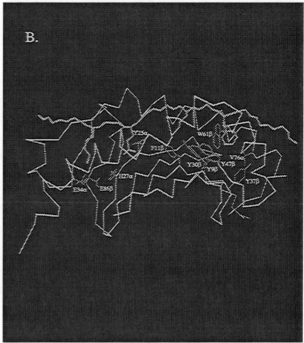

Figure 4. *Top-view and side-view images of the DQ3.2/34p complex showing anchor residues of 34p and key side chains from DQ3.2. The DQ3.2 backbone is shown in teal and DQ3.2 side chains are shown in red. The DQ3.2 side chains shown are: Glu 34α and Glu 86β in pocket A; Tyr 25α and His 27α in pocket B; Phe 11β, Tyr 30β, and Tyr 47β in pocket C; Val 76α, Tyr 9β, Tyr 30β, Tyr 37β, and Trp 61β in pocket D. The peptide is orange, with anchor residues Phe (position 4), Met (position 6), Glu (position 9), and Lys (position 11) shown.*

Construction and comparison of DQ3.1, DQ3.2 and DQ3.3 complexes with 34p

Position 57β, the single amino acid polymorphism that distinguishes DQ3.2 from DQ3.3, is located in anchor pocket D of the binding groove. Thus, any observed differences in peptide binding between DQ3.2 and DQ3.3 are due to this substitution. The primary difference predicted to arise from this Ala->Asp substitution is the probable formation of a salt bridge between Asp 57β and Arg 79α in DQ3.3. Formation of this salt bridge requires that Arg 79α adopt a side chain conformation similar to that observed for Arg 76α in the DR1 crystal structure. In our DQ3.2 model, an alternate conformation was chosen for Arg79α that orients it away from the hydrophobic environment of pocket D, and dramatically increases its solvent accessibility, as described above. This alternate conformation for Arg 79α is easily accommodated in our model structures with no backbone adjustment, and allows formation of an additional hydrogen bond with the peptide backbone, as discussed above. The main impact of this Asp-Arg salt bridge is a significant change in pocket

size. Pocket D is significantly smaller in DQ3.3 and DQ3.1 than in DQ3.2, because the Asp-Arg contact pair fills a portion of pocket D. Figures 3B and 3C show a detailed view of pocket D in the DQ3.2 and DQ3.3 models.

DQ3.1 contains four polymorphic substitutions relative to DQ3.2. As in DQ3.3, there is an Ala→Asp substitution at position 57β. There are additional polymorphisms at positions 13β (Gly→Ala), 26β (Leu→Tyr), and 45β (Gly→Glu). Position 13β is situated along the edge of pocket B between pockets B and C, and position 26β is located along the edge of pocket C, near pocket B. Both of these substitutions in DQ3.1 reduce the size and depth of the (already) shallow anchor pockets B and C. Position 45β is located on the edge of the β-sheet that forms the floor of the binding groove, and is not expected to directly impact peptide interactions within the groove (see Figure 1).

Thus, our models suggest a clear structural basis for the observed binding properties of 34p to DQ3.1, DQ3.2, and DQ3.3. Anchor pocket D is much smaller in DQ3.1 and DQ3.3, and our docking exercises suggest that peptides with a phenylalanine anchor at this position, such as 34p, should have greatly reduced binding affinities, as is observed experimentally. [17] The two additional polymorphic substitutions at positions 13β and 26β reduce the size of anchor pockets B and C in DQ3.1 relative to DQ3.2, and these changes introduce greater steric restrictions in the DQ3.1 binding groove. Again, these polymorphic substitutions would be expected to diminish binding for peptides with large anchor residues in these positions, such as the glutamate and methionine in 34p.

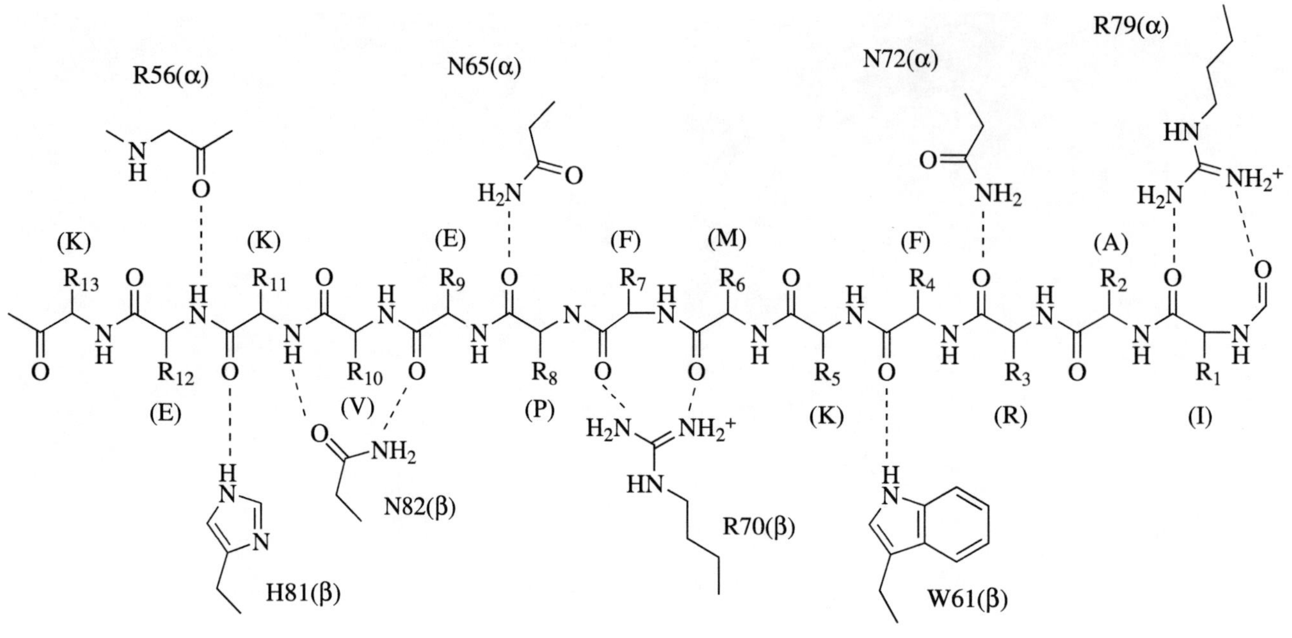

Figure 5. *Hydrogen-bonding pattern in the DQ3.2/34p complex. Side chains from DQ3.2 that form hydrogen bonds with the peptide backbone are shown and the hydrogen bonds are indicated by dashed lines. Peptide side chains are designated as R1-R13 and the residue at each position is given in parentheses.*

Discussion

The selected anchor residues from 34p provide an excellent model for the bound DQ3.2/34p complex, as shown in detail in Figure 4. This model exhibits good interactions between anchor residues and anchor pockets, appropriate spacing between anchor residues, a slightly twisted peptide backbone in extended conformation, and numerous hydrogen bonds between DQ3.2 and the peptide. This model of DQ3.2 with bound 34p suggests a novel binding mode in which the peptide orientation in the binding groove is opposite that typically observed. However, most peptide orientation data comes from crystal structures of class I complexes, in which both termini of the peptide are contained within the binding groove and contribute to specific interactions that stabilize the complex. Because the peptide termini extend beyond each end of a class II MHC binding groove, peptide binding is not restricted to the standard orientation by interactions between the peptide termini and the binding groove. Limited data on the characteristics of class II peptide anchor residues have revealed relative symmetry in the positioning and properties of anchor residues, and binding in either orientation has been suggested. [15] Src homology 3 domain (SH3) molecules also bind peptides that adopt a polyproline II conformation, and peptide binding has been observed in both orientations in SH3 protein-ligand complexes. [45, 46] It appears that

peptide side chain interactions with the SH3 binding groove determine the binding orientation in these complexes. [45] Our DQ3.2/34p model is the first proposed structure of an MHC complex with a reverse-orientation motif.

A variety of experimental data are available which lend support to our reverse-orientation binding model for peptide 34p to DQ3.2. For example, the choice of a lysine anchor residue for pocket A and phenylalanine for pocket D seems well justified, based on peptide motifs for DQ2 and DQ3.2 molecules identified in sequencing studies. [42] Since DQ3.2 and DQ2 molecules are highly homologous (91-94% sequence identity), it is not surprising that our predicted DQ3.2 peptide binding motif is quite similar to the motif determined experimentally for DQ2 molecules. The presence of a highly similar possible binding motif in the dust mite peptide antigens is also intriguing. It is quite interesting to note the sequence binding motif identified for DQ2 antigens suggest a traditional orientation in our models, while the possible motifs observed in the dust mite peptide antigens suggest a traditional binding orientation for some peptides, and a reverse-orientation binding mode for others.

Experimental studies have been performed to map the anchor residue positions in peptide 34p. [17] To assess the impact on binding to DQ3.2, single Arg substitutions were introduced at each position in 34p that is not normally a lysine or arginine residue. Our reverse-orientation model predicts that peptides containing Arg substitutions at position 4 (Phe→Arg), position 6 (Met→Arg), or position 9 (Glu→Arg) would not bind DQ3.2, since Arg would be an unfavorable anchor in each of these pockets. The experimental results indeed show that peptide binding by DQ3.2 is blocked only when an arginine substitution is introduced at positions 4, 6, or 9, as predicted by our model, and at one additional site, position 1.[17] Analysis of our model reveals the formation

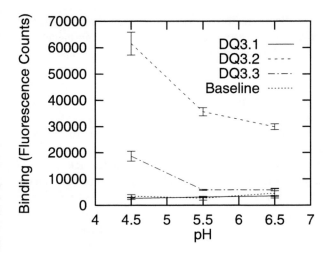

Figure 6. *Binding data for peptide 34p with DQ3.1, DQ3.2, and DQ3.3 as a function of pH. Baseline values are averages obtained for peptide 34p to the BLS1 cell line, which lacks class II MHC molecules.*

of two hydrogen bonds between Arg 79α in DQ3.2 and the backbone of the peptide at position 1. An Ile→Arg substitution at position 1 would place two Arg side chains in very close proximity, resulting in potential serious ionic and steric repulsion (see Figure 2). Replacement of Ile at position 1 with Thr, Gln, Phe, Asp, or Ala did not block peptide binding, however. [17] Since position 1 is tolerant of essentially any amino acid substitution except Arg, we propose that this position is not likely to function as an anchor residue.

A more systematic study has also been performed for positions 4, 6, and 9 of peptide 34p, using a number of amino acid substitutions. [17] The results indicate that binding is maintained only when each of these residues is replaced with a conservative substitution (e.g., Phe→Leu at position 4), or alanine, which appears to be an impartial, or permissive, anchor. Based on the data reported, position 9 seems to be least tolerant of substitution. Substitution of arginine or phenylalanine at this position abolishes binding, while threonine, leucine, and alanine mutations reduce binding substantially. These results are also consistent with the reverse-orientation model, as they imply the importance of the interaction between His27α and Glu-9 in pocket B.

We have used our model complexes to design additional experiments to further test this proposed binding mode for 34p. One of the most interesting implications of our reverse orientation model is the placement of Glu-9 from 34p in anchor pocket B, which contains a histidine residue in all three DQ molecules we have studied. At pH 4.5, both the His and Glu side chains are predicted to be predominantly charged. This suggests that peptide binding at low pH might involve formation of an ion-pair-reinforced hydrogen bond between these side chains. Our calculations also suggest that peptide binding may diminish somewhat as the pH increases to 6.5, since the histidine will begin to deprotonate noticeably at

the higher end of this range. Similar behavior should also be observed for DQ3.3, since 34p exhibits measurable, though much reduced, binding affinity for DQ3.3.

To test these model predictions, peptide 34p binding to DQ3.1, DQ3.2, and DQ3.3 was evaluated at pH 4.5, 5.5, and 6.5 (Nepom, Kwok, DeWeese, and Lybrand, unpublished results). Binding to DQ3.2 was observed to be pH-dependent, with optimal binding observed at pH 4.5, and a smooth decrease in binding as the pH is raised to 6.5, where the histidine is expected to be only ~50% protonated. These results are shown in Figure 6, and agree well with our predictions for the reverse-orientation model. As expected, 34p shows no appreciable binding to DQ3.1. The peptide does exhibit limited binding to DQ3.3, and the pH profile follows the trend observed for DQ3.2, as predicted by the models. These data are also given in Figure 6.

We also explored several alternate binding modes for 34p, to determine if any other plausible models could be generated. One alternative assumed a traditional orientation, with 34p residues 1, 4, 6, and 9 chosen as anchors (i.e., the residues identified in the Arg substitution experiments by Kwok, at al. [17]). This binding motif, however, requires a different set of interactions in the primary anchor pockets. Specifically, this binding motif positions Ile-1 in pocket A, Phe-4 in pocket B, Met-6 in pocket C, and Glu-9 in pocket D. As described above, pocket A contains charged and polar side chains; a hydrophobic side chain such as Ile would be at best impartial in this pocket. Pocket D, the other key anchor pocket of DQ3.2, contains hydrophobic, aromatic side chains, and a charged Glu side chain does not appear to be a particularly good anchor residue for this pocket. It has been proposed that Glu-9 of the peptide could form a salt bridge with Arg79α in anchor pocket D. However, this proposal is difficult to reconcile with the observed pH-dependent binding profile. If this Glu-Arg interaction were formed, we predict it would be strongest at pH 6.5, where the glutamate is fully ionized. As the pH is decreased to 4.5, the glutamate would begin to protonate, weakening the salt bridge and reducing binding affinity at least slightly. The experimentally observed pH binding profile is distinctive and shows the opposite trend (i.e., peptide binding becomes significantly stronger as the pH is lowered), consistent with our reverse-orientation model that suggests a Glu-9/His27α interaction in pocket B as discussed above. Finally, this proposed binding motif is drastically different from the motif identified by Verreck et al. for DQ2 molecules. [42] As discussed above, we expect that DQ3.2 may have a similar binding motif, due to its very high sequence similarity to DQ2 molecules. As discussed above, we expect that DQ3.2 may have a similar binding motif, due to its very high sequence similarity to DQ2 molecules, and the motif data from pool sequencing for DQ3.2. [44]

We considered other combinations of anchor residues that would generate a traditional binding orientation. Based on anchor pocket characterization, most potential combinations of anchor residues for the traditional orientation are predicted to be unfavorable. The only combination of anchor residues that fits a traditional orientation binding motif with impartial or favorable characteristics is Arg-3 for the polar and negatively charged pocket A and Val-10 for the hydrophobic pocket D. We constructed a model of the complex in this orientation. This combination of anchor residues contains i, i+7 spacing, as does our reverse-orientation model, and a similar hydrogen bonding pattern for peptide backbone with DQ side chains is observed. However, this model is inconsistent with the arginine substitution data obtained by Kwok and coworkers. [17] They found that replacement of Val-10 with Arg does not disrupt binding, as would be expected if Val were an anchor residue. Furthermore, this model requires that Phe-4, Met-6, and Glu-9 side chains project directly out of the binding groove. However, Kwok at al. observed that arginine substitution at any of these positions eliminates peptide binding to DQ3.2 completely. It is difficult to envision how arginine substitution for a residue that projects out of the binding groove could totally block peptide binding, or how arginine substitutions could be completely tolerated at all the other positions predicted by this model to project into the binding groove.

For direct comparison to the known human class II MHC-peptide crystal structures, we attempted to dock peptide 34p using both the HA and CLIP backbone alignments as templates. First, the DQ3.2 backbone atoms were aligned with the DR1 backbone atoms, and the HA coordinates were copied into the file containing the DQ3.2 model. Amino acid side chains were replaced to generate the 34p sequence. This procedure was repeated for the DR3/CLIP structure.

Since the HA and CLIP peptides have i, i+8 spacing between anchor residues, the 34p sequence was assessed for potential anchor combinations. However, 34p contains no side chains that fit the characteristic profiles needed to match anchor pockets A and D with i, i+8 spacing. The combination of Ala-2 and Val-10 appear to be the only possible combination of residues with i, i+8 spacing that would not be disruptive in pocket A or D. However, while Val is an acceptable choice for pocket D, Ala seems much less desirable for the large, polar pocket A. This motif is completely inconsistent with the arginine substitution data [17], since an Arg is not disruptive to binding at either position. In addition, this motif orients Phe-4, Met-6, and Glu-9 out of the binding groove, in contrast to the implications of the arginine substitution data, as discussed above. This motif also places the side chain of Lys-4 into pocket B, which contains His27α, another interaction predicted to be unfavorable.

One result observed by Kwok et al. is more difficult to interpret simply. Binding is enhanced when the Lys anchor at position 11 is replaced with Phe (34p11F). [17] Since our model predicts that a hydrophobic side chain such as Phe would be much less favorable than lysine in pocket A, the reverse-orientation model cannot account for the binding of peptide 34p11F. An analysis of the peptide sequence reveals two potential alternative combinations of anchor residues to explain the binding observed for the 34p11F peptide: one

that predicts a traditional orientation for binding and one that predicts an alternate reverse-orientation binding mode.

The traditional orientation model has Arg-3 and Phe-11 anchors with i, i+8 spacing. Using 1-4-6-9 spacing for anchor residues, Met-6 and Pro-8 would be the anchor residues for pockets B and C, respectively. In this binding model, favorable interactions would occur in pockets A, C and D, with an impartial or slightly unfavorable (due to the steric bulk of Met) interaction in pocket B. A model of this complex was constructed using the HA peptide backbone as a template. Eleven hydrogen bonds can be formed between the peptide backbone and DQ3.2 in this model, but no interactions are identified that might explain the pH profile observed for 34p. However, there are no experimental data to demonstrate that this modified 34p11F peptide exhibits the same pH-dependent binding profile as 34p.

The alternate reverse-orientation binding model for peptide 34p11F would involve a register shift of two residues from the reverse-orientation model for wild-type 34p. This would place Lys-13 and Met-6 anchor residues in pockets A and D, respectively. Pro-8 and Phe-11 would be anchors for pockets B and C, respectively. This would also produce favorable interactions in pockets A, C, and D, with a neutral interaction in pocket B. A model of this complex was constructed for comparison using the wild-type 34p reverse-orientation peptide backbone as a template. Ten peptide backbone hydrogen bonds are formed with DQ3.2 side chains in this complex. As for the traditional orientation model of this complex, there is no suggestion that a pH-dependent binding profile would be observed for this peptide complex either.

The proposed shift in binding register, or flip to a traditional binding mode with different anchor residues, appear to be the two most rational explanations for the extremely strong binding observed for peptide 34p11F. Because each of these models entail fundamentally different binding modes, a new series of residue substitution and pH binding profile experiments will be needed to evaluate them properly.

Conclusions

Our reverse orientation binding model provides a reasonable explanation for the interactions of a potentially diabetogenic peptide with the DQ3.2 MHC molecule, and the rather dramatic impact selected polymorphic substitutions have on peptide binding to several highly homologous DQ molecules. It includes both of the key components for peptide binding to class II MHC proteins: (1) complementary interactions between peptide anchor residues and protein anchor pockets, and (2) an extensive hydrogen bonding network between protein side chains and the peptide backbone. In addition, our model provides a reasonable explanation for the pH-dependent binding profile observed experimentally.

For peptide 34p, no peptide binding motifs are available that incorporate the typical orientation, fit the profile needed for anchor residues and observed for homologous DQ2 molecules, and are consistent with results from anchor residue mapping experiments. Thus, it seems that peptide 34p most likely binds to DQ3.2 in a reverse-orientation mode. Our modeling results also suggest that DQ molecules may bind peptides in either direction, depending on the precise nature of anchor residue interactions with the binding groove, as is observed experimentally for SH3 protein complexes. Our modeling studies also suggest that individual amino acid substitutions at certain positions in the peptide may in some cases alter the binding mode dramatically.

This is the first proposed structure of an MHC protein binding a peptide ligand in the reverse orientation. Detailed biophysical studies will be needed to confirm the exact nature and orientation of peptide binding in DQ3.2. At present, we are performing photoaffinity labeling studies to obtain definitive information for 34p orientation in the DQ3.2 binding groove.

Acknowledgments This work was supported in part by grants from the National Institutes of Health (DK41801) and the Whitaker Foundation. We also wish to thank Mary Ellen Domeier and Eric Swanson for expert technical assistance.

Supplementary Material

Full coordinates for the following models are available in Brookhaven PDB format:
    DQ3.1 (0205dg31.pdb)
    DQ3.2 (0205dg32.pdb)
    DQ3.3 (0205dg33.pdb)

References

1. Kim, S. J.; Holbeck, S. L.; Nisperos, B.; Hansen, J. A.; Maeda, H.; Nepom, G. T. *Proc Natl Acad Sci U S A* **1985**, *82*, 8139.

2. Michelsen, B.; Kastern, W.; Lernmark, Å.; Owerbach, D. *Biomed Biochim Acta* **1985**, *44*, 33.

3. Nepom, B. S.; Palmer, J.; Kim, S. J.; Hansen, J. A.; Holbeck, S. L.; Nepom, G. T. *J Exp Med* **1986**, *164*, 345.

4. Nepom, G. T.; Seyfried, C. A.; Nepom, B. S. *Pathol Immunopathol Res* **1986**, *5*, 37.

5. Nepom, G. T. *Diabetes Reviews* **1993**, *1*, 93.

6. Martin, J. M.; Trink, B.; Daneman, D.; Dosch, H.-M.; Robinson, B. *Annals of Medicine* **1991**, *23*, 447.

7. Lernmark, •.; BŠrmeier, H.; Dube, S.; Hagopian, W.; Karlsen, A.; Wassmuth, R. *Endocrinology and Metabolism Clinics of North America* **1991**, *20*, 589.

8. Yoon, J. W. *Curr. Top. Microbiol. Immunol.* **1990**, *164*, 95.

9. Stern, L. J.; Brown, J. H.; Jardetzky, T. S.; Gorga, J. C.; Urban, R. G.; Strominger, J. L.; Wiley, D. C. *Nature* **1994**, *368*, 215.

10. Sinigaglia, F.; Hammer, J. *Apmis* **1994**, *102*, 241.

11. Rotzschke, O.; Falk, K. *Curr Opin Immunol* **1994**, *6*, 45.

12. Madden, D. R. *Annu Rev Immunol* **1995**, *13*, 587-622.

13. Ghosh, P.; Amaya, M.; Mellins, E.; Wiley, D. C. *Nature* **1995**, *378*, 457.

14. Fremont, D. H.; Hendrickson, W. A.; Marrack, P.; Kappler, J. *Science* **1996**, *272*, 1001.

15. Falk, K.; Rotzschke, O.; Stevanovi'c, S.; Jung, G.; Rammensee, H. G. *Immunogenetics* **1994**, *39*, 230.

16. Johansen, B. H.; Buus, S.; Vartdal, F.; Viken, F.; Eriksen, J. A.; Thorsby, E.; Sollid, L. M. *Int. Immunol.* **1994**, *6*, 453.

17. Kwok, W. W.; Domeier, M. E.; Raymond, F. C.; Byers, P.; Nepom, G. T. *J Immunol* **1996**, *156*, 2171.

18. Goodman, J. W. In *Basic and Clinical Immunology*; 8 ed.; Stites-D-P, Terr-A-I, Parslow-T-G, Eds.; Appleton & Lange: Norwalk, CT, 1994.

19. Duquesnoy, R. J. *Clin Lab Med* **1991**, *11*, 509.

20. Marsh, S. G.; Bodmer, J. G. *Immunogenetics* **1993**, *37*, 79.

21. Kockum, I.; Wassmuth, R.; Holmberg, E.; Michelsen, B.; Lernmark, A. *Am J Hum Genet* **1993**, *53*, 150.

22. Sanjeevi, C. B.; Lybrand, T. P.; DeWeese, C.; Landin-Olsson, M.; Kockum, I.; Dahlquist, G.; Sundkvist, G.; Stenger, D.; Lemmark, A. *Diabetes* **1995**, *44*, 125.

23. Kobayashi, T.; Tamemoto, K.; Nakanishi, K.; Kato, N.; Okubo, M.; Kajio, H.; Sugimoto, T.; Murase, T.; Kosaka, K. *Diabetes Care* **1993**, *16*, 780.

24. Sanjeevi, C. B.; Falorni, A.; Kockum, I.; Hagopian, W. A.; Lernmark, Å. *Diabetic Medicine* **1996**, *13*, 209.

25. Hagopian, W. A.; Michelsen, B.; Karlsen, A. E.; Larsen, F.; Moody, A.; Grubin, C. E.; Rowe, R.; Petersen, J.; McEvoy, R.; Lernmark, A. *Diabetes* **1993**, *42*, 631.

26. Atkinson, M. A.; Bowman, M. A.; Campbell, L.; Darrow, B. L.; Kaufman, D. L.; Maclaren, N. K. *J Clin Invest* **1994**, *94*, 2125.

27. Chothia, C.; Lesk, A. M. *Embo J* **1986**, *5*, 823-6.

28. Ring, C. S.; Cohen, F. E. *Faseb J* **1993**, *7*, 783-90.

29. Sutcliffe, M. J.; Hayes, F. R.; Blundell, T. L. *Protein Eng* **1987**, *1*, 385.

30. Ponder, J. W.; Richards, F. M. *J Mol Biol* **1987**, *193*, 775.

31. Ferrin, T. E.; Huang, C. C.; Jarvis, L. E.; Langridge, R. *J Mol Graph* **1988**, *6*, 13.

32. Swanson, E. PSSHOW, Seattle, 1995.

33. Pearlman, D. A.; Case, D. A.; Caldwell, J. C.; Seibel, G. L.; Singh, U. C.; Weiner, P.; Kollman, P. A. ; University of California: San Francisco, 1991.

34. Weiner, S. J.; Kollman, P. A.; Nguyen, D.; Case, D. A. *J Comput Chem* **1986**, *7*, 230.

35. Gregoret, L. M.; Cohen, F. E. *J Mol Biol* **1990**, *211*, 959.

35. Antosiewicz, J.; McCammon, J. A.; Gilson, M. K. *J Mol Biol* **1994,** *238*, 415.

37. Antosiewicz, J.; Porschke, D. *Biophys J* **1995,** *68*, 655.

38. Davis, M. E.; Madura, J. D.; Sines, J.; Luty, B. A.; Allison, S. A.; McCammon, J. A. *Methods Enzymol* **1991,** *202*, 473.

39. Annand, R. R.; Kontoyianni, M.; Penzotti, J. E.; Dudler, T.; Lybrand, T. P.; Gelb, M. H. *Biochemistry* **1996,** *35*, 4591.

40. Cornell, W. D.; Cieplak, P.; Bayly, C. I.; Gould, I. R.; Merz, K. M.; Ferguson, D. M.; Spellmeyer, D. C.; Fox, T.; Caldwell, J. C.; Kollman, P. A. *J Am Chem Soc* **1995,** *117*, 5179.

41. Rammensee, H. G. *Curr Opin Immunol* **1995,** *7*, 85.

42. Verreck, F. A. W.; van de Poel, A.; Temijtelen, A.; Amons, R.; Drijfhout, J.-W.; Koning, F. *Eur. J. Immunol.* **1994,** *24*, 375.

43. Neeno, T.; Krco, C. J.; Harders, J.; Baisch, J.; Cheng, S.; David, C. S. *J. Immunol.* **1996,** *156*, 3191.

44. Chicz, R.M.; Lane, W.S.; Robinson, R.A.; Trucco, M.; Strominger, J.L.; Gorga, J.C. *Int. Immunol.* **1994,** *6*, 1639.

45. Feng, S.; Chen, J. K.; Yu, H.; Simon, J. A.; Schreiber, S. L. *Science* **1994,** *266*, 1241.

46. Wilson, I. A. *Science* **1996,** *272*, 973.

47. Kraulis, P. J. *Journal of Applied Crystallography* **1991,** *24*, 946.

J. Mol. Model. **1996**, 2, 217 – 226

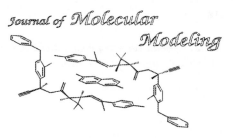

Ab Initio and Molecular Mechanics Calculations of Various Substituted Ureas – Rotational Barriers and a New Parametrization for Ureas

Thomas Strassner [1]

Institut für Organische Chemie, Universität Erlangen-Nürnberg, Henkestr. 42, D – 91054 Erlangen, Germany
(strasner@organik.uni-erlangen.de)

Received: 3 May 1996 / Accepted: 6 August 1996 / Published: 3 September 1996

Abstract

High-level *ab initio* calculations have been performed on urea, methylurea, tetramethylurea and three isomers of dimethylurea to obtain accurate rotational barriers. Results of MP2(fc)/6-31 G(d) calculations are compared to those with lower basis sets and semiempirical calculations. The MM2(87) force field has been parameterized.

Keywords: MM2, parametrization, rotational barrier, urea, *ab initio*

Introduction

Today molecular modeling is an important tool for interdisciplinary research. Different methods are applied, depending on the size of the molecules. Though the computational power is still increasing, it is not feasible to study larger systems with *ab initio* quantum-chemical methods. The applicability of *ab initio* calculations increased due to the development of new supercomputers and the accuracy of *ab initio* calculations depends only on the choosen level, but still they are only useful for model compounds. Semiempirical calculations can deal with larger molecules of up to 500 atoms, but to study natural compounds with several possible conformations they are also not useful.

Force field methods enable scientists to calculate large structures with up to 10.000 atoms, but the quality of force field calculations depends on the availability of reliable force field parameters which are consistent with earlier parameter sets to ensure that the results of previous calculations can be compared with actual computations.

Since Wöhlers synthesis in 1828 urea is known as the starting point of organic chemistry. Its structural unit can be found in molecules which are essential for many natural proc-

esses. Moreover, the $R_2N-CO-NR_2$ unit is a basic constituent of various enzymes, nucleotides and of interesting compounds with possible pharmazeutical applications. Recently published papers including a structure of a HIV-inhibitor compound show the interest in this structural unit [2]. You can find the -NH-CO-NH- fragment in many natural compounds (e.g. barbiturates, vitamine B_{13}, purines and pyrimidines) [3].

To study biological important systems force field calculations are necessary because of the size of the molecules. The importance and acceptance of molecular modelling increases, due to the efficiency and reliability of the results. These techniques are based on empirical force-field methods, which are parametrized for a wide variety of substances. Usually all parameters are developed independently, but the existing parametrization for ureas by Bowen et al. [4] changes the parameters for amides, so the consistency will be lost when the results of previous calculations of amides are compared with actual ones.

Ureas have been studied extensively in various ways, theoretically as well as experimentally. Especially the interaction between urea and water or the protein denaturation was of considerable interest [5]. Ureas are able to interact with a wide variety of compounds, they can form hydrogen bonds with electron acceptors and donators and they are known to

Table 1. *Mean values of compound classes G - M (see figure 2). Maximum deviation is given in brackets.*

	G	H	I	K	M
NC (Å)	1.37 (0.05)	1.36 (0.07)	1.37 (0.03)	1.36 (0.06)	1.36 (0.04)
CO(Å)	1.22 (0.03)	1.23 (0.03)	1.21 (0.02)	1.22 (0.03)	1.23 (0.03)
NCN (°)	107.2 (1.1)	107.7 (2.4)	116.7 (2.8)	116.5 (2.3)	113.8 (4.9)
NCNX (°)	4.1 (12.5)	4.6 (13.8)	2.9 (10.3)	3.5 (10.7)	6.8 (11.4)

form cocrystals [6]. The interactions of ureas with other molecules were studied by X-ray-analysis [7] and the question whether urea is planar or not received a lot of attention during the last years. This study cannot contribute to a discussion where results of calculations on much higher levels of theory are discussed. The intention is the development of a consistent set of reliable parameters from thorough *ab initio* studies of rotational barriers of different substituted ureas. The semi-empirical rotational barriers of ureas are well studied [8] and will be compared to the results of high-level *ab initio* calculations.

To build a parameter set for ureas the accuracy of semi-empirical calculations was not sufficient. Exact *ab initio* data concerning the rotational barriers of different substituted and unsubstituted ureas were necessary. The presented study also gave insight into the rotational barriers of different ureas, which are shown in figure 1.

Computational Details

All calculations were carried out using the GAUSSIAN 92 [9] and GAUSSIAN 94 [10] program. The standard basis sets 3-21G and 6-31G* were used in all calculations. Correlation energy was considered by Møeller-Plesset perturbation theory up to the second order including all innermost and outermost virtual orbitals. The MP2 calculations were done with frozen core. The geometries were completely optimized with a constraint on the dihedral angle ρ_1 (rotation of the NR_1R_2-group). All calculations were performed in 30 degree increments between 0 and 330 degrees, only the MP2-calcula-

tions were done from 0 to 180 degrees while the second barrier was calculated at 270 degrees.

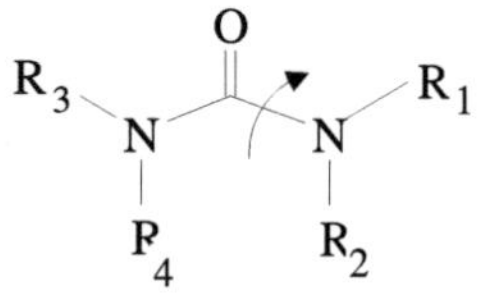

Figure 1. *Studied ureas A – F.*

A: $R_1=R_2=R_3=R_4=$ H
B: $R_1=CH_3$, $R_2=R_3=R_4=$ H
C: $R_1=R_4=CH_3$, $R_2=R_3=$ H
D: $R_1=R_4=H$, $R_2=R_3=$ CH_3
E: $R_1=R_2=CH_3$, $R_3=R_4=$ H
F: $R_1=R_2=R_3=R_4=$ CH_3

MM2 - parametrization

MM2(87) [11] with the newest parameter set was used for the parametrization. This set was announced as the final parameter set of MM2. The data of the *ab initio* rotational barriers were used to calibrate the parameters. A previous work [4] concerning the parametrization of urea compounds did not take into account the problem you encounter in the sense of consistency, if you change the values of angles and dihedral angles incorporated in amide units. This problem was

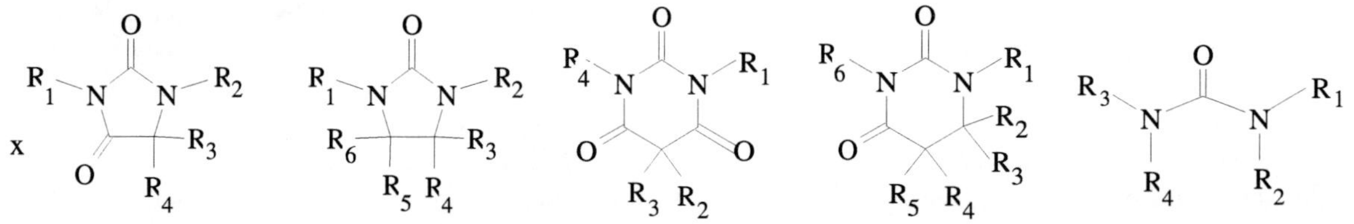

G (19)   **H (20)**   **I (36)**   **K (20)**   **L (15)**

Figure 2. *Compound classes containing a -N-CO-N- unit*

Table 2. *Energy differences of rotational isomers. All energies are given relative to the 0° torsion (3-21G: -222.73683; 6-31G*: -223.98421; MP2/6-31G*: -224.60680).*

Urea			
A	**3-21G**	**6-31G***	**MP2/6-31G***
angle	*kcal/mol*	*kcal/mol*	*kcal/mol*
0°	0	0	0
30°	+ 2.58	+ 5.17	+ 5.13
60°	+ 8.23	+ 12.07	+ 11.12
90°	+ 10.92	+ 15.50	+ 14.58
120°	+ 8.23	+ 12.07	+ 11.12
150°	+ 2.58	+ 1.32	+ 0.89
180°	0	+ 0.001	+ 1.32
210°	+ 2.58	+ 1.32	-
240°	+ 8.23	+ 6.21	-
270°	+ 10.92	+ 8.69	+ 7.64
300°	+ 8,.3	+ 6.21	-
330°	+ 2.58	+ 1.32	-
360°	0	0	0

Table 3. *Energy differences of rotational isomers. All energies are given relative to the 0° torsion (3-21G: -261.55084; 6-31G*: -263.01335; MP2/6-31G*: -263.76561)*

Methylurea			
B	3-21G	6-31G*	MP2/6-31G*
angle	*kcal/mol*	*kcal/mol*	*kcal/mol*
0°	0	0	0
30°	+ 3.18	+ 1.90	+ 5.26
60°	+ 9.39	+ 12.87	+ 12.08
90°	+ 17.88	+ 16.82	+ 15.74
120°	+ 9.70	+ 13.96	+ 12.90
150°	+ 4.74	+ 3.38	+ 2.84
180°	+ 1.72	+ 1.52	+ 1.33
210°	+ 4.75	+ 3.37	-
240°	+ 9.70	+ 7.73	-
270°	+ 12.23	+ 10.21	+ 8.96
300°	+ 9.39	+ 6.99	-
330°	+ 3.16	+ 1.90	-
360°	0.00	0.00	0

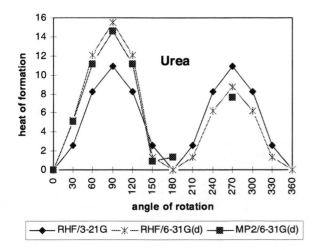
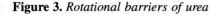

Figure 3. *Rotational barriers of urea*

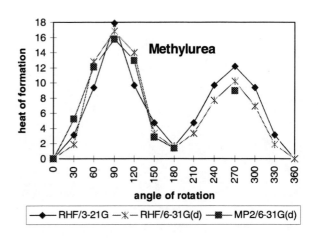

Figure 4. *Rotational barriers of methylurea.*

J. Mol. Model. **1996**, 2

Table 4. *Energy differences of rotational isomers. All energies are given relative to the 0° torsion (3-21G: -300.36468; 6-31G*: -302.04207; MP2/6-31G*: -302.92411)*

Dimethylurea Type C

C	3-21G	6-31G*	MP2/6-31G*
angle	kcal/mol	kcal/mol	kcal/mol
0°	+ 0.00	+ 0.00	+ 0.00
30°	+ 3.18	+ 1.65	+ 1.19
60°	+ 9.15	+ 12.06	+ 11.05
90°	+ 11.97	+ 15.95	+ 14.55
120°	+ 9.45	+ 13.23	+ 11.85
150°	+ 4.73	+ 3.09	+ 2.40
180°	+ 1.75	+ 1.46	+ 1.72
210°	+ 5.11	+ 3.09	-
240°	+ 9.75	+ 6.83	-
270°	+ 11.98	+ 9.04	+ 7.16
300°	+ 9.25	+ 6.79	-
330°	+ 3.18	+ 1.65	-
360°	0.00	0.00	0.000

Table 5. *Energy differences of rotational isomers. All energies are given relative to the 0° torsion (3-21G: -300.34852; 6-31G*: -302.03034; MP2/6-31G*: -302.91451)*

Dimethylurea Type D

D	3-21G	6-31G*	MP2/6-31G*
angle	kcal/mol	kcal/mol	kcal/mol
0°	+ 8.39	+ 5.91	+ 4.85
30°	+ 6.87	+ 4.88	+ 3.64
60°	+ 10.04	+ 9.00	+ 6.58
90°	+ 12.47	+ 18.21	+ 16.81
120°	+ 9.50	+ 8.03	+ 6.89
150°	+ 3.11	+ 2.23	+ 1.78
180°	+ 0.00	0.0	0.0
210°	+ 3.11	+ 2.23	-
240°	+ 9.50	+ 8.03	-
270°	+ 12.47	+ 10.58	+ 8.79
300°	+ 10.04	+ 8.12	-
330°	+ 6.87	+ 4.88	-
360°	+ 8.39	+ 5.91	+ 4.85

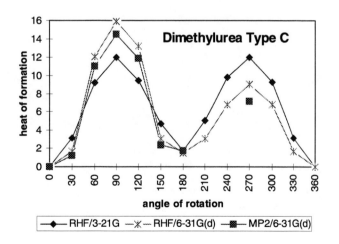

Figure 5. *Rotational barriers of dimethylurea type C*

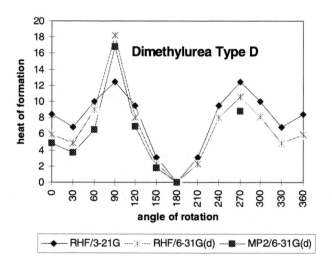

Figure 6. *Rotational barriers of dimethylurea type D*

Table 6. *Energy differences of rotational isomers. All energies are given relative to the 0° torsion (3-21G: -300.36151; 6-31G*: -302.03778; MP2/6-31G*: -302.92363)*

Dimethylurea Type E

E	3-21G	6-31G*	MP2/6-31G*
angle	kcal/mol	kcal/mol	kcal/mol
0°	+ 0.001	+ 0.01	- 0.0003
30°	+ 2.43	+ 1.18	+ 0.88
60°	+ 7.1	+ 10.71	+ 10.21
90°	+ 14.81	+ 13.74	+ 13.05
120°	+ 7.10	+ 10.71	+ 10.21
150°	+ 2.43	+ 1.18	+ 0.88
180°	+ 0.00	0	0.00
210°	+ 2.43	+1.18	-
240°	+ 7.1	+ 5.40	-
270°	+ 9.66	+ 7.61	+ 6.69
300°	+ 7.1	+ 5.40	-
330°	+ 2.43	+ 1.18	-
360°	+ 0.001	+ 0.01	- 0.0003

Table 7. *Energy differences of rotational isomers. All energies are given relative to the 0° torsion (3-21G: -377.97596; 6-31G*: -380.08382; MP2/6-31G*: -381.23600)*

Tetramethylurea

F	3-21G	6-31G*	MP2/6-31G*
angle	kcal/mol	kcal/mol	kcal/mol
0°	+ 2.52	+ 2.44	+ 2.05
30°	+ 0.000	+ 0.000	+ 0.000
60 °	+ 2.74	+ 2.68	+ 2.22
90°	+ 4.93	+ 4.75	+3.77
120°	+ 2,74	+ 2.68	+2.22
150°	- 0,001	- 0.001	+0.0001
180°	+ 2.52	+ 2.44	+ 2.05
210°	-0.001	-0.001	-
240°	+ 2.74	+ 2.68	-
270°	+ 4.93	+ 4.75	+ 3.77
300°	+ 2.74	+ 2.68	-
330°	- 0.001	- 0.0008	-
360°	+ 2.52	+ 2.439	+ 2.05

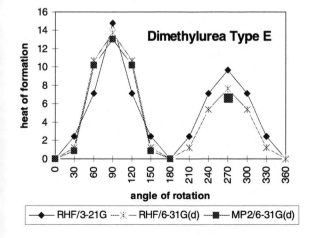

Figure 7. *Rotational barriers of dimethylurea type E*

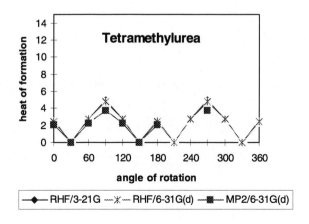

Figure 8. *Rotational barriers of tetramethylurea.*

Table 8. *Barrier heights calculated with different hamiltonians and basis sets compared to experimental values*

	A	B	C	D	E	F
RHF/3-21G	10.9	12.2	12.0	12.5	9.7	4.9
RHF/6-31G(d)	8.7	10.2	9.0	10.6	7.6	4.8
MP2(fc)/6-31G(d)	7.6	9.0	7.2	8.8	6.7	3.8
AM1	4.8	4.1	5.0	4.0	3.8	1.6
MNDO	3.2	2.9	2.1	1.8	1.6	0.8
PM3	3.4	2.5	4.6	2.4	1.0	2.0
Experiment	11.3	13.2	-	-	10.7	6.3

solved, the original MM2-parameters for amides are used and only parameters of urea units are changed. The new parameter set for urea compounds was compared to the structural information received by analyzing several structural classes (figure 2) containing urea units, which have been choosen from about 300 X-ray-structures. Many of those structures are ureas complexed by different metals, eg. Al [12], Ba [13], Co [14], Cd [15], Cu [16], Er [17], Eu [18], Fe [19], Hg [20], Ir [21], In [22], K [23], Li [24], Mg [25], Mn [26], Mo [27], Pt [28], Rb [29], Ru [30], U[31], or charged systems [32]. Those molecules often show unusual binding situations and have not been considered. The numbers in brackets give the number of considered structures, table 1 shows typical bond length, angles and dihedral angles.

Rotational barriers

Tables 2 to 7 and figures 3 to 8, respectively, show the rotational barriers of urea, methylurea, dimethylurea type C, D and E and tetramethylurea.

Comparison of *ab initio* calculations, semiempirical calculations and experimental data

Unfortunately no gas phase NMR data of the rotational barriers are available, but most of the barriers measured in solution are known. Their values are : Urea is 11.3 kcal/mol [33], methylurea for the rotation about the $NH(CH_3)$- bond is 13.2 kcal/mol [34] with a difference of 12.3 kcal/mol between the cis- and trans-isomers. Dimethylurea of type E is 10.7 kcal/ mol [34] and tetramethylurea is 6.3 kcal/mol [35]. The dimethylureas type C and D were also measured, but the only result has been that one isomer is dominant. No value for the rotational barrier was found.

Obviously there is only a small influence of the basis set. The shape of the barriers is very similar, only their height is slightly different. There are always two barriers (except tetramethylurea F) for all molecules due to the pyrimidalization of the nitrogen. The lower value of the two barriers has to be considered as rotational barrier and corre-

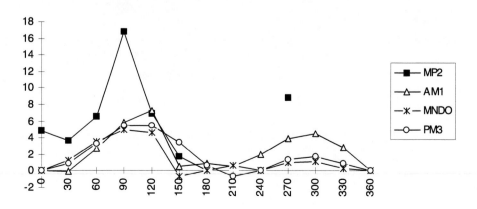

Dimethylurea Type 2 ab initio vs semiempiric

Figure 9. *Barrier of compound D calculated with MP2(fc)/6-31G(d) and AM1, MNDO, PM3*

Table 9. *Parameter set of the original MM2-code compared to the new parameter set*

	MM2			new set		
atomtype	V_1	V_2	V_3	V_1	V_2	V_3
9 3 9 3	0.8	1.3	0.0	0.8	2.0	0.0
7 3 9 3	0.6	1.3	0.0	0.0	0.0	0.0
9 3 9 28	0.0	1.0	0.0	1.0	1.0	0.0
9 3 9 1	0.0	0.0	0.01	3.0	6.0	0.0
9 1 1 9	0.0	0.0	-0.5	2.0	0.3	0.1
atomtype	k_B		angle	k_B		angle
2 1 9	0.5		109.8	0.82		109.8
2 2 9	0.5		118.0	0.52		120.0
2 9 2	0.4		107.0	0.56		110.0
3 9 2	0.6		115.0	0.6		115.0
3 9 3	0.5		125.0	0.65		120.0
9 3 9	0.5		120.0	1.0		110.0
3 9 28	0.42		119.0	0.6		119.0
0 9 28	0.010			0.8		

sponds always to the 270 degree structure, where the lone pair of the nitrogen is pointing away from the carbonyl oxygen. The 3-21G basis set calculates higher barriers than the 6-31G(d) basis set, whereas the inclusion of correlation with the MP2-calculations lowers the barriers (see table 8).

All rotational barriers were recalculated with AM1, MNDO and PM3. The results are shown in table 8. It is known for a long time that semiempirical methods have problems reproducing experimental values of rotational barriers. It is obvious that in this case of different substituted ureas this is especially true. All three semiempirical hamiltonians are unable to calculate values which are at least close to the experimental ones. Figure 9 shows the differences between the MP2-calculation, which gives the lowest barriers of all *ab initio*-methods and the barriers calculated with all three semiempirical hamiltonians.

Parameters

For a better comparison of the changes the original MM2-parameters are given together with the new set in table 9.

Comparison of rotational profiles calculated with the new parameter set / *ab initio* (MP2(fc)/6-31G(d)

Figure 10 shows the result of the comparison of the rotational barriers calculated with the new MM2-parameter set and the *ab initio* calculations.

	X-ray-data		new parameter set	
	NCN	NCNX	NCN	NCNX
G	107.2 (1.1)	4.1 (12.5)	106.1 (0.3)	0.03 (0.1)
H	107.7 (2.4)	4.6 (13.8)	106.8 (0.2)	0.7 (2.0)
I	116.7 (2.8)	2.9 (10.3)	115.8 (1.1)	9.1 (15.6)
K	116.5 (2.3)	3.5 (10.7)	115.8 (0.7)	5.0 (9.3)
L	113.8 (4.9)	6.8 (11.4)	116.7 (5.7)	3.5 (26)

Table 10. *Comparison of the angles between both nitrogens and the central carbon atom as well as the corresponding dihedral angles. The numbers in brackets mention the maximum deviation which was observed.*

	N-C	C-O	N-C-N	N-C-O
RHF/6-31G(d) [36]	1.373	1.197	114.0	123.0
MP2/6-31G(d) [37]	1.389	1.225	113.0	123.5
6-31G(d,p) [37]	1.388	1.224	113.0	123.523
MM2 (parameters from (5))	1.374	1.219	116.9	121.5
MM2 (new parameter set)	1.373	1.217	112.8	123.6

Table 11a. *Comparison of MM2-results with ab initio - calculations for urea.*

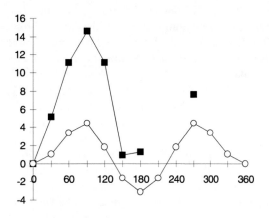

Compound A

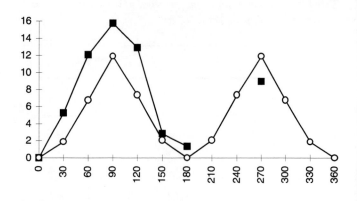

Compound B

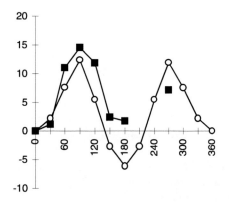

Compound C

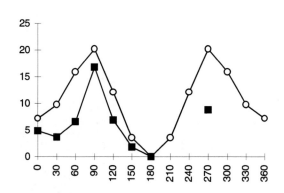

Compound D

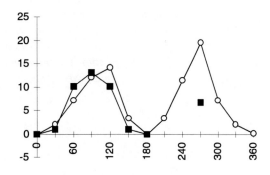

Compound E

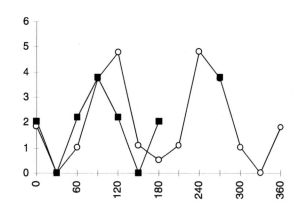

Compound F

Figure 10. *Rotational barriers calculated with the new MM2-parameter set (--o--) compared with ab initio MP2(fc)/6-31G(d) calculations (--■--). The x-axis gives the angle of rotation around the bond shown in figure 1, the y-axis energies in kcal/mol.*

Table 11b. *Comparison of MM2-results with ab initio - calculations for tetramethylurea.*

	C-N	N-C(O)	C-O	N-C-N	C-N-C-O	C-N-C-N
RHF/6-31G(d)	1.45	1.38	1.21	116.3	6.7	173.4
MM2 (parameters from (5)	1.46	1.4	1.23	124.7	-8.5	171.5
MM2 (new set)	1.46	1.4	1.23	117.9	10.9	169.1

Comparison of Xray-data with calculated structural parameters

As mentioned before several X-ray structures (G -L, see figure 2) were used for comparison. The results are shown in table 10. All calculated values are within the deviation of the Xray structures. The calculated bond length are in good agreement with experimental data.

Comparison with several examples

The parameters were used to calculate several compounds, where ab inito calculations were published or available. The results are shown in table 10 and 11. Again the force field calculations are in good agreement.

Conclusion

The MM2 force field was reparametrized for the R_2N-CO-NR_2-unit through fitting of *ab initio* and X-ray data. Molecular geometries and rotational barriers were calculated on different levels of theory. With the new parameter set MM2 was able to reproduce the results of *ab initio* calculations als well as the Xray data of several compound classes. *Ab initio* calculations of the rotational barriers of urea, methylurea, different dimethylureas and tetramethylurea allow a comparison of the influence of the basis set on the rotational barrier. The RHF/3-21G, RHF/6-31G(d) and the MP2(fc)/6-31G(d) basis sets have been employed.

References

1. Present address: Institut für Organische Chemie, Technische Universität Dresden, Mommsenstr. 13, D – 01069 Dresden, Germany. Tel.: (0351) 463-3211, Fax: (0351) 463-7030

2. (a) Adant, C.; Dupuis, M.; Bredas, J.L.; *Int. J. Quantum. Chem., Quantum Chem. Symp.* **1995**, *29*, 497. (b) Docherty, R.; Roberts, K.J.; Saunders,V.; Black, S.; Davey, R.J. *Faraday Discuss.* **1993**, *95*, 11. (c) Rasul, G.; Prakash, G.K.S.; Olah, G.A. *J. Org. Chem.* **1994**, *59*, 2552. (d) Palmer, M.H.; Blair-Fish, J.A.; *Z. Naturforsch., A: Phys.Sci.* **1994**, *49*, 146. (e) Ha, T.-K.; Puebla, C.; *Chem.Phys.* **1994**, *181*, 47. (f) Duffy, E.M.; Severance, D.L.; Jorgensen, W.L. *Isr. J. Chem.* **1993**, *33*, 323. (g) Vijay, A.; Sathyanarayana, D.N. *J.Mol.Struct.* **1993**, *295*, 245. (h) Wen, N.; Brooker, M.H. *J. Phys. Chem.* **1993**, *97*, 8608.

3. (a) Rasmussen, H.; Sletten, E. *Acta Chem. Scand.* **1973** *27*, 2757. (b) Flippen-Anderson, J.L. *Acta Cryst.* **1987**, *C43*, 2228.

4. Bowen, J.P.; Kontoyanni, M. *J. Comp. Chem.* **1992**, *13*, 657.

5. (a) Lee, I.; Kim, C.K.; Lee, B.C. *J. Phys. Org. Chem.* **1989**, *2*, 281, (b) Koizumi, M.; Tachibana, A.; Yamabe, T. *J. Mol. Struct.* **1988**, *164*, 37. (c) Orita, Y.; Pullman, A. *Theoret. Chim. Acta* **1977**, *45*, 257. (d) Kuharski, R.A.; Rossky, P.J.; *J. Am. Chem. Soc.* **1984**, *106*, 5786. (e) Cristinziano, P.; Lely, F.; Amodeo, P.; Barone, V. *Chem. Phys. Lett.* **1987**, *140*, 401. (f) Tanaka, H; Touhara, H.; Nakanishi, K.; Watanabe, N. *J. Chem.Phys.* **1984** *80*, 5170.

6. Etter, M.C.; Urbañczyk-Lipkowska, Z.; Zia-Ebrahimi, M.; Panunto, T.W. *J. Am. Chem. Soc.* **1990**, *112*, 8415.

7. (a) Lepore, U.; Castronuovo Lepore, G.; Ganis, P.; Goodman, M. *Cryst. Struct. Comm.* **1975**, *4*, 351. (b) Kashino, S.; Haisa, M. *Acta Cryst.* **1977**, *B33*, 855. (c) Brett, W.A.; Rademacher, P.; Boese, R. *Acta Cryst.* **1990**, *C46*, 880. (d) Toniolo, C.; Valle, G.; Bonora, G.M.; Crisma, M.; Moretto, V.; Izdebski, J.; Pelka, J.; Pawlak, D.; Schneider, C.H. *Int. J. Peptide Protein Res.* **1988**, *31*, 77. (e) Dannecker, W.; Kopf, J.; Rust, H. *Cryst. Struct. Commun.* **1979**, *8*, 429.

8. Strassner, T.; Feigel, M. *J. Mol. Struct. (Theochem)* **1993**, *283*, 33

9. Gaussian 92/DFT, Revision G.2, Frisch, M.J.; Trucks, G.W.; Schlegel, H.B.; Gill, P.M.W.; Johnson, B.G.; Wong, M.W.; Foresman, J.B.; Robb, M.A.; Head-Gordon, M.; Replogle, E.S.; Gomperts, R.; Andres, J.L.; Raghavachari, K.; Binkley, J.S.; Gonzalez,C.; Martin, R.L.; Fox, D.J. Defrees, D.J.; Baker, J.; Stewart, J.J.P. and Pople, J.A.; Gaussian, Inc., Pittsburgh PA, 1993.

10. Gaussian 94, Revision B.2, Frisch, M.J.; Trucks, G.W.; Schlegel, H.B.; Gill, P.M.W.; Johnson, B.G.; Robb, M.A.; Cheeseman, J.R.; Keith, T.; Petersson, G.A.; Montgomery, J.A.; Raghavachari, K.; Al-Laham, M.A.; Zakrzewski, V.G.; Ortiz, J.V.; Foresman, J.B.; Cioslowski, J.; Stefanov, B.B.; Nanayakkara, A.; Challacombe, M.; Peng, C.Y.; Ayala, P.Y.; Chen, W.; Wong, M.W.; Andres, J.L.; Replogle, E.S.; Gomperts, R.; Martin, R.L.; Fox, D.J.; Binkley, J.S.; Defrees, D.J.; Baker, J.; Stewart, J.J.P.; Head-Gordon, M.; Gonzalez, C. and Pople, J.A.; Gaussian, Inc., Pittsburgh PA, 1995.

11. (a) Allinger, N.L. *J. Am. Chem. Soc.* **1977** 99, 8127. (b) Burkert, U.; Allinger, N.L. *J. Comp. Chem.* **1982**, *3*, 40.

12. Bittner, A.; Mannig, D.; Noth, H. *Z. Naturforsch., Teil B* **1986**, *41*, 587.

13. (a) Shefter, E.; Trueblood, K.N. *Acta Cryst.* **1965**, *18*, 1067. (b) Abraham, F.; Nowogrocki, G.; Sueur, S.; Bremard, C. *Acta Cryst.* **1980**, *B36*, 799. (c) Cheng, P.-T.; Hornby, V.; Wong-Ng, W.; Nyburg, S.C.; Weinblum, D. *Acta Cryst.* **1976**, *B32*, 2251.

14. (a) Marzilli, L.G.; Epps, L.A.; Sorrell, .; Kistenmacher, T.J. *J. Am. Chem. Soc.* **1975**, *97*, 3351. (b) Kistenmacher, T.J.; Szalda, D.J. *Acta Cryst.* **1975**, *B31*, 90. (c) Kistenmacher, T.J. *Acta Cryst.* **1975** *B31*, 85. (d) Ellermann, J.; Kock, E.; Zimmermann, H.; Gomm, M. *J. Organomet. Chem.* **1988** *345*, 167.

15. Buncel, E.; Kumar, R.; Norris, A.R.; Beauchamp, A.L. *Can. J. Chem.* **1985**, *63*, 2575.

16. (a) Ravichandran, V.; Ruban, G.A.; Chacko, K.K.; Molina, M.A.R.; Rodriguez, E.C.; Salas-Peregrin, J.M.; Aoki, K.; Yamazaki, H. *J. Chem. Soc., Chem. Comm.* **1986**, 1780. (b) Sorrell, T.; Marzilli, L.G.; Kistenmacher, T.J. *J. Am. Chem. Soc.* **1976**, *98*, 2181. (c) Shieh, H.S.; Voet, D. *Acta Cryst.* **1976**, *B32*, 2354. (d) Szalda, D.J.; Kistenmacher, T.J.; Marzilli, L.G. *J. Am. Chem. Soc.* **1976**, *98*, 8371. (e) Hamelin, M. *Acta Cryst.* **1972**, *B28*, 228.

17. Mattos, M.C.; Surcouf, E.; Mornon, J.P. *Acta Cryst.* **1977**, *B33*, 1855.

18. Chieh, C.; Toogood, G.E.; Boyle, T.D.; Burgess, C.M. *Acta Cryst.* **1976**, *B32*, 1008.

19. Calogero, S.; Russo, U.; Del Pra, A. *J. Chem. Soc., Dalton Trans.* **1980**, 646.

20. Carrabine, J.A.; Sundaralingam, M. *Biochemistry* **1971**, *10*, 292.

21. Leonhard, K.; Plute, K.; Haltiwanger, R.C.; DuBois, M.R. *Inorg. Chem.* **1979**, *18*, 3246.

22. Beddoes, R.L.; Collison, D.; Mabbs, F.E.; Temperley, *J. Acta Cryst. C (Cr. Str. Comm.)* **1991**, *47*, 58.

23. (a) Parthasarathy, R.; Ohrt, J.M.; Chheda, G.B. *Acta Cryst.* **1976**, *B32*, 2648. (b) Gartland, G.L.; Gatehouse, B.M.; Craven, B.M. *Acta Cryst.* **1975**, *B31*, 203.

24. (a) Burgi, H.B.; Djuric, S.; Dobler, M.; Dunitz, J.D. *Helv. Chim. Acta* **1972**, *55*, 1771. (b) Bach, I.; Kumberger, O.; Schmidbaur, H. *Chem. Ber.* **1990**, *123*, 2267.

25. Bach, I.; Kumberger, O.; Schmidbaur, H. *Chem. Ber.* **1990**, *123*, 2267.

26. Maistralis, G.; Katsaros, N.; Memtzafos, D.; Terzis, A. *Acta Cryst.C (Cr. Str. Comm.)*, **1991**, *47*, 740.

27. Anderson, J.C.; Smith, S.C. *SYNLETT* **1990**, *2*, 107.

28. (a) Griffith, E.H.; Amma, E.L. *J. Chem. Soc., Chem. Comm.* **1979**, 322. (b) Lippert, B.; Schollhorn, H.; Thewalt, U. *J. Am. Chem. Soc.* **1986**, *108*, 6616. (c) Cramer, R.E.; Ho, D.M.; van Doorne, W.; Ibers, J.A.; Norton, T.; Kashiwagi, M. *Inorg. Chem.* **1981**, *20*, 2457.

29. Gillier, H. *Bull. Soc. Chim. Fr.* **1965**, 2373.

30. (a) Abraham, F.; Nowogrocki, G.; Sueur, S.; Bremard, C. *Acta Cryst.* **1980**, *B36*, 799. (b) Abraham, F.; Nowogrocki, G.; Sueur, S.; Bremard, C. *Acta Cryst.* **1980**, *B34*, 1466.

31. Mentzafos, D.; Katsaros, N.; Terzis, A. *Acta Cryst.C (Cr. Str. Comm.)* **1987**, *43*, 1905.

32. (a) Etter, M.C.; Kress, R.B.; Bernstein, J.; Cash, D.J. *J. Am. Chem. Soc.* **1984**, *106*, 6921. (b) Ogawa, K.; Tago, K.; Ishida, T.; Tomita, K.-I. *Acta Cryst.* **1980**, *B36*, 2095. (c) Jensen, B. *Acta Cryst.C (Cr. Str. Comm.)* **1988**, *44*, 1602. (d) Cheng, P.-T.; Hornby, V.; Wong-Ng, W.; Nyburg, S.C.; Weinblum, D. *Acta Cryst.* **1976**, *B32*, 2251. (e) Tillberg, O.; Norrestam, R. *Acta Cryst.* **1972**, *B28*, 890. (f) Karle, L.; Karle, J. *Acta Cryst.* **1971**, *B27*, 1891. (g) Macintyre, W.M.; Zirakzadeh, M. *Acta Cryst.* **1964**, *17*, 1305.

33. Stilbs, P.; Forsen, S. *J. Phys. Chem.* **1971**, *75*, 1901.

34. Martin, M.L; Filleux-Blanchard, M.L.; Martin, G.J.; Webb, G.A. *Org. Magn. Reson.* **1980**, *13*, 396.

35. (a) Stilbs, P. *Tetrahedron* **1973**, *29*, 2269. (b) Sandstrom, J.; *J. Phys. Chem.* **1967**, *71*, 2318. (c) Chan, S.O.; Reeves, L.W. *J. Am. Chem. Soc.* **1973**, *95*, 670. (d) Anet, F.A.L.; Ghiaci, M. *J. Am. Chem. Soc.* **1979**, *101*, 6857.

J. Mol. Model. **1996**, 2, 227 – 238

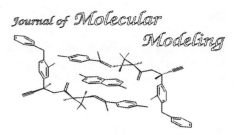

Sedimentation of Clusters of Spheres.
I. Unconstrained Systems

Konrad Hinsen[†] and Gerald Reinhard Kneller*

Institut für Theoretische Physik A, RWTH Aachen, Templergraben 55, D-52056 Aachen, Germany
(g.kneller@kfa-juelich.de)

† Present address: Institut de Biologie Structurale, Laboratoire de Dynamique Moleculaire, 41 Av. des Martyrs, F-38027 Grenoble, France (hinsen@ibs.ibs.fr)

Received: 14 May 1996 / Accepted: 1 August 1996 / Published: 4 September 1996

Abstract

We describe a numerical method for calculating hydrodynamic interactions between spherical particles efficiently and accurately, both for particles immersed in an infinite liquid and for systems with periodic boundary conditions. Our method is based on a multipole expansion in Cartesian tensors. We then show how to solve the equations of motion for translational and rotational motion of suspended particles at large Peclet numbers. As an example we study the sedimentation of an array of spheres with and without periodic boundary conditions. We also study the effect of perturbations on the stability of the trajectories.

Keywords: hydrodynamic interactions, Stokesian Dynamics simulations, colloids

Introduction

The behavior of particles suspended in a liquid has interested scientists ever since Stokes derived the drag formula for a single suspended sphere [1]. It is also important for many applications in rheology and colloid chemistry. Since all but the simplest problems require a numerical solution, numerical techniques play an important role in this field. In this article we deal with the two most important problems that any numerical simulation must address: the calculation of hydrodynamic interactions and the integration of the equations of motion.

In the treatment of hydrodynamic interactions, we limit ourselves to spherical particles and low Reynolds numbers, but aim to make the calculations both accurate and efficient.

Hydrodynamic interactions have three properties that make their numerical treatment difficult:
- They are many-body interactions, i.e. they are not pairwise additive.
- They are long-ranged, decaying as $1/R$, where R is an interparticle distance. This creates special problems for periodic systems.
- They diverge for certain types of motion when particles approach each other.

The oldest and simplest approximation, apart from neglecting hydrodynamic interactions altogether, consists of assuming pairwise additivity and describing the interactions between each pair with the Oseen tensor (see Eq. 3.3). This describes the leading $1/R$ terms correctly, but is nevertheless very inaccurate. Besides, this approximation has the funda-

* To whom correspondence should be addressed

mental problem that the diffusion matrix is not positive definite, which can only be circumvented by introducing more arbitrary approximations [2]. Improvements such as the Rotne-Prager tensor are available, but they all share the basic problem of assuming pairwise additivity and not treating long-ranged contributions correctly; they also do not take rotational motion into account. The importance of the correct inclusion of all long-ranged terms was demonstrated in [3], where sedimentation of large rigid clusters of spheres was studied. Even in such rigid structures, where short-ranged lubrication forces are irrelevant, all terms decaying as $1/R^3$ or slower must be included to prevent dramatic errors in the sedimentation coefficient.

A better approximation has been developed by Durlofsky et al. [4]. Their scheme provides correct short-distance behavior and takes the multi-body nature of hydrodynamic interactions into account; however, it still does not contain all long-range terms correctly and does not provide sufficient accuracy for many applications. The first systematic scheme that can in principle be made arbitrarily accurate was proposed by Ladd [5, 6]. Both Durlofsky et al. [4] and Ladd [7] have used their methods for dynamical simulations.

Recently, Cichocki et al. [8] presented a number of improvements that yield accurate results at a much reduced cost. We will show how these improvements can be combined with previous analytical work on hydrodynamic interactions [9, 10] and numerical techniques from the related field of electrostatic interactions [11] to construct an efficient and accurate numerical implementation that calculates hydrodynamic interactions for systems of spherical particles. This implementation is available from the CPC library [12]. It has already been used in a study of the sedimentation coefficients of conglomerates of spheres [3], which has shown very good agreement with experimental results.

We will also present an integration scheme that is suitable for accurate dynamic simulations in the Stokesian Dynamics regime, i.e. at high Peclet numbers. We include the rotational motion of the particles, which has been neglected so far. The rotational motion of suspended spheres is often interesting in itself, but its calculation becomes essential when systems with constraints, such as rigid assemblies and flexible chains, are considered. We will deal with the specific problems of constrained systems in a second paper.

To test our integrator, we study the sedimentation of a few small systems, both in unbounded and periodic geometries. Unfortunately there is little experimental data we could compare to; the sedimentation of some small clusters has been studied by Jayaweera *et al.* [13], but they do not provide enough data to allow a meaningful comparison to numerical calculations. Therefore, we must limit ourselves to demonstrating the convergence of our results with decreasing time steps.

Stokesian Dynamics regime

We consider a system of N arbitrary particles suspended in a viscous liquid which is at rest at infinity. The particles move under the influence of external forces and forces mediated by the liquid; the latter consists of deterministic and random contributions. The equations of motion for the particles are given by

$$\frac{d}{dt}[\mathbf{MV}] = \mathbf{F}_h(\mathbf{R}, \mathbf{V}) + \mathbf{F}_r(\mathbf{R}) + \mathbf{F}_{ext} \qquad (2.1)$$

In this equation, $\mathbf{R}$ is a vector containing the positions and orientations of all particles. Similarly, $\mathbf{V}$ contains the translational and angular velocities, and $\mathbf{F}$ describes forces and torques. The matrix $\mathbf{M}$ is block diagonal and contains the masses and moments of inertia of all particles.

The vector $\mathbf{F}_h$ contains the hydrodynamic forces, i.e. the deterministic forces exerted by the fluid on the particle. We assume that they are given by

$$\mathbf{F}_h = -\zeta \mathbf{V} \qquad (2.2)$$

where the matrix ζ is called *friction matrix* and depends on the viscosity of the liquid as well as on the positions and orientations of all suspended particles. The random forces $\mathbf{F}_r$ must be zero on average and fulfill the condition

$$\langle \mathbf{F}_r(0)\mathbf{F}_r(t) \rangle = kT\zeta\delta(t) \qquad (2.3)$$

which follows from the fluctuation-dissipation theorem.

Typically the time scale of observable particle motion is several orders of magnitude larger than the relaxation time of the particle momenta $\tau = m\zeta^{-1}$. In other words, the observed particle velocities, which we will denote by $\mathbf{U}$, are averages of the velocities $\mathbf{V}$ over times larger than τ. Under this condition, one can derive an expression for the displacement of the particles in a time interval Δt which is much larger than τ, but still small on the time scale a/U, where a is a typical particle size [14]. This expression is

$$\Delta \mathbf{R} = \mu \mathbf{F}_{ext}\Delta t + \nabla \cdot \mathbf{D}\Delta t + \mathbf{X} \qquad (2.4)$$

where $\mu = \zeta^{-1}$ is called the *mobility matrix* and $\mathbf{D} = kT\mu$ is called the *diffusion matrix*. $\mathbf{X}$ is a random displacement with

$$\langle \mathbf{X} \rangle = 0 \quad , \quad \langle \mathbf{XX} \rangle = 2\mathbf{D}\Delta t \qquad (2.5)$$

Simulations based on these equations are called *Brownian Dynamics* simulations.

To judge the relative importance of deterministic and the random motion, one introduces dimensionless quantities

$$\Delta \mathbf{R} = a\Delta\hat{\mathbf{R}} \quad, \quad \mathbf{D} = D_0\hat{\mathbf{D}} \quad, \quad \mathbf{U} = U\hat{\mathbf{U}}$$

$$\Delta t = \frac{a^2}{D_0}\Delta\hat{t} \quad, \quad \mu = \frac{D_0}{kT}\hat{\mu} \quad, \quad \mathbf{F}_{ext} = \frac{UkT}{D_0}\hat{\mathbf{F}}_{ext}$$

$$\nabla = \frac{1}{a}\hat{\nabla} \quad, \quad \mathbf{X} = a\hat{\mathbf{X}} \tag{2.6}$$

where a is the diameter of a typical particle, U a typical particle velocity, and D_0 a typical one-particle diffusion coefficient. Eq. 2.4 then becomes

$$\Delta\hat{\mathbf{R}} = \mathrm{Pe}\hat{\mu}\hat{\mathbf{F}}_{ext}\Delta\hat{t} + \hat{\nabla}\cdot\hat{\mathbf{D}}\Delta\hat{t} + \hat{\mathbf{X}} \tag{2.7}$$

where the dimensionless quantity $\mathrm{Pe} = Ua/D_0$ is called the *Peclet number*. At small Peclet numbers, Brownian motion dominates. At large Peclet numbers, the random displacements can be neglected. Simulations in this regime are called *Stokesian Dynamics* simulations.

From Eqs. 2.2, 2.3, and 2.4 it is evident that the effect of the liquid in which the particles are suspended is completely described by the friction matrix ζ or its inverse, the mobility matrix μ. An accurate and efficient calculation of these matrices is therefore extremely important for computer simulations of suspended particles.

Creeping flow and induced forces

Like most other approaches, our calculation of the friction and mobility matrices is based on the assumption that the liquid can be described by the so-called *creeping-flow equations*, which are valid for flow at low Reynolds numbers [15, 16]. We will also assume that the liquid is incompressible. The equations of motion for the liquid are then

$$\eta\nabla^2\mathbf{v} - \nabla p + \mathbf{f} = 0 \quad, \quad \nabla\cdot\mathbf{v} = 0 \tag{3.1}$$

where $\mathbf{v}(\mathbf{r})$ is the fluid velocity at point $\mathbf{r}$, $p(\mathbf{r})$ is the pressure, and $\mathbf{f}(\mathbf{r})$ is the force density acting on the fluid. In addition, boundary conditions at infinity and on the particle surfaces must be specified. Solutions to the creeping-flow equations for a given force density can be expressed conveniently as

$$\mathbf{v}(\mathbf{r}) = \mathbf{v}_0(\mathbf{r}) + \int d^3r' \mathbf{T}(\mathbf{r}-\mathbf{r}')\cdot\mathbf{f}(\mathbf{r}') \tag{3.2}$$

where $\mathbf{v}_0(\mathbf{r})$ is the solution for $\mathbf{f}(\mathbf{r})=0$ and the Green function $\mathbf{T}(\mathbf{r})$ for an unbounded fluid are given by the *Oseen tensor*

$$\mathbf{T}(\mathbf{r}) = \frac{1}{8\pi\eta}\left[\frac{1}{|\mathbf{r}|}\mathbf{1} + \frac{1}{|\mathbf{r}|^3}\mathbf{r}\mathbf{r}\right] \tag{3.3}$$

Finally, we must specify the boundary conditions at the surfaces of the suspended particles and at infinity. Experience has shown that real systems are best described by *stick boundary conditions*, i.e. the fluid sticks to the particle surfaces. The flow at infinity depends on the problem being studied; the two most important cases are vanishing flow and uniform shear flow.

When the particles move relative to the fluid, they exert a force density on it, which for non-permeable particles is localized on the particle surfaces. The force density $\mathbf{f}(\mathbf{r};j)$ induced on particle j at position $\mathbf{R}_j$ moving with translational velocity $\mathbf{U}_j$ and angular velocity ω_j can be written as

$$\mathbf{f}(\mathbf{r};j) = \int d^3r' \mathbf{Z}_j(\mathbf{r}-\mathbf{R}_j, \mathbf{r}'-\mathbf{R}_j)\left[\mathbf{u}_j(\mathbf{r}') - \mathbf{v}_{aj}(\mathbf{r}')\right] \tag{3.4}$$

where $\mathbf{Z}_j(\mathbf{r},\mathbf{r}')$ is a friction kernel that depends only on properties of the particle and on the boundary conditions, $\mathbf{v}_{a,j}(\mathbf{r})$ is the velocity field in absence of particle j, and $\mathbf{u}_j(\mathbf{r})$ is given by

$$\mathbf{u}_j(\mathbf{r}) = \left[\mathbf{U}_j + \omega_j\times(\mathbf{r}-\mathbf{R}_j)\right]\Theta(\mathbf{r}-\mathbf{R}_j) \tag{3.5}$$

where the step function $\Theta(\mathbf{r}-\mathbf{R}_j)$ is one inside the volume of the particle and zero outside. From Eq.3.2 it follows that the velocity field $\mathbf{v}_{a,j}$ is given by

$$\mathbf{v}_{a,j}(\mathbf{r}) = \mathbf{v}_0(\mathbf{r}) + \int d^3r' \mathbf{T}(\mathbf{r}-\mathbf{r}')\mathbf{f}(\mathbf{r}') \tag{3.6}$$

with

$$\mathbf{f}(\mathbf{r}) = \sum_i \mathbf{f}(\mathbf{r};i) \tag{3.7}$$

Calculation of the friction matrix

Eqs. 3.4 and 3.6 form an integral equation from which in principle $\mathbf{f}(\mathbf{r};i)$ can be determined for a given configuration of particles with given linear and angular velocities. Since the force and torque on particle i are related to $\mathbf{f}(\mathbf{r},i)$ by

$$\mathbf{F}_i = \int d^3r\, \mathbf{f}(\mathbf{r};i)$$

$$\mathbf{T}_i = \int d^3r(\mathbf{r}-\mathbf{R}_i)\times\mathbf{f}(\mathbf{r};i) \tag{4.1}$$

and linear and angular velocities enter via Eq.3.5, the solution of the integral equation yields the friction matrix ζ.

To find a numerical solution of the integral equation, it must be transformed into an algebraic equation, which can be done with a multipole expansion analogous to the familiar multipole expansion in electrostatic systems. As in electrostatics, the multipole expansion is guaranteed to converge only outside a spherical region containing all points where the force density does not vanish. It is therefore difficult to apply to non-spherical particles. For this reason, we will from now on restrict attention to particles of spherical shape. This is not as strong a restriction as it may seem, since many complicated shapes can be modelled by assemblies of small spherical components.

Multipole expansion

The multipole expansion for the induced force densities can be formulated in several ways; for numerical applications, the expansion in irreducible Cartesian tensors [17, 9, 10] is most convenient. The force multipole tensor of rank $p+1$ for particle j at position $\mathbf{R}_j$ is defined by

$$\mathbf{f}^{(p+1)}(j) = \frac{1}{p!} \int d^3 r \left(\mathbf{r} - \mathbf{R}_j\right)^p \mathbf{f}\left(\mathbf{r} - \mathbf{R}_j; j\right) \tag{4.2}$$

Here $\mathbf{f}(\mathbf{r};j)$ is the force density induced on particle j, and $\mathbf{r}^p$ is the p-fold tensor product of the vector $\mathbf{r}$ with itself. Similarly, we define velocity multipole tensors that describe the velocity field $\mathbf{u}(\mathbf{r}) - \mathbf{v}_0(\mathbf{r})$ around the particle by

$$\mathbf{c}^{(p+1)}(j) = \frac{1}{p!} \nabla^p \left[\mathbf{u}(\mathbf{r}) - \mathbf{v}_0(\mathbf{r})\right]_{\mathbf{r} = \mathbf{R}_j} \tag{4.3}$$

These multipole tensors can be decomposed into irreducible tensors, of which many do not give a contribution to the flow field. It has been shown in [10] that it is sufficient to consider the irreducible tensors $\mathbf{f}_{l\sigma}$ and $\mathbf{c}_{l\sigma}$, $l = 1,2,\ldots,$ which for given l and σ have $2l + 1$ independent components. These tensors are given by

$$\left(\mathbf{f}_{l0}\right)_{\gamma_1 \ldots \gamma_l} = \overbrace{f^{(l)}_{\gamma_1 \ldots \gamma_l}}$$

$$\left(\mathbf{f}_{l1}\right)_{\gamma_1 \ldots \gamma_l} = \frac{l}{l+1} \varepsilon_{\gamma_1 \lambda \mu} \overbrace{f^{(l+1)}_{\lambda \gamma_2 \ldots \gamma_l \mu}}$$

$$\left(\mathbf{f}_{l2}\right)_{\gamma_1 \ldots \gamma_l} = \frac{l(l+1)}{2(2l+1)} \delta_{\lambda\mu} \overbrace{f^{(l+2)}_{\lambda\mu\gamma_1 \ldots \gamma_l}} \tag{4.4}$$

and

$$\left(\mathbf{c}_{l0}\right)_{\gamma_1 \ldots \gamma_l} = \overbrace{c^{(l)}_{\gamma_1 \ldots \gamma_l}}$$

$$\left(\mathbf{c}_{l1}\right)_{\gamma_1 \ldots \gamma_l} = l \varepsilon_{\gamma_1 \lambda \mu} \overbrace{c^{(l+1)}_{\lambda\gamma_2 \ldots \gamma_l \mu}}$$

$$\left(\mathbf{c}_{l2}\right)_{\gamma_1 \ldots \gamma_l} = l(l+1) \delta_{\lambda\mu} \overbrace{c^{(l+2)}_{\lambda\mu\gamma_1 \ldots \gamma_l}} \tag{4.5}$$

where $\overset{\leftrightarrow}{\mathbf{a}}$ indicates the irreducible part of the tensor $\mathbf{a}$, $\varepsilon_{\alpha\lambda\mu}$ is the completely antisymmetric Levi-Cività tensor, and $\delta_{\alpha\beta}$ is the Kronecker symbol.

In this representation, the one-particle friction kernel $\mathbf{Z}_j(\mathbf{r},\mathbf{r}')$ is represented by a matrix $Z_{l\sigma\mu_1\ldots\mu_l, l'\sigma'\mu'_1\ldots\mu'_{l'}}(j)$ whose elements have been calculated for several particle models [18]. Similarly, the Oseen tensor is represented by a matrix $G_{l\sigma\mu_1\ldots\mu_l, l'\sigma'\mu'_1\ldots\mu'_{l'}}(ij)$, which expresses the flow field due to the force multipoles of particle j at the position of particle i. Expressions for its elements have been derived in [10]. The original integral equation for $\mathbf{f}(\mathbf{r};i)$ becomes a linear system of equations whose unknowns are the force multipole moments $\mathbf{f}_{l\sigma}$:

$$\tag{4.6}$$

$$\mathbf{f}_{l\sigma}(i) = \sum_{\sigma'} Z_{l\sigma\sigma'}(i) \left[\mathbf{c}_{l\sigma'}(i) - \frac{1}{l!} \sum_j \sum_{l''\sigma''} G_{l\sigma';l''\sigma''}(ij) \cdot \mathbf{f}_{l''\sigma''}(j) \right]$$

The friction matrix can be obtained by solving these equations with $\mathbf{c}_{10}(j) = \mathbf{U}_j - \mathbf{v}_0(\mathbf{R}_j)$ and $\mathbf{c}_{11}(j) = 2\omega_j - \nabla \times \mathbf{v}_0(\mathbf{r})|_{\mathbf{r}=\mathbf{R}_j}$, making use of the fact that the forces $\mathbf{F}_j$ are given by $\mathbf{f}_{10}(j)$ and the torques $\mathbf{T}_j$ by $2\mathbf{f}_{11}(j)$. A detailed description of the multipole expansion can be found in [3] and [12]. It should be noted that the friction matrix resulting from this calculation is positive definite at all levels of truncation of the multipole expansion.

The core of our numerical scheme to calculate hydrodynamic interactions is the numerical solution of Eq. 4.6, truncated to a finite number of multipole moments. Details can be found in [12], where the implementation we use is described.

Short-distance forces

Relative motion of particles at short distances creates large frictional forces, whose description by a multipole expansion requires a prohibitively large number of terms. We therefore follow the suggestion of Durlofsky and Brady [4] and incorporate the short-range forces approximately in the form

$$\zeta_L^* = \zeta_L + \sum_{i,j=1}^{N} \left(\zeta^{(2)} - \zeta_L^{(2)} \right)_{ij} \tag{4.7}$$

where ζ_L is the friction matrix as calculated according to the above description with a multipole expansion of order L, $\zeta^{(2)}$ is the exact two-particle friction matrix calculated from lubrication theory [20], and $\zeta_L^{(2)}$ is the two-particle friction matrix in order L approximation. The basic idea of this form is that the large short-range forces are localized in the region between two particles and can therefore be assumed to be pairwise additive. It is evident that ζ^*_L converges to the same value for $L \rightarrow \infty$ as ζ_L, but it does so much faster. A multipole approximation of order 3 is sufficient to calculate the friction matrix with an accuracy of about 1% [8].

Mobility calculations

In most applications of hydrodynamic interaction, such as Stokesian Dynamics simulations, it is not the friction matrix that is immediately required, but the particle velocities resulting from a given set of external forces, i.e. $\mu \cdot F_{ext}$. These velocities can be obtained by first calculating the complete friction matrix and then solving the set of equations $\zeta^*_L U = F_{ext}$ for U. Indeed this has been done by Durlofsky et al. [4] and by Ladd [5]. However, it has been shown in [8] that the velocities can be obtained directly by solving a modification of the multipole equation that leads to the friction matrix. This procedure is numerically much more efficient.

Periodic boundary conditions

The long range of the hydrodynamic interactions causes both conceptual and practical problems when periodic systems are studied. The difficulties are exactly analogous to those for the equally long-ranged Coulomb interactions, and can be solved by very similar methods. Our treatment is based on the theoretical framework developed by Felderhof [21]. We limit ourselves to elementary cells of cubic shape.

In analogy to the fact that the electrostatic potential of a periodic system is defined only if the system as a whole is neutral, the velocity field in a periodic hydrodynamic system is finite only if the total force on it vanishes. If necessary this must be enforced by adding a neutralizing homogeneous force density to the system; this is physically equivalent to applying a constant pressure gradient. In addition, the shape of the macroscopic assembly of elementary cells whose infinite limit is to be considered must be specified; we will assume it to be spherical.

Such a system can be treated much like a finite one with a different Green function [21]. The Oseen tensor $T(r)$ must be replaced by the tensor

$$T_H(r) = \frac{1}{4\pi\eta}\left[S_1(r) - \nabla\nabla S_2(r)\right] \qquad (5.1)$$

which was first introduced by Hasimoto [22]. The functions S_1 and S_2 have cubic symmetry and satisfy the equations

$$\nabla^2 S_1 = -4\pi\left[\sum_n \delta(r - nL) - \frac{1}{V}\right]$$
$$\nabla^2 S_2 = S_1 \qquad (5.2)$$

where L is the edge length of the elementary cell and V its volume. A method for the efficient calculation of S_1 and S_2, based on an analogous method for electrostatics [23], is given in [24]. It should be noted that the Hasimoto tensor (5.1) already includes the effect of the neutralizing homogeneous force density added to make the velocity field finite.

The multipole expansion in terms of irreducible Cartesian tensors that has been mentioned before must be re-derived with the new Green function. In its original form, it is valid only when the applied force density vanishes outside the particles. This assumption is violated by the addition of the neutralizing homogeneous force density. Starting from the original Taylor expansion that leads to Eq. 4.2, one finds again the formula given in [12] with the matrix $G(ij)$ replaced by

$$G_P(ij) = G_H(ij) + G'(ij) - G''(ij) \qquad (5.3)$$

Here $G_H(ij)$ is the result of evaluating Eq.(A6) from [3] or Eq. (A.13) from [12] with the Oseen tensor $T(r)$ replaced by the Hasimoto tensor $T_H(r)$. The only non-zero elements of $G'(ij)$ are

$$G'_{10\mu;12\mu'}(ij) = \frac{2}{3\eta}$$
$$G'_{11\mu;11\mu'}(ij) = -\frac{2}{3\eta}$$
$$G'_{12\mu;10\mu'}(ij) = G'_{00\mu;02\mu'}(ij) \qquad (5.4)$$
$$G'_{20\mu_1\mu_2;20\mu'_1\mu'_2}(ij) = -\frac{1}{5\eta}$$

Note that $G'(ij)$ is non-zero even for $i = j$. The matrix $G''(ij)$ is zero for $i = j$ and for $i \neq j$ given by Eq. (A6) from [3] with $T(r)$ replaced by

$$T''(r) = \frac{1}{6\eta}r^2 1 - \frac{1}{120\eta}\nabla\nabla(r^2)^2 \qquad (5.5)$$

and evaluated at $r = 0$.

For $l + l' > 4$ the elements of $G_P(ij)$ are lattice sums over the corresponding elements of $G(ij)$; in numerical calculations, they can be obtained by summing over all lattice sites within a cutoff radius. The remaining elements contain long-ranged contributions and must be calculated by evaluating the functions $S_1(r)$ and $S_2(r)$ as described in [24] and using the procedure described above.

Stokesian Dynamics

Equations of motion

In the Stokesian Dynamics regime, i.e. for high Peclet numbers (see Eq. 2.7), only the external forces need to be considered as driving forces for the particle displacements, and Eq. 2.4 becomes

$$\Delta \mathbf{R} = \mu \mathbf{F}_{ext} \Delta t \tag{6.1}$$

In principle this formula can be used to calculate particle trajectories. It is the Euler integration scheme [25] for the differential equation

$$\dot{\mathbf{R}} = \mu \mathbf{F}_{ext} \tag{6.2}$$

However, the Euler scheme is not ideal for numerical purposes. Other methods are more stable and more accurate at the same computational cost [25].

If rotational degrees of freedom are to be integrated, Eq. 6.2 must be generalized. The velocity vector replacing $\dot{\mathbf{R}}$ then contains all translational velocities, $\mathbf{V}_1,...,\mathbf{V}_N$, and all angular velocities, $\omega_1,...,\omega_N$, of the particles. Correspondingly, $\mathbf{R}$ contains the particle positions, specified by $\mathbf{R}_1,...,\mathbf{R}_N$, and the orientations, specified by $\mathbf{Q}_1,...,\mathbf{Q}_N$, where $\mathbf{Q}$ is a suitable set of angular variables. $\mathbf{F}_{ext}$ is the vector of all external forces, $\mathbf{F}_1,...,\mathbf{F}_N$, and all external torques, $\mathbf{T}_1,...,\mathbf{T}_N$. Using the above definitions, the equations of motion for Stokesian Dynamics read explicitly

$$\mathbf{V}_i = \sum_{j=1}^{N} \mu_{ij}^{tt} \mathbf{F}_j + \sum_{j=1}^{N} \mu_{ij}^{tr} \mathbf{T}_j \tag{6.3}$$

$$\omega_i = \sum_{j=1}^{N} \mu_{ij}^{rt} \mathbf{F}_j + \sum_{j=1}^{N} \mu_{ij}^{rr} \mathbf{T}_j \tag{6.4}$$

$$\dot{\mathbf{R}}_i = \mathbf{V}_i \tag{6.5}$$

$$\dot{\mathbf{Q}}_i = \mathbf{B}(\mathbf{Q}_i)\omega_i \tag{6.6}$$

where $\mu_{ij}^{tt}, \mu_{ij}^{tr} = \mu_{ji}^{rt}$, and μ_{ij}^{rr} are 3×3 submatrices of the mobility matrix μ. The linear relation (6.6) between the angular velocities and the time derivatives of the angular coordinates depends on the choice of the latter. The equations of motion (6.3) — (6.6) may be written in the compact form

$$\dot{\mathbf{R}} = \mathbf{B}\mu \mathbf{F}_{ext} \tag{6.7}$$

where $\mathbf{B}$ is a block-diagonal matrix containing unit matrices for the mapping $\mathbf{V}_i \rightarrow \dot{\mathbf{R}}_i$ and the matrices $\mathbf{B}(\mathbf{Q}_i)$ for the mapping $\omega_i \rightarrow \dot{\mathbf{Q}}_i$.

It is well known from molecular dynamics simulations that quaternion parameters are a convenient choice for the angular variables, since the resulting matrices $\mathbf{B}(\mathbf{Q}_i)$ are singularity-free [26]. A comprehensive treatise on quaternions and their relations to spatial rotations can be found in [27]. Here it is sufficient to know that rotations can be parameterized in terms of four real numbers, q_0, q_1, q_2, q_3, which are subject to the normalization condition $q_o^2 + q_1^2 + q_2^2 + q_3^2 = 1$. For quaternion parameters the relation (6.6) reads explicitly (the particle index has been dropped) [28]:

$$\begin{pmatrix} \dot{q}_0 \\ \dot{q}_1 \\ \dot{q}_2 \\ \dot{q}_3 \end{pmatrix} = \frac{1}{2} \cdot \begin{pmatrix} -q_1 & -q_2 & -q_3 \\ q_0 & q_3 & -q_2 \\ -q_3 & q_0 & q_1 \\ q_2 & -q_1 & q_0 \end{pmatrix} \begin{pmatrix} \omega_x \\ \omega_y \\ \omega_z \end{pmatrix} \tag{6.8}$$

Here the angular velocity components refer to the *laboratory-fixed* coordinate system. Eq. 6.8 is consistent with the normalization of the quaternion components, since

$$q_o \dot{q}_o + q_1 \dot{q}_1 + q_2 \dot{q}_2 + q_3 \dot{q}_3 = \frac{1}{2} \cdot \frac{d}{dt} \left(q_0^2 + q_1^2 + q_2^2 + q_3^2 \right) = 0$$

for any set of angular velocity components.

Integration of the equations of motion

Translational motion. Due to the singular behavior of the hydrodynamic interactions at short distances [20, 16] the dynamics of suspended particles can exhibit very different time scales, since relative motion for particles which are in close contact is almost completely suppressed. Differential equations describing such dynamical systems are called *stiff differential equations*. In practice so-called implicit algorithms have proven to be most useful to integrate stiff differential equations [25,29]; we find this confirmed for our case, for which we compared explicit and implicit central difference schemes for both translational and rotational motion. In the case of translation, the trajectories of particles in close contact showed unstable oscillatory relative motions, ending with unphysical overlaps, when an explicit scheme was used. This could be avoided by using the following discretized form for the translational equation of motion:

$$\dot{\mathbf{R}}_i(n) \approx \frac{\mathbf{R}_i(n+1) - \mathbf{R}_i(n-1)}{2\Delta t} = \mathbf{V}_i\left(\{\overline{\mathbf{R}}_j\}\right) \tag{6.9}$$

where $\mathbf{V}_i$ given by Eq. 6.3 and the positions $\overline{\mathbf{R}}_j$ by

$$\overline{\mathbf{R}}_j = \tfrac{1}{2}\left(\mathbf{R}_j(n) + \tfrac{1}{2}\left[\mathbf{R}_j(n+1) + \mathbf{R}_j(n-1)\right]\right) \qquad (6.10)$$

The arguments n, $n+1$, and $n-1$ are shorthands for $t = n \cdot \Delta t$ etc. Note that the $\overline{\mathbf{R}}_j$ depend on the new positions $\mathbf{R}_j(n+1)$. This leads to the following iterative update for the particle positions:

$$\mathbf{R}_i^{\nu+1}(n+1) = \mathbf{R}_i(n-1) + 2\Delta t \mathbf{V}_i\left(\left\{\overline{\mathbf{R}}_j^{\nu}\right\}\right) \qquad (6.11)$$

The superscript ν in $\overline{\mathbf{R}}_j^{\nu}$ indicates that the iterated positions $\mathbf{R}_i^{\nu}(n+1)$ are to be used in the evaluation of the positions $\overline{\mathbf{R}}_j$. We start the iteration procedure with

$$\mathbf{R}_j^0(n+1) = 2\mathbf{R}_j(n) - \mathbf{R}_j(n-1) \qquad (6.12)$$

This corresponds to setting $\overline{\mathbf{R}}_j = \mathbf{R}_j(n)$, which is the conventional central difference scheme. The iteration was stopped when the Euclidian norm of the distance between consecutive estimates for $\mathbf{R}_j(n+1)$ was below a prescribed tolerance limit $\varepsilon \cdot a$, with a being the particle radius. It should be noted that the implicit scheme described here is not essential as long as all the particle distances are sufficiently large; we found that for distances larger than $10^{-8}a$ the standard explicit scheme can be used without any problems.

Rotational motion. We found that the rotational motion, for which the singularities in the hydrodynamic interactions at short distances are weaker, could be sufficiently well integrated with a normal central difference scheme. In analogy to Eq. 6.9, the discretized form of the rotational equation of motion reads

$$\dot{\mathbf{Q}}_i(n) \approx \frac{\mathbf{Q}_i(n+1) - \mathbf{Q}_i(n-1)}{2\Delta t} = \mathbf{B}(\mathbf{Q}_i(n))\,\omega_i\left(\left\{\mathbf{R}_j(n)\right\}\right)$$

$$\qquad (6.13)$$

yielding the update formula

$$\mathbf{Q}_i(n+1) = \mathbf{Q}_i(n-1) + 2\Delta t \mathbf{B}(\mathbf{Q}_i(n))\,\omega_i\left(\left\{\mathbf{R}_j(n)\right\}\right)$$

$$\qquad (6.14)$$

for the quaternion parameters. The angular velocities ω_i are determined by Eq. 6.4. The body-fixed basis vectors $\mathbf{e}'_i$ rotating with the spheres can be updated using $\mathbf{e}'_i(n+1) = \mathbf{U}(\mathbf{Q}_i[n+1]) \cdot \mathbf{e}'_i(0)$, where $\mathbf{U}(\mathbf{Q})$ is the rotation matrix

$$\mathbf{U}(\mathbf{Q}) = $$
$$= \begin{pmatrix} q_0^2 + q_1^2 - q_2^2 - q_3^2 & 2(-q_0 q_3 + q_1 q_2) & 2(q_0 q_2 + q_1 q_3) \\ 2(q_0 q_3 + q_1 q_2) & q_0^2 + q_2^2 - q_1^2 - q_3^2 & 2(-q_0 q_1 + q_2 q_3) \\ 2(-q_0 q_2 + q_1 q_3) & 2(q_0 q_1 + q_2 q_3) & q_0^2 + q_3^2 - q_1^2 - q_2^2 \end{pmatrix}$$

Close contacts A problem one has to face with iterative integration schemes for differential equations is slow convergence. In our case this concerns the update of the particle positions (see Eq. 6.11). Slow convergence can occur in situations where the system asymptotically approaches a configuration in wich particles stick to each other (see our example below). We found that sometimes 50 iterations and more were necessary for $\varepsilon \approx 10^{-12}$, whereas for larger, but still small, particle separations only two or three iterations were necessary. Due to inaccuracies inherent in any numerical integration scheme even the unphysical situation of overlapping spheres cannot not be excluded. To solve the problem of slow convergence and unphysical particle overlaps, we removed critical close contacts by the following algorithm:

1. Find all pairs (i,j) of particles whose distance is less or equal to $2a \cdot (1 + \varepsilon)$, where a is the particle radius and ε is a tolerance limit.

2. Find all clusters of particles which have a connection *via* close contacts. Consider e.g. three particles 1,2,4 where (1,2) and (2,4) are in close contact. Then, according to the above definition, {1,2,4} form a cluster since there is a path from 1 to 4: $1 - 2 - 4$.

3. Find the centroid $\mathbf{R} = 1/N \sum_{(i)} \mathbf{R}_i$ for each cluster and scale the relative positions by $1 + 2\varepsilon$. Here N is the number of particles in a cluster. The new positions are then given by $\mathbf{R}'_i = \mathbf{R} + (1 + 2\varepsilon)(\mathbf{R}_i - \mathbf{R})$.

4. Goto 1, to check if new contacts with particles not yet involved in close contacts have been created by steps 1 to 3. If this is the case, *add* these contacts to the ones found in 1 and proceed with 2.

The procedure described above terminates after a finite number of cycles. The extreme case which can occur is that the whole system is treated as one cluster. We emphasize that the above procedure should only be used if the numerical solution fails due to inevitable accuracy problems. It is no substitute for an implicit integration scheme which can handle much closer contacts than an explicit scheme.

Accuracy. We determined the accuracy of the explicit and implicit integration schemes by simulating the sedimentation of three particles in a linear arrangement with different time steps. Here and in the following section, we use a dimensionless unit of time defined such that a single sphere in an unbounded liquid would move a distance of $2a/3$ in a time interval of length 1. Figure 1 show the differences in vertical position for the central particle. All differences are

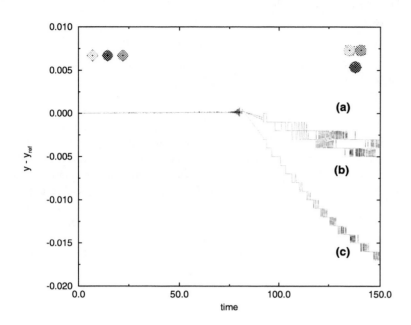

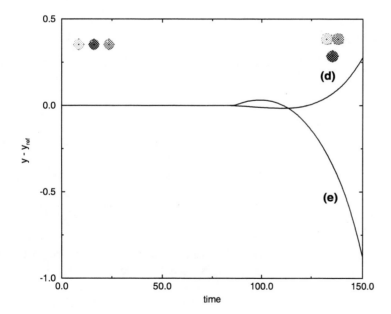

Figure 1. *The dependence of the error on the integration step. (a) Explicit/implicit integrator, $\Delta t = 0.001$. (b) Explicit/implicit integrator, $\Delta t = 0.01$. (c) Explicit/implicit integrator, $\Delta t = 0.05$. (d) Explicit integrator, $\Delta t = 0.001$. (e) Explicit integrator, $\Delta t = 0.005$.*

with respect to a reference run with a time step of $\Delta t = 0.001$ using the the implicit integrator at all times. The initial and final configurations are shown in the figures.

Figure 1, curves a – c show the results for an integration that uses the implicit integrator whenever a distance becomes smaller than 0.001 *a*. Initially, while all particle distances are large, the errors are negligible. The correction mechanism described above becomes active around $t = 80$, when the two outer spheres approach each other significantly. Nevertheless the absolute error remains small even for the larg-

est time step, $\Delta t = 0.05$. The step structure of the error is caused by the close contact elimination procedure described in the last section.

The corresponding curves (d – e) for a purely explicit integration scheme (Figure 1b) show that the error in this case already becomes very large for small step sizes (note the different scale). This demonstrates the necessity of the implicit integration scheme whenever short distances cannot be excluded.

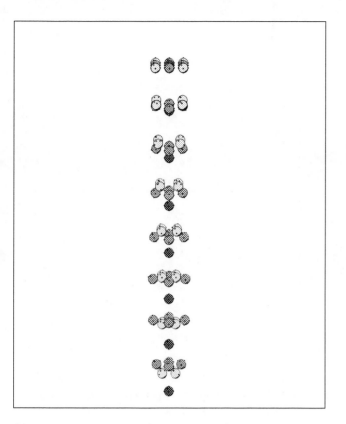

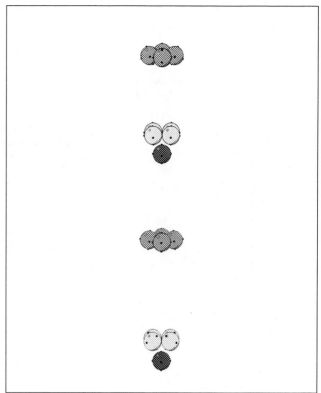

Figure 2. *Sedimentation of a planar square array of nine equal spheres in a viscous liquid under the influence of a constant gravitational force. The time separation between two consecutive configurations is $\Delta t_{frame} = 5$, corresponding to 5000 time steps of length $\Delta t = 0.001$. A single particle would move $2/3 \cdot 10^{-3}a$ during a time step, where a is the sphere's radius.*

Figure 3. *The end of the simulation whose beginning is shown in Figure 1. The two configurations correspond to t = 135 and t = 150, respectively.*

Examples

As an example we study the sedimentation of a square array of nine equal spheres in a viscous liquid under the influence of gravity — see video sequence no. Ia ('Nine equally sized spheres starting on a quadratic grid'). Here and in all following video sequences *the center of mass motion is subtracted*. The overall sedimentation of the cluster is shown on the lefthand side of the screen. The height of the frame indicates the total falling distance, and the height of the black bar corresponds approximately to the height of the screen.

Figure 2 shows the beginning of the simulation. The start configuration is shown at the top: the nine spheres are located in a plane perpendicular to the direction of gravity; the center-to-center distance between nearest neighbours is 3a, where a is the radius of a sphere. There are no interactions between the particles in addition to the hydrodynamic forces.

We ran a simulation of this system for 150,000 time steps using the methods described above. The length of each time

step was $\Delta t = 0.001$ in our units. Figure 2 shows eight configurations at time intervals of 5, beginning with the initial configuration. Figure 3 shows two configurations from the end of the simulation run, corresponding to times $t = 135$ and $t = 150$, respectively. In both pictures, we use three different colours to mark groups of spheres related by symmetry with respect to a 90° rotation around the central sphere in the initial configuration. The small dots are added to show the rotation of the spheres. The positions of the configurations in the pictures correspond to the actual distance they have moved.

In the beginning of the simulation the particles separate into three planes, the corner particles being the slowest and the center one being the fastest. This reflects the different exposure of the particles to the surrounding liquid. Then the edge particles move outward and the corner particles inward, whereupon they "dive" through the plane formed by the former edge particles. Meanwhile the center particle seems to escape from the others. In the part between Figure 2 and Figure 3 (see video sequence no. Ia), the edge particles in turn dive through the plane of the corner particles. This interchange of the planes is not periodic, however; finally the corner and center particles form a quasi-rigid cluster of five

Figure 4. *The time evolution of the rotation angles.*

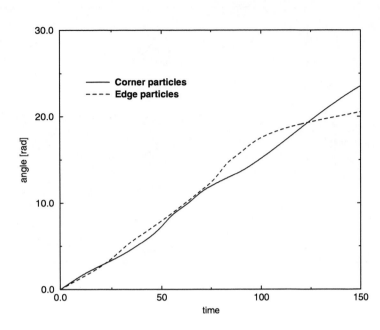

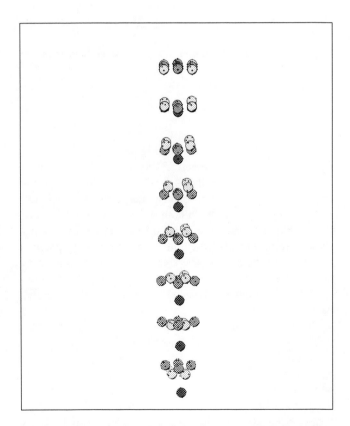

Figure 5. *The beginning of a simulation with a slightly perturbed initial configuration. The time separation between two consecutive configurations is the same as in Figure 1, i.e. Δt_{frame} = 5. The simulation time step is ten times as large as in the simulation shown in Figures 1 and 2, i.e. Δt = 0.01.*

particles falling somewhat faster than the second cluster consisting of the four former edge particles in close contact.

The angle of rotation of the corner and edge particles in the course of the simulation is shown in Figure 4. Note that the axis of rotation is constant for each particle, so that a single angle is sufficient to characterize the rotational motion.

To test the stability of the system with respect to small perturbations of the initial configuration, we ran a second simulation in which all the initial particle coordinates were randomly shifted by ± 0.1 *a*. This simulation was run with a time step of Δt = 0.01. The results are shown in video sequence no. Ib ('9 equally sized spheres starting on a slightly perturbed quadratic grid') and in Figures 5 (first part) and 6 (second part). For comparison, the total falling height indicated by the frame is the same as in the simulation shown in video sequence no. Ia. Until approximately *t* = 80 the configurations resemble those of the unperturbed simulation, but then the order is quickly destroyed.

The strong influence of the long-range terms in the hydrodynamic interactions can be seen by comparing a simulation of a system with and without periodic boundary conditions — see video sequence no. Ic ('Nine equally sized spheres starting on a quadratic grid with periodic boundary conditions'). For comparison, the total falling height indicated by the frame is again the same as in the simulation shown in video sequence no. Ia.

The nine particles now form the cubic elementary cell of a periodic system whose lattice constant *L* is 16*a* (*L* = 16 in our units). This simulation was also run with a time step of Δt = 0.01. The beginning of the simulation is shown in Figure 7. It should be noted that the time separation between

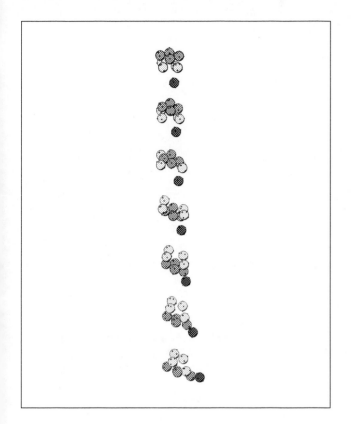

Figure 6. *The continuation of Figure 4.*

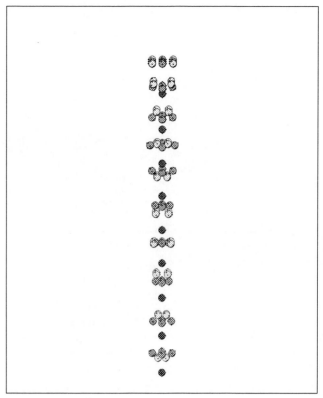

Figure 7. *The beginning of a simulation with periodic boundary conditions. Only one elementary cell is shown. The edge of the elementary cell has a length L of 16a (L = 16 in our units), where a is the radius of the spheres. Note that the time separation of two subsequent configurations is twice as much as in Figure 2, i.e. Δt_{frame} = 10. The simulation time step is Δt = 0.01.*

two consecutive configurations in this picture is twice as large as in Figure 2; the net sedimentation of the particles is much slower than in the non-periodic case. Nevertheless, the relative motion of the spheres with respect to one another is not much influenced by the presence of the periodic images.

Figure 8 shows the heat produced as the particles move downward against the friction caused by the liquid for all three simulations. The heat production is given by

$$\sum_i \mathbf{F}_i \cdot \mathbf{v}_i \tag{7.1}$$

the rotational velocities do not enter as there are no applied torques. For the periodic system, only one of the images is used. We normalize the heat production by that of nine spheres sedimenting at infinite distance. Since in our simulation the forces on all particles are identical and constant, the heat production is proportional to the velocity of the center of mass of the nine spheres. Figure 8 shows that initially the heat production increases but is modulated by an oscillation corresponding to the position interchanges of the planes of the corner and edge particles. But once the final clusters are formed, the heat production decreases monotonously. This figure also illustrates the differences between the three simu-

lations quantitatively. It can clearly be seen where the perturbed simulation starts to deviate significantly from the unperturbed one, and the smaller sedimentation speed of the periodic system is also evident.

This example shows that even very simple systems with hydrodynamic interactions can show a surprisingly complicated behavior. We are not aware of any theory that could predict more than the very first steps of our simulations.

Conclusion

We have shown how the hydrodynamic forces between spherical particles immersed in a liquid can be treated numerically, and how Stokesian Dynamics simulations can be performed on such systems. To demonstrate our method, we have performed a simulation of a simple model system.

In its current state, our method can be used to study the dynamics of colloidal suspensions, provided that the Peclet number is high enough to justify Stokesian Dynamics. It can

Figure 8. *The normalized heat production in the course of the three simulation runs.*

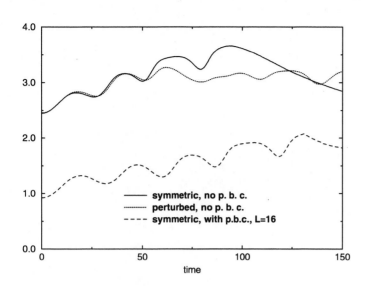

also be used for Monte-Carlo type calculations of equilibrium properties irrespective of the Peclet number. One of the most interesting applications from our point of view is the simulation of macromolecules modelled as assemblies of spheres; however, this necessitates the treatment of geometrical constraints, which we will present in a separate article.

So far we have not mentioned systems at low Peclet numbers. For such systems, all terms in Eq. 2.4 must be taken into account. This has been done in [14], but only for translational motion and the Oseen-tensor approximation for the mobility matrix. The generalization to an accurate description of finite particles remains an interesting challenge.

Acknowledgement We wish to thank the Zentralinsitut für Angewandte Mathematik of the KFA Jülich, in particular Klaudia Waschbüsch and Maik Boltes, for producing the video sequences shown in this article. The single frames were generated with the program MOLSCRIPT [30]. One of us (GRK) wishes to thank the German Space Agency for financial support.

References

1. Stokes, G.G. *Trans. Cambridge Philos. Soc.* **1851**, *9*, 8.
2. Dickinson, E. *Chem. Soc. Rev.* **1985**, *14*, 421.
3. Cichocki, B.; Hinsen, K. *Phys. Fluids* **1985**, *7*, 286.
4. Durlofsky, L; Brady, J; Bossis, G. *J. Fluid Mech.* **1987**, *180*, 21.
5. Ladd, A.J.C. *J. Chem. Phys.* **1988**, *88*, 5051.
6. Ladd, A.J.C. *J. Chem. Phys.* **1989**, *90*, 1149.
7. Ladd, A.J.C. *Phys. Fluids* **1993**, *A5*, 299.
8. Cichocki, B; Felderhof, B.U.; Hinsen, K; Wajnryb, E; Blawzdziewicz, J. *J. Chem. Phys.* **1994**, *100*, 3780.
9. Schmitz, R. *Die effektive Viskosität flüssiger Suspensionen*, PhD thesis, RWTH Aachen, **1980**.
10. Schmitz, R. *Physica*, **1980**, *A102*, 161.
11. Hinsen, K.; Felderhof, B.U. *J. Math. Phys.* **1992**, *33*, 3731.
12. Hinsen, K. *Comp. Phys. Comm.* **1995**, *88*, 327.
13. Jayaweera, K.O.L.F.; Mason, B.J.; Slack, G.W. *J. Fluid Mech.* **1964**, *20*, 121.
14. Ermak, D.L.; McCammon, J.A. *J. Chem. Phys.* **1978**, *69*, 1352.
15. Happel, J.; Brenner, H. *Low Reynolds number hydrodynamics*, Noordhoff, Leyden, **1973**.
16. Kim, S.; Karrila, S.J. *Microhydrodynamics*, Butterworth-Heinemann, Boston, **1991**.
17. Schmitz, R.; Felderhof, B.U. *Physica* **1978**, *A92*, 423.
18. Cichocki, B.; Felderhof, B.U.; Schmitz, R. *Physico-Chemical Hydrodynamics* **1988**, *10*, 383.
19. Cipriani, J; Silvi, B. *Mol. Phys.* **1982**, *45*, 259.
20. Jeffrey, D.J.; Onishi, Y. *J. Fluid Mech.* **1984**, *139*, 261.
21. Felderhof, B.U. *Physica* **1989**, *A159*, 1.
22. Hasimoto, H. *J. Fluid Mech.* **1959**, *5*, 317.
23. Cichocki, B.; Felderhof, B.U.; Hinsen, K. *Phys. Rev.* *1989*, **A39**, 5350.
24. Cichocki ,B.; Felderhof, B.U. *Physica* **1989**, *A159*, 19.
25. Press, W.H.; Teukolsky, S.A.; Vetterling, W.T.; Flannery, B.P. *Numerical Recipes in C* (second edition), Cambridge University Press, Cambridge, **1992**.
26. Evans, D.J.; Murad, S. *Mol. Phys.* **1977**, *34*, 327.
27. *Altmann*, B. *Rotations, Quaternions, and Double Groups*, Clarendon Press, Oxford, **1986**.
28. Kneller, G.; Geiger, A. *Molecular Simulation* **1989**, *3*, 283.
29. Stoer, J.; Bulirsch, R. *Introduction to numerical analysis*, Springer, Berlin, New York, **1980**.
30. Kraulis, P.J. *J. Appl. Cryst.* **1991**, *24*, 946.

J. Mol. Model. **1996**, 2, 239 – 250

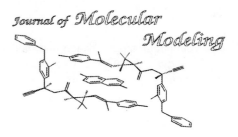

Sedimentation of Clusters of Spheres.
II. Constrained Systems

Gerald Reinhard Kneller*, Konrad Hinsen[†]

Institut fürTheoretische Physik A, RWTH Aachen, Templergraben 55, D-52056 Aachen, Germany
(g.kneller@kfa-juelich.de)

† Present address: Institut de Biologie Structurale, Laboratoire de Dynamique Moleculaire, 41 Av. des Martyrs, F-38027 Grenoble, France (hinsen@ibs.ibs.fr)

Received: 14 May 1996 / Accepted: 1 August 1996 / Published: 4 September 1996

Abstract

Starting from the *N*-body friction matrix of an unconstrained system of *N* rigid particles immersed in a viscous liquid, we derive rigorous expressions for the corresponding friction and mobility matrices of a geometrically constrained dynamical system. Our method is based on the fact that geometrical constraints in a dynamical system can be cast in the form of linear constraints for the Cartesian translational and angular velocities of its constituents. Corresponding equations of motion for Molecular Dynamics simulations have been derived recently [1]. Using the concept of generalized inverse matrices, we find the form of the constrained friction and mobility matrix in Cartesian and in reduced coordinates. We show that the equations of motion for Stokesian Dynamics can be derived from a minimum principle which is similar to Gauß' principle of least constraint in classical mechanics.

We relate our approach for deriving constrained friction and mobility matrices to Kirkwood's method where holonomic constraints acting between point-like particles are described by generalized coordinates and tensor algebra in curvilinear space.

As an application, we perform a Stokesian Dynamics simulation of sedimentation of a small model polymer consisting of five spherical monomers connected by massless sticks and joints.

Keywords: hydrodynamic interactions, geometrical constraints, macromolecules

Introduction

The dynamics of chain molecules, such as polymers and proteins, in solution has been a subject of interest in statistical physics for many years [2, 3, 4]. In order to reduce the number of degrees of freedom of these complicated macromolecules, a number of simplified mechanical models involving geometrical constraints have been considered [3, 5, 6]. The idea is to describe the preservation of molecular structures or substructures by constraints instead of including the corresponding strong intramolecular forces explicitly in the model force field, which would be a formidable task, if not an impossible one. Various schemes for computer simulations on the atomic scale (Molecular Dynamics simulations) [7, 8] and on the mesoscopic scale (Brownian Dynamics simulations) [9 – 12] have been suggested. In Brownian Dynamics simulations, the solvent in which the molecules are immersed is treated

* *To whom correspondence should be addressed*

in a continuum approximation and modelled by stochastic and hydrodynamic forces. It is well known that hydrodynamic interactions play an important role in the self-assembly of macromolecules as well as in their interactions [13, 14].

To be able to study the dynamics of realistic macromolecules one needs a simulation method which can

- handle complex chains,
- be applied to dense systems.

The method that is used almost exclusively to perform dynamical simulations of macromolecules with constraints is called SHAKE [7]. Originally, SHAKE was developed for Molecular Dynamics simulations of macromolecules, but it has also been used for Brownian dynamics simulations [11]. It works in cartesian coordinates and satisfies a set of interdependent bond constraints iteratively. The implementation in a Molecular Dynamics program is particularly simple. Later it was recognized that certain molecular geometries cannot be described by a set of bond constraints. Examples are planar rigid molecules with more than three atoms or three-dimensional molecules with more than four atoms. An extension to handle these cases has been developed [8], but is not in general use. In fact, it applies only to totally rigid molecules and cannot be used for topologically linked rigid structures as they frequently occur in macromolecules. Such systems can be treated using a method that we have developed recently [1, 15].

So far, hydrodynamic interactions in chain molecules have usually been approximated by pairwise additive interactions between 'beads' [2, 16, 11 – 13]. However, it is known that even in moderately dense systems with hydrodynamic interactions this approximation is not sufficient. Since molecular subunits in proteins and polymers are usually in close contact, the input friction matrix must describe hydrodynamic interactions in dense systems correctly. During the last few years, several authors have attacked the problem of computing friction and mobility matrices of dense systems such as colloids [17 – 19], going beyond the Oseen-Burger or Rotne-Prager descriptions [20] of hydrodynamic interactions. We use an efficient and precise method to compute the *N*-body friction and mobility matrices of spherical particles that has been published recently by Cichocki *et al.* [19]. This scheme is sufficiently accurate to describe even large closely packed assemblies of spheres; such mechanical models have already been used by Dwyer and Bloomfield to simulate the Brownian Dynamics of protein-DNA solutions [21]. We emphasize that the importance of an accurate description is not due to a desire for "correctness" at small distances; below a certain distance between particles, the assumption of length scale separation between solute and solvent breaks down. It is, however, necessary to include all long-range terms (i.e. those decaying as $1/R^3$ or slower) to obtain the correct hydrodynamic behaviour of the whole molecule or of large subunits. This has been demonstrated by Cichocki and Hinsen [22], who calculated the sedimentation coefficient of large con-

glomerates of spheres and compared with experimental data. It turns out that a level of accuracy that leads to the correct long-range terms also provides an already very good description for short-range interactions, but this fact is of little importance for simulations of macromolecules.

Here we concentrate on the aspect of describing hydrodynamic interactions in the presence of geometrical constraints. We present a rigorous and simple scheme for computing friction and mobility matrices of arbitrary chain molecules consisting of rigid constituents from the friction matrices describing the *freely moving* constituents. To derive expressions for constrained friction and mobility matrices, we start from the observation that geometrical constraints describing chain molecules can be cast in the form of linear constraints for the Cartesian translational and angular velocities of its constituents, extending an ansatz which has been employed recently to derive the equations of motion describing the classical Lagrangian mechanics of chain molecules [1]. Using the concept of generalized inverse matrices [23], we first construct a projector on the subspace of constrained Cartesian velocities. Then we derive the constrained friction and mobility matrices in full Cartesian space and in reduced space and show that the equations of motion for constrained Stokesian Dynamics can be derived from a minimum principle similar to Gauß' principle of least constraint in classical mechanics.

We also show that the relation between our constrained friction and mobility matrices in reduced space and in cartesian space can be expressed formally by appropriately defined coordinate transformations.

As an application we simulate the sedimentation of a pentamer by Stokesian Dynamics, modelling the monomers as rigid spheres linked by massless rods and joints placed between the monomers. Such a model is appropriate e.g. in the study of sedimentation of big proteins, where each monomer would represent a whole domain. The friction matrix of the unconstrained system is computed according to the scheme of Cichocki *et al.* [19, 22], using an implementation available from the CPC library [24]. To integrate the equations of motion, we employ a similar algorithm as in our previous article on Stokesian Dynamics simulations of unconstrained systems [25].

We have not yet tackled Brownian Dynamics simulations since they would require the evaluation of the divergence of the mobility matrix, $\nabla \cdot \mathbf{\mu}$ [26]. In the Oseen or Rotne-Prager approximations of hydrodynamic interactions, this term vanishes, but not for the mobility matrix computed according to [19]. The computation of $\nabla \cdot \mathbf{\mu}$ is still an unsolved problem.

Theory

Generalized inverse matrices

To derive expressions for constrained friction and mobility matrices, we will make use of *generalized inverse matrices,* which are also called *pseudoinverse matrices* or *Moore-Penrose-Inverses* [23]. The generalized inverse of an arbitrary $m \times n$-matrix $\mathbf{A}$ is generally denoted by $\mathbf{A}^+$. It is uniquely determined by the four conditions

$$\mathbf{A}\mathbf{A}^+\mathbf{A} = \mathbf{A} \tag{2.1}$$

$$\mathbf{A}^+\mathbf{A}\mathbf{A}^+ = \mathbf{A}^+ \tag{2.2}$$

$$\left[\mathbf{A}^+\mathbf{A}\right]^T = \mathbf{A}^+\mathbf{A} \tag{2.3}$$

$$\left[\mathbf{A}\mathbf{A}^+\right]^T = \mathbf{A}\mathbf{A}^+ \tag{2.4}$$

The superscript T denotes a transposition. Obviously, $\mathbf{A}^+$ equals $\mathbf{A}^{-1}$ for a quadratic non-singular matrix $\mathbf{A}$. The relations (2.1) – (2.4) express the fact that $\mathbf{A}\mathbf{A}^+$ is a projector on the column space of $\mathbf{A}$ and $\mathbf{A}^+\mathbf{A}$ is a projector on its row space.

Consider now the set of linear equations $\mathbf{A}\mathbf{x} = \mathbf{b}$. A solution exists only if the *consistency condition* $\mathbf{A}\mathbf{A}^+\mathbf{b} = \mathbf{b}$ is fulfilled. Given that this is the case, the general form of the solution reads $\mathbf{x} = \mathbf{A}^+\mathbf{b} + (1 - \mathbf{A}^+\mathbf{A})\mathbf{y}$, where $\mathbf{y}$ is an arbitrary vector of length n. The unique solution of minimum length is given by $\mathbf{x}_0 = \mathbf{A}^+\mathbf{b}$.

If $\mathbf{A}$ can be written in the form $\mathbf{A} = \mathbf{F}\mathbf{G}$, where $\mathbf{F}$ is an $m \times f$ matrix of full column rank and $\mathbf{G}$ is an $f \times n$ matrix of full row rank, there is an explicit expression for $\mathbf{A}^+$ [23]:

$$\mathbf{A}^+ = \mathbf{G}^T\left(\mathbf{G}\mathbf{G}^T\right)^{-1}\left(\mathbf{F}^T\mathbf{F}\right)^{-1}\mathbf{F}^T \tag{2.5}$$

Generalized coordinates and velocities

Linear velocity constraints. We consider a system of N rigid particles i ($i = 1 \ldots N$) immersed in a viscous liquid. Each particle is assigned a translational velocity $\mathbf{v}_i$ and an angular velocity $\boldsymbol{\omega}_i$, yielding $6N$ degrees of freedom for the unconstrained system. Throughout this paper the angular velocities refer to a laboratory-fixed reference frame. The positions and the orientations of the particles are defined by the position vectors $\mathbf{r}_i$ and sets of angular variables $\mathbf{q}_i$, respectively. Examples for angular variables are Euler angles, $\mathbf{q} = (\alpha, \beta, \gamma)$, or quaternions, $\mathbf{q} = (q_0, q_1, q_2, q_3)$ with $\mathbf{q} \cdot \mathbf{q} = 1$. In contrast to translational motion, where $\mathbf{v} = \dot{\mathbf{r}}$, the relation between the angular velocities and the time derivatives of the angular variables reads $\boldsymbol{\omega} = \mathbf{A}(\mathbf{q})\dot{\mathbf{q}}$, and the inverse relation is given by $\dot{\mathbf{q}} = \mathbf{B}(\mathbf{q})\boldsymbol{\omega}$. The form of the matrices $\mathbf{A}$ and $\mathbf{B}$ depends on the choice of angular variables. For Euler angles and similar variable sets, $\mathbf{A}$ and $\mathbf{B}$ are mutually inverse 3 $\times$ 3 matrices, of which $\mathbf{A}$ is singular for certain angles. For quaternions, $\mathbf{A}$ is a 3 $\times$ 4 matrix and $\mathbf{B}$ is a 4 $\times$ 3 matrix, with $\mathbf{A}\mathbf{B} = \mathbf{1}$ and $\mathbf{B}\mathbf{A}\mathbf{q} = \mathbf{q}$, if $\mathbf{q} \cdot \mathbf{q} = 1$. Both $\mathbf{A}$ and $\mathbf{B}$ depend

linearly on $\mathbf{q}$. Using the normalization constraint $\mathbf{q} \cdot \mathbf{q} = 1$, a non-singular relation between $\boldsymbol{\omega}$ and $\dot{\mathbf{q}}$ can be derived. This is the reason why quaternions have become popular for computer simulations [27, 28, 1].

To maintain a compact notation, we collect all translational and angular velocities into the vector $\mathbf{v} = (\mathbf{v}_1, \boldsymbol{\omega}_1, \ldots, \mathbf{v}_N, \boldsymbol{\omega}_N) = (v^1, \ldots, v^{6N})$. Correspondingly, we introduce the vector $\mathbf{x} = (\mathbf{r}_1, \mathbf{q}_1, \ldots, \mathbf{r}_N, \mathbf{q}_N) = (x^1, \ldots, x^M)$, where $M = 6N + s$ and s is the number of normalization conditions. Introducing appropriately defined supermatrices $\mathbf{A}$ and $\mathbf{B}$, the relations between the cartesian velocities and the time derivatives of the coordinates can be cast in the form

$$v^i = A^i_j\left(x^k\right)\dot{x}^j, \quad i = 1 \ldots 6N, \quad j = 1 \ldots M \tag{2.6}$$

$$\dot{x}^j = B^j_k\left(x^l\right)v^k, \quad j = 1 \ldots M, \quad k = 1 \ldots 6N \tag{2.7}$$

Here $A^i_k B^k_j = \delta^i_j$ and $B^i_k A^k_j x^j = x^i$. We use the Einstein summation rule, i.e. summation over pairwise like upper and lower indices is always assumed. To describe the conformation of a constrained dynamical system, we introduce a set of M' variables $\tilde{x}^\alpha$ that may be subject to s' constraints, such that $M' = f + s'$, where f is the number of degrees of freedom. We assume that the positions and orientations of the constituents can be written as functions of the generalized coordinates,

$$x^i = x^i\left(\tilde{x}^1, \ldots, \tilde{x}^{M'}\right) \tag{2.8}$$

Differentiating the coordinates x^i with respect to time yields

$$\dot{x}^i = \frac{\partial x^i}{\partial \tilde{x}^\alpha}\dot{\tilde{x}}^\alpha \tag{2.9}$$

where Greek indices label generalized coordinates. In analogy to (2.6) and (2.7) we assume linear relations between the $\dot{\tilde{x}}^\alpha$ and the generalized velocities $\tilde{v}^\alpha$ ($\alpha = 1 \ldots f$):

$$\tilde{v}^\alpha = \tilde{A}^\alpha_\beta\left(x^\gamma\right)\dot{\tilde{x}}^\beta, \quad \alpha = 1 \ldots f, \quad \beta = 1 \ldots M' \tag{2.10}$$

$$\dot{\tilde{x}}^\beta = \tilde{B}^\beta_\gamma\left(x^\delta\right)\tilde{v}^\gamma, \quad \beta = 1 \ldots M', \quad \gamma = 1 \ldots f \tag{2.11}$$

where $\tilde{A}^\alpha_\gamma \tilde{B}^\gamma_\beta = \delta^\alpha_\beta$ and $\tilde{B}^\alpha_\gamma \tilde{A}^\gamma_\beta \tilde{x}^\beta = \tilde{x}^\alpha$. We note that $s' = 0$, $\tilde{A}^\alpha_\gamma = \delta^\alpha_\gamma$, and $\tilde{B}^\gamma_\beta = \delta^\gamma_\beta$ for the choice $\tilde{v}^\alpha = \dot{\tilde{x}}^\alpha$. Combining relation (2.9) between the $\dot{x}^i$ and the $\dot{\tilde{x}}^\alpha$ with (2.6) and

(2.11) leads us to the following linear relation between the Cartesian velocities v^i and the generalized velocities $\tilde{v}^\alpha$:

$$v^i = A^i_j \frac{\partial x^j}{\partial \tilde{x}_\beta} \tilde{B}^\beta_\alpha \tilde{v}^\alpha \equiv C^i_\alpha \tilde{v}^\alpha, \quad i = 1...6N, \quad \alpha = 1...f \quad (2.12)$$

In contrast to the time derivatives of the coordinates, the velocities always correspond to the actual degrees of freedom. We note that in general the matrix $C = \left(C^i_\alpha \right)$ is not the Jacobian of a coordinate transformation. This is not only because C is rectangular and therefore not invertible, but also since the velocities may include angular velocities, which are non-integrable differential forms of the corresponding angular coordinates. An exception is rotation about a fixed axis.

A well known example for linear velocity constraints is the motion of a rigid body. The translational and angular velocities of the constituents are then given by

$$\mathbf{v}_i = \tilde{\mathbf{v}} + \tilde{\omega} \wedge \mathbf{r}_i \quad , \quad \omega_i = \tilde{\omega} \quad (2.13)$$

where $\tilde{\mathbf{v}}$ is the translational velocity of some reference point, and $\tilde{\omega}$ is the angular velocity of the rotational motion. The positions $\mathbf{r}_i$ refer to the reference point and $\wedge$ denotes a vector product. A generalization of the rigid-body velocity constraints to the case of topologically linked rigid bodies can be used to describe chain molecules [1]. An example will be given in the application section

Projector on the constrained Cartesian velocities. The linear velocity constraints (2.12) can be expressed in matrix form as

$$\mathbf{v}_c = C\tilde{\mathbf{v}} \quad (2.14)$$

The vector $\mathbf{v}_c$ contains the components of the constrained Cartesian velocities, whereas $\tilde{\mathbf{v}}$ comprises the reduced set of generalized velocities, $\tilde{v}^1 ... \tilde{v}^f$. The subscript c in $\mathbf{v}_c$ indicates the presence of constraints. The components of the $6N \times f$ matrix C are defined by (2.12). Using the generalized inverse of C, one can construct the projector $\wp$ on the space of constrained velocities. Multiplying Eq. (2.12) from the left by CC^+, we obtain $CC^+\mathbf{v}_c = CC^+C\tilde{\mathbf{v}} = C\tilde{\mathbf{v}} = \mathbf{v}_c$, i.e.

$$\mathbf{v}_c = \wp \mathbf{v}_c \quad , \quad \wp = CC^+ \quad (2.15)$$

The explicit form for C^+ can be found from relation (2.5). Assuming that C has full column rank, i.e. the number of its columns corresponds to the number of degrees of freedom, one can write $C = FG$, with $F = C$ and $G = 1$. Therefore

$$C^+ = \left(C^T C \right)^{-1} C^T \quad (2.16)$$

and the projector $\wp$ reads

$$\wp = C\left(C^T C \right)^{-1} C^T \quad (2.17)$$

We see from Eq. (2.16) that C^+C is equal to the unit matrix in f dimensions, where f is the number of degrees of freedom:

$$C^+C = 1_f \quad (2.18)$$

This relation can be used to express the generalized velocities in terms of the constrained Cartesian velocities. Multiplying (2.14) from the left by C^+ yields

$$\tilde{\mathbf{v}} = C^+\mathbf{v}_c = \left(C^T C \right)^{-1} C^T \mathbf{v}_c \quad (2.19)$$

which shows that C^+ can be regarded as the inverse transformation matrix with respect to C.

Friction and mobility with constraints

Constraint forces and torques. In analogy to the velocity vector $\mathbf{v}$ of the unconstrained system, we define the vector $\mathbf{f} \doteq \left(\mathbf{F}_1, \mathbf{T}_1, ..., \mathbf{F}_N, \mathbf{T}_N \right)$ containing the forces and torques acting on the particles. The equation of motion for Stokesian Dynamics reads

$$\zeta \mathbf{v} = \mathbf{f} \quad (2.20)$$

where ζ is the friction matrix and $\mathbf{f}$ is the vector of given forces and torques. The solution of (2.20) with respect to the unknown translational and angular velocities, $\mathbf{v}$, can be written formally as

$$\mathbf{v} = \mu \mathbf{f} \quad , \quad \mu = \zeta^{-1} \quad (2.21)$$

where μ is the mobility matrix.

We consider now the case of constrained motion where the velocities are subject to linear constraints. As shown above, this situation may be expressed in the form $\mathbf{v}_c = \wp \mathbf{v}_c$, where $\wp$ is a projector onto the subspace of the constrained velocities. In the following $\wp'$ denotes the projector which is orthogonal to $\wp$, i.e. $\wp' = 1 - \wp$. In the presence of constraints, Eq. 2.20 contains an additional force term comprising constraint forces and constraint torques which keep the dynamics of the system in accordance with the imposed constraints:

$$\zeta\mathbf{v}_c = \mathbf{f} + \mathbf{z} \quad , \quad \mathbf{v}_c = \wp\mathbf{v}_c \quad , \quad \mathbf{z} = \wp\mathbf{z} \tag{2.22}$$

The condition $\mathbf{z} = \wp'\mathbf{z}$ means that the constraint forces do not perform work. From a mathematical point of view, the vector $\mathbf{z}$ must be introduced to ensure the existence of a solution for $\mathbf{v}_c$. It should be noted that $\mathbf{z}$ is unknown as well and the condition $\mathbf{z} = \wp'\mathbf{z}$ ensures that both $\mathbf{v}_c$ and $\mathbf{z}$ can be obtained from the same set of linear equations, $\zeta\mathbf{v}_c = \mathbf{f} + \mathbf{z}$. Systems of linear equations like (2.22) have a unique solution. To our knowledge such systems have first been studied by Bott and Duffin in the context of electrical networks [29]. Here we use generalized inverse matrices to express the solution of (2.22). At the time when Bott and Duffin developed the theory of electrical networks, the powerful concept of generalized inverse matrices was not yet developed.

Solving for the constrained velocities. To solve Eq. 2.22, we multiply from the left by $\wp$ and make use of $\wp\mathbf{z} = 0$. With $\mathbf{v}_c = \wp\mathbf{v}_c$, the resulting equation can be cast in the form

$$\wp\zeta\,\wp\mathbf{v}_c = \wp\zeta\,\mathbf{f} \tag{2.23}$$

For the following considerations we define the *constrained friction matrix* ζ_c as

$$\zeta_c \doteq \wp\zeta\wp \tag{2.24}$$

We will now show that (2.23) has a unique solution which can be written as

$$\mathbf{v}_c = \mu_c\mathbf{f} \tag{2.25}$$

where the *constrained mobility matrix* μ_c is related to ζ_c by

$$\mu_c \doteq \zeta_c^+ \tag{2.26}$$

A solution of (2.23) exists if the consistency condition $\zeta_c\zeta_c^+\wp\mathbf{f} = \wp\mathbf{f}$ is fulfilled. This is always the case as we will see now. Using (2.17) we write $\zeta_c \doteq \wp\zeta\wp$ explicitly as

$$\zeta_c = \underbrace{\mathbf{C}(\mathbf{C}^T\mathbf{C})^{-1}}_{\mathbf{F}}(\mathbf{C}^T\zeta\mathbf{C})^{-1}\underbrace{\mathbf{C}^T}_{\mathbf{G}} \tag{2.27}$$

The pseudoinverse ζ_c^+ is obtained from relation (2.5), using a factorization $\zeta_c = \mathbf{FG}$ as indicated above. The result is

$$\zeta_c^+ = \mathbf{C}(\mathbf{C}^T\zeta\mathbf{C})^{-1}\mathbf{C}^T \tag{2.28}$$

As a prerequisite we must require that $\det(\mathbf{C}^T\zeta\mathbf{C}) \neq 0$. Since the unconstrained friction matrix, ζ, is non-singular and all column vectors in $\mathbf{C}$ are linearly independent, this is always true. From (2.27) and (2.28) we find the relations

$$\zeta_c^+\zeta_c = \zeta_c\zeta_c^+ = \wp \quad , \quad \wp\zeta_c^+ = \zeta_c^+\wp = \zeta_c^+ \tag{2.29}$$

for the generalized inverse of ζ_c. It follows immediately that $\zeta_c\zeta_c^+\wp\mathbf{f} = \wp\mathbf{f}$, showing that the consistency condition is fulfilled. The general solution for $\mathbf{v}_c$ then reads $\mathbf{v}_c = \zeta_c^+\wp\mathbf{f} + (1 - \zeta_c^+\zeta_c)\mathbf{h} = \zeta_c^+\mathbf{f} + \wp'\mathbf{h}$, where $\mathbf{h}$ is an arbitrary vector. Since we require that $\wp\mathbf{v}_c = \mathbf{v}_c$, it follows that $\mathbf{h} = 0$, i.e

$$\mathbf{v}_c = \zeta_c^+\mathbf{f} \equiv \mu_c\mathbf{f} \tag{2.30}$$

This shows that the constrained mobility matrix is indeed the generalized inverse of the constrained friction matrix, as postulated in (2.26). It remains to show that the constraint forces fulfil the condition $\wp'\mathbf{z} = \mathbf{z}$. We write

$$\mathbf{z} = \zeta\mathbf{v}_c - \mathbf{f} = (\zeta\mu_c - 1)\mathbf{f} \tag{2.31}$$

and multiply from the left by $\wp$. This yields $\wp\mathbf{z} = (\wp\zeta\mu_c - \wp)\mathbf{f}$. Since $\mu_c = \wp\mu_c$, the product $\wp\zeta\mu_c$ can be replaced by $\wp\zeta\wp\mu_c = \zeta_c\mu_c = \wp$. Therefore $\wp\mathbf{z} = (\wp - \wp)\mathbf{f} = 0$, which is equivalent to $\wp'\mathbf{z} = \mathbf{z}$.

Explicit expressions for reduced friction and mobility matrices. Expressions for the friction and mobility matrices in reduced space are obtained by writing $\mathbf{v}_c - \mu_c\mathbf{f} = 0 = \mathbf{C}\tilde{\mathbf{v}} - \mu_c\mathbf{f}$ and inserting expression (2.28) for $\mu_c = \zeta_c^+$:

$$\mathbf{C}\left[\tilde{\mathbf{v}} - (\mathbf{C}^T\zeta\mathbf{C})^{-1}\mathbf{C}^T\mathbf{f}\right] = 0 \tag{2.32}$$

Since $\mathbf{C}$ has full column rank, the vector in square brackets must be the null vector. This can be written as

$$\tilde{\mathbf{v}} = \tilde{\mu}\tilde{\mathbf{f}} \Leftrightarrow \tilde{\zeta}\tilde{\mathbf{v}} = \tilde{\mathbf{f}} \tag{2.33}$$

and defines the reduced mobility matrix $\tilde{\mu}$, the reduced friction matrix $\tilde{\zeta}$, and the reduced force vector $\tilde{\mathbf{f}}$:

$$\tilde{\mu} = \tilde{\zeta}^{-1} \quad , \quad \tilde{\zeta} = \mathbf{C}^T\zeta\mathbf{C} \quad , \quad \tilde{\mathbf{f}} = \mathbf{C}^T\mathbf{f} \tag{2.34}$$

A minimum principle. One can easily show that the equations of motion for constrained Stokesian Dynamics can be derived by minimizing

$$g(\mathbf{v}) = \frac{1}{2}\mathbf{v}\cdot\zeta\mathbf{v} - \mathbf{v}\cdot\mathbf{f} \tag{2.35}$$

with respect to $\mathbf{v}$. In the absence of constraints, the condition $\partial g(\mathbf{v})/\partial\mathbf{v} = 0$ immediately yields the equations of motion (2.20); setting $\mathbf{v} = \wp\mathbf{v}$ yields the constrained equations of motion (2.23) in Cartesian coordinates, and setting $\mathbf{v} = \mathbf{C}\tilde{\mathbf{v}}$ yields the equations of motion (2.33) if one minimizes with respect to $\tilde{\mathbf{v}}$. The principle (2.35) can be considered the equivalent of Gauß' principle of least constraint in classical mechanics [30], replacing the accelerations by velocities and the diagonal mass matrix by the friction matrix. We note that the form of $g(\mathbf{v})$ is not trivial, although it seems plausible. Consider the alternative quadratic form $g'(\mathbf{v}) = \frac{1}{2}(\zeta\mathbf{v} - \mathbf{f})^2$. Both quadratic forms give the same equations of motion for unconstrained systems but *different* equations of motion for constrained systems.

Curvilinear space

To establish a connection to Kirkwood's theory of polymer solutions [2], we will show how the relations between reduced friction and mobility matrices and their respective Cartesian counterparts can be described formally in terms of coordinate transformations between $6N$ constrained Cartesian velocities and f reduced velocities. In this framework our approach appears as an extension of Kirkwood's theory, which deals with polymers consisting of point-like beads. The correct transformation rules for the reduced friction matrix and the reduced mobility tensor are automatically obtained. In this context we briefly comment on a mistake in the Kirkwood theory which has been reported in the book by Yamakawa [4].

Basis vectors. We start by introducing a set of basis vectors $\{\mathbf{e}_1,...,\mathbf{e}_{6N}\}$ spanning the $6N$-dimensional Euclidian space, and a set of basis vectors $\{\mathbf{b}_1,...,\mathbf{b}_f\}$ spanning the reduced space of constrained velocities. The constrained velocities may then be expressed in either of the two basis sets:

$$\mathbf{v}_c = v^i_{(c)}\mathbf{e}_i = \tilde{v}^\alpha\mathbf{b}_\alpha \tag{2.36}$$

We use Latin indices to enumerate basis vectors and co-ordinates in Euclidian space and Greek indices to enumerate the corresponding quantities in reduced space. Inserting expression (2.14) for the linear velocity constraints into (2.36) shows that the basis vectors $\mathbf{b}_\alpha$ are the columns of $\mathbf{C}$. Moreover the transformation rules

$$v^i_{(c)} = C^i_\alpha\tilde{v}^\alpha \quad , \quad \mathbf{b}_\alpha = C^i_\alpha\mathbf{e}_i \tag{2.37}$$

can be read off. In addition to the basis vectors, we introduce the metric tensor whose components in reduced space read

$$\tilde{g}_{\alpha\beta} = \mathbf{b}_\alpha\cdot\mathbf{b}_\beta = \sum_i C^i_\alpha C^i_\beta \tag{2.38}$$

whereas $g_{ij} = \mathbf{e}_i\cdot\mathbf{e}_j = \delta_{ij}$ in cartesian coordinates. In general the basis vectors $\mathbf{b}_\alpha$ do not form an orthonormal basis. One can, however, define dual basis vectors, $\mathbf{b}^\alpha$, which are orthonormal to the $\mathbf{b}^\alpha$, i.e. $\mathbf{b}_\alpha\cdot\mathbf{b}^\beta = \delta^\beta_\alpha$. In terms of the metric tensor they can be expressed as

$$\mathbf{b}^\alpha = \tilde{g}^{\alpha\beta}\mathbf{b}_\beta \quad , \quad \left(\tilde{g}^{\alpha\beta}\right) = \left(\tilde{g}_{\alpha\beta}\right)^{-1} \tag{2.39}$$

Here $\left(\tilde{g}^{\alpha\beta}\right)$ denotes the matrix formed by the components of the metric tensor and $\left(\tilde{g}_{\alpha\beta}\right)^{-1}$ is the corresponding inverse matrix. We note that vector and tensor components which refer to the basis vectors $\mathbf{b}_\alpha$ are called contravariant components, and those which refer to the $\mathbf{b}^\alpha$ are called covariant components. Since $\mathbf{e}^i = \mathbf{e}_i$, co- and contravariant components in Euclidian space are the same. Consider the co- and contravariant components of the constrained velocities in reduced space, defined by $\mathbf{v}_c = \tilde{v}^\alpha\mathbf{b}_\alpha = \tilde{v}_\alpha\mathbf{b}^\alpha$. Using $\mathbf{b}_\alpha\cdot\mathbf{b}^\beta = \delta^\beta_\alpha$, one obtains the transformation rules

$$\tilde{v}^\alpha = \mathbf{b}^\alpha\cdot\mathbf{v}_c = \tilde{g}^{\alpha\beta}C^i_\beta v_{(c)i} = \left(C^+\right)^\alpha_i v^i_{(c)} \tag{2.40}$$

$$\tilde{v}_\alpha = \mathbf{b}_\alpha\cdot\mathbf{v}_c = C^i_\alpha v_{(c)i} \tag{2.41}$$

This is exactly relation (2.19) in tensor notation, since

$$\mathbf{C}^+ = \left(\mathbf{C}^T\mathbf{C}\right)^{-1}\mathbf{C}^T = \left(\tilde{g}_{\alpha\beta}\right)^{-1}\mathbf{C}^T = \left(\tilde{g}^{\alpha\beta}\right)\mathbf{C}^T .$$

Although $v_{(c)i} = v^i_{(c)}$, we distinguish between co- and contravariant Cartesian components to respect the Einstein summation rule. $\mathbf{C}^+$ replaces the normal inverse describing non-singular coordinate transformations between spaces of equal dimension. According to (2.15)) and (2.18) we have

$$C^j_\alpha\left(C^+\right)^\alpha_i = P^j_i \tag{2.42}$$

$$\left(C^+\right)^\beta_i C^i_\alpha = \delta^\beta_\alpha \tag{2.43}$$

Here $P^j_i = P_{ij} = P^{ij}$ are the Cartesian components of the projector $\wp$. Eqs. (2.40) and (2.41) also yield the familiar relations between co- and contravariant components:

$$\tilde{v}_\alpha = \tilde{g}_{\alpha\beta}\tilde{v}^\beta \quad , \quad \tilde{v}^\alpha = \tilde{g}^{\alpha\beta}\tilde{v}_\beta \tag{2.44}$$

Reduced friction and mobility matrix. Now we apply Eq. (2.34) and write the components of the reduced friction matrix and the reduced force vector as

$$\tilde{\zeta}_{\alpha\beta} = C^i_\alpha C^i_\beta \zeta_{ij} = \mathbf{b}_\alpha \cdot \zeta \mathbf{b}_\beta \tag{2.45}$$

$$\tilde{f}_\alpha = C^i_\alpha f_i = \mathbf{b}_\alpha \cdot \mathbf{f} \tag{2.46}$$

which shows that they are covariant tensor components. The components of the reduced mobility matrix are found to be contravariant tensor components:

$$\tilde{\mu}^{\alpha\beta} = \left(C^+\right)^\alpha_i \left(C^+\right)^\beta_j \mu^{ij}_{(c)} =$$
$$= \tilde{g}^{\alpha\alpha'}\tilde{g}^{\beta\beta'}C^i_{\alpha'}C^j_{\beta'}\mu_{(c)ij} = \mathbf{b}^\alpha \cdot \mu_{(c)}\mathbf{b}^\beta \tag{2.47}$$

Note that $\mu_{(c)ij} = \mu^{ij}_{(c)}$, since Cartesian tensor components refer to a Euclidian basis. Relation (2.47) follows from Eqs. (2.28) and (2.34) by writing $\mu_c = \mathbf{C}\tilde{\mu}\mathbf{C}^T$. Multiplying from the left by $\mathbf{C}^+$ and from the right by $(\mathbf{C}^+)^T$ leads to

$$\mathbf{C}^+ \mu_c \left(\mathbf{C}^+\right)^T = \left(\mathbf{C}^+\mathbf{C}\right)\tilde{\mu}\left(\mathbf{C}^+\mathbf{C}\right)^T = \tilde{\mu} \text{ , making use of } \mathbf{C}^+\mathbf{C}$$

$= \mathbf{1}$. We emphasize that the contravariant tensor $\tilde{\mu}$ must be derived from its *constrained* Cartesian counterpart, whereas the covariant tensors $\tilde{\zeta}$ and $\tilde{f}$ are derived from their *unconstrained* Cartesian counterparts. This is not surprising, since the inversion of $\tilde{\zeta}$ cannot be achieved by multiplications with $\mathbf{C}^+$. In this context we note that the Cartesian components of the friction matrix and the force vector in (2.45) and (2.46) may be replaced by the components of the corresponding constrained quantities, $\zeta_c = \wp\zeta\wp$ and $\mathbf{f}_c = \wp\mathbf{f}$, respectively. This follows from Eq. (2.34) and the identity $\wp\mathbf{C} = \mathbf{C}\mathbf{C}^+\mathbf{C} = \mathbf{C}$. We have therefore

$$\tilde{\zeta}_{\alpha\beta} = \mathbf{b}_\alpha \cdot \zeta \, \mathbf{b}_\beta = \mathbf{b}_\alpha \cdot \zeta_{(c)}\mathbf{b}_\beta \tag{2.48}$$

$$\tilde{f}_\alpha = \mathbf{b}_\alpha \cdot \mathbf{f} = \mathbf{b}_\alpha \cdot \mathbf{f}_{(c)} \tag{2.49}$$

$$\tilde{\mu}^{\alpha\beta} = \mathbf{b}^\alpha \cdot \mu_{(c)}\mathbf{b}^\beta \neq \mathbf{b}^\alpha \cdot \mu\mathbf{b}^\beta \tag{2.50}$$

A strict analogy to coordinate transformations between spaces of equal dimension can be established only if all vectors and tensors in Euclidian space are elements of the subspace defined by the constraints. In the original Kirkwood theory [2] the reduced mobility matrix was defined as $\tilde{\mu}^{\alpha\beta} = \mathbf{b}^\alpha \cdot \mu\mathbf{b}^\beta$ with $\mu = \zeta^{-1}$. As reported in Yamakawa's book [4], the mistake has been corrected by Ikeda [31] and by Kirkwood and Erpenbeck [16]. Ikeda derives an expression for the reduced mobility matrix in the Oseen approximation, starting from the correct relation $\tilde{\mu} = \tilde{\zeta}^{-1}$. Kirkwood-Erpenbeck define $\tilde{\mu}$ as the submatrix in chain space of the transformed *unconstrained* mobility matrix, using a quadratic transformation matrix for the full coordinate set describing the chain space and its complement. This is, however, still not the same as $\tilde{\mu}$, no matter how the basis vectors for chain space and its complement are chosen. We note that for point-like constituents the matrix $\mathbf{C}$ is a Jacobian, since for translational degrees of freedom $A^i_j = \delta^i_j$ and $\tilde{B}^\beta_\alpha = \delta^\beta_\alpha$ in Eq. (2.21).

Application

As an application for the computation of hydrodynamic interactions in systems with geometrical constraints, we simulate the Stokesian Dynamics of an initially stretched pentamer moving under the influence of a constant force through a viscous liquid. The system is depicted in Figure 1. It consists of five identical spherical monomers connected to massless rods. The ends of the rods are linked by joints. Each joint allows free rotation, i.e. it has three angular degrees of freedom. In the stretched conformation the distance between the spheres is $3a$, where a is the radius of the monomers, and the joints are positioned halfway between the centers of the spheres. Our model deviates in two aspects from the widely-used bead-rod models:

• The monomers have a *finite size* and therefore three translational and three rotational degrees of freedom.
• The positions of the joints do not coincide with the positions of the monomers. This picture is somewhat more realistic with respect to modelling hinges in macromolecules. We note that our method does not depend on this choice of joints; any other one could be treated as well.

For the calculation of the unconstrained friction matrix, we use an accurate scheme that is applicable even for densely packed spheres [19, 22]. The implementation we use is described in [24].

The assumptions underlying our simulation technique, i.e. clear length scale separation between solute and solvent and negligible contributions from random (Brownian) motion, mean that our monomers must be quite large, representing not small groups of atoms, but whole domains of large macromolecules. This model should not be confused with the

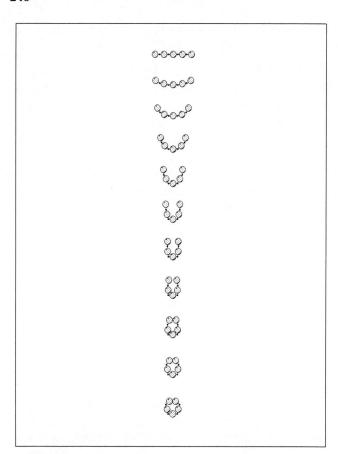

Figure 1. *Sedimentation of a pentamer. The monomers are equally sized spherical particles of radius a and the distance between two spheres in the initial configuration is 3a. The constant gravitational force **F** points from top to bottom. The figure shows 11 equidistant frames of a 75000 time step simulation with a time step of Δt = 0.001. The separation between consecutive frames is Δt$_{frame}$ = 7.5, i.e. the whole simulation is shown. Within one time step Δt a free sphere would move by 2/3 · 10^{-3} a.*

modelling of non-spherical structures (such as helices) by closely packed conglomerates of spheres [13].

Linear velocity constraints

Numbering spheres and joints from left to right, we choose the position of the first sphere, $\mathbf{r}_1$, to be the reference point for the translational motion of the chain. We obtain the following expressions for the positions of the joints and the monomers (primed quantities refer to joints):

$$\mathbf{r'}_i = \mathbf{r'}_{i-1} + \mathbf{D}(\mathbf{q}_i)(\mathbf{r'}_{0,i} - \mathbf{r'}_{0,i-1}), \quad i = 1...5 \tag{3.1}$$

$$\mathbf{r'}_0 \equiv \mathbf{r}_1 \tag{3.2}$$

$$\mathbf{r}_i = \tfrac{1}{2}(\mathbf{r'}_i + \mathbf{r'}_{i-1}), \quad i = 2,3,4 \tag{3.3}$$

$$\mathbf{r}_5 \equiv \mathbf{r'}_5 \tag{3.4}$$

We have added the fictitious joints 0 and 5 which coincide with the centers of monomers 1 and 5. The vectors with subscript 0 refer to the initial configuration and the time-dependent vectors $\mathbf{q}_i$ contain the angular parameters describing the rotation of the momomers about their respective anchor points. For practical reasons we use quaternion parameters, i.e. $\mathbf{q} = (q_0, q_1, q_2, q_3)$, where $q_0^2 + q_1^2 + q_2^2 + q_3^2 = 1$ [25, 1]. The rotation matrix $\mathbf{D}(\mathbf{q})$ in terms of quaternion parameters reads [28]

$$\tag{3.5}$$

$$\mathbf{D} = \begin{pmatrix} q_0^2 + q_1^2 - q_2^2 - q_3^2 & 2(-q_0q_3 + q_1q_2) & 2(q_oq_2 + q_1q_3) \\ 2(q_0q_3 + q_1q_2) & q_0^2 + q_2^2 - q_1^2 - q_3^2 & 2(-q_0q_1 + q_2q_3) \\ 2(-q_0q_2 + q_1q_3) & 2(q_0q_1 + q_2q_3) & q_0^2 + q_3^2 - q_1^2 - q_2^2 \end{pmatrix}$$

By differentiating the positions $\mathbf{r}_i$ with respect to time, one obtains a linear relation between the Cartesian components of the translational velocities $\mathbf{v}_i$ of the monomers and the generalized velocities, which are the translational velocity of the whole chain $\tilde{\mathbf{v}}$ and the angular velocities $\tilde{\omega}_i$ corresponding to the rotations described by $\mathbf{q}_i$. The tilde indicates the generalized velocities, which correspond to the actual degrees of freedom. An expression for the velocities $\mathbf{v}_i$ can be found by writing the time derivative of the rotation matrices $\mathbf{D}(\mathbf{q}_i)$ as $\tilde{\Omega}_i \mathbf{D}(\mathbf{q}_i)$, where the $\tilde{\Omega}_i$ are skew-symmetric matrices containing the components of the angular velocity $\tilde{\omega}_i$ of monomer i in the laboratory frame. One obtains the following expressions for the velocities of the joints and the spheres:

$$\mathbf{v'}_i = \mathbf{v'}_{i-1} + \tilde{\omega}_i \wedge (\mathbf{r'}_i - \mathbf{r'}_{i-1}), \quad i = 1...5 \tag{3.6}$$

$$\mathbf{v'}_0 = \mathbf{v}_1 = \tilde{\mathbf{v}} \tag{3.7}$$

$$\mathbf{v}_i = \mathbf{v'}_{i-1} + \tfrac{1}{2}\tilde{\omega}_i \wedge (\mathbf{r'}_i - \mathbf{r'}_{i-1}), \quad i = 2,3,4 \tag{3.8}$$

$$\mathbf{v}_5 = \mathbf{v'}_5 \tag{3.9}$$

Clearly, the angular velocities ω_i describing the rotations of the monomers about their centers are the same as those describing the rotations about the joints:

$$\omega_i = \tilde{\omega}_i \tag{3.10}$$

Therefore the relation between the 18 components of $\tilde{\mathbf{v}} = (\tilde{\mathbf{v}}, \tilde{\omega}_1,...,\tilde{\omega}_5)$ and the 30 components of $\mathbf{v} = (\mathbf{v}_1, \omega_1,...,\mathbf{v}_5, \omega_5)$ reads explicitly

$$
\begin{pmatrix}
\mathbf{v}_1 \\
\boldsymbol{\omega}_1 \\
\mathbf{v}_2 \\
\boldsymbol{\omega}_2 \\
\mathbf{v}_3 \\
\boldsymbol{\omega}_3 \\
\mathbf{v}_4 \\
\boldsymbol{\omega}_4 \\
\mathbf{v}_5 \\
\boldsymbol{\omega}_5
\end{pmatrix}
=
\begin{pmatrix}
1 & 0 & 0 & 0 & 0 & 0 \\
0 & 1 & 0 & 0 & 0 & 0 \\
1 & -\mathbf{R}'_{01} & -\frac{1}{2}\cdot\mathbf{R}'_{12} & 0 & 0 & 0 \\
0 & 0 & 1 & 0 & 0 & 0 \\
1 & -\mathbf{R}'_{01} & -\mathbf{R}'_{12} & -\frac{1}{2}\cdot\mathbf{R}'_{23} & 0 & 0 \\
0 & 0 & 0 & 1 & 0 & 0 \\
1 & -\mathbf{R}'_{01} & -\mathbf{R}'_{12} & -\mathbf{R}'_{23} & -\frac{1}{2}\cdot\mathbf{R}'_{34} & 0 \\
0 & 0 & 0 & 0 & 1 & 0 \\
1 & -\mathbf{R}'_{01} & -\mathbf{R}'_{12} & -\mathbf{R}'_{23} & -\mathbf{R}'_{34} & -\mathbf{R}'_{45} \\
0 & 0 & 0 & 0 & 0 & 1
\end{pmatrix}
\begin{pmatrix}
\mathbf{v} \\
\tilde{\boldsymbol{\omega}}_1 \\
\tilde{\boldsymbol{\omega}}_2 \\
\tilde{\boldsymbol{\omega}}_3 \\
\tilde{\boldsymbol{\omega}}_4 \\
\tilde{\boldsymbol{\omega}}_5
\end{pmatrix}
\tag{3.11}
$$

The submatrices $\mathbf{R}_{ij}$ are the skew-symmetric 3×3 matrices

$$
\mathbf{R}_{ij} = \begin{pmatrix}
0 & -z_{ij} & y_{ij} \\
z_{ij} & 0 & -x_{ij} \\
-y_{ij} & x_{ij} & 0
\end{pmatrix}
\tag{3.12}
$$

where x_{ij}, y_{ij}, and z_{ij} are the Cartesian components of $\mathbf{r}_{ij} \doteq \mathbf{r}_j - \mathbf{r}_i$.

Integrating the equations of motion

According to (2.33) and (2.34), the equations of motion can be cast in the form

$$
\tilde{\boldsymbol{\zeta}}\tilde{\mathbf{v}} = \tilde{\mathbf{f}}
\tag{3.13}
$$

where $\tilde{\boldsymbol{\zeta}} = \mathbf{C}^T\boldsymbol{\zeta}\mathbf{C}$ and $\tilde{\mathbf{f}} = \mathbf{C}^T\mathbf{f}$ are the reduced friction matrix and the reduced force vector, respectively. The constraint matrix $\mathbf{C}$ follows from (3.11) and $\boldsymbol{\zeta}$ is the friction matrix for an unconstrained system of 5 spherical particles. The latter is computed using the approach of Cichocki *et al.* [19, 22, 25, 24]. The external forces and torques in $\mathbf{f}$ are

$$
\mathbf{f} = \left(\mathbf{F}_1, \mathbf{T}_1, \ldots, \mathbf{F}_5, \mathbf{T}_5\right)
\tag{3.14}
$$

$$
\mathbf{F}_i = \left(0, 0, -F\right)
\tag{3.15}
$$

$$
\mathbf{T}_i = \left(0, 0, 0\right)
\tag{3.16}
$$

We choose the gravitational force to point towards the negative z-axis. From a numerical point of view, it is more efficient to solve (3.13) for $\tilde{\mathbf{v}}$ rather than computing $\tilde{\mathbf{v}} = \tilde{\boldsymbol{\zeta}}^{-1}\tilde{\mathbf{f}}$, which would require a full matrix inversion. The equations of motion are solved as follows:

1. For each configuration, $\{\mathbf{r}_1(n),\ldots,\mathbf{r}_5(n)\}$, compute
   (a) the unconstrained friction matrix $\boldsymbol{\zeta}$,
   (b) the constraint matrix $\mathbf{C}$,
   (c) the reduced friction matrix $\tilde{\boldsymbol{\zeta}} = \mathbf{C}^T\boldsymbol{\zeta}\mathbf{C}$,
   (d) the reduced force vector $\tilde{\mathbf{f}} = \mathbf{C}^T\mathbf{f}$.

2. Solve $\tilde{\boldsymbol{\zeta}}\tilde{\mathbf{v}} = \tilde{\mathbf{f}}$ for $\tilde{\mathbf{v}} \equiv \tilde{\mathbf{v}}(n)$.

3. Update the reference position $\mathbf{r}_1$ and the quaternion parameters $\mathbf{q}_i$ describing the orientation of the monomers according to the following central difference scheme:
   (a) $\mathbf{r}_1(n+1) = \mathbf{r}_1(n-1) + 2\Delta t \cdot \tilde{\mathbf{v}}(n)$.
   (b) $\mathbf{q}_i(n+1) = \mathbf{q}_i(n-1) + 2\Delta t \cdot \mathbf{B}[\mathbf{q}_i(n)]\tilde{\boldsymbol{\omega}}_i(n)$.

4. Update the positions of monomers 2 to 5 according to (3.1) – (3.4).

In step 3 we use the following singularity-free relation between the components of the angular velocity components and the time derivatives of the quaternion parameters:

$$
\dot{q}_\alpha = B_{\alpha j}\left(q_\beta\right)\omega_j
\tag{3.17}
$$

$$
\mathbf{B} = \frac{1}{2}\begin{pmatrix}
-q_1 & -q_2 & -q_3 \\
q_0 & q_3 & -q_2 \\
-q_3 & q_0 & q_1 \\
q_2 & -q_1 & q_0
\end{pmatrix}
$$

This relation has already been employed in molecular dynamics simulations [1, 27, 28].

Results

To simulate the sedimentation of the model pentamer described above, we performed a Stokesian Dynamics simulation of 75000 time steps of length Δt – see video sequence

Figure 2. *Comparison of the last configuration of the pentamer simulation (light grey) and the corresponding configuration of five unconnected spherical monomers of equal size (dark grey). The initial configuration for the unconstrained system was the same as for the constrained one. The dots on the unconstrained monomers indicate the presence of rotational motion.*

no. IIa ('Pentamer starting in the stretched conformation'). As in the video sequences in [25] *the center of mass motion is subtracted.* Again, the overall sedimentation of the cluster is shown on the left hand side of the screen. The height of the frame indicates the total falling distance, and the height of the black bar corresponds approximately to the height of the screen. In the simulation we used internal units with force $F = 1$, viscosity $\eta = \frac{1}{4}\pi$, particle radius $a = 1$, and time step $\Delta t = 10^{-3}$. In these units, the displacement of a single sphere in an infinite medium per dimensionless unit time is given by $\Delta r = F/(6/\pi\eta a) = 2/3$. The initial configuration of the pentamer was the stretched configuration shown in Figure 1. The monomers interact only via the background fluid – i.e no explicit interaction forces are considered. The constant driving force points from top to bottom and the time difference between consecutive configurations shown in Figure 1 is $\Delta t_{frame} = 7.5$, corresponding to 7500 time steps. For com-

parison we performed a second simulation of a system of five unconnected spherical particles of equal size – see video sequence no. IIb ('5 equally sized spheres starting in a linear configuration'). The total falling height indicated by the frame is the same as in the simulation shown in video sequence no. IIa. Apart from removing the constraints, the simulation parameters were the same as for the simulation of the pentamer. Figure 2 shows the superposition of the configurations at the end of the respective runs. The pentamer is drawn in light grey and the five unconnected spheres in dark grey. In the final configuration of the pentamer, monomers 1 and 5 touch each other. It is interesting to look at the heat production of the two systems which is defined as

$$p(\mathbf{v}) = \mathbf{v} \cdot \mathbf{f} \qquad (3.18)$$

In our example the external torques are zero and therefore

$$p = \sum_i \mathbf{F}_i \cdot \mathbf{v}_i \qquad (3.19)$$

Since the external forces are equal and constant, $\mathbf{F}_i = \mathbf{F}$, p is proportional to the average settling speed. Figure 3 shows the normalized heat production p/p_0 for the pentamer and the five unconnected spheres, where p_0 is the corresponding heat production of five spheres at infinite distance which are driven by the same force, i.e. $p_0 = 5 \cdot F^2/(6\pi\eta a) = 10/3$ in our internal units. The curve for the pentamer shows that the final configuration is reached at about $t = 60$. This can also be seen in Figure 1, where the configuration corresponding to $t = 60$ is the 9th configuration from top (3rd from bottom). Note that for $t \le 20$ the constrained and the unconstrained systems settle with approximately the same speed. The configurations in the initial phase (not shown here) are similar. Then the pentamer settles faster, reducing the friction, whereas the unconstrained system starts to lag behind and at about $t = 35$ it starts to form the separate groups (2,3,4) and (1,5) while the friction increases. As in the example we studied in [25], the heat production is neither monotonically increasing nor decreasing.

Conclusions

We have presented a rigorous method to derive the friction and mobility matrices for constrained dynamical systems consisting of rigid constituents. The method is based on the assumption that the constraints can be expressed as linear constraints for the Cartesian velocities, which is true for all situations in which the positions and orientations of the constituents can be expressed as functions of a set of generalized coordinates. We have shown that the constrained friction and mobility matrices are mutually generalized inverses

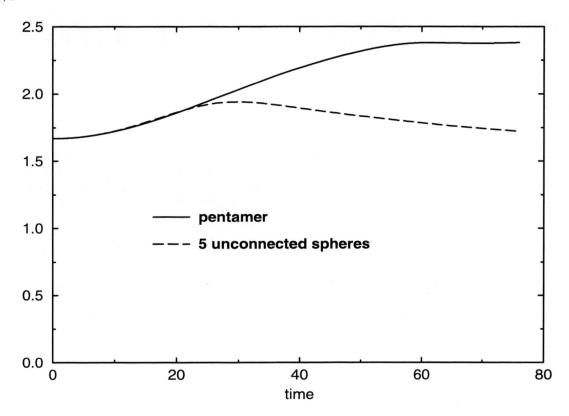

Figure 3. *Normalized heat production, p/p_0, of the pentamer and the corresponding system of five unconnected spherical monomers of equal size. The normalization factor p_0 is the heat production of five single spheres at infinite distance.*

in Cartesian velocity space, $\mu_c = \zeta_c^+$, which translates into the relation $\tilde{\mu} = \tilde{\zeta}^{-1}$ in reduced space. Explicit expressions for all relevant vectors and tensor quantities were given in both Cartesian and reduced space. The equations of motions for Stokesian Dynamics were shown to follow from a minimum principle analogous to Gauß' principle of least constraint in classical mechanics.

We have also shown that vectors and tensors in reduced space and in constrained cartesian space are formally mapped onto each other by coordinate transformations. Although this formal scheme is not of practical importance it shows that our method yields automatically the right transformation rules.

We conclude that complex constrained dynamical systems with hydrodynamic interactions can be described conveniently and correctly in the framework of generalized inverse matrices and linear velocity constraints. An efficient scheme for the computation of the divergence of the mobility matrix still needs to be developed in order to perform Brownian Dynamics simulations.

Acknowledgement. We wish to thank the Zentralinsitut für Angewandte Mathematik of the KFA Jülich, in particular Klaudia Waschbüsch and Maik Boltes, for producing the video sequences shown in this article. The single frames were generated with the program MOLSCRIPT [32]. One of us (GRK) wishes to thank the German Space Agency for financial support.

References

1.   Kneller, G.R.; Hinsen, K. *Phys. Rev. E* **1994**, *50(2)*, 1559.
2.   Kirkwood, J.G. *Macromolecules*, Auer, P.L. (ed.), series John Gamble Kirkwood Collected Works, Oppenheim, I. (Gen. Ed.), Gordon and Breach Science Publishers, New York, **1967**.
3.   Bird, R.B.; Hassager, O.; Amstrong, R.C.; Curtis, C.F. *Dynamics of Polymeric Liquids*, Vol. 2 (Kinetic Theory), John Wiley, New York, **1977**.
4.   Yamakawa, H. *Modern theory of polymer solutions*, Harper & Row, New York, **1971**.
5.   Wegener, W.A. *J. Chem. Phys.* **1982**, *76(12)*, 6425.
6.   Harvey, S.C.; Mellado, P.; Garcia de la Torre, J. *J. Chem. Phys.* **1983**, *78(4)*, 2081.
7.   Ryckaert, J.-P.; Ciccotti, G.; Berendsen, H.J.C. *J. Comp. Phys.* **1977**, *23*, 327.
8.   Ciccotti, G.; Ferrario, M; Ryckaert, J.-P. *Mol. Phys.* **1982**, *47*, 1253.
9.   Fixman, M. *J. Chem. Phys.* **1978**, *69(4)*, 1527.

10. Fixman, M. *Macromolecules* **1986**, *19*, 1195.
11. Allison, S.A.; McCammon, J.A. *Biopolymers* **1984**, *23*, 167.
12. Oettinger, H.C. *Phys. Rev. E* **1994**, *50(4)*, 2696.
13. Garcia de la Torre, J.; Bloomfield, V.A. *Quarterly Review of Biophysics* **1981**, *14(1)*, 81.
14. Brune, D.; Kim, S. *Proc. Natl. Acad. Sci. USA* **1994**, *91*, 2930.
15. Hinsen, K.; Kneller, G.R. *Phys. Rev. E* **1995**, *52(6)*, 6868.
16. Erpenbeck, J.J.; Kirkwood, J.G. *J. Chem. Phys.* **1958**, *29(4)*, 909.
17. Durlofsky, L; Brady, J.F.; Bossis, G. *J. Fluid Mech.* **1987**, *180*, 21-49.
18. Ladd, A.J.C. *J. Chem. Phys.* **1990**, *93*, 3484.
19. Cichocki, B; Felderhof, B.U.; Hinsen, K; Wajnryb, E; Blawzdziewicz, J. *J. Chem. Phys.* **1994**, *100(5)*, 3780.
20. Rotne, J.; Prager, M. *J. Chem. Phys.* **1969**, *50*, 4831.
21. Dwyer, J.D.; Bloomfield, V.A. *Biophys. J.* **1993**, *65(5)*, 1810.
22. Cichocki, B.; Hinsen, K. *Phys. Fluids* **1995**, *7*, 286.
23. Ben-Israel, S.; Greville, T.N.E. *Generalized Inverses: Theory and Application*, John Wiley, New York, **1974**.
24. Hinsen, K. *Comp. Phys. Comm.* **1995**, *88*, 327.
25. Hinsen, K. and Kneller, G.R. *J. Mol. Model.* **1996**, 2, accepted for publication.
26. Ermak, D.L.; McCammon, J.A. *J. Chem. Phys.* **1978**, *69*, 1352.
27. Evans, D.J.; Murad, S. *Mol. Phys.* **1977**, *34*, 327.
28. Allen, M.P.; Tildesley, D.J *Computer Simulation of Liquids*, Oxford University Press, Oxford, **1987**.
29. Bott, R.; Duffin, R.J. *Trans. Amer. Math. Soc.* **1953**, *74*, 99-109.
30. Gauß, C.F. *Journal für Reine und Angewandte Mathematik* **1829**, *IV*, 232; Pars, L.A. *A Treatise on Analytical Dynamics*, Heinemann, London, **1968**.
31. Ikeda, Y. *Kobayashi Rigaku Kenkyusho Hokuku* **1956**, *6*, 44, Ref. 28 in [4], p. 354.
32. Kraulis, P.J. *J. Appl. Cryst.* **1991**, *24*, 946.

J. Mol. Model. **1996**, 2, 251 – 277

Proceedings of
10th Molecular Modelling Workshop
Darmstadt, Germany
May 14 - 15, 1996

Preface

As editor of this journal and secretary of the *Molecular Modeling and Graphics Society (Deutschsprachige Sektion)*, I have pleasure in presenting this collection of abstracts and full papers from the Workshop. It contains a wealth of material on aspects of molecular modeling ranging from method development and application to multimedia data presentation.

The abstracts and full papers collected here give a flavour of the meeting. The electronic mode of publication makes it possible to give a true image of the workshop presenting – for the first time – videos and VRML-scenes.

Thanks are due to all those who have submitted contributions and the keynote speakers Paul Madden and Soeren Toxvaerd. Special thanks go to Alfons Geiger (University Dortmund), who was in charge of the organization of the scientific program and to Jürgen Brickmann and his group for the immaculate organization of the technical part of the program.

We hope that you will enjoy this collection of abstracts and full papers and are looking forward to welcoming you at the 11th Molecular Modelling Workshop 1997 in Darmstadt. The Molecular Modelling Workshops are organized under the auspices of the *Molecular Graphics and Modelling Society – Deutschsprachige Sektion.*

Tim Clark, Erlangen, September 16th, 1996

Molecular Dynamics without Effective Potentials via the Car-Parrinello Method

Paul A. Madden

Physical Chemistry Laboratory, South Parks Road, Oxford OX1 3QZ, United Kingdom (madden@vax.ox.ac.uk)

Abstract

It is now possible to perform molecular dynamics without the a priori introduction of a potential, by calculating the energy and forces on the nuclei at each instantaneous nuclear configuration from a variational prescription for the electronic energy, following methods introduced by Car and Parrinello. The method thus opens the way to the simulation of systems in which the prescription of an internuclear potential is difficult [1]. Such difficulties may arise from a variety of causes:- because chemical bonds are being broken and formed, because of polarization and other many-body phenomena, in the description of 'screening' in metallic systems..... The C-P method promotes a general solution to such problems by introducing an ab-initio prescription for the electron energetics, which should be applicable to all electronic phenomena. A high computational price paid for this generality, although some very significant calculations have been undertaken.

In order to circumvent the computational cost of a full ab-initio description of the electronic structure in large-scale applications of Car-Parrinello MD, we have developed simplified representations of the electronic structure in the C-P scheme which are applicable to particular physical systems. By sacrificing generality, it becomes possible to incorporate the relevant aspects of the electron energetics, for these particular systems, in an energy functionals which may be evaluated at a small fraction of the computational cost of a full ab-initio description whilst retaining the essential physics and accuracy of a first principles approach.

For ionic systems, the many-body aspects of polarization and dispersion interactions are included by adding additional degrees of freedom, which represent distortions of the electronic structure of an ion due to interionic interactions, to the
ionic coordinates and extending the equations of motion accordingly [2].

Short-range corrections to the asymptotic induction and dispersion terms are parameterized on the basis of ab-initio electronic structure calculations. The polarizable ion model reproduces distinctive features of short- and intermediate-range order in MCl_2 melts (where M is a group IIA or IIB metal) and gives global energy minimum structures in agreement with experiment for the MX_2 crystals.

For metals we use a density functional formalism which involves the use of the electron density as the basic variable and avoids the introduction of orbitals [3]. The form of the kinetic energy functional is chosen to incorporate several exact limits (uniform system, linear response and rapidly varying density) while the rest of the energy functional is exactly the same as in a Kohn-Sham calculation within the local density approximation. For metals the orbital-free scheme has particular advantages – the dynamics are stable and Brillouin-zone sampling is avoided. The electronic part of the algorithm scales linearly with system size and large simulation cells and long run times become possible. Good results for simple metals have been obtained.

In the talk, the general framework of the C-P method will be described and the process of 'tailoring' the electron energetics to the problem at hand will be illustrated with a number of applications.

References

1. *Molec. Phys.* **1990**, *70*, 921.
2. *J. Phys. Condens. Matt.* **1993**, *5*, 2687.
3. *J. Phys. Condens. Matt.* **1993**, *5*, 3221.

Dynamic AM1 Calculations of Polaron Migration in Polysilanes

Manfred Gröppel, Timothy Clark

Computer-Chemie-Centrum, Nägelsbachstr. 25, D-91052
Erlangen, Germany (wolfgang.roth@zfe.siemens.de)

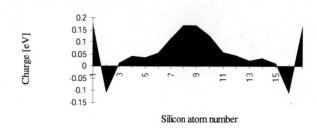

Figure 1. *Charge distribution per Si(Me)$_2$-group.*

Abstract

Polysilanes are of interest because of their unusual electronic
properties [1]. The conductivity of bulk polysilanes has been
reported to occur via hole transport [2], which can be induced by doping with strong oxidizing agents [3], electro-
oxidation [4], light [5] or heat [6]. Since there is considerable -delocalization along the backbone chain, a polysilane
molecule may be considered as a one-dimensional molecular wire, the backbone being the wire, the alkyl side groups
the insulator. We have demonstrated the ordering of
polysilanes by self-assembly on graphite. This is an experimental prerequest for future conductance measurements on
insulating substrates [7].

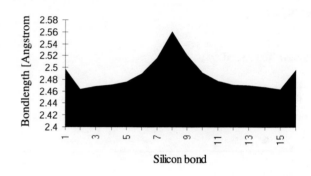

Figure 2. *Bond length distribution along the backbone chain.*

Here we focus on dynamic quantum mechanical calculations concerning the migration of an electron hole along a
single polysilane backbone. The ease of migration relates to
electrical conductivity.

The linear polymethylsilane radical cation $Si_{17}(CH_3)_{36}^{+\bullet}$
has been used in dynamic AM1 calculations to investigate
the migration of a positive charge along the backbone chain.
Both the bond length (Figure 1) and charge distributions (Figure 2) indicate polaron formation. The polarons were localized over about 6 Si-Si bonds with a significantly longer central bond. If an external field is applied to the molecule the
polaron migrates several Angstrom along the backbone. The
speed for polaron migration depend on the strength of the
electrical field (Figure 3).

Calculations were performed using the standard AM1
(Austin model 1, Dewar, 1985) method. All calculations used
the VAMP 5.6 program [8].

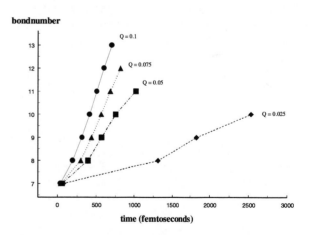

Figure 3. *Polaron-speed dependence on the electrical field-
strength.*

References

1. Miller, R.; Michl, J. *Chem. Rev.* **1989**, *89*, 1359.
2. Stolka, M.; Yuh, H.-J.; McGrane, K.; Pai, D. *J. Polym.
   Sci. Part A: Polym. Chem.* **1987**, *25*, 823.
3. Allred, A.; van Beek, D. Jr. *Polyhedron* **1991**, *10*, 1227.
4. Diaz, A.; Miller, R. *J. Electrochem. Soc.* **1985**, *132*,
   834.
5. Fujino, M. *Chem. Phys. Lett.* **1987**, *136*, 451.
6. Samuel, L.; Sanda, P.; Miller, R. *Chem. Phys. Lett.* **1989**,
   159, 227.
7. Gröppel, M.; Roth, W.; Elbel, N.; von Seggern, H. *Surf.
   Sci.* **1995**, *323*, 304.
8. Rauhut, G.; Chandrasekhar, J.; Alex, A.; Beck, B.; Sauer,
   W.; Clark, T. VAMP5.6, available from Oxford Molecular, The Magdalen Centre, Oxford Science Park,
   Sandford on Thames, Oxford OX4 4GA, United Kingdom.

J. Mol. Model. **1996**, 2

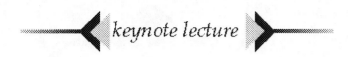

Temperature Dependence of Hydrogen Bonding in Liquid Amides

Ralf Ludwig

Physikalische Chemie, Universität Dortmund, D-44221 Dortmund, Germany
(ludwig@heineken.chemie.uni-dortmund.de)

F. Weinhold und T.C. Farrar

Department of Chemistry, University of Wisconsin-Madison, Madison, Wisconsin 53706, USA

Equilibrium and non-Equilibrium Molecular Dynamics of Complex Systems.

Soeren Toxvaerd

Department of Chemistry, H.C. Oerstedt Institute, University of Copenhagen, Denmark (tox@st.ki.ku.dk)

Abstract

Large scale molecular dynamics simulations are performed on complex systems by parallel computers. The systems investigated include a biological system with enzyme activation, open systems with chemical reactions including oscillating reactions (Volterra- Lotka) and phase separation dynamics.

Some general and technical problems concerning large scale parallel computation as well as MD of chemical reactions in open systems will be discussed.

Abstract

Temperature dependence of hydrogen bonding in neat, liquid amides is studied in a pure theoretical way. These calculations are based on standard ab initio self-consist field (SCF) methods for different molecular clusters and a quantum cluster equilibrium (QCE) model of liquids. The cluster sizes varied from one to six molecules and include linear and cyclic structures. The QCE method employs the optimized geometries, harmonic frequencies, and binding energies of each cluster to evaluate translational, rotational, vibrational and electronic partition functions by standard methods of quantum statistical thermodynamics. The condition of chemical equilibrium is used for solving the cluster populations in the canonical ensemble. The equilibrium cluster populations were determined for a large range of liquid phase temperatures. Macroscopic phase properties were then calculated by weighting the properties for each cluster by the appropriate cluster population at chosen temperatures. Quadrupole coupling constants, asymmetry parameters and geometries are sensitive probes of solution state hydrogen bonding. Therefore these properties were calculated for formamide, n-methylformamide and n-methylacetamide in the described fashion. Their temperature behaviour is in good agreement with results from NMR relaxation time and diffraction experiments.

Molecular Dynamics Simulation of the Proton Transport in Water

Roger G. Schmidt, Jürgen Brickmann

Technische Hochschule Darmstadt, Institut für Physikalische Chemie I, Petersenstr. 20, D-64287 Darmstadt, Germany (rgs@pc.chemie.th-darmstadt.de)

Abstract

The results of molecular dynamics simulations of a rigid hydronium ion $(H_3O)^+$ in liquid water are reported. The water molecules are represented by the TIP3P-model, the ion has a pyramidal structure.

From NVE simulations the equilibrium spatial and temporal distributions of the first solvation shell water molecules to the hydronium ion were ascertained. It could be shown that a nearest neighbor water molecule is most probably found within a distance of 2.6 Å between the oxygen atom of the H_3O^+-ion and the oxygen atom of the H_2O-molecule.

A two dimensional potential energy surface of the proton transfer within the $H_5O_2^+$-complex was calculated at the ab initio-level. The two coordinates of the surface were chosen to be the decisive degrees of freedom for proton transfer reactions, the oxygen-oxygen distance R_{O-O} and the position of the transfering proton. Calculation of the energy levels, the position distribution and the temporal behaviour of the proton in the double minimum potential surface together with the results of the equilibrium distributions lead to the following conclusions:

i) The proton resides in a flat double minimum potential with low barrier inbetween the oxygen atoms. Consequently, the proton transfer in water is mainly an adiabatic process with no tunneling involved.

ii) The reaction coordinate is found in the solvens coordinates rather than in the quantum coordinate of the proton. This means that the rate limiting step of the proton mobility is the association to and dissociation of a $H_5O_2^+$-complex.

On account of these findings we propose a model for treating the proton transfer process by means of a mixed MD/MC algorithm. The hydrodynamic single particle diffusion is integrated within the framework of a classical molecular dynamics simulation, the quantum mechanical proton transfer along a O—H⋯O hydrogen bond is performed by a proton jump between the donating hydronium ion and an adjacent water molecule. After parametrization the model is able to correctly simulate the diffusion coefficient of the proton in water.

pH-Dependent Protein Stability Calculation of Absolute Free Energy Differences

Michael Schaefer, Michael Sommer, Martin Karplus

Université Louis Pasteur, Institut le Bel, Laboratoire de Chimie Biophysique, 4, rue Blaise Pascal, 67000 Strasbourg (schaefer@brel.u-strasbg.fr)

Abstract

A method for calculating the absolute electrostatic free energy of a titrating system as a function of pH is proposed. Based on the theory of linked functions, the free energy is calculated by integration of the titration curve of the system. The charge-state dependent electrostatic free energy of the system is calculated using the Finite-Difference Poisson-Boltzmann program UHBD [1]. For the calculation of the titration curve, a Monte Carlo program by Beroza et al. [2] is employed.

The new approach differs from the known methods of calculating the relative pH-stability of proteins by the use of pH= ∞ as the reference for the titration curve integration, where the energy is uniquely defined by the electrostatic free energy of the unprotonated state. It thus allows for the calculation of absolute free energy differences between two conformers as a function of pH, e.g., the native and denatured states of a protein or the bound and unbound states of a protein-ligand system.

The new method is applied to the protein hen egg-white lysozyme. To account implicitly for conformational flexibility, a dielectric constant of 20 is assigned to the protein interior. The relative pH-stability of lysozyme is calculated using an extended b-structure and the "Null" model of non-interacting sites as the unfolded reference state, finding good quantitative agreement with experiment. For the extended b-structure, a net free energy difference of -1.5 kcal/mol at pH=7 is found relative to the Null model, indicating that there are significant interactions between titrating sites in the unfolded state. Good agreement between theory and experiment is shown for the pH-stability of pepsin, the dimerization of HIV-protease, and the stability of foot-and-mouth disease virus capsid.

References

1. Davis et al. *Comput. Phys. Comm.* **1991**, *62*, 187.
2. Beroza et al. *Proc. Natl. Acad. Sci. USA* **1991**, *88*, 5804.

J. Mol. Model. **1996**, 2

Molecular Dynamics Simulations of Polypeptide Chains in Internal Coordinates

Gerald R. Kneller and Konrad Hinsen,

Institut für Theoretische Physik, RWTH Aachen
(g.kneller@kfa-jülich.de)

Abstract

Using a new method for MD simulations of partially rigid macromolecules we study the influence of geometrical constraints on the dynamics of 16-polyalanin. In our approach a macromolecule is modelled as a chain of topologically linked rigid bodies where two rigid bodies can be connected in a common point or a common axis. A rigid body can have three moments of inertia or two, in case of a linear rigid body. The orientations of the rigid units are described by quaternions. This approach is well established in robot mechanics and has also been applied in MD simulations of molecular liquids consisting of completely rigid molecules. The main result of our study is that angle constraints may be imposed as long as the binding geometry of the C_{alpha} carbons is left flexible. Keeping e.g. the peptide planes rigid does not alter the essential dynamics of our model system. However, simulations in torsional angle space ('Phi-Psi-angles') lead to an unacceptable rigidification of the polypeptide backbone.

Stokesian Dynamics Simulations of Sedimenting Spheres

Gerald R. Kneller and Konrad Hinsen,

Institut für Theoretische Physik, RWTH Aachen
(g.kneller@kfa-jülich.de)

Abstract

The dynamics of particles which move in a highly viscous liquid such that all motions are overdamped and random forces can be neglected is called Stokesian Dynamics. The simplest example is a macroscopic sphere which is pushed by a constant force through a viscous medium. According to the famous Stokes law the resulting velocity is proportional to the driving force. To describe the dynamics of many-particle systems one needs to take into account hydrodynamic interactions (HI) which are mediated by the liquid. Mathematically they are described by the friction matrix which replaces the friction constant appearing in the Stokes law for a single sphere. HI are long-ranged many-body interactions. Here we present Stokesian Dynamics (SD) simulations of clusters of spheres which sediment in a viscous liquid. The friction matrix is computed by using a force multipole approach developed recently. Since the friction matrix becomes singular for particles at contact the integration of the equations of motion requires a careful handling of close contacts. To be able to describe geometrically constrained assemblies, like e.g. large multidomain proteins, we develop schemes to compute friction and mobility matrices of such constrained systems. Depending on the applied forces these matrices can be used as well for Brownian Dynamics as for Stokesian dynamics simulations.

References

1.   Kneller, G.R.; Hinsen, K. *J.Mol.Model.* **1996**, 2, 227.

Best Paper Award

Modelling of tripod Metal Compounds $RCH_2C(CH_2PR´R´´)_3ML_n$: Optimisation of Force Field Parameters by Genetic Algorithms

Johannes Hunger, Stefan Beyreuther and Gottfried Huttner

Anorganisch-Chemisches Institut, Universität Heidelberg, INF 270, D-69120 Heidelberg
(johannes.hunger@indi.aci.uni-heidelberg.de)

Abstract

The method of force field calculations has been established nearly exclusively on the basis of experimental data derived from organic compounds. In the last decade its application has been expanded to the modelling of inorganic compounds. This lead to the problem of lacking parameters, which describe the interactions of the metal center. Apart from the possibility of a more or less trial and error guess, no real method for the empirical developement of those parameters exists. Genetic algorithms [1] were found to be an efficient tool to develop missing parameters in a way highly superior to manual fitting. Often applied in Molecular Modelling for conformational search[2] they are, as far as we can see, the first time used to find local minima on the parameter hyper surface. This surface describes the caclulated potential energy of one and the same conformation varying the force field parameters. Applying an evolutionary algorithm in this way it is possible to generate a subset of selected parameters on the basis of a given set of solid state structures (Figure 1).

This is illustrated by the example of a subclass of metal complexes formed by *tripod* ligands of the type $RCH_2C(CH_2PR´R´´)_3$. A statistical analysis of the solid state conformations of *tripod* metal compounds $CH_3C(CH_2PPh_2)_3ML_n (n=2,3)$ demonstrates that the conformations observed for these templates are dominated by the inner forces while the influence of packing forces is not determining [3]. A basic condition to model compounds of the type *tripod* ML_n (n=2,3) with Molecular Mechanics on the basis of solid state structures is thus given.

One way to reduce the multiplicity of possible solutions, given by the mathematical nature of the optimisation problem, is to expand the number of underlying solid state structures. In addition this can be achieved by reference to NOE-Data.

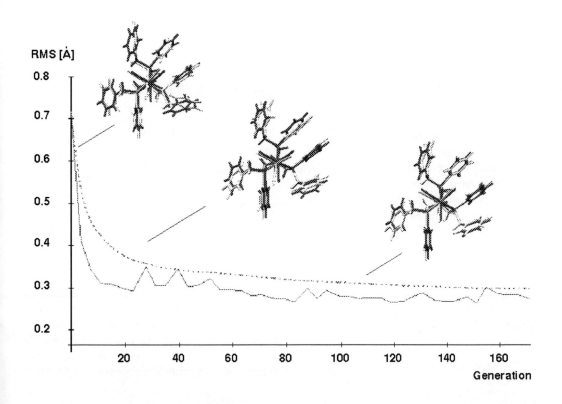

References

1. Holland, J. *Adaption in Natural and Artificial Systems*, University of Michigan **1975**; Goldberg, D. E. *Genetic Algorithms in Search, Optimization & Machine Learning*, Addison Wesley 1989
2. Brodmeier, T.; Pretsch, E. *J. Comp. Chem.* **1993**, *15*, 588.; McGarrah, D.B.; Judson, R.S. *J. Comp. Chem.* **1993**, *14*, 1387.
3. Beyreuther, S.; Hunger, J.; Huttner, G.; Mann, S.; Zsolnai, L. *Chem. Berichte* **1996**, in press.

GROMOS 96

Peter Krüger and Wilfred F. van Gunsteren

Laboratory of Physical Chemistry, Swiss Federal Institute of Technology Zurich (ETHZ) ETH Zentrum, Universitätstrasse 6, CH 8092 Zürich, Switzerland (peter@igc.phys.chem.ethz.ch)

Abstract

The program package GROMOS has been used for many years as a versatile tool to perform energy minimization, molecular and stochastic dynamics simulations including free energy calculations, distance and position restraint simulations (e.g. for NMR refinement, model-building). Most applications were in the field of biomolecular systems, but also simulations of crystals, of small molecules or of pure solvent were performed. The new version of GROMOS, called GROMOS 96, is an essentially rewritten version of GROMOS 87. It uses standard FORTRAN 77 and is therefore easy to install on machines, ranging from a PC to a supercomputer. Concerning the handling of the program a number of improvements have been made, as for instance the input/output files being now more flexible and therefore easier to read, to commentor to change. Besides an improved force field, a number of new functions were added. The main aspects are:

- Soft core potential for the non-bonded interactions
- Molecular dynamics in 4 dimensions (extended conformational search, e.g. NMR refinement, loop search etc.)
- Free energy calculation of a system simulated in 4 dimensions
- Path integral calculations
- Inclusion of a reaction field
- Time-averaged atom-atom distance and J-value restraining
- Local elevation search
- Position constraining
- Separate temperature coupling to different parts of a solute molecule
- SHAKE free energy contributions

A number of new buildings blocks (defining the topology of molecular segments) have been added and the aromatic rings of amino acid residues contain hydrogen atoms. The following solvents are available, i.e. defined in the topology and interaction function parameters set: H_2O, CH_3OH, $CHCl_3$, DMSO, CCl_4. As its previous version, GROMOS 96 will be available to academic users for a price mainly covering the processing costs. This version will be delivered with a manual, not only describing the details how to handle the input, file formats etc., but also explaining the theoretical background of the simulation technique itself and the GROMOS specific extensions.

The Puzzling Forces and Energies Occuring at Simulated Conformational Transitions

Thomas Mülders, Peter Krüger

Institut für Biochemie, Rheinisch-Westfälische Technische Hochschule Aachen, D-52057 Aachen, Germany

Wolfgang Swegat and Jürgen Schlitter

Lehrstuhl für Biophysik, Ruhr-Universität Bochum, D-44780 Bochum, Germany
(jürgen@bph.ruhr-uni-bochum.de)

Abstract

The simulation of conformational transitions in macromolecules ranging from local events to large scale changes like refolding or unfolding is a rather new field of growing interest. The common goal of all techniques applied is the search for realistic pathways including barriers and intermediates and their energetic characterization. The quantative comparison with experimental equilibrium constants and rates requires determination of free energy profiles. With a new technique also force profiles can be measured.

Like for small molecules, approximative minimum energy pathways (MEP) are sometimes determined by energy minimization (EM) to give a first idea of the actual pathway. They are characterized by the profile of minimum potential energy E_{pot}, U_o. For several reasons molecular dynamics (MD) simulations [1,2] are superior to energy minimization or even essential [3,4]. They yield broad pathways with large fluctuating energies E_{tot}, and show that the underlying true MEP has only little meaning. We shall show, however, that its energy profile U_o can be determined in a simple way by means of the Lagrangian L as $U_o \sim <-L> = <E_{pot}> - <E_{kin}>$. The expression has proved to be a good estimate which never underestimates the true value U_o [5]. By comparison of U_o as obtained by EM and MD, it is shown that the latter method caused an enormous progress in the search for favourable pathways.

The free energy profile is identical with the potential of mean force. It is therefore obtained by thermodynamical integration of a mean force $<F>$ over the reaction coordinate, which usually is a geometric variable like, for instance, the euclidian distance from the target configuration [5]. We have proved that this force is identical **in the mean** with the negative constraint force, $<F> = - <F_{constraint}>$, that must be applied in order to keep a given value of the reaction coordinate. This finding opens a practicable way for the numerical evaluation. Moreover, the fluctuations of the force can be minimized by using mass-weighted coordinates in a suitable way [6].

Recently MD simulations were successfully applied to determine force profiles and rupture forces which had been measured before by single molecule atomic force microscope experiments [7]. It is discussed why these forces are different from those mentioned above and are not related to free energy.

References

1. Jacoby, E.; Krüger, P.; Schlitter, J.; Wollmer, A. *Prot.Engin.* **1995**, in press.
2. Wroblowski, B.; Diaz, F.; Schlitter, J.; Engelborghs, Y. **1995**, in preparation.
3. Engels, M.; Jacoby, E.; Krüger, P.; Schlitter, J.; Wollmer, A. *Prot.Engin.* **1992**, *5*, 669.
4. Czerminski, R.;Elber, R. *Int.J.Quantum Chem.* **1990**, *24*, 167.
5. Schlitter, J.; Engels, M.; Krüger, P.; Jacoby, E.;Wollmer, A. *Mol.Simul.* **1993**, *10*, 291.
6. Mülders, T.; Krüger, P.; Swegat, W.; Schlitter, J. *J.Chem.Phys.* **1996**, *104*, 4869.
7. Grubmüller, H.; Heymann, B.; Tavan, P. *Science* **1996**, *271*, 997.

A Hybrid-Approach for Flexible 3D-Database Searching

Jens Sadowski

Bayer-AG, PH-R SR, Aprather Weg, D-42096 Wuppertal, Germany (ebayspk@ibmmail.com)

Abstract

A combined method for flexible 3D-database searching is presented which addresses the problem of ring flexibility. The proposed hybrid-approach combines the explicit storage of ring conformations generated by the 3D-structure generator CORINA [1] with the strength of a flexible search technique which is implemented in the 3D-database system UNITY [2]. A comparison of the hybrid approach with the original UNITY-approach was performed by using a public domain database of about 130,000 molecular structures [3].

Five pharmacophor queries taken from the literature were searched following the hybrid-approach and the original method. The hybrid approach gained on average 10-20% more hits than the original UNITY-method.

In addition, specific problems with unrealistic hit geometries produced by the original approach could be excluded.

References

1. Gasteiger, J.; Rudolph, C. and Sadowski, J. *Tetrahedron Comput. Method.* **1990**, *3*, 537; Sadowski, J. and Gasteiger, J. *Chem. Rev.* **1993**, *93*, 2567; Sadowski, J.; Gasteiger, J. and Klebe, G. *J. Chem. Inf. Comput. Sci.* **1994**, *34*, 1000; CORINA available from Oxford Molecular Ltd., Magdalen Centre, Oxford Science Park, Sandford-on-Thames, Oxford, OX4 4GA, England.
2. UNITY Version 2.4, Tripos Ass., St. Louis, MO.
3. Milne, G. W. A.; Nicklaus, M. C.; Driscoll, J. S.; Wang, S. and Zaharevitz, D. W. *J. Chem. Inf. Comput. Sci.* **1994**, *34*, 1219.

J. Mol. Model. **1996**, *2*

Free Energies for Active Site Hydration and Substrate Binding to Cytochrome P450cam Computed by Molecular Dynamics Simulation

Volkhard Helms and Rebecca Wade,

EMBL, Meyerhofstr. 1, D-69012 Heidelberg, Germany
(helms@embl-heidelberg.de)

Abstract

Solvent makes an important contribution to the thermodynamics of protein- ligand interactions. The active sites of proteins usually contain water molecules which are displaced into bulk solution when ligands bind and make a favourable entropic contribution to ligand binding. In order to calculate substrate binding free energies by computer simulation it is essential to account for the solvent. Yet, when a ligand binds on the surface of a protein, it is often difficult to identify the solvent molecules involved in the binding process. Cytochrome P450cam is therefore an ideal model system because the natural substrate, camphor, binds in a buried active site and the protein is experimentally well-characterized. Six water molecules were assigned to the electron density in the crystal structure of the ligand-free form [1]. Five of these cluster in one electron density blob and their actual number is not well defined. No water molecules were detected in the active site of the complex with camphor [2]. In order to calculate the absolute binding free energy of camphor, it is necessary to know the exact number of water molecules in the unbound state. We calculated the free energy differences for having six, seven, or eight water molecules in the active site from molecular dynamics simulations by thermodynamic integration employing a 3-step perturbation scheme. The energetic differences between the different degrees of hydration were relatively small (within 10 kJ/mol). In agreement with the crystallographic determination, we calculated that six water molecules is the most favourable arrangement. In further simulations, the six water molecules were exchanged with camphor, resulting in a computed absolute binding free energy of camphor within 5 kJ/mol of the experimental value.

References

1. Poulos, T.L.; Finzel, B.C.; Howard, A.J. *Biochemistry* **1986**, *25*, 5314.
2. Helms, V.; Wade, R.C. *Biophys.J.***1995**, *69*, 810.

Correct Ranking of the Carboxypeptidase A Inhibition Strength using mixed QM/MM calculations

Harald Lanig, Bernd Beck and Timothy Clark

Computer Chemie Centrum - Institut für Organische Chemie I, Universität Erlangen-Nürnberg, Nägelsbachstr. 25, D-91052 Erlangen, Germany
(lanig@organik.uni-erlangen.de)

Abstract

A combined QM/MM approach has been used to dock a series of seven flexible molecules into the active site of Carboxypeptidase A (CPA). In order to estimate the ranking of their inhibitor strengths (K_i), the difference between the heats of formation within the protein environment and a water cage was used. The starting structures for this investigation are the X-ray geometries of the corresponding protein/inhibitor complexes. The ligands were fully optimized within the fixed environment using our point charge model (PCM) including electrostatic polarisation and van der Waals interactions.

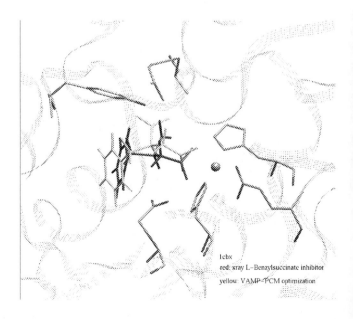

Figure 1. *X-ray structure of the active site of Carboxypeptidase A in complex with L-Benzylsuccinate inhibitor (red); VAMP-PCM optimized inhibitor (yellow).*

The figure shows an overlay of the X-ray structure of L-Benzylsuccinate (red) with the ligand calculated by VAMP-PCM (yellow).

The solvent calculations used a supermolecule approach including 28 water molecules around the inhibitor to obtain the heat of formation within the solvent. Using this approach we can reproduce the experimental inhibition strength ranking of the set of CPA ligands correctly. The following table shows the relative binding energies [kcal/mol] compared to experimental K_i values. To compare the orientation/conformation of the ligands after geometry optimisation with experimental data, the RMS deviation of the coordinates in relation to the corresponding X-ray structures is given (only heavy atoms are included).

system	binding energy [kcal/mol]	K_i [M]	RMS [Å]	Ref.
7cpa-fvf	-176.89	1.1E-14	1.65	1
6cpa-zaf	-171.88	3.0E-12	2.34	1
8cpa-agf	-144.85	7.1E-10	3.79	1
1cpa-cpm	-123.12	2.2E-7	1.29	2
1cbx-bzs	-107.15	4.5E-7	1.16	3
2ctc-lof	1.43	–	1.27	–
3cpa-gy	191.91	1.0E-4	1.58	4

References

1. Kim, H. and Lipscomb, W. N. *Biochemistry* **1991**, *30*, 8171.
2. Cappalonga, A. M.; Alexander, R. S. and Christianson, D. W. *J. Biol. Chem.* **1992**, *267*, 19192.
3. Mangani, S.; Carloni, P. and Orioli, P. *J. Mol. Biol.* **1992**, *223*, 573.
4. Bunting, J. and Myers, C. *Can. J. Chem.* **1975**, *53*, 1993.

Stabilities of and Hydrogen Shifts in Oxidized DNA Base-Pairs: *ab initio* and DFT Calculations

Michael Hutter and Timothy Clark

Computer-Chemie-Centrum, Institut für Organische Chemie, Friedrich-Alexander-Universität Erlangen-Nürnberg, Nägelsbachstraße 25, D-91052 Erlangen, Germany (hutter@organik.uni-erlangen.de)

Abstract

Ab initio (UHF/6-31G*) and density functional (Becke3LYP/D95*) calculations have been used to investigate the structures and stabilities of radical cations of the DNA base-pairs as well as possible hydrogen shifts. The calculated structures of the corresponding neutral base-pairs show excellent agreement with crystallographic data. A particularly stable radical cation is formed by guanine (which has been shown to be the most easily oxidizable base) together with cytosine, so that the calculated adiabatic ionization potential for the guanine-cytosine hydrogen-bonded complex is about 0.75 eV lower than that of guanine itself. Isodesmic reactions show that the extra stabilization enjoyed by the base-pair radical cation relative to the uncomplexed base radical cation is about 7 kcal·mol^{-1} compared to AT$^{+·}$. Calculations at UBecke3LYP/D95*//UHF/6-31G* show that shift of the central hydrogen-bonded proton at N1 of guanine to N3 of cytosine is only slightly endothermic (+1.6 kcal·mol^{-1}), while the product of the corresponding proton shift in the adenine-thymine system is unfavorable by +14.1 kcal·mol^{-1}. The two other proton exchanges in the guanine-cytosine system also result in higher energies (+8.5 and +30.7 kcal·mol^{-1}). Thus GC$^{+·}$ can be considered as a resonance structure of two tautomers with rapid proton exchange along the central low-barrier hydrogen bond. These results suggest that the guanine-cytosine radical cation represents even more of a thermodynamic sink in oxidized DNA than might be concluded from the ionization potentials of the individual bases.

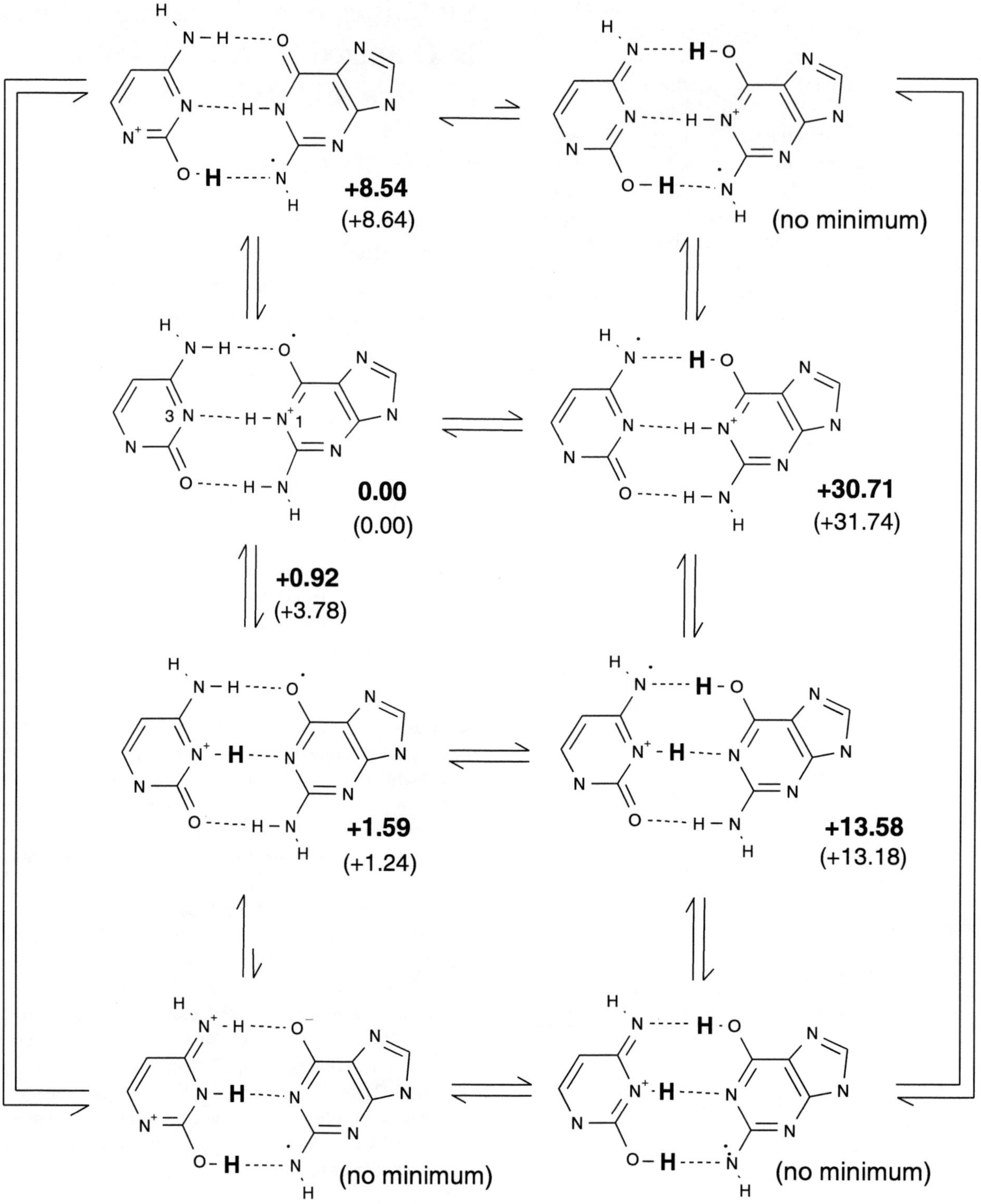

Scheme 1. *Energies of the guanine-cytosine radical cation tautomers in kcal/mol. Shifted protons are marked* **bold.**
values in brackets: UBecke3LYP/D95*//UHF/6-31G*
values in **bold***: UHF/6-31G* (incl. zero point energies)*

J. Mol. Model. **1996**, 2

Comparison of Molecular Dynamics and Energy Minimization Techniques for Pathway Determination

E. Jacoby, P. Krüger, A. Wollmer

Institut für Biochemie, Rheinisch-Westfälische Technische Hochschule Aachen, D-52057 Aachen, Germany

B. Wroblowski, F. Diaz, Y. Engelborghs

Laboratorium voor Chemische en Biologische Dynamik, Katholieke Universiteit Leuven, B-3001 Leuven, Belgium

Jürgen Schlitter

Lehrstuhl für Biophysik, Ruhr-Universität Bochum, Bochum, D-44780 Germany

(jürgen@bph.ruhr-uni-bochum.de)

Abstract

The determination of realistic pathways for conformational transitions in proteins ranging from local changes to complete unfolding is a rather new challenging task in the field of molecular simulations. As such transitions are rare events hindered by considerable barriers they are hardly seen in MD simulations except at extremely high temperature. Recently several techniques have been developped which are based either on energy minimization or modified MD simulation. Here we report studies of the extended T-R transition in the insulin hexamer and a conformational transition in alpha-chymotrypsin with both kinds of techniques. Insulin [1] was investigated using the targeted energy minimization (TEM) method [2], chymotrypsin [3] using the self penalty walk (SPW) method [4]. These techniques are based on energy minimization with different protocols for generating a pathway from a given starting structure to an given target structure. The same transitions were also investigated using targeted molecular dynamics simulation (TMD) [5,6] which differs from usual MD in that the distance from the target structure is a constraint variable which is continuously diminished during the simulation. The main features of TMD pathways are as follows:

(i) The deviation from the ficticious linear pathway is larger than with SPW. Similarity with the SPW path occurs near both ends.

(ii) They exhibit a less smooth course although they are quasi continuous with a large number of calculated points (some 10^5)

(iii) In contrast to TEM, symmetry is not maintainted on the way between the symmetric end states in the case of insulin.

(iv) Nonproductive pathways can be identified.

(v) The energies lie about 1 Megajoule/Mole below those found for energy minimization pathways.

(vi) The mean energy decreases with increasing simulation temperature.

The favourable energy behaviour has to be attributed to the fact that MD at room temperature can cross many of the numerous small barriers between conformational substates which are known to exist in proteins [7]. TMD pathways should not be interpreted as representative trajectories since they exhibit oscillations in a rather broad groove connecting the end states with intermediates. Therefore a thermostatistical evaluation, in particular a determination of free energy seems to be more appropriate, which is now in progress.

References

1. Jacoby, E.; Krüger, P.; Schlitter, J.; Wollmer, A. *Prot.Engin.* **1995** in press.
2. Engels, M.; Jacoby, E.; Krüger, P.; Schlitter, J.; Wollmer, A. *Prot.Engin.* **1992**, *5*, 669.
3. Wroblowski, B.; Diaz, F.; Schlitter, J.; Engelborghs, Y. **1995** in preparation.
4. Czerminski, R.; Elber, R. *Int.J.Quantum Chem.* **1990**, *24*, 167.
5. Schlitter, J.; Engels, M.; Krüger, P.; Jacoby, E.; Wollmer, A. *Mol.Simul.* **1993**, *10*, 291.
6. Schlitter, J.; Engels, M.; Krüger, P. *J.Mol.Graphics* **1994**, *12*, 84.
7. Elber, R.; Karplus, M. *Science* **1987**, *235*, 318.

J. Mol. Model. **1996**, *2*

Development of a Force Field for N-Oxides

Kristine M. Kast [a] and Jürgen Brickmann [a, b]

[a] Department of Physical Chemistry, Darmstadt University of Technology, Petersenstr.20, D-64287 Darmstadt, Germany (wetzel@pc.chemie.th-darmstadt.de)

[b] Darmstadt Center of Scientific Computing, Germany (brick@pc.chemie.th-darmstadt.de)

Abstract

For the development of a new set of force field parameters based on the CHARMM22 force field, *ab initio* Hartree-Fock and post-Hartree-Fock calculations at the MP2 level were carried out for pure compounds, mono- and dihydrates of aliphatic N-oxides. Structural information of the N-oxides and the position of the water molecules in the hydrates was obtained using the SCF/6-31G** basis set. Charges and dipole moments were calculated on the same level of theory.[1]

Coulomb interactions are described using "potential derived" (pd) charges, which represent the high dipole moment resulting from the wavefunction better then the charges derived from Mulliken population analysis. The Lennard-Jones parameters describing the non-bonded interactions were obtained by a simulated annealing fit [2] to the optimized structures of two N-oxide dihydrates.

Due to the high dipole moments of the N-O bonds, large H-bond energies were determined by single point *ab initio* calculations at MP2/6-311+G(2d,2p) level.1 The achieved parameters were able to reproduce the H-bond energies and geometries of optimized gas phase structures. Crystal parameters using rigid molecules are also in good agreement with experimental data.

The intramolecular force parameters were fitted to vibrational and structural data obtained from *ab initio* calculations by again using the simulated annealing approach.[3]

References

1. Kast, K. M.; Reiling, S. and Brickmann, J. in preparation.
2. Kast, S. M. and Berry, R. S. in preparation.
3. Kast, K. M.; Brickmann, J.; Kast, S. M. and Berry, R. S. in preparation.

Simulation of Small Polar Molecules on Oxide Surfaces

Dietmar Paschek and Alfons Geiger

University of Dortmund, D - 44221 Dortmund, Germany (pas@heineken.chemie.uni-dortmund.de)

Abstract

A realistic model of molecularly absorbed ammonia on the (110) surface of TiO_2 (rutile) has been studied by means of Molecular Dynamics. The simulations were carried out within a temperature range between 100 K and 450 K. The complex behaviour of the translational dynamics was attributed to the specific structure of the substrate and has been characterized as an "exchange reaction" between two distinct adsorbed states on the surface. The high lateral mobility of weakly bound 2nd layer molecules leads to rapid isotropic diffusion on the surface of micro-crystalline powders and thus serves as an explanation for the experimentally obtained singulett peak in ^{1}H-NMR spectra.

Unique Optimum Superposition of Molecules

Günter Voll,

BCMP — BioComputing & Mathematical Physics, Friedrichstr. 57, D-90408 Nürnberg, Germany

Abstract

A physically unique optimum superposition of equally large clusters of atoms or, more generally, of any data points in some d-dimensional real space can be obtained naturally in the set of inhomogeneous linear mappings. We define a quantitative measure for the similarity of (shapes of) clusters modulo linear distortions.

A Systematic Study on the Effect of Different Quantum Mechanical Methods in Combined QM/MM Potentials

Iris Antes and Walter Thiel

Organisch-Chemisches Institut, Universität Zürich, Winterthurerstr. 190, CH-8057 Zürich, Switzerland (antes@ocisgi14.unizh.ch)

Abstract

A program has recently been developed in our group which allows us to do combined QM/MM studies using semi-empirical, *ab initio* and DFT methods for the quantum mechanical part. It has a modular structure and includes an interface between the quantum mechanical programs MNDO (semiempirical methods) and CADPAC (*ab initio* and DFT methods) and the molecular simulation program AMBER. In addition to the standard electrostatic coupling model widely used [1], a classical treatment of the polarization of the molecular mechanics region by the electric field of the quantum mechanical subunit has been implemented [2-4]. To partition a molecule into a small region which can be studied quantum mechanically and a larger part which is treated in a classical way we use the concept of link atoms [1].

Calculations have been done on small organic test molecules, using the different quantum mechanical methods available in the program. We compare the effects of these methods on the combined QM/MM potential and discuss the influence of the different coupling models, the link atom treatment, and the MM charges.

References

1. Field, M. J.; Bash, P. A. and Karplus, M. *J. Comp. Chem.* **1990**, *11*, 700.
2. Thole, B. T. *Chem. Phys.* **1981**, *59*, 341.
3. Bakowies, D. Dissertation, Universität Zürich, 1994
4. Bakowies, D. and Thiel, W. *J. Phys. Chem.*, in press.

Ab Initio Calculations on Propanediol Systems

Stephan Hobohm and Rudolf Friedemann

Department of Organic Chemistry, Martin-Luther University Halle-Wittenberg Kurt-Mothes-Str. 2, D-06120 Halle (Saale) (friedemann@chemie.uni-halle.de)

Abstract

The structural and energetic properties of monomers, dimers and monohydrates of 1,2-propanediols (PDL1,2) and 1,3-propanediols (PDL1,3) were investigated in a comparative way within the program package TURBOMOLE [1]. Moreover, conformational studies on perfluoric PDL1,2 and PDL1,3 molecules were taken into account.

The *ab initio* calculations were performed on the 6-31G level. The aim of these investigations was to study the influence of intramolecular and intermolecular hydrogen bonding on the structure and stability of such diols in a systematic way. By calculations on the dimers of the chiral PDL1,2 it was tested if there is a distinct stabilization of $(PDL1,2)_2$ formed by the same (R-R) and different (R-S) enantiomers, respectively.

In the case of the perfluoric diols the influence of alternative hydrogen bonds under participation of fluorine atoms on the conformational behavior was investigated in more detail.

Finally, the *ab initio* results on the perfluoric diols were used as reference data in order to adjust fluorine parameters in force field methods.

Up to now *ab initio* calculations [2,3] and experimental studies [4] on such diols were mainly performed to investigate the hydrogen bonding of the isolated molecules. Our results on PDL1,2 and PDL1,3 monomers are compared with theoretical data and IR spectroscopic findings of other authors.

References

1. Ahlrichs, R. et al., Biosym/MSI Insight Environment and TURBOMOLE95, 1995.
2. Vázquez, S.; Mosquera, R. A.; Rios, M. A. and van Alsenoy, C. *J. Mol. Struct. (Theochem)* **1988**, *181*, 149 and **1989**, *184*, 323.
3 Bultinck, P.; Goeminne, A.; Van de Vondel, D. *J. Mol. Struct. (Theochem)* **1995**, *357*, 19.
4. Morantz, D. J. and Waite, M. S. *Spectrochim. Acta* **1971**, *A 27*, 1133.

J. Mol. Model. **1996,** *2*

Molecular Dynamics Simulations on Chiral Biamphiphilic Tetraol Clusters

Annett Fengler, Stefan Naumann, Rudolf Friedemann

Department of Chemistry, Martin Luther University Halle-Wittenberg D-06120 Halle(Saale), Germany (friedemann@chemie.uni-halle.de)

Abstract

Investigations on the structure, stability and dynamics of biamphiphilic 1,2,7,8-octanetetraol (OTL) as well as 1,2,15,16-hexadecanetetraol (HDTL) were performed within MD simulations using the program package GROMOS87 [1].

These bolaamphiphilic molecules represent a new class of recently discovered amphotropic liquid crystals [2,3] which are also interesting as model systems for the study of molecular self-organization in bilayers. First results of MD simulations on clusters up to 64 tetraol molecules in the gas phase are presented. Especially, the role of intramolecular and intermolecular hydrogen bonding as well as the function of the bolaamphiphilic head groups on the process of association were studied in more detail.

Moreover, by structural modifications in the hydrophobic part of the monomers, e. g. regarding phenyl and cyclohexyl rings as well as olefinic bonds, the formation of the energetically preferred arrangements of the cluster were investigated in a systematic way. Clusters with HDTL molecules as monomers were taken into account in order to study the hydrophobic-hydrophilic balance on the molecular self-organization.

Comparing MD simulations on (OTL)16 and (HDTL)16 clusters in aqueous solution are presented. For the visualization of the molecular dynamics results a graphics tool was created on workstations. By this tool the results of the MD runs are illustrated and analyzed in a useful way. Especially, from trajectories of the distances of the chiral carbon atoms hints on the flexibility of the hydrophobic chain were obtained.

References

1.   van Gunsteren, W. F.; Berendsen, H. J. C. *GROMOS* Library Manual, Biomos b.v., Groningen, 1987.
2.   Ringsdorf, H.; Schlarb, B.; Venzmer, J. *Angew. Chemie* **1988,** *100,* 117.
3.   Hentrich, F.; Tschierske, C.; Zaschke, H. *Angew. Chemie* **1991,** *103,* 429.

A Simple Technique for the 3-D Projection of Moved Stereoscopic Images

Stefan Naumann and Rudolf Friedemann

Department of Organic Chemistry, Martin-Luther-University Halle-Wittenberg Kurt-Mothes-Str. 2, D-06120 Halle (Saale), Germany (friedemann@chemie.uni-halle.de)

Abstract

A hard- and software development for the 3-D projection of moved stereoscopic images is presented. By an optical equipment the light from a video projector is orientated through a mirror construction as well as two polarization filters and projected on a screen. Using simple polarization glasses the 3-D projections can be viewed. In principle, the new optical equipment can be attached to any commercial video projector.

The stereoscopic image-pairs, generated by a corresponding graphics tool, can be projected both on-line from a graphics workstation and a video cartridge via the video projector. The hard- and software development enables a facil access for the projection of stereoscopic images and the 3-D viewing of moved pictures as e.g. the dynamics of molecular systems. This is of interest especially for the simulation and visualization of molecular processes in chemistry and biochemistry.

Moreover, by extentions of the graphics software any objects also under simulation of a changed viewer position can be generated as 3-D projections.

Calculation of ¹³C NMR Shifts for the Analysis of Relative Configuration in Polyketide Natural Products

Martin Stahl, Ulrich Schopfer, Gernot Frenking, Reinhard W. Hoffmann

Fachbereich Chemie, Philipps-Universität Marburg, D-35032 Marburg, Germany
(stahl@spock.chemie.uni-marburg.de)

Abstract

¹³C NMR shifts can vary strongly for different low-energy conformations as well as for different diastereomers of flexible hydrocarbon compounds [1]. The accurate calculation of ¹³C shifts can therefore be a valuable tool to analyze relative configuration, if the conformational behaviour can be modeled reliably. However, in order to be applied routinely in organic chemistry, the computational methods involved must be quite efficient.

This goal is met in a three-step procedure: Molecular mechanics calculations are used for conformational analyses (MACROMODEL 4.5[2], MM3(94)[3]) and relative energies of the conformers, chemical shifts are calculated by a density functional method on the force field geometries (SOS-DFTP/IGLO in deMon/Master[4,5]) and the shifts are then weighted according to a Boltzmann distribution. This com-

putational procedure is tested with a series of 1,3-dimethylated hydrocarbon compounds, which occurs often as segments in polyketide natural products. This yields an excellent agreement between calculated and experimental shifts. The method is then applied to elucidate the relative stereochemistry in the side chain of sambutoxine **1** and the bradykinin inhibitor **2**. It is shown that the methyl groups have relative *anti* configuration in compound **1** and *syn* configuration in **2**.

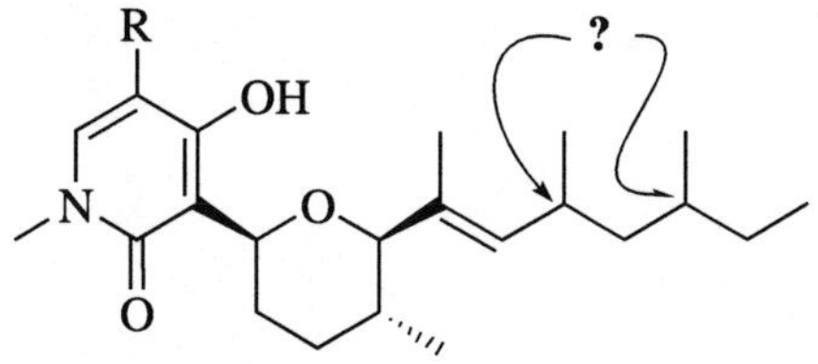

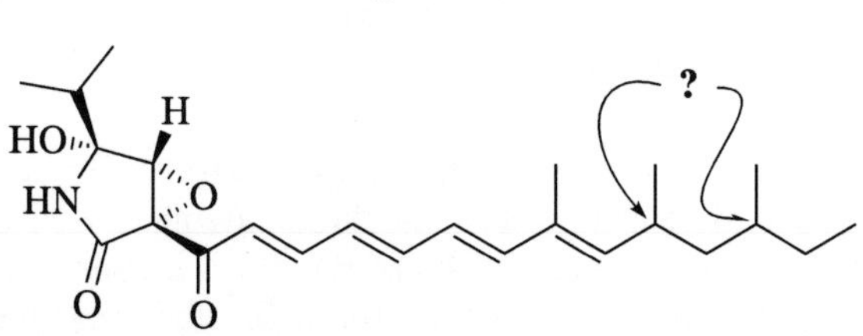

References

1. Hildebrandt, B.; Brinkmann, H.; Hoffmann, R. W.; *Chem. Ber.* **1990**, *123*, 869.
2. Macromodel 4.5, Department of Chemistry, Columbia University, New York, NY 10027.
3. Allinger, N. L.; Yuh, Y. H.; Lii, J.-H. *J. Am. Chem. Soc.* **1989**, *111*, 8551.
4. deMon 1.0, A Gaussian Density Functional Program, University of Montreal
5. Malkin, V. G.; Malkina, O. L.; Casida, M. E.; Salahub, D. R. *J. Am. Chem. Soc.* **1994**, *116*, 5898.

Binding of LMW Inhibitors to Thrombin and Trypsin Studied by Electrostatics Calculations

Michael Engels and Peter Grootenhuis

Comp. Med. Chem. Group (RK2340), Organon NV, P.O. Box 20, 5340 BH OSS, The Netherlands
(m.engels@organon.akzonobel.nl)

Abstract

The electrostatic free energy of binding has been calculated for a series of thrombin and trypsin X-ray structures in complex with LMW inhibitors. The calculations have been performed either with the UHBD or the MEAD program. Both programs use the finite difference solution to the Poisson-Boltzmann equation for determing electrostatic interactions. A modeling protocol has been optimized which also takes into account solvation entropy changes during the binding process. We demonstrate that the electrostatic energy and the empirically derived solvation entropy can reproduce the experimental binding energies almost quantitatively.

Figure 1. *Target molecules*

Rotational Barriers of Various Ureas - a New Force Field Parametrization

Thomas Straßner

Institut für Organische Chemie, Technische Universität Dresden, Mommsenstr. 13, D-01069 Dresden, Germany (strasner@organik.uni-erlangen.de)

Abstract

To build a parameter set for force field calculations of ureas the accuracy of semiempirical calculations was not sufficient. Exact ab initio data concerning the rotational barriers of different substituted and unsubstituted ureas were necessary. The presented material also gives insight into the rotational barriers of different ureas.

High-level ab initio calculations have been performed on urea, methylurea, tetramethylurea and three isomers of dimethylurea to obtain accurate rotational barriers. Results of MP2(fc)/6-31G(d) calculations are compared to those with lower basis sets and semiempirical calculations.

The MM2(87) force field was reparametrized [1] for the R_2N-CO-NR_2-unit through fitting of ab initio and Xray data. Molecular geometries and rotational barriers were calculated on different levels of theory. With the new parameter set MM2 was able to reproduce the results of ab initio calculations als well as the Xray data of several compound classes. Ab initio calculations of the rotational barriers of urea, methylurea, different dimethylureas and tetramethylurea allow a comparism of the influence of the basis set on the rotational barrier.

References

1.   Straßner, Th. *J. Mol.Model.* **1996**, 2, 217.

Structures, Energetics, and Dynamics of Transition Metal-Activated Lactone-Bridged Biaryl Complexes: a Density-Functional Study

Gerhard Bringmann and Ralf Stowasser

Institut für Organische Chemie, Universität Würzburg, Am Hubland, D-97074 Würzburg, Germany (stowasse@chemie.uni-würzburg.de)

Abstract

Axially chiral biaryls can be stereoselectively synthesized by metal-assisted cleavage of lactone-bridged helically twisted biaryl systems [1]. The stereoselective ring opening (s. Scheme 1) of the helimerizing lactone-bridged biaryls to give configuratively stable products can be achieved either by chiral metal-activated nucleophiles or by achiral nucleophiles and activated benzonaphthopyranones *e.g.* by coordination of a chiral Lewis acid to the carbonyl group.

A most promising modification of this concept is the use of planar-chiral η^6-coordinated transition metal fragments **3** [2]. Hereby three crucial questions are of interest:

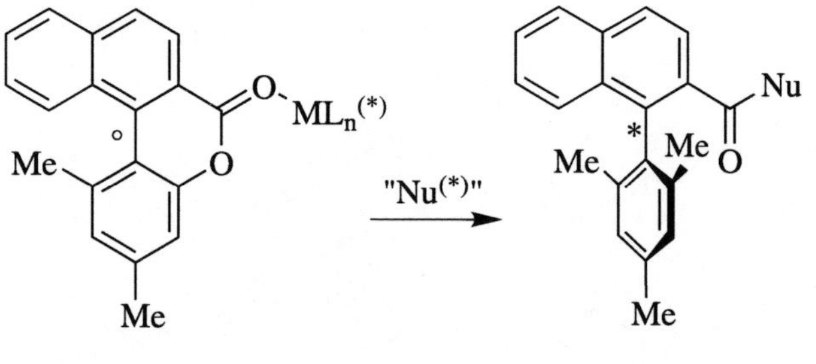

1   *(M)*-**2**

Scheme 1.   °: *stereochemically unstable*
             *: *stereochemically stable*

1.)   Which regioisomer (**I**, **III**, or **IV**) will be formed by variation of the substituents R and the metal fragment ML_n?

2.)   How is the influence of transition metal fragments on the structure and the barriers for the interconversion of the two helimers?

3.)   What kind of mechanism is involved in the ring opening reaction of **3**?

In previous work we performed *ab initio* calculations [3,4] on structure and dynamics of free and main group Lewis acid activated biaryl lactones.

Here we present first results on *ab initio* DF-calculations of the transition metal-activated biaryl lactone complexes **3**. Ground state structures and relative energies of the regioisomers of **3a** and **3b** were calculated by means of the LDA-VWN method [5]. The thermodynamically most stable regioisomers calculated agree with those found experimentally. The calculated global minimum structure parameters of the helically twisted complex **3a** match very well with those previously obtained through X-ray structure analysis [6]. Additionally, two atropisomeric minimum structures with different rotating positions of the $Cr(CO)_3$ rotor were found per regioisomer. Therefore and due to the periodically increasing steric interaction during the helimerization process, we assume a correlation of the dynamic of the $Cr(CO)_3$-group with that of the biaryl axis.

3

	ML_n	R
3a	$Cr(CO)_3$	Me
3b	$Cr(CO)_3$	tBu

References

1. Bringmann, G.; Schupp, O. *South African J. Chem.* **1994**, *47*, 83-102.
2. Bringmann, G.; Göbel, L.; Peters, K.; Peters, E.-M.; von Schnering, H. G. *Inorg. Chim. Acta.* **1994**, *222*, 255-260.
3. Bringmann, G.; Busse, H.; Dauer, U.; Güssregen, S.; Stahl, M. *Tetrahedron* **1995**, *11*, 3149–3158.
4. Bringmann, G.; Dauer, U.; Lankers, M.; Popp, J.; Posset, U.; Kiefer, W. *J. Mol. Struc.* **1995**, *349*, 431–434.
5. Vosko, S.H; Wilk, L.; Nusair, M.W. *Can. J. Phys.* **1980**, *58*, 1200-1205.
6. Bringmann, G.; Stowasser, R.; Vitt, D. *J. Organomet. Chem.*, in press.

Visualization of Local Physical Properties on Molecular Surfaces: The Mapping Function

Robert Jäger [a], Matthias Keil [a], Gerd Moeckel [a,b], Horst Vollhardt [a], Jürgen Brickmann [a,c]

[a] Department of Physical Chemistry I, Darmstadt University of Technology, Germany (jaeger@pc.chemie.th-darmstadt.de)
[b] Division of Toxicology & Cancer Risk Factors, German Cancer Research Center
[c] Darmstadt Center of Scientific Computing, Germany

Abstract

The display of local lipophilicity/hydrophobicity mapped on a molecular surface is a well established example of visualization of an intrinsic physical quality [1]. This implies that the overall hydrophobicity of a molecule can be obtained by the superposition of any defined set of atomic/fragmental contributions. However, the mapping procedure is based on the concept of molecular lipophilicity potentials (MLP) first introduced by Audry [2] in order to describe a 3D lipophilicity potential profile in the molecular environment. Crucial to the physical relevance of the lipophilicity potential is the choice of a function which governs the behavior of a physical quality with distance. While Audry chose a simple $(1+d_i)^{-1}$ term, d_i being the distance of a molecular/atomic fragment i from any point in 3D-space, several other authors proposed different functions, e.g. an exponential distance dependence

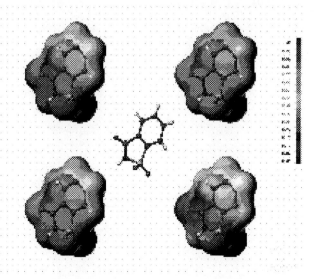

or the concept of a 'hydrophobic dipole moment' by Fauchère [3] and Eisenberg [4], respectively. The function under investigation is that developed by Brickmann and coworkers [1]. This particular function was successfully applied for predicting logP (Octanol/Water) values [6] as well as to an qualitative assessment of receptor sites by Lichenthaler et al. [7].

In this presentation the quality mapped on Connolly's *solvent accessible surface* [5] does not have any physical meaning, it merely represents the portion of an atom to the surface indicated by a predefined color code for each atom. Depending on the choice of function parameters this will result in a colored surface with areas of great atomic overlap (mixed colors) or rather single colored regions where only one atom dominately contributes to that particular surface increment. Therefore the function devises a 'contribution range' to the generated surface allowing both localization and quantification of surface dependent qualities as it has been demonstrated for *free energy surface density* (FESD) by Pixner et al. [6] The parameter dependency of the function and the consequences for the visualization is presented and illustrated.

References

1. Heiden, W., Moeckel, G., and Brickmann, J. *J. Comput.-Aided Mol. Design* **1993**, *7*, 503-514.
2. Audry, E., Dubost, J.P., Colleter, J.C., and Dallet, P. *Eur. J. Med. Chem.* **1986**, *21*, 71-72.
3. Fauchère, J.-L., Quarendon, P., and Kaetterer, L. *J. Mol. Graphics* **1988**, *6*, 203-206.
4. Eisenberg, D., Weiss, R.M., and Terwilliger, T.C. *Nature* **1982**, *299*, 371-374.
5. Connolly, M.L. *Science* **1983**, *221*, 709-713.
6. Pixner, P., Heiden, W., Merx, H., Moeckel, G., Moeller, A., and Brickmann, J. *J. Chem. Inf. Comput. Sci.* **1994**, *34(6)*, 1309-1319.
7. Lichtenthaler, F.W., Immel, S., and Kreis, U. *Starch* **1991**, *43*, 121-132.

Density Effects on the Structure of Liquid Silicon A Molecular Dynamics Study

Andreas Appelhagen and Alfons Geiger

Physikalische Chemie, Universität Dortmund, D-44221 Dortmund, Germany (geiger@heineken.chemie.uni-dortmund.de)

Abstract

The effect of a global and local density decrease on the structure of the tetrahedral bond network of liquid silicon is studied via molecular-dynamics simulations. It is found that the structural changes resemble those already obtained in molecular-dynamics models of water. Decreasing the global density leads for example to more pronounced peaks in the radial distribution function and to a smaller number of nearest neighbours up to the point where the network disrupts. In this case very large holes occur, indicated by the distribution of the volumes of the Voronoi polyhedra.

The local density decrease produced by a dissolved Lennard-Jones particle causes a similar structural enhancement in its nearest surrounding. This is indicated by more pronounced peaks in the radial distribution function of the atoms in the "solvation shell" compared to the bulk atoms. An analysis of the orientation of the Si-Si-bonds around the solute reveals that the structure of the shell can be compared to that one found in molecular-dynamics models of the hydrophobic hydration shell in water/Lennard-Jones systems.

Solvation Effects in PIMM

Ralf Schwerdtfeger and Hans Jörg Lindner

Institut für Organische Chemie, Technische Hochschule Darmstadt, Petersenstrasse 22, D-64347 Darmstadt, Germany (ralf@oc2.oc.chemie.th-darmstadt.de)

Abstract

The force field used in PIMM [1-3] is parametriced for vacuum calculations. To consider solvation effects we applied and tested several continuum models.

The most suitable method is based on the algorithm of W.C. Still et al [4] by using the generalized Born equation and calculating effective born radii for all atoms.

By using optimized contact radii for the different atom types it is possible to calculate heats of solvation in good agreement with experimental data. The continuum model we use works without any problem with our MD.

Correct trends are calculated for solvent effects such as keto enol tautomerization of acetylacetone and cyclohexandione and the rotational barriers in amides.

We are currently starting to use this approach with our MD to study small peptides in solution and the effects of solvation on the Ramachandran plots.

References

1. Lindner, H.J. *Tetrahedron* **1974**, *30*, 1127; Lindner, H.J. *Tetrahedron Lett.* **1974**, 2479.
2. Smith, A.E. and Lindner, H.J. *J. Comput.-Aided Mol. Des.* **1991**, *5*, 235.
3. Kroeker, M. PhD Thesis TH Darmstadt 1994.
4. Still, W.C.; Tempczyk, A.; Hawley, R.C. and Hendrickson, T. *J. Am. Chem. Soc.* **1990**, *112*, 6127.

A Combined Semiempirical MO/ Neural Net Technique for Estimating vicinal H,H-Coupling Constants J (H,H)$_{vic}$

Andreas Breindl and Timothy Clark

Computer-Chemie-Centrum, Institut für Organische Chemie I, Friedrich-Alexander-Universität Erlangen-Nürnberg, Nägelsbachstraße 25, D-91052 Erlangen, Germany (breindl@organik.uni-erlangen.de)

Abstract

A back-propagation neural net has been trained to estimate J (H,H) from the vic results of AM1 semiempirical MO calculations. The input descriptors include the complete electronic environment (bond orders and atomic charges) and all geometrical factors as dihedral angle, valence angles and distances for an individual H-C-C-H moiety. Our net can mainly handle rigid organic compounds, because we are not able to consider conformational effects. The resulting net estimates the vicinal H,H-coupling constants of a test set of 95 different coupling constants with a standard deviation of 0.98 Hz from the experimental values for AM1 with a maximum error of 2 Hz.

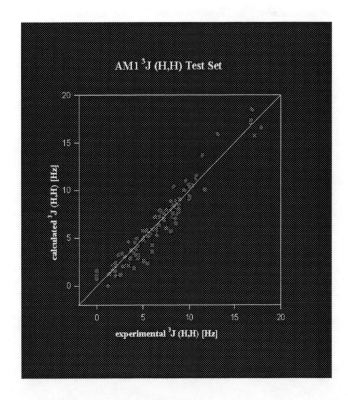

J. Mol. Model. **1996**, *2*

Homology Modeling of the Small Sialidase from *Clostridium perfringens*: Use of Laser Photo CIDNP(Chemically Induced Dynamic Nuclear Polarization) Techniques and Site-directed Mutagenesis to Ascertain the Reliability of the Modeled Structures

Hans-Christian Siebert [a,b], Emadeddin Tajkhorshid [c,d], Claus-Wilhelm von der Lieth [d], Reinhard G. Kleineidam [e], Susanne Kruse [f], Roland Schauer [f], Robert Kaptein [b], Hans-Joachim Gabius [a] and Johannes F.G. Vliegenthart [b]

[a] Institut für Physiologische Chemie der Ludwig-Maximilians-Universität, München, Germany;

[b] Bijvoet Center for Biomolecular Research, University of Utrecht, The Netherlands;

[c] School of Pharmacy, Tehran University of Medical Sciences, Tehran, Iran;

[d] DKFZ, ZentraleSpektroskopie, Heidelberg, Germany; (w.vonderlieth@dkfz-heidelberg.de)

[e] Biochemisch Laboratorium and Bioson Research Institute, University of Groningen, The Netherlands;

[f] Biochemisches Institut II, Christian-Albrechts Universität, Kiel, Germany

Abstract

Sialidases (N-acylneuraminosyl-glycohydrolases, EC 3.2.1.18) hydrolytically cleave alpha-glycosidically bound sialic acids, derivatives of the amino sugar neuraminic acid. Sialic acids are mostly found as terminal constituents of oligosaccharides, glycoproteins and glycolipids in higher animals. The sialidases, however, are widely distributed not only throughout the metazoan animals of the deuterostomate lineage, but also among protozoa, viruses, fungi and bacteria, most of which are unable to produce sialic acid by themselves. Remarkably, the enzyme is often produced by microorganisms, which live in close contact with an animal host, whereby the enzyme may serve as a pathogenicity factor, or as an important tool for nutrition [1].

X-ray structure of sialidase of *Salmonella typhimurium* was used as the template for homology modeling of *Clostridium perfringens* sialidase (CPS). Both FastA and BLAST algorithms indicate high similarity between these two enzymes. The amino acids located in the four 'Asp boxes' and those of the active site of the enzyme are highly conserved. Construction of the starting framework, fitting of the CPS backbone, addition of loop regions and missing side chains, and preliminary refinement of model were carried out using the Swiss-Model Automated Protein Modelling service [2]. The generation of hydrogen atoms and automatic assignment of partial charges of each atom were accomplished using the INSIGHTII. The structures were then submitted to an MD simulation using the CVFF force field at a temperature of 300 K with an equilibration time of 20 ps and production period of 100 ps. The ten lowest potential energy conformers were selected for further minimization and surface accessibility calculation of aromatic residues.

The function of sialidases can be studied with help of mutants constructed by site-directed mutagenesis [3]. Based on the known three-dimensional structure of the *Salmonella typhimurium* sialidase, amino acids analogous to those that seem to be important for substrate binding or catalysis, were selected for mutation in CPS. The activity of some of the mutant sialidases was strongly decreased but the Km-values were hardly changed.

The side chains of tyrosine, tryptophan and histidine are able to produce CIDNP (Chemically Induced Dynamic Nuclear Polarization) signals after laser irradiation in the presence of a suitable radical pair-generating dye [4]. The CIDNP technique has previously been used for comparative studies of non-specific and specific interaction between the lac-repressor headpiece and DNA denatured states of lysozyme as well as of glycoproteins in glycosylated and deglycosylated form or in sialylated and desialylated form in solution.

The results from CIDNP experiments with CPS and its mutant forms indicated significant changes in the pattern of surface accessibility of aromatic residues of CPS in all of the mutants, which is in complete agreement with the measured Connolly surfaces of amino acids in the modeled structures.

References

1. Schauer R. *Sialic Acids Cell Biology Monographs, 10,* Springer Verlag Wien-New York 1982.
2. Peitsch M.C. *Bio/Technology* **1995**, *13*, 658-660.
3. Roggentin, T. ; Kleineidam, R.; Schauer, R.; Roggentin, P. *Glycoconjugate* **1992**, *9*, 235-240.
4. Kaptein, R.; Dijkstra, K.; Nicolay, K. *Nature* **1978**, *274*, 293-294.

J. Mol. Model. **1996,** *2*

The Conformation of *tripod* Metal Templates in CH₃C(CH₂PPh₂)₃MLₙ: Statistical Methods, Neural Networks and Molecular Mechanics

Stefan Beyreuther, Johannes Hunger, Gottfried Huttner

Anorganisch-Chemisches Institut der Universität
Heidelberg, INF 270, 69120 Heidelberg, Germany

Tel: +49 6221 548446, Fax: +49 6221 545707
(s.beyreuther@indi.aci.uni-heidelberg.de)

Abstract

In most cases specificity in coordination chemistry is dominated by a specific choice of the ligands surrounding the metal. Homogenous catalysis and, even more so, the subset of enantioselective catalysis are clear illustrations of this statement.

Part of this specificity will arise from steric interactions. In order to be able to predict specificity a thorough knowl-edge of these steric interactions is therefore needed. One basis for acquiring such a knowledge is currently available in many cases: A considerable amount of information exists on solid state structures (determined by X-ray analysis), most of it deposited in an orderly and easily retrievable way in internationally available databases. The wealth of information contained in these data files is, as yet, far from having been systematically exploited.

The systematic analysis of all such structural data describing the solid state conformation of *tripod*Co templates (*tripod* = CH₃C(CH₂PPh₂)₃) for compounds *tripod*CoL₂ and *tripod*CoL₃ is presented (Figure1). The interest in this analysis stems from a research programme aiming at an understanding of the chemical reactivity of a subset of ligand metal templates in which the ligand is a neopentane based tripodal entity. The programme involves the three following steps: 1) synthesis of a library of neopentane based *tripod* ligands RCH₂C(CH₂X)(CH₂Y)(CH₂Z) (X, Y, Z = donor groups), 2) understanding and modelling the shape of *tripod* metal templates *tripod*M, 3) correlating the results of a catalysis mediated by *tripod*M with the shape of *tripod*M.

The synthesis of *tripod* ligands RCH₂C(CH2X)(CH2Y)(CH2Z) with three different donor groups X, Y, Z is well established including the enantioselective synthesis of ligands containing three different phosphorus donor groups. Catalysis mediated by tripod metal templates has been observed and it is thus time to approach the understanding of the shape of tripod metal templates.

An experimental basis for the development of such an understanding may be found in the X-ray data pertaining to a

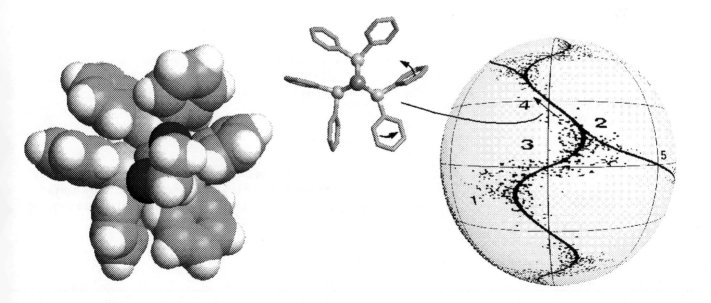

Figure 1. *Solid state structure of [CH₃C(CH₂PPh₂)₃Co(en)]²⁺ (en = 1,2-Diaminoethane), the schematic* tripod*Co template of this compound, its symbolic mapping according to the torsional positions of the phenyl groups and indicated pathways for conformational change.*

J. Mol. Model. **1996**, 2

specific tripod ligand CH3C(CH2PPh2)3. Since most of the data referring to this ligand are available for tripodCo templates the systematic analysis has been restricted to cover tripodCoL2 and tripodCoL3, thus resulting in a data basis of 82 tripodCo templates [1,2].

The traditional tools for this type of analysis [3,4], such as conformational space group scatter graphs (Figure1), principal component analysis (PCA), partial least squares (PLS) and hierarchical clustering, have been applied to the data [5]. They allow for the classification of conformations and the elucidation of pathways in conformational space. This type of information in itself forms a basis on which the construction of force field models may be built.

As a method that is not yet traditional in conformational analysis the methodology of neural networks has also been applied to the data [6,7]. It is shown that this technique reproduces the results from the traditional methods but also has merits of its own that recommend its further application in conformational analysis.

The Molecular Mechanics approach proved its ability to reconstruct the patterns obeserved in conformational space of the tripod metal templates [8].

References

1. Beyreuther, S.; Hunger, J.; Huttner, G.; Mann, S.; Zsolnai, L. *Chem. Berichte* **1996**, in the press.
2. Cambridge Structural Database: Allen, F. H.; Kennard, O. *Chemical Design Automation News* **1993**, 8, 1 & 31.
3. Bürgi, H.-B.; Dunitz, J. D. Structure Correlation. VCH: Weinheim 1994.
4. Morton, D. A. V.; Orpen, A. G. *J. Chem. Soc., Dalton Trans.* **1992**, *4*, 641.
5. SYBYL 6.2. Tripos: St. Louis 1995.
6. Kohonen, T.; Kangas, J.; Laaksonen, J. SOM_PAK, The Self-Organizing Map Program Package, Version 1.2. University of Technology: Helsinki 1992.
7. Zupan, J.; Gasteiger, J. Neural Networks for Chemists. VCH: Weinheim 1993.
8. MacroModel 5.0. Columbia University: New York 1995.

Structure-Reactivity Correlations by a Modified PIMM Force Field

Holger Kalkhof and Hans Jörg Lindner

Institut für Organische Chemie, Technische Hochschule Damstadt, Petersenstr. 22, D-64287 Darmstadt, Germany (holger@oc2.oc.chemie.th-darmstadt.de)

Abstract

In our group a force field for carbocations and anions was developed which differs from other versions of PIMM [1,2] in its special parameters for charged carbon atoms. It is appropriate to all kinds of hydrocarbons like unsubstituted and substituted aromatic cations or aliphatic carbocations. Furthermore the program has been successfully tested for aromatic carbanions.

Carbocations

The heats of formation, proton affinities, and hydrid affinities of aromatic compounds with delocalized charge (e.g. allyl-, tropyliumions) calculated with PIMM agreed well with experimental values. Applied to σ-complexes of benzoid systems or substituted allyl cations we obtained relative stabilities of these systems that are in good agreement with experimental data.

Carbanions

Proton affinities and heats of formation in a series of about 20 aromatic systems were calculated with PIMM. The results reproduced well experimental values.

Strain-reactivity correlations for the solvolysis of tertiary alkylbromides

It is well known that the solvolysis rate of a tertiary hydrocarbon is dominated by the difference of steric strain between alkylbromide and cation. It is caused by the tertiary carbon atom holding the leaving-group which changes from a tetrahedral to a planar geometry. The first force field calculations were applied to these systems by Schleyer [3]. A good strain-reactivity correlation was obtained for the solvolysis of caged hydrocarbons. Miller et al. [4,5] included also acyclic and monocyclic tertiary halides in their calculation with a modified MM2-program. Our calculations for the molecules treated by Miller with PIMM resulted in a strain-reactivity plot with a comparable scattering.

References

1. Smith, A. E.; Lindner, H. J. *J. Comput.-Aided Mol. Des.* **1991**, *5*, 235.
2. Kroeker, M. Thesis, Technische Hochschule Darmstadt, 1994.
3. Gleicher, G. J.; Schleyer, P. v. R. *J. Am. Chem. Soc.* **1967**, *89*, 582.
4. Müller, P.; Mareda, J. *J. Comput. Chem.* **1989**, *10*, 863.
5. Müller, P.; Millin, D. *Helv. Chim. Acta* **1991**, *74*, 1808.

Temperature Dependence of the Properties of Liquid Crystalline PCH5
– A molecular dynamics simulation study –

Frank Eikelschulte, Alfons Geiger

Institute of Physical Chemistry, University of Dortmund, Otto-Hahn-Strasse, D-44221 Dortmund, F. R. Germany (frank@heineken.chemie.uni-dortmund.de)

S. Yakovenko, A. Muravski

Institute of Applied Physics Problems , Kurchatova Minsk, 220064, Minsk, Belarus

Abstract

A system containing 200 molecules of the liquid crystalline substance PCH5 is studied at temperatures between 310 K and 450 K by means of molecular dynamics simulations. These investigations are aimed to study the temperature dependence of the static and dynamic properties such as order parameters, translational and orientational diffusion and pair correlation functions of a realistic model of a mesogen. Special effort was made on the electrostatic part of the interaction potential. It is found that in many cases e.g. when dealing with macroscopic parameters, the influences of the system size and simulation time limitations are seen very clearly. On the other hand single particle dynamic properties and local processes such as pair formation can be successfully studied by computer simulations.

Computer Aided Modelling of *Rhizopus Oryzae* Lipase Catalyzed Stereoselective Hydrolysis of Triglycerides

Hans-Christian Holzwarth, Jürgen Pleiss, Holger Scheib and Rolf D. Schmid

Institute of Technical Biochemistry, University of Stuttgart, Allmandring 31, D-70569 Stuttgart, Germany (jpleiss@tebio1.biologie.uni-stuttgart.de)

Abstract

Lipase from *Rhizopus oryzae* catalyzes the stereoselective hydrolysis of triacylglycerols and analogues. Enantio-preference and degree of enantiomeric excess of the product varies with the structure of the substrate: 2-O-octyl, 2-hexyl, and 2-octanoyl substituted triacylglycerols are preferentially hydrolyzed at pro-*sn*-1 with decreasing enantiomeric excess, substitution by 2-phenyl reverses the enantiopreference to pro-*sn*-3.

We have modelled the stereoselectivity of *Rhizopus oryzae* lipase by docking the tetrahedral intermediates of these substrates in both the pro-*sn*-1 and pro-*sn*-3 orientations.

The initial complexes were further relaxed by molecular dynamics simulations. In their preferred orientation the tetrahedral intermediates fit well into the binding site. In the unfavourable orientation the complex is destabilized by three effects: (1) Repulsive interaction of the *sn*-2 group with the side chain of Leu 258 and - to a minor degree - with other residues, which leads to (2) deformation of the substrate conformation and (3) destabilization of the oxyanion hole.

Our model is consistent with experimental data and explains the ranking of four different substrates. It can be used to design lipase mutants with modified enantioselectivity.

J. Mol. Model. **1996**, *2*

K411B, A Triazine-Binding Single-Chain Antibody: Structure Modelling and Hapten Docking

Sebastian Hörsch, Jürgen Pleiss, Karl Kramer$, and Rolf D. Schmid

Institute of Technical Biochemistry, University of Stuttgart, Allmandring 31, D-70569 Stuttgart, Germany (jpleiss@tebio1.biologie.uni-stuttgart.de)

§ Department of Botany at Weihenstephan, Technical University of München, D-85350 Freising

Abstract

The single-chain antibody fragment (scFv) K411B, which binds to various triazine herbicides, was cloned, sequenced and tested for relative affinities in the group of Prof. B. Hock, Freising, Germany [1,2]. As there is no experimental structure available, we have modelled the structure by homology: for both light and heavy chain, antibodies with a known structure and a high degree of sequence homology in the framework regions were found. The complementarity determining regions (CDRs) were identified, and for three of the six CDRs, canonical structure classes according to the concept of Chothia and Lesk [3] were assigned using the Kabat database [4]. The remaining CDRs were modelled by loop searching or conformational search algorithms.

Hapten docking was performed while carefully examining the cross-reactivity pattern of the scFv. Thus, the number of possible orientations of the hapten within the binding site was significantly reduced by exclusion of orientations which are not consistent with the cross-reactivity data.

This model can be used to understand the molecular basis for the specificity pattern of K411B and to predict mutants with designed specificity for biosensor applications.

References

1. Kramer, K.; Hock, B. *Food and Agricultural Immunology* **1996**, accepted for publication.
2. Giersch, T. *Journal of Agricultural and Food Chemistry* **1993**, *41*, 1006 - 1011.
3. Chothia, C.; Lesk, A. M.; Tramontano, A.; Levitt, M.; Smith-Gill, S. J.; Air, G.; Sheriff, S.; Padlan; E. A.; Davies, D.; Tulip, W. R.; Colman, P. M.; Spinalli, S.; Alzari, P. M.; Poljak, R. J. *Nature* **1989**, *342*, 877 - 883.
4. Kabat, E. A.; Wu, T. T.; Bilofsky, H.; Reid-Miller, M.; Perry, H. M.; Gottesmann, K. S. Sequences of Proteins of Immunological Interest, U.S. Dept. of Health and Human Services, Washington, DC, 1987.

Cyclofructins: Geometries, Electrostatics and Lipophilicity Patterns, and Inclusion Complexes

Stefan Immel, Guido Schmitt, and F. W. Lichtenthaler

Institute of Organic Chemistry, Technical University of Darmstadt, D-64287 Darmstadt, Germany

Abstract

Cyclofructins composed of six (α-CF, **1**) to ten (ε-CF, **5**) $\beta(1\rightarrow2)$-linked fructofuranose units (i.e. *cyclo*[D-Fru*f*β(1→2)]n with n = 6 - 10) were subjected to conformational analysis using "Monte-Carlo" simulations based on the PIMM91 [1] force field [2,3]. Far reaching similarities and identical over all conformations of the solid state geometry [4] of a cyclofructin (**1**) and its computer generated form provide information about the reliability of the computational analysis.

1 *cyclo*[D-Fru*f*β(1→2)]$_6$
2 *cyclo*[D-Fru*f*β(1→2)]$_7$
3 *cyclo*[D-Fru*f*β(1→2)]$_8$
4 *cyclo*[D-Fru*f*β(1→2)]$_9$
5 *cyclo*[D-Fru*f*β(1→2)]$_{10}$

Calculation of the molecular surfaces for the energy minimum structures establishes a disk type shape of the cyclofructins with six to eight residues, ring enlargement to nine and ten residues leads to torus shaped molecules with central cavities that conceivably allow for the formation of inclusion complexes. The color coded projection of molecular lipophilicity patterns (MLPs) [5] and electrostatic potential profiles (MEPs) onto these surfaces displays the crown ether-like properties of the disk shaped cyclofructins. The central cavities of **4** and **5** should be amenable to the formation of inclusion complexes similar to those formed by cyclodextrins.

References

1. Smith, A. E.; Lindner, H. J. *J. Comput.-Aided Mol. Des.* **1991**, *5*, 235-262.
2. Immel, S.; Lichtenthaler, F. W. *Liebigs Ann. Chem.* **1996**, 39-44.
3. Immel, S.; Lichtenthaler, F. W. to be submitted.
4. Sawada, M.; Tanaka, T.; Takai, Y.; Hanafusa, T.; Taniguchi, T.; Kawamura, M.; Uchiyama, T. *Carbohydr. Res.* **1991**, *217*, 7-17.
5. Waldherr-Teschner, M.; Goetze, T.; Heiden, W.; Knoblauch, M.; Vollhardt, H.; Brickmann, J. in *Advances in Scientific Visualization*, Post, F. H.; Hin, A. J. S. (Eds.) Springer Verlag, Heidelberg, **1992**, 58-67.

Molecular Similarities in Estrogenic Chemicals: Evaluation by Genetic Algorithms and Neural Networks

Sandra Handschuh, Markus Wagener,

Johann Gasteiger

Computer-Chemie-Centrum, Institut für Organische Chemie, Universität Erlangen-Nürnberg, Nägelsbachstraße 25, D-91052 Erlangen, Germany ({Handschuh, Wagener, Gasteiger}@eros.ccc.uni-erlangen.de)

Abstract

Applications of *Genetic Algorithms* and *Neural Networks* give rise to increased efficiency in toxicological and pharmaceutical research. In addition, new comprehensive analyses of specific receptor-substrate interactions are possible.

Natural or environmental estrogenic chemicals, such as insecticides and pesticides like p,p'-DDT, coumestrol or diethyl-4,4'-stilberol (**1**), displace the sexual hormone testosteron in male organism and prevent its binding to the androgen receptor. Blocking the natural action of testosteron causes abnormalities in male reproductive tracts. [1] The female sexual hormone estradiol (**2**) is also a well known

1 diethyl-4,4'-stilberol

2 estradiol

3 MCSS
X = O, Cl

testosteron antagonist. Thus, similarities between estradiol and thirteen environmental estrogenic chemicals have been investigated by Genetic Algorithms and Neural Networks.

Superimposing the three-dimensional structures of **2** and the estrogen-mimetics (e.g. **1**) by a Genetic Algorithm [2] reveals the parasubstituted aromatic system **3** as *MCSS (maximal common substructure)*. The MCSS of the different ligands, which bind to the androgen receptor reveals the *pharmacophore*. The latter is a prerequisite for high affinity receptor binding.

Self organizing Kohonen-networks [3] generate two-dimensional maps, which allow the comparison of molecular shapes and physicochemical surface properties, such as van-der-Waals surfaces, electrostatic and hydrogen binding potentials. These qualities are of pre-eminant importance for the evaluation of similarities between **2** and the thirteen estrogenic chemicals. The similarities are less evident for the electrostatic potentials, but much more significant for the van-der-Waals surfaces. Excellent agreement is apparent from comparisons of the hydrogen bindingpotentials of **2** and the estrogen-mimetics.

As is revealed by both, the MCSS **3** and the analysis of the molecular surfaces, the affinity to the androgen receptor is significantly determined by hydrogen bond acceptor functions (e.g. oxygen atoms) attached to the six-membered ring systems.

References

1. Kelce, R. W.; Stone, C. R.; Laws, S. C.; Gray, L. E.; Kemppainen, J. A.; Wilson, E. M. *Nature*, **1995**, *375*, 581-585.
2. Wagener, M. PhD Thesis, TU München, **1993**.
3. Gasteiger, J.; Zupan, J. *Neural Networks for Chemists: An Introduction*, VCH, Weinheim, **1993**.

J. Mol. Model. **1996**, 2, 278 – 285

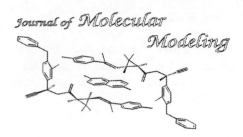

Journal of Molecular Modeling

NpH-MD-Simulations of the Elastic Moduli of Cellulose II at Room Temperatue

Johannes Ganster*

Fraunhofer Institute of Applied Polymer Research, Kantstrasse 55, D-14513 Teltow-Seehof, Germany (ganster@iap.fhg.de)

John Blackwell

Department of Macromolecular Science, Case Western Reserve University, Cleveland, Ohio 44106-7202, USA.

Received: 15 May 1996 / Accepted: 6 August 1996 / Published: 27 September 1996

Abstract

We have used molecular dynamics modeling to investigate the stucture and mechanical properties of regenerated cellulose fibres. This work is motivated by continued interest in replacing the environmentally hazardous viscose process by alternative spinning methods. An important input parameter for any realistic model of the elastic properties is the stiffness tensor of the crystalline constituent, cellulose II. Conventional molecular mechanics techniques can be used to estimate the elastic reaction of a material with respect to small external stresses or strains, i.e. the compliance and stiffness tensors, and the elastic moduli derived therefrom, at zero temperature. In order to access non-zero temperatures, it is necessary to use either the quasi-harmonic approximation for the vibrational free energy or molecular dynamics (MD) simulations. In the present work, Parrinello-Rahman constant-stress MD was performed to generate trajectories in constant particle number (N), constant external stress tensor (p or $\mathbf{t}$) and constant enthalpy H (NpH or HtN) ensemble. This was found to be less time consuming than working with isothermal conditions, as done by other authors. The fluctuations in kinetic energy and MD cell vectors were then used to calculate adiabatic elastic constants, thermal expansion coefficients and heat capacity. The isothermal elastic constants were found by applying a standard thermodynamic relation. The Young's modulus along the chain direction, E_l, was determined to be 155 GPa, whereas the values in the perpendicular directions vary between 51 and 24 GPa. These results are of the same order of magnitude as those obtained by Tashiro and Kobayashi [1] with the static (T = 0K) method, but our value of E_l is 5% lower and, unexpectedly, the lateral values are up to six times higher. A strong anisotropy is found for shear along the chains in planes containing the chain axis, the shear modulus ranging from 5 to 20 GPa. Convergence was achieved in the simulations, to the extend that the elastic constants become stationary, but significant internal stresses remain, pointing to shortcomings in the software used. Further work is necessary to resolve these problems, although the major conclusions should be unaffected.

Keywords: material modeling, molecular mechanics, MD

* *To whom correspondence should be addressed*

Introduction

The cellulose II polymorphic structure [2, 3] is character-istic of regenerated or Mercerized cellulose fibers. The mechanical properties of these materials depend on the structure of the individual molecules and their mutual interactions, to the degree to which an optimum structure has been attained [4]. One of the most simple sets of parameters describing the mechanical behavior of an array of polymer chains consists of the components of the three-dimensional small strain elastic constant or stiffness matrix **C,** and their reciprocals, i.e. the compliances, $S = C^{-1}$. Cellulose II has an approximately monoclinic $P2_1$ space group, which has 13 independent stiffness components [5]. A full experimental determination of these parameters is out of the question, but the Young's modulus of fibres can be determined by following the stress induced shifts of the X-ray diffraction peaks due to lattice deformation, provided that the relation between external load and local crystalline stress is known (or can reasonably be assumed). Estimates of E_ℓ for semicrystalline, regenerated cellulose fibers are reported in the range of 70 - 112 GPa [6-9].

Theoretical calculations employing appropriate force fields allow the determination of the whole stiffness tensor [10]. For cellulose, such calculations have been performed by Tashiro and Kobayashi [1], leading to a value for E_ℓ for cellulose II of 162 GPa [10]. A considerably lower value of 86 GPa was obtained by Kroon-Batenburg *et al.* [11], who considered only a single chain and used a markedly different force field. The main shortcoming of such calculations is their static character, i.e. the mechanical parameters are calculated for the minimum energy structure, which is effectively for 0 K.

Non-static methods for the calculation of stiffness tensors at non-zero temperatures have only recently been applied to polymer crystals. Lacks and Rutledge [12] have used the quasi-harmonic approximation for the vibrational free energy (in both the classical and quantum mechanical formulations) to allow for the thermal vibrations of the atoms, and have calculated the stiffness tensor for polyethylene at several temperatures in the range from 0 to 400 K. Gusev *et al.* [13-15] have employed Parrinello-Rahman constant-stress molecular dynamics (MD) [16] with a Nosé-Hoover thermostat [17, 18] to calculate isothermal stiffness tensors for polyethylene from the corresponding fluctuation formula [19].

In the present work, Parrinello-Rahman MD simulations have been performed for cellulose II in order to calculate the adiabatic stiffness, thermal expansion and heat capacity using the fluctuation formulae due to Ray [20]. In contrast to the work of Gusev *et al.* [13-15], we have employed less time consuming simulations utilizing a constant enthalpy H, tension **t** and particle number N (HtN) ensemble. The isothermal stiffness tensor was calculated from standard thermodynamic relations [21].

Method

According to Andersen [22], Parrinello and Rahman [16], and Ray [23, 24], MD in a scaled and augmented Hamiltonian system can be used to simulate constant stress conditions. This leads to a time-dependent MD cell with fluctuating cell vectors **a, b, c,** which are summarized in the cell matrix **h**:

$$\mathbf{h} = \begin{pmatrix} a_1 & b_1 & c_1 \\ a_2 & b_2 & c_2 \\ a_3 & b_3 & c_3 \end{pmatrix}$$

and is used to calculate the instantaneous strain tensor γ:

$$\varepsilon = \frac{1}{2}(\mathbf{h}_0'^{-1}\mathbf{h}'\mathbf{h}\mathbf{h}_0^{-1} - 1) \tag{1}$$

Here, h_0 is the average h matrix in a zero stress situation, and the prime indicates matrix transposition. The instantaneous and average volumes, V and V_0, are given by the determinants $V = \det(h)$ and $V_0 = \det(h_0)$, respectively.

The microscopic stress tensor, σ, is calculated from

$$\sigma = \frac{1}{V}\left[\sum_a \frac{\mathbf{p}_a' \circ \mathbf{p}_a}{m_a} + \sum_{a<b} \mathbf{f}_{ab}' \circ (\mathbf{r}_a - \mathbf{r}_b)\right] \tag{2}$$

where

$$\mathbf{f}_{ab} = -\frac{\partial U(\{\mathbf{r}_c - \mathbf{r}_d\})}{\partial(\mathbf{r}_a - \mathbf{r}_b)}$$

is the derivative of the potential energy function U of the system, which depends on the set $\{\mathbf{r}_c - \mathbf{r}_d\}$ of the interatomic distance vectors between atoms *c* and *d* at positions $\mathbf{r}_c$ and $\mathbf{r}_d$ in the unscaled system and $\circ$ means the dyadic product for vectors. The linear momentum $\mathbf{p}_a$ of atom *a* with mass m_a is also taken in the unscaled system, and is different from the time derivative of $\mathbf{r}_a$ [23]. For molecular systems the angle bending and torsional forces are not central, with a result that $\mathbf{f}_{ab}$ is not parallel to $\mathbf{r}_a - \mathbf{r}_b$, giving a non-symmetric instantaneous stress tensor [25, 26].

We define $\delta(ab)$ as the fluctuation of two quantities *a* and *b* according to

$$\delta(ab) = \langle ab \rangle - \langle a \rangle \langle b \rangle$$

where <...> designates the HtN ensemble average, which is equal to the trajectory average after equilibration. The adiabatic elastic constant (stiffness) tensor **C** is given by [16]

$$\delta(\varepsilon_{ij}\varepsilon_{kl}) = \frac{kT}{V} C_{ijkl}^{-1}, \tag{3}$$

where $T = 2\langle K \rangle / 3Nk$ is the instantaneous temperature and $K = \Sigma p_a^2/2m_a$ is the kinetic energy, k is Boltzmann's constant, and N is the number of atoms. The heat capacity C_p and the thermal expansion tensor, α^p (both for constant stress), can be determined [20] from

$$\delta(K^2) = \frac{3}{2} N(kT)^2 \left[1 - \frac{3Nk}{2C_p} \right], \tag{4}$$

$$\delta(\varepsilon_{ij}K) = -\frac{3}{2} N(kT)^2 \frac{1}{C_p} \alpha_{ij}^p, \tag{5}$$

allowing for calculation of the isothermal elastic stiffness, $^T\mathbf{C}$[21]

$$^T C_{ijkl}^{-1} = C_{ijkl}^{-1} + V_0 \frac{T\alpha_{ij}^p \alpha_{kl}^p}{C_p}. \tag{6}$$

To control the convergence to equilibrium, Gusev *et al.* [13] compared averaged stress-strain products (invariants) with their equilibrium values (0, 0.5 and 1) according to

$$\frac{V}{kT} \langle \sigma_{ij}\varepsilon_{kl} \rangle = \frac{\delta_{ik}\delta_{jl} + \delta_{il}\delta_{jk}}{2}, \tag{7}$$

where δ_{ij} is the Kronecker symbol: $\delta_{ij} = 1$ for $i = j$ and 0 otherwise.

Experimental

Molecular models for cellulose II were based on the X-ray work by Kolpak and Blackwell [2], which is similar to the structure also proposed by Stipanovich and Sarko [3]. The structure is described by a monoclinic unit cell (space group $P2_1$) with dimensions $a = 8.01$ Å, $b = 9.04$ Å, c (fiber axis) $= 10.36$ Å, and $\gamma = 117.1°$ [2], containing disaccharide units of two chains of opposite sense that pass through the origin and center of the *ab* projection. As described in refs. 2 and 3, the two chains have identical backbone conformations, but differ in the conformations of the CH_2OH side chains, which are *gauche-trans* (*gt*) on one chain and *trans-gauche* (*tg*) on the other. Recent work by Gessler et al. [27] has determined the structure of cellotetraose by single crystal methods, which is thought relevant to cellulose II because of the close similarity of the polymer and

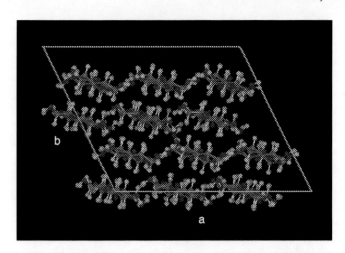

Figure 1. *Snapshot of the simulated cellulose II MD cell containing 12 cellulose chains of four anhydroglucose units, viewed along the* c *direction.*

tetramer unit cells. In the cellotetraose unit cell, the two molecules have the same CH_2OH conformations (*gt*), but have slightly different backbone conformations. We have elected to retain the published cellulose II structure as the starting model. However, there is more than sufficient freedom for rearrangement to the all-*gt* model during the dynamic modeling, and it will be seen that such a change does not in fact occur.

Construction of the molecular models, energy minimization and MD was accomplished using the Biosym molecular modeling software [28] on a Silicon Graphics Power Series 4D 220 GTXB computer. A unit cell was constructed according to the structural features described above, us-

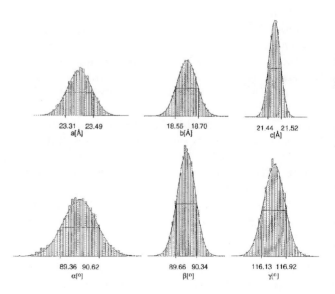

Figure 2. *Frequency distribution of MD cell lengths* a, b,c *and angles* α, β, γ *in a 100 ps HtN run. The solid lines are best fit Gaussian distributions for these data.*

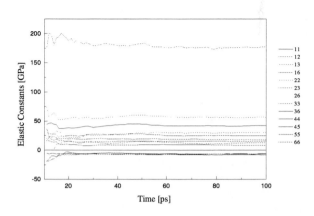
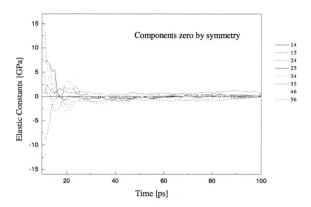

Figure 3. *Running averages of the adiabatic elastic constants (components of **C**) obtained by HtN-MD after 440 ps of relaxation. left: non-zero components; right: components predicted to be zero by symmetry.*

ing the default molecular geometry contained in the software package [28] except that the torsion angles at the glycosidic linkages, the CH_2OH conformations, and the mutual stagger of the center and corner chains along the z axis were taken from reference [2]. The force field was chosen by comparing energy minimizations with periodic boundary conditions (pbc) for the unit cell with variable cell dimensions using the CVFF [29, 30], PCFF [31] and AMBER [32-34] force fields from the DISCOVER program [28]. All three force fields produced structures similar to that in reference [2] with deviations in a, b, c and γ of less than 5 % (α and β were very close to 90°). The calculated density was always somewhat lower than the experimental value of 1.61 g/cm³, but this difference did not exceed 5 %. Even a perfect match in the static structure would not assure the quality of the force field in dynamic simulations. Since there was little difference between the results using different force fields, we used the original CVFF force field.

The MD cell consisted of three unit cells in the a direction and two each in the b and c directions. (The use of larger cells resulted in prohibitive long calculation times.) The simulations were performed with application of peri-

odic boundary conditions (pbc) in the HtN mode, these being three times faster than isothermal ensemble calculations. A time step of 1 fs was used together with a charge group-based cutoff of 9.5 Å (full interaction up to 8.5 Å) both for van der Waals and Coulombic forces, a dielectric constant of 1, default bond increments of the force field for assigning net atomic charges, a cell mass parameter of 20 atomic units, and the velocity Verlet integrator [28]. Simulations were carried through up to 540 ps. Cross terms were not included in the force field and a simple quadratic potential was used instead of a Morse term for the bond length oscillations. A starting temperature of 602 K was found appropriate to reach the target temperature of 298 K. However, a slight heating up to 303.5 K after 500 ps was observed. After some 10 ps, the average shape of the MD cell reached the final form described by the average unit cell dimensions given in *Table 1*. The relatively large c value of 10.74 Å as compared to the experimental value of 10.36 Å [2] is viewed as a defect of the force field, but is not expected to have more than a minor influence on the results of the simulations.

The quantities of interest: cell parameters, kinetic energy and instantaneous microscopic stress tensor determined by the DISCOVER program [28] were saved every 10 fs. Our coordinate system has the same orientation as that of Tashiro and Kobayashi [1], i.e. the c axis is parallel to z and the b axis lies in the z y plane. The contracted notation [5] for tensor indices (11->1, 22->2, 33->3, 23->4, 13->5, 12->6) is used for fourth rank tensors. The cell matrix **h** was constructed from the cell parameters, and $\mathbf{h}_0$ was obtained by averaging **h** over the whole run. $\boldsymbol{\varepsilon}$ was calculated according to eq. (1) for every tenth point on the trajectory.

Table 1. *Density Δ, unit cell lengths a, b, c and angles α, β, γ as determined by X-ray structure analysis [2] and the present MD calculations as well as relative deviations)*

	ρ [g/cm³]	a [Å]	b [Å]	c [Å]	α [°]	β [°]	γ [°]
X-ray [1]	1.61	8.01	9.04	10.36	90	90	117.1
MD	1.54	7.80	9.31	10.74	90	90	116.6
Δ [%]	-5	-3	3	4	-	-	-1

J. Mol. Model. **1996,** *2*

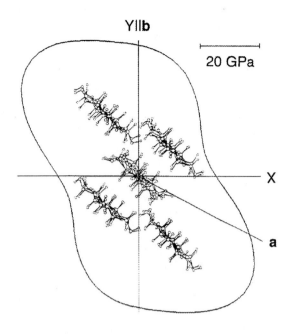

Figure 4. *Young's modulus perpendicular to the chain direction as a function of orientation. The distance from the origin to the bold line is proportional to the modulus.*

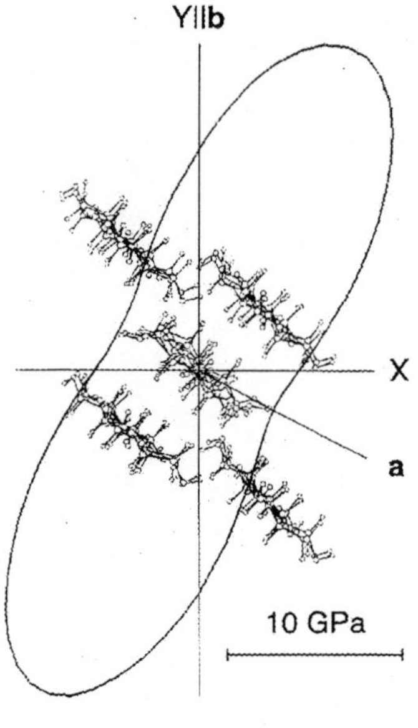

Figure 5. *Shear modulus for the planes containing the* c *axis as a function of orientation. The distance from the origin to the bold line in a given direction is proportional to the modulus for shear along* c *in the plane containing* c *and that direction.*

Results and Discussion

A snapshot of the MD cell at 300 K after 540 ps of simulation time viewed along the chain direction c is shown in Figure 1. A short animation of the trajectory of the system showed that the hydrogen bond system as described in [2] and [3] is essentially maintained. As mentioned above, the structure does not rearrange to the all-*gt* bonding scheme found in the crystal structure of cellotetraose by single crystal determination [27].

Figure 2 shows the frequency distribution of the MD cell lengths and angles over a 100 ps run (i.e. 10000 data points). It can be seen that there are good fits to the best (least squares) Gaussian distributions, indicating an appropriate choice of the cell mass parameter. As is to be expected, the covalent bonds of the polymer chains restrict the cell motion in the c direction; there is more freedom in the non bonded directions, and a shows a slightly broader distribution than b. The „weakest" unit cell angle is α, indicating a relatively easy gliding of the *020* planes past each other, which is consistent with the anisotropy of the shear modulus (see below).

Figure 3 shows running averages of the 13 non-zero and the 8 vanishing adiabatic elastic constants calculated according to eq. (3) for a 100 ps run starting after 440 ps of relaxation. The final values averaged over 100 ps after 540 ps are

Table 2. *Young's moduli E_i, Poisson ratios v_{ij} and shear moduli G_i as calculated from the stiffness constant tensor C*

E_1[GPa]	E_2[GPa]	E_c[GPa]	v_{12}	v_{13}	v_{21}
26.4	42.2	154.6	0.37	0.13	0.59

v_{23}	v_{31}	v_{32}	G_1[GPa]	G_2[GPa]	G_3[GPa]
-0.03	0.78	-0.12	13.5	12.6	5.9

$$C = \begin{pmatrix} \mathbf{42.2} & \mathbf{24.5} & \mathbf{30.0} & 0.4 & 0.2 & \mathbf{-8.1} \\ \mathbf{24.5} & \mathbf{56.7} & \mathbf{12.5} & -0.1 & 0.0 & \mathbf{-5.9} \\ \mathbf{30.3} & \mathbf{12.5} & \mathbf{177.2} & -0.3 & 0.7 & \mathbf{-6.8} \\ 0.4 & -0.1 & -0.3 & \mathbf{8.1} & \mathbf{-6.2} & -0.1 \\ 0.2 & 0.0 & 0.7 & \mathbf{-6.2} & \mathbf{17.3} & -0.1 \\ \mathbf{-8.1} & \mathbf{-5.9} & \mathbf{-6.8} & -0.1 & -0.1 & \mathbf{15.1} \end{pmatrix} GPa \cdot$$

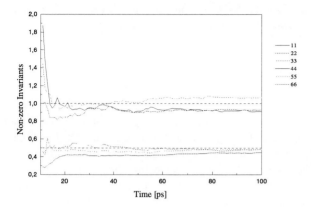

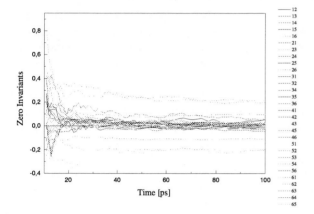

Figure 6. *Running averages of diagonal (above) and non-diagonal (below) invariants V<σ*$_{ij}$*ε*$_{kl}$*>/(kT). Relatively large deviations from zero are seen in for the 16, 61, 26, 62, 45 and 54 components.*

The directional dependence of the Young's modulus perpendicular to the chain direction E_φ([1], [5]) as calculated from **C** is visualized in Figure 4, where the distance from the origin to the bold line is proportional to the modulus in that direction. The „stiffest" direction is close to $[1\bar{1}0]$, where deformation requires stretching or compressing the hydrogen bonds, for which the modulus is 51.4 GPa. The „most compliant" direction is close to *[110]*, where a change in the O-H···O bond angle would be sufficient to allow deformation, and has a modulus of 23.9 GPa.

Figure 5 shows an analogous plot of the shear modulus $G(\varphi)$ [5] for planes containing the chain direction with respect to forces along this direction. n is the inclination angle of the shear plane with the *x* axis, i.e. G_{23} is found along the *x* axis and G_{13} along the *y* axis. The anisotropy is seen to be even more pronounced than that for E_φ. G = 20.0 GPa close to the $\bar{1}10$ plane, which is stabilized in many ways by valence forces. This is about four times greater than the value for shear close to the weaker bound *110* sheets. Our predictions correlate with the comparatively broad frequency distribution for α in Figure 2.

The running averages of the invariants according to eq. (7) are shown for the last 90 ps of the simulation in Figure 6. The values at the end of the run are

$$\frac{2V}{kT}\langle\sigma\varepsilon\rangle = \begin{pmatrix} \mathbf{0.91} & 0.05 & -0.05 & -0.01 & -0.02 & -0.20 \\ 0.06 & \mathbf{0.93} & -0.01 & -0.02 & -0.01 & 0.22 \\ -0.01 & -0.02 & \mathbf{1.07} & 0.00 & 0.01 & 0.02 \\ 0.00 & 0.03 & 0.00 & \mathbf{0.45} & 0.1 & 0.02 \\ 0.00 & 0.03 & 0.01 & -0.11 & \mathbf{0.49} & 0.01 \\ 0.20 & -0.20 & 0.01 & 0.01 & -0.02 & \mathbf{0.48} \end{pmatrix}.$$

The components predicted to be zero on the basis of symmetry are shown in non-bold face. The value of E_ℓ = 155 GPa calculated from this stiffness matrix is slightly lower than the value of 162 GPa determined by Tashiro and Kobayashi for 0 K [1]. None of our figures for the non-vanishing components are as close to zero as are those for C_{13}, C_{16}, C_{23} and C_{36} reported by Tashiro and Kobayashi [1]. Except for the stiffness in chain direction, (C_{33}), the remaining components in the present work are two to six times higher. This may be due to differences in the force field or an effect of disregarding the lack of symmetry in the microscopic stress tensor in the Biosym software [28] (see below). The stiffness components for cellulose II other than C_{33} are also comparatively high compared to those reported (5 - 9 GPa) in simulations of polyethylene and isotactic polypropylene [13, 14], most probably due to the rich hydrogen bond system maintained in the present room temperature simulations. The Young's moduli, Poisson ratios and shear moduli as calculated from the stiffness tensor **C** are given in Table 2.

Ideally, the off-diagonal terms should be zero. However, the 16, 61, 26, 62, 45 and 54 components deviate considerably from their equilibrium values of zero and show no tendency to converge further. (The same effect was noticed in earlier runs, from 240 to 340 and from 340 to 440 ps). In addition, the microscopic stress tensor output by the Biosym DISCOVER program [28] (which gives only six components, thus assuming the tensor to be symmetric) does not converge to zero after 540 ps for our **t** = 0 simulation, as shown in Figure 7. Similar problems have been encountered in a constant stress isothermal simulation of isotactic polypropylene crystals [35]. Furthermore, the average of σVh'$^{-1}$ gave non-zero values. This is not the correct behaviour, as one can see by averaging the differential equation for **h** [36]. A variation of the cell mass parameter did not lead to significantly different results, suggesting that the above effect is not due to incomplete convergence. The asymmetrical rotational components of the cell dynamics could not be studied because the cell tensor **h** is not evaluated as such, but has to be recalculated from the MD cell dimensions. However, these problems relate

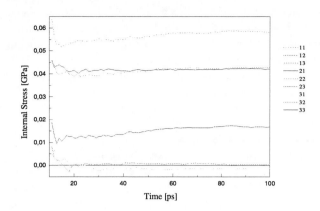

Figure 7: *Running averages of microscopic internal stresses for the simulation shown in Figure 3.*

to the rotational motion. Only fluctuations are necessary to calculate the elastic constants, and these should not be affected more than the invariants shown in Figure 6.

The final values for the specific heat and the thermal expansion are, respectively, $c_p = 3.3$ J/(gK) and

$$\alpha^p = \begin{pmatrix} 4.3 & -1.2 & 0.1 \\ -1.2 & 0.7 & 1.0 \\ 0.1 & 1.0 & 0.1 \end{pmatrix} 10^{-5} / K$$

giving a bulk expansion of $\alpha_{ii} = 5.1 \ 10^{-5}$/K and an averaged linear isotropic expansion of $\alpha = 1.7 \ 10^{-5}$/K. The specific heat is somewhat higher and the expansion coefficient lower than the values reported [37] for cellulose acetate ($c_p = 1.3$ J/(gK), $\alpha = 14 \ 10^{-5}$/K) and cellulose acetate butyrate ($c_p = 2.1$ J/(gK), $\alpha = 12.5 \ 10^{-5}$/K), but the latter are amorphous materials. A low thermal expansion must be expected for the mainly quadratic force field, especially for the bonded interactions: complete absence of anharmonic terms would result in no thermal expansion at all. Using the thermodynamic relation [20, 21]

$$\frac{c_p}{c_v} = 1 + \frac{V_0 T \alpha_{ij}^p \alpha_{kl}^p C_{ijkl}}{C_p} \qquad (8)$$

we obtain a value of 3.25 J/(gK) for c_v from the present fluctuation calculations, which is very close to that of 3.23 J/(gK) derived using the Dulong and Petit law [38]. This suggests that the errors due to low α^p and high stiffness tend to cancel out in eq. (8).

Calculating the isothermal stiffness using eq. (6) gave $^T\mathbf{C}$ values lower than the corresponding $\mathbf{C}$ components, but within the scatter of components which should be zero by symmetry, i.e. there is no significant difference between $\mathbf{C}$

and $^T\mathbf{C}$. A comparison with the constant temperature simulations (TtN ensemble) could reveal deficiencies in the force field and/or the sampling method.

Conclusions

Room temperature adiabatic elastic constants, thermal expansion coefficients and heat capacity of cellulose II have been calculated using constant stress molecular dynamics. During the simulation, the system remained close to the starting configuration, maintaining a dense hydrogen bond network very similar to that found by X-ray crystal structure analysis. Compared to other polymers, the lateral stiffness of cellulose II is high, which is most probably due to the intra- and inter- molecular hydrogen bond system. The isothermal elastic constants do not differ from the adiabatic values within the limitations of the calculation. Thermal expansion coefficients are somewhat low, whereas the specific heat is found to be almost exactly that of Dulong and Petit's rule.

Although convergence was achieved in the sense that the stiffness tensor remains constant for prolonged simulation times, terms in the averaged microscopic stress tensor differ considerably from their expected zero values and showed no tendency to converge further. It seems likely that this was caused by the use of a symmetric instantaneous stress tensor in in the commercial software used.

Acknowledgement: The support of the German Academic Exchange Service through a fellowship to J. Ganster is gratefully acknowledged.

References

1. Tashiro, K. and Kobayashi, M. *Polymer* **1991**, *32*, 1516-1526.
2. Kolpak, F.J. and Blackwell, J. *Macromolecules* **1976**, *9*, 273-278.
3. Stipanovic, A.J. and Sarko, A. *Macromolecules* **1976**, *9*, 851-857.
4. Ganster, J.; Fink, H.-P.; Fraatz, J. and Nywlt, M. *Acta Polym.* **1994**, *45*, 312-318.
5. Hearmon, R.F.S. *An Introduction to Applied Anisotropic Elasticity*, Oxford Univ. Press, 1961.
6. Mann, J. and Roldan-Gonzalez, L. *Polymer* **1962**, *3*, 549-553.
7. Sakurada, I., Ito; T. and Nakamae, K. *Macromol. Chem.* **1964**, *75*, 1-10.
8. Matsuo, M.; Sawatari, Ch.; Iwai, Y. and Ozaki, F. *Macromolecules* **1990**, *23*, 3266-3275.
9. Nishino, T.; Takano, K. and Nakamae, K. *J. Polym. Sci. B, Polym. Phys.* **1995**, *33*, 1647-1651.

10. Tashiro, K.; Kobayashi, M. and Tadokoro, H. *Macromolecules* **1978**, *11*, 908-913.

11. Kroon-Batanburg, L.M.J.; Kroon, J. and Northolt, M.G. *Polymer* **1986**, *27*, 290-292.

12. Lacks, D.J. and Rutledge, G.C. *J. Phys. Chem.* **1994**, *98*, 1222-1231.

13. Gusev, A.A.; Zehnder, M.M. and Suter, U.W. *J. Chem. Phys.* submitted for publication.

14. Gusev, A.A.; Zehnder, M.M. and Suter, U.W. presented at the ACS short Course Molecular *Modeling of Polymers*, 1.-2.7.1994, Akron, OH.

15. Gusev, A.A.; Zehnder, M.M. and Suter, U.W. *Macromolecules* **1994**, *27*, 615-616.

16. Parrinello, M. and Rahman, A. *J. Appl. Phys.* **1981**, *52*, 7182-7190.

17. Nosé, S. *Mol. Phys.* **1984**, *52*, 255-268; *J. Chem. Phys.* **1984**, *81*, 511-519.

18. Hoover, W.G. *Phys. Rev.* **1985**, *A31*, 1695-1697.

19. Parrinello, M. and Rahman, A. *J. Chem. Phys.* **1982**, *76*, 2662-2666.

20. Ray, J.R. *J. Appl. Phys.* **1982**, *53*, 6441-6443

21. Thurston, R.N. in *Physical Acoustics*, Vol. I, Part A, W.P. Mason (Ed.) Academic, New York 1964, p. 39.

22. Andersen, H.C. *J. Chem. Phys.* **1980**, *72*, 2384-2393.

23. Ray, J.R. and Rahman, A. *J. Chem. Phys.* **1984**, *80*, 4423-4428

24. Ray, J.R. *Comp. Phys. Reps.* **1988**, *8*, 109-152.

25. Nosé, S. and Klein, M.L. *Mol. Phys.* **1983**, *5*, 1055-1076.

26. Theodorou, D.N.; Boone, T.D.; Dodd, L.R. and Mansfield, K.F. *Macromol. Chem., Theory Simul.* **1993**, *2*, 191-238.

27. Gessler, K.; Krauss, N.; Steiner, Th.; Betzel, Ch.; Sandmann, C. and Saenger, W. *Science* **1994**, *266*, 1027-1029.

28. *Insight II 2.3.0, Polymer 6.0, Discover 94.0* available from Biosym Technologies, Inc. San Diego, CA, 1994.

29. Lifson, S.; Hagler, A.T. and Dauber, P. *J. Am. Chem. Soc.* **1979**, *101*, 5111-5121.

30. Hagler, A.T.; Lifson, S. and Dauber, P. *J. Am. Chem. Soc.* **1979**, *101*, 5122-5130 and 5131-5141.

31. Maple, J.R.; Hwang, H.-J.; Stockfisch, T.P.; Dinur, U.; Waldman, M.; Ewis, C.S. and Hagler, A.T. *J. Comp. Chem.* **1994**, *15*, 162-166.

32. Weiner, S.J.; Kollman, P.A.; Case, D.; Singh, U.C.; Ghio, C.; Alagona, G.; Profeta, S. Jr. and Weiner, P. *J. Am. Chem. Soc.* **1984**, *106*, 765-784.

33. Weiner, S.J.; Kollman, P.A.; Nguyen, D.T. and Case, D.A. *J. Comp. Chem.* **1986**, *7*, 230-252.

34. Homans, S.W. *Biochemistry* **1990**, *29*, 9110-9118.

35. Suter, U.W. personal communication.

36. Ray, J. *J. Chem. Phys.* **1983**, *79*, 5128-5130.

37. Encyclopedia of Poymer Science and Engineering, Vol. 16. John Wiley & Sons, New York, 1989, p. 737.

38. Becker, R. *Theory of Heat*, Springer-Verlag, New York, 1967, p. 234.

J. Mol. Model. **1996**, 2, 286 – 292

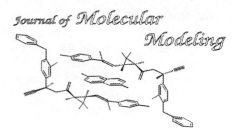

MD — Simulation of Diffusion of Methane in Zeolites of Type LTA

Siegfried Fritzsche, Martin Gaub, Reinhold Haberlandt*, Gerd Hofmann

Universität Leipzig, Fakultät für Physik und Geowissenschaften, Institut für Theoretische Physik, Augustusplatz 10-11, D-04109 Leipzig, Germany (sfri@sunstat1.exphysik.uni-leipzig.de)

Received:15 May 1996 / Accepted: 6 August 1996 / Published: 27 September 1996

Abstract

Using molecular dynamical computer simulations (MD) the dynamics of kinetic processes in zeolites will be discussed on a molecular level. Small changes in lattice parameters can cause dramatical changes in the diffusion coefficient. The presence of cations Na^+, Ca^{2+} also strongly influences the diffusion. Changes of the self-diffusivities will be discussed that appear if a vibrating lattice instead of a rigid one is used. Nonequilibrium simulations show the correlation between transport-diffusion and self-diffusion in zeolites.

Keywords: MD, zeolites, diffusion processes

Introduction

The study of diffusion processes in zeolites is of great interest because these crystals contain very regular internal surfaces and they are used for many industrial purposes [1]. Molecular dynamical computer simulations (MD) [2, 3] that give insight in the dynamics of kinetic processes in zeolites will be discussed on a molecular level. Furthermore, MD allows variations in the system parameters that are not possible in experiments. Starting from late eighties interrelations and dependencies have been examined by MD [4 – 14].

Simulations

The MD simulations were carried out using the velocity-Verlet-algorithm [2, 3] with up to 6,000,000 time steps of 5 and 10 fs, respectively. The basic MD box contains 8 up to 343 large cavities with occupation numbers varying between 1 and 11 per cavity, respectively.

For the interaction between the guest molecules CH_4 and the lattice atoms we use a Lennard-Jones (LJ) (12,6) potential

$$U(r) = 4\varepsilon \left\{ \left(\frac{\sigma}{r} \right)^{12} - \left(\frac{\sigma}{r} \right)^6 \right\} \tag{1}$$

with ε as the minimum depth of the potential energy and σ defined by $U(\sigma) = 0$, respectively. For NaCaA additional polarization energy terms due to the cations have been included. Using the conventional microcanonical MD ensemble the self-diffusivity is calculated while a more generalized non-equilibrium (NEMD) ensemble is used to determine the transport-diffusivities.

Figure 1 shows the general structure of zeolites of type LTA used for our calculations. The sodalite units form a cubic lattice with large cavities connected by so-called 'windows' consisting of eight oxygen atoms.

In figure 2 the distribution of the lattice atoms around a large cavity in the NaCaA zeolite is to be seen. Lattice atoms

* *To whom correspondence should be addressed*

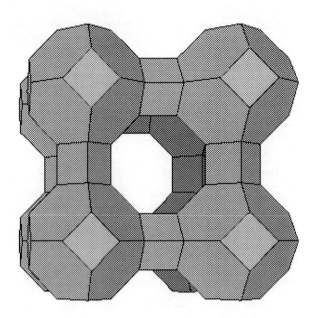

Figure 1. *General structure of zeolites of type LTA*

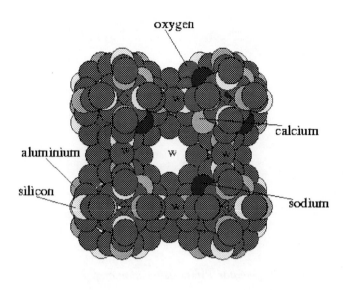

Figure 2. *View into a NaCaA - zeolite*

in front have been removed in order to see the interior of the cavity. Windows are marked by a small w.

Results

Residence times and velocity-auto-correlation functions (vacf)

We define the residence time for a given guest molecule as the time difference between subsequent passages of limiting planes. Such a plane is situated in the center of each window perpendicular to the window axis.

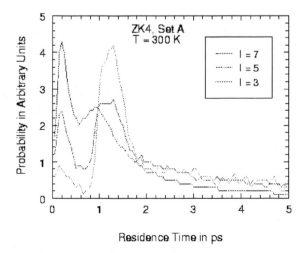

Figure 3. *Histogram of the residence time for different concentrations of guest molecules*

Figure 3 shows the probability density of the thus defined 'residence times' within the individual cavities for three different sorbate concentrations (model A). The first maximum at ~ 0.3 ps corresponds to times which are too short to allow a passage through the cavity to one of the five other windows [8].   Therefore, this maximum must be attributed to trajectories, which are reversed immediately after the molecule has passed the window. It is interesting to note that the intensity of this first maximum increases with increasing sorbate concentration. It may be concluded, therefore, that the reversal in the trajectory is mainly caused by the influence of the other adsorbate molecules. These conclusions are confirmed by minima in the velocity-auto-correlation function (vacf) (see [8]) where the time and the density dependence of these minima are in agreement with the reasoning above.

Propagator

The propagator is defined as the conditional probability density to find a particle at time t at the place $\vec{r}_0 + \vec{r}$ if it has been at time $t = 0$ at $\vec{r}_0$. For a pure random walk the propagator is a Gaussian distribution

$$P(\vec{r},t) = \left(4\pi Dt\right)^{-\frac{3}{2}} \exp\left\{-\frac{r^2}{4Dt}\right\} \qquad (2)$$

Figure 4 shows the propagator for zeolites with structural effects in correspondence to the structure of the zeolite

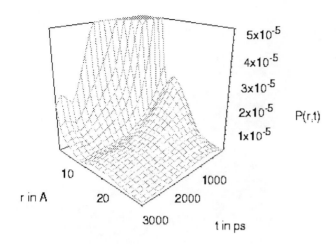

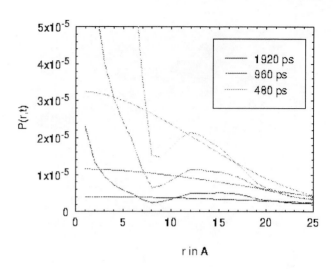

Figure 4. *Propagator* $P(\vec{r},t)$ *as function of* $\vec{r}$ *and* t *(set B,* $I = 3$, $T = 300$ K)

Figure 5. *Propagator* $P(\vec{r},t)$ *for fixed times as function of* $\vec{r}$ *compared with the ideal Gaussian-Propagators (set B,* $I = 3$, $T = 300$ K)

while in figure 5 for three special times a comparison with the ideal Gaussian behaviour is given.

The density distribution in the cation free LTA zeolite (sometimes we use for shortness the abbreviation 'ZK4' although this is somewhat inaccurate) shown in figure 6 may well be understood by the isopotential lines [8] for this zeolite. The density distributions — here demonstrated in a plane through the center of the large cavity — show a remarkable structure. Corresponding to the potential surface these distributions are different from zero practically only near the cavity wall and these densities have maxima in the window (set A) and *in front of* them (set B), respectively.

Influence of the potential parameters on D

The influence of potential parameters on the diffusion coefficients is remarkable. Even small parameter changes may cause significant changes in the diffusion processes (Figure 8).

It has been shown that the choice of the σ parameter of the Lennard-Jones potential eq. 1 for the methane - oxygen interaction has a dramatic influence not only on the value of the diffusion coefficient but also on its concentration dependence [8].

As in [8] we use the two different potentials (A, B) from the literature based on the sets of potential parameters shown in table 1.

An impression of the shape of the potential surface, especially with respect to the different behaviour in the vicinity of the window is given in figure 7 for parameter set A (left) and set B (right), respectively. Easily can be seen that the potential values are high in the center of the large cavity and, of course, at the repulsive walls. The potential has a minimum in the window in model A and in front of the window in model B where it has a saddle point in the center of the window. This threshold reduces the diffusivity in model B. From the larger value of σ in set B for the dominating oxy-

Figure 6. *Density distribution of methane in a cation free LTA (left set A, right set B)*

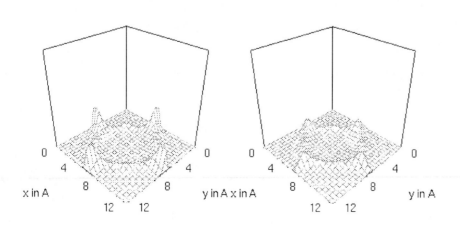

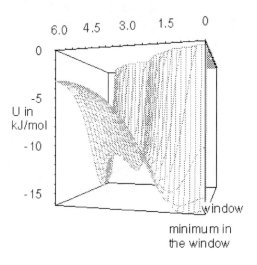

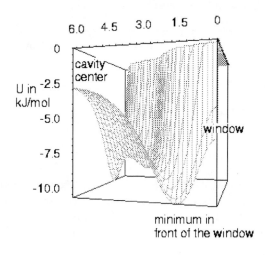

Figure 7. *Potential of cation free LTA (left: set A, right set B)*

gen - guest interaction in the zeolite a narrower window between adjacent cavities results. Especially for methane which has a similar size as the window this causes dramatical changes [8, 16].

This effect seems to be the consequence of the interrelation of two effects (reflexion at the inlet of the window, coming back after passing the window) with different density dependence [8, 16]. Fig. 8 (left) shows that the diffusion coefficient D increases with increasing mean number of guest molecules per cavity and temperature for set B while this dependence for set A demonstrated for $T = 173$ K is reversed (right). For higher loading this figure shows an interesting cross over of both curvatures which is under examination in detail.

Influence of cations on D

The dynamics even of small neutral molecules with saturated bindings is strongly influenced by the presence of exchangeable cations [9, 11]. For these investigations we have chosen the NaCaA zeolite with 4 Na and 4 Ca so that the windows — marked by w in fig. 2 — are free from cations. The unexpected (see [14]) strong effect can clearly be seen in fig. 9 and has been confirmed experimentally, meanwhile [17]. In comparison with the cation free LTA the self-diffusivity decreases up to two orders of magnitude.

It should be noted that the computational effort is much larger in this case than in our previous simulations for the cation free analogue zeolite since much longer trajectories (up to 5-10 ns) are necessary to evaluate such small D's. Additionally, the calculation of the forces resulting from the polarization energy is very time expensive although we were able to replace the full Ewald sum for our system by a corrected r space part of this sum.

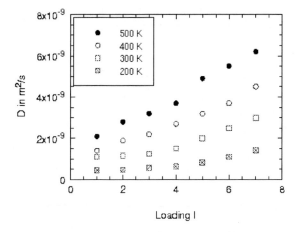

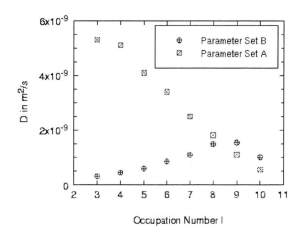

Figure 8. *The diffusion coefficient for different loadings of methane in the cation free LTA zeolite (left: different temperatures, set B; right: T = 173 K, set A and B)*

Table 1. *Potential parameters*

	σ in Å	ε in kJ/mol
CH$_4$-CH$_4$	3.817	1.232
CH$_4$-Si	2.14	0.29
CH$_4$-O (set A)	3.14	1.5
CH$_4$-O (set B)	3.46	0.81

Fig. 9 demonstrates MD results for different situations and compares these data with experiments. The agreement with experimental results from NMR measurements [18] is satisfactory in the case of set A (see fig. 9).

Influence of lattice vibrations on the diffusion coeffients

The influence of lattice vibrations on the diffusion coeffient in the cation free LTA zeolite is not very large for both of the parameter sets under consideration as can be seen in fig. 10. It should be noticed that the effect is small in the case of set B (smaller window) and larger in the case of set A. This is somewhat surprising. Suffritti and Demontis found a much stronger effect using a parameter set beyond the region of A and B [12].

Their parameters lead to a practically closed window and it appears reasonable to assume that temporary opening of windows by vibrations drastically increases *D* in such cases. All these effects are currently under examination.

Transport-diffusivity, corrected diffusivity and self-diffusivity

While the self-diffusion coefficient D_0 according to the Kubo theory may be obtained from

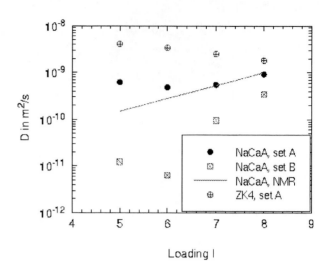

Figure 9. *Comparison D of methane in NaCaA and cation free LTA with NMR-experiments*

$$D_0 = \frac{1}{N}\sum_j \int_0^\infty dt \langle v_{zj}(0)v_{zj}(t)\rangle_0 \quad (3)$$

the so called corrected diffusion coefficient D_c includes the cross correlations between velocities of different particles.

$$D_c = \frac{1}{N}\sum_j\sum_k \int_0^\infty dt \langle v_{zj}(0)v_{zk}(t)\rangle_0 \quad (4)$$

The diffusion coefficient that appears in Fick's law usually is called transport diffusion coefficient D_T although also the other *D*'s are describing transport properties.

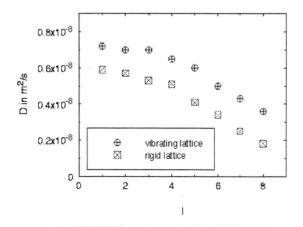

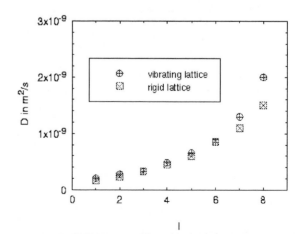

Figure 10. *Comparison of D with rigid and vibrating lattice. (left: Set A. right: Set B)*

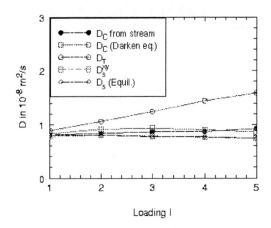

Figure 11. *Self-diffusion coefficient D_s, corrected diffusion coefficient D_c and transport-diffusion coefficient D_T from nonequilibrium-MD*

$$J = -D_T \frac{dn}{dx} \qquad (5)$$

J is the stream and n is the density. If the gradient of the chemical potential μ is interpreted as driving force for a diffusive stream then

$$J = -nB \frac{d\mu}{dx} \qquad (6)$$

where B is the mobility. According to Kubo's theory B is connected with the corrected diffusivity by the Einstein relation

$$D_c = Bk_B T \qquad (7)$$

k_B is the Boltzmann constant. Comparison of eqs. 5 and 6 together with eq. 7 leads to

$$D_T = D_c \frac{n}{k_B T} \frac{d\mu}{dn} \qquad (8)$$

which is a somewhat unusual form of the well known Darken equation. In fig. 11 these different D's obtained from equilibrium and nonequilibrium MD simulations are compared with each other. The self-diffusion coefficients are obtained from the mean square displacement. D_T results from nonequilibrium simulations in which a density gradient in six layers of cavities is created by randomly inserting particles that leave the last layer into the first layer and evaluating the stream in the intermediate region. $D_0{}^{xy}$ is obtained from the mean square displacement perpendicular to the density gradient. It turns out that the stream practically has no influence on the self-diffusion in the direction perpendicular to the stream. D_c is somewhat larger then D_0 because of the collective effects expressed by the cross correlation terms in eq. 4. D_c is obtained from D_T by the Darken equation and compared with results from annother nonequilibrium MD experiment. In this experiment a stream is produced by an external field. Measuring this stream D_c may be obtained from the Einstein relation eq. 7.

Acknowledgement We thank Professors Kärger, Pfeifer (University Leipzig), Wolfsberg (UCI Irvine) Suffritti, Demontis (University of Sassari), Brickmann (TH Darmstadt) and Theodorou (UCB) for stimulating discussions.

We are indebted to the Deutsche Forschungsgemeinschaft (SFB 294) for financial help. R.H. thanks the Fonds der Chemischen Industrie, Frankfurt, for financial support.

Finally, we are thankful for a grant of computer time by the Höchstleistungsrechenzentrum Jülich and the office of Academic Computing, University of California, Irvine.

References

1. Kärger, J.; Ruthven, D.M. *Diffusion in zeolites and other microporous solids*, Wiley, New York, **1992**.
2. Allen, M.P.; Tildesley, T.S. *Computer simulation of liquids*, Clarendon Press, Oxford, **1989**.
3. Haberlandt, R.; Fritzsche, S.; Peinel, G.; Heinzinger, K. *Molekulardynamik*, Vieweg, Wiesbaden, **1995**.
4. Yashonath, S.; Demontis, P.; Klein, M.L. *Chem. Phys. Lett.* **1988**, *153*, 551.
5. Fritzsche, S.; Haberlandt, R. Kärger, J.; Pfeifer, H.; Wolfsberg, M. *Chem. Phys. Lett.* **1990**, *171*, 109.
6. Catlow, C.R.A.; Freeman, C.M.; Vessal, B.; Tomlinson, S.M.; Leslie, M. *J. Chem. Soc. Faraday Trans.* **1991**, *87*, 1947.
7. Fritzsche, S.; Haberlandt, R.; Kärger, J.; Pfeifer, H.; Heinzinger, K. *Chem. Phys. Letters* **1992**, *198*, 283.
8. Fritzsche, S.; Haberlandt, R.; Kärger, J.; Pfeifer, H.; Heinzinger, K. *Chem. Phys.* **1993**, *174*, 229.
9. Fritzsche, S.; Haberlandt, R.; Kärger, J.; Pfeifer, H.; Waldherr-Teschner, M. *Studies in Surface Science and Catalysis*, Vol. 84, p. 2139, Elsevier Science B.V. **1994**.
10. Fritzsche, S. *Phase Transitions* **1994**, *52*, 169.
11. Fritzsche, S.; Haberlandt, R.; Kärger, J.; Pfeifer, H.; Heinzinger, K.; Wolfsberg, M. *Chem. Phys. Lett.* **1995**, *242*, 361.
12. Demontis, P.; Suffritti, G.B. *Chem. Phys. Lett.* **1994**, *223*, 355.
13. Fritzsche, S.; Haberlandt, R.; Kärger, J. *Z. phys. Chem.* **1995**, *189*, 211.
14. General Discussion during the Faraday Symposium 26 on "Molecular Transport in Confined Regions and Membranes" *J. Chem. Soc. Faraday Trans.* **1991**, *87*, 1797.

15. Demontis, P.; Fois, E.S.; Suffritti, G.B.; Quarticri, S. *J. Phys. Chem.* **1990**, *94*, 4329.

16. Fritzsche, S.; Haberlandt, R.; Hofmann, G.; Kärger, J.; Heinzinger, K.; Wolfsberg, M. *Chem. Phys. Lett.* **1996**, submitted.

17. Heink, W.; Kärger, J.; Ernst, S.; Weitkamp, J. *Zeolites* **1994**, *14*, 320.

18. Heink, W.; Kärger, J.; Pfeifer, H.; Salverda, P.; Daterna, K.P.; Nowak, A. *J. Chem. Soc. Faraday Trans.* **1992**, *88*, 515.

J.Mol.Model. (electronic publication) – ISSN 0948–5023

J. Mol. Model. **1996**, 2, 293 – 299

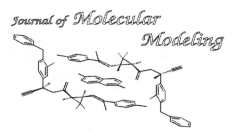

Elastic Properties of Polymer Networks

Ralf Everaers* and **Kurt Kremer**[†]

Institut für Festkörperforschung, Forschungszentrum Jülich, Postfach 1913, D-52425 Jülich, Germany

present address: Institut Charles Sadron, 6, rue Boussingault, F-67083 Strasbourg, France (ever@ics.crm.u-strasbg.fr)

† Max-Planck Institut für Polymerforschung, Postfach 3148, D-55021 Mainz, Germany

Received: 15 May 1996 / Accepted: 6 August 1996 / Published: 27 September 1996

Abstract

Many fundamental questions for the understanding of polymer melts and networks are more suitably addressed by current computer simulations than by experiments. The reason is that simulations have simultaneous access to the microscopic structure and the macroscopic behavior of well-defined model systems. The coarse-grained models used often bear little relation to actual chemical species. This is justified by the experimentally established universality of polymer dynamics and no limitation for the test and development of theories which are directed at these universal aspects. The difficulties already encountered on this level will be illustrated for entanglements between polymers which dominate the dynamic in dense systems.

For practical purposes it would, of coarse, be desirable to predict the characteristic length and time scales of experimental systems from the chemical structure of the polymer chains. Due to the extremely long relaxation times it is impossible to achieve this in brute-force simulations of truely microscopic models. Systematic coarse-graining combined with a better theoretical understanding seem to offer a practical alternative.

Keywords: polymers networks, dynamics, polymer properties, microscopic models

Introduction

Polymer networks are the basic structural element of systems as different as tire rubber and gels. They are not only technically important but also commonly found in biological systems such as the cytoskeleton. Networks of flexible macromolecules display an elastic and thermoelastic behaviour quite different from ordinary solids. [1] Crystals, metals, ceramics, or glasses can be stretched only minimally. Small deformations of the sample extend down to atomic scales and lead to an increase of the internal *energy*. Rubber-like materials reversibly sustain elongations of up to 1000% with small strain elastic moduli that are four or five orders of magnitude smaller than for other solids. Most importantly, the tension induced by a deformation is almost exclusively due to a decrease in *entropy*. As a consequence, the underlying mechanism has to be different from the case of conventional solids.

The key problem in the theory of rubber elasticity is the correct identification of the microscopic sources of this entropy change. An at least qualitative explanation was found in the 1930, when it was realized that rubber is the result of cross linking a melt of long flexible chain molecules. Such polymers adopt random coil conformations and behave as entropic springs. The classical theories of rubber elasticity [1] estimate the elastic properties of a polymer network from the elongation of the network strands. This explanation ne-

* To whom correspondence should be addressed

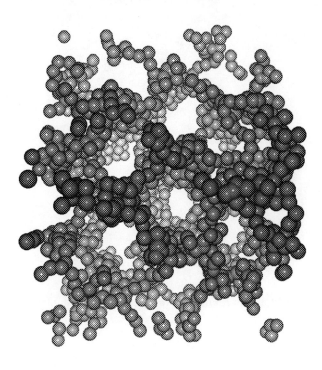

Figure 1. *Network of bead-spring polymer chains with the connectivity of a diamond lattice. The typical extension of the network strands corresponds to that of free chains in a melt. The "loose" monomers are due to the representation using periodic boundary conditions.*

glects the mutual impenetrability of the chains, although it is the reason for the viscoelastic properties of un-cross-linked polymer melts. [2] The well-known "magic putty" jumps, when formed accordingly, like an ordinary elastic ball. Left to rest on a table, it flows like thick honey. The difference between the two situations lies in the typical time scale over which the force acts on the material. This behavior is universal in the sense that beyond a chemisty-dependent minimal chain length all polymers follow the same laws.

A simple microscopic explanation of these laws is provided by the tube model [2]. Entanglements with other molecules restrict each polymer to a one-dimensional diffusion along its own coarse-grained contour (reptation). The tube is permanently reformed at its ends. To relax the tension in a sheared melt the chains have to reptate out of their original tubes. On shorter time scales the melt displays rubber elasticity with a time and chain length independent modulus; on larger times scales the melt flows like a viscous fluid with a viscosity that increases by a factor of ten when the chain length is doubled. On the whole the tube model is in good agreement with experiments. What is lacking, however, is an understanding of the relation between the parameters of the tube model and the chemical structure of the polymers. Considering how much energy is required in order to pump and to stir intermediate and final products in synthetics production the need for a microscopic understanding of entan-

glement effects becomes obvious. Here the investigation of polymer networks offers a couple of advantages. Cross-linked chains are subject to the same, in their nature topological constraints as chains in a melt without having the possiblity to free themselves by reptation. Furthermore, from a mathematical point of view entanglements are only rigorously defined between closed curves like the meshes of a network. Remarkably enough, there is no definite experimental answer to the question, if and how much entanglement effects contribute to the elasticity of polymer networks.

One reason are the great difficulties in the chemical preparton and characterisation of model networks. A randomly cross-linked melt of linear polymers has a highly irregular connectivity. Typical defects are polydispersity, dangling ends and clusters, and self loops. These imperfections are present to some degree in all experimental system and a serious complication in a study of the consequences of topology conservation. The second reason is that although it is comparatively simple to measure the macroscopic properities such as the shear modulus, experiments provide little microscopic information.

Computer simulations offer a couple of important advantages over experiments [3]. We mention the greater freedom in and control over the formation of the networks, the access to the microscopic structure and dynamics, or the possible realization of Gedankenexperiments such as the comparison of otherwise identical systems with and without topology conservation. The key, however, to the optimal use of the information available in a computer simulation is the simultaneous determination of macroscopic quantities. Unfortunately, it is particularly complicated to determine in a simulation what is easily accessible in experiments: the macroscopic elastic properties. The long relaxation times require corresponding computer resources, so that the present simulations were the first in which shear moduli could be measured reliably by investigating strained samples. Since we also have complete access to the microscopic structure and dynamics in both, the strained and the unstrained state, we are in a unique position to test statistical mechanical theories of rubber elasticity without invoking adjustable parameters.

Simulations of model networks

In this paper we summarize results of molecular dynamics simulations of model polymer networks with diamond lattice connectivity, where we have exploited all of the advantages mentioned above. Such systems, which cannot be prepared experimentally, are free of "chemical" defects in the network structure. Thus, the effect of physical knots and entanglements is isolated from other sources of quenched disorder.

The individual diamond networks are spanned across the simulation volume via periodic boundary conditions. We have chosen an average distance between connected

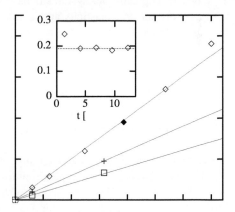

Figure 2. *Stress relaxation (inset) and stress-strain curvs for diamond networks in the common representation* $\sigma(\lambda)$ *versus* $\lambda^2 - \lambda^{-1}$: *N=44 (□), N=26 (+), (N=12) (◊). The shear modulus is given by the slope.*

crosslinks equal to the root mean square end–to–end distance of the corresponding chains in a melt. The density of a single diamond net decreases with the strand length. To reach melt density we place several of these structures in the simulation box and work with interpenetrating diamond networks (IPDN). On the length scale of the network strands the structure is similar to experimental systems, which are also locally interpenetrating with nearest neighbor crosslinks that are not connected by a network strand.

We follow two distinct strategies to isolate the entanglement effects. One is to calculate quenched averages for otherwise identical systems with different topology. In our simulations of randomly and of regularly IPDN we employ interaction potentials which ensure the mutual impenetrability of the chains, thereby preserving the topological state from the end of the preparation process. The second strategy is to calculate annealed averages over different topologies. This can either be achieved trivially by simulating non-interacting phantom chains or by using interaction potentials that allow chains to cut through each other but nevertheless preserve the monomer packing of the melt. The structure of the chains is almost identical for all systems and by comparing their behaviour we can directly access the effects of the topological constraints.

As already explained chemial details are not important for the understanding of the *universal* aspects of rubber elasticity, even though they are, of course, crucial when it comes to preparing experimental systems with as few defects as possible. For our simulations we used the same coarse-grained model as in earlier investigations of polymer melts and networks [4, 5, 6]. The network strands were modeled as freely

jointed bead spring chains of uniform length N. They were crosslinked by four-functional monomers into networks with the connectivity of a diamond lattice. There were two types of interactions, an excluded volume interaction, U_{LJ}, between all monomers and a bond potential, U_{FENE}, between chemical nearest neighbors. With ε, σ and τ as the Lennard-Jones units of energy, length and time we worked at a temperature $k_BT = 1\varepsilon$ and at a density $\rho = 0.85\,\sigma^{-3}$. The average bond length was $l = 0.97\,\sigma$ and topology was conserved. The relevant length and time scales for chains in a melt are the mean-square end–to–end distance $<R^2>$ $(N) \approx 1.7\ l^2\ N$, the melt entanglement length $N_e \approx 35$ monomers and the Rouse time $\tau_{Rouse} \approx 1.5\ N^2\ \tau$ [4]. We carried out molecular dynamics simulations, where the system was weakly coupled to a heat bath. The same samples were simulated first in the unstrained state and subsequently elongated by typically 50%. Deformations were implemented as a short sequence of small step changes at the beginning of the runs. We measured the pressure tensor, $\hat{P}$, and watched the relaxation of the normal tension $\sigma = P_{xx} - \frac{1}{2}(\ P_{yy} + P_{zz})$ to a plateau value (inset in Figure 2). The stress relaxation was completed after a period of about $2\tau_{Rouse}$ compared to our overall simulation times of the order of $10\tau_{Rouse}$. For our largest systems this is equivalent to $8\cdot10^{10}$ particle updates. Data from the initial relaxation period was discarded for the analysis of conformational properties.

By varying the interaction potentials we can investigate systems which differ only in the mutual penetrability of the chains. The simplest example are "phantom" networks, in which the excluded volume interaction is switched off ex-

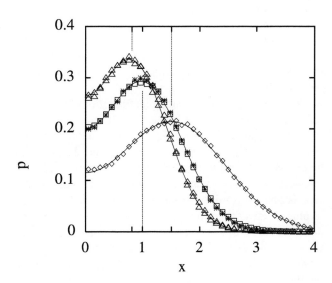

Figure 3. *End-to-end distance distribution of the network strands for N=44 in the un-strained (□) and strained state (∥ ◊ and ⊥ Δ, λ = 1.5). For comparison: data for phantom networks (*). Distances are measured in units of the lattice constant x_l of the diamond lattice.*

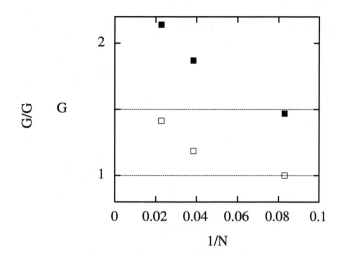

Figure 4. *Strand length dependence of the shear modulus normalized to the phantom model prediction* G_{ph}. *The filled symbols represent the measured values* G/G_{ph} *and the open symbols indicate the classical prediction* G_{class}/G_{ph}. G_{aff}/G_{ph} = 3/2 *is the upper limit for any classical theory.*

cept for nearest and next-nearest neighbors along the chains. Different strands of the network can now freely penetrate each other. What is important to note, is that the structure of the chains (and in particular $< R^2 >$) is almost identical to the original case. This is, on the one hand, useful for the preparation of relaxed initial conformations. More importantly and in contrast to experiments, the comparison allows to quantify the effect of topology conservation on the elastic properties of polymer networks.

Simulations of regularly IPDN start from intercalating conformations of strongly swollen networks with completely stretched strands. In MD runs the conformations are slowly compressed to melt density. The important point is that the topology conserving LJ interaction between all monomers is used right from the beginning.

The preparation of the initial conformations for the randomly IPDN is illustrated in the first video sequence. The networks are set up at melt density. Between the crosslinks on the diamond lattice sites we place phantom chains generated in Monte Carlo simulations with the proper end-to-end distance. After the relaxation of the lattice structure in MD runs for phantom chains, we introduce the repulsive excluded volume interaction between the monomers. This is done by slowly building up a cosine potential up to a point where the monomer distances are large enough for the LJ potential. From that point onwards the random topology is quenched and the random entanglements between meshes of the different networks become permanent.

We have investigated systems with strand lengths $N = 12, 26, 44$ corresponding to $n = 5, 7, 9$ independent, but mutually interpenetrating diamond networks. The total number of particles ranged from 8000 and to 51264 monomers. The density of elastically active strands is given by $\rho_{strand} = \rho /$

(N+1/2). Macroscopically our systems behave as ideal rubber with a purely entropic elasticity and exhibit the classical stress-strain relation (Figure 2). The measured shear moduli G for systems with conserved random topology are between 50% (N=12) to 100% (N=44) larger than in the other cases [7].

The classical theories of rubber elasticity

The most important step to an understanding of rubber elasticity is the examination of the typical conformations of a polymer chain in a melt [1]. Independent of their chemical structure all chain molecules adopt random coil conformations on large length scales. Important with regard to the elastic properties is that for a random coil those conformations are the most probable in which the chain ends are close together. An external force pulling the chain ends apart therefore forces the polymer to adopt a less probable conformation. The internal energies, on the other hand, remain practically unchanged as long as the polymer is not stretched completely. It is the loss of entropy which is responsible for the restoring forces. The force-elongation relation corresponds to that of a linear spring.

The classical theories calculate the elastic properties by treating the network strands as independent entropic springs and estimating their elongation under deformations of the sample. The predicted shear moduli are of the order of k_BT times ρ_{strand}, the density of elastically active network strands. Remarkably enough, the result (including the prefactors) is completely independent of the chemical structure of the polymers. It is, however, far from obvious that the chains may be treated as independent. The reason is the mutual impenetrability of the polymers which severely restricts the conformations accessible to individual network strands. In spite of weighty theoretical arguments [9, 10] the importance of entanglements for the elasticity of polymer networks has been disputed for decades, not least because the stress-strain relations of most experimental systems fit the predictions of the classical theories.

Test of the classical theories

The state of a network of (entropic) springs is most conveniently characterized by distribution functions $p(x)$, $p(y)$ and $p(z)$ for the Cartesian components of the spring end-to-end vectors (Figure 3). For the original unstrained state the distribution functions are identical for phantom networks and randomly IPDN and can be calculated analytically. The mean elongation is given by the bond lengths of the diamond lattice r_1. The actual end-to-end distances fluctuate around corresponding values, so that $p(x)$ etc. are given by a superposition of two Gaussian centered at $\pm x_1$.

Figure 3 illustrates the changes in the spring elongations after a deformation of the sample. While they increase par-

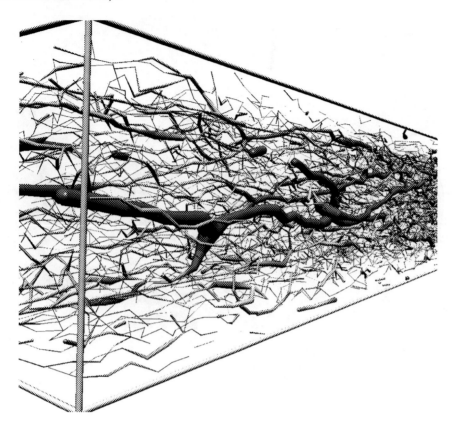

Figure 5. *Conformation of strained randomly interpenetrating diamond networks (λ=3.2). In the non-linear regime there are a few highly stretched paths (marked by thick radii and in red) where a large fraction of the induced tension is localized. The apparant interruption of the chains is due to the representation in periodic boundary conditions.*

allel to the elongation, they actually decrease in the perpendicular directions. The maxima change "affinly" with the outer dimensions of the sample: If the simulation box is stretched by a factor of two in x-direction, the same holds for the x-component of the mean elongation of the network strands. At the same time the size of the simulation box as well as the mean elongations in y- and z-direction are reduced by a factor of $\sqrt{2}$, so that the volume remains unchanged.[a] This behavior is fulfilled almost exactly by experimental systems. In our simulations the volume conservation was enforced. Regarding the width of the distributions our results confirm the crossover from phantom to affine behavior with increasing chain length which has been discussed in the framework of the most sophisticated classical theories [8]. Decisive, however, is the quantitative comparison of the classical moduli calculated from the change in the elongation of the network strands with the measured moduli. Figure 4 demonstrates that *the true change in entropy and accordingly the true modulus are by far higher than the values one obtains within the classical picture* [7]. This result shows that a theory that aims to calculate the effects of topology conservation from limitations in the crosslink fluctuations is bound to overlook relevant contributions to the total entropy change even if no further approximations are being made. The discrepancy grows with increasing chain length.

Entangled meshes

Explaining the discrepancy by including the topological constraints in a first principles statistical mechanical treatment has proven extremely complicated [9, 10, 11, 12]. Progress is often due to the study of comparatively simple models of the effects of the topological constraints. It is therefore of great importance to clarify, if a model captures the relevant physics. Simulations can play an important role in answering this question. As an example we refer the reader to the classical theories discussed above. As we have shown, it makes little sense to pursue this approach with ever more sophisticated refinements.

The reason for the increase of the modulus due to topology conservation can be illustrated by an analysis of the stress distribution in strongly stretched networks (see Figure 5 and the second video sequence). Our simulations show that in randomly IPDN with topology conservation a large part of the tension is localized on *topologically* shortest paths through the system. These paths are composed of strands as well as meshes with physical entanglements propagating the tension in the same manner as chemical cross-links. The way the chains fail to release a link is an artefact of our model. At too large stresses the connected beads at the contact point are driven so far apart that the chains can slip through each other. Thus, in contrast to a real system, the chains do not break in the process. Since the energy threshold is of the order of $70\,k_BT$ such events do not occur at small elongations.

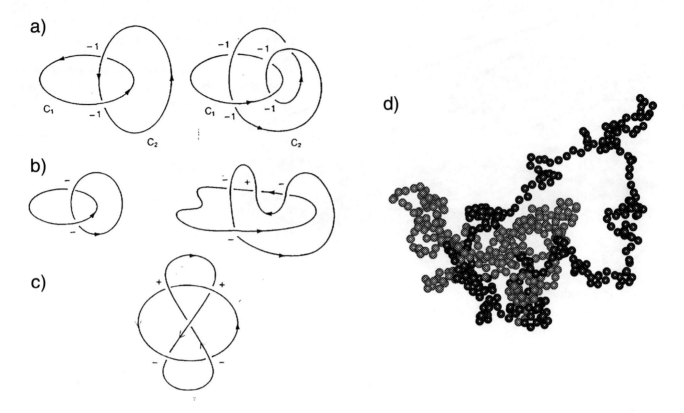

Figure 6. *a) Illustration of the index method for characterizing links. b) Invariance of the linking number under deformations. c) Trapped figure eight: a linked pair of curves with I=0. d) Example from the simulation for two linked meshes.*

The stress localization in diamond networks is completely unexpected from the point of view of the classical theory, since all network strands are equivalent. The highly arteficial regularly IPDN mimick a situation where this equivalence is preserved for a conserved topology. When these networks are stretched, all strands contribute equally to the elastic response. The high tensions we observed along a few paths and at much lower elongations in the first case, occur now homogeneously throughout the whole system when all strands are stretched to their full contour length.

A similar analysis can be performed for swollen networks (see the third video sequence).

For a single diamond network swelling reveals the regular connectivity, i.e. it leads to a state similarly to what is shown in figure 1 but with stretched network strands. For regularly IPDN the individual diamond "lattices" can move against each other. In the case of randomly IPDN, however, the entanglements lead to an aligment of the network strands and the formation of pores.

Modeling the effects of topology conservation requires two steps [9, 10]: the characterization of the entanglements and an estimate of their contribution to the restoring forces resisting a deformation of the sample. The first part is a math-

ematical problem and the subject of knot theorie, which deals with classifying single ("knots") and several ("links") closed curves embedded in three dimensional space. Usually the classification starts from a projection of the curves onto a plane. The most important tool are topological invariants, i.e. numbers or polynomials which retain their values for different projections as well as under continuous deformations of the curves which do not require opening and closing the rings in between. A topological invariant is the more suitable for the purpose of classification, the less frequent identical values are assigned to different knots or links, as they are tubulated in extensive lists.

A simple example for the approximate characterization of links is the Gauss linking number. It can be calculated by a method where all crossing points of the two curves in a projection are indexed by $\pm 1/2$ (Figure 6). The sign depends on the direction into which the tangent vector of the upper curve has to be rotated in order to coincide with the one of the bridged curve. The linking number I is defined as the sum of the indices and is a topological invariant. There are, however, a few examples (Figure 6c) for linked curves with linking number zero, so that the classification is not completely reliable.

For estimating the interaction between meshes of a network the mathematical link tables are only of limited use. On the one hand the classification is too detailed (it makes little sense to distinguish between more complex entanglements of two rings and to neglect interactions between three or more rings). On the other hand the link tables provide no information on the likelihood of the occurence of particular links for random walk-like ring polymers.

The simplest way to motivate an effective topological interaction between ring polymers is based on a center of mass distance-dependent linking probability [13]. The argument is analoguos to the treatment of a polymer chain as entropic spring. Since there are less conformations of two linked rings with a large center of mass distance, the rings behave as if connected by an entropic spring. By the same rational there is an repulsive entropic interaction between non-linked rings. In the framework of a network theory Graessley and Pearson [14] treated entanglements between meshes as additional entropic springs and predicted a contribution proportional to the entanglement density ρ_{ent} to the modulus. The prefactor depends on the interaction law and can be estimated from linking probabilities, which, in general, are accessible only through computer simulations.

It is very difficult, if not impossible, to decide in experiments, whether or not such an approach makes sense. In the framework of our simulations we determined the degree of linking for all mesh pairs in the system and estimated a topology contribution to the modulus on the basis of the model of Graessley and Pearson without invoking any adjustable parameters. The comparision with the measured values was suprisingly positive. We observed the predicted proportionallity $G - G_{ph} = 0.85\, k_B T\, \rho_{ent}$, with a prefactor of the same order of magnitude as the estimate of $1.3\, k_B T$ [15]. In view of the substancial simplifications the agreement is quite remarkable.

Summary

We presented the first simulations of model polymer networks in which shear moduli could be measured reliably by investigating strained samples. This enabled us to prove quantitatively that the classical treatment of the network strands as independent entropic springs omits important contributions to the total entropy change in deformed networks. The reason is the quenching of the topology (e.g. the state of linking of mesh pairs) during the formation network. We have estimated a topology contribution to the modulus in the framework of a simple theory based on entropic interactions between loops. The agreement with the measured values constitutes a small, but nevertheless *quantitatively* controlled step towards a topological theory of rubber elasticity. A great challenge for future work is the derivation of the empirically very successful tube model from considerations based on the topological character of the constraints. It is difficult to imagine that the necessary information for a test of such a theory could be obtained in experiments. In our view, this makes simulations an indispensible intermediate step in the understanding of the viscoelasticity of polymeric materials.

References and Footnotes

[a]. This behavior is fulfilled almost exactly by experimental systems. In our simulations the volume conservation was enforced

1. Treloar, G. *The Physics of Rubber Elasticity*, Clarendon, Oxford, 1975.

2. Doi, M. and Edwards, S.F. *The Theory of Polymer Dynamics,* Claredon Press, Oxford, 1986.

3. Everaers, R.; Kremer, K. and G.S. Grest, *Macromol. Symposia* **1995**, *93*, 53.

4. Kremer, K. and Grest, G.S. *J. Chem. Phys.* **1990**, *92*, 5057 ; **1991**, *94*, 4103 Erratum.

5. Duering, R.; Kremer, K. and Grest, G.S. *J. Chem. Phys.* **1994**, *101*, 8169.

6. Kremer, K. and Grest G.S. in *Monte Carlo and Molecular Dynamics Simulations in Polymer Science*, Binder, K. (ed.), Oxford University Press, New York, 1995.

7. Everaers, R. and Kremer, K. *Macromolecules* **1995**, *28*, 7291.

8. Erman, B. and Flory, P.J. *Macromolecules* **1982**, *15*, 800.

9. Edwards, F. *Proc. Phys. Soc.* **1967**, *91*, 513.

10. Edwards, F. *J. Phys.* **1968**, *A 1*, 15.

11. Edwards, F. and Vilgis, T.A. *Rep. Progr. Phys.* **1988**, *51*, 243.

12. Heinrich, G.; Straube, E. and Helmis, G. *Adv. Pol. Sci.* **1988**, *85*, 34.

13. Vologodskii, V.; Lukashin A.V. and Frank-Kamenetskii, M.D. *Sov. Phys.-JETP* **1975**, *40*, 932; Frank-Kamenetskii, M.D.; Lukashin, A.V. and Vologodskii, A.V. *Nature* **1975**, *258*, 398.

14. Graessley, W.W.; Pearson, D.S. *J. Chem. Phys.* **1977**, *66*, 3363.

15. Everaers, R. and Kremer, K. *Phys. Rev.* **1996**, *E 53*, R37.

J. Mol. Model. **1996**, 2, 300 – 303

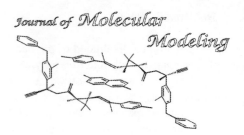

The Photodissociation/Recombination Dynamics of I₂ in an Ar Matrix: Wave Packet Propagation in a Mixed Quantum-Classical Picture

Udo Schmitt [a] and Jürgen Brickmann*

Physikalische Chemie I and Darmstädter Zentrum für Wissenschaftliches Rechnen; Technische Hochschule Darmstadt Petersenstr. 20, D-64287 Darmstadt, Germany (brick@pc.chemie.th-darmstadt.de)

[a] part of the Ph.D.-thesis of Udo Schmitt ,Technische Hochschule Darmstadt (D17).

Received: 15 May 1996 / Accepted: 12 September 1996 / Published: 27 September 1996

Introduction

The theoretical description of photodissociation processes of molecules after short time laser excitations is essentially based on the formalism of quantum mechanics, i.e. both the product formation and the energy distribution are strongly related to the time evolution of a quantum wave packet. This time evolution can be treated within fully quantum dynamical concepts with reasonable computation effort only if a very small number of degrees of freedom are considered. The situation here is very similar to that of quantum scattering theory. Only very small isolated molecules or a small number of system coordinates can be treated. Large molecules or molecules in contact with their environment are out of the scope of fully quantum dynamical calculations. In this work we present the results of a mixed quantum-classical approach wherein one relevant degree of freedom is treated quantum dynamically while all the others follow the laws of classical dynamics. The model system is a iodine molecule imbedded in a solid argon matrix. The treatment of the combined dynamics follows a discrete time-reversible propagation scheme for mixed quantum-classical dynamics which has been published recently by the present authors [1]. The results of the simulations are visualized and presented as a time sequence of images.

The time dependent SCF approximation

A considerable simplification of the dynamics of a system with many degrees of freedom can be obtained by separating the total system into two parts, the "relevant system", which is treated quantum mechanically, and the "bath" degrees of freedom, which shall be well described by classical equations of motion. A possible realization of such hybrid methods is the quantum/classical time-dependent self-consistent field (Q/C TDSCF) ansatz suggested by Gerber, Ratner and coworkers [2]. Within the Q/C TDSCF approach the total system wave function is approximated as a product of wavefunctions for each set of co-ordinates. Furthermore, it is assumed that the probability distribution of the bath degrees of freedom remains "classical" in time, i.e. the corresponding part of the wave packet remains strongly localized in co-ordinate and momentum space. The total system dynamics is then controlled by a time-dependent Schrödinger equation for the relevant system and classical equations of motion for the bath degrees of freedom. The coupling between the two sets is represented by the parametric dependency of the system Hamiltonian on the bath coordinates and by the functional dependency of the effective classical potential in the system wavefunction.

* To whom correspondence should be addressed

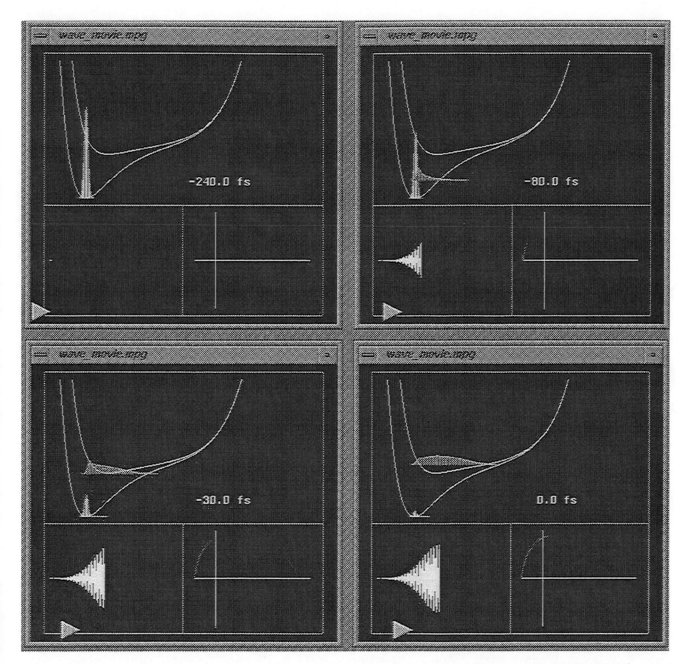

Figure 1. *Four individual scenes of the time sequences at time t = -240 fs, -80 fs, -30 fs and 0 fs.*

The discrete time propagation scheme

Intrinsic problems of the mixed mode propagation are related to the inherent time scale difference due to the large frequency disparity within both subsets. Using a single time scale for both, the quantum and classcial degrees of freedom, a very small time step is necessary for the description of the fast motion of the quantum modes. Consequently, such a procedure is time consuming and inefficient. This problem can be circumvented by the introduction of two different time steps for the quantum and classical modes,

respectively. A numerically stable multiple time step scheme has been suggested recently by the authors [1]. It is based on a reformation of the standard Q/C TDSCF equations of motion as a purely classical Hamilton-Jacobi type description by replacing the complex time-dependent coefficients representing the quantum wave function by a linear combination of formal position and momentum co-ordinates. By grouping the formal position and momentum co-ordinates of the quantum part with the co-ordinates representing the classical subset into a new formal phase space [3], the time evolution can be described by a classical Liouville equation. The total system time evolution operator can be split off by using Trotter's formula [4,5], which then allows the introduction of a second, small time step for the integration of the quantum part. For a detailed dis-

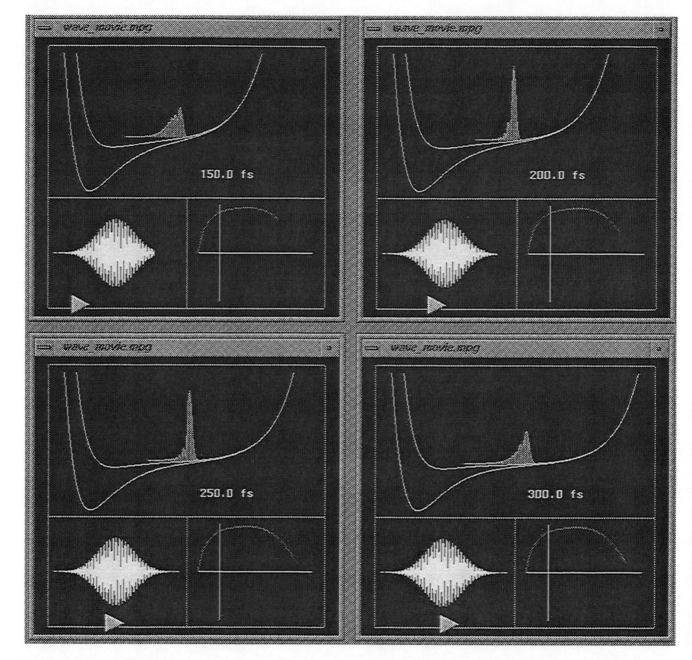

Figure 2. *Four individual scenes of the time sequences at time t = 150 fs, 200 fs, 250 fs and 300 fs.*

cussion see Ref. [1]. The resulting propagation scheme (reversible scheme, RPS) give rise to a significantly enhanced integration stability with respect to energy conservation, and rigorously accounts for time-reversibility due to the symmetric exponential factorization of the propagator.

The model system

We have applied the reversible propagation scheme (RPS) to simulate the photoexcitation process of I_2 immersed in a solid Ar matrix initiated by a femtosecond laser puls. The system serves as a prototypical model in experiment and theory for the understanding of photoinduced condensed phase chemical reactions and the accompanied phenomena like the cage effect and vibrational energy relaxation. The system/bath separation is introduced in the following manner:

The quantum subsystem is taken as the motion along the vibrational stretch co-ordinate of the I_2 molecule on two electronic surfaces, the ground state X and the excited state A, which are coupled through a transient off-diagonal interaction with a Gaussian shaped laser puls (τ=80 fs) with a carrier frequency of λ=728 nm. The wave function representing the I_2 stretch is expanded in a discrete coordinate representation using 256 equally spaced grid points for each electronic state. The two electronic states are represented by Morse potentials with appropriate parameters. The I_2 is inserted into a double-substitional site of an Ar fcc-lattice, which is modelled by 498 classical Ar atoms placed in a cubic box employing periodic boundary conditions. The Ar-Ar and I-Ar intercations are described by pairwise additive Lennard-Jones potentials. The equations of motion are integrated within the RPS using a large time step $\Delta t = 5$ fs for the classical system, and a small time step $\delta t = 0.05$ fs for the quantum part. In the classical part, the well-known velocity Verlet integrator is employed, whereas the quantum subsystem is integrated with an explicit symplectic partitioned Runge-Kutta of third order. A detailed description of the numerical implementation will be given elsewhere [6].

Wave packet visualization

The direct way of distributing the results of a dynamics calculation is the display of time sequences. Here we present two sequences, which are generated as it is shown in Figs. 1 and 2. Three different subwindows are generated: the upper part showing the electronic ground (lower curve) and the excited A state (upper curve) of the I_2 molecule potential surfaces, which are modified with respect to the gas-phase potential by the time-dependent mean-field Ar-I interaction resulting in a solvent cage repulsive wall for larger I-I separations. Furthermore, the time-dependent probability density of the I_2 stretch on the X (yellow) and the A state (purple) are depicted. A clock giving the time in femtoseconds is placed in the left part underneath the electronic states. In the lower part, the electric field amplitude representing the laser puls is shown as a function of time in the left corner, whereas in the right part the expectation values of the co-ordinate and momentum for the I_2 strectch are given in a phase space-like illustration. Therein, the time evolution is shown in a color-coded manner, starting with phase space points at $t = -250$ fs in red, and ending up with yellow at $t = 6750$ fs. The time origin is chosen as $t = -250$ fs to obtain the maximum of the laser intensity at $t = 0$ fs. The complete time evolution can be explored interactively and is attached as an electronic supplement.

Discussion

The complete time evolution of the wave packet can be studied by travelling along the time axis in both directions. Starting from a vibronic wave function respresenting the vibrational ground state on the lower electronic surface, the transient electric field intensity coupling the two electronic states induces a transition to the excited A state. Due to the strong repulsive character of the A state potential surface in the Frank-Condon regime, the resulting wave packet starts to move immediately towards larger I-I seperation. After $t \approx 60$ fs the laser pulse has almost completely populated the A state resulting in a broadened, approximately Gaussian shaped wave packet travelling towards the solvent cage wall. The collision of the wave packet with the repulsive solvent wall gives rise to a significant energy transfer to the nearest Ar atoms, which results in the solvent wall being pushed away. During that process, the wave-packet refocuses due to quantum interference of the incoming part of the wave-packet with the one already at the solvent wall reflected. While the solvent wall is moving outwards, the wave-packet, which has lost most of its kinetic energy, starts to delocalize along almost the whole I_2 stretch range. After $t \approx 600$ fs, the solvent wall starts moving backwards, kicking the delocalized wave packet, which results in an energy transfer back into the I_2 stretch vibration, causing the wave packet to travel towards shorter I-I separations. After being reflected at the inner repulsive wall of the A state, the fairly delocalized wave packet interacts with the solvent cage a second time, where the amount of energy being exchanged is noticable reduced compared to the first cage encounter.

References

1. Schmitt, U.; Brickmann, J. *Chem.Phys.* **1995**, *208*, 45.
2. (a) Gerber,R. B.; Buch, V.; Ratner, M. A. *J.Chem.Phys.* **1982**, *77*, 3022; (b) Gerber; R. B. and Ratner, M. A. *Adv.Chem.Phys.* **1988**, *70*, 97.
3. Gunkel, T.; Bär, H.-J.; Engel, M.; Yurtsever, E. and Brickmann, J. *Ber. Bunsenges. Phys. Chem.* **1994**, *98*, 1552.
4. Trotter, H.F. *Proc. Am. Math. Soc.* **1959**, *10*, 545.
5. Tuckermann, M.; Berne, B.J. and Martyna, G.J. *J. Chem. Phys.* **1992**, *97*, 1990.
6. Schmitt, U.; Brickmann, J. to be published.

J. Mol. Model. **1996**, 2, 304 – 306

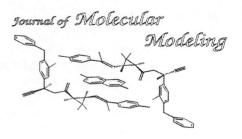

The Genetic Algorithm Applied as a Modelling Tool to Predict the Fold of Small Proteins with Different Topologies

Thomas Dandekar

Europäisches Laboratorium für Molekularbiologie, Postfach 102209, D-69012 Heidelberg, Germany
(Thomas.Dandekar@Mailserver.EMBL-heidelberg.DE)

Received: 15 May 1996 / Accepted: 6 August 1996 / Published: 27 September 1996

Abstract and Introduction

The genetic algorithm exploits the principles of natural evolution. Solution trials are evolved by mutation, recombination and selection until they achieve near optimal solutions [1].

Our own approach has now been developed [2] after a general overview on the application potential for protein structure analysis [3] to a tool to delineate the three-dimensional topology for the mainchain of small proteins [4], no matter whether they are largely helical, are mixed or β-strand rich [5].

Results on several protein examples for these different modelling tasks are presented and compared with the experimentally observed structures (RMSDs are around 4.5-5.5 Å). To start a modelling trial only the protein sequence and knowledge of its secondary structure is required. The fittest folds obtained after the evolution at the end of the simulations yield the three dimensional models of the fold. Current limitations are protein size (generally less than 100 aminoacids), number of secondary structure elements [7-8] and irregular topologies (e.g. ferridoxins).

Further, preliminary results from current simulations are illustrated. We now want to apply simple experimental or other information, which is available long before the three-dimensional structure of the protein becomes known, to refine the modelling of the protein fold and tackle also more difficult modelling examples by our tool.

Keywords: Genetic algorithm, protein structure analysis, 3D topology

Methods

To achieve protein structures close to observed starting from a population of random structures, selection of structures according to basic protein building principles is applied. They are briefly summarized in the following table (Table 1; for additional details see [2],[4],[5]) and focus around global and hydrophobic packing, avoiding clashes, stabilization of secondary structure (used as input, to avoid bias from bad prediction the DSSP assignment in many trials, but similar trials are also run using standard secondary structure prediction methods) and the build up of strands and sheets.

Results and discussion

With these criteria, we can model in our simulations the main chain topology of a number of different protein structures [5]. Table 2 gives two examples for each of the categories helical, mixed and strand.

Table 1. *Fitness function criteria*

criteria [a]	des [b]	term		specific parameters
constant [c]	C	$weight_C$		adjusted to 10% negative fitness in the first generation
clash	cl	$weight_{cl} \cdot \Sigma$ overlap [d]		$weight_{cl}$ = -500
secondary structure(ss):				
	pf	$weight_{ss} \cdot$	structural preference [e]	
	co	$weight_{ss} \cdot$	cooperativity [f]	$weight_{ss}$ = +12
tertiary structure: global scatter (gs)				
	gs	$weight_{sc} \cdot$	scatter(sc) [g]	$weight_{sc}$ = -24
hydrophobic scatter (hs)				
	hs	$weight_{hd} \cdot$	hydrophobic distribution(hd) [h]	$weight_{hd}$ = -19 hydrophobic residues include: Phe,Tyr, Met,Cys, Ile,Leu,Val,Trp
β-strand criteria [i]: hydrogen bond	hyd	$weight_{hyd} \cdot$ hydrogen bond		$weight_{hyd}$ = + 15 bondcount + betapair + bondstrand + revturn + 2 · bondsheet
sheetdir sh		$weight_{sh} \cdot$ sheetdir		$weight_{sh}$ = + 6; within 66°, reward = +1 within 35°, additional reward = +6;

[a]   The total fitness measures the quality of the structure encoded by an individual bit string. It is the sum of the general fitness terms (C + cl + pf + co + gs + hs) and the §-strand fitness (hyd + sh) plus new fitness terms exploiting experimental information currently investigated.

[b]   The term "des" refers to the abbreviated designation for the criteria involved.

[c]   The constant keeps the population of prediction trials richer since low fitness individuals may also survive.

[d]   Mainchain atom overlaps were counted.

[e]   Structural preference rewards all residue conformations encoded in a bit string which agree with the secondary structure (known or predicted) used in the trial.

[f]   Cooperativity yields a reward for any two consecutive residues in the same dihedral conformation

[g]   scatter of all residues around the center of mass

[h]   distribution of hydrophobic residues around the center of mass

The root mean square deviation in Angstrøms of topologically equivalent C_α atoms in the fittest structure relative to those observed is given. Left entry RMSD values include and right value exclude connecting loop residues. For each protein the fittest fold obtained after 10 simulation runs is given, Aa denotes the number of the amino acids in the protein; terminal loop residues were not included in the simulations. The secondary structure is sketched ("a" denotes helices; "T", turns; and "b", beta strands).

To model the topology of a wider range of proteins, in particular more complex topologies or proteins of larger size, we currently investigate additional fitness parameters which can be derived from further, for instance experimental, data available on the protein without knowing its three dimensional structure.

One such example are inclusion of disulfide bonds as a selection criterion. To implement them as a fitness parameter, different potentials and weights have to be tested and simple protein structures such as crambin act as a test fold. Figure 1 shows that a reasonable topology and slight RMSD improvement can be achieved applying this criterion even without having optimized its fitness weight (Figure 1). Another protein investigated is anemona toxin, a very irregular (not much secondary structure) protein fold (Figure2). The topology obtained by including the disulfide bonds as an additional distance constraint here is not too far from observed

Table 2. *Modelling different protein topologies*

helical proteins:

1HMD (113 Aa)	a4	4.9	3.7	hemerythrin
1ERP (37 Aa)	a3	3.5	3.3	mating pheromone

strand rich:

1BBI (71 Aa)	b6	5.7	5.1	Bowman-Birk inhibitor
1DEF (30 Aa)	bTbTT	4.5	3.5	defensin

mixed structures:

1CRN (46 Aa)	baaba	5.4	4.2	crambin
4CRO (66 Aa)	a3TbTb	6.0	5.1	lambda cro-repressor

Figure 1. *Crambin*
(A) Simulation result; the simulation result is given as a brk-file containing the main-chain atoms; RMSD to observed 5.3 Å
(B) crystal structure (1CRN.BRK)

Figure 2. *Anemona toxin*
(A) Simulation result; the simulation result is given as a brk-file containing the main-chain atoms; RMSD to observed 6.2 Å
(B) crystal structure (1ATX.BRK)

in spite of the irregularity of the protein and has an RMSD to the crystal structure of 6.2 Å.

A completely different criterion studied is for instance the formation of a protein core. This criterion can either be derived from studying and comparing the architecture of related folds or by mutagenesis data. Also with these fitness criterion a number of different folds is investigated including barnase and ubiquitin.

These and other new fitness criteria are in the moment examined in detail and in further simulations by us to allow also modelling of more complex and bigger structures.

References

1. Goldberg, D.E. *Genetic algorithms in search, optimization and machine learning.* Addison Wesley Publ., Reading, Massachusetts 1989.
2. Dandekar, T. and Argos, P. *Int.J.Biol. Macro-molecules* **1996,** *18,* 1-4.
3. Dandekar, T. and Argos, P. *Protein Engineering* **1992,** *5,* 637-645.
4. Dandekar, T. and Argos, P. *J.Mol. Biol.* **1994,** *236,* 844-861.
5. Dandekar, T. and Argos, P. *J.Mol. Biol.* **1996,** *256,* 645-660.

J. Mol. Model. **1996**, 2, 307 – 311

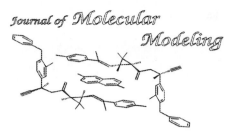

Peptides and Peptoids - A Systematic Structure Comparison

Kerstin Möhle*

Fakultät für Chemie und Mineralogie; Universität Leipzig, Talstraße 35, D-04103 Leipzig
(kerstin@quant1.chemie.uni-leipzig.de)

Hans-Jörg Hofmann

Fakultät für Biochemie, Pharmazie und Psychologie; Universität Leipzig, Talstraße 33, D-04103 Leipzig

Received: 15 May 1996 / Accepted: 6 August 1996 / Published: 27 September 1996

Abstract

A systematic analysis of the conformational space of the basic structure unit of peptoids in comparison to the corresponding peptide unit was performed based on *ab initio* MO theory and complemented by molecular mechanics (MM) and molecular dynamics (MD) calculations both in the gas phase and in aqueous solution. The calculations show three minimum conformations denoted as $C_{7\beta}$, a_D and a that do not correspond to conformers on the gas phase peptide potential energy hypersurface. The influence of aqueous solvation was estimated by means of continuum models. The MD simulations indicate the a_D form as the preferred conformation in solution both in *cis* and *trans* peptide bond orientations.

Keywords: Peptoids, peptides, *ab initio* conformational analysis, molecular dynamic simulations, peptide mimetics, solvation effect

Introduction

It is well known that peptides play an important role in different kinds of biological processes. The main disadvantage for pharmacological application is their proteolytic instability. Therefore, the development of peptide mimetics or peptide analogues with higher stability is of special interest [1-3]. Recently a new class of biopolymers was proposed - the so-called peptoids [4]. The typical structural feature of these compounds is the shift of the amino acid side-chains from the α-carbon of the amino acids to the nitrogen atom of peptide bond. This leads to oligomers or polymers of *N*-substi-

tuted glycines (NSGs). The essential advantages of these compounds are their resistance to proteases and the relatively easy synthesis with a wide variety of side chains. However, several structural consequences could be expected from these modified structures as illustrated in Figure 1 [5]:

1. The missing C_a side-chains cause the loss of chirality.
2. The hydrogen bond capacity of the peptide nitrogen responsible for the formation of the typical secondary structure elements gets lost.
3. In order to realize an equivalent orientation of the side-chains to the corresponding peptide, the peptoid chain has to be arranged in the reverse direction.

---

* *To whom correspondence should be addressed*

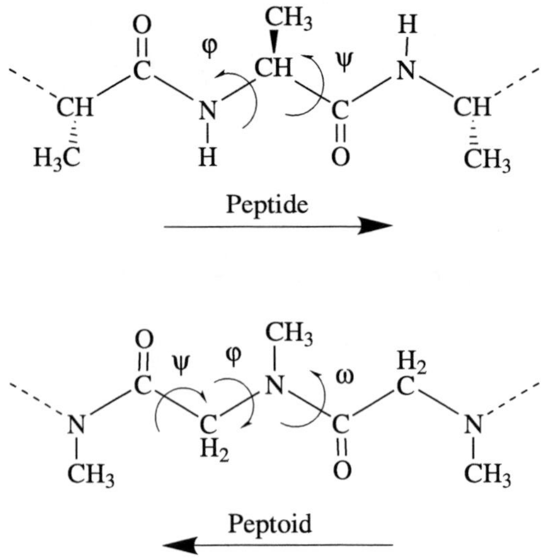

Figure 1. *Comparison between a peptide and a peptoid chain in a corresponding arrangement.*

4. In that arrangement the rotation angles determining the conformation of peptides and peptoids are not in equivalent positions.

5. The occurence of *cis* peptide bonds should increase by *N*-substitution as known for proline-containing peptide bonds.

Therefore, the examination of the conformational space of peptoids should be of special interest to show similarities and differences to the peptide structures. Recently, we presented the results of higher level *ab initio* MO calculations on the peptoid model compound *N,N*-acetylmethylglycine-*N',N'*-dimethylamide, **1** [6]. These results were compared with those calculated on the corresponding peptide analogues of glycine and alanine. In this paper, we want to complement the quantum chemical results by molecular dynamics simulations for the gas phase and aqueous solution based on empirical force fields. Thus, we intend to describe the dynamic behaviour of these compounds.

Methods

In the *ab initio* MO calculations we performed geometry optimizations for selected peptoid structures at the MP2/6-31G*, HF/6-31G* and HF/3-21G* levels. A detailed description of the methodical aspects of these calculations can be found in Ref. 6. The molecular mechanics (MM) treatments and molecular dynamics (MD) simulations were based on the CHARMm22 force field as it is part of the QUANTA 4.1 molecular modeling package [7]. In this force field all

parameters for the peptoids are available. For the MD simulations in the gas phase and in aqueous solution we used different charge models. In the gas phase charges calculated according to Gasteiger and Marsili [8] yield a potential surface in good agreement with the *ab initio* results. A greater charge polarization is considered by application of the CHARMm template charges in aqueous solution. Therefore, the increase of the dipole moments for neutral solutes in aqueous solution by about 20-30 % [9,10] is taken into account.

Starting from the different minimum conformations the MD simulations were performed at 300K in the microcanonical ensemble. The peptoid was solvated by 206 TIP3P water molecules in a cubic box of 18.9 Å length using periodic boundary conditions. A switching function based on groups truncated the nonbonded interactions between 8.0 and 9.4 Å. The SHAKE algorithm was used to constrain the XH bond lengths and consequently time steps of 1 or 2 fs in the numerical integration procedure could be employed. The whole system was heated and equilibrated for 23 ps and trajectories running for another 200 ps with the coordinates saved every 0.1 ps were used for analysis.

For the calculation of free energy differences ($\Delta\Delta A$) between different minimum conformations, we used MD simulations with holonomic internal coordinate constraints [11] and thermodynamic perturbation theory (TPT) [12,13]. Simulations of 40 ps were performed with ϕ and ψ constrained at 10 different pairs of ϕ_i and ψ_i, respectively, along a given path.

Results and discussion

First, we want to give a short review of the *ab initio* results. Our calculations showed only three minimum conformations and their symmetric counterparts on the potential hypersurface of the model peptoid independent of the basis set level (Figure 2). The torsion angles of these conformers and the energetic relations between them are summarized in Table 1. The most stable conformations are the $C_{7\beta}$ and α_D forms with comparable stability. These conformations could be interpreted as mixed structures of peptide minimum conformations. In the case of peptides, conformations with intramolecular hydrogen bonds are usually the most stable conformers on the potential hypersurfaces (e.g. the C_7 form), but this opportunity gets lost in the peptoids. Thus the $C_{7\beta}$ form represents a compromise between the C_7 and another peptide conformation, the ß form. The steric requirements of the methyl group at the peptide nitrogen lead to an extension of the original ring structure, although attractive interactions between the methyl hydrogens and carbonyl oxygen still occur. The α_D form could be derived from the fully extended peptide conformation. The dihedral ϕ has changed, but ψ still remains near 180°. The third conformer of **1** corresponds to the α helical conformation of peptides, but is distinctly more unstable than the other two conformers. This is interesting because the α form was not obtained as a minimum

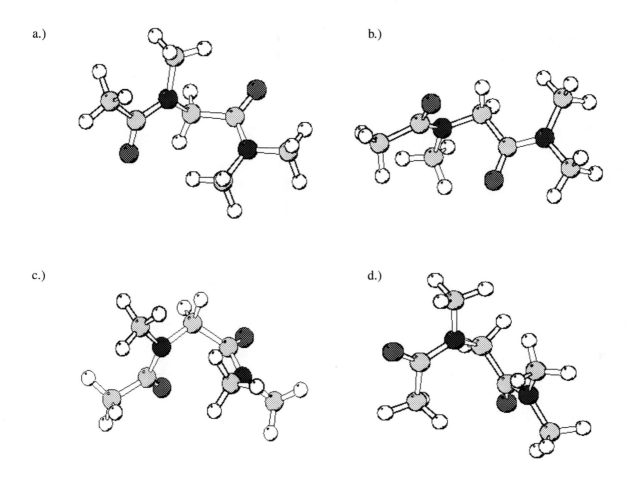

Figure 2. *Sketch of the three trans conformers (a) $C_{7\beta}$, (b) α_D, (c) α and (d) the cis α conformation of the model peptoid 1.*

structure on the gas phase potential hypersurface of peptides, but appears only in an aqueous environment [14-18].

Further calculations were performed on peptoid structures with a *cis* orientation of the peptide bond since such arrangements get more importance in tertiary amide bonds. The three conformers above-mentioned were also obtained in the *cis* peptoids without essential geometry distortions. At the different *ab initio* approximation levels the *cis* α_D form is always the most stable conformation. The *cis* $C_{7\beta}$ form is considerably destabilized, whereas the *cis* α form is even more stable than the corresponding *trans* orientation due to the additional interactions between the carbonyl oxygens and the hydrogen atoms of the different methyl groups (see Figure 2d).

The estimation of the solvation influence on the peptoid conformers by means of a quantum chemical polarizable continuum model (PCM) and a self-consistent reaction field model (SCRF) shows that all gas phase minimum structures remain stable and the α_D form is always the most stable solution conformation. Also the *trans* α helical conformation gets additional stabilization. The $C_{7\beta}$ form and the *cis* form are destabilized in solution.

The dynamic behaviour of the peptoid structure was analyzed on the basis of the MD trajectories. The results are presented as Ramachandran-like plots for the torsion angles ϕ and ψ in Figures 3a-d. Because of the missing chirality, these plots are symmetric. Whereas the symmetry of the gas phase plots is clearly visible, the time evolution of the solution trajectories is not sufficient to overcome the corresponding barriers to get into the alternative conformation range, so that these plots should be regarded as complete after symmetric reflection. When discussing the MD results, it should be remembered that the empirical force field overestimates the stability of the α helical conformation in relation to the other two conformers in the gas phase (Table 1). The gas phase trajectory plots for the *trans* (Figure 3a) and *cis* (Figure 3b) peptoid orientations indicate smaller fluctuations of the torsional angles due to larger steric restrictions in comparison to peptide analogues. The $C_{7\beta}$ and α_D forms could be considered as dynamically stable with a slight energetic preference of $C_{7\beta}$, whereas the α helical conformation disappears. Free energy calculations provide a free energy difference of $\Delta\Delta A$ ($C_{7\beta} \rightarrow \alpha_D$) = 2.9 kJ/mol. Contrary to this, the α helical conformation is the preferred one in the *cis* peptoids. In that case, the α_D form represents the second conformer and the $C_{7\beta}$ minimum disappears. The dynamics study shows that the highest energy conformers change into the more stable conformers.

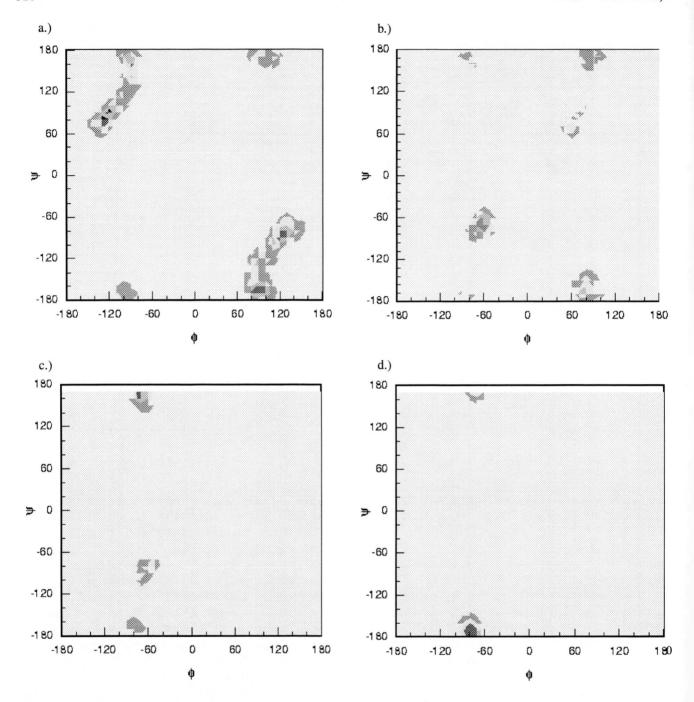

Figure 3. *Ramachandran-like plots of the gas phase trajectories for a) the trans and b) the cis peptoid and the trajectories in solution for c) the trans and d) the cis peptoid.*

energy calculations indicate a preference of the α_D form by about 6 kJ/mol over the α helical conformation.

When embedding the peptoid in a water box distinct differences occur with respect to the gas phase results as can be seen in Figures 3c and 3d. Independent of the peptide bond orientation, the α_D form is the preferred conformation in solution, whereas the $C_{7\beta}$ form disappears. Only a small amount of the *trans* α helical conformation can additionally be found in the trajectories indicating the higher probability of helical structures in aqueous solution as also found for peptides. Free

Conclusions

The results of MD simulations on the basic structure unit of peptoids reflect essential features of the *ab initio* MO conformational analysis. However, the higher energy conformers disappear on dynamic conditions and change into more stable conformers. In aqueous solution, the α_D conformer predominates and only a small amount of the α helical conformation was additionally indicated. In any case, some struc-

Table 1. *Relative energies and dihedral angles (ϕ, ψ) of the trans and cis conformers of the peptoid basic unit* **1** *at various levels of theory [a].*

	trans			cis		
	$C_{7\beta}$	α_D	α	$C_{7\beta}$	α_D	α
ΔE(MP2/6-31G*)	0.0[b](0.0)[c]	2.0 (1.2)	24.2 (24.7)	14.7 (15.5)	7.9 (7.5)	17.2 (17.9)
ϕ	-128.2	74.2	-54.7	-153.8	72.4	-62.4
ψ	77.0	-175.6	-47.2	62.5	172.1	-51.2
ΔE(HF/6-31G*)	5.4	0.0[d]	27.3	20.9	8.3	23.3
ΔG	5.9	0.0	25.5	18.0	6.6	21.1
ϕ	-128.4	79.2	-60.0	-153.2	76.9	-67.2
ψ	79.8	-174.6	-42.7	59.1	171.2	-48.7
ΔE(HF/3-21G)	0.0[e]	4.9	35.0	24.4	11.2	27.3
ϕ	-114.1	85.1	-53.3	-148.4	76.9	-57.5
ψ	96.6	-178.0	-47.3	60.6	170.3	-51.0
ΔE(CHARMm)[f]	0.0	3.2	10.5	7.3	4.9	3.1
ϕ	-120.6	94.0	-65.5	-135.9	83.4	-62.2
ψ	82.6	-164.1	-68.7	73.5	-176.4	-66.6

[a] *Energies in kJ·mol^{-1}, angles in degrees.*
[b] $E_T = -533.453074$ *a.u.*
[c] *MP2/6-31G*//HF/6-31G* single point energies in parentheses; $E_T = -533.447244$ a.u.*
[d] $E_T = -531.869079$ *a.u.*
[e] $E_T = -528.910625$ *a.u.*
[f] *Gasteiger- Marsili charges; cf. Ref. 8.*

tural differences appear in the peptoid structures when compared with the corresponding peptides which should be considered when replacing peptide units by peptoid ones.

References

1. Spatola, A.F. *Chemistry and Biochemistry of Amino Acids, Peptides and Proteins*; Vol. 7, Weinstein, B. ed., Marcel Dekker, New York, 1983; pp. 267-357.
2. Kessler, H. *Trends in Drug Research*; Claasen, V. ed., Elsevier, Amsterdam, 1990; pp 73-84.
3. Hruby, V.J.; Kazmierski, W.; Kawasaki, A.M.; Matsunaga, T.O. *Peptide Pharmaceuticals*;. Ward, D.J. ed., Open University Press, Milton Keynes, England, 1990; pp. 135-184.
4. Simon, R.J.; Kania, R.S.; Zuckermann, R.N.; Huebner, V.D.; Jewell, D.A.; Banville, S.; Ng, S.; Wang, L.; Rosenberg, S.; Marlowe, C.K.; Spellmeyer, D.C.; Tan, R.; Frankel, A.D.; Santi, D.V.; Cohen, F.E.; Bartlett, P.A. *Proc. Natl. Acad. Sci. USA* **1992**, *89*, 9367-9371.
5. Kessler, H. *Angew. Chem.* **1993**, *105*, 572-573.
6. Möhle, K; Hofmann, H.-J. *Biopolymers*, **1996**, *38*, 781-790.
7. *QUANTA 4.1 Molecular Modeling Package*; Molecular Simulations Inc., Burlington, MA, 1994.
8. Gasteiger, J.; Marsili, M. *Tetrahedron* **1980**, *36*, 3219-3222.
9. Luque, F.J.; Negre, M.J.; Orozco, M.J. *J. Phys. Chem.* **1993**, *97*, 4386-4391.
10. Gao, J.; Xia, X. *Science* **1992**, *258*, 631-635.
11. Tobias, D.J.; Brooks, C.L. III *J. Chem. Phys.* **1988**, *89*, 5115-5127.
12. Zwanzig, R. *J. Chem. Phys.* **1954**, *22*, 1420-1426.
13. Tobias, D.J.; Brooks, C.L. III *Chem. Phys. Lett.* **1987**, *142*, 472-476.
14. Rommel-Möhle, K.; Hofmann,H.-J. *J. Mol. Struct. (THEOCHEM)* **1993**, *285*, 211-219.
15. Tirado-Rives, J.; Maxwell, D.S.; Jorgensen, W.L. *J. Am. Chem. Soc.* **1993**, *115*, 11590-11593.
16. Smythe, M.L.; Huston, S.E.; Marshall, G.R. *J. Am. Chem. Soc.* **1993**,*115*, 11594-11595.
17. Shang, H.S.; Head-Gordon, T. *J. Am. Chem. Soc.* **1994**, *116*, 1528-1532.
18. Deng, Z.; Polavarapu, P.L.; Ford, S.J.; Hecht, L.; Barron, L.D.; Ewig, C.S.; Jalkanen, K. *J. Phys. Chem.* **1996**, *100*, 2025-2034.

J. Mol. Model. **1996**, *2*, 312 – 318

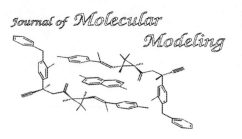

Molecular Dynamics Simulations on the Coenzyme Thiamin Diphosphate in Apoenzyme Environment

Anne von Fircks, Stefan Naumann, Rudolf Friedemann*

Department of Chemistry, Martin Luther University Halle-Wittenberg, Kurt-Mothes-Str.2, D-06120 Halle, Germany
(friedemann@chemie.uni-halle.de)

Stephan König

Department of Biochemistry, Martin Luther University Halle-Wittenberg

Received: 15 May 1996 / Accepted: 6 August 1996 / Published: 27 September 1996

Abstract

Thiamin diphosphate (ThDP) is an essential cofactor for a number of enzymes, and especially involved in the nonoxidative decarboxylation of α-keto acids by pyruvate decarboxylase (PDC). Recently the crystal structure of PDC bound ThDP has been determined. Based on these X-ray data MD simulations of the isolated coenzyme as well as of ThDP in its enzymatic environment were performed, using the GROMOS87 software package. For the ThDP-apoenzyme modelling all significant amino acid residues with a cut-off radius less than 8.5 Å from the cofactor were taken into account.

Because the activity of the coenzyme mainly depends on the formation of a specific structure, the conformational behavior of ThDP and enzyme bound ThDP were investigated within the MD simulations in more detail. Therefore, trajectories of significant structural parameters such as the ring torsion angles Φ_T and Φ_P as well as essential hydrogen bonds were analyzed by our graphics tool. Moreover, Ramachandran-like plots with respect to the torsion angles Φ_T and Φ_P were used for the illustration of preferred orientations of the two aromatic rings in ThDP.

Finally, MD simulations on ThDP analogs with less or none catalytic activity and apoenzyme mutants were included, in order to get hints of conformational effects and significant interactions in relation to cofactor-apoenzyme binding and the catalytic mechanism.

Keywords: Molecular dynamics, catalytic mechanism, enzymes

Introduction

Thiamin diphosphate (ThDP), the biological active form of vitamin B1, is an essential cofactor for a number of enzymes in the carbohydrate metabolism, where it is mainly involved in the decarboxylation of α-keto acids.

In the last years we have studied the conformational and structural properties of isolated ThDP-systems [1,2]. But recently the crystal structures of three ThDP dependent enzymes have been determined [3,4,5], which also opens new possibilities for molecular modelling.

Using the X-ray data of pyruvate decarboxylase (PDC) - the simplest of the three enzymes - we examined the influ-

---

* *To whom correspondence should be addressed*

ence of the apoenzyme environment on the conformational behavior and the catalytic activity of the cofactor.

ThDP consists of two aromatic rings and a diphosphate side chain (Fig.1). The orientation of the pyrimidine and the thiazolium ring is described by the torsion angles Φ_T and Φ_P defined by:

$$\Phi_T = \text{C2-N3-C}_{br}\text{-C5'} ,$$
$$\Phi_P = \text{N3-C}_{br}\text{-C5'-C4´}$$

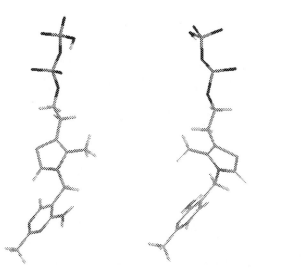

Figure 1. *ThDP system with the torsion angles Φ_T and Φ_P*

Based on this torsion angles three basic conformations were introduced in literature [6] (Fig.2).

F ($\Phi_T = 0°$, $\Phi_P = ±90°$) and S conformers ($\Phi_T = ±100°$, $\Phi_P = ±150°$) are found in the crystal structures of isolated ThDP systems, while the V-conformation ($\Phi_T = ±90°$, $\Phi_P = \mp 90°$) was assumed to be the biological active form of the cofactor, which was verified by the X-ray structure of enzyme bound ThDP. In apoenzyme bound ThDP torsion angles of $\Phi_T = 95,5°$ and $\Phi_P = -69,9°$ were found [5].

The enzyme catalyzes the decarboxylation of pyruvate to acetaldehyde and CO_2. For the activity of the cofactor Mg^{2+} ions are also required. The holoenzyme is a tetramer, consisting of four identical subunits, whereby each subunit contains a cofactor. A subunit is formed by three domains α, β, γ (Fig.3).

The cofactors are located in clefts between the α and γ domains of two different subunits (Fig.4), whereby the α domain is mainly involved in binding the pyrimidine ring, while the γ domain interacts with the diphosphate side chain.

The diphosphate side chain, is tightly bound to the enzyme by hydrogen bonds and the octahedral coordinated Mg^{2+}-ion, which assists in anchoring the diphosphate to the protein. The binding of the pyrimidine ring by hydrogen bonds to the amino acid residues GLU 51, ILE 476 and GLY 413 is important for the mechanism [5] (Fig.5).

Beside ThDP a number of ThDP analogs (Fig.6) with different catalytic activity synthesized by Schellenberger et al. [7] were taken into account.

The N1'- and N3'-ThDP analogs differ from ThDP by the substitution of the nitrogen atoms N1' and N3' with a C-H group respectively, and show less or none activity.

A substitution of the 6'-H atom by a methyl group results in the 6'-Me-ThDP-analog which is also inactive.

These experimental findings are hints, that the N1'atom plays an important role in the catalytic mechanism of decarboxylation [8].

The mechanism of decarboxylation of α-keto acids by ThDP is a subject of intensive studies since a long time.

In the mechanism suggested in 1958 by Breslow [9] (Fig.7) the formation of an ylide (**1**) by deprotonation of the C2 atom of the thiazolium ring is assumed. The addition of the ylide to the substrate gives the 2-lactyl-ThDP intermediate (**2**). Its decarboxylation leads to a α-carbanion structure (**3**) and finally to the elimination of acetaldehyde.

Especially from the recently solved X-ray structure of PDC a refined mechanism for the deprotonation of the C2 atom was supported [5]. The N1'atom is protonated by the amino acid residue GLU 51. This protonation increases the acidity of the 4'-NH_2 group and favours the formation of the imino structure. In the active V-like conformation the 4'-N position and the C2 atom are neighbouring. Therefore the

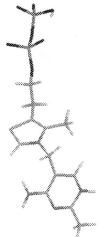

Figure 2. *Basic conformations of ThDP from left to right: F-conformation S-conformation V-conformation*

J. Mol. Model. **1996,** 2

Table1 *Occurence of the hydrogen bonds HN1'--- GLU 51 in %*

System	N1'-H --- OE1	N1'-H --- OE2	OE1 --- N1'-H --- OE2
N1'-H -ThDP	84,5	43	19,6
N1'-H-pyridyl-ThDP	65,9	62,8	18,5
N1'-H-6'-MeThDP	551	48,7	15,2

deprotonation of the C2 atom by interaction with the imino group should be possible.

For that reason we have performed MD-simulations on isolated and enzyme bound ThDP and its analogs, in order to investigate the conformational behavior and to get hints about the catalytic activity of the cofactors as well as mechanistic aspects .

Methods

The calculations were performed on a SGI workstation. For the MD simulations we used the GROMOS 87 software package [10]. Atomic net charges of the cofactor were calculated by PM3 [11] and adapted to the GROMOS 87 charges. During the calculations the N1' was protonated and for the isolated ThDP-systems Mg^{2+} were taken into account as well. Classical MD studies with NVP ensemble were performed at a temperature of 300 K and simulation periods of 500 ps for the isolated systems and 150 ps for the enzyme bound systems were used. Time steps of 1 fs were regarded. The starting structures are based on the pdb-file of the X-ray structure of the enzyme.

For the MD simulations of the cofactor in the apoenzyme environment we considered all amino acid residues located in a cut off radius of 8.5 Å from the ThDP molecule. Due to the long calculation times required by this model, amino acids described in the literature as significant were taken into account, only [5] (Fig.8). In order to simulate the rather rigid structure of a complete protein environment the C and N atoms of the protein backbone have been localized with position restraining while the calculations.

Results and Discussion

In order to get information about the conformational dynamics of the isolated and enzyme bound ThDP-systems the behavior of the torsion angles Φ_T and Φ_P during the MD run

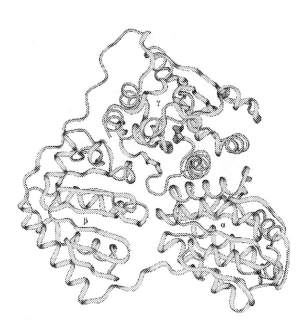

Figure 3. *Domains of the apoenzyme monomer (image available as PDB and inventor file)*

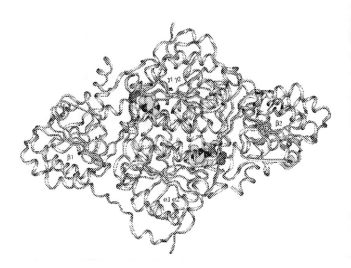

Figure 4. *Ribbon drawing of the PDC-Dimer (image available as PDB and inventor file)*

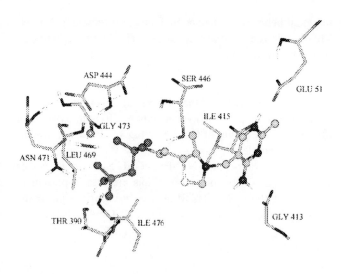

Figure 5. *Significant coenzyme-apoenzyme interactions (image available as PDB file or VRML scene)*

is considered in more detail. The preferred orientations of the two aromatic rings are indicated by fluctuations of these angles resulting from the corresponding trajectories, which are illustrated by Ramachandran-like plots. The position of the V-conformation is specified in the plots as a reference value.

By these Ramachandran-like plots we wanted to investigate if there are significant differences in the conformational behavior of ThDP and its analogs and consequently hints for a different catalytic activity by conformational reasons.

N1'-pyridyl-ThDP

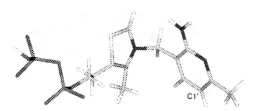

N3'-pyridyl-ThDP

6'-Me-ThDP

Figure 6. *ThDP analogs*

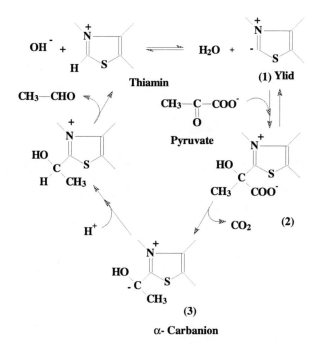

Figure 7. *Breslow mechanism of thiamin catalysis*

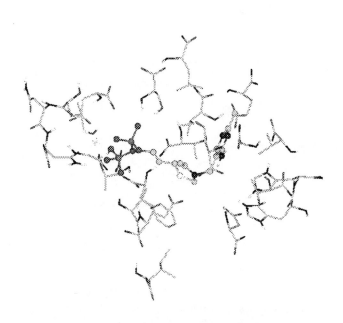

Figure 8. *Apoenzyme environment used for MD simulations (image available as PDB file and VRML scene)*

1. Isolated ThDP-Systems

The diagramms (Fig.9) illustrate that in the isolated ThDP-Systems V-like conformations are generally energetically preferred. There are no hints for a different catalytic activity by steric effects.

2. Apoenzyme bound ThDP-Systems

From the Ramachandran plots (Fig.10) it is obvious, that in case of the N3'-pyridyl-ThDP, which shows no catalytic activity, V-like-conformations are less stable within the MD run. The active N1'-H-pyridyl-ThDP shows similar confor-

mational behavior in the apoenzyme environment as N1'-H-ThDP. It is remarkable that in the case of the apoenzyme bound inactive 6'-Me-N1'-H-ThDP beside V-like conformations other conformers are also found.

In our MD simulations on coenzyme-apoenzyme interactions we have assumed that the N1'atom is protonated and the formation of possible hydrogen bonds to both oxygens of the carboxylate group of GLU 51 (Fig.11) has been investigated in more detail.

The trajectories of the N1'-H --- GLU 51 hydrogen bonds show, that these specific interactions are less important in the case of the 6'-Me-ThDP analog in comparision to the N1'-pyridyl-ThDP one as well as ThDP. Obviously, the for-

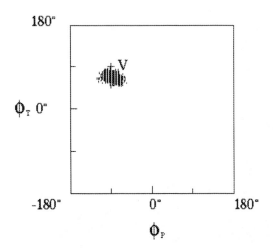

N1'-H- Mg^{2+}-ThDP

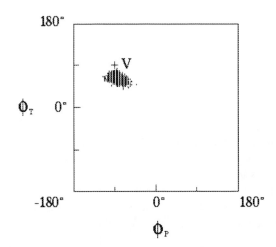

N1'-H-6'-Me-Mg^{2+}-ThDP

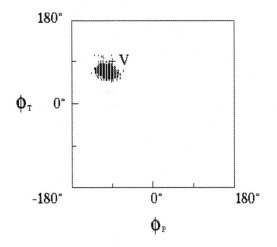

N1'-H-pyridyl-Mg^{2+}-ThDP

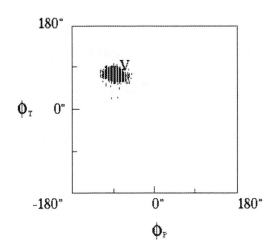

N3'-pyridyl-Mg^{2+}-ThDP

Figure 9. *Ramachandran-like plots of isolated ThDP-systems*

mation of the hydrogen bonds is hindered by the additional methyl group in the 6'-position by sterical reasons. For the N3'-pyridyl-ThDP analog the formation of a hydrogen bond to GLU 51 is excluded.

For a better illustration of these findings we have calculated the corresponding histograms from the trajectories of the distances N1'-H---OE1 and N1'-H---OE2 (Fig.11) with respect to the three ThDP systems. The histograms indicate the frequency distributions of the possible two hydrogen bonds as functions of the distances N1'-H---OE1 and N1'-H---OE2.

The histogram of N1'-H-ThDP shows an asymmetric frequency distribution of the two hydrogen bonds. The formation of the OE1---HN1' hydrogen bonding in comparision to the OE2---HN1' one is significantly preferred (Fig12). The frequency distributions of both hydrogen bonds in the N1'-H-pyridyl-ThDP system are comparable but show lower absolute values for the maxima related to N1'-H-ThDP. The lowest tendency to form hydrogen bonds to GLU 51 was found for the 6'-Me-N1'-H-ThDP system.

Similar results were also obtained within the special analyzing option for hydrogen bonds in the GROMOS Program. This procedure calculates the percental occurence of hydrogen bonds during the MD-run.

The results for the both hydrogen bonds to GLU 51 for the considered systems are illustrated in Tab.1.

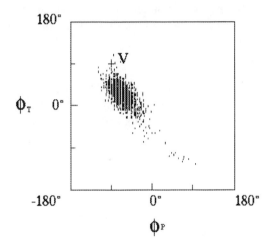

N1'-H-ThDP-Apoenzyme

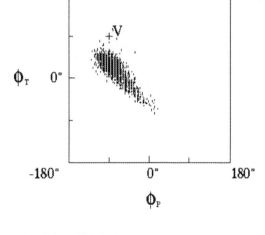

N1'-H-6'-Me-ThDP-Apoenzyme

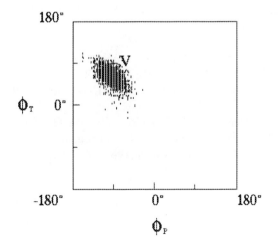

N1'-H-pyridyl-ThDP-Apoenzyme

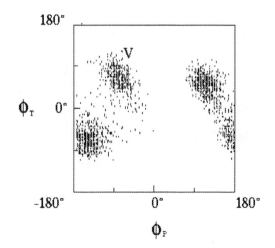

N3'-pyridyl-ThDP-Apoenzyme

Figure 10. *Ramachandran-like plots of enzyme bound ThDP-systems*

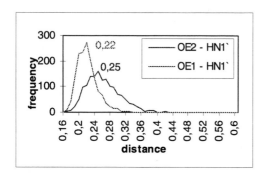

Figure 11. *Schematic drawing of the hydrogen bond N1'-H---GLU 51*

N1'-H-ThDP

N1'-H-pyridyl-ThDP

N1'-H-6'-Me-ThDP

Figure 12. *Histograms of hydrogen bonding distances HN1'--- GLU 51 for enzyme bound ThDP-systems*

A tendency between the frequency distributions of the hydrogen bonding and the catalytic activity of the ThDP systems was found. Nevertheless the inactivity of the 6'-Me-ThDP analog can not be explained with this simulation. A possible reason for this failure could be seen in the limited included amino acid residues within our calculations.

Our MD simulations on coenzyme-apoenzyme interaction support the assumption, that the hydrogen bonding of the cofactor to the amino acid residue GLU 51 has a key function in the formation of the active V-like conformation. Moreover the hydrogen bond is necessary to explain the mechanistic pathway of the ThDP catalysis with respect to the cleavage of the C2-H bond of the thiazolium ring to form an ylide like structure.

For a deeper understanding of mechanistic aspects of the ThDP catalysis further simulations including coenzyme-apoenzyme-substrate interaction are required in order to investigate the possible function of the substrate as an activator in the catalytic process.

Acknowledgements: The authors are grateful to the Deutsche Forschungsgemeinschaft and Fonds der Chemischen Industrie for the financial support of this work and the access to new computer facilities.

References

1. Friedemann, R.; Breitkopf, C. *Bioorganic Chemistry* **1994**, *22*, 119-127.
2. Friedemann, R.; Breitkopf, C. *Int. J. Quant. Chem.* **1996**, *57*, 943-948.
3. Lindqvist, Y.; Schneider, G. *EMBO J.* **1992**, *7*, 2373-2379.
4. Muller, Y.A.; Schulz, G. *Science* **1993**, *259*, 965 -967.
5. Dyda, F. et al. *Biochemistry* **1993**, *32*, 165-6170.
6. Pletcher, J.; Sax, M. *J. Am. Chem. Soc.* **1972**, *94*, 3998-4005.
7. Neef, H.; Golbik, R. *Liebigs Annal. Chem.* **1990**, 913.
8. Schellenberger, A. *Chem. Ber.* **1990**, *123*, 1489-1494.
9. Breslow, R. *J. Am. Chem. Soc.* **1958**, *80*, 3719-3726.
10. van Gunsteren, W. F.; Berendsen, H. J. C. GROMOS Library Manual, Biomos b.v., Groningen, 1987
11. Stewart, J.J.P. *J. Comp. Chem.* **1989**, *10*, 209.

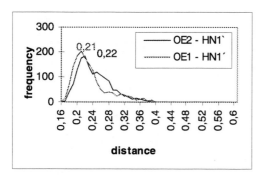
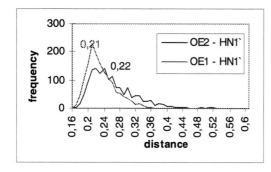

J. Mol. Model. **1996**, 2, 319 – 326

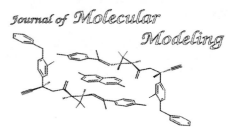

© Springer-Verlag 1996

Gibbs-Ensemble Molecular Dynamics:
A New Method for Simulations Involving Particle Exchange

Reinhard Hentschke*, Tim Bast, Ewald Aydt, and Michael Kotelyanskii

Max-Planck-Institut für Polymerforschung, Postfach 3148, D - 55021 Mainz, Germany
(hentschk@mpip-mainz.mpg.de)

Received: 15 May 1996 / Accepted: 6 August 1996 / Published: 27 September 1996

Abstract

We discuss a novel simulation method suitable for simulating phenomena involving particle exchange. The method is a molecular dynamics version of the Gibbs-Ensemble Monte Carlo technique, which has been developed some years ago for the direct simulation of phase equilibria in fluid systems. The idea is to have two separate simulation boxes, which can exchange particles or molecules in a thermodynamically consistent fashion. We discuss the general idea of the Gibbs-Ensemble Molecular Dynamics technique and present examples for different simple atomic and molecular fluids. Specifically we will discuss Gibbs-Ensemble Molecular Dynamics simulations of gas-liquid and liquid-solid equilibria in Lennard-Jones systems and in hexane as well as an application of the method to adsorption.

Keywords: Gibbs-Ensemble MD, Monte Carlo, particle exchange, adsorption

Introduction

There are numerous phenomena, which can be studied via computer simulation, where it is advantageous or even necessary to vary the particle number during the simulation. Three simple examples may serve to illustrate this point. The forces between two planar surfaces at close proximity can be related to the microscopic interactions on the molecular scale using the so called surface force apparatus [1]. A computer simulation of a surface force apparatus would have to include not only the solid surface-to-solid surface interface. Usually there are molecules adsorbed or tethered to the surfaces as well as the molecules of a liquid filling the intermediate space. In addition there has to be a reservoir, where those liquid molecules can be 'deposited', which are squeezed out from

between the surfaces as the inter-surface separation is reduced. Thus, such simulations are plagued by the comparatively large portion of the system devoted to the 'interface' between the slit and the reservoir, which causes unwanted artifacts. A second example is the simulation of osmotic phenomena. Here the computer simulation would have to include both the osmotic cell as well as the corresponding bulk solvent surrounding it. One may for instance think of simulating the swelling of a polymer network in contact with a solvent. Because molecular simulations still contain rarely more than a few thousand atoms, the inclusion of a large interfacial region may render such a simulation impossible (for realistic systems). Finally, we may think about the adsorption equilibrium between a porous bulk solid (e.g., zeolites) and a gas. Here the comparatively small external surface area of the solid contributes little to the overall adsorbing surface area and even may be ill de-

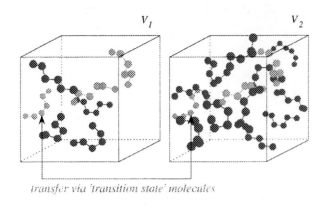

Figure 1. *Sketch of two simulation volumes or boxes with the respective volumes V_1 and V_2 exchanging molecules.*

fined. Again one faces a situation, where in principle it is sufficient to model a small number of unit cells of the porous bulk solid in which gas particles are inserted or eliminated according to the chemical potential of the surrounding gas. Thus, these examples illustrate the motivation for simulating open systems in which the number of molecules is no longer constant.

In the 1970's the Monte Carlo methodology was extended to include such open systems through the development of the grand canonical Monte Carlo (GCMC) method [2–4]. An extension of the Molecular Dynamics method to open systems (GCMD) was not developed until about 20 years later [5, 6]. However, in most applications a certain deficiency of both the GCMC and GCMD methods is the need for specifying or independently calculating the chemical potential. Even though the latter in principle can be determined, based on the Widom potential-distribution theory [7, 8], from NVT Monte Carlo [9] or Molecular Dynamics [10, 11], this adds another complicated calculation. A method which combines these two aspects into a single simulation run was developed by Panagiotopoulos [12] in the context of phase coexistence in one-component fluids. The Gibbs-Ensemble Monte Carlo (GEMC) technique does not require the reference chemical potential as input. In GEMC two separate simulation volumes can exchange particles or molecules so that in equilibrium, for instance, the temperature, the pressure and the chemical potential in the boxes are the same (cf. figure 1). In this fashion a single simulation run of a one-component fluid at a particular subcritical temperature and pressure yields the pure gas and the pure liquid at coexistence, where each one occupies its respective simulation volume. However, other thermodynamic conditions, for instance requiring different pressures in the two simulation volumes, are also easy to implement [13]. A drawback of the MC based methods usually is the low acceptance rate for particle creation or insertion in a dense phase. That Molecular Dynamics may

be better suited for dense systems was first suggested by Cagin [14], who also discusses the idea of a Molecular Dynamics version of the Gibbs-Ensemble method. Shortly afterwards the GEMD method was indeed realized via two independent approaches [15–17].

It is one of these approaches, i.e., [16, 17], on which we focus in the following. First we discuss the general idea as well as the specific implementation of the method. We then illustrate its performance using a number of examples, i.e., the gas-liquid-solid phase behavior of Lennard-Jones systems as well as the gas-liquid phase behavior of a realistic molecular system, which here is n-hexane. Finally, we show an application of the method to adsorption in a zeolite, which pertains to the last of the examples discussed above.

The GEMD method

The GEMC method originally was introduced in [12] to simulate the liquid-gas coexistence of a single-component fluid system. In this case one has two physically separated volumes or simulation boxes, whose combined volume and combined particle number is constant (as a consequence of the phase rule). Both boxes are at the same temperature and pressure. In addition they can exchange particles in a way such that the chemical potential also is the same in both boxes. For a specified temperature, and with the proper choice of the total volume, the system may phase separate so that there will be the pure liquid in one box and the pure coexisting gas in the other. This can happen below the critical temperature, when the average density corresponds to the thermodynamically unstable states below the liquid-vapor coexistence line. In the following we discuss an analogous scheme for a molecular system within the MD framework.

The potential

In conventional MD one numerically solves the equations of motion for a system of N particles contained in a box of volume V having the total potential energy U. Here we consider a molecular system consisting of N atoms in M molecules, and we write the total potential energy U as a sum over inter- and intra-molecular interactions, i.e.

$$U = U_{inter} + U_{intra} = \sum_{\substack{i<j \\ \alpha \in i, \beta \in j}} \Phi\left(\vec{r}_{\alpha\beta}\right) + \sum_{i} U_{intra}\left(\left\{\vec{r}_{\alpha\in i}\right\}\right) \tag{1}$$

The first term, U_{inter}, is the sum over atom-atom pair potentials $\Phi\left(\vec{r}_{\alpha\beta}\right)$, where $\vec{r}_{\alpha\beta} = \vec{r}_\alpha - \vec{r}_\beta$, and $\vec{r}_a$ and $\vec{r}_\beta$ are the position vectors of the atoms labeled α and β belonging to two different molecules i and j. The second potential energy term, U_{intra}, encompasses all interactions within a molecule, i.e., bond stretching, valence angle and torsional distortions as well as intra-molecular non-bonded interactions.

In order to simulate a variable number of molecules in each of the two boxes we introduce an additional degree of freedom ξ_i for every molecule i. ξ_i can vary between 1 and 0, where $\xi_i=1$ means that the molecule i is in box 1, whereas $\xi_i=0$ means that it is in box 2. For $1> \xi_i > 0$ the molecule is in a 'transition state', where it is 'felt' in both boxes. Thus, we rewrite the inter-molecular potential energy of the system as a function of the coordinates and the ξ_i as

$$U_{inter}\left(\{\vec{r}_\alpha\},\{\xi_i\},V_1,V_2\right)=$$

$$\sum_{\substack{i<j \\ \alpha\in i,\beta\in j}} \left[\Phi\left(\vec{r}_{\alpha\beta},V_1\right)\xi_i\xi_j + \right.$$

$$\left. +\Phi\left(\vec{r}_{\alpha\beta},V_2\right)\left(1-\xi_i\right)\left(1-\xi_j\right)\right]+\sum_i g\left(\xi_i\right) \quad (2)$$

$$=U_1+U_2+\sum_i g\left(\xi_i\right)$$

where V_1 and V_2 are the volumes of the two boxes. The first two terms, U_1 and U_2, represent the inter-molecular potential energies of the first and the second box, respectively. Notice that as soon as we apply periodic boundary conditions and the inter-particle interactions are calculated involving the particle's closest images, the distance between them, and therefore the inter-molecular potential energy, is a function of the box dimensions (or the volume if the shape of the box is kept fixed).

At equilibrium the number of unphysical but necessary transition state molecules should be small in comparison to the total number of molecules. What exactly 'small' means we will define below. However, to satisfy this condition it often is necessary to introduce an additional potential function $g(\xi_i) > 0$, which is equal to zero only at $\xi_i=0$ and at $\xi_i=1$. Here we use

$$g\left(\xi_i\right)=\begin{cases} \omega\left[\tanh\left(u\xi_i\right)+\tanh\left(u\left(1-\xi_i\right)\right)-1\right] & , \quad 0\leq\xi_i\leq1 \\ \infty & , \quad otherwise \end{cases} \quad (3)$$

This additional potential introduces a barrier of height w and steepness u between the states corresponding to the 'real' particles or molecules, i.e., particles or molecules which are entirely in one or the other of the two boxes, making the transition state unfavorable.

The GEMD equations of motion

In the following we present the equations of motion for our GEMD method. In the case of liquid gas-coexistence in a one-component system the temperature, the pressure, and the chemical potential, even though the latter two are not explicitly specified, must be equal in the two phases and thus in the two boxes. Similar to the GEMC method this can be achieved if every change of the volume of one of the boxes is accompanied by an opposite but equal change of the volume of the other box. Thus, the total volume of the two boxes is constant, while the individual volumes are variable. The GEMD equations of motion describing this case are

$$\vec{p}_\alpha = m_\alpha\dot{\vec{r}}_\alpha$$

$$\dot{\vec{p}}_\alpha = -\frac{\partial U}{\partial \vec{r}_\alpha}-\eta\vec{p}_\alpha$$

$$\dot{\eta} = \frac{1}{Q_T}\left[\sum_\alpha \frac{\vec{p}_\alpha^2}{m_\alpha}-Xk_BT\right]$$

$$p_{\xi_i} = m_{\xi_i}\dot{\xi}_i$$

$$(4)$$

$$\dot{p}_{\xi_i} = -\frac{\partial U}{\partial \xi_i} =$$

$$=-\sum_{\substack{i(\neq j) \\ \alpha\in i,\beta\in j}}\left[\Phi\left(\vec{r}_{\alpha\beta},V_1\right)\xi_j-\Phi\left(\vec{r}_{\alpha\beta},V_2\right)\left(1-\xi_j\right)\right]-\sum_i\frac{\partial}{\partial\xi_i}g\left(\xi_i\right)$$

$$p_{V_1} = Q_P\dot{V}_1$$

$$\dot{p}_{V_1} = P_1^e-P_2^e$$

Here $\vec{p}_\alpha$ and p_{ξ_i} are the momenta conjugate to the cartesian coordinates $\vec{r}_\alpha$ and the transfer coordinate x_i, respectively. h is an additional degree of freedom, and Q_T is a parameter governing the temperature relaxation. The first three equations describe the evolution of a system coupled to an external heat bath with the temperature T [18], where X is a coefficient, which is equal to the number of degrees of freedom coupled to the thermostat. If there would be only one box, this would correspond to the well known NVT simulation algorithm, with the temperature being controlled via a Nose-Hoover thermostat. The next two equations govern the evolution of the x_i, i.e., the transfer of the molecules between the boxes. The last two equations are the equations of motion of the box volume V_1, where p_{V_1} is a momentum variable conjugate to V_1, and Q_P is a parameter governing the volume relaxation. Thus, the volume changes are controlled by the difference between the instantaneous values of the 'external' pressures P_1^e and P_2^e in the two boxes. Here, for each box, we employ the (single-box) constant-pressure MD algorithm proposed in reference [19], i.e., if again there would be only one box, say the first box, the equations $p_{V_1} = Q_P \dot{V}_1$ and $\dot{p}_{V_1} = P_1^e - P$ would be identical to the equations of motion derived for the box volume in reference [19] (cf. also equation (16) in reference [20]), where P is just the preassigned pressure, which here is replaced by the instantaneous external pressure P_2^e in the second box. Here

$$P_1^e = \sum_{\substack{i>j \\ \alpha \in i, \beta \in j}} \frac{1}{3V_1} \left(\vec{\nabla}_{\vec{r}_{\alpha\beta}} \Phi\left(\vec{r}_{\alpha\beta}, V_1\right) \xi_i \xi_j \vec{R}_{1,n} \right) \tag{5}$$

where $\vec{R}_{1,n} = V_1^{1/3}\left(n_x, n_y n_z\right)$ is a vector, which maps the separation of the coordinates $\vec{r}_\alpha$ and $\vec{r}_\beta$ into the proper distance between the atoms α and β according to the minimum image convention assuming a cubic box. In this algorithm the $\vec{r}_\alpha$ are not scaled by the box dimensions and they are not mapped back into the box according to the boundary conditions whenever a particle leaves the primary box and enters one of the surrounding image boxes. The coupling to the volume fluctuations rather happens through the positions of the image particles. Note that the summations involving the inter-particle or inter-molecular interactions in the equations (1), (2), (4), and (5) are meant to implicitly include the interactions between the primary particles or molecules with their (nearest) images. Thus some care has to be taken in the case of the interaction of a particle or molecule with it's own images. Note also that a detailed discussion including a formal justification of the above equations of motion is given in [16, 17].

The above equations of motion are appropriate for the simulation of two-phase coexistence in a one-component system, and we will discuss two examples below. In order to simulate two-phase coexistence in a two-component system, the last two equations in (4) are replaced by

$$p_{V_k} = Q_P \dot{V}_k$$
$$\dot{p}_{V_k} = P_k^e - P \tag{6}$$

where $k = 1,2$. The GEMD equations of motion appropriate for the examples discussed in the introduction are even more simple. In these cases the last two equations in (4) are replaced by

$$p_{V_1} = Q_p \dot{V}_1$$
$$\dot{p}_{V_1} = P_1^e - P \tag{7}$$

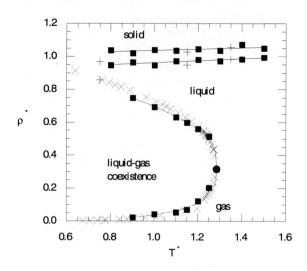

Figure 2. *Phase diagram of the Lennard-Jones system in terms of the LJ density, ρ*, and the LJ temperature, T*. The solid squares indicate the GEMD results for the location of the phase boundaries. The solid lines are fits to the GEMD results as explained in the text. The crosses and the plusses are experimental data points for methane and argon, respectively.*

Here P is the pressure in the reference system represented by box 2, i.e., the pressure in the liquid outside the force apparatus or in the reference solution in contact with the swollen network or the pressure of the gas in contact with the porous solid.

Some technical aspects

If a long-range cutoff distance r_{cut} is used when calculating the non-bonded interactions, then the corresponding continuum corrections modify $\dot{p}_{\xi_i} = -\partial U / \partial \xi_i$ in equation (4), where we must add $-\partial \left(U_1^{cor} + U_2^{cor}\right) / \partial \xi_i$ to the right hand side. In the case of Lennard-Jones non-bonded interactions, the usual long-range corrections become

$$U_1^{cor} \approx \sum_{\nu\mu} \frac{1}{2} 4\pi \frac{1}{V_1} \sum_i \xi_i \sum_j \xi_j \int_{r_{cut}}^{\infty} \Phi_{\nu\mu}(r) r^2 dr \tag{8}$$

and U_2^{cor} is given by an analogous expression in which ξ_i is replaced by $(1 - \xi_i)$ and V_1 is replaced by V_2. The indexes ν and μ in equation (8) label distinct atom types in the interacting molecules i and j. The only other change affects P_k^e (where $k = 1,2$). Here we must add the correction

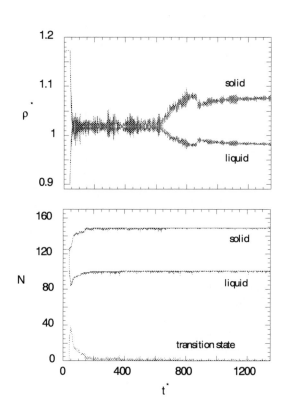

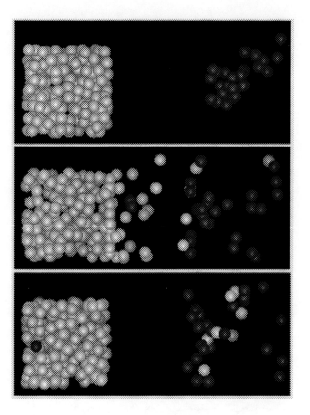

Figure 3. *The upper panel shows the time evolution of ρ^* for the two boxes at $T^*=1.4$ in the case of liquid-solid coexistence. The lower panel shows the corresponding evolution of the number of particles N in the solid (upper curve), in the liquid (middle curve) and in the transition state (bottom curve).*

Figure 4. *The three panels illustrate the particle exchange between the two boxes for one temperature at different times towards the beginning as well as towards the end of the simulation run. Here the ξ_i-values of the transition state particles are used to scale their cartesian position relative to the shifted positions of the boxes, which in reality are superimposed. The color coding distinguishes particles, which initially are in different boxes (top). Subsequently the transfer is allowed. Early on there are more particles in the transfer state (middle) than at later times (bottom)*

$$P_1^{e,cor} \approx -\sum_{\nu\mu} \frac{2\pi}{3} \frac{1}{V_1^2} \sum_i \xi_i \sum_j \xi_j \int_{r_{cut}}^{\infty} \frac{d\Phi_{\nu\mu}(r)}{dr} r^3 dr \qquad (9)$$

where $P_2^{e,cor}$ is also given by (9), however, again with the above replacements for ξ_i and V_1.

The GEMD method discussed here also requires a short-range cutoff r'_{cut}. This is because two atoms α and β belonging to the respective molecules i and j may 'collide' due to the strong divergence of $\Phi(\vec{r}_{\alpha\beta})$ as $\vec{r}_{\alpha\beta} \to 0$ even though according to their ξ-values, i.e., $\xi_i \approx 0$ and $\xi_j \approx 1$, they would belong to different boxes. Thus, the value for r'_{cut} should meet two requirements. It should be sufficiently large to minimize the effect of unphysical ghost collisions. In addition it should not affect the interactions between atoms belonging to the same box. In the Lennard-Jones systems described below, the optimal value for r'_{cut} can be found via a series of independent NVT simulations (in the relevant temperature and density range) using increasing values for r'_{cut}. The best value for r'_{cut} lies just below the onset of noticeable effects on the average pressure and

other bulk properties. In the following examples we use $r'_{cut} = 0.88$ (in Lennard-Jones units). It should be noted that currently we assume a constant non-bonded potential below the short-range or 'ghost' cutoff distance. However, other functional forms, which approach a finite value at $\vec{r}_{\alpha\beta} = 0$, are possibly better alternatives.

The equations of motion (4) do not depend on explicit threshold ξ_i-values according to which a particle is counted as being in box 1, in box 2, or in the transition state. Nevertheless the transition state is an artificial state, whose population should be small in comparison to the population of the real states represented by the particles considered to belong to either of the two boxes. In other words, for long times each particle should be either 'close to 0' or 'close to 1' most of the time. 'Close' means that the long time deviation of the particle's ξ_i-values from 0 or 1 has a negligible effect on the bulk properties of the systems rep-

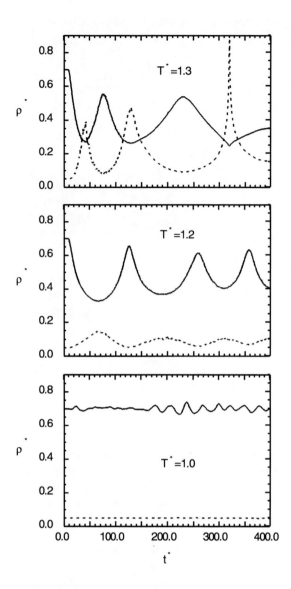

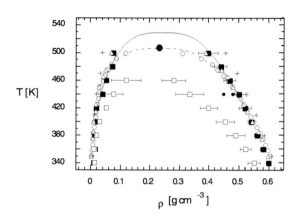

Figure 6. *Liquid-gas coexistence curve for hexane. Hollow circles: experimental data; hollow triangles: GEMC result obtained with a 13.8 Å cutoff including long-range corrections. The large solid circle corresponds to the experimental critical point. Hollow squares: GEMD using a cut and shifted LJ potential, were the long-range cutoff is at 10 Å; solid squares: using a 15 Å cut-off instead; small solid circles (at T=440 K): liquid densities for a 12 Å and a 14 Å cutoff, respectively; plusses: result obtained for a 10 Å cutoff including long-range corrections. The error bars indicate standard deviations. These are omitted for the plusses in order to not obscure the figure. The lines (solid: GEMD; dashed: experiment) are again fits as described in the text to figure 2.*

Figure 5. *This series of panels each shows the evolution of the densities (gas: dashed line; liquid: solid line) at different temperatures below and close to the critical point for the case of gas-liquid coexistence in the LJ system. Note that the spike-like feature in the upper panel is due to a numerical constraint on the maximum fluctuation of the volume.*

Finally we like to mention that the stability of the numerical solution of (4) is markedly improved if the Berendsen thermostat [21] is used instead of the Nose-Hoover thermostat. Even though only for the latter one can mathematically justify that the algorithm does produce the proper ensemble averages, the obtained results were the same in both cases within the statistical error [17]. An additional numerical improvement of the algorithm can be achived by also thermalizing the transfer degrees of freedom by a direct coupling to the Berendsen thermostat. This and some related aspects are discussed in detail in [22].

Application to phase coexistence in Lennard-Jones and molecular systems

Our first application is the gas-liquid-solid phase diagram of the Lennard-Jones (LJ) system, using the inter-particle potential $4\varepsilon[(\sigma/r)^{12} - (\sigma/r)^6]$ including long-range continuum corrections [22]. The GEMD density-temperature phase diagram is shown in figure 2. The solid boundary of the gas-liquid coexistence density is a fit based on the law of rectilinear diameters, $(\rho_l + \rho_g)/2 = \rho_c + C_1 (1-T/T_c)^\beta$, together with the power law $\rho_l - \rho_g = 2C_2 (1-T/T_c)^\beta$ using the

resented by the two boxes. That this is indeed the case is illustrated by the examples below. However, for bookkeeping purposes, we sometimes distinguish between particles belonging to box 1 or to the transition state or to box 2 according to whether their ξ_i-values are in the range between 1 and $1 - 10^{-4}$ or between $1 - 10^{-4}$ and 10^{-4} or between 10^{-4} and 0. It should be noted in this context that we use the Verlet algorithm to integrate the equations of motion. A special modification, described in detail in reference [17], can be used to handle the 'hard walls' at 0 and 1 along the ξ_i-direction.

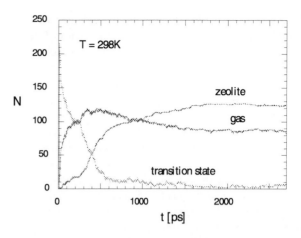

Figure 8. *Time evolution of the number of particles N in the zeolite, in the gas box, and in the transition state.*

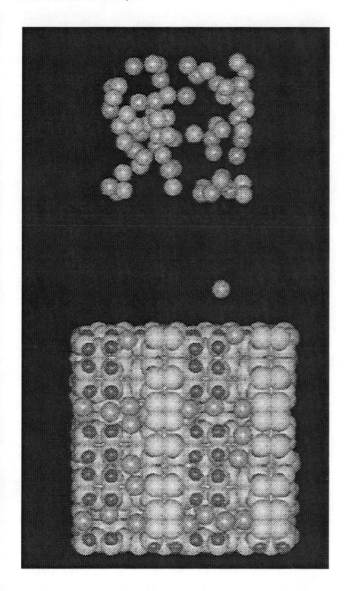

Figure 7. *Using an analogous representation as in figure 4 this figure shows a snapshot of the model zeolite (bottom) containing methane molecules (shown as blue spheres) diffusing along channels in the zeolite. The methane molecules can transfer between the gas phase (top) and the cavities in the solid.*

critical density, ρ_c, the critical temperature, T_c, and C_1 as well as C_2 as fit parameters. For β we use the 3D Ising value of ≈ 0.325. The experimental data for methane, shown for comparison, are taken from [23]. The data are converted to LJ units via $\varepsilon/k_B = 148.7$ K and $\sigma = 3.79$Å based on a fit to the experimental 2nd virial coefficient. Figure 2 also includes the liquid-solid coexistence, showing that the GEMD method works well at high densities. Here the solid lines are simply least-squares fits to the GEMD results. The experimental data are for argon, again using the LJ parameters obtained via the 2nd virial coefficient to convert to LJ units [24]. The scatter of the GEMD results in this case

is due to the rather small system containing a total of 250 particles. The deviation of the simulation data from the experimental results near the triple point is possibly due to the oversimplified form of the potential (cf. [24] and a reference therein). Figure 3 shows the evolution of the particle density for the case of liquid-solid coexistence together with the corresponding instantaneous particle numbers corresponding to the liquid, the solid, and the transition state. Notice that the number of particles in the transition state quickly decreases to a level which is low compared to the number of particles in the boxes. The three snapshots shown in figure 4, which correspond to one particular temperature ($k_BT/\varepsilon = 1.0$), present a pictorial illustration of the particle transfer at different times along the system's trajectory. Figure 5, finally, shows the gas-liquid density evolution for a series of temperatures below and close to the critical temperature. Notice how the densities in the two boxes become virtually indistinguishable near the critical temperature as they should.

We also want to include the analogous phase diagram for gas-liquid coexistence in a molecular system. Figure 6 shows the corresponding result for hexane in a system containing a total of 144 molecules. The details of the model potential and the specific parameters used in this calculation are described elsewhere [17]. The agreement between GEMD and GEMC [25, 26] as well as with the experiment [27] is quite good, depending of course on the accuracy of the interaction potential. Notice that the position of the coexistence region in the T-ρ-plane is quite sensitive to the choice of r_{cut} and the inclusion of long-range corrections.

Application to adsorption

Here we briefly want to mention some preliminary results obtained for the physisorption of methane in a zeolite using the GEMD approach. For the sake of comparison we chose silicalite I (ZSM5), which has been studied via GCMC by Goodbody et al. [28]. In this case the atoms of the solid do not transfer between the boxes, one of which contains eight ZSM5 unit cells and the other corresponds to the gas phase. Again the details of the model parameters will be given elsewhere [29].

Figure 7 shows a simulation snapshot, which illustrates the methane transfer between the gas phase and the cavities in the solid. The actual evolution of the number of particles in the two boxes as a function of time is shown in figure 8 at a gas pressure of 70 bar and a temperature of 298 K. Notice that the gas box is kept at constant pressure, whereas the zeolite box is kept at constant volume. This situation therefore corresponds to the set of equations of motion, where the last two equations in (4) are replaced by the equations (7).

Conclusion

The above examples show that the GEMD method is useful for simulations in which a single large molecular system can be replaced by two smaller systems, which exchange molecules at constant chemical potential. The specific advantage of the two-box approach is that the chemical potential in the reference system must not be determined separately. An advantage that the GEMD method has over the GEMC method is that it provides direct dynamic information in the two boxes, like the self-diffusion coefficient in the two coexisting phases [17]. Another important point is that the GEMD method performs quite well in dense systems.

References

1. Israelachvili, J. *Intermolecular and Surface Forces*; Academic Press: London, 2nd ed., 1992.
2. Norman, G. E.; Filinov, V. S. *High Temp. Res. USSR* **1969**, *7*, 216.
3. Adams, D. *Mol. Phys.* **1974**, *28*, 1241-1252.
4. Adams, D. Mol. Phys. 1975, 29, 307-311.
5. Cagin, T.; Pettitt, B. M. *Mol. Simul.* **1991**, *6*, 5-26.
6. Cagin, T.; Pettitt, B. M. *Mol. Phys.* **1991**, *72*, 169-175.
7. Widom, B. *J. Chem. Phys.* **1963**, *39*, 2808-2812.
8. Widom, B. *J. Phys. Chem.* **1982**, *86*, 869-872.
9. Shing, K. S.; Gubbins, K. E. *Mol. Phys.* **1983**, *49*, 1121.
10. Romano, S.; Singer, K. *Mol. Phys.* **1979**, *37*, 1765-1772.
11. Powles, J. G.; Evans, W. A. B.; Quirke, N. *Mol. Phys.* **1982**, *46*, 1347-1370.
12. Panagiotopoulos, A. Z. *Mol. Phys.* **1987**, *61*, 813-826.
13. Panagiotopoulos, A. Z. *Mol. Simul.* **1992**, *9*, 1-23.
14. Cagin, T. In *Computer Aided Innovation of New Materials II*; Doyama, M., Kihara, J., Tanaka, M., Yamamoto, R., Eds.; Elsevier Science Publishers, 1993; pages 255-259.
15. Palmer, B. J.; Lo, C. *J. Chem. Phys.* **1994**, *101*, 10899-10907.
16. Kotelyanskii, M. J.; Hentschke, R. *Phys. Rev.* **1995**, *51*, 5116-5119.
17. Kotelyanskii, M. J.; Hentschke, R. *Mol. Simul.* **1996**, *17*, 95-112.
18. Hoover, W. G. *Phys. Rev.* **1985**, *31*, 1695-1697.
19. Winkler, R. G.; Morawitz, H.; Yoon, D. Y. *Mol. Phys.* **1992**, *75*, 669-688.
20. Winkler, R. G.; Hentschke, R. *J. Chem. Phys.* **1993**, *99*, 5405-5417.
21. Berendsen, H. J. C.; Postma, J. P. M.; van Gunsteren, W. F.; DiNola, A.; Haak, J. R. *J. Chem. Phys.* **1984**, *81*, 3684-3690.
22. Bast, T.; Kotelyanskii, M.; Hentschke, R. in preparation, 1996.
23. Grigor, A. F.; Steele, W. A. *J. Chem. Phys.* **1968**, *48*, 1032-1037.
24. Hansen, J.-P.; Verlet, L. *Phys. Rev.* **1969**, *184*, 151-161.
25. Siepmann, J. I.; Karaborni, S.; Smit, B. *Nature* **1993**, *365*, 330-332.
26. Computer simulations of vapour-liquid phase equilibria of n-alkanes. Smit, B.; Karaborni, S.; Siepmann, J. I. preprint, 1995.
27. Smith, B. D.; Srivastava, R. Thermodynamics Data for Pure Compounds: Hydrocarbons and Ketones; Elsevier: Amsterdam, 1986.
28. Goodbody, S. J.; Watanabe, K.; MacGowan, D.; Walton, J. P. R. B.; Quirke, N. *J. Chem. Soc. Faraday Trans.* **1991**, *87*, 1951-1958.
29. Aydt, E. M.; Bast, T.; Hentschke, R. in preparation, 1996.

J. Mol. Model. **1996**, 2, 327 – 329

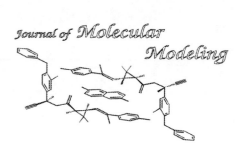

Interaction Energies and Dynamics of Alkali and Alkaline-Earth Cations in Quadruplex-DNA-Structures

Jutta Töhl, Wolfgang Eimer

Universität Bielefeld, Physikalische Chemie I, D - 33615 Bielefeld, Germany. (jutta@chep122.uni-bielefeld.de)

Received: 15 May 1996 / Accepted: 6 August 1996 / Published: 27 September 1996

Abstract

We have investigated the nonbonded interaction energies and dynamical properties of different types of cations in quadruplex DNA structures using the GROMOS force field [1]. Quadruplex structures consist of planar guanine-quartets stacking together and causing the formation of a chan-

nel, large enought to enclose several cations (Figure 1). In recent years many experimental studies have indicated a prefered formation of this unusually stabel complexes with K^+-ions. However, the high selectivity of this cation has not yet been understood [2].

To determine the most stable coordination sites and the mobility of cations, we have calculated the pair potential energy of alkali and alkaline-earth cations along the heli-

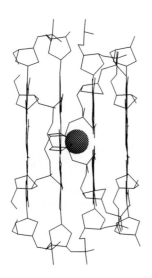 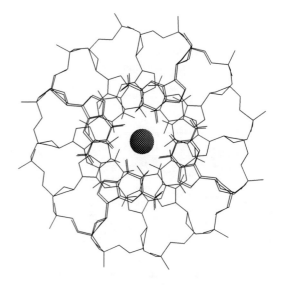

Figure 1. *Wireframe presentations of the d[GGGG]₄ quadruplex structure. View perpendicular to the helical axis (left) and in the helical axis (right).*

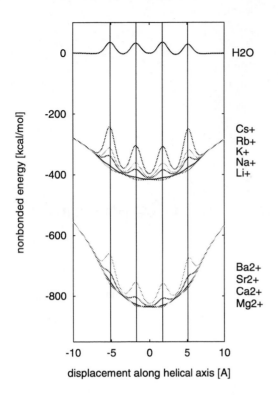

Figure 2. *Pairwise interaction of alkali, alkaline-earth metal ions, and water with all DNA-atoms, along the helical axis.*

cal axis of a model quadruplex structure (Figure 2). Our force field calculations indicate that small ions like Li$^+$, Na$^+$, Mg^{2+} and Ca^{2+} are free to move throught the channel. In contrast, for K$^+$ and larger ions a high potential barrier appears, located in the plane of the tetramer unit. These findings are in agreement with data from X-ray crystallography, indicating that K$^+$ cations are located between two planes while Na$^+$ ions also can occupy coordination sites in the G-quartet plane.

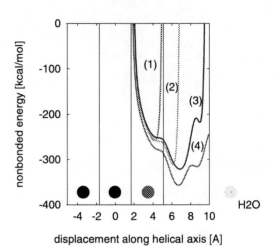

Figure 3. *Pairwise interaction of K$^+$ (red sphere) of different water configurations (1 to 4). Two additional potassium cations in the neighbouring cages were considered (black spheres). Simulation-period: ca. 2ps.)*

Considering solvent atoms in our calculations leads to the observation that a cation at the end of the quadruplex strongly interacts with one water molecule located near the entrance of the cage. Snapshots taken at different times of the MD simulation provide configurations which differ mainly in the position of this complexing water molecule. Moving away from the entrance of the cage causes a significant decrease of the potential barrier for K$^+$ and smaller 'cage cations' (Figure 3). For the larger ions the potential barrier is much higher than the thermal energy (not shown), preventing the cations from leaving and entering the cage.

This conclusion is in agreement with results from our MD simulations. We followed the dynamics of different cations. While K$^+$ is able to leave as well as to re-migrate into the channel (movie I), this was not observed for other

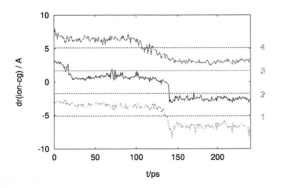

Figure 4. *z-component of the ion fluctuation about the center of geometry. The horizontal lines indicates the tetrameric planes.*

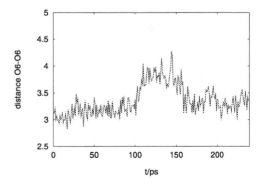

Figure 5. *Averages distance of O6 atoms in the fourth plane*

types of cations. Figure 4 shows the time history of the positional fluctuations of potassium along the helical axis, and in Figure 5 we have monitored the distance between the O6-oxygen atoms of the outer G-quartet. It becomes clearly evident that the cation movement through the planes is correlated with the dynamic behaviour of the tetrameric planes. When the K^+-ion penetrates the tetrameric unit to enter the quadruplex, the O6-O6-distance - a measure of size of the hole of the plane - increases. After a while the cation has reached the cage position and the G-quartet contracts to the initial value. That means the tetrameric planes perform a kind of breathing motion.

For lithium ions we find a much higher mobility of the cation within the quadruplex channel. Two of three ions are leaving the cage instantaniously (not shown).

Another indication of the experimentally observed much weaker complexation tendency of quadruplexes with lithium is the change in the distances of planes (a measure for the cage size) with time. While in the case of potassium the distance of planes is nearly the same for all three cages, for lithium the central occupied cage is much smaller than the cage in the starting structure, indicating that the DNA structure has to adjust its conformation to the cation size. On the other side the outer unoccupied cages are much greater and less stable. Due to this cation induced quadruplex deformation we observe an unwinding of the DNA-structure in the presens of lithium ions at longer simulation periods (400 ps).

Movie

The movie shows the dynamics of the model quadruplex d$[GGGG]_4$ and the potassium ions (yellow) over a simulation period of 11 ps. Two water molecules at the ends of the channel are considered (dots indicates the van der Waals radii). At the end of the movie one ion is leaving the channel (see Figure 4), while the water molecule is displaced from the channel exit. Then the ion, which occupies the central cage jumps into the empty outer cage.

References

1. van Gunsteren, W.F.; Berendsen, H.J.C. Groningen Molecular Simulation (GROMOS) Library Manual; University of Groningen, Groningen **1986**.
2. Williamson, J.R. *Ann. Rev. Biophys. Biomol. Struct.* **1994**, *23*, 703 - 730.

J. Mol. Model. **1996**, 2, 330 – 340

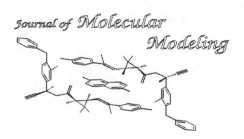

Molecular Dynamics Simulation of a Micellar System

Tim Bast* and Reinhard Hentschke

Max-Planck-Institut für Polymerforschung, Postfach 3148, 55021 Mainz, Germany (bast@mpip-mainz.mpg.de)

Received: 15 May 1996 / Accepted: 6 August 1996 / Published: 27 September 1996

Abstract

We present the results of an atomistic molecular dynamics simulation based on the AMBER/OPLS force field applied to segments of isolated one-dimensional micelles, 2,3,6,7,10,11-Hexa-(1,4,7-Trioxaoctyl)-Triphenylene, in aqueous solution using the SPC/E water model. The quantities which we study include the intra-micellar monomer structure, e.g., the equilibrium monomer-monomer separation along the micelle, the micelle-water interface, which yields the effective micellar diameter, and the flexibility of the micelle in terms of its persistence length as a function of temperature. In addition, we determine the micelle size distribution at low concentration via the free enthalpy gain per monomer-monomer contact using a hydration shell model in combination with thermodynamic integration. Finally, we locate the isotropic-to-nematic transition by using our results as input for an analytical model.

Keywords: MD simulation, micellar system, force field, water models

Introduction

Many amphiphilic molecules in aqueous solution spontaneously self-assemble into labile anisotropic aggregates which exhibit complex liquid-crystalline phase behavior (e.g. [1]). Over the past decade the theoretical modeling of the interplay between self-assembly and lyotropic phase behavior in these systems has been addressed by various authors (e.g. [2] (review article), [3 – 6]). Common to these models is that they focus on rod-shaped micelles (some exceptions are discussed in [2]) and that they are extensions of Onsager's excluded volume theory for the isotropic-to-nematic transition in dilute systems of slender rods. The theoretical models usually depend on a number of a priori unknown parameters such as the monomer shape and size, the flexibility of the micelle expressed in terms

of its persistence length (for a rod-like micelle), or the monomer-monomer contact enthalpy within a micelle, etc. These parameters are either adjusted by fitting the model results to the experiment or deduced from independent experimental information. In the present study we want to use the molecular dynamics technique to obtain these parameters via computer simulations.

Today's advanced computer technology allows the modeling of simplified micelles and a fair number of such studies can be found in the literature [7 – 11]. Nevertheless, atomistic molecular modeling on workstation computers is still limited to a few thousand atoms and to the nano-second time scale. This usually prohibits the full simulation of micellar assembly for realistic molecules, and of course the modeling of the phase behavior of such systems. An alternative possibility is to model a small portion of a micelle only using the proper boundary conditions to

* To whom correspondence should be addressed

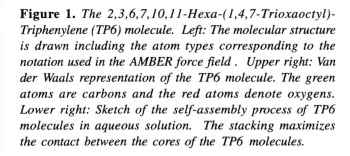

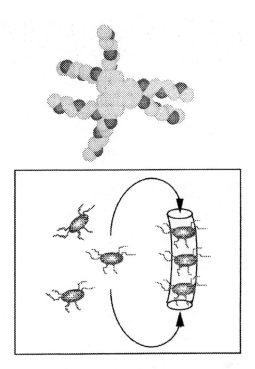

Figure 1. *The 2,3,6,7,10,11-Hexa-(1,4,7-Trioxaoctyl)-Triphenylene (TP6) molecule. Left: The molecular structure is drawn including the atom types corresponding to the notation used in the AMBER force field . Upper right: Van der Waals representation of the TP6 molecule. The green atoms are carbons and the red atoms denote oxygens. Lower right: Sketch of the self-assembly process of TP6 molecules in aqueous solution. The stacking maximizes the contact between the cores of the TP6 molecules.*

extract local information, which then can serve as input to, for instance, analytical theories. This is what we do here. The specific system which we study is 2,3,6,7,10,11-Hexa-(1,4,7-Trioxaoctyl)-Triphenylene (TP6) in water (cf. Figure 1). The monomer is a disk-like molecule consisting of an aromatic core surrounded by hydrophilic side-chains at its periphery. Above the critical micelle concentration the monomers stack, the reason for this is one aspect of this work, and thus reversibly aggregate to form one-dimensional or rod-like micelles (see Figure 1). This system has been studied extensively by Boden and coworkers [12 – 17] (and references therein), which makes it an interesting model system for the above concept for combining theory and simulation.

In the present work we simulate a section of a one-dimensional micelle representing a segment within an isolated, long micelle as well as a complete short micelle both solvated by molecular water. We study the side chain-water interface and deduce the effective diameter of the micelle. In our simulation the length of the simulated micellar segment is allowed to fluctuate freely subject to the

external hydrostatic pressure. This allows to determine the equilibrium monomer-monomer separation along the micelle's axis, which is important for relating the aggregation number of a micelle to its length. Using a previously developed approach [18] we use the conformation statistics of the simulated segments to construct long micelles, for which we obtain the persistence length for a number of different temperatures. Furthermore, we study the micelle size distribution via the free enthalpy gain per monomer-monomer contact within a micelle using a thermodynamic integration technique. Finally, we use the results obtained in this study as input parameters for a previously developed analytical model to locate the isotropic-to-nematic transition.

Methodology

In molecular dynamics (MD) simulations the Newtonian equations of motion

$$\frac{d^2}{dt^2}\vec{r}_i = -\frac{1}{m_i}\vec{\nabla}V(\vec{r}_1,...,\vec{r}_N) \quad ; \quad (i=1...N) \tag{1}$$

are integrated numerically for all N atoms of the system located at positions $\vec{r}_i$ and having masses m_i. Here we employ the program package AMBER 4.0 (Assisted Model Building with Energy Refinement) [19] to compute the corresponding trajectory. The potential V used in AMBER 4.0 consists of five different contributions, i.e.

$$V(\vec{r}_1,...,\vec{r}_N) = \sum_{bonds\ i} k_b^{(i)}\left(b_i - b_0^{(i)}\right)^2$$

$$+ \sum_{angles\ i} k_\alpha^{(i)}\left(\alpha_i - \alpha_0^{(i)}\right)^2$$

$$+ \sum_{dihedrals\ i} k_d^{(i)}\left[1 + cos\left(n^{(i)}\phi_i - \gamma^{(i)}\right)\right]$$

$$+ \sum_{atom\ pairs\ ij} \left(\frac{A_{ij}}{r_{ij}^{12}} - \frac{B_{ij}}{r_{ij}^6}\right) \qquad (2)$$

$$+ \sum_{atom\ pairs\ ij} \frac{q_i q_j}{r_{ij}}$$

The first two terms model all bond length b_i and valence angle α_i deformations in terms of harmonic potentials. Here we use the SHAKE algorithm to constrain the bond lengths to their equilibrium values $b_0^{(i)}$ [20]. The third term together with the 1-4 non-bonded interactions (cf. below) approximates the torsional potential variations in terms of the torsion angle ϕ_i. The fourth and fifth term describe non-bonded interactions, i.e. Lennard-Jones and Coulomb pair-interactions. The summation over atom pairs includes all atom pairs separated by three (1-4 interactions) or more bonds in the same molecule or atom pairs, where each atom belongs to a different molecule. Note that

$r_{ij} = \left|\vec{r}_i - \vec{r}_j\right|$. The non-bonded interactions are only calculated within a residue-based cutoff radius R_{cut}. This means that if two atoms belonging to two different residues are closer than R_{cut}, then all pair-interactions between the two residues are calculated. In the following each water and each TP6 molecule constitute one residue. Note also that in addition the 1-4 interactions are scaled by a factor of ½ [19]. The Lennard-Jones parameters $A_{ij} = \varepsilon_{ij}\sigma_{ij}^{12}$ and

$B_{ij} = 2\varepsilon_{ij}\sigma_{ij}^6$ for the mixed interactions are obtained via the Lorentz-Berthelot mixing rules $\sigma_{ij} = \sigma_i + \sigma_j$ (where a factor of ½ is absorbed into σ_i and σ_j) and $\varepsilon_{ij} = \sqrt{\varepsilon_i\varepsilon_j}$ [21]. The TP6 molecule is simulated using the united atom representation, whereas the water molecules are described via the SPC/E model [22]. The details of the parameterization and the numerical values of the parameters are given elsewhere [23].

We integrate the equations of motion (1) using the leapfrog verlet algorithm [21] and we apply periodic boundaries using the minimum image convention to calculate the non-bonded interactions. During the simulations temperature and pressure in the simulation box are controlled by

the weak coupling velocity scaling thermostat and barostat of Berendsen et al. [24]. However, the pressure is computed for each dimension separately, which is necessary due to the uniaxial anisotropy induced by the presence of the micelle. Then the atom positions are scaled for each dimension separately according to the respective pressures. This anisotropic pressure scaling is necessary because of two different effects. First, water molecules penetrate into the side chains of the micelle and the simulation box thus shrinks. With the pressure scaling perpendicular to the micellar axis we achieve the proper water bulk density away from the micelle (as discussed below in the context of Figure 4). Second, the monomer-monomer separation is a priori unknown. By applying a constant pressure along the micelle any stretch or compression of the micelle is al-

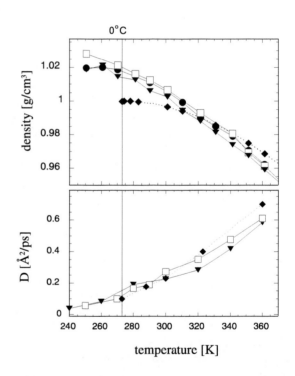

Figure 2. *Test of the SPC/E water model. Upper panel: Simulated water density vs. temperature for different systems, i.e. two different volumes containing 125 (triangles) and 216 (circles and squares) water molecules respectively. For the case of 125 molecules the cutoff radius is R_{cut}=7.5 Å. whereas for the 216 molecules system we compare two different cutoff radii, i.e. R_{cut} = 9 Å (squares) and R_{cut} = 7.5 Å (circles). The experimental values (diamonds) are taken from [25]. Lower panel: Self-diffusion coefficients D of the water molecules vs. temperature for two different box sizes, i.e. 125 molecules using R_{cut} = 7.5 Å (triangles) and 216 molecules using R_{cut} = 9 Å (squares). The experimental values (diamonds) are taken from [26].*

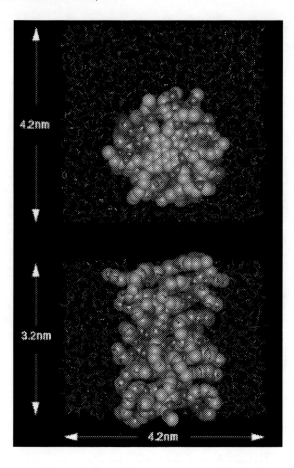

Figure 3. *Snapshot of the simulation box at T = 300 K containing a stack of eight TP6 molecules in van der Waals representation surrounded by water molecules in stick representation. The upper panel shows a view along the micelle's axis, whereas the lower panel shows a side view of the simulation box. Note that due to the application of periodic boundary conditions we model a segment of eight monomers within a virtually infinite micelle.*

lowed to relax, and the average monomer-monomer separation can fluctuate around its equilibrium value.

Test of the water model

Figure 2 summarizes the results of several 0.5ns constant pressure test simulations of pure SPC/E water [22] for different box sizes (i.e., 125 and 216 water molecules) at various temperatures ranging from 250K to 360K. Figure 2(a) shows the simulated density as a function of temperature in comparison to the experiment [25]. The deviation of the liquid densities from their experimental values is always less than 2%. For 216 water molecules and at $T > 300$K the deviation is even less than 1%. Also, for 216 water molecules the figure includes the results for two different values of the cutoff radius, i.e. $R_{cut} = 7.5$ Å and $R_{cut} = 9$ Å. The

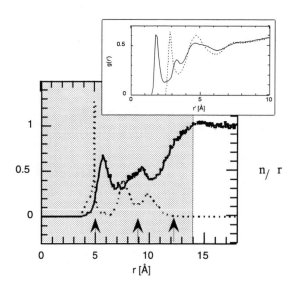

Figure 4. *The radial density distribution of the water oxygen atoms ρ (normalized to its experimental bulk value ρ_0, solid line) and the radial distribution of the side chain oxygen atoms $\Delta n/\Delta r$ (in arbitrary units, dotted line) at T = 300 K. Note that r is the radial distance from the micelle's backbone. The arrows indicate the radial positions of the oxygen atoms in a fully extended side chain. From the ρ/ρ_0 curve we estimate a radius of the micelle of 14 Å. The respective area of the plot is shaded. Inset: Side Chain oxygen-water hydrogen pair distribution function g(r') (solid line) and the corresponding side chain oxygen-water oxygen g(r') (dotted line).*

reduction of the cutoff radius leads to a very minor decrease in density due to the neglect of attractive interactions. Figure 2(b) shows the corresponding temperature dependence of the self-diffusion coefficient, D, in comparison to the experiment [26]. For liquid water the self-diffusion coefficients are in very good agreement with the experimental data, especially in the temperature range which is of interest in this work ($T=280...300$K). Notice, that we do not observe any indication of freezing in the simulated temperature range below 273K. The O-O pair correlation function, which is not shown here, is very similar to that obtained in Monte Carlo studies [27]. The positions of the first three peeks of the experimental curve are reproduced to within 0.1 Å. Only the height of the first peak is somewhat exaggerated (roughly by 30 %).

The TP6/water system

For the various simulations in this work which include TP6 the starting configurations are prepared as follows. Stacks consisting of six and eight TP6 molecules are assembled by first building and energy-minimizing a single TP6 mol-

$$\langle \vec{u}_i \cdot \vec{u}_{i \ n} \rangle_i$$

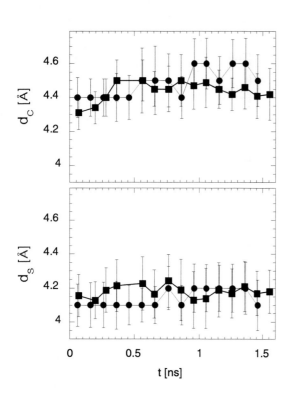

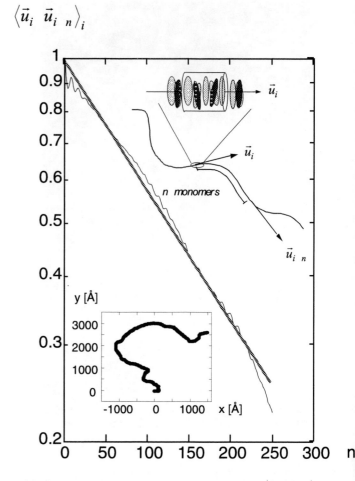

Figure 5. *Upper panel: Center of mass separation d_c between adjacent monomers in the 8-micelle vs. simulation time t. Lower panel: Corresponding stacking distance d_s vs. simulation time t. The stacking distance is the center of mass separation projected onto the backbone of the micelle. Circles: T = 280 K; squares: T = 300 K. All values are obtained by averaging over all monomer-monomer contact sites within the micelle as well as over 100 ps timeslices along the trajectory. The error bars denote the standard error assuming that the contact sites are independent of each other.*

Figure 6. *Orientation correlation function $\langle \vec{u}_n \cdot \vec{u}_{i+n} \rangle_i$ vs. distance n. The $\vec{u}_i$ and $\vec{u}_{i+n}$ are tangent vectors along the micelle's contour at monomers i and i+n. The average is taken over i. The smooth straight line is an exponential fit to the simulation result for T = 300 K. Upper right insert: Definition of a tangent vector along the micelle's contour. Lower left insert: x,y-Projection of the contour of one large micelle.*

ecule with the program INSIGHT II using the Discover version 3.2 force field [28]. The TP6 molecule is then replicated employing a 4 Å repeat distance perpendicular to the triphenylene plane. Subsequently, the stacks are surrounded by water molecules placed on a simple cubic lattice. The box with the 6-mer contains 2500 water molecules whereas the box with the 8-mer contains 1827 water molecules. Note that both systems are prepared large enough to include bulk behavior of the water at large radial separations from the micelle. In the case of the 6-micelle this is also true along the axial direction. Notice also that we apply a constant pressure of 1bar to the box. In the case of the 8-micelle the axial box dimension coincides with eight

times the monomer-monomer separation. The latter, however, is flexible due to the constant pressure condition. Note that because of periodic boundary conditions the 8-mer corresponds to a segment of a virtually infinite micelle. In a similar fashion a third system is prepared which contains two isolated TP6 molecules solvated in 2004 water molecules.

The results reported below are for MD simulations at three different temperatures (T = 280, 290 and 300 K). After a short 10 ps MD run using a 1fs time step the remainder of each simulation is carried out with a 4fs time step. The total length of each run is between 0.6 and 1.5ns during which the atom positions are stored every 0.2 ps. As an example, Figure 3 shows an instantaneous configuration during the simulation of the 8-mer. The eight monomers in

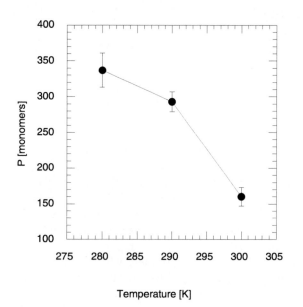

Figure 7. *Average persistence length, P, for the constructed long micelles vs. temperature T for three different temperatures.*

Table 1. *Mean values for the center of mass separation d_C and the stacking distance d_S between two adjacent monomers for the three different temperatures.*

T [K]	$<d_C>$ [Å]	$<d_S>$ [Å]
280	4.5 ± 0.14	4.2 ± 0.12
290	4.5 ± 0.19	4.3 ± 0.20
300	4.4 ± 0.10	4.2 ± 0.13

Comparing the two radial distributions it can be seen that the radial density of the water oxygens shows local minima at the preferred positions of the side chain oxygen distribution peaks. The oxygen atoms in the water molecule as well as in the side chains carry negative partial charges. Therefore, a repulsive Coulomb force acts between these atoms. This force explains the minima mentioned above. Nevertheless, there appear to exist hydrogen bonds between the neighboring water molecules and the side chain oxygen atoms. This can be seen from the pronounced first peak at ≈ 1.9 Å in the side chain oxygen – water hydrogen pair distribution function shown in the inset of Figure 4.

Monomer-monomer stacking distance

The monomer-monomer distance in the micellar aggregates is important in converting aggregation numbers into the actual lengths of the micelles. These lengths play an important role in the analytical description of the phase behavior of TP6 in aqueous solution. Here we study two differently defined monomer-monomer separations. In the first case we determine the separation of the centers of mass d_C of adjacent triphenylene cores. In the second case we compute the projection of the previously determined separation d_C onto the backbone in order to obtain the stacking distance d_S along the backbone. In this study we define the backbone of the micelle as the principal axis of inertia corresponding to the direction along the micelle.

Figure 5 shows d_C and d_S averaged over the monomer-monomer contacts within the periodic 8-micelle as a function of simulation time t for two temperatures, i.e. $T = 280$ K and $T = 300$ K, which bracket the temperature range investigated here. The average over all contacts for the separations d_C and d_S are given in Table 1. On average d_C exceeds d_S by 0.2 to 0.3 Å. However, we do not see any change of the mean distances with temperature exceeding the error margin. Thus, the magnitude of the error, where we assume that different contacts are statistically independent, is certainly an upper limit for the temperature dependence of d_C and d_S. It is worth noting that the averages for d_C and d_S in Table 1 are rather close to the experimentally determined ring-ring separation of between ≈ 4.2 Å at $T = 280$ K and 5 Å at $T = 300$ K [17].

the box form a stack with the core regions in contact. Notice that the water molecules penetrate into the side chains but do not enter the core region.

The side chain-water interface

In Figure 4 the radial density distributions of the water oxygen atoms and the radial distribution of the side chain oxygen atoms are shown. Note that the water oxygen density is normalized to its experimental bulk value. The results plotted in Figure 4 are taken from the trajectory at $T = 300$ K. The radial distributions extracted from the other two trajectories do not show significant deviations from the curves shown in Figure 4. The radial distribution function of the water molecules shows two peaks at a distance of $r = 6$ Å and $r = 9.5$ Å. In this range the average density is approximately half that of the bulk value, which is approached in the range 11 Å $< r <$ 14 Å. From this we estimate a micelle diameter of about 28 Å. Notice that an appreciable onset of the water distribution occurs at $r = 4$ Å, i.e. these water molecules are right at the periphery of the triphenylene core. However, the water molecules do not penetrate into the core region between the monomers.

The corresponding radial distribution of the side chain oxygen atoms shows three peaks which can be identified with the three oxygen atoms in each side chain. The first peak is sharp, whereas the broadness and the inward shift of the two outer peaks reflects the conformational freedom of the side chains. Notice that the arrows in Figure 4 denote the radial oxygen positions in a fully extended side chain.

Note that we compare the results obtained for the infinite micelle with those for the micelle consisting of six monomers only. Within the fluctuations we do not observe any difference neither in d_C nor in d_S between the short and the infinite micelle.

Micellar flexibility

To obtain the flexibility of the rod-like micelle we estimate its persistence length, *P*. *P* is defined by the correlation between the tangent vectors along the micelle (see upper inset of Figure 6). For not too small separations the correlation for two tangent vectors $\vec{u}_i$ and $\vec{u}_{i+n}$ separated by *n* monomers is given by [29]

$$\langle \vec{u}_i \cdot \vec{u}_{i+n} \rangle_i = e^{-n/P} \qquad (3)$$

In order to compute the persistence length via this equation a large micelle is required to compute the average in (3). Here we use an adaptation of a method, which we have developed previously to compute *P* for a helical polypeptide (Poly-γ-benzyl-L-glutamate) [18]. The method is a build-up procedure by which a long micelle is constructed by chaining together instantaneous conformations taken from a simulation of a short segment, i.e. the 8-micelle in the present case.

The persistence length is then calculated using equation (3) by averaging the dot products of the tangent vectors $\vec{u}_i$ and $\vec{u}_{i+n}$ along the micelle. The direction of each tangent vector is defined as the moment of inertia axis along a cylindrical segment of the micelle consisting of five monomers centered at the origin of the tangent vector (cf. inset in the upper right of Figure 6). As a measure of the uncertainty of the persistence length to be computed we calculate the standard deviation of *P* based on ten different long micelles each constructed as described by choosing at random ten different initial segments. Figure 6 shows an exponential fit to $\langle \vec{u}_n \cdot \vec{u}_{i+n} \rangle_i$ for one of the ten micelles at *T* = 300 K. The lower left inset of Figure 6 shows an example micelle in terms of its contour's *x,y*-projection.

In Figure 7 the mean value of the persistence length is shown for three different temperatures. It can be seen that the persistence length decreases with increasing temperature. Thus, the micelles become more flexible with rising temperature. The error bars denote the standard error as explained above.

Thermodynamics of micellar assembly

To good approximation the chemical potential μ_n of a one-dimensional n-micelle in dilute solution can be written as

$$\mu_n = n\tilde{\mu}_n^0 + RT \ln\left(\frac{1}{n} X_n\right) + RT \ln\gamma - \alpha RT(n-1) \qquad (4)$$

An extensive discussion of this equation can be found in the article by A. Ben-Shaul and W.M. Gelbart in reference [1]. In the first term $\tilde{\mu}_n^0$ is a standard average chemical potential per monomer in an n-micelle, excluding the interaction between the monomers, which is treated separately. The second term is the contribution due to the solute mixing entropy. Note that here X_n refers to the mole fraction of monomers within micelles of aggregation number n. In particular $n = 1$ corresponds to the free monomers. The third term describes the interaction between the solute particles. However, here we are merely interested in the limit of no interaction, where the activity coefficient $g = 1$ so that this term vanishes (At high concentrations, however, excluded volume effects as well as entropic repulsion between the persistent-flexible micelles due to their confinement must be included). The last term constitutes a simple one-parameter model of monomer-monomer interaction within a one-dimensional *n*-micelle, where *n*-1 is the number of contacts between monomers. The quantity –a is the molar free enthalpy per monomer-monomer contact and RT, i.e.

$$\alpha = -\frac{\Delta G_{contact}}{RT} \qquad (5)$$

which in general depends on the type of solvent, temperature, pressure etc. and usually also on *n*, because monomers near the end of a micelle feel a different environment compared to those in the bulk of the micelle. Here, however, we will be exclusively dealing with the bulk value of α, which is independent of *n*. Using equation (4) in conjunction with the equilibrium condition $\mu_n = n\mu_1$ yields the micellar size distribution

$$X_n = nX_1^n e^{\alpha(n-1)+n\left(\tilde{\mu}_1^0 - \tilde{\mu}_n^0\right)/RT} \qquad (6)$$

under the given conditions. In the following we will be concerned with the calculation of α or rather $\Delta G_{contact}$, which, as equation (6) shows, is an important ingredient for the calculation of the micellar size distribution. Also, the analysis of the experimental data in reference [17] indicates that for the present system the approximation $\tilde{\mu}_1^0 = \tilde{\mu}_n^0$ seems to be justified, and that the size distribution is mainly determined by α alone.

In the following we apply the simulated thermodynamic integration approach (cf. e.g. [30]), which allows to evaluate $\Delta G_{contact}$ via the relation

$$\Delta G_{contact} = \frac{1}{n-1} \int_{\infty}^{\lambda_{eq}} \left\langle \frac{\partial H_c}{\partial \lambda} \right\rangle_{\lambda} d\lambda \qquad (7)$$

Here $H_c(\lambda)$ is the Hamiltonian of the system with a fixed monomer-monomer separation λ. $\langle ... \rangle_{\lambda}$ denotes an ensemble average at the same fixed λ. The numerical integration of the right hand sight of equation (7) requires a series of MD simulations to evaluate the ensemble average at several distinct values of λ. However, the method is computationally too expensive for the complete system containing the micelle and the solvent. Therefore we calculate $\Delta G_{contact}$ along the alternative route (double lined arrows) in the following thermodynamic cycle

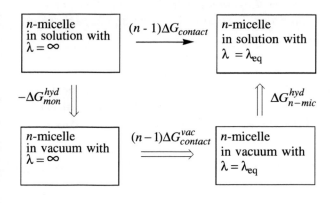

(8)

i.e.

$$\left(n-1\right)\Delta G_{contact} =$$
$$-n\Delta G_{mon}^{hyd} + \left(n-1\right)\Delta G_{contact}^{vac} + \Delta G_{n-mic}^{hyd} \qquad (9)$$

Here ΔG_{mon}^{hyd} and ΔG_{n-mic}^{hyd} are the free enthalpies of hydration of the monomer and the n-micelle, respectively, whereas $\Delta G_{contact}^{vac}$ is $\Delta G_{contact}$ in vacuum. Note that $\Delta G_{contact}^{vac}$ requires substantially less computational effort, because without solvent the number of interactions is greatly reduced.

To obtain the hydration free enthalpies of the micelle ΔG_{n-mic}^{hyd} from our trajectories we use the hydration shell model developed by Scheraga and co-workers [31, 32]. It is based on the assumption that ΔG^{hyd} can be split up in contributions proportional to the average water accessible volume $< V_{(i)k} >$ of the individual atoms or groups. Making the simplifying assumption that the weighting of

the sidechain conformations in solution is the same as in vacuum it can be written

$$\Delta G^{hyd} \approx \sum_k \Delta g_k^{hyd} \left\langle V_{(i)k} \right\rangle \qquad (10)$$

Δg_k^{hyd} is the hydration free enthalpy density of the kth atom or group. Note that these densities are independent of the molecule's conformation. The $< V_{(i)k} >$ can be obtained directly from our simulation trajectory by Monte Carlo integration. The hydration free enthalpies of the monomer ΔG_{mon}^{hyd} are obtained in a similar fashion except that the trajectories are taken from three additional simulations of two isolated TP6 molecules in water.

Note that the righthand side of equation (7) is the reversible work necessary to separate a monomer from the micelle. Here we evaluate the righthand side of (7) in vacuum by calculating

$$\Delta G_{contact}^{vac} = \int_{\infty}^{\lambda_{eq}} \left\langle f_{ax}^{vac} \right\rangle_{\lambda} d\lambda \qquad (11)$$

where $\left\langle f_{ax}^{vac} \right\rangle_{\lambda}$ is the time averaged force along the axis of the micelle in vacuum acting on a monomer, which is pulled off the end of the micelle. The micelle is approximated by a trimer for which we perform molecular dynamics simulations (as discussed above) constraining the positions of the carbon atoms of the central hexagon in each monomer so that these hexagons in neighboring monomers

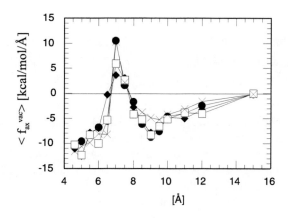

Figure 8. *Time averaged contact force* $\left\langle f_{ax}^{vac} \right\rangle_{\lambda}$ *as a function of λ for three different temperatures T = 280 K (squares), 290 K (diamonds), and 300 K (circles). For comparison, the result of a similar calculation for T = 300 K only with a different charge model (AM1 charges) is also shown (crosses). All points are averages over MD trajectories of 0.5 ns.*

Table 2. *Free enthalpies calculated along the alternate path in (9), and* $\Delta G_{contact}$ *and* α *calculated according to (9) and (5) respectively in the limit for large n.*

ΔG [kcal/mol]	$T = 280$ K	$T = 290$ K	$T = 300$ K
ΔG_{mon}^{hyd}	-47.8	-47.6	-48.4
$\frac{1}{n}\Delta G_{n-mic}^{hyd}$	-29.3	-29.4	-29.8
$\Delta G_{contact}^{vac}$	-43	-38	-35
$\Delta G_{contact}$	-24.5	-19.8	-16.4
α	43.9	34.3	27.4

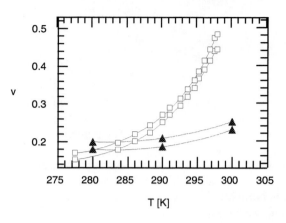

Figure 9. *The isotropic-to-nematic phase coexistence region in terms of temperature T and solute volume fraction* υ *obtained in this study (triangles) and experimentally (squares).*

are eclipsed and parallel. All other atoms of the triphenylene core and in particular the side chains remain unconstrained. Initially, we set the distance of the triphenylene to the equilibrium separations discussed above. Subsequently, the monomer is pulled away along the axis of the micelle, and $\left\langle f_{ax}^{vac}\right\rangle_{\lambda}$ is calculated as a function of λ.

Table 2 lists the hydration contribution to $\Delta G_{contact}$ according to equation (10). Thus in the indicated temperature range the hydration contribution to $\Delta G_{contact}$, $-\Delta G_{mon}^{hyd} + \frac{1}{n}\Delta G_{n-mic}^{hyd}$, is close to 18.5kcal/mol which means that the hydration disfavors micelle formation. Table 2 also contains the values of $\Delta G_{contact}^{vac}$ calculated according to (11) – the corresponding $\left\langle f_{ax}^{vac}\right\rangle_{\lambda}$ vs. λ curves are shown in Figure 8. The values for $\Delta G_{contact}^{vac}$ obtained in this fashion range from –43 kcal/mol for $T = 280$ K to –35 kcal/mol for $T = 300$ K. Note that the decrease in $\Delta G_{contact}^{vac}$ with decreasing temperature is physically reasonable, because the entropic repulsion (due to spatial confinement) between the sidechains should become less as temperature decreases. The resulting values for $\Delta G_{contact}$ yield an α (cf. Table 2) which is between 43.9 for $T = 280$ K and 27.4 for $T = 300$ K, and thus our α is about two to three times larger than the experimental value quoted above.

One source of error in the present calculation is the neglect of the polarization contribution to ΔG^{hyd} in equation (10). The parameterization is strongly model dependent and unfortunately cannot be carried over to the model used in this work. We also probe the sensitivity of $\left\langle f_{ax}^{vac}\right\rangle_{\lambda}$ to the charge model, because the partial charges are a possi-

ble source of considerable error in every force field calculation. Figure 8 shows the integrand on the righthand side of (11) for $T = 280$, 290, 300 K. All three curves show a similar behavior with a positive (repulsive) maximum at $\lambda \approx 7$ Å. We test the sensitivity of $\left\langle f_{ax}^{vac}\right\rangle_{\lambda}$ on the partial charges by comparing the result for the partial charges of the Q-equilibration algorithm, which we use here, to the result for AM1 charges (which on average are 2.6 × smaller). Overall we find little difference in the $\left\langle f_{ax}^{vac}\right\rangle_{\lambda}$ vs. λ curves, and thus the corresponding areas under the curves are close, and we would estimate the error at about 10%. Thus, if the error for ΔG_{mon}^{hyd}, $\frac{1}{n}\Delta G_{n-mic}^{hyd}$ and $\Delta G_{contact}^{vac}$ in Table 2 is 10%, then the resulting error for α is already about 25%, becaused $\Delta G_{contact}^{vac}$ essentially is calculated as the difference between significantly larger numbers.

Isotropic-to-nematic phase transition

In [33] a theoretical model for the phase behavior including the isotropic-to-nematic phase transition for a system of reversibly assembling long, rod-like, flexible, monodisperse, linear aggregates is given. The phase behavior is driven by a subtle interplay between aggregate flexibility, the inter-aggregate steric interactions, and the internal energy of the aggregates. The orientation free energy in the model depends on the angular distribution of the aggregate's contours and on their length. The flexibility of the aggregates therein is characterized by their persistence length. For the contribution due to the excluded volume an extension of Parson's decoupling approximation (cf. [34]) is used. This expression simplifies to Onsager's trial function description [35] of long, rigid rods for small volume fractions of the solute. Finally, the internal free energy of the aggregate is described via a monomer-monomer contact free energy analogous to the last term in equation (4).

Figure 9 shows a preliminary calculation, where we have converted the theoretical isotropic-nematic phase diagram of Figure 3 in reference [33] from the P-υ-plane to the T-υ-plane, using the simulation result of Figure 7. Here υ is the solute (TP6) volume fraction. The comparison with the experimental transition [16] is rather satisfactory at the low temperatures, but for higher temperatures the deviation is considerable. We mainly attribute this to the independence of α on temperature and concentration assumed in [33], and, in addition, to the neglect of polydispersity in the analytical model.

Conclusion

In this paper we have studied the structure and thermodynamics of a rod-like micelle using molecular dynamics simulations. From the radial water density profile and the radial positions of the side chain atoms we estimate the effective diameter of the TP6 micelle in water to be 28 Å. For the average monomer-monomer center of mass and stacking distance within the micelle we obtain 4.5 Å and 4.2 Å respectively for all three temperatures considered here. These average values are in good agreement with the experimentally determined ring-ring separations. Long TP6 micelles are flexible, and we obtain persistence lengths between $\approx$ 340 monomers at 280 K and $\approx$ 160 monomers at 300 K. It is worth mentioning that the temperature dependence of the phase behavior of micellar systems like the one studied here enters mainly through two quantities - the persistence length and the contact free enthalpy. Here we have shown that the temperature dependence of the first is rather strong in the case of TP6 in water, i.e. P is reduced by a factor of $\approx$ 2 between T = 280 K and 300 K. Thus, the decrease of the persistence length with increasing temperature is substantial. As mentioned before, the effective micelle diameter as well as the persistence length are key quantities in the theoretical description of the phase behavior of the TP6/H$_2$O-system, i.e. excluded volume interactions are dependent on the micelle's geometry (aspect ratio), whereas the orientation distribution of the micelle's contour and the attendant entropy contribution depend on the persistence length and on the micelle's contour length which in turn depends on the monomer-monomer separation. The present simulation shows how, even for a large system (in terms of the monomer size), these quantities can be calculated including microscopic detail. Furthermore, we have considered the monomer-

monomer contact free enthalpy which largely governs the micellar size distribution at low concentration. We base this calculation on a combination of our simulation trajectories with a hydration shell model due to Scheraga and co-workers. Even though our simulation results of α = 27.4 43.9 overestimate the experimental findings they allow to compare the various contributions to the monomer-monomer contact free enthalpy. We find that the main contribution to α is due to the competition between the monomer contact potential energy (which favors micellation) and the loss of free enthalpy of hydration when a monomer-monomer contact is formed (which disfavors micellation). Finally, we used our simulation results as an input for an previously developed theoretical model for the phase behavior of reversibly assembling, rod-like, flexible aggregates. In this preliminary calculation we have found quantitative agreement for T = 280 K and qualitative agreement for the two higher temperatures.

References

1. Micelles, Membranes, Microemulsions and Monolayers, Gelbart, W. M.; Ben-Shaul, A. and Roux, D. (eds.) Springer, New York, **1994**.
2. Taylor, M. P. and Herzfeld, J. *J. Phys.: Condens. Matter* **1993**, *5*, 2651.
3. McMullen, W. E.; Gelbart, W. M. and Ben-Shaul, A. *J. Chem. Phys.* **1985**, *82*, 5616.
4. Odijk, T. *J. Physique* **1987**, *48*, 125.
5. Hentschke, R. *Liq. Cryst.* **1991**, *10*, 691.
6. van der Schoot, P. and Cates, M. E. *Europhys. Lett.* **1994**, *25*, 515.
7. Böcker, J.; Brickmann, J. and Bopp, P. *J. Phys. Chem.* **1994**, *98*, 712.
8. Rector, D. R.; van Swol, F. and Henderson, J. R. *Mol. Phys.* **1994**, *82*, 1009.
9. Laaksonen, L. and Rosenholm, J. B. *Chem. Phys. Lett.* **1993**, *215*, 429.
10. Shelley, J. C.; Sprik, M. and Klein, M. L. *Langmuir* **1992**, *9*, 916.
11. Karaborni, S.; Esselink, K.; Hilbers, P. A. J.; Smit, B.; Karthäuser, J.; van Os, N. M. and Zana, R. *Science* **1994**, *266*, 254.
12. Boden, N.; Bushby, R. J. and Hardy, C. *J.Phys. Lett. Paris* **1985**, *46*, L 325.
13. Boden, N.; Bushby, R. J.; Hardy, C. and Sixl, F. *Chem. Phys. Lett.* **1986**, *123*, 359.
14. Boden, N.; Bushby, R. J.; Jolly, K. W.; Holmes, M. and Sixl, F. *Mol. Cryst. Liq. Cryst.* **1987**, *152*, 37.
15. Ferris, L. M. *Ph.D. thesis*, University of Leeds, **1989**.
16. Edwards, P. J. B. *Ph.D. thesis*, University of Leeds, **1993**.
17. Hubbard, J. and Boden, N. privat communication.
18. Helfrich, J.; Hentschke, R. and Apel, U. M. *Macromolecules* **1994**, *27*, 472.

19. Pearlman, D. A.; Case, D. A.; Caldwell, J. C.; Seibel, G. L.; Singh, U. C.;Weiner, P. K. and Kollman, P. A. AMBER 4.0, Molecular Dynamics Simulation Package, University of California, San Francisco, **1991**.

20. Ryckaert, J. P.; Cicotti, G. and Berendsen, H. J. C. *J. comput. Phys.* **1977**, *23*, 327.

21. Allen, M. P. and Tildesley, D. J. *Computer Simulations of Liquids*. Oxford University Press, Oxford, **1987**.

22. Berendsen, H. J. C.; Grigera, J. R. and Straatsma, T. P. *J. Phys. Chem.* **1987**, *91*, 6269.

23. Bast, T. and Hentschke, R. *J. Phys. Chem.* **1996**, in print.

24. Berendsen, H. J. C.; Postma, J. P. M.; van Gunsteren, W. F.; DiNola, A. and Haak, J. R. *J. Chem. Phys.* **1984**, *81*, 3684.

25. Riddick, J. A.; Bunger, W. B. and Sakano, T. K. *Organic Solvents*, John Wiley & Sons, New York, **1986**.

26. Landolt-Börnstein, Vol. 5: Eigenschaften der Materie in ihren Aggregatzuständen, Schäfer, K. (ed.), Springer, Berlin, Heidelberg, New York, **1969**.

27. Jorgensen, W.; Chandrasekhar, J.; Madura, J.; Impey, R. and Klein, M. *J. Chem. Phys.* **1983**, *79*, 926.

28. INSIGHT, Vers. 2.0, Molecular Dynamics Simulation Software, Biosym Technologies, **1989**.

29. Landau, L. D. and Lifshitz, E. M. *Statistische Physik, Teil 1*, Akademie Verlag, Berlin, **1979**, Vol. 5.

30. van Gunsteren, W. F. and Berendsen, H. J. C. *J. Comp.-Aid. Mol. Des.* **1987**, *1*, 171.

31. Kang, Y. K.; Nemethy, G. and Scheraga, H. A. *J. Phys. Chem.* **1987**, *91*, 4105.

32. Kang, Y. K.; Nemethy, G. and Scheraga, H. A. *J. Phys. Chem.* **1987**, *91*, 4110.

33. Hentschke, R.; Edwards, P. J. B.; Boden, N. and Bushby, R. J. *Macromol. Symp.* **1994**, *81*, 361.

34. Lee, S. *J. Chem. Phys.* **1987**, *87*, 4972.

35. Onsager, L. *Ann. N. Y. Acad. Sci.* **1949**, *51*, 627.

J. Mol. Model. **1996**, 2, 341 – 350

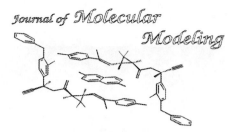

Journal of Molecular
Modeling

© Springer-Verlag 1996

Reparametrisation of Force Constants in MOPAC 6.0/7.0 for Better Description of the Activation Barrier of Peptide Bond Rotations

Olaf Ludwig, Heiko Schinke and Wolfgang Brandt*

Fachbereich Biochemie/Biotechnologie, Martin-Luther-Universität Halle-Wittenberg, Kurt-Mothes-Str. 3, D-06099 Halle/
Saale, Germany (brandt@biochemtech.uni-halle.de)

Received: 15 May 1996 / Accepted: 6 August 1996 / Published: 27 September 1996

Abstract

We have reparametrised the force constants in AM1 and PM3 for better description of activation barriers of peptide bond rotations. A new keyword MMOP was introduced for special recognition of peptide bonds preceding a proline residue or other N-dialkyl substituted amides. The bug in the original MOPAC was corrected where in the case of amides where the nitrogen atom is linked to two hydrogens the force field correction term is counted twice. The new parametrisation of the force constants for peptide bond rotations leads to more realistic rotational barriers of peptide bond rotations. The PM3 optimised pyrrolidine ring in proline adopts no longer a pyramidalised nitrogen atom but a more sp^2-hybridised flat peptide bond geometry.

Keywords: proline, MOPAC, AM1, PM3, force field correction, peptide bond, rotational barrier.

Introduction

The peptide bonds in peptides or proteins usually adopt a *trans* conformation. From peptide bonds preceding a proline residue, however, about 10% occur as a *cis* isomer in native proteins [1-3].

It has been shown that *cis-trans* isomerisation reactions can be the rate limiting step in protein folding processes [4-30]. Furthermore, *trans-cis* isomerisation has been suggested as essential step in enzyme catalysis mechanisms in proline specific proteases in particular for dipeptidyl peptidase IV [31].

The investigation of these mechanisms by means of theoretical calculations requires a correct description of the activation barrier for torsion of the peptide bond. Ab initio methods can be used for the calculation of such activation barriers for relatively small molecules only [32].

Semiempirical methods such as AM1 and PM3 [33-36] do not reflect the mesomery stabilisation of the peptide bond correctly (see Scheme 1) [36, 37].

Scheme 1: *Mesomery stabilisation of the peptide bond*

Therefore, additional force field correction terms (E = $k \cdot \sin^2(w)$) (Keyword MMOK) have been introduced in MOPAC for better description of the activation barrier for peptide bond isomerisation.

Despite these artificial improvements of the methods several problems remain by employing these methods for inves-

* *To whom correspondence should be addressed*

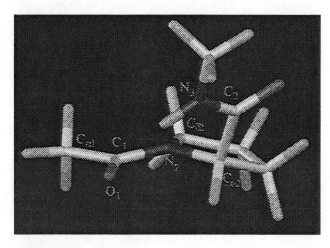

Figure 1: *Trans conformation of N-acetylproline-methylamide with marked atoms to define the considered atoms for peptide bond rotations:* $\zeta = C_{\alpha 1}\text{-}O_1\text{-}C_{\delta 2}\text{-}C_{\alpha 2}$; $\eta = C_1\text{-}C_{\alpha 2}\text{-}N_2\text{-}C_{\alpha 2}$; $\omega = C_{a1}\text{-}C_1\text{-}N_2\text{-}C_{\alpha 2}$; $\phi = C_1\text{-}N_2\text{-}C_{\alpha 2}\text{-}C_2$; $\psi = N_2\text{-}C_{\alpha 2}\text{-}C_2\text{-}N_3$; $\chi = C_{\alpha 2}\text{-}C_{\beta 2}\text{-}C_{\gamma 2}\text{-}C_{\delta 2}$

tigations in *trans-cis* or *cis-trans* isomerisation processes: (i) The keyword MMOK leads to the identification of peptide bonds possessing an amide group with at least one N-H bond. Thus, the additional force field correction is not applied for peptide bonds preceding a proline residue or other N-dialkyl substituted compounds (e.g. N,N-dimethyl acetamide). (ii) The conformation of the pyrrolidine ring in proline residues differs considerably from X-ray structures. That is, a pyramidal structure of the nitrogen (in particular with PM3 optimisation) and a too flat pyrrolidine ring conformation. (iii) Since the application of the keyword MMOK leads to the recognition of amide bonds when at least one nitrogen is linked to a hydrogen atom the force field correction term is counted twice for amides with two N-H bonds such as acetyl amide or peptidyl-amides. (iv) The activation barriers for rotation of the peptide bonds are too small even if the keyword MMOK is applied (see Table 7).

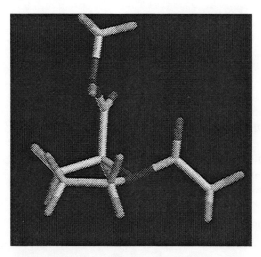

anti-endo

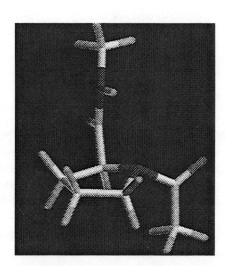

anti-exo

Figure 2: *The four possible transition states of peptide bond rotations.*

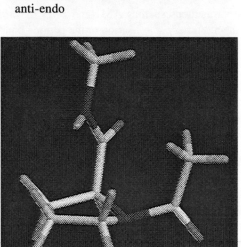

syn-endo

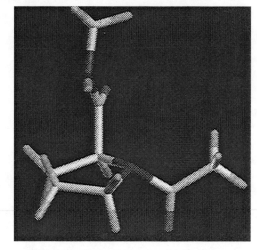

syn-exo

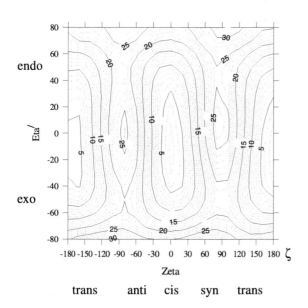

Figure 3. *Conformational η´-ζ rotation map for N-acetyl-(S)-proline-methylamide.*

η´ = 180° - η in correspondence to [32].

Methods

The semiempirical methods AM1 and PM3 embedded in MOPAC 6.0/7.0 were used to calculate and adjust the activation barriers for torsion of peptide bonds.

The rotation of a peptide bond is usually described by the dihedral angle ω ($C_{\alpha 1}$-C_1-N_2-$C_{\alpha 2}$) (see Figure 1). However, it has been shown by *Feigel et al. 1993* [37] and *Fischer et al. 1994* [32] that a simple twisting of ω leads to incorrect results for the energy height of the activation barrier (see Table 1). The degrees of freedom are not only the torsion ω but also the out-of-plane deformation of the amide nitrogen (pyramidalisation). For this reason the virtual dihedral angles ζ ($C_{\alpha 1}$-O_1-$C_{\delta 2}$-$C_{\alpha 2}$) and η (C_1- $C_{\alpha 2}$-N_2- $C_{\alpha 2}$) (see Fig-

Table 1. *Comparison of barriers for trans-cis isomerisation in N-Acetyl-(S)-proline-methylamide calculated with AM1 and PM3*

barrier	AM1		PM3	
	ω-plot	η–ζ-plot	ω-plot	η–ζ-plot
anti	22.5	21.4	23.9	19.7
syn	22.4	18.5	21.7	18.6

energies in kcal/mol obtained with MMOP and HTYP (see below)

ure 1) were used (see also [32]). The dihedral angle ζ describes the peptide bond and adopts a value of 0° for a *cis* and 180° for a *trans* conformation.

The pyramidalisation of the peptide bond nitrogen atom depends on the dihedral angle η which is 0° for a plane sp²-nitrogen and 120° or -120°, respectively, for a pyramidal sp³-nitrogen atom.

In this way all four possible transition states can be recorded. On the one hand, these four transition states result from two possible orientations of the lone pair of the nitrogen which is in periplanar (endo) or antiperiplanar (exo) position to the carbonyl group and on the other hand from two alternatives in the twisting of the peptide bond (syn or anti, respectively). Accordingly to *Fischer et al. 1994* [32] these transition states are named anti-endo ζ -80°, η 120°, anti-exo ζ -80°, η -120°, syn-endo ζ 80°, η 120° and syn-exo ζ 80°, η -120° (see Figure 2 and Tables 2 to 6).

All semiempirical calculations were performed using the eigenvector following method (keyword EF or TS, respectively) by setting the SCF cut off criterium to SCFCRT = 1.D-12 and that one of the gradient to GNORM = 0.1. A grid search was carried out for ζ in 20° increments from 180° to -180° and for η from 120° to -120° in steps of 10°.

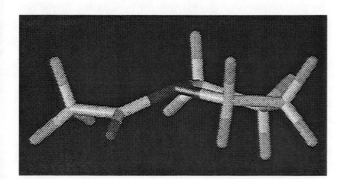

Figure 4a. *PM3 optimised conformation (pyramidal nitrogen) of N-acetyl-pyrrolidine without a force field correction of the peptide bond.*

Figure 4b. *PM3 optimised conformation (planar nitrogen) of N-acetyl-pyrrolidine with a force field correction (MMOP/ HTYP) of the peptide bond.*

Table 2. *Ground and transition states of acetamide calculated with AM1 and PM3*

variable	keywords	trans	cis	anti-endo	anti-exo	syn-endo	syn-exo
energy	**NOMM**	-50.7	-50.7	-41.1	-37.1	-37.1	-41.1
[kcal/mol]		-51.0	-51.0	-45.5	-42.9	-42.9	-45.5
	MMOK	-50.7	-50.7	-31.8	-27.6	-27.6	-31.8
		-49.3	-49.4	-25.3	-22.7	-22.7	-25.2
	MMOK/HTYP	-50.7	-50.7	-32.7	-28.6	-28.6	-32.7
		-49.3	-49.3	-31.3	-28.8	-28.8	-31.3
ζ [°]	**NOMM**	-179.9	-0.1	-86.0	-80.7	80.6	87.8
		-172.1	6.1	-85.9	-80.8	81.4	86.1
	MMOK	-179.9	-0.1	-86.0	-80.7	80.8	86.3
		-179.9	-0.1	-85.7	-80.9	80.6	86.7
	MMOK/HTYP	-179.9	-0.1	-86.1	-81.0	80.9	86.1
		-179.4	-0.1	-86.4	-81.1	81.1	86.4
η [°]	**NOMM**	179.9	179.9	116.8	-117.3	117.3	-117.3
		133.7	-134.3	120.4	-119.6	119.6	-120.5
	MMOK	179.6	179.9	116.6	-117.4	117.4	-116.7
		179.6	179.6	112.7	-111.1	111.3	-112.6
	MMOK/HTYP	179.9	179.9	117.1	-118.0	118.0	-122.7
		179.3	176.7	114.8	-113.4	113.4	-123.9
ω [°]	**NOMM**	179.9	-0.1	-122.6	-57.6	57.3	124.5
		165.1	31.4	-119.9	-59.5	60.2	120.0
	MMOK	179.9	-0.1	-122.7	-57.5	57.6	123.2
		179.9	-0.1	-124.6	-53.8	53.6	125.8
	MMOK/HTYP	-179.9	-0.1	-122.7	-58.2	58.2	122.7
		179.3	-2.0	-123.9	-55.9	55.9	123.9

For the investigation of the influence of the pyrrolidine ring conformation to peptide bond rotation a grid search for ζ and χ in 10° increments ranging from -40° to 40° was performed. The influence of the C-terminal amide group was estimated by varying ζ and ψ in 20° increments. For all grid searches the keywords EF and DMAX = 0.1 were used.

The determination of the transition states was performed by using the highest energy values for ζ where for η the lowest energy was obtained. These conformations were optimised in direction to a transition state (TS, DMAX = 0.1). In such case where no transition state could be calculated the NLLSQ minimiser was applied. The resulting transition states were proved by the eigen values (one negative) of the hessian matrix using FORCE.

Results and discussions

The investigations of the barriers for peptide bond rotations were carried out for the following compounds: acetamide, N-methylacetamide, N,N-dimethylacetamide, N-acetyl-pyrrolidine and N-acetyl-(S)-proline-methylamide.

For the determination of the transition state of rotation around a peptide bond it is insufficient to twist the dihedral angle ω [32, 37]. In this case the resulting barriers of rotation are too high (see Table 1). The reason for this is an incorrect consideration of the nitrogen inversion. Therefore, we calculated for all investigated compounds η-ζ maps to determine the correct transtition states. The conformational map for η-ζ rotation is shown for example for N-acetyl-(S)-proline-methylamide in Figure 3.

The comparison of the calculated activation barriers is listed in Table 1. These results are based on the modifica-

Table 3. *Ground and transition states of N-methyl-acetamide calculated with AM1 and PM3*

variable	keywords	trans	cis	anti-endo	anti-exo	syn-endo	syn-exo
energy	**NOMM**	-47.3	-47.1	-38.4	-35.0	-35.0	-38.4
[kcal/mol]		-51.4	-51.9	-46.9	-44.6	-44.6	-46.9
	MMOK	-47.3	-47.1	-33.3	-29.7	-29.7	-33.3
		-50.0	-50.7	-36.0	-33.7	-33.7	-36.0
	MMOK/HTYP	-47.3	-47.1	-29.3	-25.6	-25.6	-29.3
		-49.8	-50.7	-32.1	-29.9	-29.9	-32.2
ζ [°]	**NOMM**	-180.0	0.2	-92.4	-80.6	80.5	92.6
		-175.2	-12.9	-95.8	-79.2	79.0	96.1
	MMOK	180.0	-179.1	-93.1	-80.4	80.4	92.8
		178.7	2.9	-95.3	-79.9	79.8	95.4
	MMOK/HTYP	180.0	0.5	-93.2	-80.3	80.2	93.3
		-179.1	0.5	-95.5	-80.0	80.0	95.6
η [°]	**NOMM**	179.9	-178.8	124.1	-128.1	128.2	-124.3
		-136.1	138.2	123.7	-123.7	123.7	-123.7
	MMOK	-180.0	-175.9	121.1	-123.8	124.1	-121.2
		-153.6	-157.6	118.7	-118.2	118.2	-118.8
	MMOK/HTYP	-180.0	-177.3	119.2	-121.1	121.2	-119.2
		163.2	-174.2	117.1	-116.3	116.3	-117.1
ω [°]	**NOMM**	180.0	0.9	-118.2	-64.0	64.0	118.4
		-166.3	-33.7	-121.3	-60.3	60.0	121.6
	MMOK	180.0	2.9	-120.6	-61.9	62.0	119.8
		-170.7	14.4	-122.8	-58.2	58.0	122.9
	MMOK/HTYP	180.0	2.0	-121.1	-60.5	60.5	121.2
		174.1	3.6	-123.8	-57.2	57.2	123.8

tions within MOPAC which will be described in the following.

Since the amide bonds in N,N-dialkylamides are not recognised for a force field correction within the original MOPAC-program we have modified the source code of MOPAC slightly and introduced a new keyword MMOP. With this keyword the detection of peptide bonds without an N-H group is possible and a force field term for the twisting of such peptide bonds can be added. In general, a peptide bond is recognised by measuring the distances between the atoms of this bond (C_1, O_1, N_2, H). MMOP works in the same manner as MMOK except searching for two N-C bonds included in a peptide bond.

With this keyword the bug in the original MOPAC where in the case of two N-H bonds the additional force field term is applied twice has also been corrected. The results obtained with this modifications are summarised in Tables 2-6 for all investigated molecules. In each table the upper rows indicate AM1 and the other corresponding rows the PM3 results (highlighted in red).

In spite of these improvements of MOPAC the calculated barriers for the twisting of the peptide bond are in each case to small in comparison with experimentally measured values or results from ab initio calculations (see Table 7). The constants k of the force field term are stored in the array HTYPE[2] for AM1 and HTYPE[3] for PM3. It seems that

Table 4. *Ground and transition states of N,N-dimethyl-acetamide calculated with AM1 and PM3*

variable	keywords	trans	cis	anti-endo	anti-exo	syn-endo	syn-exo
energy	NOMM	-41.4	-41.3	-33.6	-30.7	-30.7	-33.6
[kcal/mol]		-52.3	-52.3	-49.0	-47.1	-47.1	-49.0
	MMOP	-41.3	-41.3	-28.2	-25.1	-25.1	-28.2
		-51.1	-51.1	-37.4	-35.4	-35.4	-37.4
	MMOP/HTYP	-41.3	-41.3	-24.0	20.8	-20.8	-24.0
		-50.9	-50.9	-33.3	-31.2	-31.2	-33.3
ζ [°]	NOMM	179.6	0.0	-85.4	-79.1	78.9	86.1
		-159.8	16.6	-85.9	-79.2	79.0	86.2
	MMOP	179.7	-0.4	-85.8	-78.9	78.8	85.8
		177.4	-2.0	-86.4	-79.0	79.0	86.3
	MMOP/HTYP	-179.1	-0.3	-86.0	-78.6	78.6	86.0
		-179.1	-1.3	-86.5	-78.9	78.8	86.5
η [°]	NOMM	-171.6	-171.4	129.7	-133.4	133.4	-129.2
		141.6	-141.6	129.6	-130.3	130.3	-129.4
	MMOP	-174.3	-178.1	126.7	-130.1	130.1	-126.5
		157.5	-157.9	125.5	-126.2	126.2	-125.5
	MMOP/HTYP	163.2	178.1	124.9	-128.0	128.0	-124.8
		163.2	162.9	124.2	-124.9	124.8	-124.2
ω [°]	NOMM	-176.7	5.0	-114.9	-67.1	66.9	115.6
		-179.2	39.1	-115.0	-65.4	65.2	115.4
	MMOP	-177.7	-1.5	-116.4	-65.4	65.3	116.5
		-172.0	-8.1	-117.4	-63.1	63.1	117.3
	MMOP/HTYP	174.1	-1.5	-117.4	-64.2	64.1	117.5
		174.1	-6.7	-118.2	-62.3	62.2	118.1

the parametrised value for N-methylacetamide of 14.0 kcal/mol is too low in comparison to experimental values (18.3 to 18.9 kcal/mol, see Table 7). To enlarge the activation barriers these constants were adjusted for better correlation with experimental values. As the result of systematic conformational studies the AM1 HTYPE[2] array was enlarged from 3.3191 to 5.9864 and that one for PM3 (HTYPE[3]) from 7.1853 to 9.8526. These new constants will be applied if the new keyword HTYP is used in the MOPAC calculations. The results of calculated barriers for rotation of a peptide bond are summarised in Table 7 and compared with the results obtained by using the original parameters (MMOK/MMOP without HTYP) as well as with experimental results.

The comparison of the calculated isomerisation energies with experimental results (Table 7) is not without its problems. The experimental values listed in Table 7 result from different methods and measurements in distinct solutions and at various temperatures. Depending on the methods used and applied evaluation differences in the isomerisation energies between 0.7 kcal/mol [40] and 2.8 kcal/mol [48] can be obtained. This is also indicated by the values listed in particular for N, N-dimethylacetamide. The free activation enthalpies for the isomerisation range from 15.6 kcal/mol to 21 kcal/mol.

Furthermore, the comparison between the experimental free enthalpies and the calculated heat of formation differ-

Table 5. *Ground and transition states of N-acetyl-pyrrolidine calculated with AM1 and PM3*

variable	keywords	trans	cis	anti-endo	anti-exo	syn-endo	syn-exo
energy	NOMM	-46.0	-46.0	-38.4	-35.0	-35.0	-38.4
[kcal/mol]		-57.3	-57.3	-53.5	-51.0	-51.0	-52.5
	MMOP	-46.0	-46.0	-33.2	-29.7	-29.7	-33.2
		-55.9	-55.9	-41.4	-39.9	-39.9	-41.4
	MMOP/HTYP	-46.0	-46.0	-29.1	-25.6	-25.6	-29.1
		-55.8	-55.8	-37.3	-35.9	-35.9	-37.3
ζ [°]	NOMM	177.0	-2.2	-86.1	-78.9	79.6	86.1
		-167.4	-7.4	-86.3	-78.4	77.7	86.4
	MMOP	172.6	-7.4	-86.3	-77.4	80.2	86.3
		178.5	-1.1	-86.3	-77.4	80.2	86.3
	MMOP/HTYP	179.9	0.1	-86.4	-77.5	79.6	86.4
		-179.9	-0.1	-86.9	-79.0	79.0	86.9
η [°]	NOMM	-162.1	160.1	127.9	-130.4	130.5	-128.1
		145.0	145.0	129.3	-128.3	128.2	-129.2
	MMOP	-175.0	175.0	126.0	-126.6	126.6	-126.2
		-165.0	164.4	125.2	-124.5	124.0	-125.4
	MMOP/HTYP	-179.4	179.9	124.6	-124.6	124.6	-124.8
		177.8	177.9	124.1	-123.5	123.5	-124.2
ω [°]	NOMM	-173.7	-14.4	-117.4	-63.4	64.1	117.3
		173.1	-28.9	-116.8	-61.2	60.5	117.1
	MMOP	176.1	-10.9	-118.4	-63.3	63.1	118.3
		-174.1	-11.2	-119.3	-60.3	60.3	119.2
	MMOP/HTYP	179.7	0.0	-119.1	-59.7	61.7	119.0
		179.0	-1.4	-119.9	-59.7	59.7	119.9
χ [°]	NOMM	-1.6	-1.7	-0.3	-0.6	-0.0	-0.1
		-19.0	19.2	-12.8	-1.3	1.5	1.3
	MMOP	-1.3	-1.0	-0.4	-3.3	-2.9	-0.2
		15.1	-16.5	-0.8	0.0	0.3	-1.3
	MMOP/HTYP	-0.1	-0.7	-0.3	-2.5	-2.1	-0.3
		16.1	-15.6	-0.5	0.0	0.2	-0.1

Table 6: *Ground and transition states of N-acetyl-(S)-proline-methylamide calculated with AM1 and PM3*

variable	keywords	trans	cis	anti-endo	anti-exo	syn-endo	syn-exo
energy	NOMM	-81.7	-79.9	-70.0	-69.6	-65.5	-72.2
[kcal/mol]		-93.3	-93.7	-86.0	-87.6	-82.6	-89.2
	MMOP	-81.5	-79.7	-64.5	-64.2	-59.6	-67.1
		-92.4	-92.3	-74.2	-76.6	-70.5	-77.8
	MMOP/HTYP	-81.4	-79.6	-60.0	-59.9	-55.1	-62.9
		-92.1	-91.8	-69.7	-72.4	-66.1	-73.5
ζ [°]	NOMM	178.2	2.1	-96.3	-71.6	82.0	88.3
		177.4	1.0	-97.3	-79.2	73.3	88.0
	MMOP	178.5	0.2	-94.1	-73.9	81.4	87.6
		178.9	0.1	-90.7	-79.3	79.3	87.5
	MMOP/HTYP	178.5	-0.3	-93.0	-74.8	81.0	87.5
		179.4	-0.2	-90.2	-79.6	80.3	87.6
η [°]	NOMM	-158.4	-154.9	139.6	-135.2	145.0	-128.5
		-146.3	-142.2	136.9	-126.3	141.4	-130.0
	MMOP	-164.8	-165.8	134.5	-131.2	140.8	-126.6
		-164.4	-152.2	131.0	-123.2	134.4	-126.4
	MMOP/HTYP	-167.1	-170.7	131.8	-128.7	137.4	-125.2
		-171.4	-156.9	129.3	-122.0	132.3	-125.2
ω [°]	NOMM	-170.7	17.3	-124.4	-56.4	71.9	119.4
		-166.0	24.2	-126.3	-62.5	59.3	118.2
	MMOP	-173.4	8.8	-124.1	-57.8	69.2	119.5
		-173.3	17.2	-121.2	-61.4	63.7	119.1
	MMOP/HTYP	-174.5	5.1	-124.0	-58.0	67.1	119.8
		-176.2	13.9	-121.4	-61.2	63.2	119.7
ϕ [°]	NOMM	-88.3	-88.0	-34.7	-97.1	-48.4	-107.4
		-99.1	-104.4	-38.0	-92.6	-46.8	-106.4
	MMOP	-84.9	-80.5	-30.5	-100.0	-45.5	-108.3
		-85.1	-98.1	-32.9	-95.4	-41.3	-109.1
	MMOP/HTYP	-82.9	-77.3	-29.4	-102.1	-43.4	-108.5
		-81.0	-94.6	-32.2	-92.8	-40.6	-109.0
ψ [°]	NOMM	63.6	-51.9	-47.9	25.5	-64.6	-46.8
		93.7	-55.9	-63.1	-44.7	-71.9	-59.7
	MMOP	64.0	-53.0	-44.6	28.8	-65.2	-46.5
		100.3	-56.5	-69.1	-45.1	-75.6	-59.6
	MMOP/HTYP	65.0	-52.5	-55.5	28.6	-67.3	-45.9
		100.4	-54.5	-69.5	-42.8	-74.6	-58.4
χ [°]	NOMM	-4.6	0.6	-5.4	0.9	-9.6	2.6
		-20.9	-17.9	-12.5	-7.9	-10.7	-20.1
	MMOP	-5.9	-3.6	-7.3	3.5	-10.4	3.8
		-18.5	-15.8	-11.9	-8.1	-9.6	-20.4
	MMOP/HTYP	-6.1	-4.1	-8.6	5.6	-9.8	5.0
		-18.0	-15.5	-10.9	-2.2	-9.0	-20.7

Table 7: *Barriers of rotation (in kcal/mol) for trans-cis isomerisations of the peptide bond*

Compound	NOMM	MMOK MMOP	HTYP	experimental $\Delta G^{\#}$ [kcal/mol]	Ref.
acetamide	9.6	18.9	18.0	16.7-17.3	[38]
	5.5	24.1	18.0		
N-methylacetamide	8.9	14.0	18.0	18.3-18.9	[39]
	4.5	14.0	17.6		
N, N-dimethylacetamide	7.8	13.1	17.3	18.2-18.6	[40]
				18.3 (366 K)	[41]
				17.4-20.3	[42]
				21.0 (401 K)	[43]
	3.3	13.7	17.6	15.6 (gas phase)	[44]
N-acetylpyrrolidine	7.6	12.8	16.9	16.4-17.1 [a] (303-343 K)	[45]
	3.8	14.5	18.5	16.4 [b] (358 K)	[46]
N-acetyl-(S)-proline-methylamide	9.4	14.4	18.5	18.7-20.7 [c] (333K)	[47]
	4.1	14.6	18.6	17.9 [d]	[32]

Enthalpies (differences in heats of formation) in kcal/mol (standard conditions 298 K, 1 atm)

[a] *N-acetyl-4-methylpiperidine and N-acetylmorpholine,*
[b] *N-acetyl-4-methylpiperidine,*
[c] *N-acetylprolinemethylester,*
[d] *ab inito 6-31G**

ences requires the estimation of the entropy contributions. The measured activation entropies of peptide bond isomerisations are positive for peptides but for simple amides negative activation entropies have also been reported [40]. The calculations of free enthalpy differences gave deviations smaller than 0.5 kcal/mol in comparison to the obtained enthalpy differences with both AM1 and PM3. Since obviously solvents do influence the isomerisation energies [38,39,42,43,47-52] which are not considered in these calculations we list the enthalpies of the barriers for peptide bond isomerisations only (see Tables 1 and 7).

Despite these problems, the agreement between the theoretical and experimental values of the peptide bond isomerisation energies is more satisfactoring than obtained with the original MOPAC force field correction constants.

The introduction of a force field term for peptide bonds preceding a proline residue leads not only to an improved description of the peptide bond rotation but also of the ge-

ometry of the nitrogen bonds inside the five membered ring (see Figure 4). Without a force field correction the pyrrolidine nitrogen atom adopts a pyramidal structure in particular with PM3 calculations which is indicated by the dihedral angle η 145° (PM3) in Tables 5 and 6 for *trans* and *cis* conformations (see Figure 4a). The introduction of MMOP causes a more realistic planar geometry ($\eta \approx 177°$) of such a peptide bond (comp. Figure 4a and 4b).

The puckering of the pyrrolidine ring in proline is a function of the dihedral angle χ. In comparison to the X-ray structure of acetyl-proline-methylamide ($\chi = -36.2°$) [53] the obtained values of the dihedral angle χ by AM1 and PM3 calculations are too small (see Tables 5 and 6). Local minima for $\chi \approx 20°$ or $-20°$ possess a relative energy 0,1 to 0,6 kcal/mol higher than for $\chi = 0°$. These minima could be found only with NLLSQ and not by using TS as a keyword. The determined two transition states of each χ-ζ map are conformationally and energetically identical with the transition states obtained by the η-ζ maps.

The fact that AM1 and PM3 calculate five membered rings too flat is known from the literature and is due to approaches within the semiempirical methods [54]. The energetical minimum of the H-H interaction is about 2.1 Å for both methods. Since the potential hypersurface of five membered rings is very flat and the distance of diaxial protons is shortened from a twisted to a planar conformation from 2.5 Å to 2.3 Å the H-H repulsion is a minimum in a plane conformation.

These effects could not be corrected by the force field correction for proline peptide bonds.

References

1. Dyson, H.J.; Rance, M.; Houghten, R.A.; Lerner, R.A.; Wright, P.E. *J.Mol.Biol.* **1988**, 201, 161.
2. Steward, D.E.; Sarkar, A.; Wampler, J.E. *J.Mol.Biol.* **1990**, 214, 253.
3. Frömmel, C.; Preissner, R. *FEBS Lett.* **1990**, 277, 159.
4. Brandts, J.F.; Halvorson, H.R.; Brenman, M. *Biochemistry* **1975**, 14, 4953.
5. Schmid, F.X.; Baldwin, R.L. *Proc.Batl.Acad.Sci.U.S.A.* **1978**, 75, 4764.
6. Schmid, F.X.; Baldwin, R.L. *J.Mol.Biol.* **1979**, 133, 185.
7. Harding, M.W.; Galat, A.; Uehling, D.E.; Schreiber, S.L. *Nature* **1989**, 341, 758.
8. Siekierka, J.J.; Hung, S.H.Y., Poe, M.; Lin, C.S.; Sigal, N.H. *Nature* **1989**, 341, 755.
9. Fischer, G.; Wittmann-Liebold, B.; Lang, K.; Kiehaber, T.; Schmid, F. *Nature* **1989**, 337, 476.
10. Drakenberg, T.; Forsen, S. *J.Phys.Chem.* **1970**, 74, 1.
11. Roques, B.P.; Garbay-Jaureguiberry, C.; Combrisson, S.; Oberlin. R. *Biopolymers* **1977**, 16, 937.
12. Cheng, H.N.; Bovey, F.A. *Biopolymers* **1977**, 16, 1465.
13. Bächinger, H.P. *J.Biol.Chem.* **1987**, 262(35), 17144.
14. Lang, K.; Schmid, F.X.; Fischer, G. *Nature* **1987**, 329, 268.
15. Tropschug, M.; Wachter, E.; Mayer, S.; Schönbrunner, E.R.; Schmid, F.X. *Nature* **1990**, 346, 674.
16. Schmid, F.X.; Mayr, L.M.; Muecke, M.; Schönbrunner, E.R. *Adv.Protein Chem.* **1993**, 44, 25.
17. Gething, M.J.; Sambrook, J. *Nature* **1992**, 355, 33.
18. Fischer, G.; Schmid, F.X. *Biochemistry* **1990**, *29*, 2205.
19. Jayaramann, T.; Brillantes, A.M.; Timerman, A.P.; Fleischer, S.; Erdjument-Bromage, H., Tempst, P.; Marks, A.R. *J.Biol.Chem.* **1992**, 267, 9474.
20. Fischer, S. **1992** In Curvilinear reaction coordinates of conformational change in macromolecules. Application to rotamase catalysis, Ph.D. Thesis, Harvard University.
21. Fischer, S.; Michnick, S.; Karplus, M. *Biochemistry* **1993**, 32, 13820.
22. Schmid, F.X. *Annu.Rev.Biophys.Biomol.Struct.* **1993**, *22*, 123.
23. Hurle, M.R.; Anderson, S.; Kuntz, I.D. *Protein Eng.* **1991**, 4, 451.
24. Wood, L.C.; White, T.B.; Ramdas, L.; Nall, B.T. *Biochemistry* **1988**, 17, 8562.
25. Kiefhaber, T.; Grunert, H.P.; Hahn, U.; F.X. Schmid *Biochemistry* **1990**, 29, 6475.
26. Kiefhaber, T; Schmid, F.X. *J.Mol.Biol.* **1992**, 224, 231.
27. Mayr, L.; Landt, O.; Hahn, U.; Schmid, F.X. *J.Mol.Biol.* **1993**, 231, 897.
28. Kelley, R.; Richards, F.M. *Biochemistry* **1987**, 26, 6765.
29. Texter, F.L.; Spencer, D.B.; Rosenstein, R.; Matthews, C.R. *Biochemistry* **1992**, 31, 5687.
30. Fischer, G.; Bang, H.; Mech, C. *Biomed.Biochim.Acta* **1984**, *43*, 1101.
31. Brandt, W.; Ludwig, O.; Thondorf , I.; Barth, A.; *Europ.J.Biochem.* **1996**, 236, 109
32. Fischer, S.; Dunbrack Jr, R.L.; Karplus, M.; *J.Am.Chem.Soc.* **1994**, 116, 11931
33. Dewar, M.S.; Zoebisch, E.G.; Healy, E.F.;.Stewart, J.J.P.; *J.Am.Chem.Soc.* **1985**, 109, 3902
34. Stewart, J. J. P.; *J.Comp.Chem.* **1989**, 10, 209
35. Stewart, J. J. P.; *J.Comp.Chem.* **1989**, 10, 221
36. Stewart, J.J.P. "Manual MOPAC (Seventh Edition)" **1993**
37. Feigel, M.; Strassner, T.; *J.Mol.Struct.* **1993**, 283, 33
38. Drakenberg, T.; *Tetrahedron Lett.* **1972**, 18, 1743
39. Drakenberg, T.; Forsen, S.; *Chem.Commun.* **1971**, 1404
40. Neuman Jr., R.C.; Jonas, V.; *J.Am.Chem.Soc.* **1968**, 90, 1970
41. Montaudo, G.; Maravigna, P.; Caccamese, S.; Librando, V.; *J.Org.Chem.* **1974**, 39, 2806
42. Neuman Jr., R.C.; Jonas, V.; *J.Org.Chem.* **1974**, 39, 929
43. Rabinowitz, M.; Pines, A.; *J.Am.Chem.Soc.* **1969**, 91, 1585
44. Feigel, M.; *J.Phys.Chem.* **1983**, 87, 3054
45. Hirsch, J.A.; Augustine, R.L.; Koletar, G.; Wolf, H.G.; *J.Org.Chem.* **1975**, 24, 3547
46. Lynch, D.M.; Cole, W.; *J.Org.Chem.* **1966**, 31, 3337
47. Eberhardt, E.S.; Loh, S.N.; Hinck, A.P.; Raines, R.T.; *J.Am.Chem.Soc.* **1992**, 114, 5437
48. Neuman Jr., R.C.; Jonas, V.; *J.Org.Chem.* **1974**, 39, 925
49. Stewart, W.E.; Siddall, T.H.; *Chem.Rev.* **1970**, 70, 517
50. Calzolari, A.; F.Conti, F.; *J.Chem.Soc.(B)* **1970**, 555
51. Eberhardt, E.S.; Loh, S.N.; Raines, R.T.; *Tetrahedron Lett.* **1993**, 34, 3055
52. Duffy, E.M.; Severance, D.L.; Jorgensen, W.L.; *J.Am.Chem.Soc.* **1992**, 114, 7535
53. DeTar, D.F.; Luthra, N.P.; *J.Am.Chem.Soc.* **1977**, 99, 1232
54. Ferguson, D.M.; Gould, I.R.; Glauser, W.A.; Schroeder, S.; Kollman, P.A.; *J.Comp.Chem.* **1992**, 13, 525

J. Mol. Model. **1996**, 2, 351 – 353

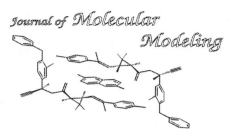

Role of Aromatic Amino Acids in Carbohydrate Binding. Laser Photo CIDNP(Chemically Induced Dynamic Nuclear Polarisation) and Molecular Modeling Study of Hevein-domain Containing Lectins

Hans-Christian Siebert [a,b], Claus-Wilhelm von der Lieth* [c], Robert Kaptein [b], Ukun M.S. Soedjanaatmadja [d], Johannes F.G. Vliegenthart [b], Christine S.Wright [e] and Hans-Joachim Gabius [a]

[a] Institut für Physiologische Chemie der Ludwig-Maxi-milians-Universität, München, Germany;

[b] Bijvoet Center for Biomolecular Research, University of Utrecht, The Netherlands;

[c] German Cancer Research Center, ZentraleSpektroskopie, Im Neuenheimer Feld 280, D - 69120 Heidelberg; Germany
(W.vonderlieth@dkfz-heidelberg.de)

[d] Biochemisch Laboratorium and Bioson Research Institute, University of Groningen, The Netherlands;

[e] Department of Biochemistry and Molecular Biophysics, Virginia Commonwealth University, USA

Received: 15 May 1996 / Accepted: 6 August 1996 / Published: 27 September 1996

Introduction

The recognition of carbohydrate determinants by lectins is involved in mediation of intercellular binding and elicitation of biosignalling processes [1,2]. Structural aspects of this glycobiological interplay can in principle be disclosed by multidimensional nuclear magnetic resonance (NMR) experiments and/or X-ray crystallography. However, these techniques can suffer from a number of inherent restrictions, for example molecular mass in multidimensional NMR studies and in case of X-ray crystallography a lack of information about the dynamic behaviour of molecules in solution. The side chains of tyrosine, tryptophan and histidine often involved in carbohydrate recognition are able to produce CIDNP signals [3] after laser irradiation in the presence of a suitable radical pair-generating dye. Elicitation of such a response in protein simplifies accessibility of the respective groups to the light-absorbing dye. This technique allows in close combination with molecular modeling tools to monitor on an atomic level the effect of ligand binding to a receptor if CIDNP-reactive amino acids are affected.

Experimentally, the shape and intensity of CIDNP signals are determined in the absence and in the presence of specific glycoligands. When the carbohydrate ligand is bound,

CIDNP signals of side chain protons of tyrosine, tryptophan or histidine residues are altered, e. g. they can be broadened and of reduced intensity or disappear completely.

To determine the actual value of this method in glycosciences, several well-characterised N-acetyl-glucosamine-binding lectins from different species with increasing molecular complexity are selected as models. In detail, laser photo CIDNP and molecular modeling experiments of hevein pseudo-hevein, Urtica dioica agglutinin (UDA), wheat germ agglutinin (WGA) and its B domain (domB) have been carried out in the absence and in the presence of their specific ligands. Hevein (43 amino acid residues, PDB entry 1HEV) and pseudo-hevein (43 amino acid residues, modelled by homology from PDB entry 1HEV) are small lectins of rubber trees (Hevea brasiliensis) with ragged C-terminal sequences. The amino acid sequences of the two proteins differ only in six positions from each other. Among these differences the exchange of Trp 21 present in hevein to Tyr 21 in pseudo-hevein is especially notable in the context of CIDNP experiments.

Urtica dioica agglutinin (UDA) from stinging nettle rhizomes is a member of the chitin-binding family of plant lectins whose structure consists of two hevein domains. It has a single chain of 89 amino acids, constituting two binding sites with different affinities for oligomers of N-acetylglucosamine.

---

* *To whom correspondence should be addressed*

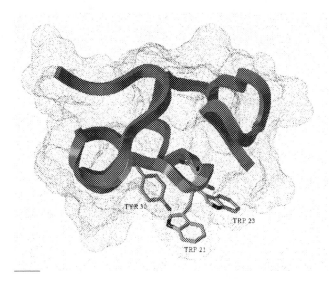

Figure 1: *The backbone of hevein (PDB-entry 1HEV) is drawn in magenta. The aromatic amino acids TRP-21, TRP-23 and TYR 30 are displayed in green. The water accessible surface is indicated with a dotted surface.*

The isoforms of wheat germ agglutinin (WGA PDB entry 7WGA), which are built up from four hevein domains can also bind N-acetylglucosamine. Their structures with a subunit molecular mass of 16 kDa and aggregation of two chains display a considerably increased complexity in retention to the other lectins of this family. The carbohydrate-specific unit, namely the B-domain, was cloned enabling to logically extend the list of test substances.

The lectins where no X-ray or NMR-derived structures were available were modelled using standard homology modelling techniques (ProMod). All lectins are relaxed by MD-simulations in aqueous solution using the CVFF force field.. The surface accessibility of the aromatic amino acids is calculated for a certain number of minimised structures taken from MD-simulations. The comparisons of the calculated surface accessibility with the CIDNP signals showed a good agreement.

CIDNP has to be considered as an excellent tool for protein-carbohydrate interaction studies in solution supplementing the commonly used panel of X-ray-crystallography, high-resolution multidimensional NMR and molecular modeling.

Computational details

The starting structures of the lectins were either taken from the Brookhaven Protein Data Bank or generated using knowledge-based homology modeling technique. The generation of the H-atoms, automatic assignment of partial charges for each atom of the molecule were accomplished using the INSIGHTII software. The center of mass of the lectin was placed in the center of a 30x30x30 Å large water box. A simple point charge model for the water was used allowing the atoms of the water molecules to vibrate. Periodic boundary conditions were applied using a double cutoff of 10 and 12 Å. Each molecular ensemble was submitted to a molecular dynamics simulation using the CVFF force field at a temperature of 300 K, with an equilibration time of 20 ps, a production period of 100 ps, an integration step of 1 fs and a dielectric constant of 1. Each 250 steps of integration a conformation was stored. From the 400 individual conformations of the production time the ten with the lowest potential energy were automatically selected and submitted to a energy minimization of the complete ensemble using the conjugate gradient method. Delineation of possible conforma-

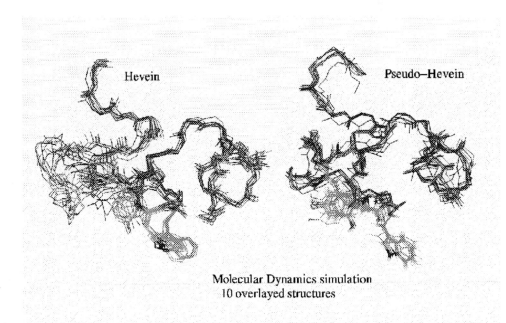

Figure 2: *10 overlayed conformation taken from a molecular dynamics simulation (see computational details) of hevein and pseudo hevein. All selected structures were minimized.*

tions for domB was achieved using different approaches. I) singling out the domB from the X-ray structure of the complete lectin (PDB-entry 7WGA). II) kowledge-based homology modeling techniques using the ProMod tools [4]

a) based on 1HEV as 3D template and

b) based on 1WGC,2WGC,1WGT and 1HEV as 3D template.

The MD calculation were performed on an IBM-SP2 parallel machine using four processors. The handling of the molecules and the analysis of the MD results were accomplished with various modules of the INSIGHTII software on an Silicon Graphics Indigo^2 worksatation.

References

1. Sharon and Lis, *Trends Biochem. Sci.* **1993**, *6*, 221-226.
2. Gabius and Gabius (eds.) Glycosciences: Status and Persepctives, Chapman&Hall, Weinheim, 1996.
3. Kaptein et.al. *Nature* **1978**, *274*, 293-294.
4. Peitch, M.C. *Bio/Technology* **1995**, *13*, 658 - 660.

J. Mol. Model. **1996**, 2, 354 – 357

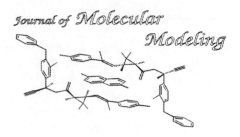

Theoretical Study of the Water Exchange Reaction on Divalent Zinc Ion using Density Functional Theory

Michael Hartmann [a,b], Timothy Clark [b] and Rudi van Eldik [a]

[a] Institut für Anorganische Chemie, Universität Erlangen-Nürnberg, D-91058 Erlangen, Germany

[b] Computer-Chemie-Centrum, Institut für Organische Chemie I, Universität Erlangen-Nürnberg, D-91052 Erlangen, Germany (hartmann@organik.uni-erlangen.de)

Received: 15 May 1996 / Accepted: 6 August 1996 / Published: 27 September 1996

Abstract

Recent *ab initio* studies reported in the literature have challenged the mechanistic assignments made on the basis of volume of activation data [1,2]. In addition to that *ab initio* molecular orbital calculations on hydrated zinc(II)-ions were used to elucidate the general role of this ion in metalloproteins [3]. Due to our interest in both inorganic reaction mechanisms and enzymatic catalysis we started a systematic investigation of solvent exchange processes on divalent zinc-ion using density functional calculations. Our investigations cover aqua complexes of the general form $[Zn(H_2O)_n]^{2+} \cdot mH_2O$ with n=3-6 and m=0-2, where n and m represent the number of water molecules in the coordination and solvation sphere, respectively.

The complexes $[Zn(H_2O)_5]^{2+} \cdot 2H_2O$ and $[Zn(H_2O)_4]^{2+} \cdot 2H_2O$ turnend out to be the most stable zinc complexes with seven and six water molecules, respectively. This implies that a heptacoordinated zinc(II) complex, where all water molecules are located in the co-ordination sphere, should be energetically highly unfavorable and that $[Zn(H_2O)_6]^{2+}$ can quite readily push two coordinated water molecules into the solvation sphere. For the pentaqua complex $[Zn(H_2O)_5]^{2+}$ only one water molecule is easily lost to the solvation sphere, which makes the $[Zn(H2O)_4]^{2+} \cdot H_2O$ complex the most favorable in order to consider the limiting dissociative and associative water exchange process of hexacoordinated zinc(II). The dehydration and hydration energies using the most stable zinc(II) complexes $[Zn(H_2O)_4]^{2+} \cdot 2H_2O$, $[Zn(H_2O)_5]^{2+} \cdot 2H_2O$ and $[Zn(H_2O)_4]^{2+} \cdot H_2O$ were calculated to be 24.1 and -21.0 kcal/mol, respectively.

Keywords: *ab inito*, DFT, zinc complexes, hydration energies

Introduction

The energetics and dynamics of the hydration of transition metal ions form the basis of many fundamental chemical and biochemical processes and have received the attention of many experimentalists in recent years. The ap-plication of *ab initio* molecular orbital and SCF calculations reported in the literature has challenged the mechanistic assignments made on the basis of volume of activation data [1,2]. In addition to that *ab initio* molecular orbital calculations on hydratcd Zn^{2+} ions were used to elucidate the general role of this ion in metalloproteins [3]. In this respect, the application of density functional theory

Table 1. *Total molecular energies (au) and zero-point energies (kcal/mol) of* $[Zn(H_2O)_n]^{2+} \cdot mH_2O$

n+m [a]	n	m	B3LYP/SHA1 [b]	ZPE [c]	B3LYP/SHA2 [d]	B3LYP/SHA3 [e]
7	6	1	-2313.239 035	107.0	-2314.205 809	-2314.163 070
7	5	2	-2313.250 312	107.3	-2314.215 864	-2314.171 175
6	6	0	-2236.850 392	89.9	-2237.710 038	-2237 672 247
6	5	1	-2236.856 859	91.5	-2237.713 126	-2237.674 948
6	4	2	-2236.862 967	92.4	-2237.716 919	-2237.678 007
5	5	0	-2160.460 229	75.1	-2161.207 801	-2161.175 511
5	4	1	-2160.464 991	76.5	-2161.210 592	-2161.178 306
5	3	2	-2160.443 825	76.8	-2161.189 460	-2161.157 477
H₂O			-76.341 524	13.0	-76.461 570	-76. 456 609
Zn(II)			-1778.187 233		-1778.375 868	-1778.375 868

[a] n = number of coordinated water molecules,
m = number of closely associated water molecules in the solvation sphere
[b] B3LYP/SHA1=B3LYP/SHA1//B3LYP/SHA1
[c] calculated at B3LYP/SHA1//B3LYP/SHA1
[d] B3LYP/SHA2= B3LYP/SHA2//B3LYP/SHA1
[e] B3LYP/SHA3 = B3LYP/SHA3//B3LYP/SHA1

(DFT) to transition metal complexes opens new opportunities to describe the hydration of metal ions, as recently shown for the stepwise hydration of Fe^{3+} [4]. Due to our interest in both inorganic reaction mechanisms and enzymatic catalysis we started a systematic investigation of solvent exchange processes on divalent zinc ion using density functional calculations. Our investigations cover aqua complexes of the general form $[Zn(H_2O)_n]^{2+} \cdot mH_2O$ with n=3-6 and m=0-2, where n and m represent the number of water molecules in the coordination and solvation sphere, respectively. First we studied the stepwise hydration of Zn^{2+} ions. In order to include the limiting dissociative and associative substitution modes of the octahedral hexaqua cation, we next focussed on coordination numbers between 5 and 7. Our findings enable us to comment on related results reported in the literature for the hydration of Zn^{2+} ions [1-3] and to emphasize fundamental requirements that must be fulfilled in order to draw mechanistic conclusions from such calculations.

Methods

Calculations were carried out on a CONVEX SPP-1000 computer and on HP-735 PA-RISC workstations at the Computer-Chemie-Centrum Erlangen, using GAUSSIAN 94 [5] and the B3LYP hybrid functional [6,7]. The optimized split valence basis set of Ahlrichs et al. [8], augmented with p- and d-type polarization functions, was used for geometry optimizations and frequency analyses. The latter procedure verified that all structures presented in this work are indeed local minima with respect to their potential energy surface. The final adopted zinc-basis set was of the form [14s9p5d/5s3p3d], whereas the oxygen contraction pattern was [7s4p1d/3s2p1d]. The hydrogen [4s/2s] set remained unchanged. For convenience, this combined basis set will be further referred to as SHA1. For the refinement of the total energy, single point calculations were performed using two modified double zeta Schäfer-Horn-Ahlrichs basis sets for zinc, oxygen and hydrogen [8]. Both types of combinations have two additional p- and d-type polarization functions for zinc forming a [14s11p5d2f/8s7p3d2f] basis set and two d-type polarization functions for oxygen giving a [10s6p2d/6s3p2d] basis set. However they differ in the hydrogen and oxygen parameter sets. The first combination, referred to as SHA2, includes a p-type polarization function on hydrogen forming a [5s1p/3s1p] basis set, whereas the second parameter set, abbreviated as SHA3, contains a diffuse s function for hydrogen giving a [6s/4s] basis set and an additional diffuse sp function for oxygen.

Results and Discussion

Total molecular energies of all structures of the type $[Zn(H_2O)_n]^{2+} \cdot mH_2O$, where n and m represent the water molecules in the coordination and solvation sphere, respectively, are summarized in Table 1. In the case of heptahydrated zinc(II) ions, the most stable structures, in contrast to those published by another group [1,2], have at least one water molecule in the solvation sphere. Although

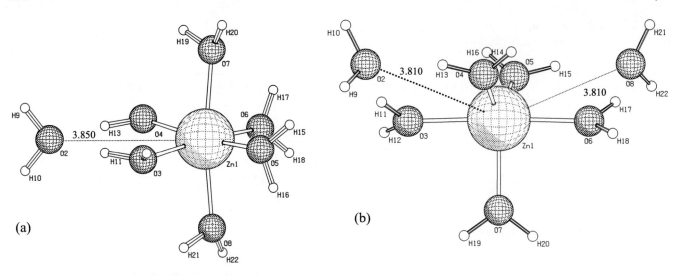

Figure 1. *Molecular geometries from B3LYP/SHA1//B3LYP/ SHA1 calculations. Bond lenghts given in Angstrom. (a) [Zn(H₂O)₆]²⁺·H₂O; PG=C₂.*

(b) [Zn (H₂O)₅]²⁺·2H₂O; PG=C₁

Glusker et al.[3] mentioned this behavior, they did not present any structural data or total energies for heptahydrated Zn^{2+} ions. However, in order to discuss the water exchange behavior of Zn^{2+} ions, these solvent complexes are essential, especially when searching for a suitable transition state structure. Calculations on the B3LYP/ SHA1//B3LYP/SHA1 level show that $[Zn(H_2O)_6]^{2+} \cdot H_2O$ and $[Zn(H_2O)_5]^{2+} \cdot 2H_2O$ are more stable than the corresponding hexaqua complex with one separated water molecule (Figure 1). Taking into account the zero point energy correction (ZPE), these differences were calculated to be 25.5 kcal/mol and 32.3 kcal/mol, respectively, thus

making $[Zn(H_2O)_5]^{2+} \cdot 2H_2O$ the most favourable heptaqua complex in the gas phase. The same trend can be observed when going to the hexaqua complexes, i.e. $[Zn(H_2O)_4]^{2+} \cdot 2H_2O$ is 2.9 kcal/mol more stable than $[Zn(H_2O)_5]^{2+} \cdot H_2O$, and 5.4 kcal/mol lower in energy than $[Zn(H_2O)_6]^{2+}$. This implies that in the gas phase hexaqua Zn^{2+} can push two coordinated water molecules into the solvation sphere. This behavior changes when going to the pentaqua complexes. Here only one water molecule is easily lost to the solvation sphere, which makes the $[Zn(H_2O)_4]^{2+} \cdot H_2O$ complex the most favorable one. The energy difference between this complex and $[Zn(H_2O)_5]^{2+}$ was calculated to be 1.6 kcal/mol, compared to 13.6 kcal/mol between this complex and $[Zn(H_2O)_3]^{2+} \cdot 2H_2O$.

Inclusion of additional polarization and diffuse functions in the SHA2 and SHA3 basis sets, does not change the above conclusions. The tendency of Zn^{2+} ions with more than four water molecules in the coordination sphere to release their fifth and sixth waters into the solvation sphere

Table 2. *Calculated relative stabilities (kcal/mol) for the movement of water molecules from the coordination sphere into the solvation sphere and hydration energies (kcal/mol) for the reaction of the most stable aqua complexes of the zinc(II)-ion corrected for zero-point energies.*

n+m [a]	(n,m) → (o,p) [b]	B3LYP/SHA1 [a]	B3LYP/SHA2 [a]	B3LYP/SHA3 [a]
7	(6,1) → (5,2)	-6.8	-6.0	-4.8
6	(6,0) → (5,1)	-2.5	-0.3	-0.1
6	(5,1) → (4,2)	-2.9	-1.5	-1.0
5	(5,0) → (4,1)	-1.6	-0.4	-0.4
5	(4,1) → (3,2)	13.6	13.6	13.4
6 +H₂O	(4,2) +H₂O → (5,2)	-26.9	-21.6	-21.0
6 - H₂O	(4,2) - H₂O → (4,1)	32.5	25.2	24.1

[a]   see footnote to Table1

[b]   (n,m)=[Zn(H₂O)ₙ]²⁺·mH₂O, (o,p) = [Zn(H₂O)ₒ]²⁺·pH₂O

is maintained. The calculated energies are in good agreement with those published by Glusker et al.[3]. However, the trend of releasing water molecules into the solvation sphere is more pronounced and is most conspicuous when dealing with the heptaqua Zn^+ ions. Here the energy difference between the most stable complex $[Zn(H_2O)_5]^{2+} \cdot 2H_2O$ and $[Zn(H_2O)_6]^{2+} \cdot H_2O$ on the B3LYP/ SHA3//B3LYP/ SHA1 level is approximately 5 times higher than the analogous difference between $[Zn(H_2O)_5]^{2+} \cdot H_2O$ and $[Zn(H_2O)_4]^{2+} \cdot 2H_2O$ and even 50 times higher than the energy difference between $[Zn(H_2O)_6]^{2}$ and $[Zn(H_2O)_5]^{2+} \cdot H_2O$ (Table 2). This, however, indicates that a heptacoordinated zinc(II) complex, where all water molecules are located in the coordination sphere, should be energetically highly unfavorable. Thus a mechanistic assignment based only on the complexes $[Zn(H_2O)_7]^{2+}$, $[Zn(H_2O)_6]^{2+}$ and $[Zn(H_2O)_5]^{2+}$ as used by others [1,2] should be considered with care.

In order to cover the limiting associative and dissociative water exchange processes of hexacoordinated Zn^{2+} and to calculate energies for hydration and dehydration, respectively, it is necessary to consider the most stable gas phase complexes obtained by the DFT calculations. Scheme 1 shows the reaction of the most stable hexacoordinated zinc(II) complex $[Zn(H_2O)_4]^{2+} \cdot 2H_2O$ with water, yielding the most stable heptacoordinated aqua species $[Zn(H_2O)_5]^{2+} \cdot 2H_2O$. In the same way, the dissociation of water results in a pentaqua complex which is best described as $[Zn(H_2O)_4]^{2+} \cdot H_2O$.

$$[Zn(H_2O)_4]^{2+} \cdot 2H_2O + H_2O \rightarrow [Zn(H_2O)_5]^{2+} \cdot 2H_2O$$

$$[Zn(H_2O)_4]^{2+} \cdot H_2O + H_2O \rightarrow [Zn(H_2O)_4]^{2+} \cdot 2H_2O$$

Scheme 1

The energies for the stepwise release of water molecules from the coordination sphere into the solvation sphere and the reaction energies for the hydration and dissociation of water, respectively, are also included in Table 2.

References

1. Akesson, R.; Petterson, L. G. M.; Sandström, M. *J. Phys.Chem.* **1993**, *97*, 3765-3774.
2. Akesson, R.; Petterson, L. G. M.; Sandström, M.; Wahlgren, U. *J. Am. Chem. Soc.* **1994**, *116*, 8705-8713.
3. Bock, C.W.; Kaufman, A.; Glusker, J. P. *J. Am. Chem. Soc.* **1995**, *117*, 3754-3765.
4. Ricca, A; Bauschlicher, Jr., W. *J. Phys. Chem.* **1995**, *99*, 9003-9007.
5. Frisch, M. J.: Trucks, G. W.; Schlegel, H. B.; Gill P. M. W.; Johnson, B. G.; Robb, M. A.; Cheeseman, J. R.; Keith, T.; Petersson, G. A.; Montgomery, J. A.; Raghavachari, K.; Al-Laham, M. A.; Zakrzewski, V. G.; Ortiz, J. V.; Foresman, J. B.; Cioslowski, J.; Stefanov, B. B.; Nanayakkara A.; Challacombe M.; Peng, C.Y.; Ayala P. Y.; Chen, W.; Wong, M. W.; Andres, J. L.; Replogle, E. S.; Gomperts, R.; Martin, R. L.; Fox, D. J. ; Binkley, J. S.; Defrees D. J.; Baker, J.; Stewart, J. P. ; Head-Gordon, M.; Gonzalez, C.; Pople, J. A. *GAUSSIAN 94, Revision B.2;* Gaussian Inc.: Pittsburgh, PA, 1995.
6. Becke, A. D. *J. Chem. Phys.* **1993**, *98,* 5648-5652.
7. Lee, C.; Yang, W.; Parr, R. G. *Phys. Rev.* **1988**, *B37*, 785-789.
8. Schäfer, A.; Horn, H.; Ahlrichs, R. *J. Chem. Phys.* **1992**, *97*, 2571-2577.

J. Mol. Model. **1996**, 2, 358 – 361

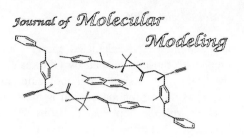

Investigation of the Uncatalyzed Hydration of CO_2 and First Approximations to the Active Site of Carbonic Anhydrase – A Combined *Ab initio* and DFT Study –

Michael Hartmann [a,b], Timothy Clark [b] and Rudi van Eldik [a]

[a] Institut für Anorganische Chemie, Universität Erlangen-Nürnberg, D-91058 Erlangen, Germany.

[b] Computer-Chemie-Centrum, Institut für Organische Chemie I, Universität Erlangen-Nürnberg, D-91052 Erlangen, Germany (hartmann@organik.uni-erlangen.de)

Received: 15 May 1996 / Accepted: 6 August 1996 / Published: 27 September 1996

Introduction

Carbonic anhydrase (CA) is found in both plants and animals and the only known biological function of this ubiquitous enzyme is to catalyze the interconversion of CO_2 and HCO_3^-. In aqueous solution in the absence of CA the reactions

$$H_2O \ + \ CO_2 \ \rightarrow \ H_2CO_3 \qquad (1)$$

$$OH^- \ + \ CO_2 \ \rightarrow \ HCO_3^- \ + \ H^+ \qquad (2)$$

have activation energies of 17.7 kcal/mol (forward) and 14.6 kcal/mol (reverse) for reaction 1 and 13.1 kcal/mol (forward) for reaction 2 [1]. The rate constants for the forward reactions of reaction 1 and 2 are 0.035 s^{-1} and 8500 $s^{-1}M^{-1}$, respectively. In comparison to the uncatalyzed reaction human CA enhances the hydration of CO_2 by a factor of about 10^7. This however implies that the enzyme reduces the energy of activation by about 10 kcal/mol. Although the gas-phase reaction of carbondioxide with water and with hydroxide ion have been studied earlier by others [1-6] we found it necessary to apply density functional theory to the nonenzymatic hydration of CO2 and to compare these results with our ab initio calculations. The active site environment of CA shows that the zinc atom is bound to three histidine imidazole groups with the fourth coordination site occupied by either water or hydroxide ion. In order to understand the central role of the zinc ion in the catalytic mechanism of CA, the next step in our investigation involves the approximation of the active site of this enzyme. Starting with the simplest model, namely the $[ZnOH_2]^{2+}$ / $[ZnOH]^+$ system we also considered the analogous triammine and trisimidazole complexes $[Zn(NH_3)_3OH_2]^{2+}/[Zn(NH_3)_3OH]^+$, $[Zn(Im)_3OH_2]^{2+}/[Zn(Im)_3OH]^+$. The optimized structures and energies for the latter two systems and the first steps in the reaction of $[ZnOH]^+$ with CO_2 are presented in this poster.

Methods

Calculations were carried out on a CONVEX SPP-1000 computer and on HP-735 PA-RISC workstations at the Computer-Chemie-Centrum Erlangen, using GAUSSIAN 94 [7] and the B3LYP hybrid functional [8,9]. The optimized split valence basis set of Ahlrichs et al. [10], augmented with p- and d-type polarization functions, was used for geometry optimizations and frequency analyses. The latter procedure verified that all structures presented in this work are indeed local minima with respect to their potential energy surface. The final adopted zinc-basis set was of the form [14s9p5d/5s3p3d], whereas the oxygen contraction pattern was [7s4p1d/3s2p1d]. The hydrogen [4s/2s] set remained unchanged. For convenience, this type of model chemistry will be further referred to as B3LYP/SHA1. The Moeller-Plesset correlation energy corrected calculations cover geometry optimizations as well as frequency calculations on the MP2(FC) level of theory in conjunction with

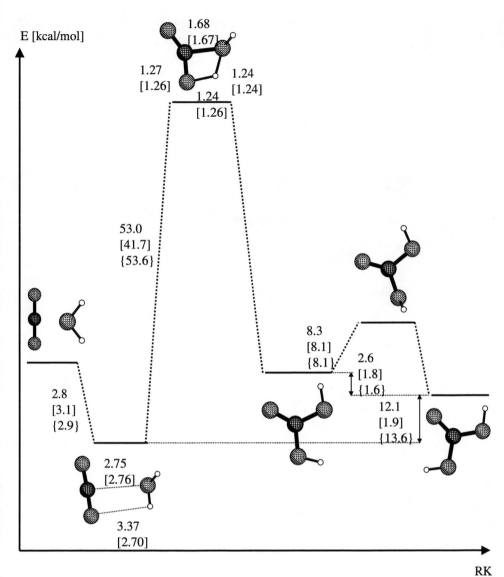

Figure 1. *Energy profile of the reaction of carbondioxide and water. The levels of theory are signed with:*
MP2(FC)/6-31+G(d,p)//
MP2(FC)/6-31+G(d,p),
{MP4(SDTQ)/6-31+G(d,p)//
MP2(FC)/6-31+G(d,p)} and
[B3LYP/SHA1//B3LYP/SHA1].

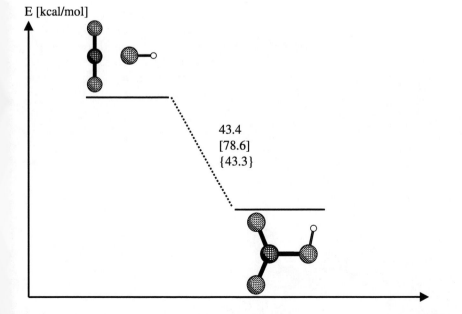

Figure 2. *Energy profile of the reaction of carbondioxide and hydroxide ion. The levels of theory are signed with:*
MP2(FC)/6-31+G(d,p)//
MP2(FC)/6-31+G(d,p),
{MP4(SDTQ)/6-31+G(d,p)//
MP2(FC)/6-31+G(d,p)} and
[B3LYP/SHA1//B3LYP/SHA1].

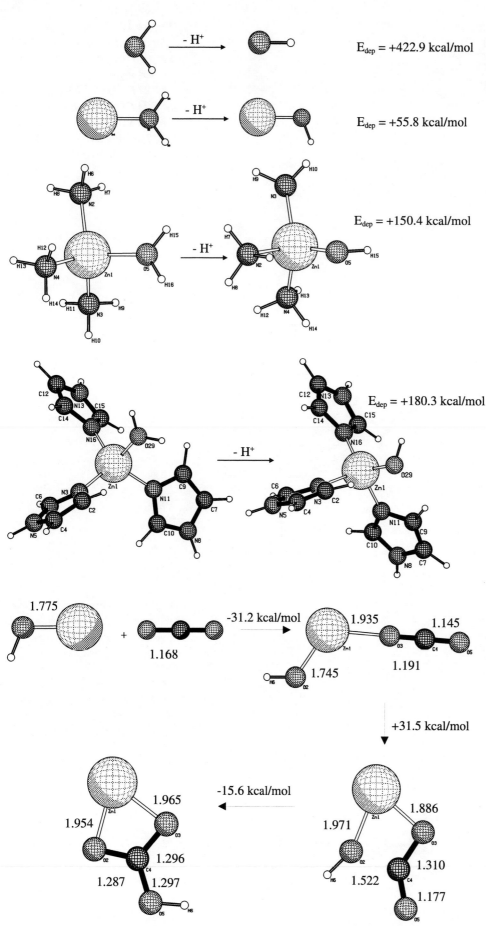

Figure 3. *Energies of deprotonation [kcal/mol] for selected model complexes of the active site of carbonic anhydrase based on B3LYP/SHA1//B3LYP/SHA1 calculations*

Figure 4. *Suggestion for the first steps in the reaction of [ZnOH]⁻ and CO2 calculated on the B3LYP/SHA1//B3LYP/SHA1 level of theory. Bond lenghts are given in Angstrom*

a 6-31+G(d,p) basis set. For the refinement of the total energies, single point calculations were performed using MP4(SDTQ)/6-31+G(d,p) calculations. All of the energy values given in the figures are corrected by means of zero-point energies (ZPE).

Results and Discussion

The energy barrier for the reaction of CO_2 with H_2O (Figure 1) varies with respect to the computational technique from 53.6 kcal/mol (MP4/6-31+G(d,p)//MP2/6-31+G(d,p)) to 41.7 kcal/mol (B3LYP/SHA1//B3LYP/ SHA1). The deviation from the experimental energy of activation measured in solution (17.7 kcal/mol) can be explained by means of solvent interactions, which have an important contribution to the barriers. However, in comparison to semiempirical and *ab initio* results found in the literature [1-4], density functional calculations give better values for the energy barrier. The overall reaction yielding the trans-trans isomer of the carbonic acid is predicted to be endothermic with reaction energies ranging from 13.6 kcal/mol (MP4/6-31+G(d,p)//MP2/6-31+G(d,p)) to 1.9 kcal/mol (B3LYP/SHA1// B3LYP/SHA1). Again density functional theoretical calculations resemble the experimental value of 4.8 kcal/mol better than any other computational technique described in the literature. For the analogous gas-phase hydration of CO_2 by OH^- instead of H_2O (Figure 2), we found, as have other workers [1-4], that this reaction is activationless, while the experimental energy of activation in solution is 13.1 kcal/mol. Almost all of this error is again associated with the underestimation of the solvent contribution. The calculated reaction energies vary from -43.3 kcal/mol mol (MP4/6-31+G(d,p)//MP2/6-31+G(d,p)) to -78.6 kcal/mol (B3LYP/ SHA1//B3LYP/ SHA1). Now, MP4 calculations in conjunction with a diffuse basis set give results remarkably close to the experimental value of -49 kcal/mol, while density functional theory yields energies that are in the range of semiemprical (AM1) ones [2].

Theoretical and experimental work have shown that the deprotonation of zinc-bound water is more favorable than either the deprotonation of zinc-bound imidazolium or that of imidazole itself [2]. Hence it seems that the role of zinc is to reduce the pK_a of bound H_2O by electrostatic interaction, thus allowing the hydroxyl group to retain significant nucleophilic character to attack CO_2. This effect can be shown by looking at the deprotonation energies of a selection of model complexes (Figure 3). For the extreme cases, i.e the totally uncatalyzed and catalyzed deprotonation, respectively, these energies were calculated to be +422.9 kcal/mol and +55.8 kcal/mol (B3LYP/SHA1/ /B3LYP/SHA1). On approaching the active site environment of CA by means of complexation of the zinc with ammonia and imidazole, we obtained energy values between these limits. For the σ-donor ligand we obtained +150.4 kcal/mol while the σ-acceptor ligand yields +180.3 kcal/mol. In contrast to this, the calculations for the first steps in the reaction of $[ZnOH]^-$ with CO_2, show that the direct complexation of carbondioxide is more favorable than the nucleophilic attack of the zinc-bound hydroxide ion. However due to the saturation of the coordiantion sphere with other ligands, we expect this behaviour to be changed when dealing with other model complexes like the triammine and trisimidazole complexes.

References

1. Liang, J.-Y.; Lipscomb, W. N. *J. Am. Chem. Soc.* **1986**, *108*, 5051.

2. Merz, K. M.; Hoffmann, R.; Dewar, M. J. S. *J. Am. Chem. Soc.* **1989**, *111*, 5936.

3. Jönsson, B.; Karlsröm, G.; Wennerström, H.; Forsen, S.; Roos, B.; Almlof, J. *J. Am. Chem.Soc.* **1977**, *99*, 4628.

4. Jönsson, B.; Karlsröm, G.; Wennerström, H. *J. Am. Chem.Soc.* **1978**, *100*, 1658.

5. Nguyen, M. T.; Ha, T. K. . *J. Am. Chem.Soc.* **1984**, *106*, 599.

6. Williams, J. O.; van alsenoy, C.; Schafer, L. *J. Mol. Struct.* **1981**, *76*, 109.

7. Frisch, M. J.: Trucks. G. W.; Schlegel, H. B.; Gill P. M. W.; Johnson, B. G.; Robb, M. A.; Cheeseman, J. R.; Keith, T.; Petersson, G. A.; Montgomery, J. A.; Raghavachari, K.; Al-Laham, M. A.; Zakrzewski, V. G.; Ortiz, J. V.; Foresman, J. B.; Cioslowski, J.; Stefanov, B. B.; Nanayakkara A.; Challacombe M.; Peng, C.Y.; Ayala P. Y.; Chen, W.; Wong, M. W.; Andres, J. L.; Replogle, E. S.; Gomperts, R.; Martin, R. L.; Fox, D. J. ; Binkley, J. S.; Defrees D. J.; Baker, J.; Stewart, J. P. ; Head-Gordon, M.; Gonzalez, C.; Pople, J. A. *GAUSSIAN 94, Revision B.2;* Gaussian Inc.: Pittsburgh, PA, 1995.

8. Becke, A. D. *J. Chem. Phys.* **1993**, *98*, 5648-5652.

9. Lee, C.; Yang, W.; Parr, R. G. Phys. Rev. B37, 1988, 785-789.

10. Schäfer, A.; Horn, H.; Ahlrichs, R. J. Chem. Phys. 1992, 97, 2571-2577.

J. Mol. Model. **1996**, 2, 362 – 369

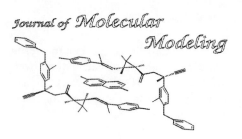

Computer-Aided Receptor Modelling of Human Opioid Receptors: (*Mu, Kappa & Delta*).

Bahram Habibi-Nezhad*, Mahindokht Hanifian [a], Massoud Mahmoudian [b]

Deutsches Krebsforschungszentrum (DKFZ), Molekulare Biophysik (810), D-69120 Heidelberg, Germany. (b.habibinezhad@dkfz-heidelberg.de)

[a] Dept. Med. Chem., Faculty of Pharmacy, Tehran University of Med. Sciences, 14155-6451, Tehran, Iran. (42bahram@rose.ipm.ac.ir)

[b] Dept. Pharmacology, Faculty of Medicine, Iran University of Medical Sciences, 14155-6183, Tehran, Iran. (masmah73@rose.ipm.ac.ir)

Received: 15 May 1996 / Accepted: 6 August 1996 / Published: 27 September 1996

Abstract

Opioid receptors (OPRs) are important agents in the central nervous system (CNS) function. These receptors belong to "G-Protein Coupled Receptors (GPCRs)" which have structural similarity with the *BACTERIORHODOPSIN* (bR). Because of receptor location in the membrane, three dimensional (3D) structure of GPCRs are unknown. The Computer-Aided Receptor Modelling on the basis of amino acid sequence, accompanied by the experimental results is a useful method to understanding the structure and mechanism of these receptors.

In this study we tried to modell three types of Human Opioid Receptors; *Mu, Kappa* and *Delta*. We applied several methods to predict secondary structure (such as Hydropathicity Plot) of opioid receptors and also determined the possible regions of transmembrane helices (TMHs). Results were confirmed by inclusion of other human GPCRs sequence in multiple alignment methods. Then similarity between these receptors and bR were calculated on the basis of parameters such as Mutation Matrix and Secondary Structure Scale. After calculation and refinment of geometric coordinates of atoms located in helices by computerized mutation method (on the basis of 3D structure of bR, as a template) these data were corrected and optimized using Molecular Mechanics Calculations (AMBER Force Field). We used Morphin, Naloxone, Ethylketazocine (EKC) and SKF-10047 as general/specific ligand for these receptors. We optimized conformation of ligands by Quantum Mechanical Semiemprical Calculations (MOPAC). In final step we tried to dock ligands into the receptor cavity with attention to Mutagenesis Data and Structure-Activity Relationships (SAR) information.

Our results show that in Delat receptors 'ASP-96' in TMH-II is important to binding of agonists and antagonists. In *Mu* receptors charged amino acid residues in TMH-II (ASP-116), TMH-III (ASP-149) and TMH-VI (HIS-299) interact with agonists. In Kappa receptors TMH-VI (GLU-297) and TMH-II (ASP-106) play a major role in interaction with antagonists. All of the mentioned residues are located in or near the inner cavity of receptors. With attention to results we suggest that other sites of receptors (such as loops and terminals) may be interact with ligands.

Keywords: Opioid Receptors, Morphine, Modelling, G-Protein Coupled Receptors, Bacteriorhodopsin.

* *To whom correspondence should be addressed*

Introduction

Opioid receptors are most identified with the analgesic properties of opiate drugs [1]. Their agonists can modify pain in virtually every test of spinal and supraspinal analgesia and they perominently implicated in mechanism of opiate-induced reward and reinforcement [2]. Opioid drugs are the principle agents used for treating sever pain. Their value is reflected by the effort expended on producing new compounds and understanding their mechanism of action and effects.

The cloning of cDNAs encoding a number of OPRs [3-5] has demonstrate that the three most prevalent OPR subtype *Kappa, Mu* and *Delta*, all belong to the family of Rhodopsin-like receptors within the superfamily of GPCRs [6]. A number of observation suggests that all GPCRs evolved from a common ancestor.

Because of location GPCRs in membrane, The 3D structure of these receptors are unknown. There is now enough evidence to generate reasonable 3D models of GPCRs using "*Computer-Aided Molecular/Receptor Modelling*". But 3D interpretation is still very speculative, so the projection map can not be used directly as a modelling template. However, it clearly exhibits a 7-helix bundle, and a number of further criteria [7] suggests that the GPCRs are structurally anlogous to *BACTERIORHODOPSIN* although sequence homology is not detectable [8]. Despite of sequence homology with GPCRs, the parallel between the overall 3D structural patterns is striking. The 3D structure of helices in bR was revealed by electron cryomicroscopy [9].

Attemps to build a GPCR model vary in their degree of adherence to the bR structure. In several models the overall topology, i.e. the 7-helix bundle, was incorporated, and this information was supplemented by general structural features of membrane proteins and by experimental data [10-11]. On the other hand, these models are based on the assumption of structural analogy and, hence adhere more closely to the bR structure [12-13].

Beacause of human OPRs are the ultimate targets of therapeutic opiods drugs, it is particularly important to have models of these receptors. We report here our investigation of human OPRs primary sequence homology and alignment, prediction of secondary structure and the construction of 3D models for *Mu, Kappa* and *Delta* human opioid receptors with their general ligands using bR as a template.

Methods

As a first step in the construction of the GPCR 3D models, exhaustive primary sequence comparison and hydropathicity analysis were required. The following GPCR sequences were analyzed: Human *Mu* (*OPRM_HUMAN*), *Kappa* (*OPRK_HUMAN*) and *Delta* (*OPRD_HUMAN*).

The alignment was performed with the method of Needleman-Wunsch [14] and Lipman-Pearson [15] using the Dayhoff Similarity Table [16] for amino acids as implemented in the HUSAR [17] and MULTALIN [18] softwares. To obtain an optimal alignment, we used several Gap Penalty and finally the comparison was refined manually (Fig. 1).

The prediction of secondary structure of OPRs was performed with the several methods:

- Kyte-Doolittle Parameter Set [19].
- Goldman-Engelman-Steitz Parameter Set [20].
- Garnier Scale (GOR Method) [21].
- Manual Refinment (GPCRs Overall Topology).

The refined model of bR was obtained from Brookhaven Protein Databank (entry *1BRD*) and the primary structure of bR and human OPRs (*Mu, Kappa* and *Delta*) were obtained from Swiss-Prot databank (entry *P02945, P35372, P41145* and *P41143*, respectively).

Due to the conformational flexibility of the extra- and intracellular loop region, we have only attempt to model the transmembrane helices (TMHs) of the OPRs. In order to obtain a Homology-Based Model of the TMHs of the OPRs the following protocol was followed:

i) The sequences was aligned with that of bR as described above. (Fig. 2).
ii) The backbone of bR (1BRD) was used as a template for the positioning of the TMHs of the OPRs.
iii) The side chains were adjusted to adopt likely positions.

The receptors were optimized by the AMBER, Ver.4.1 [22] force fields using molecular mechanics calculations with the „Kollman All Atoms" parameter set (unconstrained pathway) in the following way:

Step I) The single helices were minimized for 1000 steps using the conjugate gradient minimizer.
Step II) The transmembrane part the receptor models was constructed and again minimized for 2000 steps.

A nonbonded cutoff of 8 Å was used. To account to some extent for the membrane environment, a distance-dependent dielectric constant of 5 and 1-4 non bonded interactions of 0.5 were chosen. Conjugate gradient minimization used until the RMS energy gradient was acheived a value below 0.1 Kcal/mol·$Å^2$. The N-terminus was capped with an Acetamido group, and C-terminus with a Carboxamido group.

We used some compounds as a general/selective or specific ligands:

- For Mu : Morphine as a agonist and Naloxone as a antagonist (Fig. 3 and 4, respectively).

J. Mol. Model. **1996,** 2

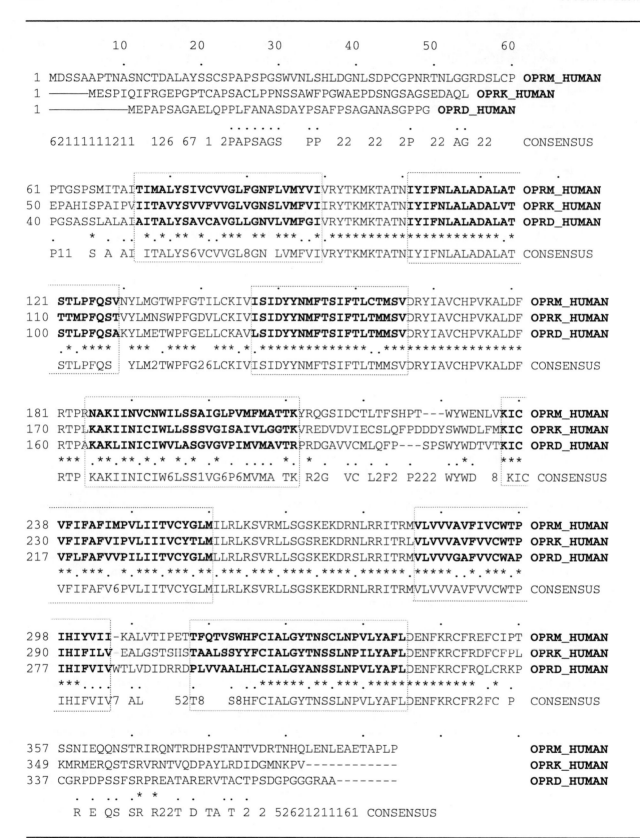

```
              10          20          30          40          50          60
               .           .           .           .           .           .
  1 MDSSAAPTNASNCTDALAYSSCSPAPSPGSWVNLSHLDGNLSDPCGPNRTNLGGRDSLCP   OPRM_HUMAN
  1 ——————MESPIQIFRGEPGPTCAPSACLPPNSSAWFPGWAEPDSNGSAGSEDAQL        OPRK_HUMAN
  1 ——————————MEPAPSAGAELQPPLFANASDAYPSAFPSAGANASGPPG            OPRD_HUMAN
                        .......          ..    .    .    .    ..
    62111111211  126 67 1 2PAPSAGS    PP   22   22   2P   22 AG 22   CONSENSUS


 61 PTGSPSMITAI TIMALYSIVCVVGLFGNFLVMYVI VRYTKMKTATN IYIFNLALADALAT  OPRM_HUMAN
 50 EPAHISPAIPV IITAVYSVVFVVGLVGNSLVMFVI IRYTKMKTATN IYIFNLALADALVT  OPRK_HUMAN
 40 PGSASSLALAI AITALYSAVCAVGLLGNVLVMFGI VRYTKMKTATN IYIFNLALADALAT  OPRD_HUMAN
       *  .. * *.*.** *.*** ** ***..* .********************.*
    P11   S A AI ITALYS6VCVVGL8GN LVMFVI VRYTKMKTATN IYIFNLALADALAT CONSENSUS


121 STLPFQSV NYLMGTWPFGTILCKIV ISIDYYNMFTSIFTLCTMSV DRYIAVCHPVKALDF  OPRM_HUMAN
110 TTMPFQST VYLMNSWPFGDVLCKIV ISIDYYNMFTSIFTLTMMSV DRYIAVCHPVKALDF  OPRK_HUMAN
100 STLPFQSA KYLMETWPFGELLCKAV LSIDYYNMFTSIFTLTMMSV DRYIAVCHPVKALDF  OPRD_HUMAN
     *.****   *** .****   ***..* .************* .***************
    STLPFQS   YLM2TWPFG26LCKIV ISIDYYNMFTSIFTLTMMSV DRYIAVCHPVKALDF CONSENSUS


181 RTPR NAKIINVCNWILSSAIGLPVMFMATTK YRQGSIDCTLTFSHPT---WYWENLV KIC  OPRM_HUMAN
170 RTPL KAKIINICIWLLSSSVGISAIVLGGTK VREDVDVIECSLQFPDDDYSWWDLFM KIC  OPRK_HUMAN
160 RTPA KAKLINICIWVLASGVGVPIMVMAVTR PRDGAVVCMLQFP---SPSWYWDTVT KIC  OPRD_HUMAN
    *** .**.**.* *.* .* * .* * *.  R2G   VC L2F2 P222 WYWD  8 KIC
    RTP  KAKIINICIW6LSS1VG6P6MVMA TK R2G   VC L2F2 P222 WYWD  8 KIC CONSENSUS


238 VFIFAFIMPVLIITVCYGLM ILRLKSVRMLSGSKEKDRNLRRITRM VLVVVAVFIVCWTP  OPRM_HUMAN
230 VFIFAFVIPVLIIIVCYTLM ILRLKSVRLLSGSREKDRNLRRITRL VLVVVAVFVVCWTP  OPRK_HUMAN
217 VFLFAFVVPILIITVCYGLM LLRLRSVRLLSGSKEKDRSLRRITRM VLVVVGAFVVCWAP  OPRD_HUMAN
    ** .***. * .*** *** ** .*** .*** .**** .**** ****** .*.***.*
    VFIFAFV6PVLIITVCYGLM ILRLKSVRLLSGSKEKDRNLRRITRM VLVVVAVFVVCWTP CONSENSUS


298 IHIYVII-KALVTIPET TFQTVSWHFCIALGYTNSCLNPVLYAFL DENFKRCFREFCIPT  OPRM_HUMAN
290 IHIFILV EALGSTSIIS TAALSSYYFCIALGYTNSSLNPILYAFL DENFKRCFRDFCFPL  OPRK_HUMAN
277 IHIFVIV WTLVDIDRRD PLVVAALHLCIALGYANSSLNPVLYAFL DENFKRCFRQLCRKP  OPRD_HUMAN
    ***...       . AL    52T8    S8HFCIALGYTNSSLNPVLYAFL DENFKRCFR2FC P
    IHIFVIV7 AL    52T8    S8HFCIALGYTNSSLNPVLYAFL DENFKRCFR2FC P CONSENSUS


357 SSNIEQQNSTRIRQNTRDHPSTANTVDRTNHQLENLEAETAPLP        OPRM_HUMAN
349 KMRMERQSTSRVRNTVQDPAYLRDIDGMNKPV------------        OPRK_HUMAN
337 CGRPDPSSFSRPREATARERVTACTPSDGPGGGRAA--------        OPRD_HUMAN
      .  .. .   * *
      R E QS SR R22T D TA T 2 2 52621211161 CONSENSUS
```

Figure 1. *The Multiple Alignment of Human Opioid Receptors. Blocks Showed Our Predicted Trans Membrane Helices Region.*

Figure 2 (next page). *The Alignments of Helices Between Bacteriorhodopsin* (BACR_HALHA) *and Human Opioid Receptors.*

- **HELIX   I:**

BACR_HALHA_I	21	PEWIWLALGTALMGLGTLYFLVKGM	45
OPRM_HUMAN_I	81	ITIMALYSIVCVVGLFGNFLVMYVI	105
OPRK_HUMAN_I	60	VIITAVYSVVFVVGLVGNSLVMFVI	84
OPRD_HUMAN_I	50	IAITALYSAVCAVGLLGNVLVMFGI	74

- **HELIX II:**

BACR_HALHA_II	51	DAKKFYAITTLVPAIAFTMYLSMLL	69
OPRM_HUMAN_II	106	.NIYIFNLALADALATSTL......	123
OPRK_HUMAN_II	96	..IYIFNLALADALVTTTMPFQST.	117
OPRD_HUMAN_II	85	.NIYIFNLALADALATSTL......	102

- **HELIX III:**

BACR_HALHA_III	87	EQNPIYWARYADWLFTTPLLLL	108
OPRM_HUMAN_III	146	.ISIDYYNMFTSIFTLCTMSV.	165
OPRK_HUMAN_III	133	IVISIDYYNMFTSIFTLTMMSV	154
OPRD_HUMAN_III	125	.LSIDYYNMFTSIFTLTMMSV.	144

- **HELIX IV:**

BACR_HALHA_IV	119	GTILALVGADGIMIGTGLVGAL	140
OPRM_HUMAN_IV	196	..LSSAIGLPVMFMATTK....	211
OPRK_HUMAN_IV	174	KAKIINICIWLLSSSVGISAIV	195
OPRD_HUMAN_IV	175	.....LASGVGVPIMVMAVTR.	190

- **HELIX   V:**

BACR_HALHA_V	149	VWWAISTAAMLYILYVLFFGFT	170
OPRM_HUMAN_V	235	.KICVFIFAFIMPVLIITVCYG	256
OPRK_HUMAN_V	227	.KICVFIFAFVIPVLIIIVCYT	247
OPRD_HUMAN_V	215	.KICVFLFAFVVPILIITVCYG	235

- **HELIX VI:**

BACR_HALHA_VI	179	EVASTFKVLRNVTVVLWSAYPVVWLI	204
OPRM_HUMAN_VI	283	MVLVVVAVFIVCWTPIHIYVIIK...	305
OPRK_HUMAN_VI	276	VLVVVAVFVVCWTPIHIFILVEAL..	299
OPRD_HUMAN_VI	262	.MVLVVVGAFVVCWAPIHIFVIVW..	284

- **HELIX VII:**

BACR_HALHA_VII	215	NIETLLFMVLDVSAKVGFGLILLR	238
OPRM_HUMAN_VII	314	.TFQTVSWHFCIALGYTN......	330
OPRK_HUMAN_VII	312	YYFCIALGYTNSSLNPILYAFL..	333
OPRD_HUMAN_VII	294	...PLVVAALHLCIALGYAN....	

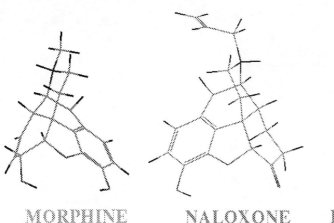
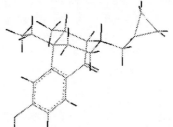

MORPHINE   NALOXONE   ETHYLKETAZOCINE   SKF-10047

Figure 3 - 6. *The Structure of General / Specific Ligands for Opioid Receptors.*

- For Kappa: Ethylketazocine (EKC) as a agonist and SKF-10047 as a antagonist (Fig. 5 and 6, respectively).
- For Delta: SKF-10047 as a antagonist.

A systematic search for obtaining best conformation of ligands was performed by the INSIGHT II package and finally geometry and charges distribution of suitable conformation of ligands was calculated by the MOPAC package [23], using PM3 and AM1 [24] hamiltonian. (Figure 3 – 6)

Data about probable binding sites and important residues involved in interaction of ligand-receptor obtained from mutagenesis experiments. With attention to these data and structure-activity relationships studies of ligands, the selective and/or specific ligands were manually and rigidly docked into their putative binding sites. The docking procedure was repeated several times with different initial orientations of the side chains and of the ligand in order to obtain the best possible inteaction complexes. Charges for the ligands were imported from the MOPAC output files.

The drug-receptor complexes were optimized by molecular mechanics calculation (AMBER force fields, 4000 Steps, Conjugate Gradient, Cutoff = 8 Å and Gradient less than 0.1 Kcal/mol·Å^2). Final geometry was acheived for ligands and receptors (Figures 7, 8, 9 and 10)

The primary interactive modelling, display and file generation was acheived with molecular modelling package; WhatIf, Ver.3.0 [25] and finally display, systematic search of ligands conformation and file generation was acheived with molecular modelling package INSIGHT II, Ver.2.9/3.1 [26]. All calculations were performed on Silicon Graphics and SP2 computers.

Results and Discussion

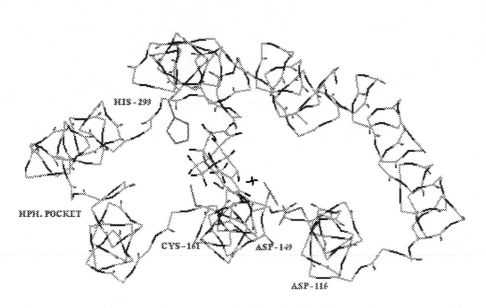

Figure 7. *Three-Dimensional Views of Ligands-Opioid Receptors Complexes. Backbone of human Mu receptor, active site and morphine*

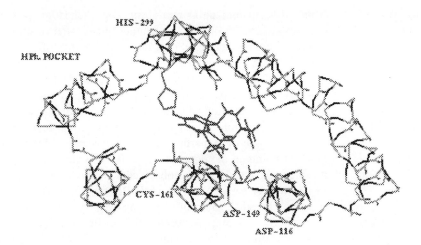

Figure 8. *Three-Dimensional Views of Ligands-Opioid Receptors Complexes. Backbone of human, Mu receptor active site and naxolone.*

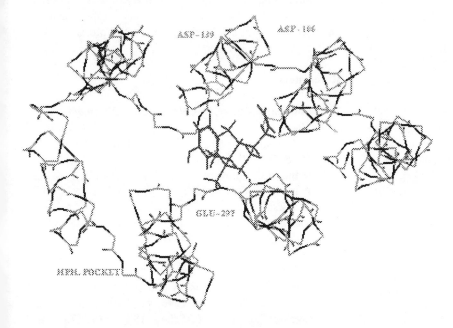

Figure 9. *Backbone of human Kappa receptor, active site and ethylketazocine.*

We briefly explain the results obtained through this study:

1.) In TMH-I, the motif GXXGN occures in OPRs, rather than the GN motif present in biogenetic amine. In all sequences the last five C-terminal residues of this helix are frequently occupied by basic residues, indicating the end of trans membrane domain. Such basic residues may serve as "Membrane Anchors".

2.) In TMH-II, the residue preceding the conserved Leu in the LXXXD motif is a conserved Serine for the biogeneic amine receptors, but is an Asn in the OPRs. The PRO is consistently spaced by seven residues from the 'ASP' in the OPRs (eight residues in biogenic amine receptors).

3.) The conserved 'ASP' in TMH-II was shown to be a Sodium-Dependent Allosteric Regulatory Site in the OPRs.

4.) The DRY motif, characteristic of the third transmembrane domain in GPCRs, is supposedly important for coupling of the OPRs to G-proteins unit and not for ligand binding.

5.) In *Kappa* receptor one or more of the first six positions in the N-terminal sequence of TMH-IV are generally occupied by LYS or ARG residues. Again these residues could well serve as "Membrane Anchor". This condition has not been found in *Mu* and *Delta* receptors.

6.) The CXXP motif and WXP motif in TMH-VI has been found in all of the OPRs.

7.) Results showed that the hydrophobic side of each helix was facing the lipid face and the hydrophilic side of each helix was facing either another helix or the pore formed by the putative bundle.

8.) The assembly of helices maintained a clockwise order, when seen from the intracellular side. Non of the helices were intersecting.

9.) Five potential glycosylation sites are present on the extracellular N-terminal amino acid sequence.

10.) The *Mu* receptor has about 60% amino acid identity to the *Kappa* and 65% to *Delta* receptors.

11.) The *Kappa* receptor has about 59% amino acid identity to the *Delta* and 61% to *Mu* receptors.

12.) 'ASP' in TMH-II and TMH-III,is generally seen as the main anchoring point for agonist and antagonist binding. Residues in other TMH domain (TMH-V) appear to be involved in determining the selectivity of these receptors for their agonists, e.g. Serine interacts with the agonist hydroxyl moiety. Our finding suggest that the induction of a conformational change in 'ASP' by an agonist could be a genertal and crucial step in OPRs stimulation. On the other hand conformational changes in intracellular loops to rearrangment of the seven-helix bundle, the so called „ARG Switch" [27], or ligand-mediated "Proton Transfer" mechanims [28].

13.) Docking showed that the amino group of Morphine to be within 3 to 4 Å of the 'ASP' residue (in TMH-III) and the aromatic ring to be in close (less than 5 Å) proximity of TRP residues of the TMH-IV and TMH-VI.

14.) It appears that the ligand is in contact with only four of the TMH-III, TMH-IV, TMH-V and TMH-VI, suggesting that these helices are responsible for binding selectivity. Probabely 'HIS' in TMH-VI is important in interaction with ligands.

15.) Our results showed that following residue are important in the drug-receptor interactions:

• In **Mu** Receptor:
  - 'ASP-116' in TMH-II.
  - 'ASP-149' in TMH-III.
  - Hydrophobic residues in TMH-VII, TMH-VI and TMH-I as a hydrophobic pocket.
  - 'HIS-299' in TMH-VI.
  - 'CYS-161' in TMH-IV.

• In **Kappa** Receptor:
  - 'ASP-106' in TMH-II.
  - 'ASP-139' in TMH-III.
  - Hydrophobic residues in TMH-VII, TMH-VI and TMH-I as a hydrophobic pocket.
  - 'GLU-297' in TMH-VI.

• In **Delta** Receptor:
  - 'ASP-96' in TMH-II.
  - 'ASP-129' in TMH-III.
  - Hydrophobic residues in TMH-VII, TMH-VI and TMH-I as a hydrophobic pocket.

Conclusion

In the present study we have combined results from site-specific mutagenesis studies on the OPRs with findings from different molecular modelling approaches such as conformational analysis, pharmacophore fitting and receptor docking studies. From our findings the followin conclusions emerge:

• The different conformations of 'ASP' in TMH-III and TMH-II observed in our modelling studies upon agonists or antagonists binding indicate that these Aspartic Acids may play a key role in receptor simulation. Upon binding of agonists one of Aspartic Acids changes its conformation and points in the direction of TMH-V, which contains residues responsible for the observed selectivity, i.e. LYS. In this way Aspartic Acids in TMH-III and TMH-II is assigned a crucial function in triggering GPCRs stimulation.

• There still remain much work to be done on the characteization of OPRs ligand recognition domaine, loop building (important for peptide ligands), G-protein coupling mechanisms, and receptor correlates for opioid tolerance and dependence.

• Modelling of GPCRs has been become an important tool in understanding drug-receptor interactions and in the development of new ligands for these receptors.

References

1. Porreca, F.; Burks, T.F. *Handb. Exp. Pharmacol.* **1993**, *104*, 21-51.

2. Di Chira, G.; Imperato, A. *Proc. Natl. Acad. Sci. U.S.A.* **1988**, *85*, 5274-5278.

3. Wang, J.B.; Johnson, P. S.; Persico, A.M.; Hawkins, A.L.; Griffin, C. A.; Uhl, G.R. *FEBS Lett.* **1994**, *338*, 217-222.

4. Mansson, E.; Bare, L.; Yang, D. *Biochem Biophys. Res. Commun.* **1994**, *202*, 1431-1437.

5. Warner, R.L.; Johnston, C.; Hamiton, R.; Skolnick, M.H.; Wilson, O.B. *Life Sci.* **1994**, *54*, 481-490.

6. Hjorth, S.A.; Thirstrup, K.; Grandy, D.K.; Schwartz, T.W. *Mol. Pharmacol.* **1995**, *47*, 1089-1094.

7. Ovchinnikov, Y.A. *FEBS Lett.* **1982**, *148*, 179-191.

8. Fryxcll, K.J.; Mcycrowitz, E.M. *J. Mol. Evol.* **1991**, *33*, 367-378.

9. Henderson, R.; Baldwin, J.M.; Ceska, T.A. *J. Mol. Biol.* **1990**, *213*, 899-929.

10. Baldwin, J.M. *EMBO J.* **1993**, *12*, 1693-1703.

11. Huss, K.M.; Lybrand, T.P. *J. Mol. Biol.* **1992**, *225*, 859-871.

12. Yamamoto, Y.; Kamiya, K.; Terao, S. *J. Med. Chem.* **1993**, *36*, 820-825.

13. Brann, M.R.; Klimkowski, V.J.; Ellis, J. *Life Sci.* **1993**, *52*, 405-412.

14. Needleman, S.B.; Wunsch, C.D. *J. Mol. Biol.* **1970**, *48*, 443-453.

15. Lipman, D.J.; Pearson, W.R. *Science* **1985**, *227*, 1435-1441.

16. Dayhoff. H.O.; Schwartz, R.M.; Orcut, B.C. in *Atlas of Protein Sequence and Structure,* Dayhoff, H.O. (Ed.) **1978**, NBFR,5, Suppl. 3.

17. Singer, M.; Glatting, K.H.; Ritter, O.; Suhai, S. *Computer Methods and Programs in Biomedicine* **1995**, *46*, 131-141.

18. Caron, P.R.; Grossman, L. *Nucl. Acids Res.* **1988**, *16*, 10881 -10890.

19. Kyte, J.; Doolittle, R.F. *J. Mol. Biol.* **1982**, *157*, 105-132.

20. Engelman, D.M.; Steitz, T.A.; Goldman, A. *Annu. Rev. Biophys. Chem.* **1986**, *15*, 321-353.

21. Garnier, J.; Osguthorpe, D.J.; Robson, B.; *J. Mol. Biol.* **1978**, *120*, 97-120.

22. Weiner, S.J.; Kollman, P.A.; Nguyen, D.T.; Case, D.A. *J. Comput. Chem.* **1986**, *7*, 230.

23. Stewart, J.J.P, MOPAC; A General Molecular Orbital Package, QCPE BULL **1983**, *3*, 43.

24. Dewar, M.J.S.; Zoebisch, E.G.; Healy, E.F.; Stewart, J.J.P. *J. Am. Chem. Soc.* **1985**, *107*, 3902-2909.

25. Vriend, G. *J. Mol. Graph.* **1990**, *8*, 52-56.

26. Biosym Technologies, INSIGHT II Molecular Modeling Software, 1993.

27. Olivera, L.; Paiva, A.C.M.; Sander, C.; Vriend, G. *TIPS* **1994**, *15*, 170-172.

28. Nederkoorn, P.H.J.; Timmerman, H.; Donne-Op den Kelder, G.M. *TIPS* **1995**, *16*, 156-160.

J. Mol. Model. **1996**, 2, 370 – 372

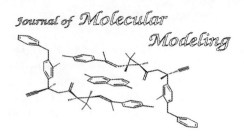

VRML in Cancer Research
Local Molecular Properties of the p53 Tumor Suppressor Protein-DNA Interface

Gerd Moeckel [a,b], Matthias Keil [a], Bertold Spiegelhalder [b], Jürgen Brickmann [a,c]

[a] Department of Physical Chemistry, Darmstadt University of Technology, Germany
(moeckel@pc.chemie.th-darmstadt.de)
[b] Division of Toxicology & Cancer Risk Factors, German Cancer Research Center, Germany
[c] Darmstadt Center for Scientific Computing , Darmstadt University of Technology, Germany

Received: 15 May 1996 / Accepted: 6 August 1996 / Published: 27 September 1996

Abstract

The three-dimensional structure information of the p53 core domain-DNA complex is presented. The Virtual Reality Modeling Language (VRML), a new concept of information transfer is used for this biochemical application. VRML provides an object oriented method for the three-dimensional description of molecular models [1, 2].
This VRML-WWW pages contain the 3D structure of the p53 core domain-DNA complex crystallized by Pavletich and coworkers [3], now available from the Brookhaven Protein Databank, and the p53-DNA binding region with related biochemical information like hydrophilicity/lipophilicity and local electrostatic partial charges.
These investigations are done in a close cooperation with H. Bartsch and his group at the Division of Toxicology & Cancer Risk Factor Prevention of the German Cancer Research Center in Heidelberg, Germany.

Keywords: VRML, WWW, biochemical application, p53 protein-DNA interface

p53 core domain-DNA complex

In the first 3D-VRML-application (figure 1), the structure of the p53 core domain-DNA complex is represented in a wire model according to the crystallographic structure determination of Nikola Pavletich and coworkers [3]. DNA is colored blue, the three p53 core domain molecules are magenta (c1), yellow (c2), and red (c3). The zink ions are represented as white spheres.

In the next 3D-VRML-applicationon (figure 2) the DNA binding p53 core domain2 molecule is represented as a yel-

low wire model (left), a yellow wire model with ribbons along the backbone (middle), and with the six red colored mutation hotspots (right).

The *solvent accessible surface* of the p53-DNA binding region is presented in the third 3D-VRML-application (figure 3). The molecular surfaces of the p53 core domain2 molecule and the DNA is colored according to local hydrophilic (blue) - lipophilic (red) properties.

The *solvent accessible surface* of the p53-DNA binding region is presented in the last 3D-VRML-application (figure 4). The molecular surfaces of the p53 core domain2 mol-

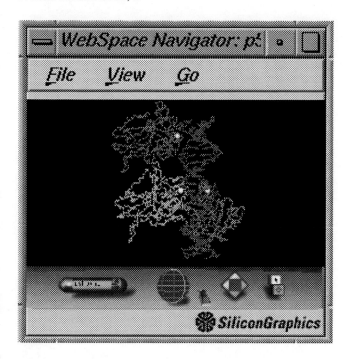

Figure 1. *Complete p53-DNA complex - Wireframe*

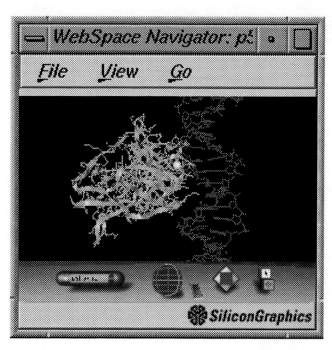

Figure 2. *p53 core domain2-DNA complex:Wireframe, Ribbon and mutation hotspots*

ecule and the DNA is colored according to partial charge distribution from negative (red) to positive (green).

Details

P53-DNA Interface

The p53 tumor suppressor protein controls the cell cycle checkpoint responsible for maintaining the integrity of the genome. When DNA is damaged, the p53 level increases and the cell cycle is stopped at the G1/S phase to allow DNA repair followed by normal cell growth or induction of apoptosis. The p53 protein is frequently altered by mutations in almost all types of cancer. Up to now more than 5000 tumor-specific p53 mutations (85% missense mutations) have been identified and are compiled in a database available at the European Bioinformatics Institute (EBI, Cambridge UK; http://www.ebi.ac.uk/pub/databases/p53/) [4]. The majority of these mutations are found in the central p53 core region containing the DNA binding domain and the zinc binding domain.

The 3D structure of the p53 protein-DNA complex has been crystallized [3] and this allows investigation into the influence of mutations on DNA binding and protein functions. The molecular properties of the p53 tumor suppressor protein-DNA complex available from the Brookhaven Protein Database (PDB) are analyzed using the molecular modeling package SYBYL/MOLCAD [1]. The investigations are based on Connolly's concept of molecular surfaces. These solvent-accessible surfaces show the three-dimensional size

and topography of the molecules. Additionally the surfaces are used as maps for a color coded representation of local molecular properties like hydrophilicity/lipophilicity [5] or the electrostatic partial charges calculated with CHARMM. Important to the stability of the protein-DNA complex and to the biological activity as well is the significant amount of water molecules found in the contact region of the complex. The embedded water molecules form hydrogen bonds to both DNA and protein and thus contribute substantially to the protein-DNA interface.

The topographical analysis of this p53 protein-DNA complex shows, that the five mutation hotspots, constituting more than 25% of the database mutations [4], are located either in the zinc binding region or in the protein-DNA interface. ARG248 and ARG273 especially which are most frequently mutated in human tumors bind directly to DNA. The local hydrophobic/hydrophilic properties of the p53 core domain protein are similar to the surrounding DNA interface while the electrostatic charges are complementary leading to intermolecular attraction. Point mutations at these hotspots lead to disturbances of the partly water-bridged DNA binding.

VRML Application

The three-dimensional structure information of this p53 core domain-DNA complex is presented in the authors World Wide Web pages (Darmstadt University of Technology, Physical Chemistry I, http://www.pc.chemie.th-darmstadt.de/vrml/ p53). A summary of this 3D structure data is delivered as a

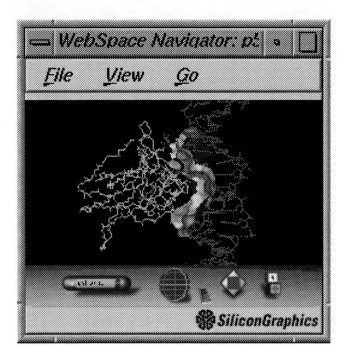

Figure 3. *Hydrophilic/lipophilic properties of the p53-DNA interface: Dots, lines and solid.*

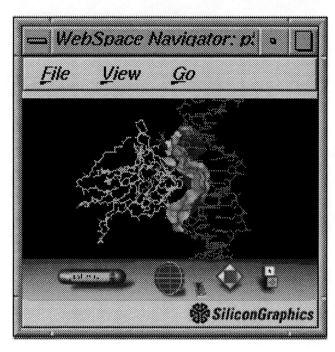

Figure 4. *Electrostatic properties of the p53-DNA interface: Dots, lines and solid.*

HTML/VRML supplement with this issue. For this information transfer of biochemical 3D data, a new concept, the Virtual Reality Modeling Language (VRML) is used. VRML provides an object oriented method for the description of molecular models [2]. The language is based on a subset of the Open Inventor File Format. This subset was extended with networking capabilities, such as WWW hyperlinks. With this feature, VRML is an equivalent to the HyperText Markup Language (HTML). Like HTML files describe the layout of 2D-text pages to be displayed by WWW browers, VRML files describe the layout of 3D scenarios. Therefore VRML is an excellent tool for visualization of 3D molecular scenarios. It is used to prepare platform-independent files to visualize the output of all kinds of 3D information commonly generated with molecular modeling technologies [6, 7]. Our VRML-supplement contains the 3D structure data of the p53-DNA complex and the p53-DNA binding region combined with local molecular properties. This example shows an aspect of VRML capabilities to deliver important 3D structure related biochemical information in science.

References

1. Brickmann, J.; Goetze, T.; Heiden, W.; Moeckel, G.; Reiling, S.; Vollhardt, H. and Zachmann, C.-D. in *Data Visualisation in Molecular Science: Tools for Insight and Innovation*, Addison-Wesley Publishing Company Inc., **1995**.

2. Vollhardt, H.; Henn, C.; Teschner, M. and Brickmann, J. *J. Mol. Graphics* **1995**, *13*, 368.

3. Cho, Y.; Gorina, S. and Pavletich, N. *Science* **1994**, *265*, 346.

4. Hollstein, M.; Shomer, B.; Greenblatt, M.; Soussi, T.; Hovig, E.; Montessano, R. and Harris, C. *Nucl. Acid Res.* **1996**, *24*, 141

5. Moeckel, G.; Spiegelhalder, B.; Bartsch, H. and Brickmann, J. *Medizinische Forschung -Ärztliches Handeln*, MMV Medizin Verlag, **1995**.

6. Brickmann, J.; Heiden, W.; Marhoefer, R.; Moeckel, G.; Sachs, W.; Vollhardt, H. and Zachmann, C.-D. *Proceedings of the IV International Conference on Chemical Structures, Noordwijkerhout, The Netherlands, June 2.-6. 1996*, submitted for publication.

7. Brickmann, J. and Vollhardt,H. *Trends in Biotechnology* **1996**, *14*, 167

J. Mol. Model. **1996**, 2, 373 – 375

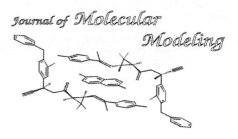

Ab initio Calculation of the Conformations and Vibrational Spectra of 2-Phenylbutane

Christian Borsdorf*, Thomas Dorfmüller

University of Bielefeld, Fakulty of Chemistry, Physical Chemistry 1, 33615 Bielefeld, Germany
(chris@chep118.uni-bielefeld.de)

Received: 15 May 1996 / Accepted: 6 August 1996 / Published: 27 September 1996

Abstract

Molecules with internal degrees of rotation are of particular interest to understand the behavior of synthetic or natural polymers. With the extremely rapid increase in floating point performance of modern computers we are able to calculate a subspace of the Born-Oppenheimer hypersphere for quite large molecules with *ab initio* methods. We choose

2-Phenylbutane as an elementary model for Polystyrene (PS), because it possesses two internal degrees of freedom, that are highly relevant for PS. These are the rotation of the benzene ring and the chain like motion of the ethylgroup.

We minimized the energy of one conformation with respect to all internal coordinates and used different basis function sets. From that the standard gaussian type basis 3-21G elaborates as an optimum function set regarding computing time and accuracy. With the help of multidimensional search

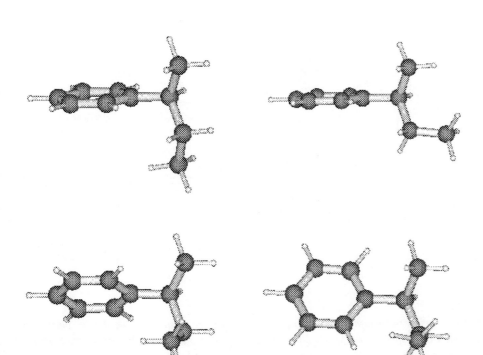

Figure 1. *Conformations of 2-Phenylbutane calculated with GAUSSIAN92 at 6-31G.*
a: T (ΔE=0.0 kJ/mol)
b: G⁺ (ΔE=3.7 kJ/mol)
c: G⁻ (ΔE=7.9 kJ/mol)
d: G₂⁻ (ΔE=19.9 kJ/mol)

PDB-files available:
0373f01a.pdb
0373f01b.pdb
0373f01c.pdb
0373f01d.pdb

* To whom correspondence should be addressed

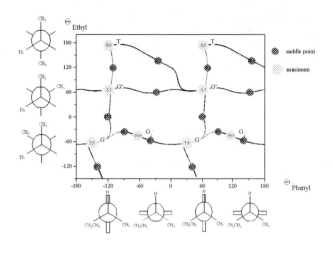

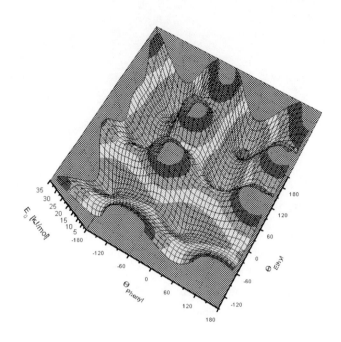

Figure 2. *Changes of the internal torsion angles along the IRC between the calculated minimum and saddle point structures. The Newman projections for different torsion angles are shown to give an idea of the conformational structure within the plot.*

Figure 3. *3 dimensional representation of the Born-Oppenheimer hypersphere for 2-Phenylbutane. The contour is calculated from the IRC coordinates shown in fig. II.*

algorithms, that are available in the GAMESS [1] and GAUSSIAN92 [2] program packages, we were able to calculate 4 different conformations of the 2-Phenylbutane that are energetically relevant (figures 1a-d). In a more extensive search we show that it is possible to calculate the saddle point structures between the minima without restricting the mini-

mization to any coordinate. The calculated intrinsic reaction coordinate (IRC) between the saddle points and the minima provide a detailed view in the Born-Oppenheimer hypersphere of 2-Phenylbutane (figure 2). A three dimensional representation of this energy landscape in dependence of the two internal rotational degrees of freedom is calculated that gives a further inside into the flexibility of the studied molecule (figure 3). Our investigation proofs that the assumption of rigid rotors in molecules with more than 1 internal degree of rotation is - especially for PS - a crude approximation.

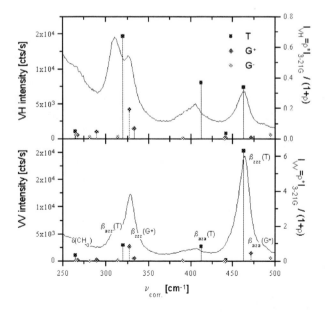

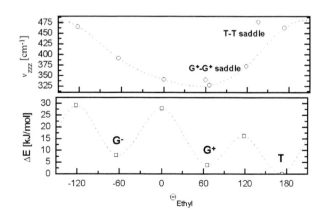

Figure 4. *Calculated Raman active vibrations in the region of CCC bending modes. The polarized and depolarized intensities are in agreement with the experimental data, that are also shown in the figure.*

Figure 5. *The calculated frequencies of the 'in phase' CCC bending mode ν_{zzz} are showing a large variation with the torsion angle of the butane chain.*

We verify the level of energy relaxation in the calculated conformations by performing a normal coordinate analysis at the minimum and saddle point structures. The calculated polarized and depolarized Raman spectra are in agreement with our experimental data (figure 4). We characterized the measured bandstructure of the strong symmetric ring breathing vibration as a superposition of the vibrations of different conformers. The coupling of the two CCC-bending modes of the chain leads to a normal vibration whose frequency is highly dependent on the conformation of the chain (figure 5), which can be used to monitor the internal rotation of 2-Phenylbutane with Raman spectroscopy [3].

The analysis of the normal vibrations of 2-Phenylbutane with ab initio calculations give a quantitative view of the complex vibrational Raman spectra. With the help of these investigation we are able to show in an Raman experiment the freezing of an internal mode of 2-Phenylbutane during the glass transition [4].

References

1. Schmidt, M. W.; Baldridge, K. K.; Boatz, J. A.; Elbert, S. T.; Gordon, M. S.; Jensen, J. H.; Koseki, S.; Matsunaga, N.; Nguyen, K. A.; Su, S.; Windus, T. L.; Dupuis, M.; Montgomery, Jr., J. A. *J. Comp. Chem.* **1993**, *14*, 1347-1363.

2. Gaussian 92/DFT, Revision G.2, Frisch, M. J.; Trucks, G. W.; Schlegel, H. B.; Gill, P. M. W.; Johnson, B. G.; Wong, M. W.; Foresman, J. B.; Robb, M. A.; Head-Gordon, M.; Replogle, E. S.; Gomperts, R.; Andres, J. L.; Raghavachari, K.; Binkley, J. S.; Gonzales, C.; Martin, R. L.; Fox, D. J.; Defrees, D. J.; Baker, J.; Stewart, J. J. P.; Pople, J. A. Gaussian, Inc., Pittsburgh PA, 1993.

3. Borsdorf, Ch. Thesis, Universität Bielefeld 1995.

4. Borsdorf, Ch.; Dorfmüller, Th. to be published.

J. Mol. Model. **1996**, 2, 376 – 378

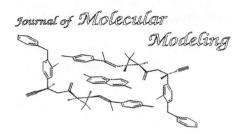

The PIMM Force Field - Recent Developments

Martin Kroeker and Hans Jörg Lindner*

Institut für Organische Chemie. TH Darmstadt, Petersenstr. 22; D-64287 Darmstadt, Germany
(martin@oc2.oc.chemie.th-darmstadt.de; lindner@oc1.oc.chemie.th-darmstadt.de)

Received: 15 May 1996 / Accepted: 6 August 1996 / Published: 27 September 1996

Introduction

Even though the steady advances in computing power and algorithms continue to increase the competition by semiempirical and *ab initio* methods, they also serve to open new realms for modern force fields. Among the key advantages of the classical molecular mechanics approach is their ability to handle systems with a large number of atoms and degrees of freedom. A second and increasingly important feature lies in the fact that they can often provide a simple but efficient means of modeling metal complexes.

In recent years, our PIMM force field [1] for organic molecules has found successful application in a number of problems in organic and bio-organic chemistry. Examples range from peptide complexes of alkaline earths cations to porphyrin complexes of the transition metals. In its traditional field, the conformational analysis of organic molecules, satisfying results have been reported especially for carbohydrates [2]. Besides conventional energy minimization techniques and systematic permutations, molecular dynamics calculations have gained increasing importance.

In its current form, the program fills the niche between the small-molecule force fields and the protein modeling packages. Even on the pc platform, systems of up to 2000 atoms can readily be studied without simplifications. This entails not only an all-atom representation, but also a semiempirical SCF calculation of the pi electrons and a Hessian matrix-based optimization algorithm. Current projects address the specific continuum effects in condensed phases, i.e. a model representation of solvent and packing effects.

Parameters

Metal complexes can in theory be modeled by two conceptually different approaches. While some authors favor an explicit parametrization of metal- ligand bonds, we have achieved good results based on electrostatic and van der Waals interactions alone. At first glance, the bonded approach offers the advantage of prescribing exact bond angles and torsional angles in addition to distances. On the other hand, this requires a great number of parameters and will usually lead to a very stiff model, where e.g. the number of ligands is defined by the starting structure.

The flexible non-bonded model used in PIMM naturally benefits greatly from the strong contribution of charge interactions in the force field. Its disadvantages become apparent only in cases where orbital interactions have a dominating influence on the 3D structure of a complex. Results for porphyrin complexes, metallocenes and carbonyl complexes generally display good agreement between calculated and experimental geometries.

Catering especially to the growing interest in coordination chemistry, the parameter set has been extended over the past year to include several new classes of compounds. Besides the ferrocenes, this now includes zirconocenes as well as chromium carbonyls. Parameters for covalently bound boron have been added to allow modelling of pyrazolylborate complexes and tetraalkylborate salts. As the parameter set is extended mainly in conjunction with

* *To whom correspondence should be addressed*

specific projects, the current set as depicted in figure 1 does not reflect a fundamental limitation of the program.

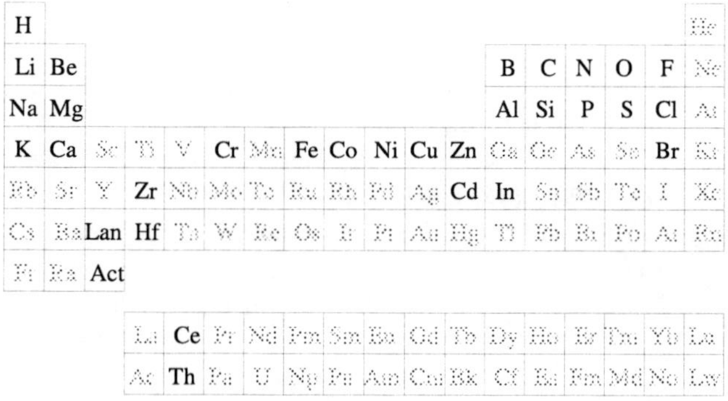

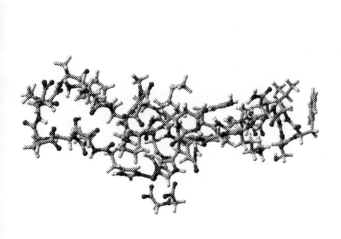

Figure 1: *Chemical elements available in the current parameter set*

Algorithms

Until recently, the molecular dynamics code in PIMM constituted one of its major weaknesses. As a relatively recent addition originally intended for the generation of starting conformations rather than true MD, it contained a number of shortcomings that impaired its performance. The revised algorithm has shown vastly improved performance in calculations of cyclodextrins and peptides. Conformational analysis of larger molecules has been facilitated by a Metropolis Monte Carlo algorithm. Irrespective of the type of starting coordinates, this can be used to vary torsional angles as well as the cartesian coordinates of individual atoms. For cyclic systems, variation of a torsional angle is automatically transformed into the appropriate puckering or folding motion.

The concomitant changes to the torsional angle driver have also removed the previous restrictions in the systematic variation of torsional angles. Thus it is no longer necessary to include a torsional angle in the Z-matrix in order to permute it, and substituents can have arbitrary reference atoms. Periodic boundary conditions have been introduced to allow the realistic modeling of molecular crystals. While the full crystallographic symmetry can be taken into account, it is currently necessary for the asymmetric unit to contain a complete molecule. Although this precludes the study of covalently linked polymers, hydrogen bonding networks are already treated correctly.

Although the code has primarily been tested in energy minimization runs, it can be used as well for molecular dynamics and monte carlo simulations. The results obtained so far generally indicate a good agreement of experimental and calculated heats of formation, though in some cases coupled with a small contraction of the unit cell.

Sample Applications

These snapshots of current research projects are meant to serve only as illustrations of the versatility of the program. In-depth discussions of the individual topics will be published elsewhere.

A. Mutational study of zinc binding sites

Based on previous modeling work with calcium-binding proteins, the goal of this project is the prediction of zinc

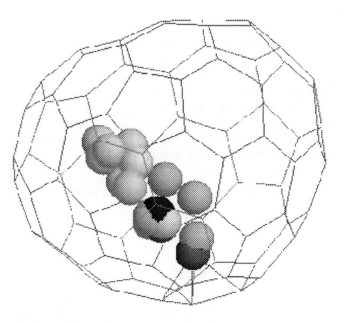

Figure 2: *Peptide fragment encompassing the Zn binding region*

Figure 3: *Trajectory of Fe color-coded by binding energy*

binding affinities in active site mutants of carbonic anhydrase A2 (CA2) [3] and a chimeric protein [4] composed of the retinol binding protein (RBP) and the zinc site of CA2. Preliminary calculations on protein fragments as depicted in figure 2 yield a good correlation between experimental binding constants and calculated binding energies.

B. Fullerene complexes

Molecular dynamics and monte carlo calculations are being employed to locate energetically favourable positions for divalent iron ions in the cage structure of C82. The results obtained with PIMM appear to be in good agreement with spectroscopic data and *ab initio* calculations that used a minimal basis set (figure 3).

C. Porphyrin complexes

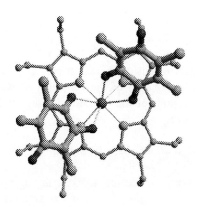

Figure 4: *Model structure of Fe(OEP)(hquin)2*

The structure of a novel zirconium complex [5] with axial hydroxyquinone ligands was modeled by combined MM and MD techniques, allowing the interpretation of spectroscopic features even though no crystals suitable for an x-ray analysis could be obtained.

References

1. Lindner, H.J. *Tetrahedron* **1972**, *30*, 1127; Smith, A.E. and Lindner, H.J. *J. Comput.-Aided Mol. Des.* **1991**, *5*, 235; Kroeker, M. PhD Thesis, Technische Hochschule Darmstadt 1994.
2. Immel, S.; Lichtenthaler, F. W. *Liebigs Ann.* **1996**, *27* and references cited therein.
3. Kiefer, L.L.; Paterno, S.A.; Fierke, C.A. *J. Am. Chem. Soc.* **1995**, *117*, 6831.
4. Müller, H. N.; Skerra, A. *Biochemistry* **1994**, *33*, 14126.
5. Eberle, M.; Buchler, J.W. unpublished results.

J. Mol. Model. **1996**, 2, 379 – 382

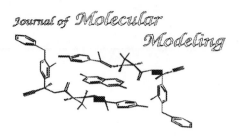

Quadrupole Relaxation of the $^7Li^+$ Ion in Dilute Aqueous Solution Determined by Experimental and Theoretical Methods

Ralf Baumert*, Ralf Ludwig and Alfons Geiger

Physical Chemistry, University of Dortmund, 44221 Dortmund, Germany (bau@heineken.chemie.uni-dortmund.de)

Received: 15 May 1996 / Accepted: 6 August 1996 / Published: 27 September 1996

Introduction

A combination of molecular dynamics simulations (MD), *ab initio* selfconsistent field (SCF) calculations and nuclear magnetic resonance relaxation time experiments (NMR) is a powerful battery of techniques to investigate the molecular origins of the nuclear quadrupole relaxation mechanism for $^7Li^+$ ions in dilute aqueous solution.

NMR relaxation time experiments for $^7Li^+$ in dilute aqueous solutions can be performed only in the extreme narrowing region. Any quantitative analysis of the measured relaxation rates suffers from the fact that there is no strict separation of the quadrupole coupling constant and the correlation time of the electric field gradient fluctuations at the $^7Li^+$ ion site.

Molecular dynamics simulations allow the calculation of the electric field gradient time correlation function (EFG-TCF). The electric field gradient and the correlation time of the $^7Li^+$ ion can be determined separately and the obtained relaxation rates are compared with experimental data from NMR. The molecular dynamics simulations also provide detailed information about the structure and dynamics of the hydration shell which can not be seen in the experiment. The EFG-TCF is mainly due to the water molecules in the first hydration shell and shows a complex time dependence not describable with a mono-exponential behaviour.

Ab initio self-consistent field (SCF) calculations on molecular clusters from the MD simulation may be a good check for the existence of strong polarizability or many-body effects which are not taken into account in pure electrostatic MD potentials. The MD simulations were used to generate snapshots of the $^7Li^+$ in aqueous solution (Figure 1) representing the pair correlation function of the whole system. The electric field gradient at the $^7Li^+$ nucleus was then calculated by *ab initio* methods for these configurations and averaged to obtain the quadrupole coupling constant.

NMR relaxation time measurements

The $^7Li^+$ magnetic relaxation time measurements were performed at 360 MHz with the 180 - τ - 90 pulse sequence using a BRUKER AM 360 spectrometer. All measurements were performed on a LiCl solution of the concentration C* = 0.555 $\overline{m}$ ($\overline{m}$ = aquamolality scale, i.e. moles salt/55.5 moles water) which is the same used in the MD studies.

Figure 1. *Snapshot from a Molecular Dynamics (MD) Simulation of $^7Li^+$ in water for the first hydration shell.*

* *To whom correspondence should be addressed*

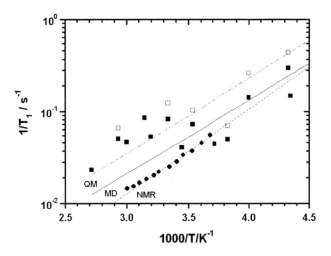

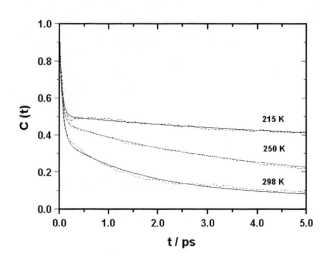

Figure 2. *Temperature dependence of lithium relaxation rates from NMR relaxation time experiments (dashed line and filled circles), Molecular Dynamics Simulations (solid line and filled squares) and ab initio calculations on MD configurations (dashed-dotted line and open squares).*

Figure 3. *Normalized time correlation functions for the electric field gradient (efg) fluctuations shown for three different temperatures. The solid lines represent the fitted functions for a threefold exponential approach.*

Beside the desired electric quadrupolar also dipolar interaction contributes to the $^7Li^+$ relaxation rate. A combination of measurements in D_2O and H_2O as well as a correction for the dynamic isotope effect allow the separation of the pure quadrupolar interaction of $^7Li^+$ in water. Our obtained quadrupolar relaxation rates are shown in figure 2. They are comparable with earlier results by Mazitov et. al [1] at a concentration of about 0.2 $\bar{m}$. From nmr measurements only, no further conclusion can be drawn. Any quantitative analysis suffers from the fact that there is no strict separation of the quadrupole coupling constant and the correlation time of the electric field gradient fluctuation at the $^7Li^+$ ion site.

$$NMR : \left(\frac{1}{T_1}\right)_{Li} = \frac{3}{40} \frac{(2I+3)}{I^2(2I-1)} \left[\frac{eQ}{\hbar}(1+\gamma_\infty)\right]^2 \bar{V}_{zz}^2 \tau_c$$

Molecular Dynamics simulation

The incontestable advantage of MD compared to experiment and theory is the detailed insight into the system at the molecular level. Therefore quadrupolar relaxation of any free nucleus with spin $I > \frac{1}{2}$ can be examined by MD, since the strength of the electric field gradient (EFG) is strongly distance dependent and hence only *those* molecules neighbouring the relaxing nucleus have to be taken into account.

Two different mechanisms are discussed as source of the fluctuating field gradient. On the one hand the so called *electronic* model by Deverell [2] supposes that the fluc-

tuations are due to distortions of the solute electronic cloud by collisions with solvent molecules. On the other hand Hertz [3, 4] and Valiev [5, 6] assume that the *electrostatic* effect of the solvent point dipoles and their thermal motions let arise the fluctuating EFG. Hertz's electrostatic approach has been utilized very successfully to reproduce experimental relaxation rates of different nuclei in a variety of solvents [7]. Consequently this model has already found application in combination with MD-Simulations [8 - 12] because this offers the facility to calculate the mean square amplitude $< V_{zz}^2 >$ and the correlation time τ_c separately.

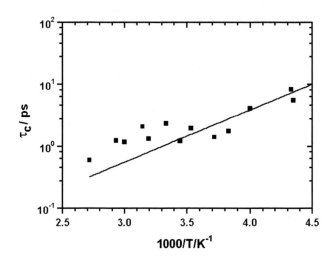

Figure 4. *Temperature dependence of the correlation time τ_c of the efg time correlation function (first hydration shell) from MD simulations.*

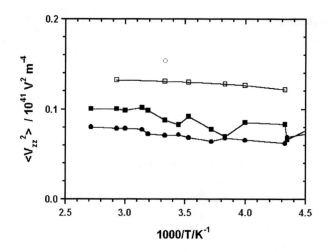

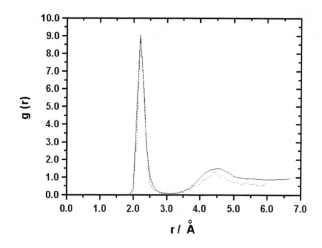

Figure 5. *Electric field gradient fluctuations at the $^7Li^+$ position obtained by Molecular Dynamics Simulations (filled symbols) and by ab initio calculations on selected clusters from the MD study (open symbols). The squares represent the efg arised from the first hydration shell, the rings show the efg for the full system. For the ab initio calculations only the second hydration shell was included. The efg from the MD simulation is already multiplied by the Sternheimer antishielding factor to make the results comparable with the ab initio data.*

Figure 6. *Radial distribution function, $g_{(r)}$, for the oxygen atom relative to lithium (solid line). The dashed line represent the $g_{(r)}$ of one hundred selected MD configurations used for the ab initio calculations.*

$$\text{MD}: \left(\frac{1}{T_1}\right)_{Li} = \frac{3}{40}\frac{(2I+3)}{I^2(2I-1)}\left[\frac{eQ}{\hbar}(1+\gamma_\infty)\right]^2 V_{zz}^2\tau_c$$

In order to access the NMR-relaxation rates in aqueous solution we performed 15 NVE MD Simulations in the temperature range of 215 K $\leq$ T $\leq$ 350 K. With respect to the small system size (100 water molecules + one ion) pbc of a truncated octahedron were applied. To ensure energy conservation Steinhauser-tapering [13] was used with a cut-off radius of 7.868 Å. Newton's equations of motion were solved by the leap-frog algorithm using a time step of 1 *fs*. The density has been kept constant at ρ = 0.997 $g \cdot cm^{-3}$. For the water-water interaction the simple point charge (SPC) model was chosen. The water-ion interaction was obtained by applying Lorentz-Berthelot mixing-rules taking the parameter set for $^7Li^+$ of Palinkas et al. [14].

During the initial equilibration run ($\geq$ 1 *ns* for T $\leq$ 280 *K*) the temperature was adjusted by the Berendsen-thermostat with a time constant of 0.1 *ps*. The production runs were performed over 400000 steps (0.4 ns) calculating the EFG-tensor generated by either the whole system or only the first hydration shell at each step. Afterwards < V_{zz}^2 > (Figure 5) and the EFG time correlation function (tcf) (Figure 3) up to 20 *ps* were obtained. The latter show a complex time behaviour that can be described well by a threefold exponential approach. To yield the correlation time τ_c the tcf is integrated up to infinity (Figure 4).

These results - in combination with the Sternheimer factor [15] - reveal access to the relaxation rate.

Ab Initio calculations

Equilibrated snapshots of the molecular configurations from MD-simulations were taken at intervals of 4000 steps (4 ps) for six temperatures. In all snapshots a $^7Li^+$ cation is surrounded by a first (4-6 molecules) and a second (30-35 molecules) hydration shell of water molecules. One hundred clusters selected in this way are already able to represent the pair correlation function g_{Li-O} of the whole system as shown in figure 6. For these clusters *ab initio* cal-

Table 1. *Comparison of electric field gradient fluctuations at the $^7Li^+$ position directly obtained from MD simulations and from ab initio calculations on MD configurations at 300 K. The efg are listed for MD runs with different system sizes (100, 218 and 400 water molecules). For each simulation the data for the first hydration shell and the total system are shown. In the ab initio calculations the efg are calculated including the second hydration shell instead of the full system (results given in 10^{41} V^2m^{-4}).*

Method	100		218		400	
	1.HS	2. HS	1.HS	2. HS	1.HS	2. HS
MD	0.0873	0.0705	0.0873	0.0911	0.0858	0.0863
QM	0.1305	0.1535	0.1228	0.1487	0.1101	0.1295

culations were performed on a 6-31+G* basis set [16, 17], Hehre on the Hartree-Fock level of theory to obtain the electric field gradient at the $^7Li^+$ nucleus. In a first set the electric field gradient at $^7Li^+$ caused by the first hydration shell is calculated, in a second set of calculations the influence of the first as well as the second hydration shell is studied. The mean values of the electric field gradient for all chosen clusters are used for comparison with the results obtained directly from Molecular Dynamics Simulations [17, 18] (Figure 5). At 300 K we also investigated whether the efg caused by the first or the first plus the second hydration shell are dependent on the system size in the MD simulations (100, 218 and 400 water molecules). The results obtained by the MD study and the ab initio calculations are listed in Table 1.

$$QM : \left(\frac{1}{T_1}\right)_{Li} = \frac{3}{40}\frac{(2I+3)}{I^2(2I-1)}\left[\frac{eQ}{\hbar}(1+\gamma_\infty)\right]^2 \overline{V_{zz}^2}\,\tau_c$$

Conclusions

• The pure solvent contribution to the relaxation of lithium in water was studied by NMR relaxation time experiments, Molecular Dynamics Simulations and *ab initio* calculations on MD configurations.

• The electric field gradient correlation functions for all temperatures show a complex time behaviour and can be fitted by a threefold exponential approach.

• The static part $\overline{V}_{zz}^2$ is a function of the arrangements of the water molecules around lithium. Although the first hydration shell dominates the efg, the influence of outersphere waters is significant for all temperatures.

· The differences of the efg's obtained directly from MD simulations and from *ab initio* calculations can be explained by the Sternheimer approximation.

· All electric field gradient fluctuations caused by the first hydration shell as well as the full configuration decrease slightly with decreasing temperature. At lower temperatures the system becomes more structured and the effects of different water molecules will tend to cancel each other.

· In contrast to $\overline{V}_{zz}^2$, the correlation time τ_c show great fluctuations with temperature which may require longer simulation runs for better statistics. On the other hand the time correlation functions look reasonable and the simulation runs took already 400 ps.

· In spite of the strong fluctuations with temperature, the activation energies of about 17.6 kJ/mol (MD) and 15.74 kJ/mol (QM) for the lithium relaxation rates are in reasonable agreement with the experimental result of 15.34 kJ/mol. In all cases simple exponential fits for the data were applied although the experimental rates do not strictly follow an Arrhenius-type equation.

References

1. Mazitov, R.; Müller, K.J.; Hertz, H.G. *J. Phys. Chem. NF* **1984**, *140*, 55.
2. Deverell, C. *Mol. Phys.* **1969**, *16*, 491.
3. Hertz, H.G. *Z. Elektrochem.* **1961**, *65*, 20.
4. Hertz, H.G. *Ber. Bunsenges. Phys. Chem.* **1973**, *77*, 531.
5. Valiev, K.A. *Sov. Phys. JETP* **1959**, *37* , 77.
6. Valiev, K.A.; Khabibullin, B.M. *Russ. J. Phys. Chem.* **1961**, *35*, 1118.
7. Weingärtner, H.; Hertz , H.G. *Ber. Bunsenges. Phys. Chem.* **1977**, *81*, 1204.
8. Luhmer, M.; van Belle, D.; Reisse, J.; Odelius, M.; Kowalewski, J.; Laaksonen, A. *J. Chem. Phys.* **1993**, *98*, 2.
9. Schnitker, J.; Geiger, A. *Z. Phys. Chem. NF* **1987**, *155*, 29.
10. Roberts, J.E.; Schnitker, J. *J. Phys. Chem.* **1993**, *97*, 5410.
11. Dejaegere, A.; Luhmer, M.; Stien, M.L.; Reisse, J. *J. Magn. Reson.* **1991**, *91*, 362.
12. Engström, S.; Jönsson, B.; Jönsson, B. *J. Magn. Reson.* **1982**, *50*, 1.
13. Steinhauser, O. *Mol. Phys.* **1982**, *45*, 335.
14. Palinkas, G.; Palinkas, W.O.; Riede, K.; Heinzinger, *Z. Naturforsch.* **1977**, *32a* , 1137.
15. Sternheimer, R.M. *Phys.Rev.* **1966**, *146* , 140.
16. Gaussian 92, Revision A, M. J. Frisch, G. W. Trucks, M. Head-Gordon, P.M.W. Gill, M. W. Wong, J.B. Foresman, B. G. Johnson, H. B. Schlegel, M.A. Robb, E.S. Replogle, R. Gomperts, J.L. Andres, K. Raghavachari, J.S. Binkley, C. Gonzalez, R.L. Martin, D.J. Fox, D.J. Defrees, J. Baker, J.J.P. Steward, and J. A. Pople, Gaussian, Inc., Pittsburgh, PA, 1992.
17. Eggenberger, R.; Gerber, S.; Huber, H.; Searles, D. and Welker, M. *J. Chem. Phys.* **1992**, *97*, 5898.
18. Eggenberger, R.; Gerber, S.; Huber, H.; Searles, D. and Welker, M. *Mol. Phys.* **1993**, *80*, 1177.

J. Mol. Model. **1996**, 2, 383 – 385

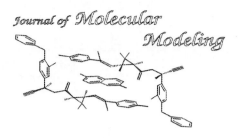

Exploring the Conformational Behavior of Rigid Porphyrin-Quinone Systems by High-Temperature MD Simulations and Temperature-Dependent 1H-NMR Experiments

Martin Frank*, Katrin Peraus and Heinz A. Staab

Max-Planck-Institut für medizinische Forschung, Abteilung Organische Chemie, Jahnstr. 29, 69120 Heidelberg, Germany (M.Frank@dkfz-heidelberg.de)

Received: 15 May 1996 / Accepted: 6 August 1996 / Published: 27 September 1996

Abstract

Photoinduced electron transfer reactions play an important role in the primary step of the biological photosynthesis process. In an attempt to understand better the mechanism of the charge separation organic donor-acceptor molecules containing porphyrins and quinones were designed as photosynthesis models. In order to study the structure dependence of the photoinduced electron transfer twofold and fourfold bridged porphyrin-quinone systems with increasing donor-acceptor distance were synthesized (Figure 1) [1, 2, 3]. It was assumed that in these molecules the porphyrin and quinone should be linked in a rigid and well-defined orientation. To verify this assumption the conformational behavior of these systems was

studied by high-temperature MD simulations in combination with conformational analysis of selected minimized structures [4, 5].

As an example we describe the dynamical behavior of the quadruply bridged porphyrin-quinone with naphthyl groups as aryl spacer ("naphthyl cage") (Figure 2) which was investigated by the following generally used method:

The initial structure used for MD simulation was constructed with the builder modul of INSIGHT II [6]. To stimulate the dynamics in such rigid molecules as the "porphyrin-quinone cages" during a simulation time of 1000 ps we had

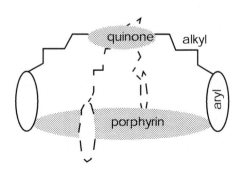

Figure 1. *Schematic representation of the twofold and fourfold bridged porphyrin-quinone systems (aryl spacer: phenyl, naphthyl, biphenylenyl, anthryl; alkyl spacer: tetramethylene).*

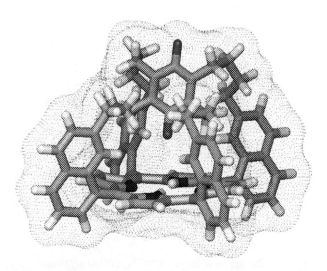

Figure 2. *Calculated structure of the "naphthyl cage" with lowest potential energy.*

* To whom correspondence should be addressed

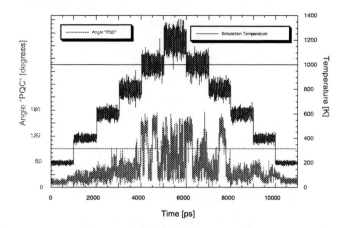

Figure 3. *Simulated heating and annealing (equilibration time = 50 ps, simulation time for each temperature step = 1000 ps, temperature intervals = 200 K, maximum temperature = 1200 K). Definition of angle "PQC" see Figure 4.*

to choose 1000 K as simulation temperature (Figure 3). The CFF91-forcefield [7] was used because it had been proven to reproduce well the Xray-structures of our porphyrin-quinone systems [4]. To analyse the dynamic behavior we defined internal coordinates by which the characteristic motions of the molecule can be described efficiently ("characteristical coordinates") (Figure 4). These "characteristical coordinates" were evaluated by statistical methods (frequency plots) (Figure 5). In addition we carried out conformational analysis on selected minimized structures with respect to the same internal coordinates (Figure 6). In contrast to the basic assumption we found that the "naphthyl cage" is not completely rigid. There is a characteristic dynamic process in the "naphthyl

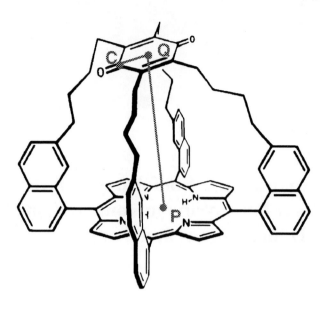

Figure 4. *Definition of angle "PQC" which is the "characteristical coordinate" for the motion illustrated in Figure 7 (P: center of porphyrin, Q: center of quinone, C: carbon atom of quinone carbonyl group).*

cage" (Figure 7). Nevertheless, the conformational mobility of the fourfold bridged porphyrin-quinone system is significantly decreased compared to the twofold bridged porphyrin-quinone system [4]. These calculated results exhibit good agreement with experimental results derived from temperature dependent ^{1}H-NMR experiments [3].

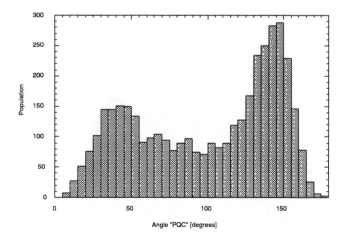

Figure 5. *Histogram of angle "PQC" (1000 ps MD simulation, history = 0.25 ps, simulation temperature = 1000 K, sampling interval = 5 degrees)*

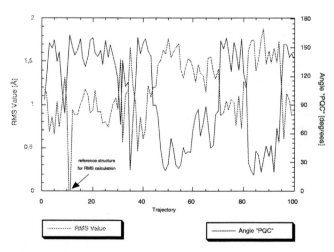

Figure 6: *Relationship between RMS value (superimposition of all heavy atoms, reference structure: 11) and angle "PQC" of 100 minimized structures selected in regular intervals from the MD trajectory. It is evident that there is a direct correlation between the change of the RMS value and the change of the angle "PQC". Therefore the motion which is described by the change of the angle "PQC" should be the only significant motion which leads to a conformational change of the molecule.*

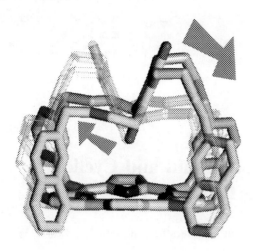

Figure 7. *Characteristic dynamic process of the "naphthyl cage". The plane of the quinone moiety alternates between two positions with an effective rotation of about 120 degrees so that the carbonyl which points up in one conformer points down in the other conformer. The arrows indicate schematically the movement of the quinone oxygens.*

Keywords: High-Temperature MD Simulations, Conformational Analysis, Porphyrin-Quinone Systems, Photosynthesis Models

References

1. Staab, H.A.; Feurer, A.; Hauck, R. *Angew. Chem.* **1994**, *106*, 2542 - 2545.
2. Staab, H.A.; Döhling, A.; Voigt, P.; Dembach, M. *Tetrahedron Lett.* **1994**, *41*, 7617 - 7620.
3. Peraus, K. *Dissertation*, Universität Heidelberg, **1996** .
4. Frank, M. *Diplomarbeit*, Universität Heidelberg, **1995**
5. Dernbach, M. *Dissertation*, Universität Heidelberg, **1993**
6. INSIGHT II Version 2.3.5, Biosym Technology Inc., San Diego, USA, **1994**
7. DISCOVER Version 2.9.5, Biosym Technology Inc., San Diego, USA , **1994**

J. Mol. Model. **1996**, 2, 386 – 389

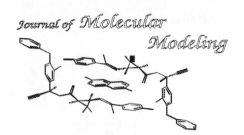

Molecular Modeling Studies on Novel Open-chain and Cyclic Thia Compounds and their Ag(I) and Hg(II) Complexes

Torsten Krueger*, Karsten Gloe, Holger Stephan, Bernd Habermann [a], Kerstin Hollmann [a], Edwin Weber[b]

Institute of Inorganic Chemistry, Technical University Dresden, D-01062 Dresden, Germany (TK@cach05.chm.tu-dresden.de)

[a] Institute of Organic Chemistry, University Leipzig, D-04103 Leipzig, Germany

[b] Institute of Organic Chemistry, Technical University Freiberg, D-09596 Freiberg, Germany

Received: 15 May 1996 / Accepted: 6 August 1996 / Published: 27 September 1996

Introduction

Selective and effective binding of metal ions by multi-functional structure units play an important role in biology, analytics and technique [1]. Therefore the design of novel macrocyclic and structure related open-chain ligands for cations is a main topic of investigations. Solvent extraction studies offer good possibilities to find structure-complexation relationships [2].

In this paper molecular modeling calculations of some novel sulfur containing compounds and their Ag(I) and Hg(II) complexeas are demonstrated and used for the interpretation of the complexation and extraction behaviour of this ligands.

Computational Procedure

The structure of novel thia compounds (type **1 – 3**) and their complexes has been determined by force field, semiempirical

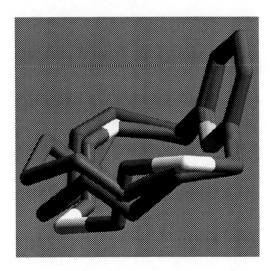

Figure 1. *Ligand 3 (X=N): Three calculated minimum geometries (PM3) of the thermodynamic most stable chair conformation (the gray coloured lowest conformation is in very good agreement with the X-ray structure)*

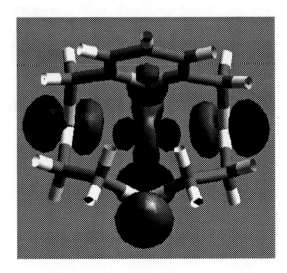

Figure 2. *HOMO alpha orbitals of the 1:1 silver(I) complex with ligand 3 (X=N) using ADF*

---

* *To whom correspondence should be addressed*

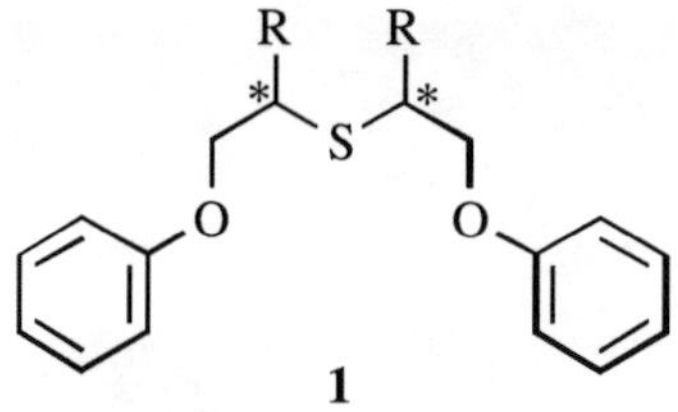

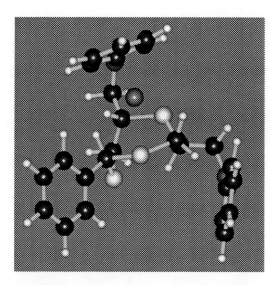

R = -CH$_3$; -CH$_2$-O-CH$_3$; -CH$_2$-O-(C$_2$H$_4$-O-)$_n$H

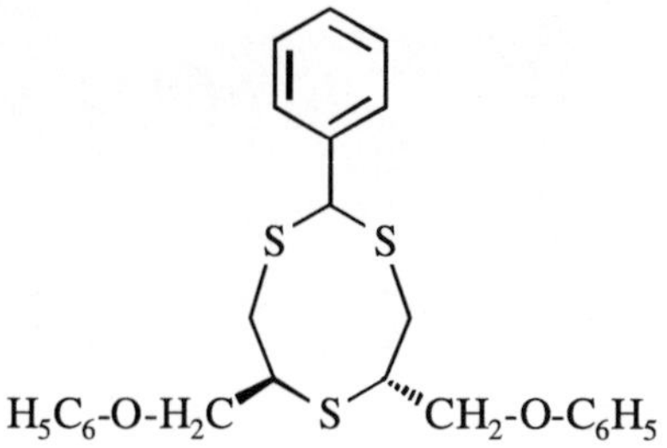

1

H$_5$C$_6$-O-H$_2$C⟩⟩⟩⟩CH$_2$-O-C$_6$H$_5$

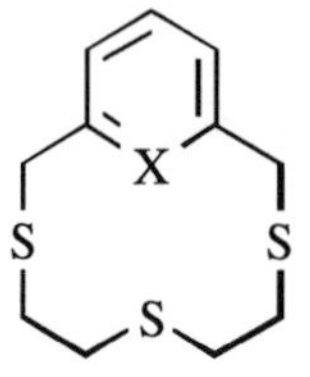

2

X = CH; N

3

Scheme 1 *Investigated compounds.*

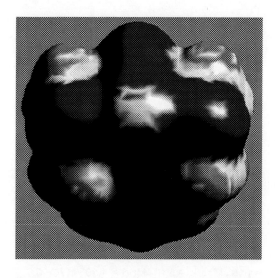

Figure 3. *Electron density surface with electrostatic potential of the 1:1 silver(I) complex with ligand* **3** *(X=N) using ADF*

and ab initio calculations and molecular dynamics simulations. The usefulness of the applied computer methods are proved by comparison of calculated and X-ray structure results [4, 5].

The initial model of the molecular structure were acieved using the UNIVERSAL FORCE FIELD of the CERIUS^2 1.6 (BIOSYM/MSI) software package. The semiempirical calculations were performed with MOPAC 6.0 [6] using the PM3 hamiltonian (ligands and Hg(II) complexes) and the HYPERCHEM 4.5 (HyperCube) program package using the ZINDO/1 method (Ag(I) complexes). The molecules under

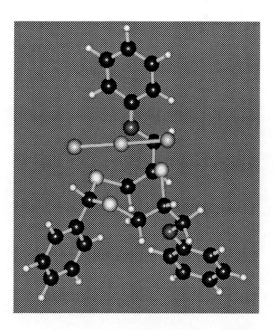

Figure 4. *ZINDO/1 calculated structure of 1:1 Ag(I) complex with ligand* **2**

Figure 5. *PM3 calculated structure of 1:1 HgCl$_2$ complex with ligand* **2**

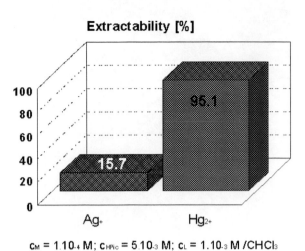

Extractability [%]

$c_M = 1.10_{-4}$ M; $c_{HPic} = 5.10_{-3}$ M; $c_L = 1.10_{-3}$ M /CHCl$_3$

Figure 6. *Solvent extraction studies of silver picrate and mercury chloride with ligand* **2**

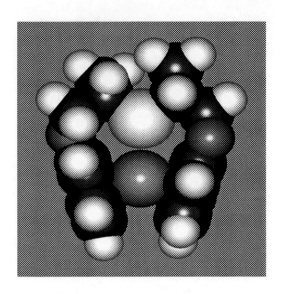

Figure 7. *ZINDO/1 calculated structure of 1:1 silver(I) complex with ligand* **1** *(R=CH$_3$)*

study were completely geometry optimized at the SCF level until a gradient norm less than 0.0001 kcal·mol^{-1}·Å^{-1}. The ab initio calculations were carried out with GAUSSIAN 94 (Gaussian Inc.) using the 3-21G, 4-31G, 6-31G, 6-31G*, LanL2MB and LanL2DZ basis sets at the Hartree-Fock level. The DFT calculations were done with ADF (Amsterdam Density Functional) using the II(DZ) and IV(DZ) large core basis sets at the CGA method.

Discusssion

The investigated thia compounds **1 – 3** give varying extractabilities towards Ag(I) and Hg(II) in dependence on the structure of the ligands. Compound **3** with X = CH gives a high

selectivity for Ag(I) over Hg(II), whereas **3** with X = N (structure see Fig. 1) extracts both Ag(I) and Hg(II) in a comparable order of magnitude.

These considerations are clearly confirmed by the calculated structures of Ag(I) and Hg(II) complexes. In case of **3** with X = N the structures of Ag(I) and Hg(II) complexes are very similar; the metal ions are coordinated by three S donor atoms and the pyridin nitrogen atom (see Figs. 2, 3). In contrast for **3** with X = CH only Ag(I) gives a bonding to all three sulfur atoms and, in addition, to the aromatic π-system. It is interesting that the order of selectivity is reversed in the case of **2** (Fig. 6). The HgCl$_2$ species extracted by **2** is best coordinated by all S atoms and one oxygen of the

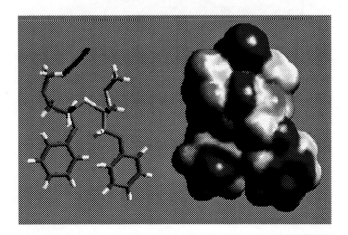

Figure 8. *Electron density surface with electrostatic potential of the 1:1 HgCl$_2$ complex with ligand* **1** *(R=CH$_2$-O-CH$_3$) using PM3 (MOPAC 6.0)*

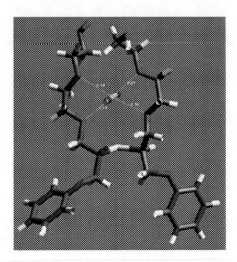

Figure 9. *PM3 calculated structure of 1:1 HgCl$_2$ complex with ligand* **1** *(R=CH$_2$-O-(C$_2$H$_4$-O-)$_2$H)*

phenoxy substituent (Fig.5). In contrast Ag(I) is bonded by two sulfur and the aromatic π-system of the benzylic group only (Fig. 4). The extraction behaviour of type **1** compounds is characterized generally by a preference of Ag(I). So **1** with R = CH_3 can form a typical intramolecular sandwich structure of high hydrophobicity (Fig. 7). The highest extractabilita for Ag(I) is achieved using **1** with R = CH_2-O-$(C_2H_4O-)_2$H and is obviously caused by a bonding pattern including sulfur and three ether oxygen atoms. In contrast the same ligand with $HgCl_2$ gives an interaction with four oxygen donor atoms only (Fig. 9). The best extractant of type **1** for $HgCl_2$ is given with R = CH_2-O-CH_3. According to our calculations the mercury is complexed by sulfur and the oxygens of two methoxy groups (Fig. 8). But also in this case the Ag(I) extraction is significantly better, because of the shorter bonding distances.

It can be shown that molecular modeling leads to a deeper insight in the phenomena of structure-property relations and stimulates the further development of efficient and selective complexing agents.

References

1. Lehn, J.-M. *Supramolecular Chemistry*, VCH, Weinheim, **1995**

2. Gloe, K.; Heitzsch, O.; Stephan, H.; Weber, E. *Solvent Extr. Ion Exch.* **1994**, *12(3)*, 475

3. Gloe, K.; Mühl, P.; Beyer, J. *Z. Chem.* **1988**, *28(1)*, 1.

4. Krüger, T.; Gloe, K.; Habermann, B.; Mühlstädt, M.; Hollmann, K. *Z. amor. allg. Chem.* in press.

5. Krüger, T.; Gloe, K.; Hollmann, K.; Weber, E. *J. Comput. Chem.* in preparation.

6. Stewart, J. J. P.; Frank, J. Seiler, Research Laboratory, U.S. Air Force Academy, Colorado 80840; QCPE 455

J. Mol. Model. **1996**, 2, 390 – 398

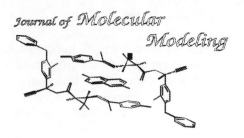

Anionic Tetrahedral Complexes as Serine Protease Inhibitors

Michael Shokhen[‡]* and Dorit Arad[$]*

The George S. Wise Faculty of Life Sciences, Tel-Aviv University, Ramat Aviv, Israel 39040, Tel: 972-3-6408723, FAX: 972-3-6409407

[‡] Department of Biochemistry (shokhen@etgar.tau.ac.il)

[$] Department of Molecular Biology and Biotechnology (dorit@argaman.tau.ac.il)

Received: 4 April 1996 / Accepted: 10 September 1996 / Published: 11 October 1996

Abstract

Potent inhibitors of proteases are constantly sought because of their potential as new therapeutic lead compounds. In this paper we report a simple computational methodology for obtaining new ideas for functional groups that may act as effective inhibitors. We relate this study to serine proteases. We have analyzed all of the factors that operate in the enzyme–substrate interactions and govern the free energy for the transformation of the Michaelis complex (*MC*) to the anionic covalent tetrahedral complex (*TC*). The free energy of this transformation ($\Delta\Delta G_{MC\text{-}TC}$) is the quantitative criterion that differentiates between the catalytic and inhibitory processes in proteases. The catalytic *TC* is shifted upwards ($\Delta\Delta G_{MC\text{-}TC} > 0$) relative to the *MC* in the free energy profile of the reaction, whereas the inhibitory tetrahedral species is shifted downward ($\Delta\Delta G_{MC\text{-}TC} < 0$). Therefore, the more stable the *TC*, the more effective it should be as an inhibitor. We conclude that the dominant contribution to the superstabilization of an anionic *TC* for transition state analog inhibitors originates from the formation of a σ-covalent bond between the reactive centers of the enzyme and its inhibitor. This energetic effect is a quantitative value obtained in *ab initio* calculations and provides an estimate as to whether a functional group is feasible as potent inhibitor or not. To support our methodology, we describe several examples where good agreement is shown between modeled *ab initio* quantum chemical calculations and experimental results extracted from the literature.

Keywords: Protease, methodology of inhibitor design, MO ab initio calculations, free energy profile.

Introduction

Serine proteases play an important role in regulating a wide variety of biological activities. Specific protease inhibitors thus serve as targets for many therapeutic applications [1-3].

Transition state analog inhibitors were first introduced by Wolfenden [4], who suggested that stable analogs of the transition state (*TS*) structure of any enzyme-substrate catalytic process should inhibit enzyme activity. The TS analog concept has been widely used over the last two decades for designing *TS*-analog protease inhibitors [5-8]. The *TS*-ana-

* *To whom correspondence should be addressed*

log inhibitor is a chemical structure that contains a functional group (for example aldehyde or ketone), which, in contrast to the native substrate, cannot be transformed further by the chemical machinery of the enzyme. The common feature of *TS*-analog inhibitors is that they mimic the shape of the native substrate in its *TS* state [5-8]. In this paper we analyze a specific type of TS analogs in which the inhibitory effect results from the intrinsic stability of the covalent complex, forming a stable charged species, rather than an intermediate on the reaction potential surface.

The first step in hydrolysis catalyzed by serine proteases is the conversion of the non-covalent enzyme-substrate (E-S) Michaelis complex (*MC*) to an anionic covalent tetrahedral complex (*TC*) – *MC* ⇒ *TC* [9], where the new covalent bond is formed between the attacking nucleophilic atom in the active site of the enzyme and the electrophilic center of the substrate. On the reaction potential surface of enzymatic catalysis, the *TC* is located in a very shallow local minimum, close to the saddle point corresponding to the *TS* [10-14] and cannot accumulate [15]. Consequently, according to Hammond's postulate [16], the structure of the *TC* in the reaction with a native substrate should be close to that of the *TS*. The reactive center of native substrates of proteases contains a carbonyl group. We suggest that an appropriate structural variation of the reaction center of a native substrate can lead to an anionic superstable *TC*, which will be lower in energy than the *MC* or the products. Thus, an anionic inhibitor is located in the bottom of a potential well rather than on the top of the hill on the free energy profile of enzymatic reaction path. This type of inhibition implements thermodynamic control on the enzymatic catalysis. Anionic *TC*s have been detected experimentally by direct NMR measurements [17, 18]. Stabilization of the tetrahedral intermediate has been established to be an important factor for increasing the in-vitro potency of inhibitors [18]. Substitution of the location alpha to the reaction center carbonyl with electron withdrawing substituents as fluorine, trifluormethyl and heterocyclic substituted aromatic rings, contribute to *TS* stabilization, and derivatives bearing these substituents are found to be potent inhibitors [18]. The effect of substituents on chemical reactivity in general is also highlighted in a number of papers [19-22], and good correlation is found between calculated values (semi-empirical and *ab initio*) and experimental values. However, a theory that can analyze and quantify the experimental values and build an easy to follow general concept for obtaining new derivatives that may provide novel charged superstable inhibitors is missing. We now report one approach to such a theory.

Which forces contribute to the formation of a superstable *TC* in inhibitory processes in contrast to the unstable *TS* formed during the catalysis of native substrates? Wolfenden's theory [4], predicting the principal ability of the *TS*-analog inhibitors neither assumes the specific nature of the attractive forces involved nor requires that the enzyme be rigid or flexible. Opinions in the literature that favor the notion that in enzymatic catalysis noncovalent forces are responsible for

stabilizing either the *TS* or the *TC* arose from the following concepts: (a) formation of hydrogen bonds with the oxyanion hole (reviewed by Ménard, & Storer [23]); (b) electrostatic stabilization [18]; (c) environmental effects of the active site [12,25,26]; and (d) hydrophobic effects [27].

In the present work we have estimated the relative contributions from different effects (proton transfer, hydrogen bond formation and etc.) accompanying the *MC* ⇒ *TC* transformation to the total free energy of this process. By analyzing factors that control the *MC* ⇒ *TC* step in catalysis, we conclude that the formation of a new covalent σ-bond during the nucleophilic attack of the protease nucleophilic center on the carbonyl group or its analog makes the main contribution to the energy stabilization of the *TC*. The superstabilization

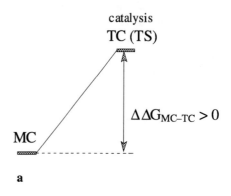

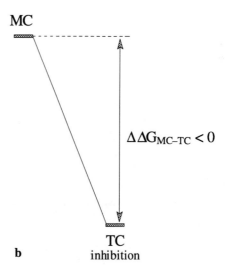

Figure 1. *(a) For a normal catalytic process, a tetrahedral complex (TC) is the transition state–the saddle point on a reaction free-energy profile. The TC is shifted upward from the energy of the Michaelis complex (MC).*

(b) For the inhibition of proteases, the TC is a thermodynamically very stable structure corresponding to the potential minimum on the reaction pass, which is shifted downward from the energy of the MC.

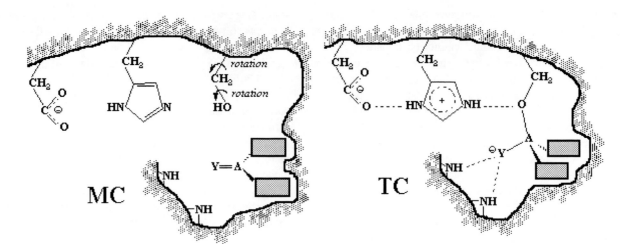

Figure 2. *Generalized schematic presentation of the structural change in the active site of a serine protease during the conversion of MC ⇒ TC. The rectangles designate the subsites of a substrate (inhibitor) belonging to the recognition site (RS) which are bound to the enzyme by noncovalent interactions. The two NH groups depict an oxyanion hole. Formation of the new covalent bond, O-A, in a TC between the reactive centers results in the creation of ring skeleton, which contains a Ser residue, an inhibitor molecule, and part of the backbone of the enzyme. Hydrogen bonds in the TC are depicted by dotted lines to stress that they are more rigid than in the MC. Additional binding between the enzyme and the inhibitor results in a loss of degrees of internal rotation.*

of the *TC* depends on the strength of the new forming σ-bond. This contribution can easily be estimated by model calculations, and thus provide a tool for a semi-quantitative choice of substituents needed in order to stabilize the *TC*. We demonstrate that simple routine quantum-chemical calculations of a small model system–the analog of the *TC* can be applied successfully for the design of new classes of protease inhibitors. Using examples of known inhibitors from the literature, we show that the effect of inhibition results in the formation of the superstable anionic *TC*. Our model calculations on these systems show excellent agreement between the predicted values and the experimental results.

Models and Methods

Structural analysis

In a normal catalytic pathway that corresponds to the nucleophilic attack of the enzyme nucleophile on the substrate carbonyl, which already locates as the *MC* in the active site, the *TC* lies on a hill higher in energy than the *MC*. If we succeed in stabilizing the *TC* to such an extent that it lies in a potential energy well lower than the *MC*, we achieve inhibition.

We define an unstable *TC* as the *TS* corresponding to the saddle point on a normal proteolytic reaction path. Conversely, we define here an inhibitory tetrahedral species as a thermodynamically stable enzyme-inhibitor complex that corresponds to the potential minimum on the reaction path. The free energy effect of the transformation $MC \Rightarrow TC$, $\Delta\Delta G_{MC\text{-}TC}$, is the quantitative criterion that differentiates between the catalytic and inhibitory processes. The catalytic *TC* is shifted upwards ($\Delta\Delta G_{MC\text{-}TC} > 0$) relative to the *MC* in the energy profile of the reaction (Fig. 1a), whereas the inhibitory tetrahedral species is shifted downward ($\Delta\Delta G_{MC\text{-}TC} < 0$), see Fig. 1b. Therefore, the more stable the *TC*, the more effective it should be as an inhibitor.

We limit our consideration of E-S interactions to the region of enzyme active site (*AS*) only, so that the Gibbs free-energy effect of the $MC \Rightarrow TC$ transformation can be approximated by the value of $\Delta\Delta G(AS)$ only.

The *AS* is divided into two structural regions: the catalytic (reactive) site (*CS*), where the chemical interaction between the enzyme and substrate occurs, accompanied by the formation and cleavage of covalent bonds, and the recognition site (*RS*), which binds the substrate by non-covalent interactions. No chemical transformations occur in the *RS*. Usually the *CS* fragment is much smaller than the *RS*.

$$AS = RS \cup CS \qquad (1)$$

The model introduced here for the subdivision of an E-S complex into its component parts is a trivial application of the approach that is widely used in molecular mechanics and quantum chemistry-structural-additivity analysis of molecular total energy. We assume that the *RS* fragment of an E-S complex conserves its 3D molecular structure in the $MC \Rightarrow TC$ path: $\Delta G_{MC} (RS) \approx \Delta G_{TC} (RS)$. Hence, we can use Eq. 1, to rewrite the expression for the Gibbs free-energy difference between the $MC \Rightarrow TC$ [$\Delta\Delta G(AS)$] for a reaction profile of the same substrate (or inhibitor):

$$\Delta\Delta G (AS) = \Delta G_{TC} (AS) - \Delta G_{MC} (AS) \approx \Delta\Delta G(CS) =$$

$$= \Delta\Delta H(CS) - T\Delta\Delta S_{vib} \ (CS) \qquad (2)$$

where we have neglected the contribution from the *RS* region to $\Delta\Delta G(AS)$ in Eq. 3. The overall translational and rotational entropy, $T\Delta\Delta S (AS)_{tr,rot}$, does not contribute to $\Delta\Delta G(AS)$ because in both the *MC* and the *TC* states the substrate is bound to the enzyme in one complex. $S (AS)_{tr,rot}$ is determined by the mass and geometrical parameters of the complex [28], where the enzyme, which does not change its shape in the *MC*$\Rightarrow$*TC* transformation, dominates, so that $\Delta\Delta S(AS)_{tr,rot} = 0$.

The E-S interactions in the *RS* ("recognition site") provide the driving force for aligning the substrate (or inhibitor) in the active site of an enzyme, which strongly influences the efficacy of enzymatic catalysis or inhibition. The chemical transformation of the substrate (or inhibitor) in the reaction path *MC*$\Rightarrow$*TC* can proceed in the initial stage of the *MC* only if all the atoms of the CS are positioned in very specific, optimal interatomic distances for the desired chemical process. The *RS* fragment contributes the dominant part of the noncovalent binding between the enzyme and the substrate. The alignment can be characterized quantitatively by the values of $\Delta G_{TC} (AS)$ and $\Delta G_{MC} (AS)$. Although the noncovalent E-S interactions in the *RS* strongly influence the values of $\Delta G_{TC} (AS)$ and $\Delta G_{MC} (AS)$, according to Eq. 2, such influence is canceled in the value of their Gibbs free-energy difference, $\Delta\Delta G(AS)$: $\Delta\Delta G(RS) \approx 0$. Thus, we conclude that the free energy contribution from the CS [$\Delta\Delta G(CS)$] determines the overall difference between the E-S binding energy during the transformation , the noncovalent E-S interactions in the *RS* play a minor role here.

In a related paper, Menger suggested the "Split-Site" model [29], where an active site is divided into a "binding region" and a "reaction region". The theoretical findings of Menger [29] are based on his idea that only *destabilizing* interactions between the enzyme and the substrate occur in the CS region (Menger's reactive site). Menger also accepted that the 3D structure of the "binding site" (analogous to the *RS* in our analysis) remains constant during enzymatic transformations. He postulated that (a) in the reactive site (*CS*), destabilization of the *TS* is much stronger than that of the *MC* and (b) in the binding site (*RS*), only noncovalent interactions can stabilize E-S complexes. We believe that covalent binding in the *CS* contribute to the stabilization effect. The quantitative picture is demonstrated by calculations.

Free energy components in the CS

In this section we discuss the contribution to $\Delta\Delta G(CS)$ from different processes accompanying the *MC*$\Rightarrow$*TC* transformation. For simplicity we designate $\Delta G(CS)$ instead of $\Delta\Delta G(CS)$.

Reactivity center. The dominant process that occurs in the transformation of *MC*$\Rightarrow$*TC* is the nucleophilic attack of the catalytic site of the enzyme on the electrophilic center of a substrate or an inhibitor occurring in the *CS*. In this analysis

we separate a fragment in the *CS* – the "reactivity center", defined as the fragment that contains the reactive centers of the enzyme and substrate and the atoms of their valent surrounding. Most structural changes of the substrate in the reaction step are located in the reactivity center. ΔG_{rc} expresses the contribution of the reactivity center to the value of $\Delta\Delta G(CS)$.

The electrophilic center of the native substrate is either an amide or an ester carbonyl. The object of the nucleophilic attack in an inhibitor is a polarized double bond of the electrophilic site, $A=Y$, where A is an electrophilic center and Y is usually an electronegative substituent [6, 30]. The nucleophile atom D ($D = O$ or S) forms the new bond D-A, which is partially covalent and partially ionic [31]. The transformation is accompanied by an $sp_2 \Rightarrow sp_3$ rehybridization, which causes pyramidalization at the A center and $A=Y$ bond elongation. The ΔG_{rc} builds up mainly from a considerable stabilization energy, which results from the formation of a new covalent σ-bond D-A, (see Fig. 2). This strong stabilization effect is partially compensated by the destabilizing energy resulting from reducing the π character of the A=Y bond to a single bond. The net result of the transformation, however, is always $\Delta G_{rc} < 0$.

Proton Transfer. Serine proteases facilitate nucleophilic attack on a substrate by transferring a proton from the nucleophile [9]. The proton transfer occurs simultaneously with *TC* formation in reactions catalyzed by serine proteases. The process contributes [designated as $\Delta G_{pt} (CS)$] $\Delta G(CS)$, of the transformation to the total free energy effect. Warshel & Russell [12], using empirical procedures, estimated that the energy change involved in the proton transfer Ser-His in trypsin is 14 kcal/mol, so $\Delta G_{pt} (CS) > 0$.

Environmental Effects. It has been established that the main source of *TS* stabilization lies in the environmental effects in the active site of the enzyme (designated as $\Delta G_{env} (CS)$). The oxyanion binding site in serine proteases is one example of a widely studied environmental factor [23]. Such factors are often considered the main sources of anionic *TC* stabilization [32,33]. The structure of chymotrypsin-trifluoromethylketone inhibitor complexes provides an experimental demonstration of the contribution of environmental effects [34]. The negatively charged oxygen atom attached to the tetrahedral carbon of a hemiketal adduct (*TC*) is hydrogen-bonded to Ser195 and Gly193 amides in the oxyanion hole. In the oxyanion hole, the hydrogen bonding for a negatively charged oxygen atom in the ionized hemiketal in the *TC* is stronger than the carbonyl oxygen binding of the neutral *MC* [12, 35,36]. The superstabilization of the anionic *TC* relative to the *MC* is ascribed to this differential binding strength.

McMurray, and Dyckeys [37] studied trypsin inhibition using the series of model peptide ketones Lys-Ala-LysCH$_2$X. From a Hammett plot of -log K_i vs. σ_I; the authors concluded that the strength of binding of the hemiketals to the AS of the enzyme increases with the electron-withdrawing ability of the varied substituent, X. The tightest binding was determined for the fluoromethylketone X = F. In our analysis, *ab initio* calculations were used to obtain a quantitative estimate on the energetic contribution of this effect. We considered the small molecules $(CF_3)HCO$ and $(NH_2)HCO$ to compare the CF_3 substituent with NH_2, which is the model for a native amide substrate. To simulate the hydrogen bonds in the oxyanion hole, we used one water molecule. The energy of hydrogen bonds was calculated by Gaussian 92 [38] at the 6-31+G*/3-21G level according to the following reaction equations:

$$MC: \ (NH_2)HCO + HOH \Rightarrow (NH_2)HCO\text{—}HOH,$$
$$E_{MC} \ (NH_2) = -4.0 \ \text{kcal/mol} \tag{3a}$$

$$TC: \ HO(NH_2)HCO^- + HOH \Rightarrow HO(NH_2)HCO^-\text{–}HOH,$$
$$E_{TC} \ (NH_2) = -12.4 \ \text{kcal/mol} \tag{3b}$$

$$MC: \ (CF_3)HCO + HOH \Rightarrow (CF_3)HCO\text{—}HOH,$$
$$E_{MC} \ (CF_3) = -2.6 \ \text{kcal/mol} \tag{3c}$$

$$TC: \ HO(CF_3)HCO^- + HOH \Rightarrow HO(CF_3)HCO^-\text{—}HOH,$$
$$E_{TC} \ (CF_3) = -14.7 \ \text{kcal/mol} \tag{3d}$$

Our simple model for hydrogen bonds in the oxyanion hole provides an estimate of tighter binding for the *TC*:

$$\Delta E(NH_2) = E_{TC}(NH_2) - E_{MC}(NH_2) = -8.4 \ \text{kcal/mol} \tag{4a}$$

$$\Delta E(CF_3) = E_{TC}(CF_3) - E_{MC}(CF_3) = -12.1 \ \text{kcal/mol} \tag{4b}$$

$$\Delta\Delta E = \Delta E(CF_3) - \Delta E(NH_2) = -3.7 \ \text{kcal/mol} \tag{4c}$$

The values are quantitatively close to experimental values [36] and to the theoretical estimations of Warshel et al. [12, 39, 40]. We can conclude that the hydrogen bonds in the oxyanion hole contribute not only to the additional stabilization of a *TC* but also to the superstabilization of a *TC* when *TS*-analog inhibitors with a strong electron-withdrawal substituent are involved. The value of the effect is small, however, when compared to the gain in energy that is due to the formation of a covalent bond in a *TC* (see the "Validity" section in Results and Discussion below).

Hwang and Warshel [39] proposed that electrostatic effects are key factors in serine protease catalysis. Enzymes provide the proper environment of polar-group dipoles to complement the changes in charge distribution from the *MC* to the *TS* [12,39]. The estimated energy value of *TS* stabilization is -7 kcal/mol [12]. Another theory claims that in water solution, substrate-enzyme binding is driven by a hydrophobic effect [27]. The most reasonable conclusion is that the environmental effects cause the *TC* to be more stable than the $MC[\Delta G_{env} \ (CS) < 0]$.

Vibrational Entropy. The last component of free energy considered here is the contribution of vibrational entropy ΔS_{vib} *(CS)*. In a *TC* the nucleophilic center of the enzyme is covalently bound to the electrophile. Topologically, the formation of this bond may be considered a ring closure (see Fig. 2), meaning a loss of at least two degrees of internal rotation. The entropy difference between linear and cyclic compounds gives an entropy loss of 2.7 to 4.3 kcal/mol per internal rotation [41]. In addition, the skeleton of hydrogen bonds should be more rigid in the *TC* than in the *MC*. Such phenomena generally reduce the integral vibrational entropy of the *TC* relative to the *MC*. A quantitative estimate of the value of ΔS_{vib} *(CS)* can be calculated only by a sophisticated computational procedure because the system under consideration is extremely complex. For a qualitative picture, however, we may accept that ΔS_{vib} *(CS)* < 0.

In summary, we can write the following qualitative expression reflecting the additivity of all the components of the main contributors to the $\Delta G(CS)$:

$$\Delta G(CS) = [\Delta G_{rc} < 0] - [T\Delta S_{vib} \ (CS) < 0] +$$

$$+ [\Delta G_{pt} \ (CS) > 0] + [\Delta G_{env} \ (CS) < 0] + Rest \tag{5}$$

where *Rest* = the free energy contribution from all other effects. We assume that the *Rest* effects are minor and principally do not change the general energetic picture.

Results and Discussion

Equation 5 provides the guideline for design of superstable anionic *TC*s because it summarizes the trends of influence of the main factors governing the thermodynamic stability of the *TC* in the *CS* region of the E-S complex. We concluded that the first term (ΔG_{rc}), which relates to the formation of a covalent σ-bond (D-A), should be the dominant component of $\Delta G(CS)$. The results of our computations in the following section support this conclusion. Actually, we only have one tool to design an effective inhibitor forming the superstable *TC*–the variation of the molecular structure of the desired candidate. Therefore, one must query which of the possible contributors to $\Delta G(CS)$ (reactivity center interactions, environmental effects, proton transfer, and vibrational entropy) are the most susceptible to these variations. The answer is that the variation of the *CS* fragment of a substrate influences only ΔG_{rc}, and the other terms in Eq. 5 are not relevant. Indeed, any change in ΔS_{vib} *(CS)* relating to the loss of degrees of internal rotation on the amino-acid residues of the enzyme is specific to a concrete type of enzyme (Fig. 2). Basic catalysis that accelerates reactivity of its nucleophilic group is also an intrinsic feature of an enzyme (for example the catalytic triad in serine proteases). Because catalysis proceeds between nonvariable structural fragments (the amino-acid residues of enzyme), the value of ΔG_{pt} *(CS)* is independent of substrate variation. On the other hand, the value of $\Delta G_{env}(CS)$ depends on the variation in the molecular structure of the substrate. Because $\Delta G_{env}(CS)$ involves only noncovalent interactions, which are one or two orders of magnitude lower than the energy contribution from covalent forces, ΔG_{rc} makes the dominant contribution to $\Delta G(CS)$. We conclude, therefore, that the formation of a new, covalent D-A bond between the reactive centers of the enzymatic nucleophile and its substrate (inhibitor) is the main source of the tighter binding of a *TC* compared with a *MC*.

Methodology of design

The results of the thermodynamic analysis summarized in Eq. 5 and the refinement of the most susceptible region for the variations in the inhibitor structure lead us to the conclusion that the reaction center, which is a very small fragment of the real *TC*, can be used to model the relative stability between *MC* and *TC*.

The dominant structural factor – the new covalent bond formed during the $MC \Rightarrow TC$ transformation determines the choice of the computational method. This property–the formation of a covalent bond can be separated from the rest of the factors, and followed quantitatively. Only quantum chemical calculations can correctly take into consideration the covalent interactions and the electronic effects of substituents on the stability of the *TC*. The system is small enough to allow the use of *ab-initio* quantum mechanical calculations with large basis sets.

The design of the most potent inhibitor in a series for any serine protease can be reformulated as the task of the choice in a series of inhibitors. According to our model, a potential protease inhibitor I_k is simulated by the simplest molecule A_k, which contains the reactivity center of the I_k. The enzyme's nucleophilic site is also simulated by the simplest molecule or anion D.

We take into account only the values of $\Delta G_{rc}(k)$ for each inhibitor I_k. The modeled reaction scheme of $MC \Rightarrow TC$ can be presented as follows:

$$D + A_k \Rightarrow DA_k \qquad (6)$$

For the nucleophile D conserved in the series, the stability of the DA_k depends only on the electronic features of the atom of the electrophilic center in A_k and its covalently bound substituents. Because we calculate the actual total energies of molecules, we calculate the values of $\Delta E_{rc}(k)$ rather than those of $\Delta G_{rc}(k)$. This approach is based on the suggestion of Dewar [42] that the variation in the entropy component of the free energy of reaction for a reaction series of analogous reagents should be insignificant. When designing potential inhibitors, the species A_k that form the most stable DA_k in the series should be the optimal candidates to construct the reactivity center of inhibitors.

Validity of the model

In this section we shall present examples to prove our model for the design of functional groups that can act as serine protease inhibitors. Our guideline was to design small molecules that mimic the reactivity centers of virtual substrates and inhibitors. All examples were selected from the literature to show that we can predict experimentally observed tendencies for the influence of the substituents on the catalytic/inhibition constants. Quantum-chemical calculations have shown that neutral nucleophiles as H_2O and CH_3OH cannot mediate nucleophilic addition to the carbonyl group without supporting basic or acidic catalysis [43-45]. Therefore, OH^- anion was used as the simplest model for the serine nucleophile of the enzyme. Consequently, all the modeled products of nucleophilic addition are anions, as illustrated by reaction schemes in Figs. 1 to 3.

We examined the following two types of carbonyl group derivatives: (1) variation of substituents at the conserved electrophilic center, and (2) variation of the atom of the electrophilic center. The calculated stabilities of the anionic *TC*s are presented in Table 1. All quantum mechanical calculations were performed using the standard Gaussian 92 program [38]. Our general position was to simplify the computational procedure, keeping in the same time the level of accuracy of the molecular calculations that provides reproduction of the principal chemical picture of the observed experimental reaction and structural features of reagents and products. This route is convenient for practical application for computer assisted drug design, where a large series of lead compounds needs to be calculated in a reasonable time. In our computational experiments we found that semi-empirical quantum chemical procedures like AM1 or PM3 (see the review article by Stewart about these methods [46]) failed to reproduce geometries and energies of anionic *TC* structures. In some cases, especially for the anionic *TC*s containing the second row elements, semi-empirical methods gave unrealistic geometries and energies for the $MC \Rightarrow TC$ transformation. For example, the high level *ab initio* calculations demonstrated the inability of the HS^- anion to form *TC* with formaldehyde [31, 47]. In contrast, semi-empirical procedures predicted the formation of a stable *TC*. Thus, we used the non expensive variant of *ab initio* calculations, which reproduce reasonable molecular geometries and avoids basis set superposition error (BSSE) in the reaction energy estimations [48]. Molecular geometries were optimized in the 3-21G* basis set and then total energies were recalculated in the basis set 6-31+G*. This is a commonly used computational approach that gives good results for anionic species [49].

The first set of calculations include a series of carbonyl derivatives where the substituents examined have different electron-donor and electron-acceptor properties. Figure 3 shows the list of varied substituents and the general scheme of *TC* formation for this series.

Table 1. *Ab initio 6-31+G*/3-21G* calculated stabilization energies E_{st} of Tetrahedral Complexes*

Substrate	E_{st} kcal/mol
H(NH$_2$)C=O	-22.7
H(OH)C=O	-32.6
H(CF$_3$)C=O	-59.9
H(F)C=O	-50.5
H(CN)C=O	-62.2
HB(OH)$_2$	-53.2
F(OH)$_2$P=O	-61.7 [a]

[a] *The E_{st} for the F(OH)$_2$P=O was estimated from the reaction equation:*
$$HO^- + F(OH)_2P=O \Rightarrow (HO)_2PO_2^- + HF$$

The first example provides a detailed analysis of standard free energy for a-chymotrypsin-catalyzed hydrolysis of N-acetyl-L-tryptophan methyl ester and N-acetyl-L-tryptophan amide. Bender et al. [50] showed that the activation barrier for the acylation step measured from the *MC* is 13.5 kcal/mol for the ester substrate and 19.6 kcal/mol for the amide substrate. For the deacylation step, the activation barrier of 17.8 kcal/mol is the same for both cases because their acylenzymes are identical. The physical meaning is that the rate-determining step is acylation for hydrolyzing amides and deacylation for hydrolyzing esters. The free energy of stabilization of the *MC* compared with that of the separated enzyme and substrate is similar for both substrates:

-4 kcal/mol for the ester and -3 kcal/mol for the amide. Hence, the molecular structure of the acylation step *TS* is about 6 kcal/mol lower in energy for the ester than for the amide, which causes the barrier difference. The calculations agree well with the experimental result: the model *TC* is about 10 kcal/mol less stable for an NH$_2$ substituent than for an OH substituent (see Table 1 and Figure 3).

The second example relates to the polyfluoroketones. That these moieties are strong *TS*-analog inhibitors for many members of the serine hydrolase class is well established [32, 33, 51]. The new bond forming between the electrophilic center of a carbonyl substrate and an attacking nucleophile in a *TC* should be more sensitive to the variation of the substituents than the hydrogen bonds discussed previously. In the model calculations the molecules CF$_3$CHO and NH$_2$CHO that mimic the reactivity center of substrates were subjected to nucleophilic addition of an HO$^-$ anion. The reaction scheme presented in Figure 3 and the estimated values of the *TC* stability shown in Table 1 demonstrate that when compared with the NH$_2$ substituent, the CF$_3$ substituent superstabilizes the *TC* by -37.2 kcal/mol. This value is higher by one order-of-magnitude than the energetic effect of hydrogen bonds (environmental effect) considered above (see Eq. 4c). Therefore, the results confirm our hypothesis that the energetic contribution from the covalent bond between the reactive centers plays the leading role in superstabilizing the anionic *TC* for *TS*-analog protease inhibitors. To strengthen our case, we calculated additional model molecules with strong electron-withdrawal substituents attached to the carbonyl carbon – a fluorine substituent [FCHO] and cyano group [(CN)CHO]. The results of the calculated *TC* stability, presented in Table 1, show that superstabilization relative to the reference substituent, NH$_2$, is -27.8 kcal/mol for the fluorine and -39.5 kcal/mol for the cyano substituent. Thus, we can predict that peptidyl aldehyde derivatives containing F(CO)$^-$ and NC(CO)$^-$ fragments should be effective *TS*-analog inhibitors for serine proteases.

The following examples concern the variation of the atom of the electrophilic center of the substrate from a carbonyl carbon atom to the boron and phosphorous atoms. At nanomolar concentrations, peptide boronic acids are very

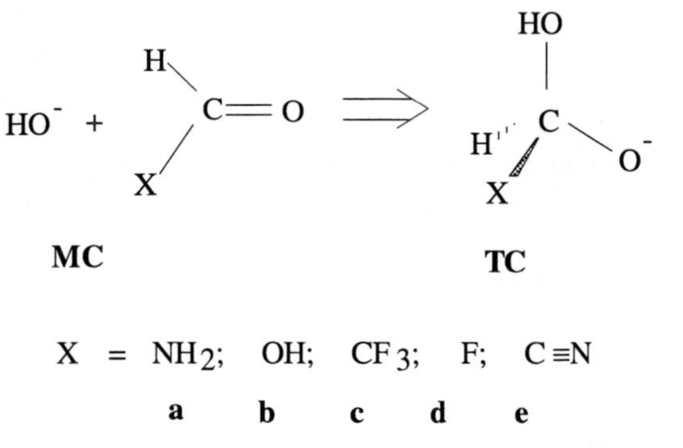

Figure 3. *Scheme of TC formation by the series of carbonyl derivatives where the varied substituents have different electron-donor and electron-acceptor properties.*

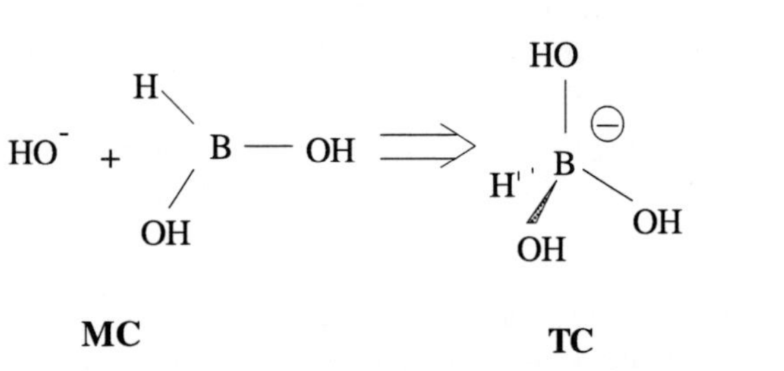

Figure 4. *Scheme of TC formation by the HB(OH)$_2$ – the simplest molecule simulating a boronic acid moiety of the inhibitor.*

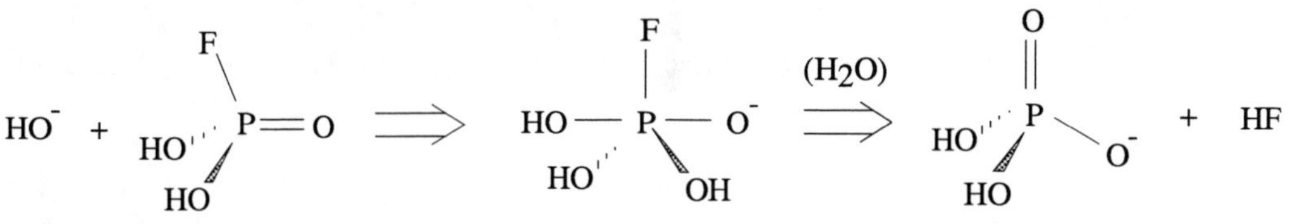

MC   "aged" TC

Figure 5. *The reaction center of DFP is modeled by* $F(HO)_2PO$. *Here TC is the aged tetrahedral anion,* $O(HO)_2PO^-$.

effective *TS*-analog inhibitors in a wide range of serine proteases [52, 53]. The active-site serine forms a covalent tetrahedral adduct *TC* with the boronic acid moiety of the inhibitor. We used the simplest molecule [$HB(OH)_2$] that forms a *TC* [$HB(OH)_3^-$] with OH$^-$ anion (see the reaction scheme in Figure 4 and the stabilization energy of a *TC* in Table 1).

Many serine proteases are inhibited by organophosphorus compounds [8]. Diisopropylphosphofluoridate (DFP), which stoichiometrically and irreversibly inactivates serine proteases [54, 55] is used extensively as a diagnostic test for the presence of the active serine in an enzyme. Phosphorylated serine hydrolases are susceptible to an "aging" reaction, which involves hydrolysis of one phosphate bond [56]. The tetrahedral structure of an aged enzyme-inhibitor complex has a covalent bond between the serine oxygen and phosphorous atom [57]. In our calculations, the reactivity center of DFP was modeled by $F(HO)_2PO$. The aged product of the nucleophilic addition of the HO$^-$ anion is the tetrahedral anion, $O(HO)_2PO^-$, as shown in Figure 5.

The calculated stabilization energies of boronic and phosphorous *TC*s are presented in Table 1. The superstabilization energies of these *TC*s relative to the *TC* of the amide ($HONH_2CO^-$) are -30.5 kcal/mol for boronic and -39.0 kcal/mol for phosphorous inhibitor. As for the carbonyl-based inhibitors, the *TC*s are more stable with the inhibitor than with the native substrate, and indeed, are found as selective inhibitors.

All energy values calculated here – the superstabilization of the inhibitor *TC*s relative to the *TC* for the modeled substrate – emerge from the formation of a s-covalent bond between the reactive centers. The mean energetic value of these effects is -30 kcal/mol, exceeding by one order of magnitude the relevant value of the hydrogen bonds in the oxyanion hole, which we estimated in Eq. 4c as -3.7 kcal/mol. This result confirms our theoretical prediction that the dominant contribution to the superstabilization of a *TC* for *TS*-analog inhibitors originates from the formation of a σ-covalent bond between the reactive centers.

Summary

We analyzed the mechanism of serine protease inhibition in a single catalytic step–the chemical transformation of *MC* ⇒ *TC*. We demonstrated that the dominant contribution to the superstabilization of a *TC* originates from the formation of a σ-covalent bond between the reactive centers in the E-S complex. We suggest a simple computational methodology for designing functional groups that can serve as the reactivity center for new classes of serine protease inhibitors–superstable anionic *TC*. To support our theoretical findings, we described several examples where excellent agreement exists between model *ab initio* quantum chemical calculations and well-known experimental results extracted from the literature.

Acknowledgments We are grateful to Mr. Yehuda Bronitsky from Ormat Industries for supporting the work of Dr. M. Shokhen. We also thank the Israeli Ministry for Absorption of New Immigrants for the Giladi scholarship and the Ministry of Science and Education for their help in supporting in this research.

References

1. Tapparelli, C; Metternich, R.; Ehrhardt, C.; Cook, N. S. *Trends Pharmacol. Sci.* **1993**, *14*, 366-376.
2. Wilhelm, O.; Reuning, U.; Jaenicke, F.; Schmitt, M.; Graeff, H. *Onkologie* **1994**, *17*, 358-366.
3. Claeson, G. *Blood Coagulation & Fibrinolysis* **1994**, *5*, 411-436.
4. Wolfenden, R. *Acc. Chem. Res.* **1972**, *5*, 10-18.
5. Wolfenden, R. *Ann. Rev. Biophys. Bioeng.* **1976**, *5*, 271-306.
6. Tomasselli, A. G.; Howe, W. J.; Sawyer T. K.; Wlodawer, A.; Heinrikson, R. L. *Chimicaoggi- Chemistry Today* **1991**, *9*, 6-27.
7. Dunn, B. M. *Advances in Detailed Reaction Mechanisms*; Vol.2; JAI Press, Inc, 1992; pp. 213-241.
8. Powers, J. C.; Harper, J. W. In *Research Monographs in Cell and Tissue Physiology*; Vol.12; *Proteinase Inhibitors* Barrett, A. J.; Salvesen, G., (Eds.); Elsevier: Amsterdam, 1988; pp. 55-152

9.   Pòlgar, L. *Biol. Chem. Hoppe-Seyler* **1990**, *371, Suppl.,* 327-331.

10. Bender, M. L.; Kezdy, F. J.; Gunter, C. R. *J. Am. Chem. Soc.* **1964**, *86,* 3714-3721.

11. Longo, E.; Stamato, F. M. L.; Ferreira, R.; Tapia, O. *J. Theor. Biol.* **1985**, *112,* 783-798.

12. Warshel, A.; Russel, S. *J. Am. Chem. Soc.* **1986**, *108,* 6569-6579.

13. Daggett, V.; Schroder, S.; Kollman, P. *J. Am. Chem. Soc.* **1991**, *113,* 8926-8935.

14. Mulholand, A. J.; Grant, G. H.; Richards, W. G. *Protein Eng.* **1993**, *6,* 133-147.

15. Mackenzie, N. E.; Malthouse, J. P.; Scott, A. I. *Science* **1984**, *225,* 883-889.

16. Hammond, G. S. *J. Am. Chem. Soc.* **1955**, *77,* 334.

17. Abeles, R. H.; Alston, T. A. *J. Biol. Chem.* **1990**, *265,* 16705-16708.

18. Edwards, P.D.; Wolanin, D.J.; Andisik, D.W; Davis, M.W. *J. Med. Chem.* **1995**, *38,* 76–85.

19. Jursic, B.S.; Zdravkovski, Z. *Theochem* **1994**, *303,* 177.

20. Cork, D.G.; Hayashi, N. *Bull.Chem.Soc.Jpn.* **1993**, *66,* 1583.

21. Gilliom, R.D. *J.Compt.Chem.* **1985**, *6,* 437.

22. Pietrzycki, W.; Tomasik, P. *J.Chem.Phys.* **1993**, *90,* 2055.

23. Ménard, R.; Storer, A. C. *Biol. Chem. Hoppe-Seyler* **1992**, *373,* 393-400.

24. Warshel, A.; Naray-Szabo, G., Sussman, F.; Hwang, J.-K. *Biochemistry* **1989**, *28,* 3629-3637.

25. Dewar, M. J. S.; Storch, D. M. *Proc. Natl. Acad. Sci. USA* **1985**, *82,* 2225-2229.

26. Dewar, M. J. S. *Enzyme* **1986**, *36,* 8-20.

27. Wescott, C. R.; Klibanov, A. M. *Biochim. Biophys. Acta* **1994**, *1206,* 1-9.

28. Pitzer, K.S.; Brewer, L. *Thermodynamics*; McGraw-Hill: New York, 1961.

29. Menger, F. M. *Biochemistry* **1992**, *31,* 5368-5373.

30. Demuth, H-U. *J. Enzyme Inhibition,* **1990**, *3,* 249-278.

31. Shokhen, M.; Arad, D. *J. Mol. Model.* **1996**, *2,* 399-409.

32. Liang, T.-C.; Abeles, R. H. *Biochemistry* **1987**, *26,* 7603-7608.

33. Brady, K.; Liang, T.-C.; Abeles, R. H. *Biochemistry* **1989**, *28,* 9066-9070.

34. Brady, K.; Wei, A.; Ringe, D.; Abeles, R. H. Biochemistry **1990**, 29, 7600-7607.

35. Carter, P.; Wells, J. A. *Proteins: Struct., Funct., Genet.* **1990**, *7,* 335 - 342.

36. Braxton, S.; Wells, J. A. *J. Biol. Chem.* **1991**, *266,* 11797-11800.

37. McMurray, J. S.; Dyckes, D. F. *Biochemistry* **1986**, *25,* 2298-2301.

38. Frisch, M.J.; Head-Gordon, M.; Trucks,G.W.; Head-Gordon, M.; Gill, P.M.W.; Wong, M.W.; Foresman, J. B.; Johnson, B. G.; Schlegel, H. B.; Robb, M. A.; Replogle, E. S.; Gomperts, R.; Andres, J. L.; Raghavachari, K.; Binkley, J. S.; Gonzalez, C.; Pople, J. A. *Gaussian 92, Revision A,* Gaussian, Inc., Pittsburgh PA, 1992.

39. Hwang, J.-K.; Warshel, A. *Biochemistry* **1987**, *26,* 2669-2673.

40. Warshel, A.; Aqvist, J. *Annu. Rev. Biophys. Biophys. Chem.* **1991**, *20,* 267-284.

41. Page, M. I. *Chem. Soc. Revs.* **1973**, *2,* 295-323.

42. Dewar, M. J. S. *The Molecular Orbital Theory of Organic Chemistry*; McGraw-Hill: New York, 1969.

43. Williams, I. H.; Spangler, D.; Femec, D. A.; Maggiora, G. M.; Schowen, R. L. *J. Am. Chem. Soc.* **1983**, *105,* 31-40.

44. Williams, I. H. *J. Am. Chem. Soc.* **1987**, *109,* 6299-6307.

45. Madura, J. D.; Jorgensen W. I. *J. Am. Chem. Soc.* **1986**, *108,* 2517-2527.

46. Stewart, J.J.P. in *Reviews of Computational Chemistry,* Lipkovitz, K.; Boyd, D.B., Eds. VCM Publishers, New York, 1990, 45 - 81.

47. Howard, A. E. ; Kollman, P. A. *J. Am. Chem. Soc.* **1988**, 110, 7195 - 7200.

48. Feller, D.; Davidson, E.R. in *Reviews of Computational Chemistry,* Lipkovitz, K.; Boyd, D.B., Eds. VCH Publishers, New York, 1990, 1 – 43.

49. Here, J.W. Practical Strategies for Electronic Structure Calculations, Wavefunction (Irvine, California, USA) 1995.

50. Bender, M. L.; Kezdy, F. J.; Gunter, C. R. *J. Am. Chem. Soc.* **1964**, *86,* 3714-3721.

51. Abeles, R. H.; Alston, T. A. *J. Biol. Chem.* **1990**, *265,* 16705-16708.

52. Kettner, C.; Shenvi, A. B. *J. Biol. Chem.* **1984**, *259,* 15106-15114.

53. Bone, R.; Shenvi, A. B.; Kettner, C. A.; Agard, D. A. *Biochemistry* **1987**, *26,* 7609-7614.

54. Jansen, E. J.; Nutting, M. D. F.; Jang, R.; Ball, A. K. *J. Biol. Chem.* **1949**, *179,* 189-199.

55. Jansen, E. J.; Nutting, M. D. F.; Ball, A. K. *J. Biol. Chem.* **1949**, *179,* 201-204.

56. Bender, M. L; Wedler, F. C. *J. Am. Chem. Soc.* **1972**, *94,* 2101-2109.

57. Harel, M.; Su, C.-T.; Frolow, F.; Ashani, Y.; Silman, I.; Sussman, J. L. *J. Mol. Biol.* **1991**, *221,*909-918.

J. Mol. Model. **1996**, 2, 399 – 409

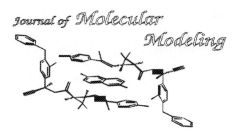

The Source for the Difference Between Sulfhydryl and Hydroxyl Anions in Their Nucleophilic Addition Reaction to a Carbonyl Group: A DFT Approach.

Michael Shokhen [‡][*] **and Dorit Arad** [$][*]

 The George S. Wise Faculty of Life Sciences, Tel-Aviv University, Ramat Aviv, Israel 39040, Tel: 972-3-6408723 FAX: 972-3-6409407

[‡] Department of Biochemistry (shokhen@etgar.tau.ac.il)

[$] Department of Molecular Biology and Biotechnology (dorit@argaman.tau.ac.il)

Received: 4 April 1996 / Accepted: 10 September 1996 / Published: 11 October 1996

Abstract

Ab initio calculations show that sulfhydryl anion has a significantly lower potential than the hydroxide anion for stabilizing the products of its attack on carbonyl moieties – the tetrahedral complexes (*TC*). In this paper we analyze the factors that contribute to this phenomenon. Quantum mechanical MO *ab initio* calculations were used for studies of two reaction series, one for the attack of hydroxyl and one for the attack of sulfhydryl anion on different carbonyl compounds and their analogs. All of the anionic *TC*s formed by HS⁻ are characterized by higher charge transfer, but are significantly less stable than the relevant *TC* of HO⁻. To explain the phenomenon we used a simple qualitative model based on Density Functional Theory (DFT). The crucial role of the occupied valence MOs is demonstrated in the process of electronegativity equalization between the donor and acceptor fragments in the final *TC* product. The sulfhydryl anion has significantly lower potential to stabilize *TC* products in comparison with the hydroxide anion because of the larger extent of electron back-donation from the electrophile's $HOMO_A$ to the nucleophile's $LUMO_D$. This electron back-donation thus reduces the stability of the anionic *TC* in the case of HS⁻ and may account for the calculational results. Applications of this work to enzyme reactions help in understanding the differences in mechanisms of serine and cysteine proteases and may be used to guide the design of inhibitors for these enzymes. In perspective, the back-donation phenomenon discussed here may be applied to the study of electron transfer processes involving oxidation-reduction enzymes.

Keywords: Nucleophilic addition to carbonyl group, Sulfhydryl anion, Hydroxyl anion, Tetrahedral Intermediate, Tetrahedral Complex, Density Functional Theory, MO *ab-initio* calculations.

---

* To whom correspondence should be addressed

Introduction

Nucleophilic addition of an anionic nucleophile to a carbonyl group accompanied by $sp^2 \Rightarrow sp^3$ rehybridization of the atomic orbitals (AOs) on an electrophilic center results in the formation of a tetrahedral product, which in most cases is a reactive intermediate [1,2]. Hydrolysis of amides and esters are representative reactions of this type [3-5], and are the key process of the enzymatic catalysis provided by proteases [6]. The widely accepted terminology for the tetrahedral product is Tetrahedral Complex *(TC)* [6]. Howard and Kollman [7] showed that in its attack on a carbonyl species in the gas phase, a sulfhydryl nucleophile (SH⁻ or SR⁻) does not form a stable covalent tetrahedral adduct. We address this paper to the question of why a sulfur nucleophile behaves differently than oxygen during nucleophilic attack on a carbonyl center. This understanding is important in relation to differentiation between hydrolytic enzyme mechanisms in serine and cysteine proteases [8]. The knowledge gained will provide clues for designing selective, mechanistic-based inhibitors for serine and cysteine proteases. Several attempts have been described to explain the different behavior of the hydroxyl and sulfhydryl nucleophiles. Cardy et al [9] applied the model of Heilbronner [10], who attempted to use molecular orbital (MO) theory to resolve this problem. However, the authors, considered only the occupied orbitals and neglected the charge-transfer process connected with the unoccupied MOs of the acceptor. The chemical nature of *TC* is predominantly covalent. Creation of a *TC* is accompanied by considerable donor-acceptor charge transfer, as will be discussed in detail below. A different interpretation [11] of the relative stability of carbanions and their isoelectronic analogs was based on hyperconjugation [12] and the anomeric effect, [11] where only frontier orbitals were considered. Neither approach alone can provide a comprehensive understanding of the reactions of hydroxyl and sulfhydryl anions with carbonyl species.

In this paper, we combine certain qualitative aspects of the molecular orbital approach and density functional theory (DFT) to analyze and explain the factors governing the *ab initio* calculated stability of *TC* products that are formed during nucleophilic attack on a carbonyl group by hydroxyl or sulfhydryl anions.

Results and Discussion

Two reaction series were calculated, one for the attack of hydroxyl and one for the attack of sulfhydryl on different carbonyl compounds and their analogs. Performing the calculations on a gas-phase model eliminates the effects of the media, so that intrinsic reactivity can be determined solely by the electronic and structural nature of the nucleophile and the carbonyl substrate. All structures were fully optimized at the HF/6-31+G*// HF/6-31+G* level, using the Gaussian 92 program [13].

Several geometrical and electronical changes are associated with the formation of a stable *TC*.

(a) The $sp^2 \Rightarrow sp^3$ rehybridization of the atomic orbitals (AOs) on an electrophilic center causes its pyramidalization and elongation of covalent bonds.

(b) The newly formed bond between the reactive centers (which is partially covalent and partially ionic) is usually longer and consequently weaker than an equivalent ordinary covalent bond [14], (see a comparison between calculated and corresponding experimental values of bond lengths in Table 1).

(c) A significant amount of charge transfer occurs from the nucleophile to the electrophile (shown in Table 1). The general reaction scheme for *TC* formation is presented in Fig. 1.

Energies

The reaction energy (E_{react}) for the formation of the tetrahedral covalent product – *TC* was calculated using eq 1.

$$E_{react} = E_{DA} - (E_D^o + E_A^o) \qquad (1)$$

where E_D^o and E_A^o are the total energies of the isolated donor and acceptor molecules, and E_{DA} is the total energy of the nucleophilic addition product, the *TC*.

The original observations of Howard and Kollman [7] demonstrated at the *ab initio* MP2-FC/6-31*G//RHF/4-31G level of calculations that the hydrosulfide anion cannot form tetrahedral covalent adducts with formamide and formaldehyde. Our geometry optimizations using a higher level basis set including diffuse functions agree with their results. We found that for several substrates– formaldehyde $CH_2=O$, cyclopropenone $(CH)_2C=O$ and urea $(NH_2)_2C=O$ there is no minimum on the reaction potential surface corresponding to a *TC*. All tetrahedral complexes formed by HS⁻ are characterized by larger values of charge transfer, (see *ab initio* calculated values in Table 1) but are significantly less stable

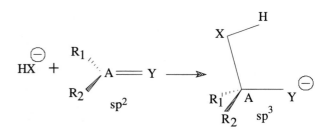

Figure 1. *The principal scheme of the formation of a TC from reagents. HX⁻ is the anion (HO⁻ or HS⁻), and R1 and R2 denote the variable substituents at the electrophilic center A on the carbonyl–like group. The varied atom A here is C or Si, and Y is O or S. The reaction is accompanied by the $sp^2 \Rightarrow sp^3$ rehybridization of the electrophilic center A.*

Table 1. *Electronic and geometric characteristics of a stable TC.*

no.	Tetrahedral complex (TC)	Values of charge-transfer			Charges on reactive centers				X–A [a,b]
		ab initio Mulliken	*ab initio* NAO	eq. 2	Mulliken		NAO		
					X	A	X	A	
1	HO–CH$_2$–O$^-$	0.6417	0.5563	0.3720	-0.8337	0.2381	-0.9257	0.3584	1.470
2	HO–CF$_2$–O$^-$	0.6053	0.6348	0.3883	-0.8893	1.3604	-0.8623	1.3537	1.396
3	HO–CH$_2$–S$^-$	0.7168	0.6261	0.4124	-0.7728	-0.0734	-0.8708	-0.2104	1.428
4	HO–(CH)$_2$C–O$^-$	0.6865	0.6107	0.3733	-0.8002	0.4608	-0.8818	0.6563	1.438
5	HO–(NH$_2$)$_2$C–O$^-$	0.6618	0.5823	0.2683	-0.8283	0.6777	-0.9073	0.9485	1.459
6	HO–(NH$_2$)$_2$C–S$^-$	0.7827	0.6401	0.3801	-0.7229	0.1011	-0.8654	0.4687	1.421
7	HO–SiH$_2$–O$^-$	0.4639	0.2914	0.4207	-1.0296	1.1723	-1.2062	1.9172	1.705
8	HO–SiH$_2$–S$^-$	0.5670	0.3140	0.4346	-0.9354	0.8547	-1.1943	1.4206	1.692
9	HS–CF$_2$–O$^-$	0.8109	0.8634	0.3623	-0.2650	0.7802	-0.2492	1.0459	1.915
10	HS–CH$_2$–S$^-$	0.8813	0.9634	0.3832	-0.2286	-0.4190	-0.1702	-0.7127	1.861
11	HS–(NH$_2$)$_2$C–S$^-$	0.8746	0.9669	0.3428	-0.2160	-0.0346	-0.1640	0.1308	1.897
12	HS–SiH$_2$–O$^-$	0.5673	0.4821	0.3959	-0.5042	0.8339	-0.6445	1.5987	2.248
13	HS–SiH$_2$–S$^-$	0.6420	0.55731	0.4087	-0.4476	0.6188	-0.5814	1.0107	2.225

[a] *The experimental values [14] of bond lengths for the relevant ordinary covalent bonds in Å are: O–C (1.42), O–Si (1.63), S–C (1.81), S–Si (2.14).*

[b] *The bond lengths in Å for the relevant ordinary covalent bonds in neutral species ab initio calculated in this work with full geometry optimization in the 6-31+G* basis set are O–C (1.382) in HOCH$_2$OH; O–Si (1.637) in HOSiH$_2$OH; S–C (1.803) in HSCH$_2$OH; S–Si (2.134) in HSSiH$_2$OH.*

than the relevant tetrahedral products of HO$^-$, as is shown in Table 2.

Density Functional Theory

Our strategy emerged from analyzing the MOs of the *TC* and combining the results with simple concepts from DFT that relate to donor-acceptor (D-A) interactions [15,16]. The charge transfer energy ΔE_{CT} can be expressed either as a function of electronegativity and hardness or as the energy gap between the donor-acceptor frontier orbitals [15,16]. Hence, a quantitative functional relation between the stabilization energy E_{react} resulting from charge transfer and an expression of the energies of HOMO-LUMO orbitals of the reactants can be obtained. The concept and the novelty of our approach lies in a surprising result, presented in the following section, which we obtained by straight forward merging of two DFT equations. The result is that back donation of

electrons occur from the electrophile's HOMO$_A$ to the nucleophile's LUMO$_D$. The extent of back donation relates on the nucleophile's type and affects the stability of the tetrahedral product.

The value of the donor-acceptor charge transfer ΔN_{CT} for the simplest DFT model, using a fixed external potential, is presented in eq 2 [15,16].

$$\Delta N_{CT} = (\chi_A - \chi_D) / 2(\eta_A + \eta_D) \qquad (2)$$

The energy-change term ΔE_{CT} contributed by the charge transfer can be calculated according to eq 3 [15,16].

$$\Delta E_{CT} = (E_A - E_A^0) + (E_D - E_D^0) =$$

$$-1/4 (\chi_A - \chi_D)^2 / (\eta_A + \eta_D) \qquad (3)$$

where χ_D, η_D and χ_A, η_A represent the absolute electronegativity and hardness for the donor D and the acceptor A. E_D^0, E_A^0 are the respective ground-state energies of the isolated donor and acceptor, and E_D, E_A are the respective energies of the donor and acceptor in the new valence states corresponding to the complex, D:A. Equation 2 shows that (a) the differences in electronegativity drive the electron transfer, and (b) the sum of the absolute hardness inhibits electron transfer.

Table 2. *Reaction energies E_{react} of the TCs and the values of energy gaps α and β between the frontier MOs of the reagents.*

no.	Tetrahedral complex (TC)	E_{react} [kcal/mol]	α a.u.	β a.u
1	$HO–CH_2–O^-$	–35.2	0.1792	0.8608
2	$HO–CF_2–O^-$	–69.8	0.1697	0.9772
3	$HO–CH_2–S^-$	–69.6	0.1389	0.7701
4	$HO–(CH)_2C–O^-$	–24.6	0.1618	0.8107
5	$HO–(NH_2)_2C–O^-$	–19.4	0.1666	0.8323
6	$HO–(NH_2)_2C–S^-$	–41.4	0.1518	0.7359
7	$HO–SiH_2–O^-$	–111.9	0.1258	0.8638
8	$HO–SiH_2–S^-$	–118.7	0.1118	0.7802
9	$HS–CF_2–O^-$	–9.8	0.1606	0.8203
10	$HS–CH_2–S^-$	–22.2	0.1298	0.6132
11	$HS–(NH_2)_2C–S^-$	15.4	0.1427	0.5790
12	$HS–SiH_2–O^-$	–56.2	0.1167	0.7069
13	$HS–SiH_2–S^-$	–62.0	0.1027	0.6233

The following operational definitions of χ and η are useful for calculating several experimental values [15,16]:

$$\chi = (I + A) / 2 \qquad \eta = (I - A) / 2 \qquad (4)$$

where I is the ionization potential, and A is the electron affinity of the system. I and A can be extracted directly from the output of quantum mechanical calculations.

According to Koopman's theorem [17], eq. 4 may be rewritten in terms of the energies of frontier orbitals (HOMO and LUMO) [18], where the frontier orbitals are given by:

$$-\varepsilon_{HOMO} = I \quad \text{and} \quad -\varepsilon_{LUMO} = A.$$

$$\chi = -(\varepsilon_{HOMO} + \varepsilon_{LUMO}) / 2; \quad \eta = (\varepsilon_{LUMO} - \varepsilon_{HOMO}) / 2 \qquad (5)$$

The values of χ and η estimated in eq 5, however, are only rough estimates compared with the values in eq 4. Therefore, before using eq 5 for further applications, we compared the values of χ and η derived by both eq 4 and eq 5 for our series of calculated reagents. The results are presented in Table 4. I and A used in eq 4 were determined according to eq 6 [15,16].

$$I = E(N – 1) – E(N); \qquad A = E(N) – E(N + 1) \qquad (6)$$

$E(N)$ is the total energy of a starting system with N electrons and charge q; $E(N-1)$ and $E(N+1)$ are the total energies of the same species going (on ionization) from the charge q to the charge $q+1$ or to the charge $q-1$ (accepting one additional electron), respectively. The calculations were done with a frozen geometry for the initial system (vertical I and A values). All energetic values were estimated with second order Møller-Plesset correlation corrections implemented in the standard Gaussian 92 program [13], using restricted Hartree-Fock calculations MP2//RHF/6-31+G* for closed shell - and unrestricted MP2//UHF/6-31+G* for open shell species. As can be seen from Table 4, the two sets of values for χ and η are well correlated, so we may use eq 5 for our development. Hence, a quantitative function of the relation between the value of ΔE_{CT} resulting from charge transfer and an expression of the energies of HOMO-LUMO orbitals of the reactants can be obtained.

The value of ΔE_{CT}, derived from eq 3, is always negative and roughly approximates the stabilization energy E_{react} of TC, as calculated by eq 1. E_{react} comprises several components: charge-transfer energy, the energy of formation of the covalent bond between the reactive centers, and the nuclear repulsion energy.

We use eq 3 as a working tool for a qualitative analysis of how the intrinsic properties of distinct isolated chemical species D and A (energies of their frontier MOs) determine the

Table 3. *Values of ε_{HOMO}, ε_{LUMO}, I, A (in a.u.) calculated for the reagents.*

no.	Comp.	ε_{HOMO}	ε_{LUMO}	I [a]	A [a]
1	HO^-	–0.10333	0.41779	0.0589	–0.3902
2	HS^-	–0.09416	0.26090	0.0660	–0.2444
3	$CH_2=O$	–0.44295	0.07593	0.4046	–0.0577
4	$CF_2=O$	–0.55939	0.06644	0.5047	–0.0536
5	$CH_2=S$	–0.35227	0.03559	0.3373	–0.0110
6	$(CH)_2C=O$	–0.39286	0.05850	0.3593	–0.0498
7	$(NH_2)_2C=O$	–0.41446	0.06333	0.3762	–0.1722
8	$(NH_2)_2C=S$	–0.31813	0.04853	0.2911	–0.0340
9	$SiH_2=O$	–0.44595	0.02247	0.4075	–0.0098
10	$SiH_2=S$	–0.36239	0.00849	0.3396	0.0079

[a] The values of I and A are calculated by eqs. 6, see details in the text.

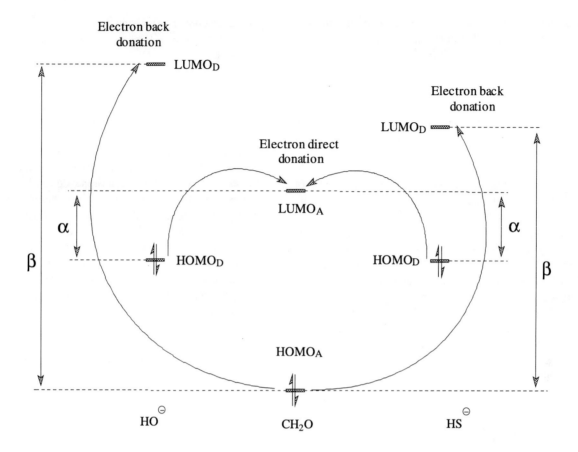

Figure 2. *A scheme describing the directions of charge transfer by the frontier orbitals of the reagents. Direct charge transfer occurs from the HOMO$_D$-nucleophile to LUMO$_A$-electrophile, some extent of charge transfer occurs in the opposite direction from the HOMO$_A$-electrophile to LUMO$_D$-nucleophile. The amount of electron back-donation is an important factor in determining the stability of the TC. The difference in orbital levels between SH$^-$ and OH$^-$ is reflected by the a and b parameters, and determines the value of the ΔE_{CT} affecting the TC stabilization.*

stability of combined system *D:A* (*TC*). By combining eqs. 2 and 3 with eq 5, we can obtain an expression for *DN* (eq 7).

$$\Delta N_{CT} = (\beta - \alpha) / 2(\beta + \alpha) \qquad (7)$$

The expression for ΔE_{CT} is directly derived from ΔN_{CT} in eq 8.

$$\Delta E_{CT} = -1/8 \, (\beta - \alpha) \qquad (8)$$

where $\alpha = (\varepsilon^A_{LUMO} - \varepsilon^D_{HOMO})$, $\beta = (\varepsilon^D_{LUMO} - \varepsilon^A_{HOMO})$; and ε^D_{LUMO}, ε^D_{HOMO} and ε^A_{LUMO}, ε^A_{HOMO} are the respective frontier orbital energies for the initial donor *D* and acceptor *A*.

The values for α and β are positive, and because $\beta > \alpha$, the calculated value of ΔE_{CT} in eq 8 is always negative. According to perturbational MO (PMO) theory, the stability of

D:A is inversely proportional to the value of the energy gap $\varepsilon^A_{LUMO} - \varepsilon^D_{HOMO}$ (α in eqs 7 and 8) [19]. The same result is obtained by DFT, where the stability of *D:A* increases as the value of *a* decreases (see eq. 8 and Fig. 2).

Equation 8, in contrast to the well known nonlinear Eq. 3, has a very simple mathematical form – it demonstrates the linear dependence of the value of ΔE_{CT} on the gaps between energy levels of reagents' frontier MOs – α and β. This is a big advantage for practical applications. It is important to emphasize that eq 8 provides a new interpretation of frontier orbital control of the *D:A* stability because it takes into consideration the simultaneous charge transfer in the opposite direction (LUMO$_D$ $\Leftarrow$ HOMO$_A$ from the acceptor's HOMO to the donor's LUMO) as well. Such "back-donation" of electrons decreases the stability of the *TC*. Therefore, the stability of the *TC* becomes higher as the energy gap $\varepsilon^D_{LUMO} - \varepsilon^A_{HOMO}$ (β) increases because of the back-electron transfer from the acceptor to the donor (D $\Leftarrow$ A) is smaller. Thus, the result that emerges from the mathematical development described above (Eq. 7-8), is that electron back donation from the HOMO$_A$ of the carbonyl group to the LUMO$_D$ of the nucleophile is not a negligible component according to DFT, and because of this it has a role in destabilizing the products. This conclusion has a general implication because, to the best of our knowledge, prior interpretations [20] of donor-acceptor interactions in terms of frontier orbitals have considered only the direct charge transfer (HOMO$_D$ $\Rightarrow$ LUMO$_A$). The validity of this development is shown by checking the linear correlation between the energy gaps of the frontier MO's - values of α and β discussed above, and the E_{react} for the stable *TC*'s calculated by *ab initio* (eq. 9).

$$E_{react} = 1702.46 \ \alpha - 333.91 \ \beta - 35.17, \qquad (9)$$

R = 0.937, F = 35.8, Standard Error = ± 14.9 kcal/mol, observations = 13. The E_{react} values are presented in kcal/mol and α and β – in the a.u.

The signs of the coefficients reflect the trend that would indicate in which direction α and β influence the value of E_{react}. The signs of the coefficients for E_{react} turn out to be the same as the signs for the α and β parameters established for ΔE_{CT} in eq 8. Consequently, the *ab initio* calculated *TC* stability (E_{react}) and the donor-acceptor charge transfer energies (ΔE_{CT}), determined from the simple qualitative model based on DFT, are governed by the same electronic factors. The conclusion is that the charge transfer process (ΔE_{CT}) is the main energy component that reflects the different abilities of SH⁻ and OH⁻ to stabilize the anionic *TC* species. Figure 3 presents a graphical comparison of the E_{react} values calculated by *ab initio* and estimated in Eq. 9.

The orbital energies of the $HOMO_D$ orbitals of SH⁻ and OH⁻, which reflect the value of the direct charge transfer, are very close (see Table 3). The energy of the SH⁻ $HOMO_D$ is even slightly higher (by 0. 011 a.u.) than that of OH⁻. Because the difference is negligible, however, the stability of the *TC*s formed by HO⁻ and HS⁻ are expected to be similar. On the other hand, the orbital energy difference between the $LUMO_D$ orbitals of SH⁻ and OH⁻ is very large: the SH⁻ $LUMO_D$ is 0. 16 a.u. lower (6-31+G*) than that of OH⁻. The contribution of the $LUMO_D \Leftarrow HOMO_A$ back donation significantly increases as the energy gap between the relevant orbitals decreases (see eqs. 8 and 9). Because the energy gap is much smaller for the sulfur nucleophile (see Table 3), the main reason for destabilization of the *TC* by SH⁻ is the in-

creased amount of back donation involved in this type of attack. Stable *TC*s will be obtained when the electrophilic group has a low-lying $HOMO_A$ orbital (as in CF₂O).

In the present work we have analyzed only the energy levels of the interacting reagent's MOs, because we were interested exclusively in the process of charge transfer between nucleophile and electrophile during the formation of a *TC*. We demonstrated that the significance of this phenomena for the stabilization of a tetrahedral product emerges only if one takes into consideration mutual directions of the charge transfer: donor-acceptor and acceptor-donor. In such a case the general trend in product stability for the relevant reaction can be well predicted in a simple linear correlation equation. Nevertheless, we would like to stress that for a better understanding of the full picture and to improve quantitative estimations of the E_{react} values in any linear correlation like Eq.9, one must consider also the overlap interactions between AOs of reactive centers of reagents creating the covalent bond in a *TC* product (or/and electron populations on these atoms). The carbonyl like bond A=Y of a reagent due to the more electronegative atom Y is polarized. As a result, the orbital lobe on the atom A in the doubly occupied π_A MO decreases. This causes the reduction of the value of overlap integrals between the relevant AOs of the reactive centers and the formation of a weaker covalent bond, which is the reason of the general elongation of this bond in *TC*s observed here in comparison with the equivalent ordinary covalent bonds (see Table 1). According to the Hard and Soft Acids and Bases (HSAB) principle, overlap interactions play a much more significant role for the soft base HS⁻ than for the hard HO⁻ [16]. Our *ab initio* calculated values of E_{react} confirm this statement. Table 2 demonstrates that HS⁻ stabilizes the anionic *TC* much better for the soft substrate H₂CS (-22.2 kcal/mol) with a weakly polarized bond C=S, than for the hard F₂CO (-9.8 kcal/mol), where the central carbonyl carbon is surrounded by three extremely electronegative substituents

ΔE_{react} (ab initio)-Series 1 vs. ΔE_{react} (regression)-Series 2

Figure 3. *Graphical comparison of the* E_{react} *values calculated by ab initio and estimated in Eq. 9.*

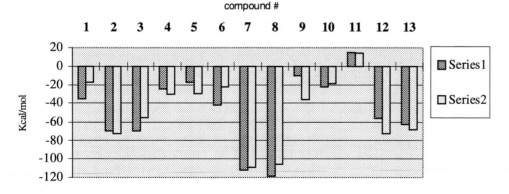

– F and O. In contrast, the relevant values of E_{react} are indistinguishable for the hard base HO⁻, where the charge transfer due to the interaction of frontier reagent's MOs dominates (see Eqs. 7 and 8, and Table 1 for the values of ΔN_{CT} calculated by Eq. 2). The values are -69.6 kcal/mol for H_2CS, and -69.8 kcal/mol for F_2CO. Thus, the general tendency observed here– much more stable *TC*s are formed for the HO⁻ series – is regulated by the charge transfer process.

An open question remains, however. As emerges from the qualitative approach of DFT, the values of ΔE_{CT} and ΔN_{CT} are proportional (eqs. 2 and 3). Consequently, because they are more stable for OH⁻ than for SH⁻, *TC*s derived from a hydroxy nucleophile attack are expected to have a larger charge transfer value than those derived by an SH⁻ attack. Equation 2 indeed confirmed this conclusion for all the calculated series (see Table 1). The charge transfer values derived from *ab-initio* calculations, on the other hand, show the opposite trend: Table 1 shows that in all calculated structures, the charge transfer values, as extracted from the Mulliken population analysis [21] and the Natural Atomic Orbitals (NAO) population analysis [22] (both of which are implemented in the GAUSSIAN-92), are higher for the less-stabilized sulfur nucleophile. Thus, a discrepancy occurs between the accurate MO *ab-initio* and the simplified model of donor-acceptor interactions [15,16] based on DFT used here.

Participation of inner orbitals in the interaction.

To explain this discrepancy we should examine the orbital picture of D-A interactions. To derive eqs. 7 and 8, we used Koopman's theorem, which provides a way to estimate of electronegativity and hardness through the energies of frontier orbitals. The frontier-orbital approach characterizes mainly the *global* intermolecular donor-acceptor charge transfer process, related to the reacting molecules as a whole. On the other hand, as shown by Parr et al. [23], all molecular orbitals of the reagents, including their inner shells, should be involved in the process of interatomic charge transfer between the reactive centers and their orbital electronegativities should be equalized. Therefore, the frontier-orbital approach only partially reflects the picture of electron-density redistribution in a *TC*. To obtain a more complete picture, the doubly occupied, valence-shell orbitals should also be taken into account. We must take into consideration the *local* electron redistribution between the nucleophilic and electrophilic centers in the *TC*, where the major role belongs to the reagents' inner occupied orbitals. Summarizing the process of electronegativty equalization between all MO's of reagents that combine into the product, we can identify two channels for electron density redistribution between reagents: *global* – originating in the reagent's frontier MOs interactions, and *local*, caused mainly by the electronegativity equalization between AOs of reactive centers. Several electronegativity equalization formulations have been proposed for calculating the partial charges of atoms in molecules [24]. For our qualitative picture, however, we shall use only the basic idea that the electron flow on a polar bond is directed from the atom with low χ to the atom with high χ and that the value of the charge transfer is proportional to the difference in their electronegativities. Yet, such a scheme is convenient for a qualitative systematization of the dominant factors influencing the process of charge transfer. The *local* channel of electron-density redistribution can be characterized by the absolute atomic electronegativities [25] of the reactive centers. Thus, the value of the total charge transfer ΔN_{CT} may be expressed through its components ΔN_{global} and ΔN_{local} in symbolic form as shown in eq 10.

$$\Delta N_{CT} = \Delta N_{global} + sign[\chi_{at}(A) - \chi_{at}(D)]\, \Delta N_{local} \qquad (10)$$

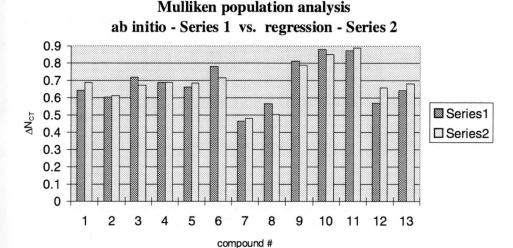

**Mulliken population analysis
ab initio - Series 1 vs. regression - Series 2**

Figure 4. *Graphical comparison of the values ΔN_{CT} (ab initio) derived from the Mulliken population analysis with linear regression estimation by eq. 11.*

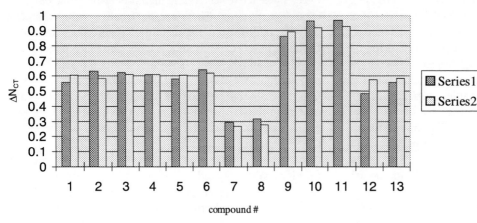

Figure 5. *Graphical comparison of the values ΔN_{CT} (ab initio) derived from the NAO population analysis with linear regression estimation by eq. 12.*

where $\chi_{at}(A)$ and $\chi_{at}(D)$ are the respective absolute atomic electronegativities [25] of the D and A reactive centers. To check the validity of eq 10, we performed a multiple regression analysis that correlates calculated *ab initio* values of ΔN_{CT} with the values of α and β characterizing the frontier orbital control on the global charge transfer (ΔN_{global}) and the difference in absolute atomic electronegativities of reaction centers – $[\chi_A - \chi_D]$, which depends mainly on the features of occupied inner and valence MOs and characterizes the local charge transfer – ΔN_{local}. We used two sets of *ab initio* calculated charge transfer values ΔN_{CT} derived from Mulliken (eq. 11) and NAO (Eq.12) population analysis (see Table 1).

$$\Delta N_{CT} \ (ab\ initio) =$$

$$1.661 \ \alpha - 0.548 \ \beta + 0.080 \ [\chi_A - \chi_D] + 0.966 \qquad (11)$$

Multiple R = 0.931; F=19.4; Standard Error = ±0.053; Number of observations = 13.

$$\Delta N_{CT} \ (ab\ initio) =$$

$$0.206 \ \alpha - 0.148 \ \beta + 0.218 \ [\chi_A - \chi_D] + 0.973 \qquad (12)$$

Multiple R = 0.979; F= 68.3; Standard Error = ± 0.050; Number of observations = 13.

Eqs. 11 and 12 show good correlations between ΔN_{CT} (*ab initio*) and the reagents' electronic structure parameters characterizing ΔN_{global} and the ΔN_{local}. Thus, we may conclude that our idea, expressed in eq. 10, about the two operating channels for the electronic density redistribution between combining reagents looks reasonable. This resolves the discrepancy posed above.

At the end of this discussion we would like to stress that the NAO variant of population analysis demonstrates much

better applicability (compare the regression quality in Eqs. 11 and 12) for the description of the donor-acceptor charge transfer processes in comparison with the traditional Mulliken approach. The graphical comparison of the values ΔN_{CT} (*ab initio*) derived from the Mulliken and NAO population analysis with linear regression estimations by eqs. 11 and 12 are presented in Figures 4 and 5, respectively.

The analysis of the computational results for the charge distribution on the reactive centers for different electrophiles reacted with SH⁻ and OH⁻ is described below. The results illustrate the validity of eq. 10. For the calculated series of *TC*s formed by an OH⁻ nucleophile, we observe that the ΔN_{local} charge transfer on the newly formed bond is always directed

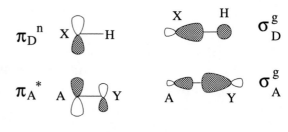

Figure 6. *Examples of the lobal structure of the MOs of the reagents participating in the overlap interactions between the reactive centers – nucleophilic X and electrophilic A. The overlap has two symmetry types: σ (for the HOMO_D and LUMO_A) and π (for the doubly occupied σ_D^g and σ_A^g MOs of the reagents' valence shells) in the local coordinates of the newly forming bond X–A in the TC.*

Table 4. *Values of* χ, η *[a] calculated for the reagents.*

no.	Comp.	$\chi = 1/2(I + A)$	$\chi = -(\varepsilon_{HOMO} + \varepsilon_{LUMO})/2$	$\eta = 1/2(I - A)$	$\eta = (\varepsilon_{LUMO} - \varepsilon_{HOMO})/2$
1	HO$^-$	−0.1657	−0.1573	0.2246	0.2606
2	HS$^-$	−0.0892	−0.0834	0.1552	0.1776
3	CH$_2$=O	0.1735	0.1836	0.2312	0.2595
4	CF$_2$=O	0.2256	0.2465	0.2792	0.3129
5	CH$_2$=S	0.1632	0.1584	0.1742	0.1940
6	(CH)$_2$C=O	0.1548	0.1672	0.2046	0.2257
7	(NH$_2$)$_2$C=O	0.1020	0.1756	0.2742	0.2389
8	(NH$_2$)$_2$C=S	0.1286	0.1348	0.1626	0.1833
9	SiH$_2$=O	0.1989	0.2118	0.2087	0.2343
10	SiH$_2$=S	0.1738	0.1770	0.1659	0.1855

[a] All values of χ and η (in a.u.) are calculated with the molecular geometries optimized at 6–31+G* level.

from the electrophilic center to the attacking, highly electronegative oxygen atom (absolute atomic electronegativity = 7. 54 [25]). For the parallel set of *TC*s formed by HS$^-$, the local interatomic charge transfer ΔN_{local} should be either absent or directed in the opposite direction – from sulfur to carbon, depending on the valent surrounding of the electrophilic center. The reason is that sulfur has an absolute atomic electronegativity that is slightly lower than that of carbon [6. 22 (S) vs. 6. 27 (C)] [25]. During the formation of a *TC* with carbon electrophiles, the total *ab-initio* calculated charge transfer ΔN_{CT} is expected to be considerably larger for HS$^-$ than for HO$^-$. It is found that for the carbon electrophile reactions, the charge on the nucleophilic atom on the *TC* (oxygen or sulfur) is always more negative on the oxygen derivative than on the sulfur derivative (see values of the NAO analysis in Table 1). The charge on oxygen in compound **2** is -0.8623 compared with -0.2492 on sulfur in compound **9**. In principle, in the case of the low electronegative silicon atom (4. 77) [25] as the electrophile center, the charge transfer picture should differ from carbon. Indeed, the observed values of the negative charge for both oxygen and sulfur nucleophiles in *TC*s is significantly smaller for carbon electrophiles than for silicon electrophiles (Table 1) (see compounds **10** and **13** for sulfur and compounds **1** and **7** for oxygen).

It is interesting to speculate about the "geometrical" aspect of the electron charge transfer between reagents. Because of a local symmetry, at least two orthogonal types of overlap interactions, σ and π, occur in a new bond forming between the orbitals of the reactive centers. Thus, the global and local redistribution of electrons between the donor and the acceptor should be expressed through two parallel channels of orbital interactions, with a different local symmetry type that depends on the bond between the reactive centers. One channel (global), shown in eq 2, is characterized by the total electronegativity and hardness of the reagents, which are the global parameters of interacting molecules. Eq. 7 demonstrates that the electron flow occurs through the interaction of the HOMO and LUMO pairs of the donor and acceptor. Because of the geometrical features of *TC* products (see Fig. 1), where the donor fragment X-H makes a frontal attack on the bond of the acceptor A-Y, the direct donor-acceptor charge transfer channel occurring through the overlap interaction of HOMO$_D$ (π_D^n) and LUMO$_A$ (π_A^*) is of σ-symmetry in the local coordinates of the bond between *D* and *A*. The schematic lobal diagram is presented in Fig. 6.

Fig. 6 also demonstrates an example of the π-symmetry overlap interaction between the occupied σ_D^g and σ_A^g MOs of the reagents that are responsible for the covalent bonds in the donor X-H and the acceptor A=Y fragments, respectively. If the frontier orbitals operate through a σ channel ΔN_s, then logically, the alternative channel of the local electron density redistribution (ΔN_{local}) between reactive centers should be of local π-symmetry. This channel belongs to the inner orbitals of the reagents. The separation of the frontier-MO and inner-MO interactions of the reagents into different symmetrical types is an oversimplification used here only for transparency of a qualitative model.

Summary

A simplified model of donor-acceptor charge transfer processes based on the DFT was applied to the explanation of the *ab initio* calculations. The sulfhydryl anion has a significantly lower potential to stabilize *TC* products than the hydroxide anion because of the larger extent of electron back donation from the HOMO$_A$ to the LUMO$_D$ of the sulfhydryl anion.

The importance of the electron back-donation effect is a new result, derived by straightforward manipulations with DFT equations. Carbonyl derivatives, which less back-donation are predicted to form fairly stable *TC*s, even with sulfur. For example, fluorine substituted derivatives (e.g. compound **9**, table 2) are calculated to give a stable tetrahedral anionic species.

We can identify two channels for the electron density redistribution between reagents: *global* – originated from the reagent's frontier MOs interactions, and *local*, caused mainly by the electronegativity equalization between AOs of reactive centers. The direction of the *local* flow depends on the differences between the atomic electronegativities of the reaction centers in the *TC* species. If the atomic electronegativities of the reactive centers are close, as in the case of the HS⁻ nucleophile that attacks carbon electrophiles, the charge transfer through the *local* channel is practically absent. If the effective atomic electronegativity of an electrophilic center is lower in comparison with a nucleophilic center, the directions of electron flows through the *global* and *local* channels are opposite. For example, the values of the donor-acceptor charge transfer are reduced for all *TC*s in the HO⁻ series in comparison with the HS⁻ series. The charge-transfer through the *global* channel is strongly dominated over the *local* channel, so in all stable *TC*s one can always observe the net charge transfer from a donor to an acceptor.

The knowledge of the rules governing the stability of *TC* has a direct application for the design of inhibitors for serine and cysteine proteases. In perspective, the back donation phenomenon discussed here, may be applied to the study of electron transfer processes involving oxidation- reduction enzymes.

Acknowledgments The Israeli Ministry for Absorption of New Immigrants and the Ministry for Science and Education are acknowledged for their support of Dr. M. Shokhen. The authors thank Dr. Virginia Buchner for editing the manuscript.

References

1. Lowry, T. H.; Richardson, K. S. *Mechanisms and Theory in Organic Chemistry*; Harper and Row: New York, **1976**, 402 - 461.

2. March, J. *Advanced Organic Chemistry;* Wiley & Sons: New York, **1992**, 378 - 383.

3. Guthrie, J. P. *J. Am. Chem. Soc.* **1973**, *95*, 6999 - 7003.

4. Jencks, W. P. *Accounts Chem. Res.* **1980**, *13*, 161- 169.

5. Jencks, W. P. *Acc. Chem. Res.* **1976**, *9*, 425 - 432.

6. (a) Fersht, A. *Enzyme Structure and Mechanism*; W. H. Freeman and Company: New York, **1988**, 405 - 426. (b) Polgar, L.; Halasz, P. *Biochem. J.* **1982**, *207*, 1 - 10. (c) McMurray, J. S.; Dyckes, D. F. *Biochemistry* **1986**, *25*, 2298 - 2301. (d) Imperiali, B.; Abeles, R. H. *Biochemistry* **1986**, *25*, 3760 - 3767. (e) Brady, K.; Wei, A.; Ringe, D.; Abeles, R. H. *Biochemistry* **1990**, *29*, 7600 - 7607. (f) Weiner, S. J.; Seibel, G. L.; Kollman, P. A. *Proc. Natl. Acad. Sci.* **1986**, *83*, 649 -653. (g) Stamato, F. M. L. G.; Longo, E.; Ferreira, R.; Tapia, O. *J. Theor. Biol.* **1986**, *118*, 45 - 59. (h) Warshel, A.; Russell, S. *J. Am. Chem. Soc.* **1986**, *108*, 6569 - 6579. (i) Warshel, A.; Naray-szabo, G.; Sussman, F. , Hwang, J. -K. *Biochemistry*, **1989**, *28*, 3629 - 3637. (j) Schroder, S.; Daggett, V.; Kollman, P. A. *J. Am. Chem. Soc.* **1991**, *113*, 8922 - 8925. (k) Duncan, G. D.; Huber, C. P.; Welsh, W. J. *J. Am. Chem. Soc.* **1992**, *114*, 5784 - 5794.

7. Howard, A. E.; Kollman, P. A. *J. Am. Chem. Soc.* **1988**, *110*, 7195 - 7200.

8. (a) Arad, D.; Langridge, R.; Kollman, P. A. *J. Am. Chem. Soc.* **1990**, *112*, 491 - 502. (b) Arad, D.; Kreisberg, R.; Shokhen, M. *J. Chem. Inf. Comput. Sci.* **1992**, *33*, 345 - 349.

9. Cardy, H.; Loudet, M.; Policard, J.; Olliver, J.; Poquet, E. *Chem. Phys.* **1989**, *131*, 227- 243.

10. Heilbronner, E.; Schmelzer, A. *Helv. Chim. Acta* **1975**, *58*, 936 - 967.

11. (a) Hoffmann, R.; Radom, L.; Pople, J.; Schleyer, P. R.; Hehre, W. J.; Salem, L. *J. Am. Chem. Soc.* **1972**, *94*, 6221- 6223. (b) David, S.; Eisenstein, O.; Hehre, W. J.; Salem, L.; Hoffmann, R. *J. Am. Chem. Soc.* **1973**, *95*, 3806 - 3807. (c) Schleyer, P. R.; Kos, A. J. *Tetrahedron* **1983**, *39*, 1141-1150. (d) Apeloig, Y.; Karni, M. *J. Chem. Soc. Perkin Trans. II* **1988**, 625 - 636.

12. (a) Dewar, M. J. S. *Hyperconjugation*; Ronald Press: New York, **1962**. (b) de la Mare, P. B. D. *Pure Appl. Chem.* **1984**, *56*, 1755 - 1766.

13. Frish, M. J.; Head-Gordon, M.; Trucks,G. W.; Head-Gordon, M.; Gill, P. M. W.; Wong, M. W.; Foresman, J. B.; Johnson, B. G.; Schlegel, H. B.; Robb, M. A.; Replogle, E. S.; Gomperts, R.; Andres, J. L.; Raghavachari, K.; Binkley, J. S.; Gonzalez, C.; Pople, J. A. *Gaussian 92, Revision A*, Gaussian, Inc. , Pittsburgh PA, 1992.

14. *CRC Handbook of Chemistry and Physics*; ed. by Lide, D. R.; CRC Press: Boca Raton, **1993**, 9-15 – 9-41.

15. Parr, R. G.; Yang, W. *Density-Functional Theory of Atoms and Molecules;* Oxford University Press: New York, 1989.

16. Parr, R. G.; Pearson, R. G. *J. Am. Chem. Soc.* **1983**, *105*, 7512-7516.

17. Koopmans, T. C. *Physica.* **1934**, *1*, 104-106.

18. (a) Pearson, R. G. *Acc. Chem. Res.* **1993**, *26*, 250-255. (b) Zhou, Z; Parr, R. G. *J. Am. Chem. Soc.* **1990**, *112*, 5720-5724.

19. Dewar, M. J. S. *The Molecular Orbital Theory of Organic Chemistry*; McGraw-Hill: New York, **1969**.

20. (a) Fuigimoto, H. : Fukui, K. *Intermolecular Interactions and Chemical Reactivity; in Chemical Reactivity and Reactions Paths*, Klopman, G. (Ed.); Wiley: New York, **1974**; 23 - 54. (b) Klopman, G. *ibid* 57 - 166. (c) Shaik, S. S. *J. Am. Chem. Soc.* **1981**, *103*, 3692 - 3701.

(d) Shaik, S. S. *Prog. Phys. Org. Chem.* **1985**, *15*, 197. (e) Pross A. *Adv. Phys. Org. Chem.* **1985**, *21*, 99 - 196.

21. (a) Mulliken, R. S. *J. Chem. Phys.* **1955**, *23*, 1833. (b) Mulliken, R. S. *J. Chem. Phys.* **1962**, *36*, 3428.

22. Reed, A.E.; Curtiss, L.A.; Weinhold, F. *Chem. Rev.* **1988**, *88*, 899.

23. (a) Parr, R. G.; Donnelly, R. A.; Palke, W. E. *J. Chem. Phys.* **1978**, *68*, 3801- 3807. (b) Donnelly, R. A.; Parr, R. G. *J. Chem. Phys.* **1978**, *69*, 4431- 4439. (c) Liu, G. -H.; Parr, R. G. *J. Am. Chem. Soc.* **1995**, *117*, 3179 - 3188.

24. (a) Sanderson, R. T. *Chemical Bonds and Bond Energy,* 2nd edn.; Academic Press: New York, **1976**. (b) Ray, N. K.; Samuels, L.; Parr, R. G. *J. Chem. Phys.* **1979**, *70*, 3680 - 3684. (c) Politzer, P.; Weinstein, H. *J. Chem. Phys.* **1979**, *71*, 4218 -4220. (d) Gasteiger, J.; Marsili, M. *Tetrahedron*, **1980**, 36, pp 3219 - 3228. (e) Parr, R. G.; Bartolotti, L. J. *J. Am. Chem. Soc.* **1982**, *104*, 3801 - 3803. (f) Nalewajski, R. F.; Koninski, M. *J. Phys. Chem.* **1984**, *88*, 6234 - 6240.

25. Pearson, R. G. *Inorg Chem.* **1988**, 27, pp 734 - 740.

J. Mol. Model. **1996**, 2, 410 – 416

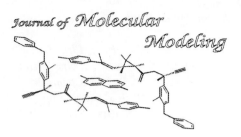

Molecular Modeling of Host-Guest Inclusion Compounds: Calculations and Practical Application to Chemical Sensors

Franz L. Dickert*, Hubert Reif and Helen Stathopulos

Institute of Analytical Chemistry, University of Vienna, Währinger Straße 38, A-1090 Wien, Austria
(fdickert@olivin.anc.univie.ac.at)

Received: 17 May 1996 / Accepted: 19 September 1996 / Published: 31 October 1996

Abstract

Molecular modeling by force-field methods is a straightforward highly convenient tool for the calculation of host-guest systems comprising a large number of atoms. [1, 2, 3] As a practical application we use the calculated data to interpret and predict the potential of mass-sensitive chemical sensors that utilize coating materials on the basis of supramolecular host compounds. Inclusion complexes of modified β-cyclodextrins and tetraazaparacyclophanes with fluorochlorinated anesthetics were calculated and compared to experimental data.

Keywords: MM3 force field, host-guest-chemistry, β-cyclodextrin, paracyclophanes, fluorochlorinated hydrocarbons, mass-sensitive detection, prediction of sensor effects.

Introduction

Supramolecular host molecules are innovative synthetic materials that engulf analyte molecules by enzyme-analogue molecular recognition mechanisms,[4] both in the liquid[5] and the gas phase [6]. When they are used as coatings on mass-sensitive devices, such as the quartz microbalance (QMB) or the surface acoustic wave (SAW) resonator, the detection of halogenated or aromatic solvents in the vapour phase is possible down to a few ppm.[7,8] If such a transducer is integrated as the frequency determining element into an oscillator circuit, every change of mass due to analyte incorporation is measured as a frequency shift Δf.[9, 10] Both methylated β-cyclodextrin **1** and the series of tetraazaparacyclophanes **2a - 2d** (Figure 1) are highly capable of incorporating solvent molecules without distinct geometry or functionality by a host-guest mechanism.[11, 12, 13] Conse-

quently such coated QMB- or SAW-oscillators offer a favorable method for the detection of anesthetics belonging to the group of halogenated hydrocarbons in concentrations relevant to medical applications. Molecular modeling by the MM3 force-field enables an efficient analyte incorporation to be predicted.[14, 15] The theoretical determination of the stabilization enthalpies of the complexes formed helps the selection of a promising coating material. To promote knowledge on the inclusion mechanism, the geometries of the complexes were examined.

In this paper we present the MM3-calculations of host-guest interactions between methylated β-cyclodextrin **1** or tetraazaparacyclophanes **2a - 2d** as host compounds and gaseous anesthetics, such as the halogenated hydrocarbons halothane, enfluran, isoflurane, and sevoflurane and we demonstrate the practical significance of those theoretical data for the development of mass-sensitive chemical sensors.

---

* *To whom correspondence should be addressed*

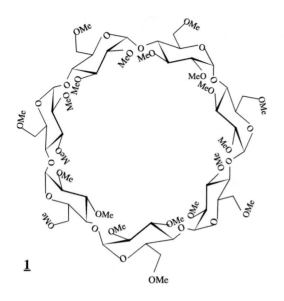

1

2a -2d

Figure 1. *Structure of the established host-compounds;*
1: TMBCD, 2: R = -CH$_3$, [m,n] : 2a: m = 3, n = 4, 2b: m = 4,
n = 4, 2c: m = 4, n = 5, 2d: m = 5, n = 5.

Computational Procedure

Molecular Modeling

The host-guest stabilization enthalpy ΔH_f(host/guest) is given
as the difference between the heat of formation of the host-
guest complex ΔH_f (complex) and the sum of the heats of
formation of the isolated host ΔH_f (host) and the free guest
ΔH_f (guest) (Equation 1).

$$\Delta H_f \ (host / guest) =$$
$$\Delta H_f \ (complex) - [\Delta H_f \ (host) + \Delta H_f \ (guest)] \qquad (1)$$

These data were computed by Allingers MM3 program.[16, 17, 18, 19] The input files were created with the
HyperChem4.0 MM+ force field and transformed to the MM3
format with the help of a self-made converting software. After optimization by the MM3 force-field the output data were
used for calculating the energy contributions (Equation 1)
and with the back-converted files the optimized host-guest
geometries were visualized. Since the MM3 force field lacks
of some parameters for the calculation of fluorinated hydrocarbons these were generated by a parametrization procedure demonstrated in the following for the dihedral angle F-
C(sp^3)-O-C(sp^3) of the MM3-type atoms 11-1-6-1. Accord-

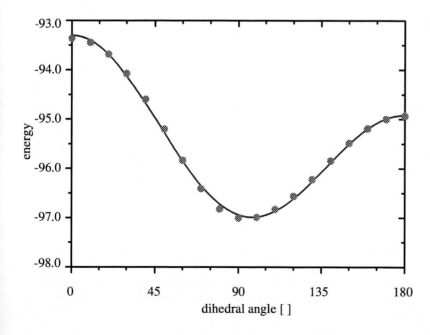

Figure 2. *Fit function of the dihedral angle with*
the calculated parameters (blue line) and semi-
empirically calculated energies (dots).

J. Mol. Model. **1996**, 2

Table 1. *Dipole-dipole interactions of anesthetics and TMBCD ($\mu_{TMBCD} = 1.90$)*

anesthetic		$\mu_{anesthetic}$	$angle_{\mu(host/guest)}$	dipole-dipole-stabilization [kcal/mol]
halothane **3a**	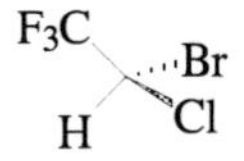	1.64	69°	-0.24
isoflurane **3b**	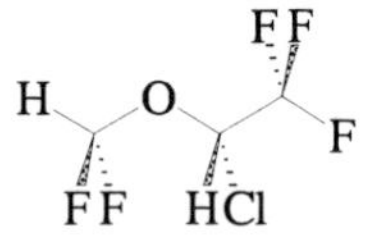	2.17	54°	-0.82
enfluran **3c**	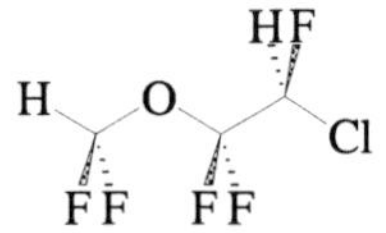	2.34	104°	-1.58
sevoflurane **3d**	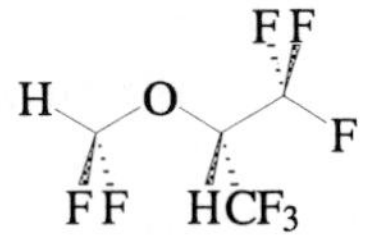	3.31	131°	-4.05

ing to Equation 2 which gives the energy of torsion at a dihedral angle ω in a Fourier series expansion in cosine the parameters V_1, V_2 and V_3 are needed in the MM3-program.

$$E_T = \frac{V_1}{2} (1 + \cos \omega) +$$
$$+ \frac{V_2}{2} (1 - \cos 2\omega) + \frac{V_3}{2} (1 + \cos 3\omega) \qquad (2)$$

The complete energy is the sum of all torsional energies that contribute to this rotation. The parameters V_1, V_2 and V_3 are determined *via* a complete rotational profile of monofluorodimethylether as model substance that was semiempirically calculated by MOPAC. The complete rotational energy E(rot) was corrected for the torsion energies of the remaining participating groups. The resulting calculated profile of the torsion energy was used for estimating the parameters of the F-C(sp³)-O-C(sp³) torsional angle with the help of PLS. In Figure 2 the semiempirically calculated energies and the fit function of the dihedral angle making use of the determined parameters $V_1 = 1.589$, $V_2 = -2.826$ and $V_3 = 0.031$ for the F-C(sp³)-O-C(sp³)-type angle are shown.

Experimental

Measurements

For the mass-sensitive measurements, QMB devices consisting of an AT-cut quartz with gold electrodes of 5.5 mm in diameter were covered with sensitive layers in a thickness up to one hundred nanometers. The QMBs work at a resonance frequency of 10 MHz and were measured with a Keithley 775A frequency counter at a resolution of ± 0.1 Hz. All data were computer-aided on-line interpreted. To eliminate ambient parameters such as varying temperature or humidity that lead to signal drifting, the measurements were performed under thermostated conditions in a differential quartz setup with an uncoated quartz as internal reference.

Chemicals

The synthesis of the tetraazaparacyclophanes has been described previously.[20, 21, 22] The methylated β-cyclodextrin TMBCD can be prepared according to literature.[23] A methylated and polymeric linked β-cyclodextrin was synthesized by the reaction with epichlorohydrin[24] which was followed by methylation.

interaction type [a]	halothane	enfluran	isoflurane	sevoflurane
compression	-0.18	-0.05	-0.11	-0.24
bending	-0.33	-0.31	+0.10	-0.94
bend-bend	-0.02	+0.10	+0.03	+0.04
vdW 1,4	+0.25	-0.18	-0.17	-0.19
vdW other	-14.19	-12.27	-11.61	-11.40
torsional	-0.31	+1.66	+0.01	+2.52
dipole-dipole	-0.24	-1.58	-0.82	-4.05
overall stabilization H_f [b]	-17.67	-14.96	-14.90	-16.53

Table 2. *[a] Energetic contributions of host-guest interactions [kcal/mol] and [b]overall stabilization enthalpies ΔH_f(host/guest) [kcal/mol] for TMBCD*

Results and Discussion

We pre-calculated the inclusion of the mixed halogenated anesthetics halothane **3a**, isoflurane **3b** and enfluran **3c** and the fluorinated hydrocarbon sevoflurane **3d** (Table 1) by compounds **1** and **2a - d**. The analytes show an increasing dipole moment in the order mentioned and host-guest inclusion should be mainly on the basis of electrostatic and dispersive interaction forces.

Methylated β-cyclodextrin (TMBCD) **1** is an already synthetically optimized cone-shaped host compound with a hydrophobic inner surface and all three hydroxyls methylated. Thus, even with varying humidity the sensor material shows no distinct interaction with water molecules, while all established anesthetics are completely incorporated due to shape and stereoelectronic recognition. In Figure 3 the side-cut view of the incorporation complex of sevoflurane reveals how the analyte joins closely to the lipophilic inner walls of the cav-ity. This excellent nestling is the reason why all four host-guest-complexes show high stabilization enthalpies ΔH_f in the range of -14.9 kcal/mol up to about -17.7 kcal/mol. The energetic contributions of several host-guest complexes and the overall energetic stabilizations which represent the heats of complex stabilization ΔH_f are listed in Table 2. The major part of the complex stabilization are van der Waals (vdW) forces. Contributions of this nature are split into intramolecular 1-4-interactions and the sum of all other van der Waals increments. Compared to the latter, the 1-4 van der Waals contributions are extraordinarily small for the ether compounds and destabilizing for halothane, indicating that the close neighbourhood of the atoms does not change them electronically both in the host and the guest compound. Therefore the inclusion mechanism seems to be similar to a key lock principle without distinct conformational adaption.

The second stabilization type worth mentioning is of electrostatic nature. In Table 1 the calculated data for dipole-

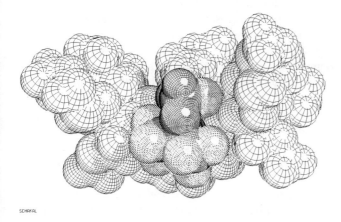

Figure 3. *Host-guest complex of TMBCD **1** and sevoflurane; wide grids: host, narrow grids: guest; blue: oxygen, black: carbon, green: fluorine.*

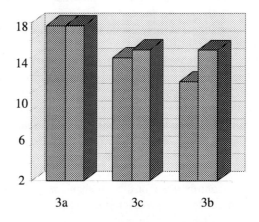

Figure 4. *Predicted values $-\Delta H_f$ (host/guest) [kcal/mol] ▦ and experimentally measured normalized logarithmic sensor effect at 1000 ppm analyte concentration ▪.*

Table 3. *Stabilization enthalpies of monocyclic tetraazacyclophanes and anesthetics*

anesthetic	tetraazaparacyclophane [m,n] **2**			
	[3,4] **2a**	[4,4] **2b**	[4,5] **2c**	[5,5] **2d**
halothane	-15.20	-13.45	-13.54	-13.04
enfluran	-14.61	-14.77	-13.36	-13.85
isoflurane	-15.53	-13.56	-14.31	-14.68
sevoflurane	-16.03	-17.79	-14.53	-15.45

Table 4. *Energetic contributions of several host-guest interactions [kcal/mol]*
[a] halothane
[b] sevoflurane

interaction type [a]	[3,4] **2a**	[4,4] **2b**	[4,5] **2c**	[5,5] **2d**
compression	-0.22	-0.16	-0.20	-0.21
bending	-0.67	+0.74	-1.02	-0.40
bend-bend	0.00	-0.01	-0.06	-0.02
vdW 1,4	-0.10	+0.11	+0.08	+0.01
vdW other	-11.88	-11.25	-10.87	-9.63
torsional	+1.07	-0.22	+1.03	-0.12
dipole-dipole	-1.62	-0.73	-0.56	-0.90

interaction type [b]	[3,4] **2a**	[4,4] **2b**	[5,5] **2d**
compression	-0.44	-0.31	-0.20
bending	-2.17	-0.59	-1.02
bend-bend	-0.07	-0.04	-0.06
vdW 1,4	-0.59	+0.53	+0.08
vdW other	-11.54	-11.76	-10.87
torsional	+2.54	-0.31	+1.03
dipole-dipole	-2.34	-2.73	-0.56

dipole interactions are presented. The amounts of these stabilizing forces depend on the sum of the dipole moments of the isolated compounds as well as on their orientation. The highest electrostatic stabilization enthalpy is calculated for the inclusion of sevoflurane with -4.05 kcal/mol, since this anesthetic has the highest dipole moment of $\mu = 3.31$ D in the series and the angle of interaction of 131° is nearest to the optimum 180° value. At the other end of the series for the TMBCD/halothane complex the lowest dipole-dipole stabilization results with a value of -0.24 kcal/mol because of the small dipole moment of the free halothane molecule and the quite disadvantageous angle of 69° between host and guest. In conclusion, the strongly polar sevoflurane molecule shows the highest electrostatic stabilization while the overall host-guest stabilization enthalpy of the TMBCD/halothane complex is calculated to be the most stable. According to these results the sensor behavior of a methylated β-cyclodextrin polymer, which was linked by epichlorohydrin, to the series of anesthetics was tested. The experimentally measured sensor effects at 1000 ppm analyte concentration were compared to the theoretical prediction. Since the sensor signal is determined by intracavitative incorporation, the stabilization enthalpy was found to correlate to the sensor effect. As demonstrated in Figure 4 the procedure established is a preferential method for the investigation of sensor behavior in the forefield of synthetic or experimental work.

Subsequently azaparacyclophanes of the [m,n]-type **2** in Figure 1 that differ in their inner diameter were theoretically examined according to their analyte engulfing abilities. Their tendency to form host-guest complexes can be compared to that of the β-cyclodextrin, although these host molecules have aromatic biphenylether moieties, but due to the chemical nature of the system no other interactions than electrostatic or van der Waals bonding play an important role. In Table 3 the stabilization enthalpies for the series of **2a** - **2d** are listed. The energetic contributions for the halothane or sevoflurane incorporation into tetraazaparacyclophanes [m,n] with a successively increasing number of bridging carbon atoms in the compounds **2a** - **2d** are presented in Table 4. The inner contact surface is the limiting factor for the development of an effective incorporation geometry and therefore for a strong

stabilization. Again in this systems mainly van der Waals and dipole-dipole forces are responsible for the host-guest inclusion. The small halothane molecule is intimately adapted to the [3,4]-paracyclophane with the smallest cavity size. With increasing diameter of the host the intermolecular van der Waals stabilization decreases from -11.88 kcal/mol to -9.63 kcal/mol for the largest [5,5]-compound **2d** and the same is observed in overall host-guest stabilization enthalpies. In a similar way the inclusion of sevoflurane can be interpreted. Here the spacy analyte is optimally incorporated into the [4,4]-host **2b** with a medium diameter. For this sevoflurane/[4,4]-complex the van der Waals stabilization has a maximum with -11,76 kcal/mol which is reduced by 1,89 kcal/mol up to the largest [5,5] complex.

As demonstrated, the driving force for host-guest inclusion of the halogenated anesthetics by cyclodextrins or tetraazaparacyclophanes are mainly shape recognition and an optimum fit to the inner surface of the host-structure is important. For the prediction of sensor effects of the tetraazaparacyclophanes we calculated the inclusion complexes with aromatic and chlorinated solvents, too. Here chloroform, benzene and the bulky analyte toluene show the best fit with the [5,5] host **2d**, while due to their diameters the complete incorporation in smaller cavities is hindered and because the larger more flexible hosts adopt an unfavorable

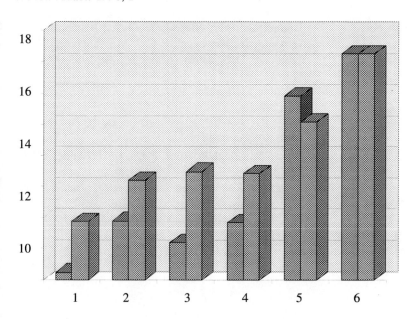

Figure 5. *Predicted values* $-\Delta H_f$ *(host/guest) [kcal/mol]* ▦ *and experimentally determined lnK* ■ *for complexation formation of* **2** *(R = $-C_5H_{11}$, [5,5]) and various analytes; 1) methylene-chloride, 2) chloroform, 3) carbontetrachloride 4) benzene, 5) toluene, 6) tetrachloroethylene.*

conformation during complexation.[14] The stabilization enthalpies ΔH_f (host/guest) for some aromatic analytes as well as halogenated hydrocarbons and the [5,5] compound are set against the logarithm of the experimentally obtained equilibrium constant K of the host-guest formation in Figure 5. Equation 3 gives the relationship between mass-normalized sensor effect, equilibrium constant K and stabilization enthalpy ΔH_f. For a first consideration the entropic terms are assumed to be constant since the condensation entropies of the solvents should follow the rule of Pictet and Trouton.

$$\ln (sensor\ effect) \sim lnK = -\frac{\Delta H_f}{RT} + \frac{\Delta S_f}{R} \qquad (3)$$

Altogether, for the widespread range of chemically different analytes and sensor materials with force-field methods host-guest interactions can be modeled involving van der Waals, dipole-dipole or even π-bondings with sufficient accuracy.[25, 26, 27, 28, 29]

Conclusion

Molecular modeling is a favorable tool for the interpretation of host-guest inclusion phenomena for systems with up to one thousand atoms. A promising application is the adaption of theoretical data for the prediction of sensor responses of mass-sensitive devices that make use of host-guest-chemistry as the detection principle. In this way computational chemistry helps the sophisticated design of sensitive chemical coatings.

Acknowledgement We gratefully thank Dr. Castor, SIEMENS ELEMA, Sweden, for the supply of the anesthetics halothane, enfluran and isoflurane.

References

1. Cornell, W. D.; Cieplak, P.; Bayly, C. I.; Gould, I. R.; Merz, K. M.; Ferguson, D. M. Spellmeyer, D. C.; Fox, T.; Caldwell, J. W.; Kollman, P. A. *J. Am. Chem. Soc.* **1995**, *117*, 5179-5197.
2. Immel, S.; Brickmann, J.; Lichtenthaler, F. W. *Liebigs Analen* **1995**, *6*, 929-942.
3. Klein, C. T.; Kohler, G.; Mayer, B.; Mraz, K.; Reiter, S.; Viernstein, H.; Wolschann, P. *J. Inclusion Phenom. Mol. Recogn.* **1995**, *22*, 15-32.
4. Vögtle, F.; Sieger, H.; Müller, W. M. In *Topics in Current Chemistry 98*; Boschke, F. L. Ed.; Springer Verlag: Berlin, 1981; pp 107-161.
5. Saenger, W. *Angew. Chem.* **1980**, *92*, 343-361.
6. Schierbaum, K. D.; Gerlach, A.; Göpel, W.; Müller, W. M.; Vögtle, F.; Dominik, A.; Roth, H. J. *Fres. J. Anal. Chem.* **1994**, *349*, 372-379.
7. Cammann, K.; Lemke, U.; Rohen, A.; Sander, J.; Wilken, H.; Winter, B. *Angew. Chem., Int. Ed. Engl.* **1991**, *30*, 516-538.
8. Dickert, F. L.; Haunschild, A. *Adv. Mater.* **1993**, *5*, 887-895.
9. Hauptmann, P. *Sens. Actuators A* **1991**, *25-27*, 371-377.
10. Sauerbrey, G. *Z. Phys.* **1959**, *155*, 206-222.
11. Lucklum, R.; Henning, P.; Hauptmann, P.; Schierbaum, K. D.; Vaihinger, S.; Göpel, W. *Sens. Actuators A* **1991**, *25-27*, 705-710.
12. Dickert, F. L.; Bauer, P. *Adv. Mater.* **1991**, *3*, 436-438.
13. Dickert, F. L.; Haunschild, A.; Kuschow, V.; Stathopulos, H. *Anal. Chem.* **1996**, *68*, 1058-1061.
14. Dickert, F. L.; Reif, M.; Reif, H. *Fresenius J Anal Chem* **1995**, *352*, 620-624.
15. Dickert, F. L.; Haunschild, A.; Maune, V. *Sens. Actuators B* **1993**, *12*, 169-173.

16. Allinger, N. L.; Yuh, Y. H.; Lii, J. H. *J. Am. Chem. Soc.* **1989**, *111*, 8551-8566.

17. Allinger, N. L.; Schmitz, L. R.; Motoc, I.; Bender, C.; Labanowski, J. K. *J. Am. Chem. Soc.* **1992**, *114,* 2880-2883.

18. Allinger, N. L.; Rahman, M.; Lii, J. H. *J. Am. Chem. Soc.* **1990**, *112*, 8293-8307.

19. Allinger, N. L.; Schmitz, L. R. *J. Am. Chem. Soc.* **1990**, *112*, 8307-8315.

20. Soga, T.; Odashima, K.; Koga, K. *Tetrahedron Lett.* **1980**, *21*, 4351-4354.

21. Snyder, H. R.; Heckert, H. *J. Am. Chem. Soc.* **1952**, *74*, 2006-2009.

22. Stetter, H.; Roos, E. E. *Chem. Ber.* **1955**, *88*, 1390-1395.

23. Cramer, F.; Mackensen, G.; Sensse, K. *Chem. Ber.* **1968**, *102*, 494-508.

24. Xu, W.; Wang, Y.; Shen, S.; Li, Y.; Xia, S.; Zhang, Y. *Chin. J. Polym. Sic.* **1989**, *7*, 16-22.

25. Timmermann, P.; Nierop, K. G. A.; Brinks, E. A.; Verboom, W.; Vanveggel, F. C. J. M.; Vanhoorn, W. P.; Reinhoudt, D. N. *Angew. Chem. Int. Ed.* **1995**, *34*, 132-143.

26. Fathallah, M.; Fotiadu, F.; Jaime, C. *J. Org. Chem.* **1994**, *59*, 1288-1293.

27. Perez, F.; Jaime, C.; Sanchez-Ruiz, X. *J. Org. Chem.* **1995**, *60*, 3840-3845.

28. Linert, W.; Margl, P.; Renz, F. *Chemical Physics* **1992**, *161*, 327-338.

29. Bekers, O.; Kettenes-van den Bosch, J. J.; van Helden, S. P.; Seijkens, D.; Beijnen, J. H.; Bult, A.; Underberg, W. J. M. *J. Inclusion Phenom. Mol. Recogn.* **1991**, *11*, 185-193.

J. Mol. Model. **1996**, 2, 417 – 426

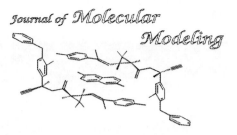

Computational Analysis of Cysteine Substitutions Modelled on the α- and β-domains of Cd₅,Zn₂-Metallothionein 2

Núria Romero[1], Mercè Capdevila[1], Pilar González-Duarte[1], Baldomero Oliva[2,*]

1) Departament de Química

2) Institut de Biologia Fonamental, Universitat Autònoma de Barcelona, 08193 Bellaterra, Barcelona, Spain,
Phone: +34-3-5812807, Fax: +34-3-5812011 (baldo@pug.uab.es)

Received: 8 July 1996 / Accepted: 23 October 1996 / Published: 5 November 1996

Abstract

Several mutant forms of rat liver Cd₅,Zn₂-metallothionein 2 (Cd₅,Zn₂-MT 2) [1] have been computationally modelled and analysed. All terminal cysteines (5, 13, 19, 21, 26, 29, 33, 36, 41, 48, 57 and 59, Figure 1) have been independently substituted by three other co-ordinating amino-acids (aspartate, glutamate and histidine), and the side-chains of the mutated residues have been modelled to co-ordinate the seven metal ions while minimizing the conformational variations with respect to the wild type protein. We have compared the ability of the putative mutant forms to maintain the MT binding properties. Substitution by aspartate residue best preserves the 3D MT structure. In addition, the mutations C5H plus C21H/E/D show neighbouring impairments that prevent their simultaneous substitution. Although replacement of cysteine by aspartate is feasible in all cases, to our knowledge there is no example of aspartate and cysteine residues co-ordinating to the same zinc atom. Accordingly, the use of histidine or glutamate instead of aspartate cannot be ruled out. The mutant forms in the β-domain of Cd₅,Zn₂-MT 2 have yielded more neighbouring contacts than those in the α-domain, which is corroborated by the accessible surface areas [2] of the sulfur atoms [3] in the native form.

Keywords: Protein modeling, metalloprotein, metallothionein
Abbreviations: MT = metallothionein; Cd₅,Zn₂-MT = Cadmium, Zinc-metallothionein, RMSD = Root Mean Square Deviation; PDB = Protein Data Bank; FEP = Free Energy Perturbation; CnX = mutant form of cysteine n (n = residue number) substituted by X (X = H, E or D, with H = histidine, E = glutamate, D = aspartate); CnX/Y = mutant forms CnX and CnY.

Introduction

Our current interest in metallothioneins (MT) and their α- and β- fragments [3, 4] stimulated us to study the possibility of substituting some cysteine residues to avoid the oxidation problems associated with the apo forms while maintaining the binding abilities of these proteins. Previous studies have either substituted the bridging [5, 6] or terminal cysteines [5, 6, 7] by low co-ordinating residues such as Ser, Tyr or Ala. These mutations altered domain structure by disrupting the normally tight protein clusters.

With the purpose of affecting in a lesser degree the 3D-structure of MT we have only chosen the terminal cysteines

* To whom correspondence should be addressed

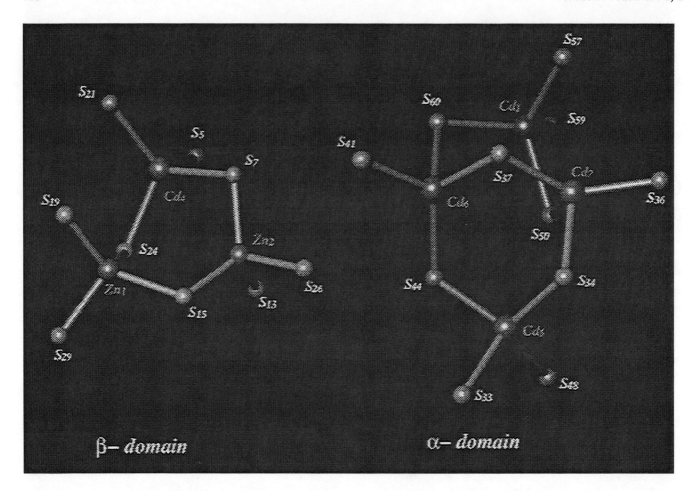

Figure 1. α *and* β *clusters that make up the two-domain structure found for rat liver* Cd$_5$Zn$_2$*-MT 2. Data taken from [1].*

as possible residues to be substituted. Among the 20 natural amino-acids, the residues with significant binding capacity are cysteine, histidine, aspartic and glutamic [8]. Combinations of these residues constitute the most common co-ordination sites in metallobiomolecules of known crystallographic structure [9] and particularly in Zn metalloproteins [10]. According to these data, we have studied the effect of replacing terminal Cys residues by His, Asp and Glu by means of a molecular modelling approach based on the X-ray structure of rat liver Cd$_5$Zn$_2$-MT 2.

Modelling the positions of side-chains, given that of the backbone, is difficult. Nevertheless, it has been tackled by different authors [11 , 12 , 13] who have succeeded with relatively high accuracy (80%) [14]. To overcome this difficulty, the simplest energy functions, van der Waals and Lennard Jones potentials, have given excellent results for buried residues [15]. However, these simple approaches have failed for totally or partially exposed residues, for which it is necessary to add empirical solvation terms [16 , 17]. Other procedures emphasize an optimal choice of conformation for a particular side-chain by local energy analysis [18], consid-

eration of local backbone conformation [19] or evaluation of neighbouring 3D homology [20]. The weakest point of most side-chain packing methods is the effect of main chain displacements [21].

The ability to co-ordinate metal ions by a wild type protein or any of its putative mutant types requires the study of the conformational space of the apoprotein in a solvated environment in the presence of the metal ions. In accordance with the aim of this study, the co-ordination of the metal ions by MT was assayed by a Molecular Dynamics simulation of the apoprotein solvated with water molecules and Zn^{2+} ions. This did not lead to satisfactory results. The Cd$_5$Zn$_2$-MT 2 structure is due to the co-ordination of the metal ions to the protein, which involves folding. Molecular Dynamics simulation of this system requires too much time to reach the adequate folding. Hence, given the difficulty associated with this method we decided to carry out the replacement of the Cys residues and corresponding structural analysis without including energetic terms. Thus, it was assumed that the mutant types reproduce the binding abilities of native mammalian MTs. The previous assumption entails the local displacement of the side-chain by application of a geometrical refinement with simple energy functions in order to assess the correct geometry. To analyse the ability of the mutant forms to preserve the native structure, individual substitutions of each terminal Cys by Asp, Glu or His were consid-

Table 1. *Statistical analysis of the Zn(II) co-ordination in a set of proteins at high resolution (<2.5Å) from the PDB.*

Glutamate is found to coordinate to the metal ions either by means of one or both carboxylic oxygen atoms, whilst aspartate coordinates only with one.

	Zn-O_1 Distance (Å)		Zn-O_2 Distance (Å)		Zn-N Distance (Å)	
	Average	Range	Average	Range	Average	Range
Glu [u]	2.04	1.91-2.17	2.82	2.60-2.94	—	—
Glu [b]	2.15	2.07-2.18	2.30	2.12-2.48	—	—
Asp [u]	2.00	1.91-2.08	3.30	2.96-3.47	—	—
His	—	—	—	—	2.07	1.91-2.17

[u] unidentate ligand
[b] bidentate ligand

Table 2. *Analysis of the single substitutions of terminal cysteine residues by histidine, glutamate and aspartate in the β-domain of rat liver Cd_5,Zn_2-MT 2.*
The analysis of the single replacements is based on the conformational variations of the putative mutant form with respect to the native type. The bond distance variation, Δd, corresponds to the difference |ML - MS|, where ML and MS represent the distances between the metal ion and the donor atom (L= N or O). Δτ = angle (LMS), is the bond angle variation. RMSD, denotes the root mean square deviation of the backbone atoms of the mutant type with respect to the wild type. Neighbours, represents the number of atoms at a distance of less than 3 Å from any of the atoms of the substituted residue in the putative mutant form.

ered first and then the effect of combined substitutions was taken into account. The best residue is that which introduces fewest steric hindrances and thus causes the least rearrangement of the backbone while maintaining the co-ordination of the metal ions.

A Free Energy Perturbation study (FEP) would be advantageous for evaluating the goodness of the model [22]. However, this is a highly time consuming process and would also involve a previous step of Quantum Mechanics Optimisation of the metal co-ordination in order to obtain the correct forcefield of the system. Finally, the same hypothesis, which assumes that the putative mutant form is able to co-ordinate to the metal ions with the same conformational arrangement as that found in Cd_5,Zn_2-MT 2, should have been taken into account in the FEP calculation. Therefore, we did not consider the calculation of the free energy strictly necessary and

		Cys 5	Cys 13	Cys 19	Cys 21	Cys 26	Cys 29
His	Δd (Å)	0.22	0.22	0.14	0.50	0.47	0.26
	Δτ (deg)	14.9	15.4	26.2	23.7	32.0	7.7
	RMSD (Å)	[a]	0.06	0.03	0.01	0.00	0.09
	Neighbours	2	1	5	3	4	3
Glu	Δd (Å)	0.13	0.10	0.09-0.19	0.01	0.50	0.01
	Δτ (deg)	16.4	0.6	[b]	5.6	21.0	7.4
	RMSD (Å)	[a]	0.17	0.15	0.17	0.02	0.04
	Neighbours	3	2	4	0	0	1
Asp	Δd(Å)	0.71	0.24	0.00	0.08-0.06	0.15-0.06	0.04
	Δτ (deg)	24.0	1.2	5.8	[b]	[b]	2.4
	RMSD(Å)	[a]	0.16	0.19	0.00	0.00	0.04
	Neighbours	0	0	3	2	1	0

[a] RMSD values for the mutant forms of Cys 5 are not considered because the large flexibility of the N-terminal tail of the protein.

[b] Mutation of Cys by Asp or Glu with both carboxylic oxygen atoms co-ordinating the metal ion.

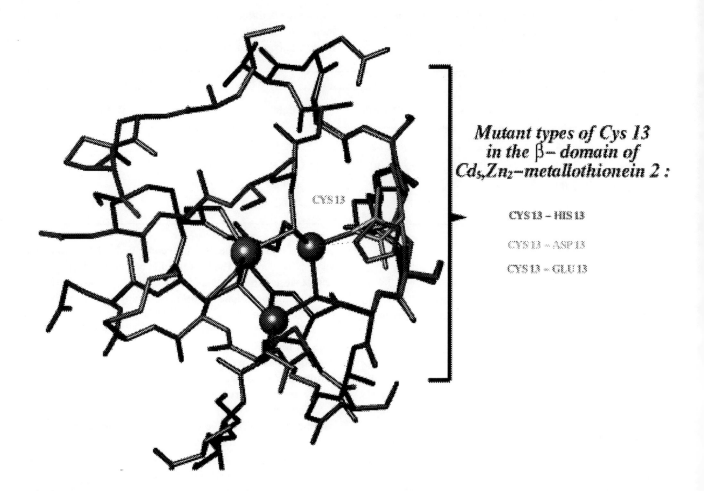

*Mutant types of Cys 13
in the β– domain of
Cd₅,Zn₂–metallothionein 2 :*

CYS 13 – HIS 13

CYS 13 – ASP 13

CYS 13 – GLU 13

Figure 2. *Mutant types C13H (red), C13D (yellow), and C13E (green) are compared to the wild type form of Cd₅,Zn₂-metallothionein 2. Zn(II) and Cd(II) metal ions are depicted as compact spheres (orange) and their bonds to the protein in blue. The deviation of the imidazol ring of His 13 with respect to the position of the native cysteine initially bound to Zn is identified by a dashed orange line.*

we limited the analysis of the mutant forms to the study of their structural impairments.

Methods

Several mutations were performed computationally upon the α- and β-domain of the X-ray crystallographic structure of rat liver Cd₅,Zn₂-metallothionein 2 [1]. All terminal cysteines (5, 13, 19, 21, 26, 29, 33, 36, 41, 48, 57 and 59) were replaced by aspartate, glutamate or histidine, independently. The substitution was performed with the program TURBO-FRODO [23] on a Silicon Graphics workstation. The cysteine residues were subsequently replaced by the residues chosen and the backbone conformation was maintained. Meanwhile,

the side-chains were rotated to approach the donor atom of the residue to the corresponding metal ion (Zn or Cd). A window of five residues centred on the substituted one was geometrically optimised to avoid conformational deformations. In most substitutions it was considered that aspartate and glutamate behaved as unidentate ligands. However, they can also bind to the metal centre *via* both carboxylic oxygen atoms causing an increase in the co-ordination number of the metal. This was only presumed when the unidentate behaviour led to significant distortions in the structure.

Each mutation was analysed in terms of variations in the native conformation and in the co-ordination geometry about the metal ion, which were estimated by the following differences between the native and the putative mutant forms: 1) metal ion-donor atom bond distance; 2) angle with vertex on the metal ion formed by the donor atom of the mutated residue (O_δ for aspartate, O_ε for glutamate, and N_ε for histidine) and the sulfur atom of the cysteine of the native form; 3) backbone conformation, calculated by the RMSD [24] of the protein backbone; and 4) steric hindrances reckoned by the number of atoms in the neighbourhood of the mutated residue (less than 3 Å distance) impairing the conformational stabilisation.

Table 3. *Variation of the neighbouring contacts for combined substitutions of terminal cysteines in the β-domain of rat liver Cd₅,Zn₂-MT 2.*
Number of new contacts at a distance shorter than 3 Å obtained by the combined replacements of two cysteines with respect to their independent single substitution effect. The total number of contacts is obtained by adding those of Table 2 for the putative mutation considered.

	His 21	Glu 21	Asp 21	His 29	Glu 29	Asp 29	His 26	Glu 26	Asp 26
His 5	16	15	11	0	0	0	0	0	0
Glu 5	-1	0	0	0	0	0	0	0	0
Asp 5	1	1	1	0	0	0	0	0	0
His 19	0	0	0	2	0	0	0	0	0
Glu 19	0	0	0	1	0	0	0	0	0
Asp 19	0	0	0	0	0	0	0	0	0
His 13	0	0	0	0	0	0	-2	2	0
Glu 13	0	0	0	0	0	0	-2	2	0
Asp 13	0	0	0	0	0	0	-1	0	0

Table 4. *Analysis of the single substitutions of terminal cysteine residues by histidine, glutamate and aspartate in the α-domain of rat liver Cd₅,Zn₂-MT 2.*
The analysis of the single replacement is based on the conformational variations of the putative mutant form with respect to the native type. The bond distance variation, Δd, corresponds to the difference |ML - MS|, where ML and MS represent the distances between metal ion and the donor atom (L= N or O). Δτ = angle (LMS), is the bond angle variation. RMSD, denotes the root mean square deviation of the backbone atoms of the mutant type with respect to the wild type. Neighbours, represents the number of atoms at a distance of less than 3 Å from any of the atoms of the substituted residue in the putative mutant form.

Results and Discussion

The zinc-protein structures at high resolution (< 2.5 Å) found in the PDB [25] were used to check the metal co-ordination by histidines, glutamates and aspartates. 24 binding sites were selected and analysed statistically. 67% were constituted by zinc ions tetrahedrally co-ordinated by one glutamate, two histidines and either a water molecule or a carbonyl oxygen belonging to the protein backbone. Consideration of the Zn-O distances less than 2.5 Å (Table 1) allowed division of the previous percentage in two groups. Accordingly, in 21% of the cases the glutamate co-ordinated by means of the two carboxylic oxygen atoms while in the remaining 46% it behaved as an unidentate ligand. Additionally, in 17% of the

		Cys 33	Cys 36	Cys 41	Cys 48	Cys 57	Cys 59
His	Δd (Å)	0.16	0.05	0.22	0.14	0.09	0.07
	Δτ (deg)	6.2	3.2	1.0	4.1	2.5	4.8
	RMSD (Å)	0.02	0.19	0.22	0.08	0.13	0.04
	Neighbours	0	1	0	0	1	0
Glu	Δd (Å)	0.17	0.03	0.17	0.01	0.05	0.19
	Δτ (deg)	8.2	4.2	1.2	4.9	0.9	3.4
	RMSD (Å)	0.07	0.05	0.16	0.11	0.20	0.07
	Neighbours	0	0	0	3	2	0
Asp	Δd (Å)	0.22	0.25	0.10	0.09	0.23	0.10
	Δτ (deg)	10.5	1.0	0.6	1.5	3.2	1.3
	RMSD (Å)	0.03	0.04	0.15	0.06	0.15	0.04
	Neighbours	0	0	0	0	1	0

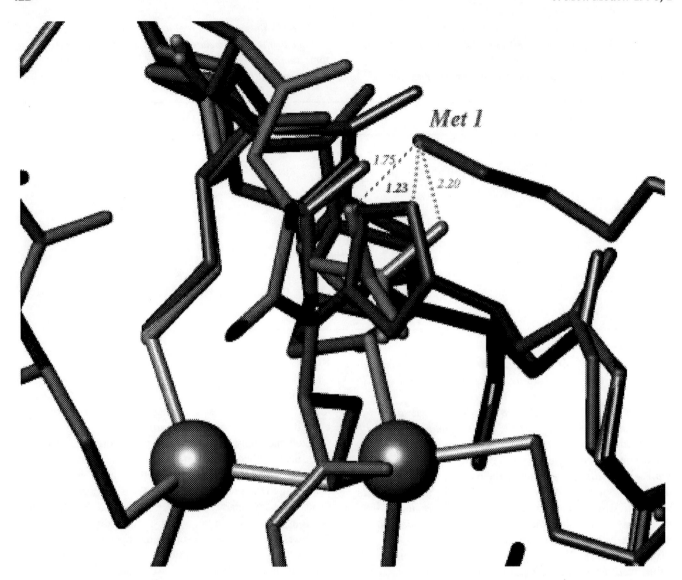

Figure 3. *Mutant types C19H (red), C19D (yellow), and C19E (green) are shown and compared to the wild type form of Cd$_5$,Zn$_2$-metallothionein 2. Zn(II) and Cd(II) metal ions are depicted as compact spheres (orange) and their bonds to the protein in blue. Closest distances between neighbouring residues are depicted (His19-Met1: 1.23; Glu19-Met1: 1.75; Asp19-Met1: 2.20).*

total examples found, the binding sites were constituted by three histidines and one aspartate, where the latter co-ordinated only by means of one oxygen. The lower and higher limits found for the distances between the metal ion and the donor atom (Table 1) were used to evaluate the co-ordination capabilities of the mutant forms.

Analysis of the single substitutions in the β-domain

Results on the single replacements of terminal cysteines in the β-domain of Cd$_5$,Zn$_2$-MT 2 are summarized in Table 2.

The RMSD [24] of the mutant forms C5D, C5E and C5H were discarded because this residue is in the N-terminal region, where there is a large conformational flexibility of the backbone. However, for the rest of substitutions the maintenance of the backbone conformation with respect to its native form cannot be neglected. Large changes on the conformation would affect the properties of the protein and therefore must be avoided (conformers deviating about 2.4 Å may preserve only 85% of the native contacts occurring between residues located at 8 Å [26]). Nevertheless, none of the studied replacements involved an RMSD value [24] greater than 0.20 Å.

The steric hindrances between the replaced side-chains and the rest of the protein were significant in number for the substitution by glutamate and histidine because of their volume. Consideration of mutants of the β-domain showed that cysteine 13 allows the greater space to accommodate the side-chain of the new residue (Figure 2). This result agrees with the accessible surface area of the sulfur atom of cysteine 13 in the native form (20 Å^2), which has the largest area of all

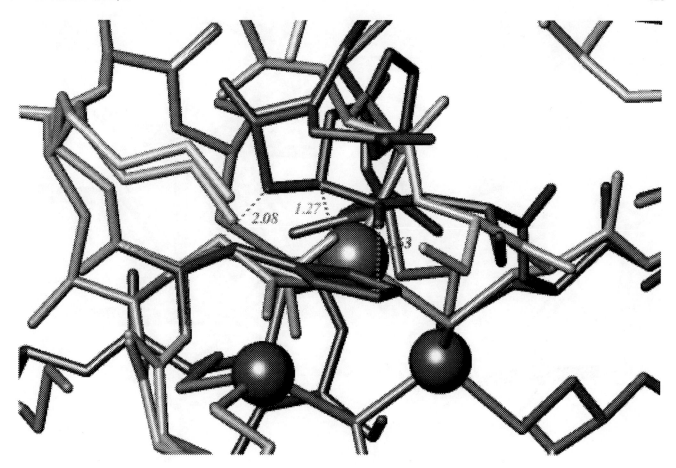

Figure 4. *Interaction between side-chains of the mutant form of C5H and the mutant forms of C21H/E/D (C5H and C21H in red, C21E in green and C21D in yellow). The closest neighbouring contacts are shown. Zn(II) and Cd(II) metal ions are depicted as compact spheres (orange) and their bonds to the protein in blue.*

the S-Cys atoms in the β-domain [3]. The substitution of the remaining cysteines (5, 19, 21, 26 and 29) by His, Asp or Glu produced several neighbouring contacts. In these cases, the percentage of accessible surface areas of the S-Cys atoms in the native form is less than 10%. Replacement of cysteine 19 (Figure 3), which has the lowest sulfur accessible surface area, led to the largest number of neighbouring collisions (a total of 12 contacts) and therefore constitutes the least favoured cysteine to be substituted.

Analysis of combined substitutions in the β-domain

It was assumed that only those residues that co-ordinate the same metal ion (Cys5-Cys21; Cys19-Cys29, and Cys13-Cys26, Figure 1) show a co-operative effect when substituted and thus this effect was disregarded for long range interactions. Table 3 gives the variation of contacts as a consequence of the combined substitutions of these pairs of cysteines, taking as reference the independent single replacement effect.

In those cases where two or more Cys residues were co-ordinated to different metal ions, the distances between the residues were long enough in 3D space so that the final result of the substitution could be taken as the addition of their independent effects.

From Table 3 and Figure 4 it can be inferred that the mutations C5H plus C21H/E/D give rise to a high number of contacts and are therefore forbidden. Except for these cases, the rest of combined substitutions could still be accepted as plausible forms for Zn or Cd binding. Note that the combined substitution of cysteines may lead to a decrease in the number of neighbouring contacts. This is the case for mutations C26H plus C13H/E/D (Figure 5).

Experimental studies on the metal binding properties of several β-MT related peptides, where some Cys residues have been substituted by His or Asp, are now in progress [27]. Preliminary results indicate that Co(II), Zn(II) and Cd(II) co-ordinate tetrahedrally to $S_2(Cys)N_2(His)$ binding sites. In contrast, co-ordination is not observed when Cys and Asp ligands are simultaneously present and chemically able to participate in metal binding.

Analysis of the substitutions in the a-domain

Analysis of the single replacements in the α-domain is summarized in Table 4. According to the RMSD values [24], the substitutions of cysteines 36 and 41 by histidine and cysteine

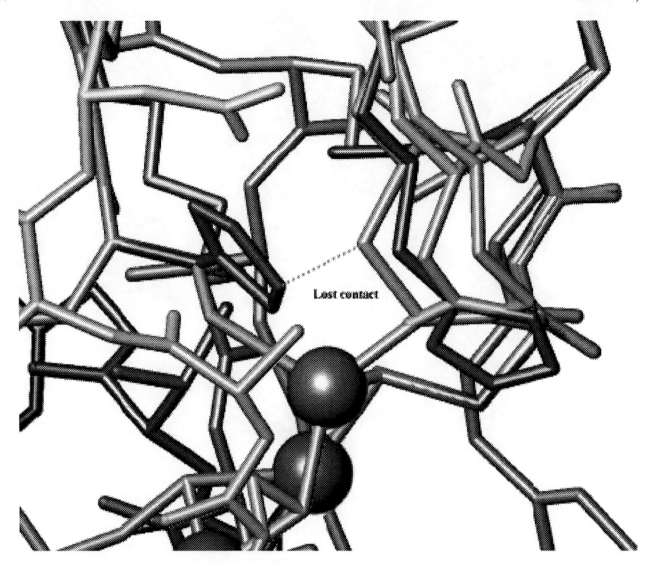

Figure 5. *Interaction between side-chains of the mutant form of C26H and the mutant forms of C13H/E/D (C26H and C13H in red, C13E in green and C13D in yellow). The closest neighbouring contacts are shown. Zn(II) and Cd(II) metal ions are depicted as compact spheres (orange) and their bonds to the protein in blue.*

57 by glutamate gave rise to the greatest changes in the backbone. However, they are smaller than 0.22 Å, and thus any considered substitution does not cause significant changes in the backbone conformation. In addition, the geometrical restraints imposed by the co-ordination of the metal ions are fully accomplished by all single substitutions.

The neighbouring contacts found for each mutant type agrees with calculations for the sulfur accessible surface areas for cysteines 33, 36, 41, 57 and 59 [3]. Cysteine 57 shows the smallest sulfur accessible surface area on the α-domain (3.45 Å^2), while cysteines 33, 41 and 59 present larger accessible surface areas than those found in the β-domain (about

20 Å^2). Consequently, the substitution of cysteine 57 by histidine, glutamate or aspartate gives rise to steric hindrances to accommodate the replaced side chain in the protein (Figure 6).

In order to assess whether the steric effects could preclude Zn$_4$-αMT cluster formation, the C57H mutant form of mouse α-MT fragment was cloned in *E. coli* and the peptide was recovered from cell cultures. Preliminary circular dichroism studies on metal binding indicate that Zn$_4$ and Cd$_4$ cluster formation occurs, probably with a similar architecture to that found for the native domain [28].

If considering simultaneous substitutions, only two pairs of terminal cysteines in the α-domain are linked to the same metal ion (Figure 1). The only mutant type with co-operative effect, unless long distance interactions were considered, was found for the substitution of cysteine 57 by aspartate and cysteine 59 by glutamate appearing an additional contact in the neighbourhood of aspartate 57.

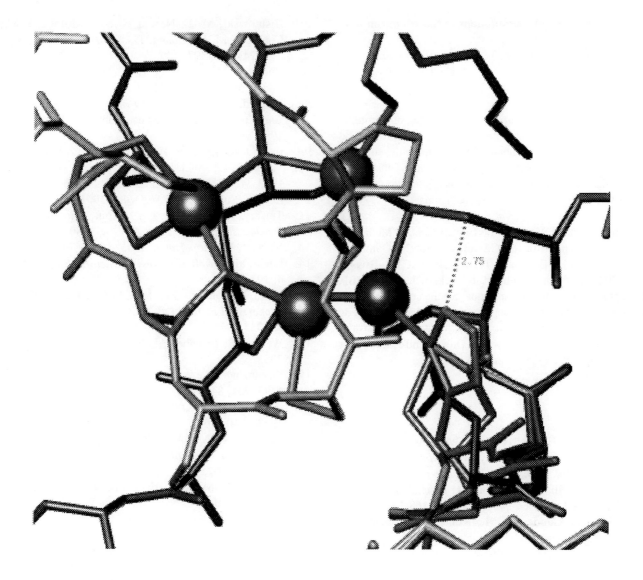

Figure 6. *Representation of main contacts found for the replaced side-chains of the mutant forms C57D (red), C57E (yellow), and C57H (green) within its neighbourhood. Zn(II) and Cd(II) metal ions are depicted as compact spheres (orange) and their bonds to the protein in blue.*

Conclusions

The conformational parameters analysed (RMSD, bond distance and bond angle variations and neighbouring contacts) have indicated that all single substitutions of terminal cysteines in rat liver Cd_5,Zn_2-metallothionein 2 by His, Asp or Glu are feasible. However, substitution of those cysteines highly buried has led to difficult accommodation of the replaced side-chain, especially with glutamate and histidine due to their large side-chain volume. The possible hindrances have been evaluated by the number of neighbouring contacts. In general, the putative mutant forms of the β-domain have shown a greater number of neighbouring side-chain contacts if compared with the α-domain and therefore the latter appears to be preferable for terminal Cys substitutions. Replacement of cysteine 19 is the most unfavourable case because it gives rise to the largest number of neighbouring collisions. The remaining mutations are viable with small perturbations in the native protein structure.

Aspartate has shown to be the best substitute of terminal cysteines of Cd_5,Zn_2-metallothionein 2 if compared with histidine and glutamate, the two latter showing similar conformational impairments. It should be noted that aspartate is not found in combination with cysteine co-ordinating to the same metal ion in the database of proteins [25]. More detailed conclusions in respect to a particular position in the mutated MT sequence are summarized below.

Single substitutions in the β-domain can be categorized in three main groups according to the side-chain neighbouring collisions:

Cys 5 and Cys 13 :	Asp > His > Glu
Cys 19 and Cys 29 :	Asp > Glu > His
Cys 21 and Cys 26 :	Glu > Asp > His

Analogously, single substitutions in the α-domain have led to:

Cys 33, Cys 41 and Cys 59 :   Asp $\approx$ His $\approx$ Glu
Cys 48 and Cys 57 :   Asp $\approx$ His > Glu
Cys 36 :   Asp $\approx$ Glu > His

The simultaneous substitution of more than one terminal cysteine co-ordinating the same metal ion has afforded good results for all possible combinations except for the substitution of cysteine 5 by histidine, which precludes substitution of cysteine 21 by histidine, aspartate or glutamate residues, owing to significant neighbouring collisions.

Acknowledgements We thank the Departament de Medi Ambient de la Generalitat de Catalunya and the Comisión Interministerial de Ciencia y Tecnología (CICYT, BIO92-0591-C02), for financial support. BO acknowledges the IBF, Universitat Autònoma de Barcelona, for the computer facilities. MC is indebted to the Departament de Medi Ambient de la Generalitat de Catalunya for a post-doctoral scholarship. Authors also want to thank Drs M. Orozco and J. Luque for their MD studies on the apoprotein.

References

1. Robbins, A. H.; McRee, D. E.; Williamson, M.; Collet, S. A.; Xoung, N. H.; Furey, W. F.; Wang, B. C.; Stout, C. D. *J. Mol. Biol.* **1991**, *221*, 1269-1293.

2. Silla, E.; Villar, F.; Nilsson, O.; Pascual-Ahuir, J. L.; Tapia, O. *J. Mol. Graphics.* **1990**, *8*, 168-172.

3. Cols, N.; Capdevila, M.; Romero, N.; Oliva, B.; González-Duarte, P.; González-Duarte, R.; Atrian, S. *Eur. J. Biochem.*, submitted.

4. a. Capdevila, M.; Romero, N.; Cols, N.; González-Duarte, R.; Atrian, S.; González-Duarte, P. *Eur. J. Biochem.*, submitted. b. Capdevila, M.; Romero, N.; González-Duarte, P.; Cols, N.; Atrian, S.; González-Duarte, R.; Stillman, M.J. *An. Quim. Int. Ed.* **1996**, *92*, 199-201. c. Atrian, S., Capdevila, M., Cols, N., González-Duarte, R., González-Duarte, P., Romero, N., Stillman, M.J. *J. Inorg. Biochem.* **1995**, *59*, 103.

5. Cismowski, M. J.; Huang, P. C. *Biochemistry* **1991**, *30*, 6626-6632.

6. Cismowski, M. J.; Narula, S. S.; Armitage, I. M.; Chernaik, M. L.; Huang, P. C. *J. Biol. Chem.* **1991**, *266*, 24390-24397.

7. Chernaik, M. L.; Huang, P. C. *Proc. Natl. Acad. Sci. Usa.* **1991**, *88*, 3024-3028.

8. Lippard, S.J., Berg, J.M. In *Principles of Bioinorganic Chemistry*; University Science Books, Mill Valley, CA, 1994.

9. Ibers, J. A., Holm, R. H. *Science* **1980**, *209*, 223-235.

10. Vallee, B. L.; Auld, D. S. *Proc. Natl. Acad. Sci. USA* **1993**, *90*, 2715-2718 and references cited whithin.

11. Lee, C.; Subbiah, S. *J. Mol. Biol.* **1991**, *217*, 373-388.

12. Tuffery, P.; Echtebest, C.; Hazout, S.; Laver, R. *J. Biomol. Struct. Dyn.* **1991**, *8*, 1267-1289.

13. Vasquez, M. *Biopolymers* **1995**, *36*, 53-70.

14. Levitt, M. *Current Opinion in Structural Biology* **1996**, *6*, 193-194.

15. Holm, L.; Sander, C. *Proteins* **1992**, *14*, 213-223.

16. Cregut, D.; Liautard, J. P.; Chiche, L. *Protein Eng.* **1994**, *7*, 1333-1344.

17. Wesson, L; Eisenberg, D. *Protein Sci.* **1992**, *1*, 227-235.

18. Eisenmenger, F.; Argos, P.; Abagyan, R. *J. Mol. Biol.* **1993**, *231*, 849-860.

19. Dunbrack, R. L.; Karplus, M. *Nature Struct. Biol.* **1994**, *1*, 334-340.

20. Laughton, C. A. *J. Mol. Biol.* **1994**, *235*, 1088-1097.

21. Vasquez, M. *Current Opinion in Structural Biology*, **1996**, *6*, 217-221.

22. Yun-yu, S.; Mark, A. E.; Cun-xin, W.; Fuhua, H.; Berendsen, H. J. C.; van Gunsteren, W. F. *Protein Eng.* **1993**, *6*, 289-295.

23. Roussel, A.; Inisan, A. G.; Knoops-Mouthy, E. In *Turbo-Frodo Manual v 5.0.*; Biographics, Technopole de Chateaux-Gombert, Marseille, France, 1991.

24. Havel, T. F.; Wütrich, K. *J. Mol. Biol.* **1985** , *182*, 281-294.

25. Bernstein, F. C.; Koetzle, T. F.; Williams, G. J. B.; Meyer, E. F.; Brice, M. D.; Rodgers, J. R.; Kennard, O.; Shimanouchi, T.; Tasumi, M. *J. Mol. Biol.* **1977,** *112*, 535-542.

26. Park, B. H.; Levitt, M. *J. Mol. Biol.*, **1995**, *249*, 493-507.

27. Unpublished results.

28. Unpublished results.

J. Mol. Model. **1996**, 2, 427 – 445

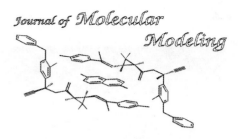

A Molecular Dynamics Study of the Repressor/Operator(OR1,OR3) Complexes from Bacteriophage 434

Laurent David and Martin J. Field

Laboratoire de Dynamique Moléculaire, Institut de Biologie Structurale – Jean-Pierre Ebel, 41, Avenue des Martyrs, 38027 Grenoble Cedex 1, France (ldavid@ibs.fr and mjfield@ibs.fr)

Received: 21 August 1996 / Accepted: 2 October 1996 / Published: 11 November 1996

Abstract

We have performed three molecular dynamics simulations using the CHARMM molecular modeling program to study the repressor protein from bacteriophage 434 complexed with DNA operators of two different sequences. Two approaches to the modeling of the solvent were used. In the first method, applied to the R1-69/OR1 truncated complex, water molecules were included explicitly in conjunction with a stochastic boundary force to solvate the complex. In the second approach, used for simulations of the R1-69/OR1 and the R1-69/OR3 complexes, the solvent was omitted and implicitly represented by using a distance-dependent dielectric constant and a scaling of the charges on the exposed residues. The simulation with the model which explicitly includes the solvent serves as a validation of the simulations using a simpler solvent representation. In our discussion of the results we focus upon the important interactions between the DNA binding motif of the 434 repressor (motif helix turn helix) and the operators and how the structures of the complexes change with time.

Keywords: Protein-DNA complex, repressor protein, bacteriophage 434, molecular dynamics, solvation models

Introduction

The regulation of gene activity in cells from a wide variety of organisms has been the subject of intense study. Much of the basis for this regulation is due to the specificity of the interaction of various proteins with DNA. Early studies which gave insight into these effects, were based primarily upon biochemical techniques [1, 2, 3]. More recent investigations have made increasing use of structural data obtained from X-ray crystallography.

Systems which have been a focus of investigation are the λ, P22 and 434 bacteriophages [4, 5, 6] in which the regulation is mediated by two proteins, cro and a repressor, which bind to the DNA. Three structures of the 434 repressor complexed to DNA are available from the Protein Data Bank [7]. They are the R1-69/OR1 [8], the R1-69/OR2 [9] and the R1-69/OR3 [10] complexes. These structures have all been determined at 2.5 Å resolution. Their availability provides a very powerful resource for investigating and understanding the crucial interactions that occur between the protein and different sequences of DNA. However, in spite of the high resolution structures, the relative affinities of the 434 repressor for its different operators are not fully understood. Thus, a comparison of the complexes reveals structural differences which do not fully explain the variation in the affinities.

Biochemical studies based on the existing structures have clarified the effect of some specific DNA bases or protein residues in the complexation and highlight some major contacts. To support the first crystallographic low resolution

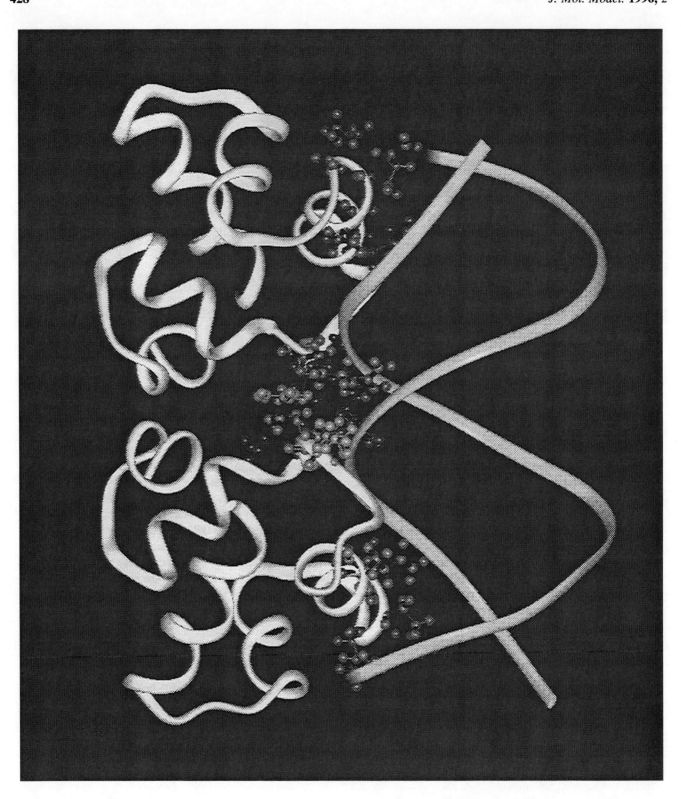

Figure 1. *An image of the protein-DNA complex. The DNA is green, the protein backbone yellow and the atoms of the amino acids which interact with the DNA strand are shown with a ball and stick representation.*

model of the complex [11], the ethylation interference method [12, 13] was used to determine protein-DNA contacts between the 434-repressor and OR1. This study showed that the truncated 434 repressor protein, called R1-69, had the same interaction at the phosphate level with the OR1 operator as with the entire protein.

Table 1. *Calculated pK$_a$ values of the ionizable groups in the protein. The values in the left half of the protein are listed first and then the right half.*

residue	N-term	arg 5	lys 7	lys 9	arg 10	glu 19	lys 23	glu 32
pK$_a$	5.5	12.9	10.6	10.2	12.9	3.8	10.7	3.6
residue	glu 35	lys 38	lys 40	arg 41	arg 43	glu 47	asp 57	C-term.
pK$_a$	2.1	10.2	9.7	12.9	11.5	3.7	3.6	3.2
residue	N-term.	arg 5	lys 7	lys 9	arg 10	glu 19	lys 23	glu 32
pK$_a$	5.5	12.9	10.3	10.4	12.9	4.0	9.6	2.9
residue	glu 35	lys 38	lys 40	arg 41	arg 43	glu 47	asp 57	C-term
pK$_a$	2.3	11.1	9.8	12.5	11.4	2.4	3.4	3.6

In more recent work, Koudelka and Lam [14] and Bell and Koudelka [15, 16] have worked on differentiating the affinity of the 434 repressor for the OR1 and the OR3 operators. They made mutants of the OR1 operator to transform it, step by step, to OR3 [15]. Bell and Koudelka [16] showed a relation between the intrinsic structure of the operator and the affinity of the operator for the 434 repressor by mutating residues suspected of contributing to the specificity of the interactions between R1-69 and OR1 [14, 17]. Another study of the dependence of the DNA structure on its composition by time-resolved fluorescence polarization anisotropy revealed that the deformation in the DNA depended only slightly on its sequence [18]. As a follow up to this work, Koudelka and Carlson [19] showed that there is a relationship between the intrinsic twist of an operator and its affinity for repressor.

Molecular dynamics simulation is an important tool that can provide information characterizing the specific interactions that occur in complex formation. In this paper, we present the results of three molecular dynamics simulations which were performed with the aim of understanding the specificity of the interactions between R1-69 and its different operators. The first simulation is a 0.5 nanosecond (ns) molecular dynamics simulation of the explicitly hydrated complex R1-69/OR1. The objective was to analyze the different interactions which occur between the protein and a part of the DNA and to show the effect of water on this complexation. The second part of the study is a comparison of two 0.5 ns simulations, carried out on the R1-69/OR1 and the R1-69/OR3 complexes. These simulations use a simplified representation of the electrostatic interactions which allows us to simulate the complete complexes whereas in the first simulation we truncated the operator sequence.

The outline of this paper is as follows. In section the next section, we detail the techniques used to perform and analyze the simulations. we than present a discussion of the results followed by our conclusion.

Methods

The initial structures for all three simulations were obtained from the Brookhaven Protein Data Bank (PDB) [7]. These were the crystal structure of Aggarwal and co-workers for the R1-69/OR1 complex [8] and the structure of Rodgers and co-workers for the R1-69/OR3 complex [10]. The preparation of both structures for simulation is described in the following sections.

Determination of the Protonation States of the Ionizable Groups in R1-69

For the simulations, it was necessary to choose the protonation states of the various ionizable amino acid groups in the protein that were appropriate for the conditions used in the crystallization of the complex (pH=6 for R1-69/OR1 and pH=5.5 for R1-69/OR3). Four types of residue were considered to have variable protonation states – arginine, aspartate, glutamate and lysine as well as the C and N-terminal ends of the protein chain (there are no histidines in the protein).

Recent progress in theoretical techniques has shown how to choose the protonation states of amino acid residues to agree with experiment [20, 21, 22]. The new methods are based upon the resolution of the Poisson-Boltzmann equation which determines the change in electrostatic potential at each residue due to changes in the protonation state of the protein. These changes are directly related to the pK$_a$ of the ionizable groups.

We used these methods to determine the charges of ionizable groups for the simulation of the R1-69 protein. Our calculations combined the cluster method described by Gilson [23] with the electrostatic free energy calculation method described by Antosiewicz et al. [22]. The Poisson-Boltzmann equation was solved using the package UHBD [24]. A Richards probe accessible surface definition with a dielectric constant of 80 for water and 20 for the protein was used. Calculations were performed with a probe sphere radius of 1.4 Å, a Stern layer of 2.0 Å and an ionic strength of 200 mM for the solvent surrounding the complex.

In Table 1, the different pK_a values obtained for the ionizable groups in the uncomplexed R1-69 repressor protein are listed. No calculations of the protein complexed with the OR1 operator were performed because the parameters (charges and radius) [24] have not been optimized for DNA. The pK_a values of the acidic groups are all less than 4 and so they are all charged at pH 5.5 and 6 (the pH of crystallization of the complexes). We protonate the N-termini, which have the same probability of being charged as uncharged. Similarly, the basic groups are all protonated under the same conditions because they all have pK_a greater than 9. Because of the approximate two-fold symmetry, the pK_as are very similar for both subunits of the protein.

Parameters and Protocols

The molecular modeling program CHARMM23 (version F2) was used for all the simulations [25] along with the most recent all-atom CHARMM force field (version 22) [26]. For both types of simulation, we used the Verlet algorithm with a 1 fs time step for the integration of Newton's equations. All simulations were run for 0.5 ns ($\equiv$ 500 picoseconds (ps)) and coordinate sets were saved each 0.1 ps giving 5000 structures for each trajectory. The initial velocities were assigned to the atoms from a Maxwell-Boltzmann distribution with a temperature of 300 K.

Molecular Dynamics Simulation with Explicit Solvent. We took the crystal structure of R1-69/OR1 [8] and deleted the three bases at the ends of each piece of DNA to obtain a

complex, R1-69/OR1, close to the one described by Anderson et al. [11]. Seventeen sodium ions were added to the system. Each position was chosen so that it lay between 3 and 4 Å from the phosphorus atom of a phosphate group along the bisector of the two free oxygen atoms [27]. The resulting system had a total charge of -3 comprised as follows: -28 (OR1), + 8 (R1-69), +17 (Na$^+$). To hydrate the system, a sphere with a radius of 30 Å containing water molecules of type TIP3P [28] was superimposed upon the complex. All water molecules closer than 2.8 Å from a protein or a DNA atom, or 2 Å from a Na$^+$ ion, were deleted.

The system was subdivided into two regions, a sphere of radius 27 Å where molecular dynamics was performed, and a shell between 27 and 30 Å where Langevin dynamics was performed. To mimic the effect of the environment outside the sphere, a stochastic boundary approximation was used which adds a term to the energy function that approximates the mean-field interaction due to the solvent [29].

One hundred steps of minimization, using a conjugate gradient algorithm, were performed on the system with all non-water molecules kept fixed. To fill up any holes created by the reorganization of the water molecules, a sphere of water was superposed with the system and all the water molecules of the sphere closer than 2 Å from an atom of the system were removed. The remaining water molecules were then added to the system. This "two step" cycle of minimization and superposition was repeated three times, until a correct hydration of the system was deemed to have occurred. The final system had 12765 atoms. The cutoff distance for the calculation of the non-bonded (electrostatic and Lennard-Jones) interactions was taken to be 13 Å and a switching function was applied between 12 and 13 Å. All the atoms of the protein and the DNA were free to move during the simulation.

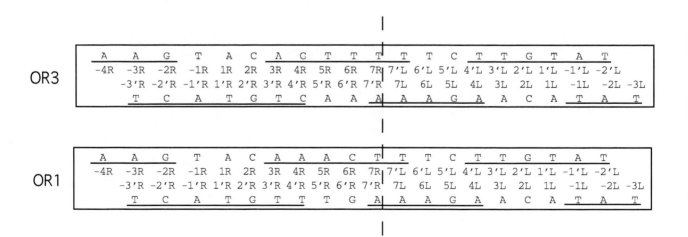

Figure 2. *Sequence of bases for the DNA operators OR1 and OR3. The first sequence represents the OR3 operator and the second the OR1 operator. The numbering of the DNA is the crystallographic numbering with the operator consisting of right (R) and left (L) parts. The bases whose exposed charges were scaled by a factor of 0.3 for the simulations with the implicit solvent model are underlined.*

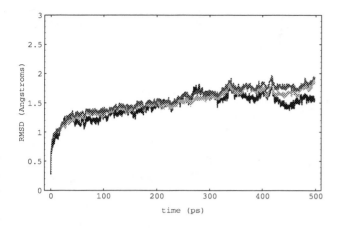

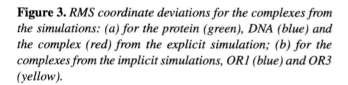

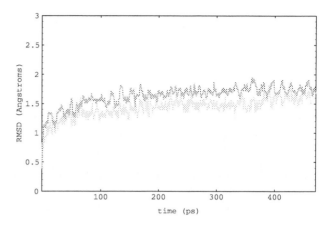

Figure 3. *RMS coordinate deviations for the complexes from the simulations: (a) for the protein (green), DNA (blue) and the complex (red) from the explicit simulation; (b) for the complexes from the implicit simulations, OR1 (blue) and OR3 (yellow).*

Molecular Dynamics Simulation with Implicit Solvent. To run the simulations with an implicit model for the solvent, we modified our treatment of the electrostatic interactions used in the previous simulation. On the basis of previous studies, the charges of the charged protein residues pointing toward the solvent were reduced [30]. For the DNA, we reduced the charges of the sugar-phosphate backbone which were not protected by the protein. This reduction was done by multiplying each charge by 0.3 and mimics two major physical effects – the reorientation of the water molecules, which has a shielding effect, and the effect of the counterion layer, which is not explicitly simulated. We used a van der Waals radius representation of the atoms in the complex to help us choose if the residue or base charges should be scaled. The sequence of the bases in the operators and a list of those bases whose charges were scaled as a result of this procedure are shown in Figure 2. It is to be noted that although the complexes were not explicitly solvated in the simulations, the crystallographic waters (44 for OR1 and 40 for OR3) **were** included in the simulation as these have important structural roles.

The structures were minimized until convergence, which was taken to be when the root mean square gradient fell below 0.1 kcal mol^{-1}·Å^{-1}. The non-bonded interaction cutoff distance was 15 Å. The extremities of the DNA were fixed during the entire simulation (last 3 bases). A distance-dependent dielectric was used so that the electrostatic interactions were calculated using the following expression

$$E_{elec} = \sum_{i<j<N} \frac{q_i q_j}{\varepsilon r_{ij}^2} \qquad (1)$$

ε was set to 1 for all simulations. Guenot and Kollmann have obtained a reasonable representation of various dynamical properties, including the RMS coordinate deviations and fluctuations of proteins, by using an equivalent model [31].

Each simulation was preceded by 20 ps of heating and 30 ps of equilibration (in which the temperature of the system was constrained to be between 290 and 310 K) before data collection began.

Results and Discussion

In this section, we describe the results of all three simulations. Firstly we present a general analysis of the results for all the simulations before we compare the differences between them in detail.

Behaviour of the Dynamics. To characterize the general behaviour of the dynamics, we calculated the root mean square coordinate deviations (RMSD) between the starting structures of each simulation and the structures from the dynamics trajectories. All the structures collected were first superimposed upon the initial structure using the algorithm due to Kabsch [32]. The RMS coordinate differences were then calculated between the starting and the reoriented structures. The results for the three dynamics simulations are shown in Figure 3. The RMSD between the crystallographic structures and the equilibrated structures (structures at t=0 in the dynamics) are, 0.7Å for the OR1 complex with explicit water, 1.8 Å for the OR1 complex with implicit water and 1.4 Å for the OR3 complex respectively.

For the simulation with explicit water we show the RMSD value for the operator, the repressor and the full complex. The RMSD values for the DNA are slightly inferior to the other RMSD values. At about 300 ps the three curves seem to reach a plateau value of about 1.6 Å for the protein and slightly higher for the full complex. The RMSD curves for the simulations with the implicit solvent model reach their stable values more quickly than those for the explicit solvent. For both implicit simulations the RMSD values are rela-

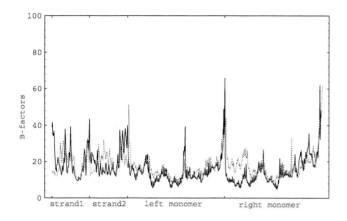
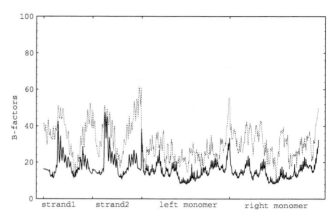

Figure 4. *The B-factors ($Å^2$) for the atoms of the complex for each of the three simulations: (a) the calculated B-factors for the OR1 complex are superposed with the results for the explicit simulation in black and the implicit simulation in red; (b) the calculated B-factors for the OR3 complex from the implicit simulation (black) are compared to the experimentally obtained ones (red).*

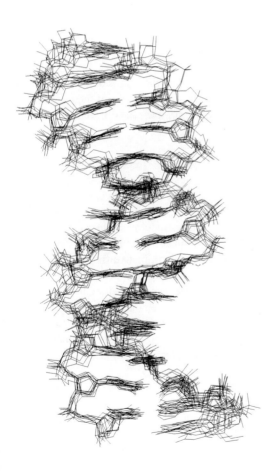

Figure 5. *Superposition of nine structures of the OR1 operator taken from the simulation with explicit solvent. All hydrogens are deleted.*

tively stable throughout the length of the simulation and stay around 1.6 Å for both complexes.

The mean coordinate fluctuations for the atoms in the complexes are shown in Figure 4. The fluctuations were computed for the backbone heavy atoms (N, CA, C and O) of the protein and for the P, O'x and C'x atoms of the DNA. The values are multiplied by 26.32 (= 8 π^2 / 3) so as to compare them directly with the crystallographic B-factors [33].

There is a reasonable correlation between the fluctuations calculated from the simulation with the OR3 operator and the implicit solvent model and the crystallographic B-factors. The calculated fluctuations are smaller than the experimental values (by about a factor of 2) but the higher mobility regions correspond in both sets of data. For the simulations with the OR1 operator a comparison with the B factors is more difficult as the latter values are very different from those for the OR3 complex. However, a comparison between the two simulations of this complex (one with implicit and one with explicit solvent) shows very similar behaviour with the largest fluctuations in roughly the same regions of the operator and the repressor. The results obtained by Arnold and Ornstein, in their study of different solvent models [34], showed greater values with the implicit simulations than with the explicit ones. A comparison of the fluctuations in the OR1 and OR3 complexes shows that while the fluctuations in the DNA are similar the fluctuations in the C-terminal regions of the protein chains are much larger.

Structure of the Hydrated R1-69/OR1 Complex

The DNA Structure. We superimposed 9 OR1 operator structures obtained at 50 ps intervals in the range 100 – 500 ps. They are displayed in Figure 5. The structure is well maintained even at the ends where there are no constraints. In contrast, structures from the first 100 ps do not superimpose very well due, in large part, to a shortening in the DNA strands which takes 100 ps to complete.

To obtain a clearer idea of structural changes, the widths of the minor and major grooves were calculated at various points along the operator. These were estimated as the minimum distance between two phosphates of each strand (a more

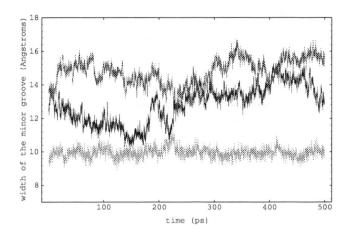

Figure 6. *The width of the minor groove of the OR1 operator DNA taken from the simulation with explicit solvent (green for the center, and blue and red for the extremities of the operator).*

rigorous analysis using the program Curves (see below) gives the same results). The major groove width is approximately constant throughout the simulation along the entire sequence. Its mean width is approximately 17.5 Å which compares with the value for a B-DNA strand of about 17.3 Å.

The minor groove shows more variation. This is shown in Figure 6. For B-DNA the average minor groove width is 11.5 Å while for the crystallographic R1-69/OR1 structure it is 8.8 Å in the middle of the operator and 14 Å at its ends. This

minor groove compression at the center of OR1 is due to a curvature in the DNA produced by interactions between the sugar phosphate backbone and helices 3 and 4 of the repressor [8]. During the simulation the minor groove width at the center of the operator stays constant with a mean value between 9 and 10 Å. In contrast, the width of the grooves at the ends of the chain varies much more, although its average value is about 15 Å.

Parameters characterizing the local DNA structure were calculated with the Curves program [35] for 11 structures taken at 50 ps intervals. These parameters are given in Tables 2a, 2b and 2c. The average helical twist is 34.5° and it differs by less than 3 % from the crystallographic average. The average rise per base pair is 3.25 Å and is the same as the one in the crystallographic structure. The average propeller twist angles show the same deformation of the DNA during the simulation as those in the crystallographic model, particularly the one due to the non-Watson-Crick hydrogen bond formed between O4 of T7R and N6 of A7L and the one between N2 of G5L and O2 of T4'L. The sugar puckers are predominantly in the forms C1'exo and C2'endo except for the bases T4'R, G6'R, A4L and T3'L. These sugars are in contact with the protein and consequently more likely to be subject to conformational changes.

The R1-69 Structure. The 434 repressor has a strong homology with the λ [36] phage repressor at the DNA-binding helix turn helix motif (HTH). The RMS coordinate deviations (calculated using the Cα atoms of the motif residues only) between both subunits of the crystallographic structures of the 434 repressor and the λ repressor vary between 0.48 Å and 0.59Å (depending on which monomers of each dimers are paired).

Table 2(a). *Angle variables of the DNA during the explicit simulation.*

Base Pair	Buckle		Propel		Base Pair	Rise		Twist	
	mean	var.	mean	var.		mean	var.	mean	var.
A2-T14	5.59	5.84	-2.72	5.81	A2-C3	-3.05	.23	-35.28	2.45
C3-G13	19.61	9.33	-17.55	4.85	C3-A4	-3.85	.39	-26.49	2.45
A4-T12	-3.10	7.17	-12.27	8.34	A4-A5	-2.95	.23	-34.67	2.07
A5-T11	4.35	5.78	-12.73	6.43	A5-A6	-3.19	.20	-34.60	3.25
A6-T10	2.25	6.95	-16.42	3.74	A6-C7	-3.20	.20	-38.94	4.08
C7-G9	3.87	9.18	-11.94	7.81	C7-T8	-3.27	.32	-33.00	3.24
T8-A8	6.58	10.13	-22.74	7.19	T8-T9	-3.31	.32	-35.79	2.38
T9-A7	-.32	5.23	-28.06	6.71	T9-T10	-3.22	.24	-38.55	3.63
T10-A6	-9.77	6.88	-24.97	4.59	T10-C11	-3.37	.22	-43.17	3.57
C11-G5	-17.95	7.36	-14.38	5.26	C11-T12	-3.12	.18	-29.83	2.53
T12-A4	-9.97	6.07	-17.47	4.43	T12-T13	-3.20	.17	-38.56	3.67
T13-A3	-8.54	7.91	-2.76	4.73	T13-G14	-3.31	.35	-25.41	4.85
G14-C2	-2.02	7.50	-1.49	11.77					

Table 2(b). *Sugar pucker conformations of the DNA during the explicit simulation. The number of structures (out of 11 total) at each conformation are listed.*

C4'exo	C3'endo	O1'endo	C2'endo	C1'exo			C1'exo	C2'endo	O1'endo	C3'endo	C4'exo
						T1	6	4	1	0	0
						A2	4	3	3	1	0
1	1	4	0	5	G13	C3	3	8	0	0	0
0	0	1	3	7	T12	A4	4	1	6	0	0
2	1	1	3	4	T11	A5	7	4	0	0	0
0	0	0	8	3	T10	A6	5	6	0	0	0
4	0	6	0	1	G9	C7	9	2	0	0	0
0	0	0	4	7	A8	T8	4	7	0	0	0
0	2	2	4	3	A7	T9	2	2	7	0	0
0	0	0	11	0	A6	T10	6	5	0	0	0
0	0	0	8	3	G5	C11	4	7	0	0	0
3	2	1	2	3	A4	T12	8	1	2	0	0
0	0	1	2	8	A3	T13	0	1	1	7	2
0	0	2	6	3	C2						
0	0	2	2	6	A1						

Table 2(c). *Non-Watson-Crick hydrogen bonds for the OR1 and OR3 operator sequences for the simulations with explicit and implicit solvent models. Only non-Watson-Crick bonds which occur in about 20% or more of the structures are listed. d_{min} is the minimum distance between the non-hydrogen atoms that define the bond during the simulation.*

Base (atom)	Base (atom)	OR1 explicit		OR1 implicit		OR3 implicit	
		d_{min}	Freq.	d_{min}	Freq.	d_{min}	Freq.
A1R(N6)	G2'R(O6)	2.7	19.7	2.6	29.8	2.7	18.5
A3R(N1)	G2'R(N1)	2.9	23.9	2.9	11.0		
A4R(N6)G(N2)	T5'R(O4)T3'R(O2)	2.7	22.6	2.7	19.2	2.6	61.5
G4R(O6)	A5'R(N6)					2.6	66.3
A5R(N6)T(O4)	G6'R(O6)A6'R(N6)	2.6	65.3	2.5	66.2	2.6	44.0
C6R(N4)T(O4)	A7'R(N6)			2.8	38.2	2.7	16.9
T7R(O2)	G6'R(N2)	2.6	77.1	2.6	87.0		
T7R(O4)	A7L(N6)	2.6	58.8	2.6	40.6	2.6	57.2
T7'L(O4)	A6L(N6)	2.6	56.7	2.7	12.0	2.7	17.8
T6'L(O4)	G5L(O6)	2.5	76.8	2.6	44.7		
C5'L(N4)	A4L(N6)	2.8	39.9	2.9	4.1	2.8	31.0
T4'L(O2)	G5L(N2)	2.6	68.6	2.6	73.7	2.6	70.9
G2'L(N1)	A3L(N1)	2.8	33.1	2.8	14.8	2.8	29.9

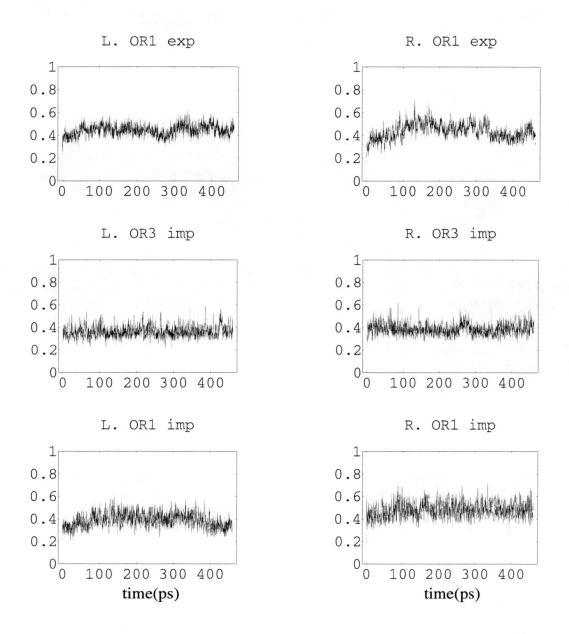

Figure 7. *The RMS coordinate deviations of the Cα atoms in the repressor helix-turn-helix motif for each of the three simulations.*

The calculated RMS coordinate deviations between the initial and intermediate trajectory structures during the simulations for this motif are shown in Figure 7 for the Cα atoms. It can be seen that, compared to the total protein RMS coordinate deviations, the HTH motif deviations are smaller and reach a plateau much more rapidly. If we compare the RMS coordinate deviations taking account of all the atoms for this motif, we obtain the same variations but with a slightly higher value.

Behaviour of the Sodium Ions. The self diffusion coefficient, D, for the sodium ions, was calculated using the Einstein relation:

$$D = \frac{1}{6} \lim_{t \to \infty} \frac{\partial}{\partial t} \left\langle \left(R(t) - R(0) \right)^2 \right\rangle \qquad (2)$$

Figure 8 shows a plot of the right hand side of equation 2 for the Na$^+$ ions in the explicit simulation. The diffusion coefficient is equal to $0.18(\pm 0.01) \times 10^{-9} \cdot m^2 \cdot s^{-1}$. This is lower than the values calculated by other workers. For example, van Gunsteren et al. [37] determined values in the range [0, 5 × 10^{-9}] $m^2 \cdot s^{-1}$ for a B-strand of DNA in water and Norbert and Nilsson [38] found a value of $1.3 \times 10^{-9} \cdot m^2 \cdot s^{-1}$ for a dinucleotide in water. In our simulation, the sodium ions are all located near a phosphate, and they stayed in the vicinity of these groups during the simulation. After 500 ps, only two

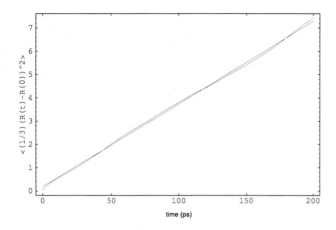

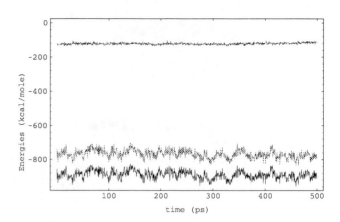

Figure 8. *A plot of the time-dependent mean square displacement (Å^2) for the sodium cations from the simulation with the explicit solvent model.*

Figure 9. *The non-bonded interaction energies between the protein and the DNA throughout the simulation using the explicit water model. The van der Waals energy is in blue, the electrostatic is in red and the total non-bonded energy is in black.*

ions are located at a distance greater than 10 Å from a phosphate. This, together with the fact that the volume of the simulation system is relatively small, helps to explain the low value of the diffusion coefficient.

DNA-Protein Contacts. Crystallographic Data: Analysis of the crystallographic complex R1-69/OR1 reveals that the R1-69 homodimer binds OR1 by an interaction of helix 3 with the major groove and by contacting the sugar phosphate backbone with two NH_2 groups at the N-terminal ends of helices 2 (Asn16,Gln17) and 4 (Arg43). In the following, we only consider the right half of the complex as the left side is the same by symmetry. A1R(N6,N7) has a double hydrogen bond with Gln28, and G2'R(O4,O6) with NE2 of Gln29. Base pair 3 does not have polar contacts with R1-69, in spite of the presence of the side chain of Glu32 which points toward it. However, a van der Waals pocket is created between Thr27, Gln29 and T3'R. The fourth base pair (A-T) has a hydrogen bond with Gln33 (T4'R O4) and a van der Waals contact with Gln29 and Ser30. The other hydrogen bonds between DNA and R1-69 are with a phosphate group–T1R with Arg10 and Gln17, the NH groups of Lys40 and Arg41 with G6'R and the NH of Arg43 with T5'R. A1R has a hydrogen bond with Gln17 and Asn36, and G2R with Asn16. These last three interactions are not detected by the ethylation interference experiment [8]. There are many van der Waals contacts between the bases T5'R, G6'R, T-1R and A1R and R1-69.

Simulation Results: The electrostatic and van der Waals energies between OR1 and R1-69 are shown in Figure 9. The van der Waals energy lies between about -100 and -140 kcal·mol⁻¹ throughout the simulation whilst the electrostatic interaction energy is about six times larger in absolute value and fluctuates much more.

A major part of the electrostatic interaction is represented by the hydrogen bond energy. The criterion that we used for

the existence of a hydrogen bond was to say that a bond existed if the distance between the donor (D) and the acceptor (A) was less than 3.2 Å. In the case of multiple hydrogen bonds involving the same donor or acceptor, we calculated the angle D-H··· A and kept the bond with the largest angle (i.e. nearest 180°). In Table 3, we list the hydrogen bonds between the DNA and the protein which appeared in more than 10% of the structures of a trajectory. Most of the hydrogen bonds that can be seen in the crystallographic model are conserved during the simulation. The two hydrogen bonds between Lys40L (resp. Lys40R) with O1P from 6'L (or 6'R)

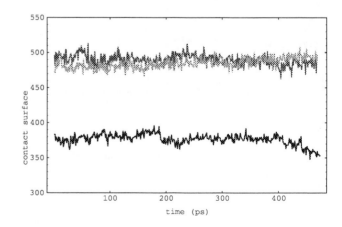

Figure 10. *The contact surface between the protein and the DNA (Å^2) for the three simulations: OR1 explicit solvent (blue); OR1 implicit solvent (red); OR3 implicit solvent (green).*

Table 3: *The hydrogen bonds between R1-69 and OR1 in the simulation with explicit water. R1 is the first strand of DNA and R2 is its complement. SL is the left subunit and SR the right. The base and residue numberings are the same as in the crystallographic model [8]. d_min is the minimum distance between the non-hydrogen atoms that define the bond during the simulation. An asterisk next to a bond means that the bond exists in the crystallographic structure.*

R1		SL				R1		SR			
Base	atom	residue	atom	d_{min}	Freq	Base	atom	residue	atom	d_{min}	Freq
7'L*	O3'	R43	NH2	2.76	49.	1R*	O5'	Q17	OE1	2.52	10.2
6'L*	O1P	K40	N	2.57	34.6	1R*	O1P	N36	ND2	2.51	47.3
6'L*	O2P	R41	O	2.68	21.4	1R*	N7	Q28	NE2	2.72	68.9
6'L*	O2P	K40	N	2.63	37.7	1R*	N6	Q28	OE1	2.56	79.8
6'L*	O2P	R41	N	2.50	99.8	1R	N6	Q29	NE2	2.83	18.7
5'L*	O2P	R43	N	2.53	94.9	2R*	O1P	K38	NZ	2.45	99.5
5'L	O2P	F44	N	2.63	34.5	6'L*	O2P	R43	NH2	2.46	95.9
5'L	O1P	S30	OG	2.47	62.4						
5'L	N4	Q33	OE1	2.57	83.9						
4'L*	O4	Q33	NE2	2.57	85.7						
3'L	O4	Q29	NE2	2.66	21.6						
2'L*	O6	Q29	NE2	2.55	70.4						

R2		SL				R2		SR			
Base	atom	residue	atom	d_{min}	Freq	Base	atom	residue	atom	d_{min}	Freq
1L*	N7	Q28	NE2	2.69	59.8	6'R	O5'	K40	N	2.79	14.4
1L*	N6	Q28	OE1	2.50	91.0	6'R*	O1P	K40	N	2.63	17.5
1L	N6	Q29	NE2	2.83	10.5	6'R*	O2P	K40	N	2.61	10.3
6'R*	O3'	R43	NH2	2.70	11.	6'R*	O2P	R41	N	2.50	98.6
6'R*	O2P	R43	NE	2.63	15.8	6'R*	O2P	R41	O	2.63	28.7
						5'R*	O2P	R43	N	2.48	99.8
						4'R*	O4	Q33	NE2	2.63	65.0
						4'R	O1P	T26	OG1	2.40	82.8
						2'R*	O6	Q29	NE2	2.53	97.4
						1'R	O4	Q29	NE2	2.62	29.4

are broken quickly, although we find that the hydrogen bond on the left half of the site is found in the average structure for the simulation, in agreement with the crystallographic structure.

Bases -2L and -1L are not in our model and the phosphate group of T-1R is not represented, so there is no possibility of van der Waals contacts here. Moreover, as no constraints were applied to the DNA strand during the dynamics, the hydrogen bonds at the end of the operator with Asn16, Gln17 and Asn36 disappear early in the simulation. Two hydrogen bonds not described in the crystallographic model, however, are observed. These are C5'L (O1P) with Ser30L

(Oγ) and C5'L (N4) with Gln33L (Oε1). In the crystallographic model [8] they underline the proximity of the Oε1 atom and the C5 atom (3.6 Å) of C5'L. In the right half, these hydrogen bonds cannot occur because of the methyl group of the thymine that replaces the cytosine.

Another way to characterize the interactions between the repressor and the operator is by looking at the contact surface between them. This surface was calculated for the structures from the three simulations and is displayed in Figure 10. We compute these surfaces without taking account of the water molecules using the following expression

$$S_{contact} = \frac{1}{2}\left(S_{ADN} + S_{protein} - S_{complex}\right) \qquad (3)$$

where S is the van der Waals surface calculated with a probe of radius 1.6 Å. The contact surface values for the crystallographic complexes are 362 Å^2 for the R1-69/OR1 truncated complex used in the explicit simulation, 420 Å^2 for the entire R1-69/OR1 complex and 439 Å^2 for the R1-69/OR3 crystallographic complex. The difference between the contact surfaces of the truncated and the crystallographic structures at time t=0 shows the importance of the bases that are not explicitly part of the operator sequence. In the explicit simulation, the contact surface is around 370 Å^2, close to the crystallographic value. It decreases slowly at the end of the simulation. This may be explained by the lack of the van der Waals contacts at the end of the DNA. In the implicit simulation, the surface contacts seem more stable, although there are more of them than in the crystallographic structure. This is due to a lack of water molecules, which is compensated by a closer contact between the DNA and the protein.

Interactions with Water Molecules. Forty four water molecules are present in the crystallographic structure of the complex between the repressor and the OR1 sequence. They contribute to multiple bridging interactions between the protein and DNA through hydrogen bonds. Table 4 lists those bridges observed in the simulation with lifetimes of more than 30 ps. Those between Ser30L and the phosphates of 5'L and 4'L are conserved in the simulation. The three bridges between Gln33, the DNA (C5'L and A3R) and a water molecule observed in the crystallographic structure are slightly modified in the simulation. The lifetimes are short (around 35 ps) for A4R(N6)–(Oε_2) in the right site while in the left site Gln33 contacts A3L(N6) and A4L(N6) for a large portion of the simulation. The hydrogen bonds mediated by a water mol-

Table 4. *Indirect hydrogen bonds between R1-69 and OR1 in the simulation with explicit solvent. The notation used is the same as in Table 3. <T> and n are the mean lifetime of the bridge (in ps) and the number of different intermediate water molecules which create a bridge between the protein and the DNA.*

R1		SS3				R1		SS4			
Base	atom	residue	atom	<T>	n	Base	atom	residue	atom	<T>	n
T4'L	O1P	T26	OG1	61	1	A1R	O1P	Q17	NE2	126	1
T4'L	O1P	T27	OG1	193	1	A1R	O1P	E32	O	83	1
T4'L	O1P	S30	OG	42	1	A1R	O1P	E32	OE1	80	2
C5'L	O5'	T26	OG1	117	1	A1R	O1P	E35	OE2	39	1
C5'L	O5'	S30	OG	151	1	A1R	O1P	N36	ND2	99	1
C5'L	O1P	S30	O	180	1	C2R	N4	E32	OE1	301	1
C5'L	O1P	S30	OG	32	2	A4R	N6	Q33	OE2	34	1
C5'L	O1P	T39	OG1	397	1						
T6'L	O1P	K38	O	126	2						

R2		SS3				R2		SS4			
Base	atom	residue	atom	<T>	n	Base	atom	residue	atom	<T>	n
C2L	O1P	E32	OE1	168	3	G2'R	N7	Q29	NE2	93	1
C2L	N4	E32	OE2	134	1	T4'R	O1P	T26	OG1	72	1
C2L	O1P	N36	ND2	53	1	T5'R	O1P	S30	O	94	1
A3L	N7	E32	OE2	114	1	T5'R	O1P	T39	OG1	399	1
A3L	N6	E32	OE2	282	1	T5'R	O3'	R43	NH1	33	1
A3L	N6	Q33	NE2	42	1	G6'R	O2P	R41	NH2	43	1
A4L	N6	Q33	OE1	97	1	G6'R	O1P	K40	NZ	46	2
A7'R	O4'	R43	NH1	48	1	G6'R	O1P	K38	O	41	3
A7'R	N3	R43	NH2	111	1	G6'R	O2P	R41	NE	167	2
A7'R	N3	R43	NH1	37	2	G6'R	O3'	R43	NH2	151	1

ecule between Glu35L–A1L and between Gln29L–C2L are found in the left half site in the crystallographic structure, but are located uniquely in the right half site of our simulation.

The following interactions exist in the simulation but not in the crystallographic structure. In the left half there are many indirect contacts between Ser30, Thr26 and Thr27 which do not occur in the right half. In contrast, Lys40 and Arg41 interact with the base A7'R in the right half but not in the consensus left half. Four main interactions are conserved in both halves: Glu32R with C2R and A1R, Lys38R (O) with G6'R and Thr39R with T5'R. The greater number of indirect interactions in our simulations is due to the larger number of water molecules in the simulation than in the crystallographic model and is evidence of the importance of these types of interaction for stabilizing the complex.

Comparison of the Different Simulations

The structures from the three simulations were compared using the programs Curve and Pcurve [35, 39]. The structures of the DNA and the protein are described using the helical parameters of Sklenar et al. [39] for the protein, and those given at an EMBO workshop for the DNA [40].

The global dynamical behaviour of the simulations with the implicit solvent model have already been discussed in a previous section. In Figure 3, we saw that the RMS coordinate deviations for the implicit solvent model reach a plateau after about 200 ps whose values are of the same order of magnitude as those obtained by Falsafi and Reich on their study of B-DNA [41]. The DNA's RMSD values are smaller than the protein's in the implicit model as is also the case for the explicit simulation. The RMSD values for the HTH DNA-binding motif show a variation in its structure which is slightly greater than the one from the explicit model. This fact can be explained by the lack of solvating water molecules which creates a reorganization of some hydrogen bonds.

Hydrogen Bonds. An analysis of the hydrogen-bond bridges between the protein and the DNA permits an understanding of some of the structural changes in the different simulations (see Figure 11).

There are two different interactions between R1/69 and the bases 4' and 5' in both halves of the operator OR1. The interaction in the left half is hydrophilic but hydrophobic in the right half. The hydrogen bonds between Ser30L and the base 5'L oscillate between being direct and mediated by a water molecule in both OR1 simulations, but this interaction is always mediated by a water molecule in the crystallographic structure and in the OR3 simulation. The residues Thr26L and Thr27L have direct and water-mediated hydrogen bonds in the implicit simulation with OR1 and only water-mediated contacts in the other simulations. In the right site, hydrophobic contacts replace the multiple hydrogen bonds which occur between the protein, the water and the DNA. These produce a contact between Thr26R and the base 4'R

that allows the formation of a hydrogen bond. We find in both simulations with OR1 that there is a bridging interaction between T4'R(O1P)-Thr26R which is not found in the crystallographic model. This interaction is also found in the simulation of the OR3 complex. In the left half site the interactions between Thr26 and Thr27 and the DNA are either direct or mediated by a water molecule in the implicit simulation with OR1. In the other two simulations, these contacts are always via a bridging water.

The hydrogen bond between Gln33R and T4'R in both OR1 simulations does not exist in the OR3 simulation, although it was described in the crystallographic complex [10]. The water bridges Q33(NE2)–A3R and Q33(OE1)–A5'R (between R1-69 and OR3) replace these direct and specific contacts. This is the major difference between the implicit OR1 and OR3 complexes.

The interaction between Lys38L(NZ), in the implicit simulations, disappears in the explicit one. Interactions with the side chains of Arg43 are slightly different in both simulations of OR1, although the OR3 arrangement is identical to the one found in the OR1 explicit simulation.

Interactions at the end of the DNA cannot be compared with the explicit model because it was truncated. Asn16R has a weak hydrogen bond at the beginning of the simulation which disappears in the OR3 simulation. This interaction does not exist in the OR1 simulation, although it was present in the crystallographic model but not defined by the ethylation experiment [8]. The side chains of Gln17 and Asn36 have some hydrogen bonds with the DNA despite the fact that the ethylation experiment gave no indication of such an interaction. Many hydrogen bonds are created with water molecules in the three models, although some of these interactions are replaced by direct contacts between protein and DNA in the simulations with implicit solvent models. These crucial interactions are made with Glu32L(R) and Asp36L with A1L(R) and C2L and correspond to the major part of the new hydrogen bonds created in the simulations with the implicit solvent model.

Comparison of the Three Simulations with the Curve Program. The DNA Structures: The Curves program [35] calculates the global and local geometrical properties of DNA using a comprehensive set of parameters. We computed these parameters for the structures of the OR1 and OR3 operators obtained every ten picoseconds from the three simulations, leading to more than 160,000 parameters. We represent these data using a curve, dial and window representation that was inspired by the work of Swaminathan et al. [42]. To reduce the very large amount of data, we superposed all graphs obtained for OR1 in the implicit simulation, with, in a first step, those obtained for OR3 and in a second step, with those obtained for the explicit simulation. For a useful schematic representation of the helicoidal parameters, readers are referred to the paper by Lavery and Sklenar [43].

The first comparison concerns the geometry of the minor and major grooves, and the bending of the two operators.

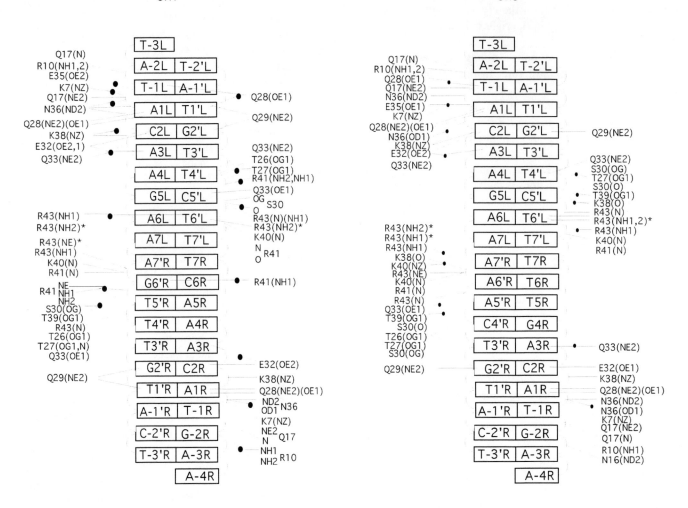

Figure 11. *Schematic representation of the different hydrogen bonds between the R1/69 protein and the DNA for the implicit simulations.*

The work of Fujimoto and Schurr [18], which measured DNA torsion constants from time-resolved fluorescence polarization anisotropy, did not show a large correlation between the sequence of the operator and its structure. Koudelka and Carlson [19] showed that there is a relationship between the sequence of the central bases of an operator and its intrinsic twist. A theoretical study of multiple sets of different sequences of DNA made by Poncin et al. [43] improved the understanding of the role of base sequence on the DNA conformation using the Curves program. They showed that many substates exist which depend on the sequence of the DNA.

The structure of the DNA is strongly linked to the state of its grooves. In Figure 12, density plots are used to show the differences in the widths of the major and minor grooves between the OR1 and OR3 sequences during the implicit simulations and between the OR1 (OR3) implicit and OR1 explicit simulations [44]. The darker the color, the lower the

density, so the density is positive or negative if the color is white or black respectively.

There is a strong correlation between the depth and the width of the grooves and so we only describe the width of the major and of the minor grooves. The greatest variation is seen around the mutation triplet (AAC) for the implicit simulations. The width of the minor groove is larger for OR3 than for OR1 throughout the simulation. The depth of the major groove at the site A3R–A4R is also greater for OR3 than for OR1, which accounts for the larger accessibility of the base 4'R in OR1 to the protein than in OR3. It also explains the disappearance of the direct hydrogen bond between Q33R (NE2) and T4'R in the simulation with OR3. This is replaced by an indirect water-mediated interaction between Q33R (NE2) and T3'R. The greatest variation in the DNA structures between the explicit and implicit OR1 simulations is also found around the triplet (AAC). The major groove's width is greater for the explicit than for the implicit simulation making the base 4'R more accessible in the former.

The global curvature of DNA can be evaluated by comparing both the end-to-end distance, and the helix axis path length described by Ravishanker et al. [45]. This calculation

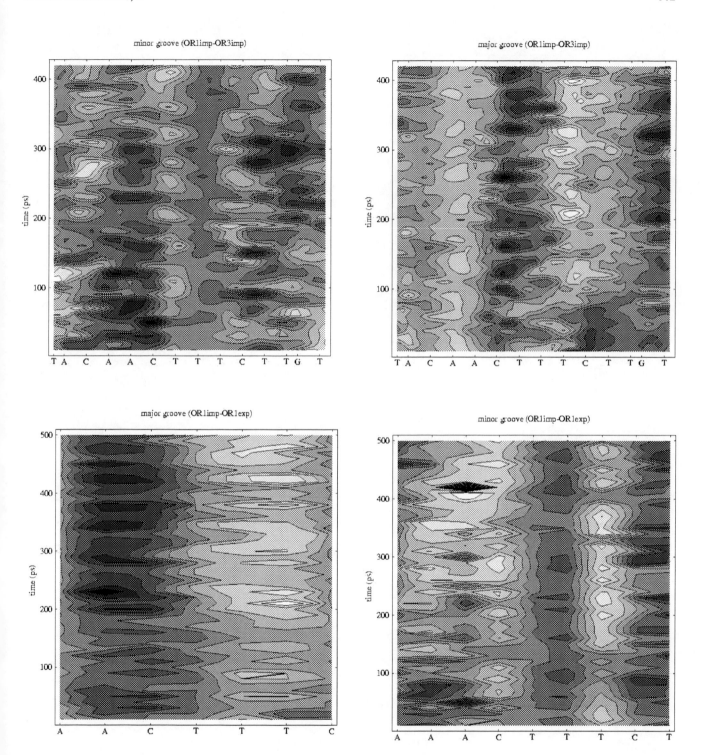

Figure 12. *(continues next page) Density plots to illustrate the difference in the width of the minor and major grooves for the OR1 and OR3 sequences from the simulations with the three models. A zero difference is denoted by a medium grey colour, a positive difference is darker and a negative difference lighter.*

was done only for the implicit simulations as these both have the full 20 base pairs of DNA. The curves showing these quantities reveal that there is no difference between the two global curvatures. The maximum variation for both ratios, of helix axis path length and end-to-end distance, between OR1 and OR3 in both implicit simulations is lower than 2 %. Because of their similarity these data are not shown.

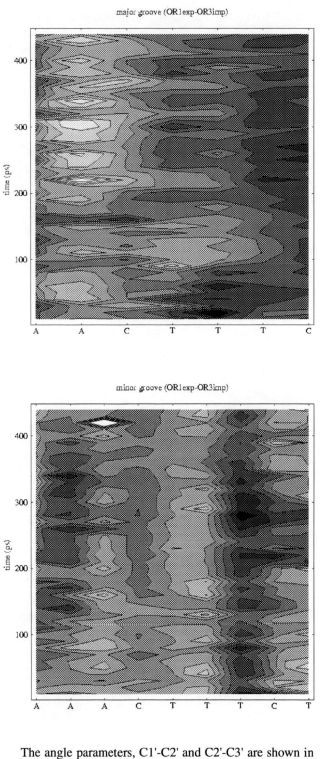

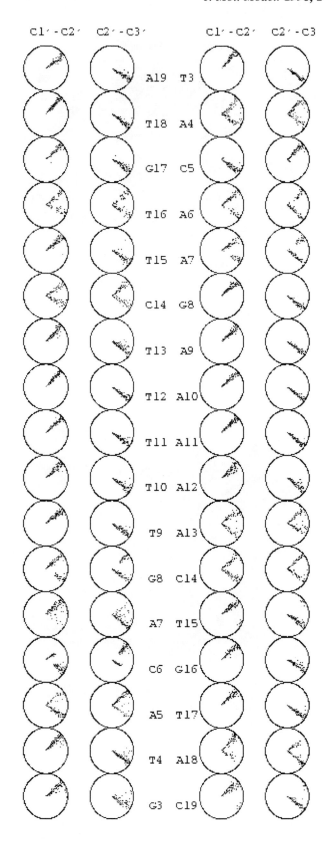

The angle parameters, C1'-C2' and C2'-C3' are shown in Figure 13 for the implicit simulations. They are different for OR1 and OR3 in both the consensus left and the non-consensus right halves of the site. We obtain the same variations with the torsion parameters α, β, γ, ε and ζ – these differences exist between both halves. They are mainly located near the regions A3L, A4L and A3R, C4'R.

A comparison of the global axis curvature parameters (not shown) from both implicit simulations shows differences for

Figure 13. *Analysis of the DNA structure: dihedral angles of the sugar for both implicit simulations using the Curve program (OR1 in red and OR3 in green).*

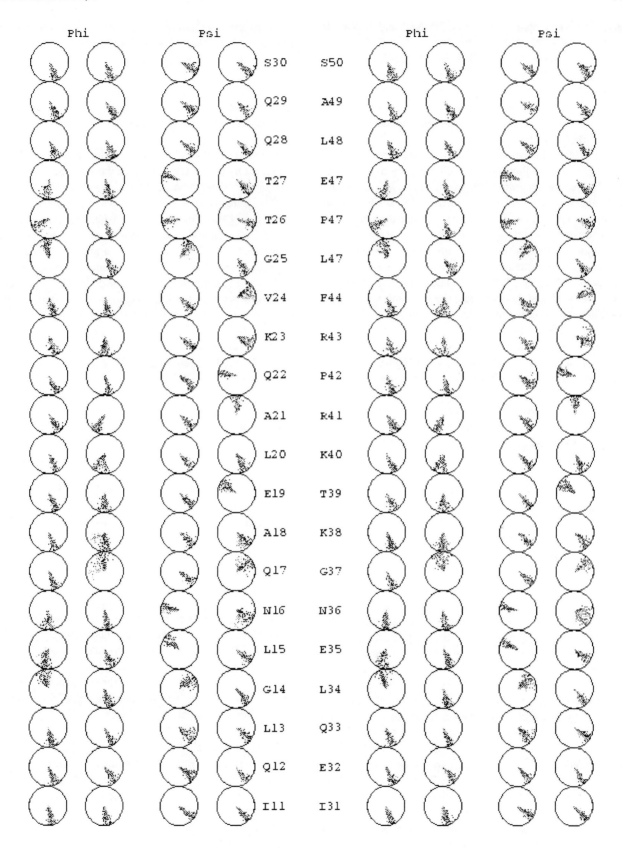

Figure 14. *The angles Φ and Ψ for the R1/69 homodimer for the three simulations using the PCurve program ((OR1 implicit solvent in black, OR3 implicit solvent in red and OR1 explicit solvent in green).*

the angle parameters to the bases C2R-A3R and A3R-G4R, for the axis X-displacement parameter of T7R-T7'R and for the axis tip parameter for A3R-G4R and T6'L-C5'L. The variation in the other parameters are too scattered to see trends between the values of both simulations. There are no significant changes for the inter-base pair parameters between the three simulations and the small variations that do appear are not localized at a specific spot.

The Protein Structure: An analysis of the three structures of R1-69 in the implicit and explicit simulations was made with the Pcurve program [39]. In the same way as for the DNA analysis, the protein structure is described using helicoïdal parameters. These parameters are displayed with dials and windows and superposed for the three simulations. The major changes between the simulations for the global peptide-axis parameters occur for the residues L13, G14 and E35, N36, G37, K38, R43, F44. There are two groups of residues (13–15 and 35–44) where the differences are important. These form turns linking helix 1 with helix 2 and helices 2 and 3. These variations exist in both halves of the site, but they are greater in the non-consensus right site. The values of the angles Φ and Ψ are shown in Figure 14. The Ω angle parameters are not shown since they change very little. Both halves of the site for the OR3 simulation have the same profile, which means that the effect of the mutation does not change the structure of the protein backbone. The greatest variations for these angles are located in both the turns described previously.

Conclusions

Three simulations have been performed of the 434 repressor protein complexed to two different operator sequences of DNA, OR1 and OR3. The major goal of this work was to improve the understanding of the interaction between R1/69 and its different operators by using simulations to obtain dynamical views of the complexes. All the simulations display reasonable behaviour despite the simplifications of the shortening of DNA for the explicit simulation and a use of a "distance-dependent dielectric" for the implicit simulations.

The first point to be underlined deals with the local structure of the DNA, which can be analyzed by characterizing the topology of its grooves. One of the major factors influencing the different specificity of the protein for the two operators is that the minor groove of the OR1 operator at the AAC triplet is shallower and the major groove is narrower than the corresponding grooves at the GTT triplet of the OR3 operator. This effect appears to be due to the closer contact that occurs between the base and the sugar-phosphate backbone in the OR1 sequence and affects the accessibility of the bases for interactions with the protein.

Huang et al. [17] showed that the helical motif Gln28-Gln29-Ser30-X31-X32-Gln33 is the unique sequence which gives a specific binding to 434 repressor for OR1. The strong preference for Gln amino acids is explained by their ability to hydrogen bond and to form van der Waals contacts with DNA. The important roles of Ser30L and Ser30R is surprising. In the crystallographic structure they have only water mediated contacts with DNA while in both OR1 simulations Ser30L has direct hydrogen bonds with DNA. The residue Thr27 is not important [17] and can be replaced by others without affecting the specificity. In the explicit simulation, both Thr27s have no direct hydrogen bonds with DNA. In the implicit simulation, the contacts that are observed can be explained by the reorganization of the side chains due to the lack of additional solvating water.

There are no significant changes in the base-base parameters during the simulation. The backbone parameters are very different in both consensus sites of OR1 and OR3. This shows that the DNA structure has a strong dependence on external parameters, such as its sequence or a complexed molecule. For the protein, the major differences between the structures occur in the turn regions – the remainder of the backbone stays essentially similar – and implies that the protein can be well approximated as a relatively rigid entity outside of the turns.

An implicit representation of the solvent, which uses a distance dependent dielectric with a reduction in the accessible charges, is an inexpensive method with which to model the solvation of macromolecules. It gives results which compare favourably with those obtained from a simulation with explicit waters. For example, the RMS coordinate displacement for both implicit and explicit simulations reaches a plateau around 1.8 Å and the heavy atom fluctuations of the R1-69/OR1 complexes are very similar in both types of simulations. Of course, an implicit model has some limitations. For example, the absence of water results in a partial reorganization of the DNA and the repressor, although this can be avoided by adding some explicit water molecules at the DNA-protein interface in addition to the ones defined crystallographically. In conclusion, for the simulations performed here, it appears that the implicit representation of solvent gives results of reasonable accuracy while enabling simulations to be performed with much less computational expense. The approximations introduced by the use of the implicit model also seem to be less drastic than the one resulting from the truncation of the three bases at the edge of the DNA which was necessary in the simulations with the explicit solvent model.

Acknowledgements The authors would like to thank Dr. Patricia Amara for her comments on the manuscript and the Institut de Biologie Structurale – Jean-Pierre Ebel, the Commissariat à l'Energie Atomique and the Centre National de la Recherche Scientifique for support of this work.

References

1. Wharton, R.P.; Brown, E.L.; Ptashne, M. *Cell* **1984,** *38*, 361.

2. Koudelka, G.B.; Harbury, P.; Harrison, S.C.; Ptashne, M. *Proc. Natl. Acad. Sci. USA* **1988**, *85*, 4633.

3. Koudelka, G.B.; Harrison S.C.; Ptashne, M. *Nature* **1987**, *326*, 886.

4. Bushman, F.D. *J. Mol. Biol.* **1993**, *230*, 28.

5. Harrison, S.C.; Aggarwal, A.K. *Annu. Rev. Biochem.* **1990**, *59*, 933.

6. Harrison, S.C. *Nature* **1991**, *353*, 715.

7. Bernstein, F.C.; Koetzle, T.F; Williams, G.J.B.; Mayer, E.F.; Brice, J.M.D.; Rodgers, J.R.; Kennard, O.; Shimanouchi T.; Tasumi, M. *J. Mol. Biol.* **1977**, *112*, 535.

8. Aggarwal, A.K.; Rodgers, D.W.; Drottar, M.; Ptashne, M.; Harrison, S.C. *Science* **1988**, *242*, 899.

9. Shimon, L.J.W.; Harrison, S.C. *J. Mol. Biol.* **1993**, *232*, 826.

10. Rodgers, D.W.; Harrison, S.C. *Structure* **1993**, *1*, 227.

11. Anderson, J.E.; Ptashne, M.; Harrison, S.C. *Nature* **1987**, *326*, 846.

12. Bushman, F.D.; Anderson, J.E.; Harrison, S.C.; Ptashne, M. *Nature* **1985**, *316*, 651.

13. Bushman, F.D.; Ptashne, M. *Proc. Natl. Acad. Sci. USA* **1983**, *83*, 9353.

14. Koudelka, G.B.; Lam, C-Y. *J. Biol. Chem.* **1993**, *268*, 23812.

15. Bell, A.C.; Koudelka, G.B. *J. Biol. Chem.* **1995**, *270*, 1205.

16. Bell, A.C.; Koudelka, G.B. *J. Mol. Biol.* **1994**, *234*, 542.

17. Huang, L.; Sera, T.; Schultz, P.G. *Proc. Natl. Acad. Sci. USA* **1994**, *91*, 3969.

18. Fujimoto, B.S.; Schurr, M. *Nature* **1990**, *344*, 175.

19. Koudelka, G.B.; Carlson, P. *Nature* **1992**, *355*, 175.

20. Bashford, D.; Karplus, M. *Biochem.* **1990**, 29, 10219.

21. Bashford, D.; Karplus, M. J. Phys. Chem. 1991, *95*, 9556.

22. Antosiewicz, J.; McCammon, J.A.; Gilson, M.K. *J. Mol. Biol.* **1994**, *238*, 415.

23. Gilson, M.K. *Proteins: Structure, Function, and Genetics* **1993**, *15*, 266.

24. Davis, M.E.; Madura, J.D.; Luty, B.A.; McCammon, J.A. *Comp. Phys. Comm.* **1991**, *62*, 187.

25. Brooks, B.R.; Bruccoleri, R.E.; Olafson, B.D.; States, D.J.; Swaminathan, S.; Karplus, M. *J. Comp. Chem.* **1983**, *4*, 187.

26. MacKerell Jr., A. *Developmental Residue Topology File for Proteins Using All Hydrogens.*

27. Seibel, G.L.; Singh, U.C.; Kollman, P.A. *Proc. Natl. Acad. Sci. USA* **1985**, *82*, 6537.

28. Jorgensen, W.L.; Chandraseklar, J.; Madura, J.D.; Impey, R.W.; Klein, M.L. *J. Chem. Phys.* **1983**, *79*, 926.

29. Brooks III, C.L.; Karplus, M. *J. Chem. Phys.* **1983**, *79*, 6312

30. Mouawad, L.; Perahia, D. *J. Mol. Biol.* **1996**, *258*, 393.

31. Guenot, J.; Kollman, P.A. *Protein Sci.* **1992**, *1*, 1185.

32. Kabsch, W. *Acta Crystallog. Sect. A* **1978**, *34*, 827.

33. McCammon, J.A.; Harvey, S.C. *Dynamics of Proteins and Nucleic Acids.* Cambridge University Press, Cambridge, 1987.

34. Arnold, G.E.; Ornstein, R.L. *Proteins: Structure, Function, and Genetics* **1994**, *18*, 19.

35. Lavery, R.; Sklenar, H. *J. Biomol. Str. Dynam.* **1988**, *6*, 63.

36. Pabo, C.O.; Aggarwal, A.K.; Jordan, S.R.; Beamer, L.J.; Obeysekare, U.R.; Harrison, S.C. *Science* **1990**, *247*, 1210.

37. van Gunsteren, W.F.; Berendsen, H.J.C.; Geurtsen, R.G.; Zwinderman, H.R.J. *Ann. N.Y. Acad. Sci.* **1986**, *482*, 287.

38. Norberg, J.; Nilsson, L. *Chem. Phys. Lett.* **1994**, *224*, 219.

39. Sklenar, H.; Etchebest, C.; Lavery, R. *Proteins: Structure, Function, and Genetics* **1989**, *6*, 46.

40. Dickerson, R.E.; Bansal, M.; Calladine, C.R.; Diekmann, S.; Hunter, W.N.; Kennard, O.; Lavery, R.; Nelson, H.C.M.; Olson, W.K.; Saenger, W.; Shakked, Z.; Sklenar, H.; Soumpasis, D.M.; Tung, C-S.; von Kitzing, E.; Wang, A.H-J.; Zhurkin, V.B. *EMBO J.* **1989**, *8*, 1; *J. Biomol. Struct. Dynam.* **1989**, *6*, 627; *J. Mol. Biol.* **1989**, *205*, 787.

41. Falsafi, S.; Reich, N.O. *Biopolymers* **1993**, *33*, 459.

42. Swaminathan, S.; Ravishanker, G.; Beveridge, D.L.; Lavery, R.; Etchebest, C.; Sklenar, H. *Protein: Structure, Function, and Genetics* **1990**, *8*, 179.

43. Poncin, M.; Hartmann, B.; Lavery, R. *J. Mol. Biol.* **1992**, *226*, 775.

44. Stofer, E.; Lavery, R. *Biopolymers* **1994**, *34*, 337.

45. Ravishanker, G.; Swaminathan, S.; Beveridge, D.L.; Lavery, R.; Sklenar, H. *J. Biomol. Str. Dynam.* **1989**, *4*, 669.

46. Lavery, R.; Sklenar, H. *J. Biomol. Str. Dynam.* **1989**, *6*, 655.

J. Mol. Model. **1996**, 2, 446 – 455

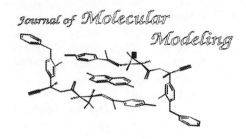

© Springer-Verlag 1996

Knowledge-based Homology Modeling and Experimental Determination of Amino Acid Side Chain Accessibility by the Laser Photo CIDNP (Chemically Induced Dynamic Nuclear Polarization) Approach in Solution: Lessons from the Small Sialidase of *Clostridium perfringens*

Hans-Christian Siebert[1,2], Emadeddin Tajkhorshid[3,4], Claus-Wilhelm von der Lieth[3], Reinhard G. Kleineidam[5,6], Susanne Kruse[6], Roland Schauer[6], Robert Kaptein[2], Hans-Joachim Gabius[1], and Johannes F. G. Vliegenthart[2,*]

[1] Institut für Physiologische Chemie, Tierärztliche Fakultät, Ludwig-Maximilians-Universität, Veterinärstr. 13, D-80539 München, Germany

[2] Bijvoet Center for Biomolecular Research, University of Utrecht, P.O. Box 80075, NL-3508 TB Utrecht, The Netherlands
Tel: ++31 30 2532184, Fax: ++31 30 2540980 (vlieg@cc.ruu.nl)

[3] Zentrale Spektroskopie, Deutsches Krebsforschungszentrum, Im Neuenheimer Feld 280, D-69120 Heidelberg, Germany

[4] Department of Medicinal Chemistry, School of Pharmacy, Tehran University of Medical Sciences, P.O. Box 14155/6451, Tehran, Iran

[5] Biochemisch Laboratorium, University of Groningen, Nijenborgh 4, NL-9747 AG Groningen, The Netherlands

[6] Biochemisches Institut, Christian-Albrechts-Universität, Olshausenstraße 40, D-24098 Kiel, Germany

Received: 9 September 1996 / Accepted:11 November 1996 / Published: 22 November 1996

Abstract

The success of knowledge-based homology modelling is critically dependent on the predictive potency of the program structure-based calculations, which attempt to translate homologous sequences into three-dimensional structures, and on the actual relevance of the crystal structure for the protein topology. As quality control, experimental data for selected parameters of the protein's conformation are required. Using the crystal structure of the sialidase of *Salmonella typhimurium* as framework for model building of the homologous enzyme from *Clostridium perfringens*, a set of energy-minimised conformers is derived. These proteins present e.g. Tyr, Trp and His residues with an assessable area on the surface, since the side chains of these amino acid residues are responsive to chemically induced dynamic nuclear polarization (CIDNP), monitored by NMR. Hence, as first lesson, a comparative analysis for model-derived and experimentally determined values can be performed. The second lesson of this study concerns the notable impact of single amino acid substitutions (Tyr/Phe, Cys/Ser) on the surface accessibility of the CIDNP-reactive amino acid side chains in mutant forms of the sialidase. Corroborating the predictions from the theoretical calculations, the spectra of the engineered mutants reveal marked and non-uniform alterations. Thus, the effect of apparently rather conservative amino acid substitutions on a distinct conformational aspect of this protein, even at distant sites, should not be underestimated.

Keywords: Sialidase, NMR, protein modelling, molecular dynamics

---

* *To whom correspondence should be addressed*

Introduction

Since the role of oligosaccharides in cellular glycoconjugates as information-storing structures mediating diverse physiological functions is well appreciated [1-3], efforts are warranted to describe in detail the properties of enzymes which are used for deliberately altering the sequence of the carbohydrate chains. Sialidases (N-acylneuraminosyl-glycohydrolases, EC 3.2.1.18) hydrolytically cleave the linkage between α–glycosidically bound *N*-acylneuraminic acid derivatives and the penultimate sugar (Fig. 1). N-acetylneuraminic acid and a wide variety of variants are often found as terminal constituents of oligosaccharide chains in glycoproteins and glycolipids serving as recognition or masking sites [4-8]. They are removed from the glycoconjugates by sialidases which are widely distributed not only throughout metazoan animals of the deuterostomate lineage, but also among viruses, bacteria and protozoa, many of which are unable to produce sialic acids [4]. Remarkably, the enzyme is frequently produced by microorganisms, which live in close contact with an animal host, whereby the enzyme may serve as a pathogenicity factor or as an important tool for processing of nutrients [4]. To gain insight into the structure/ function relationship the knowledge about the evolutionary pathway of sialidase phylogenesis and a comparison of the primary structure of this protein obtained from different sources can be instructive. Indeed, the alignment of the sequences of eight bacte-

rial sialidases indicates that they are most probably derived from a common ancestor by divergent evolution [9]. The complete conservation of 25 amino acid positions in these enzymes implies their involvement in a common functional and/ or structural role to be delineated by detailed inspection of mechanistic/ conformational aspects. At present, tertiary structures of three bacterial sialidases, namely from *Salmonella typhimurium* [10, 11], *Vibrio cholerae* [12] and *Micromonospora viridifaciens* [13], have been elucidated. Only the data set for the sialidase of *Salmonella typhimurium* [10] is stored in the Brookhaven data bank. Based on the mentioned evolutionary relationship within this family, the available structural information is employed as a framework to concoct a knowledge-based model of the small sialidase of *Clostridium perfringens* with a molecular mass of 43 kDa. Since regions of homology will most likely be centered in the core structures of related enzymes, the large sialidase of *Clostridium perfringens* with a molecular mass of 73 kDa is not considered in this study.

At least two sources of error may reduce the level of accuracy for the resulting model, namely the precision of the calculated predictions by modelling and the occurrence of differences between crystallographic and solution structures. Since the conditions for crystallization and the crystal packing order may affect the conformation of the protein, the solution structure will not necessarily be identical. Therefore, these inherent uncertainties within the modelling approach starting from data sets of the protein crystal call for collecting experimentally derived information. The actual solution structures are used to improve the reliability of the model, which is a demanding task for special NMR spectroscopic methods. Techniques which provide information on the positioning of distinct classes of amino acid side chains would be of assistance to contribute to address the defined issue.

Such a technique with focus on the side chains of tyrosine, tryptophan and histidine has been developed, taking advantage of a radical pair-generating dye [14, 15]. Upon laser irradiation a CIDNP (chemically induced dynamic nuclear polarization) radical reaction takes place in the presence of e. g. a flavin derivative [14, 15]. In glycosciences, this special technique has recently been instrumental to prove the occurrence of conformational changes after desialylation of human serum amyloid P component with only one invariant carbohydrate chain [16]. Moreover, the involvement of aromatic residues in the architecture of carbohydrate recognition domains in N-acetylglucosamine-binding lectins in solution has been documented with this approach [17]. These studies encourage to employ this technique as a quality control for knowledge-based homology modelled structures. In detail, the concomitant calculation of surface accessibilities of the respective types of side chains, which is possible by performing computer-assisted Connolly surface area assessment in the three-dimensional model [18, 19], enables a detailed comparison of these data with the obtained CIDNP results. To evaluate the potential of the combination of these two independent techniques, we herein report their applica-

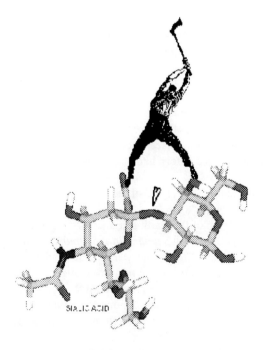

Figure 1. *Sialidases hydrolytically cleave the a-glycosidic linkage of glycosidically bound sialic acids which are mostly found as terminal constituents of oligosaccharide chains in glycoproteins and glycolipids. The subterminal sugar is galactose.*

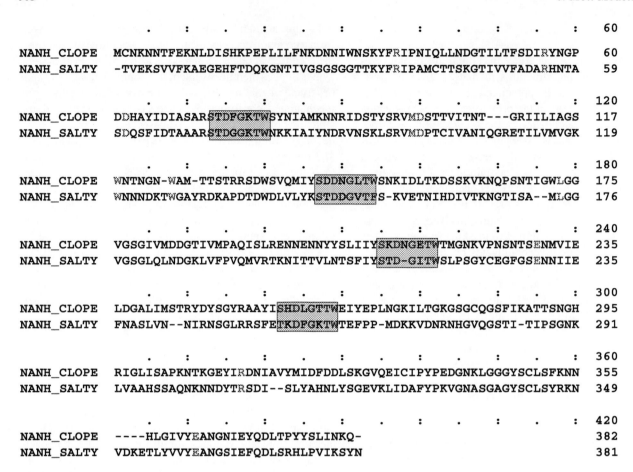

Figure 2. *Sequence alignmnent of* Clostridium perfringens *and* Salmonella typhimurium *sialidases. Four Asp boxes and active site amino acids are shown in blue and red, respectively. (-) is used to present the gaps.*

tion with respect to the *Clostridium perfringens* small sialidase. As an internal control of both the sensitivity of the method and the validity of the interpretation, the experimental examination of deliberately engineered enzyme mutants has been included in the study. These calculations and the experimental data are also pertinent to answer the question to which extent introduction of single-site mutations can affect the monitored conformational parameter.

Materials and Methods

A. Knowledge-based homology modelling

The framework for modelling of the *Clostridium perfringens* small sialidase structure is provided by the crystallographic data for the sialidase of *Salmonella typhimurium* [entries 1 SIL and 1 SIM in the Brookhaven Protein Databank]. Both FastA [20] and BLAST algorithms [21] were employed to assess the scoring level of homology. The next steps of computa-

tions were performed at ExPASy Molecular Biology Server (Geneva University Hospital, University of Geneva, Geneva, Switzerland), using SwissModel Automated Protein Modelling Server (running at the Geneva Biomedical Research Institute, Glaxo Wellcome Research and Development S.A., Switzerland) which which makes use of ProMod (PROtein MODelling tool) [22-24]. The program is accessible through internet browsers like Netscape and e-mail. The alignments were introduced separately into the optimized mode of the program.

The program is comprised of the following model building steps :

I. Construction of the starting framework from the three-dimensional structure of *Salmonella typhimurium* sialidase.

II. Fitting of the *Clostridium perfringens* sialidase backbone onto this framework using primary sequence alignment optimized for three-dimensional similarity.

III. Reconstruction of loop regions from their 'stems' by structural homology searches through the Protein Databank, as described previously [25].

IV. Rebuilding of missing side chains using a library of allowed rotamers [26].

V. Optimization of bond geometries and compensation of unfavorable non-bonded contacts by 30 steps of steepest descent followed by 500 steps of conjugate gradient minimization using CHARMM (Chemistry at HARvard

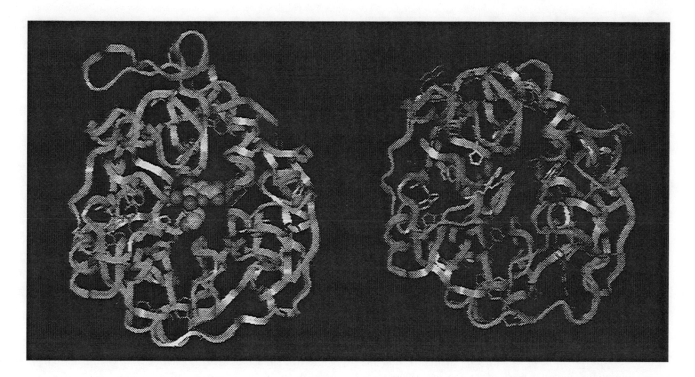

Figure 3. *Homology modelling-derived structure of the small sialidase of Clostridium perfringens (right) on the basis of the crystal structure of Salmonella typhimurium (left). The first 30 amino acids, which have not been incorporated in the modelling procedure, are shown in magenta. (tyrosine residues in red, tryptophan residues in green and histidine residues in blue).*

Macromolecular Mechanics) program [27] with the PARAM19 parameter set.

VI. Analysis of the resulting structure using the 3-D profile matching procedure [28].

The generation of hydrogen atoms, automatic assignment of partial charges of each atom and the solvation of the molecule were accomplished using the InsightII software. A molecular dynamics simulation was carried out for each molecule using the CVFF (Consistent Valence Force Field) at a temperature of 300 K with an equilibration time of 20 ps, a production period of 100 ps and an integration step of 1 fs. A cut-off distance of 10 Å and a dielectric constant of $\varepsilon = 4.0$ were used in all calculations.

Surface accessibilities of side chains of distinct amino acids of interest were calculated. The assessment of 'surface' command implemented in InsightII pinpoints the accessible exterior part of the relevant portions of the molecule by smoothening the van der Waals surface with a test sphere that displays the average radius of the solvent water (1.5 Å). The dot density of the spheres which represent the Connolly surface area was generally set to a value of 1. This dot density guarantees a sufficient distribution of an ensemble of calculated coordinate values which represent the surface area.

Based on the computationally generated model, two tyrosine residues with distinct differences in surface presentation were chosen for substitution with phenylalanine by site-directed mutagenesis. As a further control to evaluate the impact of Cys/Ser-substitution on the calculated/measured parameter, such a mutant was selected.

B. Site-directed mutagenesis, purification and characterization of recombinant enzymes

Site-directed mutagenesis, protein purification and enzymatic assays were performed, as described in detail elsewhere [29,30]. Substitutions are assumed to keep a potential impact on conformational aspects minimal. In detail, the following mutants were generated:

mutant 1: Y336F; mutant 2: Y347F; mutant 3: C349S.

Table 1. *Kinetic parameters for the cleavage of methyl-umbelliferyl-N-acetylneuraminic acid by the* Clostridium perfringens *wild-type sialidase and its three mutants with single amino acid substitution.*

	K_M (mM)	V_{max} (%)
wild type	0.17	100
Y336F	0.19	11
Y347F	0.30	0.009
C349S	0.16	37

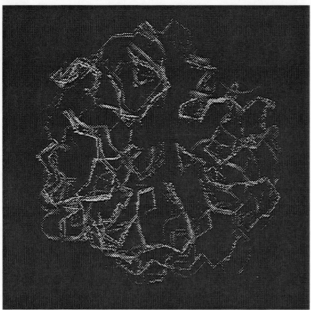

Figure 4. *Superposition of the starting structure and the energy-minimized conformations of ten structures obtained by molecular dynamics simulation of Clostridium perfringens small sialidase.*

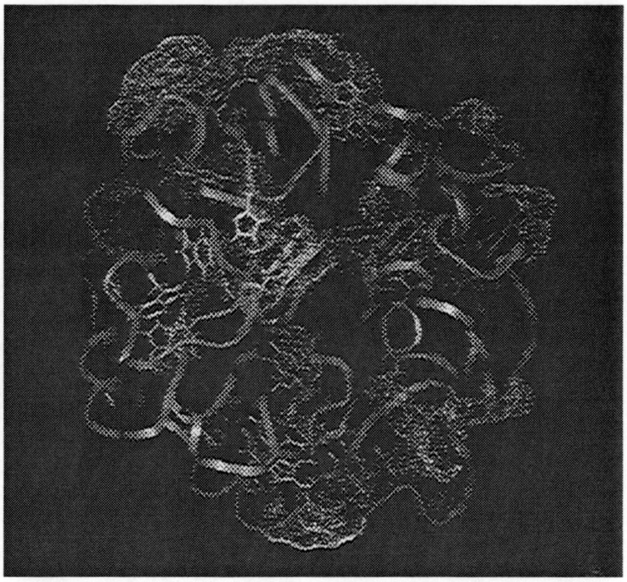

Figure 5. *The Connolly surface areas of aromatic amino acid side chains for the wild-type Clostridium perfringens small sialidase (tyrosine residues in red, tryptophan residues in green and histidine residues in blue).*

C. CIDNP Method

CIDNP experiments were performed at 360 MHz on a Bruker AM-360 NMR spectrometer, as described in detail elsewhere [15-17]. CIDNP was induced by using flavin I mononucleotide as radical pair-generating dye. Briefly, the light of a continuous-wave argon ion laser (Spectra Physics, Mountain View, USA) that operates in the multiline mode with principal wavelengths of 488.0 and 514.5 nm was directed to the sample by an optical fiber and chopped by a mechanical shutter to avoid harmful sample heating. Its operation was controlled directly by the spectrometer. Typical operation conditions were: 1 s presaturation pulse for water suppression, 0.5 s light pulse (5 W), 5 ms RF pulse (90 degree flip angle), 1 s acquisition time, and 5 s delay. A number of 16 or 32 light scans gave an adequate signal-to-noise ratio for the tested samples. The CIDNP experiments were carried out at pH 6.5 in 10 mM phosphate buffer. The CIDNP effect caused by the Tyr-residues corresponds to a spin-density distribution of the intermediate phenoxy radical with strong negative signals of the ε_1, ε_2 protons and less intense positive signals for the δ_1, δ_2 protons. The CIDNP signals of tryptophan are generated by an intermediate radical with strong spin density at the δ_1, ε_3 and h_2 positions of the indole unit and very small spin density at the ζ_2 and ζ_3 positions, which all lead to positive CIDNP signals. The CIDNP difference spectra of histidine displays positive singlet signals for proton ε_1 and proton δ_2 [15].

Results and Discussion

Application of both FastA [20] and BLAST [21] algorithms with default set of functions implemented in PROMOD yield relatively high scores of similarity between the two bacterial enzymes (Fasta optimised score of 558; BLAST score of 202 and Poison probability P(N) of 4.3e-21). This result is illustrated by the alignment of the sequences of these two enzymes, as shown in Fig. 2. Owing to an insufficient extent of similarity, the N-terminal stretch of amino acid residues and two C-terminal amino acids of the small *Clostridium perfringens* sialidase have been excluded from further modelling (Fig. 3). The lack of conservation is judged to be an indicator that these parts are not crucial for functional aspects common to both enzymes. Notably, the alignment conserves the amino acids which are located in the four Asp boxes and the active site of the enzyme. Considering the extent of similarity and the homology score as meaningful indicators for the decision to perform knowledge-based modelling, these results justify to proceed to the calculations, implementing the program steps described in materials and methods. Modelling was performed with the wild-type enzyme and the mutants derived from single amino acid substitutions. The individual sequence alterations in mutant enzymes primarily affected their V_{max} values, as shown in Table 1. The K_M-values for the substrate methylumbelliferyl-N-acetylneuraminic acid is increased for the Y347F sialidase and is only slightly affected for the other two mutant en-

Table 2. *Surface accessibilities (Å^2) of CIDNP-reactive amino acid side chains of wild-type Clostridium perfringens sialidase and its three mutants*

		wild type		Y336F		Y347F		C349S	
	residue	area	SD	area	SD	area	SD	area	SD
Tyr	35	24.7	2.0	12.2	13.3	13.6	4.3	16.0	5.4
	57	73.2	9.4	70.2	7.1	65.7	8.0	74.4	9.2
	65	54.3	6.6	49.7	9.5	49.2	15.0	55.7	4.5
	82	63.9	6.4	43.3	4.5	65.6	5.5	61.6	6.2
	95	58.7	1.8	56.1	4.6	87.6	8.4	56.9	8.9
	141	19.1	2.5	29.1	5.7	29.9	5.4	26.5	5.4
	203	61.5	4.8	63.0	8.7	41.5	5.8	59.3	5.2
	204	41.4	4.7	4.5	5.1	0.6	1.9	22.3	2.7
	209	23.4	9.1	26.1	9.7	34.1	14.0	1.8	3.0
	246	0.0	0.0	0.0	0.0	1.8	2.3	0.0	0.0
	248	25.1	5.3	30.4	4.2	15.6	4.9	17.7	4.0
	251	64.2	4.2	52.7	3.9	74.7	13.2	67.7	4.7
	255	0.0	0.0	2.1	3.4	0.9	2.7	34.9	5.9
	267	17.0	3.7	68.6	5.9	79.6	6.0	34.5	5.6
	310	56.3	5.2	101.2	3.9	34.4	13.5	49.4	10.6
	318	25.4	3.3	22.6	3.4	51.1	18.0	24.6	11.0
	336	64.8	6.3	****	****	122.0	8.7	96.5	7.0
	347	0.0	0.0	0.0	0.0	****	****	1.1	3.4
	361	36.6	3.6	27.3	6.6	36.7	9.0	22.1	6.1
	369	40.9	1.0	3.9	3.7	3.1	3.5	14.0	6.1
	375	38.8	4.5	91.6	7.1	98.8	9.8	5.3	4.0
	376	8.3	8.7	48.8	6.3	71.9	9.6	74.7	5.1
Trp	80	47.4	7.5	26.6	8.9	44.9	7.0	35.1	6.9
	118	0.0	0.0	0.2	0.8	2.8	2.5	0.0	0.0
	124	59.9	6.1	85.4	7.4	38.5	13.0	106.8	6.1
	135	0.0	0.0	14.0	10.9	18.4	4.9	13.8	2.7
	149	55.3	3.4	67.2	5.2	45.0	5.2	50.4	4.3
	172	3.4	4.1	14.9	4.2	0.0	0.0	10.6	9.8
	217	9.4	10.7	1.8	4.0	20.8	4.5	36.4	18.1
	264	4.3	4.7	19.9	9.1	16.0	4.5	13.8	6.5
His	63	35.9	4.4	27.9	4.4	33.2	5.0	1.8	3.2
	258	64.8	10.9	83.3	6.2	84.3	10.9	83.9	14.9
	285	77.1	14.3	75.3	8.6	80.5	6.5	87.5	10.5
	356	17.4	9.3	22.8	7.9	7.6	5.7	13.3	5.8

Table 3. *Calculated surface accessibility changes (Å^2) of CIDNP-reactive amino acid side chains of three Clostridium perfringens sialidase mutants in comparison to the respective residues of the modelling-derived wild-type structure*

	residue	wild type	Δ-area Y336F	Δ-area Y347F	Δ-area C349S
Tyr	35	24.7	-12.5	-11.1	-8.7
	57	73.2	-3.0	-7.5	+1.2
	65	54.3	-4.6	-5.1	+1.4
	82	63.9	-20.6	+1.7	-2.3
	95	58.7	-2.6	+28.9	-1.8
	141	19.1	+10.0	+10.8	+7.4
	203	61.5	+1.5	-20.0	-2.2
	204	41.4	-36.9	-40.8	-19.1
	209	23.4	+2.7	+10.7	-21.6
	246	0.0	+0.0	+1.8	+0.0
	248	25.1	+5.3	-9.5	-7.4
	251	64.2	-11.5	+10.5	+3.5
	255	0.0	+2.1	+0.9	+34.9
	267	17.0	+51.6	+62.6	+17.5
	310	56.3	+44.9	-21.9	-6.9
	318	25.4	-2.8	+25.7	-0.8
	336	64.8	****	+57.2	+31.7
	347	0.0	0.0	****	+1.1
	361	36.6	-9.3	+0.1	-14.5
	369	40.9	-37.0	-37.8	-26.9
	375	38.8	+52.8	+60.0	-33.5
	376	8.3	+40.5	+63.6	+66.4
Trp	80	47.4	-20.8	-2.5	-12.3
	118	0.0	+0.2	+2.8	0.0
	124	59.9	+25.5	-21.4	+46.9
	135	0.0	+14.0	+18.4	+13.8
	149	55.3	+11.9	-10.3	-4.9
	172	3.4	+11.5	-3.4	+7.2
	217	9.4	-7.6	+11.4	+27.0
	264	4.3	+15.6	+11.7	+9.5
His	63	35.9	-8.0	-2.7	-34.1
	258	64.8	+18.5	+19.5	+19.1
	285	77.1	-1.8	+3.4	+10.4
	356	17.4	+5.4	-9.8	-4.1

zymes (Table 1). The maximal rate of cleavage, however, is decreased in all mutants, i. e. to a level of 37 % (C349S), 11 % (Y336F) and 0.009 % (Y347F).

The obtained four modelled structures (wild type and three mutants) were subjected to the molecular dynamics simulation and further minimization prior to the measurement of surface accessibilities of the three kinds of CIDNP-reactive amino acid side chains. The applied value for the dielectric constant guarantees a reasonable consideration of the solvent effect (e.g. dumping hydrogen bond forces), as already shown e.g. in the cases of the oligosaccharide chain of the ganglioside 9-O-acetyl-GD1a and a galectin-binding disaccharide (Galb1-2Galb1-R) [31-35]. A production period of 100 ps and an integration step of 1 fs was necessary to produce reliable trajectories within each simulation. Following a series of 250 integrations, the actual conformational parameters were stored. Based on 400 acquired conformations within the production time, those ten data sets with the lowest potential energy level were automatically selected and processed for energy minimization of the complete ensemble using the conjugate gradient method. The starting structure with the indication for spatial flexibility, as inferred by the calculations, is shown in Fig. 4.

As already outlined, model building is based on the assumption that sequence similarities of homologous proteins will translate into equivalent three-dimensional structures. Moreover, their elucidation by X-ray crystallography anchoring knowledge-based homology modelling is assumed to be of relevance for structural aspects in solution. Since the obtained structural models will yield detailed predictions on distinct conformational aspects such as Connolly surface areas of side chains, the degree of validity of such theoretical conclusions should be assessable by an adequate experimental approach such as laser photo CIDNP-technique. By individually calculating the surface accessibilities of the principally reactive residues on the basis of the modelling-derived data set, compiled in Table 2 and exemplarily illustrated for the wild-type enzyme in Fig. 5, testable assumptions for the shape of the CIDNP-spectra are envisaged. It is remarkable that the calculations for the three mutant enzymes unveil obvious differences in this conformational aspect (Table 3). Attributing a similar essential role in the catalytic mechanism for Tyr347, as inferred for the sialidase of *Salmonella typhimurium* [10], these data suggest that Phe347 is less suitable for this role. Furthermore, substitutions may trigger conformational changes affecting the active-site topology. However, evident changes in surface accessibility are not only seen for the Y347F mutant but also for the other Tyr/Phe- and the Cys/Ser-substitutions. Therefore, these substitutions do not seem to be close to neutral for this modelling derived surface accessibilities, which is tested experimentally in the next step. The analysis of the CIDNP-spectra will allow to infer the degree of validity for these theoretically attained conclusions.

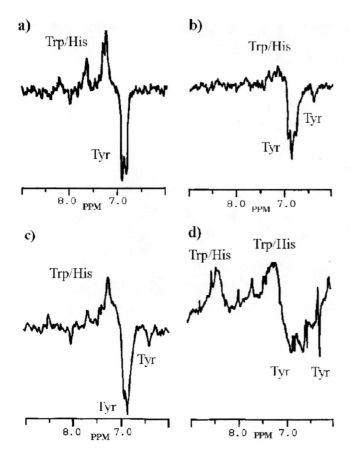

Figure 6. *Laser photo CIDNP difference spectra (aromatic part) of the Clostridium perfringens small sialidase: a. Wild-type enzyme; b. Mutant 1 (Y336F); c. Mutant 2 (Y347F); d. Mutant 3 (C349S).*

First of all, signals for the targeted side chains should be observable for the wild-type enzyme (Table 2). Therefore, *Clostridium perfringens* small sialidase and the three mutant forms with single-site substitutions are suitable for a CIDNP-NMR study directed towards the surface-exposed aromatic amino acids, tyrosine, tryptophan and histidine as sensitive sensors. According to our CIDNP-studies on hevein-like lectins one can expect to monitor a CIDNP-signal, when the surface accessibility area reaches a level of above 80 Å^2 [17]. Residues in the range between 60 and 80 Å^2 should be considered to be potentially feasible to generate or contribute to the intensity of the CIDNP signals. When more than one side chain of the same type is reactive, the overall signal will be shaped by the contributions of the individual moieties. Since five Tyr-residues are predicted to have a high surface exposition (Table 2), we can presume that the Tyr-signal in the spectrum of the wild-type enzyme, shown in Fig. 6a, is generated by an overlap. The CIDNP spectra of mutant 1 (Y336F) and mutant 2 (Y347F) display no major intensity changes and the occurrence of a new small Tyr-signal, as shown in Fig.

6b, c. These results point to alterations in the surface accessibility of a distinct Tyr-residue by conformational rearrangements induced by the amino acid replacement. In mutant 1 (Y336F) a tyrosine residue with a remarkably large surface accessibility area is substituted by a CIDNP-inert phenylalanine. The detailed comparison of the data sets reveals that this substitution is predicted to affect the positioning of several side chains, i.e. Tyr82, Tyr204, Tyr267, Tyr310, Tyr375 and Trp124 (Table 2, Table 3). A similar situation occurs for mutant 2 (Y347F). In this case a completely buried Tyr-residue has been chosen for the replacement, the resulting pattern of changes being non-identical. The impact on several residues hampers unambiguous interpretation of the spectra.

The new small Tyr-signal may be attributed to an increase of surface accessibility areas in both mutants Y336F and Y347F (Fig. 7). Such a common area–increase has been predicted for Tyr267 and even more distinct for Tyr375, as seen in Table 3. The residue Tyr267 with an surface area of 17.0 Å^2 in the wild type displays an area enlargement of 51.6 Å^2 for mutant 1 (Y336F) and 62.6 Å^2 for mutant 2 (Y347F). Remarkably, the corresponding Tyr-residue of mutant 3 (C349S) has an area increment of only 17.5 Å^2 and shall not give a contribution to a CIDNP-signal as similarly inferred for Tyr375 (Table 3). The surface area of this residue increases from 38.8 Å^2 to about 52.8 Å^2 in mutant 1 and about 60 Å^2 in mutant 2 which raises the expectation for a clearly visible Tyr-signal.

The detailed analysis of alterations in surface area above the relevant threshold value suggests that the overall Tyr-signal intensity should not vary markedly, as actually can be seen in Fig. 6a-c. Among the group of tyrosine residues, increases and compensatory reduction (or vice versa) are fairly balanced. However, the shape of Tyr-signals in the CIDNP-spectrum of the C349S mutant vary (Fig. 6d). The surface accessibility area for Tyr375, e.g. in mutant 3, decreases about 33.5 Å^2 (Table 3).

Relative to the interpretation of the tyrosine part of the CIDNP spectra, the analysis of the Trp/His-relevant part is aggravated by signal overlap to which weak Tyr signals of the δ_1, δ_2 protons can even contribute [17].

Likewise, the CIDNP spectrum of mutant 3 shows peculiarities with two rather strong Trp/His-signals one at 8.4 ppm and the other enhanced one at 7.2 ppm (Fig. 6d). The enhancement in Trp/His signal intensity for mutant 3 can be reconciled with an increase above the threshold value for clearly visible CIDNP signals (80 Å^2) as occurring for His285. The surface accessibility area of this residue enlarges from 77.1 Å^2 to 87.5 Å^2. Furthermore, Trp124 has a very high surface accessibility area of 106.8 Å^2 only in mutant 3. Although an unambiguous assignment for the altered Trp/His signals of mutant 3 is not possible, we have nonetheless found evidence that a Cys/Ser replacement indeed leads to distinct changes in the spectrum, which are reflected in the calculations (Table 2, Table 3).

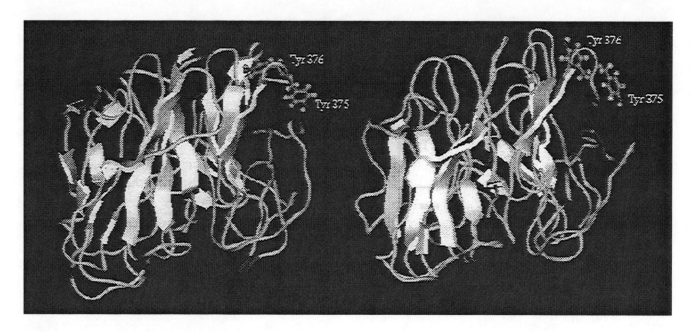

Figure 7. *Side views of the three-dimensional model structures of the Y336F (left) and Y347F (right) mutants of* Clostridium perfringens *small sialidase. Only side chains of Tyr375 and Tyr376, which have apparently enlarged their surface accessibility area in both mutants, are shown. Coloring is on the basis of secondary structure; yellow for sheets, green for random coils and blue for turns.*

Conclusion

Starting from the crystal structure of a bacterial sialidase, knowledge-based homology modelling produces a set of energy-minimized conformations for the small sialidase of *Clostridium perfringens*. Although the number of CIDNP-reactive amino acids is too large to unequivocally assign signals to defined residues, what has been possible for hevein [17], the signal intensity can be reconciled with the model-derived expectations. To obtain clear-cut conclusions about the validity of the assumed solution structure, a more restricted size of the group of surface-presented amino acids is indispensable, defining the limits of applicability of the method. In addition to the description of certain properties of the wild-type enzyme the impact of introducing amino acid substitutions by site-directed mutagenesis has been theoretically delineated and experimentally tested with respect to CIDNP-reactive side chains. Intriguingly, the changes in surface accessibilities of widely separated residues affected by a Tyr/Phe- or a Cys/Ser-substitution intimate that disparate conformational changes of mutant enzymes relative to the wild-type form should not be underestimated. Extending the combination of computational and NMR techniques, as illustrated in recent reports and reviews with emphasis on sugar-binding proteins [31-35], it will technically amenable to figure

out reliably, whether such rather global alterations can affect carbohydrate binding properties.

Acknowledgment. This work was supported by the Human Capital and Mobility Program of the European Community, the German Research Council (Grant Scha 202/ 19-2), the Fonds der Chemischen Industrie, the Verein zur Förderung des biologisch-technischen Fortschritts in der Medizin e. V. and the Sialic Acid Society e. V. (Kiel).

References

1. Gabius H.-J. and Gabius S. (eds.) (1993) *Lectins and Glycobiology* , Springer Verlag,Heidelberg-New York.
2. Gabius H.-J., Kayser K. and Gabius S. (1995) *Naturwissenschaften* **82**, 533-543.
3. Gabius H.-J. and Gabius S. (eds.) (1997) *Glycosciences: Status and Perspectives* , Chapman & Hall, Weinheim.
4. Schauer R. (1982) *Adv. Carbohydr. Chem. Biochem.* **40**, 131-234.
5. Schauer R. (1982) *Sialic Acids: Chemistry, Metabolism and Function*, Springer Verlag, Wien-New York.
6. Corfield A. (1992) *Glycobiology* **2**, 509-521.
7. Schauer R., Kelm S., Reuter G., Roggentin P. and Shaw L. (1995) in *Biology of the Sialic Acids* (Rosenberg A., ed.) pp. 7-67, Plenum Press, New York.
8. Reuter G. and Gabius H.-J. (1996) *Biol. Chem. Hoppe-Seyler*, **377**, 325-342.
9. Roggentin P., Schauer R., Hoyer L. L. and Vimr E. R. (1993) *Mol. Microbiol.* **9**, 915-921.
10. Crennell S. J., Garman E. F., Laver W. G., Vimr E. R. and Taylor G. L. (1993) *Proc. Natl. Acad. Sci. USA* **90**, 9852-9856.

11. Crennell S. J., Garman E. F., Philippon C., Vasella, A., Laver W. G., Vimr E. R. and Taylor G. L. (1996) *J. Mol. Biol.* **259**, 264-280.

12. Crennell S. J., Garman E. F., Laver W. G., Vimr E. R. and Taylor G. L. (1994) *Structure* **2**, 535-544.

13. Gaskell A., Crennell S. J., and Taylor G. (1995) *Structure* **3**, 1197-1205.

14. Kaptein R., Dijkstra K. and Nicolay K. (1978) *Nature* **274**, 293-294.

15. Kaptein R. (1982) in *Biological Magnetic Resonance* **4** (Berliner, L. J., ed.) pp. 145-191, Plenum Press, New York.

16. Siebert H.-C., André S., Reuter G., Gabius H.-J., Kaptein R. and Vliegenthart J. F. G. (1995) *FEBS Lett.* **371**, 13-16.

17. Siebert H.-C., von der Lieth C.-W., Kaptein R.,Soedjanaatmadja U. M. S., Vliegenthart J. F. G., Wright C. S. and Gabius H.-J. (1996) *J. Mol. Model.* **2**, 351-353.

18. Connolly M. L. (1983) *J. Appl. Cryst.* **16**, 548-558.

19. Connolly M. L. (1983) *Science* **221**, 709-713.

20. Pearson W. R. and Lipman D. J. (1988) *Proc. Natl. Acad. Sci. USA* **85**, 2444-2448.

21. Altschul S. F., Gish W., Miller W., Myers E. W. and Lipman D. J. (1990) *J. Mol. Biol.* **215**, 403-410.

22. Peitsch M. C. and Jongeneel C. V. (1993) *Int. Immunol.* **5**, 233-238.

23. Peitsch M. C. (1995) *Bio/Technology* **13**, 658-660.

24. Peitsch M. C. (1996) *Biochem. Soc. Trans.* **24**, 274-279.

25. Greer J. (1990) *Proteins* **7**, 317-334.

26. Ponder J. W. and Richards F. M. (1987) *J. Mol. Biol.* **193**, 775-791.

27. Brooks B. R., Bruccoleri R. E., Olafson B. D., States D. J., Swaminathan S. and Karplus M. (1983) *J. Comp. Chem.* **4**, 187-217.

28. Luthy R., Bowie J. U. and Eisenberg D. (1992) *Nature* **356**, 83-85.

29. Roggentin T., Kleineidam R. G., Schauer R. and Roggentin P. (1992) *Glycoconjugate J.* **9**, 235-240.

30. Kruse S., Kleineidam R. G., Roggentin P. and Schauer R. (1996) *Protein Express. Purific.* **7**, 415-422.

31. Siebert H.-C., von der Lieth C.-W., Dong X., Reuter G., Schauer R., Gabius H.-J. and Vliegenthart J. F. G. (1996) *Glycobiology*, in press.

32. Siebert H.-C., Gilleron M., Kaltner H., von der Lieth C.-W., Kozár T., Bovin N. V., Korchagina E. Y., Vliegenthart J. F. G. and Gabius H.-J. (1996) *Biochem. Biophys. Res. Commun.* **219**, 205-212.

33. Siebert H.-C., Kaptein R. and Vliegenthart J. F. G. (1993) in *Lectins and Glycobiology* (Gabius H.-J. and Gabius S., eds.) pp. 105-116, Springer Verlag, Heidelberg-New York.

34. Siebert H.-C., von der Lieth C.-W., Gilleron M., Reuter G., Wittmann J., Vliegenthart J. F.G. and Gabius H.-J. (1997) in *Glycosciences: Status and Perspectives* (Gabius H.-J. & Gabius S., eds.) Chapman & Hall, Weinheim, pp. 291-310.

35. von der Lieth C.-W., Kozár T., Hull W. E. (1996) *J. Mol. Struct. (Theochem)*, in press.

J. Mol. Model. **1996**, *2*, 456 - 466

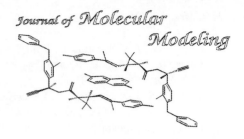

Novel Hoogsteen-like Bases for Recognition of the C-G Base Pair by DNA Triplex Formation

Jeffrey H. Rothman[†]**, W. Graham Richards***

Physical and Theoretical Chemistry Laboratory, Oxford University, South Parks Road, Oxford, UK OX1 3QZ
(jhr@vax.ox.ac.uk)

† Present address: Department of Chemistry, Havemeyer Hall, Box 3154, New York, NY 10027-6948, USA

Received: 21 June 1996 / Accepted: 28 October 1996 / Published: 27 November 1996

Abstract

Effective sequence-specific recognition of duplex DNA is possible by triplex formation with natural oligonucleotides via Hoogsteen H-bonding. However, triplex formation is in practice limited to pyrimidine oligonucleotides that bind duplex A-T or G-C base pair DNA sequences specifically at homopurine sites in the major groove as T·A-T and C[+]·G-C triplets. Here we report the successful modelling of novel unnatural nucleosides that recognize the C-G DNA base pair by Hoogsteen-like major groove interaction. These novel Hoogsteen nucleotides are examined within model A-type and B-type conformation triplex structures since the DNA triplex can be considered to incorporate A-type and/or B-type configurational properties. Using the same deoxyribose-phosphodiester and base-deoxyribose dihedral angle configuration, a triplet comprised of a C-G base pair and the novel Hoogsteen nucleotide, Y2, replaces the central T·A-T triplet in the triplex. The presence of any structural or energetic perturbations due to the central triplet in the energy-minimized triplex is assessed with respect to the unmodified energy minimized $(T·A-T)_{11}$ starting structures. Incorporation of this novel triplet into both A-type and B-type natural triplex structures provokes minimal change in the configuration of the central and adjacent triplets.

Keywords: DNA Triplex, Hoogsteen, molecular recognition

Introduction

The ever increasing knowledge of gene sequences has made DNA a suitable drug target. It has become worthwhile to consider sequence selective ligands such as DNA triple helix forming oligonucleotides (TFOs), as one of the more promising routes. Sequence-specific recognition of duplex DNA by triplex formation is induced by major groove Hoogsteen H-bonding to the duplex Watson-Crick base pairs. Oligodeoxynucleotide-implemented triple helix formation also furnishes one of the most versatile methods for sequence-specific recognition of double helical DNA [1,2]. The ability to target a broad scope of DNA sequences, its high stabilities, and single-base mismatch sensitivity make this a powerful method for binding exclusive sites within large segments of duplex DNA. Since the base sequence of a 17-mer oligonucleotide is statistically unique in the sequence of the

* *To whom correspondence should be addressed*

human genome, extremely selective intervention ought to be possible [3]. However, that approach is severely limited if we restrict attention to natural nucleotides since triplex formation is limited to pyrimidine TFOs binding duplex A-T or G-C base pair DNA sequences specifically at homopurine sites in the major groove parallel to the homopurine strand as T·A-T or C⁺·G-C triplets. Helix-coil transition melting temperature, UV mixing curve, and ¹H NMR experiments all give credence to these homopolymeric structures [4-7]. However, the construction of homopolymeric triplex structures structurally and configurationally analogous to T·A-T and C⁺·G-C via TFOs binding in the major groove of a T-A or C-G duplex parallel to the homopyrimidine strand have yet to be experimentally confirmed.

As a preliminary stage, triplets composed of novel Hoogsteen nucleosides designed to bind in the major groove of T-A or C-G base pairs should show structural stability enclosed within a known stable triplex structure. Its appropriateness is then determined by scrutiny of any possible configurational perturbations imposed upon adjacent structure by the triplet and the configuration of the test triplet itself. Previous molecular modelling studies [8-10] have involved a proposed series of novel unnatural nucleosides which demonstrate selective binding to the major groove of a T-A base pair in the center of a T·A-T triplex. In this study base design is targeted for recognition of the major groove of a C-G base pair.

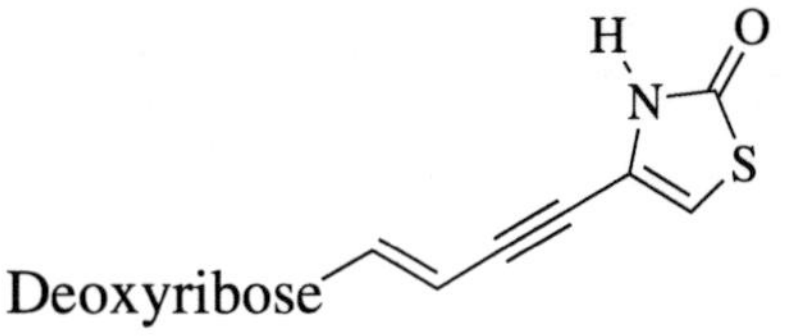

Application in DNA duplex recognition

Depending entirely upon the recognition inherent in the T·A-T and C⁺·G-C triplets, the usage of natural TFOs has been successful in mimicing repressors and the construction of artificial restriction enzymes [11-13]. For example, TFOs have been successful in accomplishing single or double site specific cleavage of yeast and human chromosomal DNA [14]. Even at micromolar TFO concentrations sequence-specific inhibition of DNA binding proteins such as prokaryotic modifying enzymes and eukaryotic transcription factor have been successful [15]. TFOs have also been shown to be useful as competitors for DNA-binding proteins and as site-specific DNA damage or cleavage reagents [12,13,16]. As illustrated by the suppression of human *c-myc* gene transcription with nanomolar TFO concentrations, suppression of gene expression via triplex formation also has potential [17]. Additionally, successful suppression of transcription has also been implemented by blocking the promoter region thereby inhibiting the binding of the eukaryotic transcription factor [18]. In light of the adaptability portrayed in these examples, this method demonstrates potential to be a universal solution for DNA recognition.

Much effort has gone into the design of nonnatural nucleotide bases for TFOs, especially those that aim to bind specifically to T-A or C-G base pairs via Hoogsteen H-bonding with the same parallel orientational geometry as the known T·A-T and C⁺·G-C natural triplexes [19-24]. Previ-

Figure 1a. *Hoogsteen base X3*

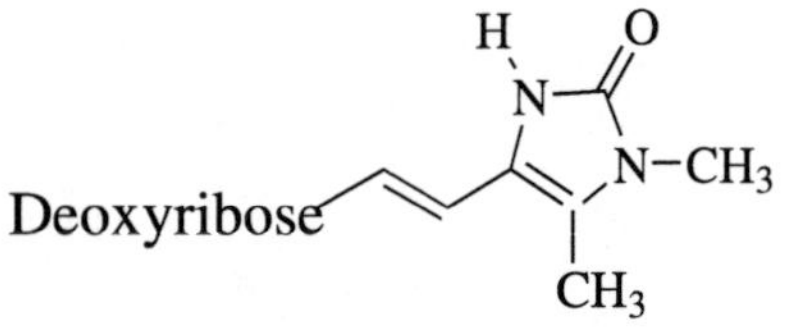

Figure 1b. *Hoogsteen base X5*

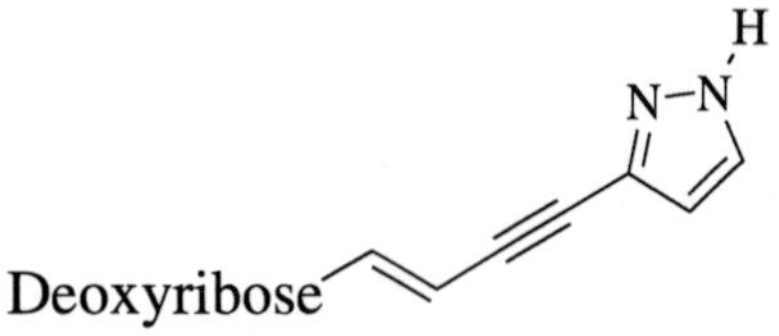

Figure 2a. *Hoogsteen base Y1*

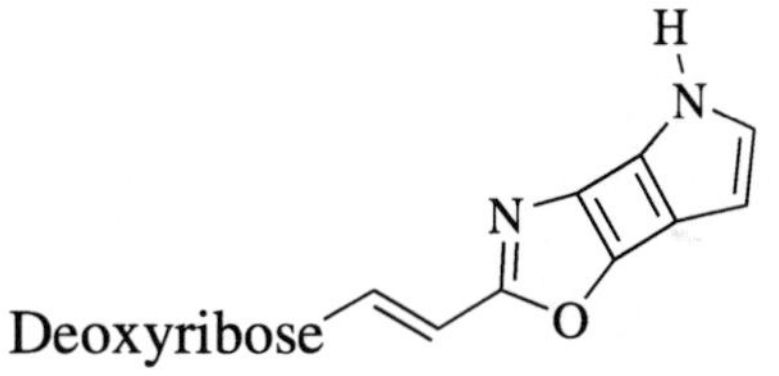

Figure 2b. *Hoogsteen base Y2*

ous energy minimization modelling studies [8,10] of Hoogsteen-like bases, X3 (Figure 1a), and X5, (Figure 1b) have shown successful specific binding of the T-A major groove within an 11mer T·A-T triplex in the A-type and B-type configuration, respectively. These energy minimization studies demonstrate minimal or no structural perturbation to the adjacent triplets with respect to the energy minimized control (T·A-T)₁₁ model. Further molecular dynamics studies of this (T·A-T)₅(A·T-X)(T·A-T)₅ explicitly solvated system with counter ions showed similar structural root mean square deviation behavior in both configurations to that of its related (T·A-T)₁₁ triplex configuration [9].

In the present study a Hoogsteen base is designed to target the C-G major groove. Favorable stacking and Hoogsteen interaction energies, and a comparable minimized phosphodiester backbone and nucleotide geometry to that of the known natural triplets are necessary requirements for a successful Hoogsteen base. Ligand construction within con-

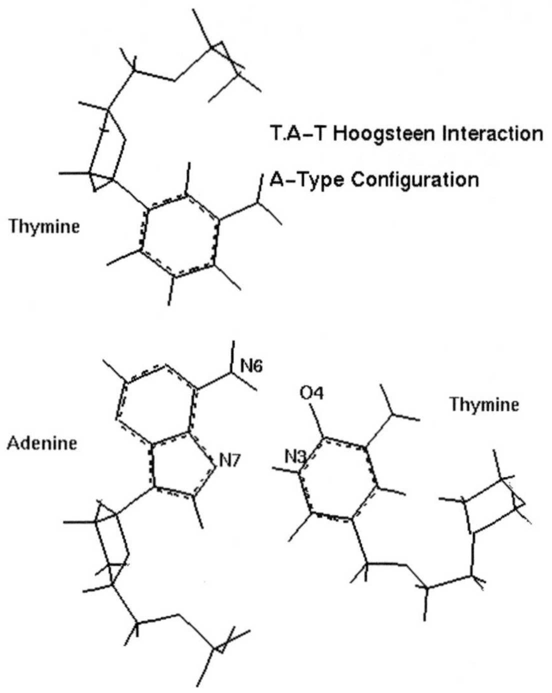

J. Mol. Model. **1996,** *2*

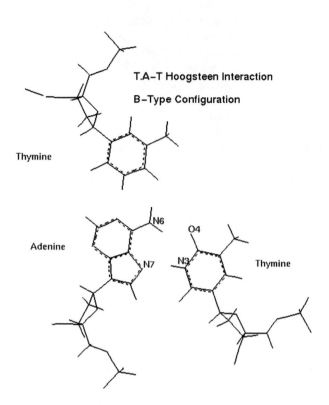

Figure 3a. *Energy minimized T.A-T triplet configuration within the (T.A-T)$_{11}$ triplex from the A-type configuration*

Figure 3b. *Energy minimized T.A-T triplet configuration within the (T.A-T)$_{11}$ triplex from the B-type configuration.*

formationally accommodating targets which need not remain near to their given geometry for optimal binding interactions is not easily accommodated by current algorithm-based ligand design routines. Nonnatural bases may bind with comparable energetics with respect to known stable triplets, but are allowed conformational latitude without significant energetic penalty to the host triplex. This is best portrayed by the detectable binding of guanine in the G·T-A triplet by sequence specific binding-cleavage methods [21], while NMR evidence indicates significantly distorted triplex geometry [25]. Due to this Hoogsteen triplet energy degeneracy any subsequent distortions imposed upon the adjacent triplets must be monitored for interaction energy decreases and maintenance of comparable structural geometry. Here we report the successful modelling of the novel nucleoside, Y2, (Figure 2b) cyclobuta [1,2-d] (*E*) 2-(1-(2-deoxy-b-D-ribofuranosyl)-1-ethene) oxazole [4,3-b] pyrrole, as the Y2·C-G triplet within a T·A-T triplex.

Design Premise

Development of the nonnatural bases from the C1' position of the central deoxyribose of the Hoogsteen strand (strand III) were directed by van der Waals boundaries and potential H-bonding sites of the central bases of the Watson-Crick

strands (strands I and II). Due to the planar base stacking constraints the constructed nonnatural bases are limited to aromatic structures. In order to achieve maximal interaction energy and specificity the Hoogsteen base design plan includes H-bonding with both the cytosine of strand II and guanine of strand I rather than only its neighboring strand II base, which is the circumstance for the natural T·A-T triplet (Figures 3a,b). Positioning the aryl portion of Y2 by an ethenyl linker accomplishes this H-bonding that spans these Watson-Crick nucleotide bases (Figure 4). By spacing appropriately the oxazole N1 and pyrrole 1NH H-bonding components of Y2, repulsive cross interference is minimized [26] between N1 and Gua-O6 and between 1N-H Cyt-4NH$_2$. Initial modelling studies demonstrated that close proximity between H-bond donor and acceptor moieties such as in a pyrazole type Hoogsteen base, Y1, (Figure 2a) causes repulsive interactions between the same H-bond moieties of the Watson-Crick nucleoside bases (Figure 5a,b). For this reason Y2 was developed to allow maximum separation of the adjacent H-bonding components, N: and N-H, and place them in favorable positions to interact with the Cyt-4NH$_2$ (strand II) and Gua-O6 (strand I), respectively. The Y2·C-G triplet demonstrates comparable stacking and Hoogsteen interbase interaction energies and geometries with respect to those of a T·A-T triplet in the center of a (T·A-T)$_{11}$ triplex.

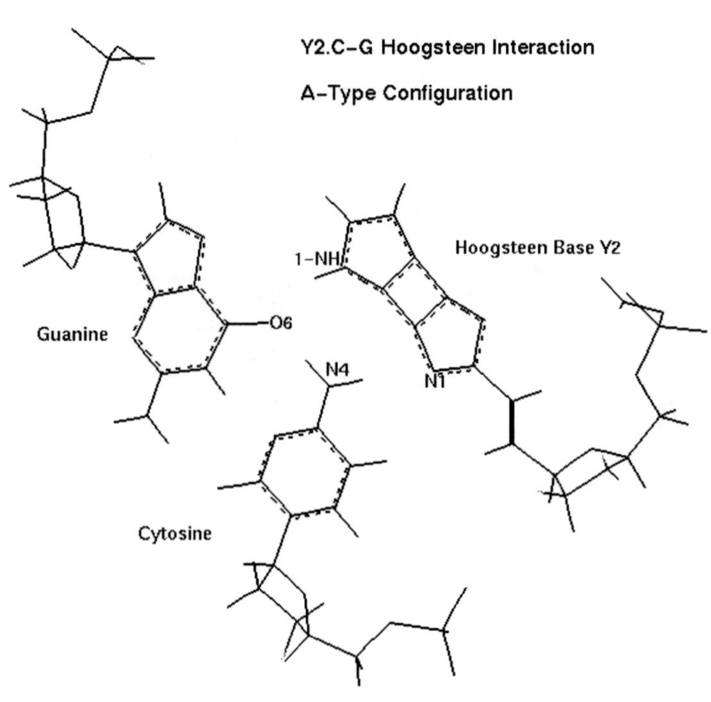

Figure 4a. *Energy minimized Y2.C-G triplet configuration within the (T.A-T)$_5$-(Y2.C-G)-(T.A-T)$_5$ triplex from the A-type configuration.*

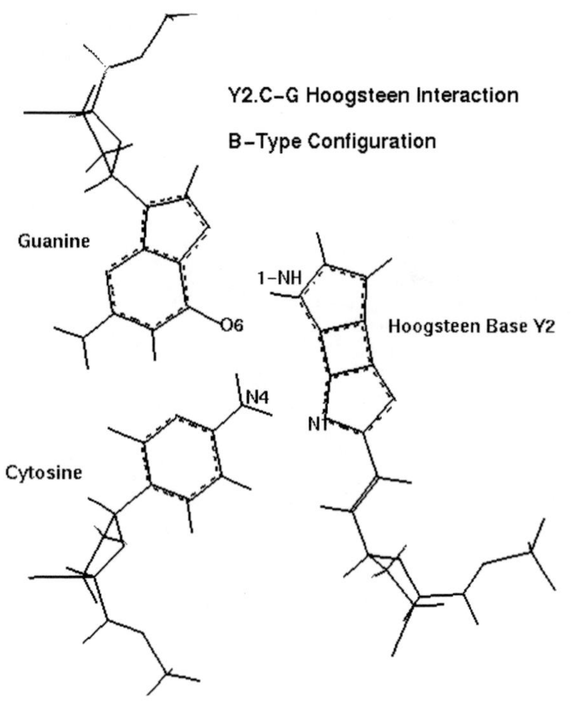

Figure 4b. *Energy minimized Y2.C-G triplet configuration within the (T.A-T)$_5$-(Y2.C-G)-(T.A-T)$_5$ triplex from the B-type configuration.*

Triplex conformations

The Arnott fibre diffraction data [27] provided the first DNA triplex models. Configurations of T·AT triplex derived from these results are considered to be similar to A-DNA with C3'-endo ribose puckers. Results from NMR NOE experiments are consistent with the fibre diffraction data indicating C3'-endo configuration on all three strands of T·AT triplex [28]. At present date, there are still no reports of crystal diffraction studies of DNA triplex. However, recent solution IR spectroscopy studies [29] of T·AT triplex and solution NMR studies [30] of various triplex oligonucleotide systems suggest that many of the nucleotide residues have ribose puckers nearer to a C2'-endo configuration which would be in better agreement with a B-DNA type conformation. Consequently, in light of these new findings a more appropriate model T·AT triplex has been devised [31,32] in which all three strands have the same phosphodiester geometry and C2'-endo ribose pucker that characterizes B-type DNA geometry. More insight into this dilemma may be gained from recent molecular dynamics simulations [33] of the T·AT DNA triplexes from both the A-type and B-type starting configurations. These simulations show similar trajectories which converge to a structure that is structurally equidistant from, but not very similar to either of the initial A-type or B-type

structures. Although it may be difficult to produce an accurate representation of a time-averaged helical unit from these dynamics simulations, the resultant convergence to similar structure suggests available configurational pathways between A-type and B-type DNA triplex conformations. Unfortunately, overwhelming proof or rejection of the preference of one type of conformation over the other is not evident. In light of this situation it would be prudent to design Hoogsteen bases that are viable for each general triplex conformation.

Methods

For test purposes, the A-type conformation host undecamer T·A-T DNA triplex, (T·A-T)$_{11}$, was constructed from the Arnott fibre diffraction model [27] where the triplet step height of 3.26Å and a turn angle of 30.0° were used. This is related to placement of a Hoogsteen binding pyrimidine strand into the major groove of an A-DNA duplex parallel to the purine strand. The B-type conformation (T·A-T)$_{11}$ DNA triplex was constructed from the T·AT triplex structure proposed by Sasisekharan [31] in which all three strands have the same ribophosphodiester geometry, nevertheless the step height and turn angle are equivalent to those of the A-type

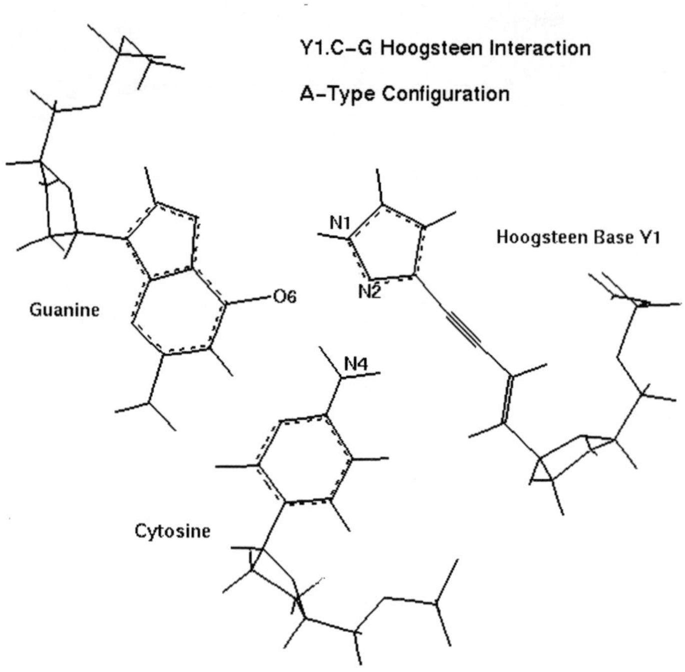

Figure 5a. *Energy minimized Y1.C-G triplet configuration within the (T.A-T)$_5$-(Y1.C-G)-(T.A-T)$_5$ triplex from the A-type configuration.*

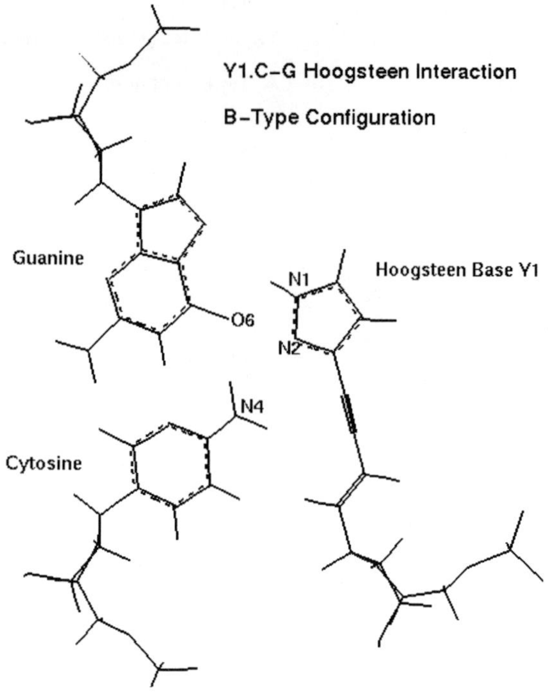

Figure 5b. *Energy minimized Y1.C-G triplet configuration within the (T.A-T)$_5$-(Y1.C-G)-(T.A-T)$_5$ triplex from the B-type configuration.*

conformation. In order to create the test triplexes, the center triplet of each host undecamer was then modified to create the binding region for testing nonnatural base candidates by replacing the A-T nucleotide bases of the Watson-Crick strand with C-G and replacing the nucleotide base of the Hoogsteen strand with the nonnatural bases to be tested, retaining the strand specific deoxyribose-base dihedral angles.

The AMBER force field does not contain all of the parameters required for the proposed nucleoside bases, Y1 and Y2. For purposes of calculating the atomic partial charges of these proposed bases their geometries were determined with MOPAC [34] using the AM1 hamiltonian. The charges of the MOPAC determined structure were then obtained from GAUSSIAN90 [35] with an RHF/STO-3G basis set, using the CHELPG [36] method, and scaled to fit the AMBER 4.0 force field for natural nucleotides. Analyses of the proposed bases within the host triplex were performed with the AMBER [37] suite of programs. Energy minimizations of these triplexes were performed via 100 steps steepest descent and 1000 steps of conjugate gradient method with a nonbonded cutoff distance of 8Å, and a linear distance dependent dielectric to model the implicit water solvation [38].

Results

Charge compatibility of proposed unnatural nucleoside bases with AMBER

Since there are no pyrazole constructs in the AMBER force field, comparison of the Y1 fitted charges is not possible. However, comparison of the fitted charges on the significantly electrostatic interacting portions of Y2 show similar values to those of analogous 5-membered heterocyclic substructures in AMBER. The fitted charges of the pyrrole N-H (N: -0.131, H: 0.246) in Y2 are similar to that of the imidazole (N: -0.142, H: 0.228) portion of histidine in AMBER. The fitted charge of the oxazole N: (N: -0.536) portion of Y2 is also similar to that of the imidazole portion of histidine (N: -0.502), and the other 5-membered aromatic imine substructures of guanine and adenine (both N: -0.543) in AMBER. In fact these charge values of Y2 are also comparable to the similarily calculated charge values of imidazole N-H (N: -0.245, H: 0.232) and N: (N: -0.550). The apparently comparable charges calculated by the two procedures for imidazole and related substructures indicates a reasonable compatibility between the fitted charges of the unnatural Y2 base and charges on similar structures in the AMBER force field.

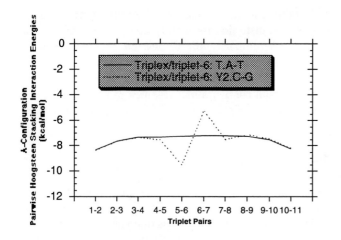

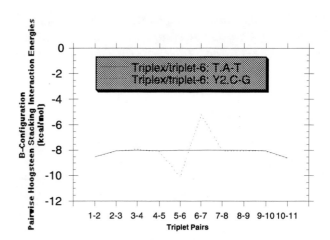

Figure 6a. *Pairwise Hoogsteen Stacking interaction energies for triplexes with the central triplets: T·A-T, Y2·C-G in the A-type conformation.*

Figure 6b. *Pairwise Hoogsteen Stacking interaction energies for triplexes with the central triplets: T·A-T, Y2·C-G in the B-type conformation.*

Analysis of Control Structure

The structural and energetic behavior of the Y2·C-G triplet is compared to that of a T·A-T triplet as their structures are energy minimized within the center of a (T·A-T)$_{11}$ triplex. The C$^+$·G-C triplet was not chosen as a basis for comparison over that of the T·A-T in triplet-6 due to the inconsistencies in comparisons as the protonated form. Before analyzing the energetic perturbation effects on the modified triplex (T·A-T)$_5$(Y2·C-G)(T·A-T)$_5$ due to replacement of the central triplet within the (T·A-T)$_{11}$ triplex in both the A-type and B-type configurations, the naturally occuring inherent perturbations in the energy minimized (T·A-T)$_{11}$ triplex model itself must be studied. End effects concerning the end triplets are evident mainly due to the lack of base stacking interactions on their outer surface. Not only is this manifested in the weaker total interaction energies of the end bases in

the Hoogsteen strand, but also the configuration of the bases of the end triplets demonstrate increased buckle angle, thus placing the center of the triplet closer to their adjacent triplet. This distortion appears to allow a more favorable stacking energy between Hoogsteen bases, and is propagated in a decreasing manner towards the center of the triplex for A-type and B-type conformations as demonstrated by the pairwise stacking energies in Figures 6a,b. This distortion also allows a closer and more favorable H-bonding distance between the Hoogsteen base and the Watson-Crick bases as demonstrated by slightly increased Hoogsteen interaction energies with the nucleosides of its triplet nearer to the ends of the triplex as demonstrated in Figures 8a,b. Aside from end effects due to the lack of outlying base stacking interactions, energies between bases of the Watson-Crick strands (I-II) and their base-stacking interaction energies (Figures 6a,b) appear consistent throughout the (T·A-T)$_{11}$ triplex in

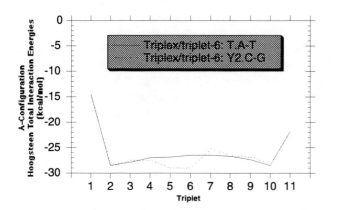

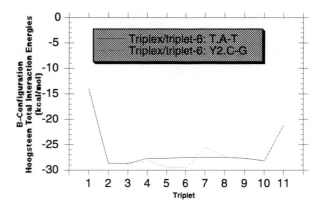

Figure 7a. *Interaction energies for each Hoogsteen base with the rest of the triplex with central triplets: T·A-T, Y2·C-G for the A-type conformation.*

Figure 7b. *Interaction energies for each Hoogsteen base with the rest of the triplex with central triplets: T·A-T, Y2·C-G for the B-type conformation.*

Table 1. *Deoxyribose-phosphodiester dihedral angles (degrees) for A-configuration triplex structures: Triplex-0: (T·A-T)$_{11}$ and Triplex-1: (T·A-T)$_5$-(Y2·C-G)-(T·A-T)$_5$ (see page 464)*

Strand-I Triplet-6 Backbone Dihedrals

Dihedrals	Triplex-0	Triplex-1
P-O3'	-64.7	-64.0
O3'-C3'	-169.1	-168.1
C3'-C4'	79.2	78.1
C4'-C5'	61.6	63.4
C5'-O5'	173.7	173.1
O5'-P	-73.0	-71.3
C1'-Base	29.9	23.3

Strand-II Triplet-6 Backbone Dihedrals

Dihedrals	Triplex-0	Triplex-1
P-O3'	-72.9	-62.05
O3'-C3'	172.2	-167.4
C3'-C4'	63.3	80.6
C4'-C5'	78.2	64.8
C5'-O5'	-169.2	171.1
O5'-P	-63.8	68.0
C1'-Base	25.2	28.5

Strand-III Triplet-6 Backbone Dihedrals

Dihedrals	Triplex-0	Triplex-1
P-O3'	-73.4	-63.3
O3'-C3'	174.4	-163.8
C3'-C4'	61.2	81.4
C4'-C5'	79.8	60.9
C5'-O5'	-167.9	173.7
O5'-P	-65.3	-72.42
C1'-Base	31.5	39.93

both configurations and relatively insensitive to terminal configurational distortion for both types of triplex starting conformations. Similarly, the Hoogsteen interaction energies between Hoogsteen bases (strand III) and the Watson-Crick bases (strands I,II) are also consistent throughout the neat triplex in both configurations (Figures 8a,b). Consequently, the central nine triplets remain energetically and structurally consistent.

A-configuration results

Upon examination of the Y2·C-G triplet within the energy minimized (T·AT)$_5$(Y2·C-G)(T·AT)$_5$ triplex starting in the A-configuration, the total interaction energy of the unnatural Hoogsteen nucleoside Y2 (-28.94 kcals) is 109% that of the analogous value in the control (T·AT)$_{11}$ triplex (-26.57 kcals). This is primarily due to the Y2 nucleoside having favorable interactions in recognition of both of the Watson-Crick nucleosides in its triplet whereas the Hoogsteen base in a natural T·AT and C$^+$·GC triplet only recognize the strand II Watson-Crick base. Accordingly, the Hoogsteen interaction energy of the Y2 nucleoside (-5.59 kcals) with the C-G Watson-Crick nucleosides is 112% that of the analogous value in the control (T·AT)$_{11}$ triplex (-5.00 kcals). Although the Y2 base accomplishes two H-bonds just like a natural Hoogsteen base, it orients one to each Watson-Crick base whereas a natural Hoogsteen base orients both to one. The improvement in the Y2 Hoogsteen interaction in comparison to a natural Hoogsteen interaction lies in the added favorable van der Waals contact which is available due to its proximal orientation to both Watson-Crick bases. The Hoogsteen H-bonding distances of 2.80Å and 3.24Å are within range of viable H-bonding interactions between Y2-1NH — Gua-O6, and Y2-N1 — Cyt-4NH$_2$, respectively. Analogous H-bonding distances in the A-configuration for a T·AT central triplet are 2.90Å and 2.88Å for the Thy-O4 — Ade-6NH$_2$ and Thy-3NH — Ade-N7 interactions, respectively. The Y2 H-bonding distance to Cyt-4NH$_2$ is somewhat attenuated with respect to the equilibrium distance, and this subsequent difference in H-bonding strength is reflected in the Y2 -2.92 kcal interaction energy with the strand II cytosine and the -2.67 kcal interaction energy with the strand I guanine.

Additionally the p-stacking interaction energy of the Y2 Hoogsteen base (-14.75 kcals) with its adjacent Hoogsteen bases are 102% that of the analogous value in the control (T·AT)$_{11}$ triplex (-14.45 kcals). However, due to the orientation of the Y2 base within its triplet plane with respect to the analogous Hoogsteen base in a central T·AT triplet, the 3' adjacent thymine base (triplet 7) receives less p-overlap and the 5' adjacent thymine base (triplet 5) receives more p-overlap than their analogous adjacent bases in the control (T·AT)$_{11}$ triplex. Their p-interaction energies, -5.23 kcals and -9.53 kcals, 5' and 3' respectively, reflect the disparity in p-overlap above and below the Y2 base in comparison to the -7.26 kcals and -7.19 kcals p-interaction energies of the analogous central triplet Hoogsteen base in the control (T·AT)$_{11}$ triplex.

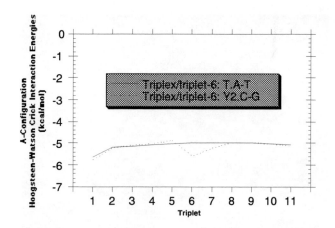

Figure 8a. *Interaction energies between the Hoogsteen base and its accompanying Watson-Crick base pairs for triplexes with central triplets: T·A-T, Y2·C-G for the A-type conformation.*

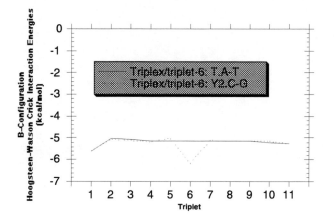

Figure 8b. *Interaction energies between the Hoogsteen base and its accompanying Watson-Crick base pairs for triplexes with central triplets: T·A-T, Y2·C-G for the B-type conformation.*

However, this does not cause any detrimental structural asymmetries as shown by the phosphodiester and c-dihedrals of the energy minimized $(T·AT)_5(Y2·C-G)(T·AT)_5$ triplex structure as compared to those of the energy minimized $(T·AT)_{11}$ control triplex (Table 1, Figure 9a). Consequently, the Hoogsteen nucleoside stacking interaction energy, Hoogsteen nucleoside total interaction energy profiles, and Hoogsteen-Watson Crick energy profiles also closely follow that of the $(T·AT)_{11}$ control triplex (Figures 6a,7a,8a). The Watson-Crick interaction energy profile closely follows that of the control $(T·AT)_{11}$ triplex except for the stronger interaction energy of C-G in triplet-6 due to its three H-bond interaction in contrast to two for the A-T pair. This is in accord with the dihedral results (Table 1) which show almost no structural perturbation to the Watson-Crick strands (I-II).

Table 2. *Deoxyribose-phosphodiester dihedral angles (degrees) for B-configuration triplex structures: Triplex-0: $(T.A-T)_{11}$ and Triplex-1: $(T·A-T)_5-(Y2·C-G)-(T·A-T)_5$ (see page 464)*

Strand-I Triplet-6 Backbone Dihedrals

Dihedrals	Triplex-0	Triplex-1
P-O3'	-86.0	-86.2
O3'-C3'	-175.8	-175.8
C3'-C4'	103.4	106.2
C4'-C5'	58.6	57.5
C5'-O5'	171.5	172.2
O5'-P	-66.3	-65.7
C1'-Base	42.1	40.3

Strand-II Triplet-6 Backbone Dihedrals

Dihedrals	Triplex-0	Triplex-1
P-O3'	-87.2	-86.3
O3'-C3'	-176.3	-176.2
C3'-C4'	105.7	104.2
C4'-C5'	57.6	57.8
C5'-O5'	172.9	172.0
O5'-P	-66.2	-66.7
C1'-Base	41.6	40.7

Strand-III Triplet-6 Backbone Dihedrals

Dihedrals	Triplex-0	Triplex-1
P-O3'	-86.6	-89.1
O3'-C3'	-176.2	-176.9
C3'-C4'	104.7	113.0
C4'-C5'	58.4	57.3
C5'-O5'	172.6	174.1
O5'-P	-66.9	-66.3
C1'-Base	41.1	50.4

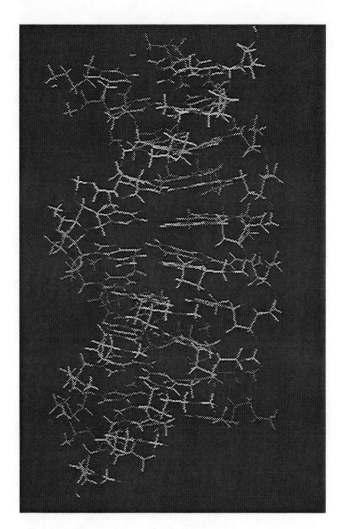

Figure 9a. *Energy minimized (T·A-T)$_5$-(Y2·C-G)-(T·A-T)$_5$ A-type triplex configuration. 3D-structure as PDB-file included.*

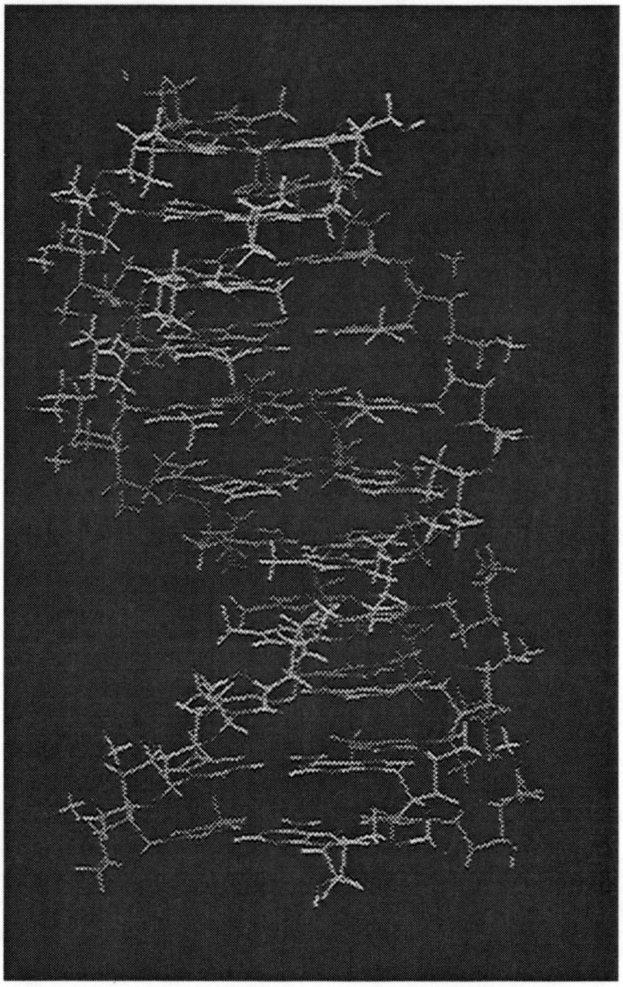

Figure 9b. *Energy minimized (T·A-T)$_5$-(Y2·C-G)-(T·A-T)$_5$ B-type triplex configuration. 3D-structure as PDB-file included.*

B-configuration results

With respect to energetic trends the Y2·C-G triplet within the energy minimized (T·AT)$_5$(Y2·C-G)(T·AT)$_5$ triplex performs similarily in a starting B-configuration in comparison to the results for the starting A-configuration. While the total interaction energy of the unnatural Hoogsteen nucleoside Y2 (-29.58 kcals) is 108% that of the analogous value in the control (T·AT)$_{11}$ triplex (-27.52 kcals), its Hoogsteen interaction energy with the Watson-Crick bases (-6.17 kcals) is 120% of the analogous interaction energy in the central triplet of the control (T·AT)$_{11}$ triplex (-5.14 kcals) in the starting B-configuration. The Hoogsteen H-bonding distances of 2.81Å and 2.92Å are within range of viable H-bonding interactions between Y2-1NH — Gua-O6, and Y2-N1 — Cyt-4NH$_2$, respectively. In comparison, the analogous H-bonding distances in the B-configuration for a T·AT central triplet are a very near 2.87Å and 2.92Å for the Thy-O4 — Ade-6NH$_2$ and Thy-3NH — Ade-N7 interactions, respectively. The similar Y2 interaction energies to the strand II cytosine

(-3.12 kcals) and the strand I guanine (-3.05 kcals) reflect the likeness in the H-bonding distance to each Watson-Crick base.

The planar orientation of the Y2 base also causes a similar disparity in the Hoogsteen stacking interaction energies with its adjacent Hoogsteen bases in the B-configuration. A biased increased p-overlap with the 5'-adjacent Hoogsteen base (-10.08 kcals) and decreased overlap with the 3'-adjacent Hoogsteen base (-5.22 kcals) compared to the p-overlap energies (-8.02 kcals) and (-7.99 kcals) of the analogous 5' and 3' adjacent Hoogsteen bases in the control (T·AT)$_{11}$ triplex is as evident in the B-configuration as in the A-configuration. However, the total p-stacking interaction energy for the Y2 base (-15.30 kcals) is 95.6 % of the total stacking interaction for the analogous Hoogsteen base (-16.01 kcals) in the control (T·AT)$_{11}$ triplex whereas the total p-stacking energy for its 5'-adjacent Hoogsteen base (triplet-5) (-17.09 kcals) is 107% and the 3'-adjacent Hoogsteen base (triplet-7) (-12.76 kcals) is 80% of the total p-stacking interaction for the analogous Hoogsteen base in the control (T·AT)$_{11}$

triplex. This disparity in pstacking interaction energies is noticeable in the Hoogsteen p-stacking interaction profile (Figure 6b) as well as the Hoogsteen nucleoside total interaction energy profile (Figure 7b). However, this does not cause any detrimental structural asymmetries as shown by the phosphodiester and c-dihedrals of the energy minimized $(T \cdot AT)_5(Y2 \cdot C\text{-}G)(T \cdot AT)_5$ triplex structure as compared to those of the energy minimized $(T \cdot AT)_{11}$ control triplex (Table 2, Figure 9b). Just like the A-configuration triplex results, the Hoogsteen nucleoside stacking interaction energy, Hoogsteen nucleoside total interaction energy profiles, and Hoogsteen-Watson Crick energy profiles in the B-configuration closely follow that of its $(T \cdot AT)_{11}$ control triplex (Figures 6b,7b,8b). The Watson-Crick interaction energy profile closely follows that of the control $(T \cdot AT)_{11}$ triplex except for the stronger interaction energy of C-G in triplet-6 due to its three H-bond interaction in contrast to two for the A-T pair in accord with the dihedral results (Table 2).

Conclusion

The emulation of a known deoxyribose-phosphodiester triplex configuration with novel base arrangements shows promise. Previous modelling work has demonstrated the viability of unnatural Hoogsteen bases that specifically recognize the T-A major groove [8,10]. Molecular dynamics studies further confirmed the stability of these triplets with respect to T·A-T triplet [9]. With respect to the presented results, the Y2 base and its triplet constructs in both the A-type and B-type configurations appear to be structurally and energetically viable within naturally occuring triplex structures. Although the B-configuration is presently believed to be most representative of the triplex solution structure, elements of A-type geometry are not necessarily obviated. The use of the T·A-T triplet as a standard for comparison is merely a guide and not a goal. The Y2·C-G triplet studies performed here are merely a starting point for dynamic evaluation which allows a more rigorous assessment of the stability of the anticipated configuration. Having a similar stable triplet configuration to that found in nature is not a guarantee for configurational stability during dynamics simulation. However, it is likely to be a prerequisite.

With respect to molecular modelling exercises, much caution is required to assess the correctness and precision of the results. For the purposes of this study the calculated interaction energies are only utilized for comparative purposes and obviously do not represent free energies of binding. In this manner systematic inaccuracies are reduced. However, more formal modelling of solvation with respect to explicit waters and counter-ions and the application of molecular dynamics to these systems are the next obvious steps to follow this preliminary study. The final test must be synthesis and binding energy studies, but the simulations should provide encouragement for experimental work.

Acknowledgement: J. H. R. is supported by a Hitchings-Elion postdoctoral fellowship from The Burroughs-Wellcome Fund.

References

1. Strobel, S.A.; Moser, H.E.; Dervan, P.B. *J. Am. Chem. Soc.* **1988**, *110*, 7929-9.
2. Povsic, T.J.; Dervan, P.B. *J. Am. Chem. Soc.* **1989**, *111*, 3059-61.
3. Uhlmann, E.; Peyman, A. *Chem. Rev.* **1990**, *90*, 544-579.
4. Pilch, D.S.; Brousseau, R.; Shafer, R.H. *Nucl. Acids Res.* **1990**, *18*, 5743-50.
5. Pilch, D.S.; Levenson, C.; Shafer, R.H. *Proc. Natl. Acad. Sci. U.S.A.* **1990**, *87*, 1942-6.
6. Pilch, D.S.; Levenson, C.; Shafer, R.H. *Biochemistry* **1991**, *30*, 6081-7.
7. Plum, E.G.; Park, Y.W.; Singleton, S.F.; Dervan, P.B.; Breslauer, K.J. *Proc. Natl. Acad. Sci. U.S.A.* **1990**, *87*, 9436-40.
8. Rothman, J.H.; Richards, W.G. *J. Chem. Soc. Chem. Comm.* **1995**, 1589-90.
9. Rothman, J.H.; Richards, W.G. *Molecular Simulation* **1995**, *18*, 13-42.
10. Rothman, J.H.; Richards, W.G. *Biopolymers* in press.
11. Moser, H.E.; Dervan, P.B. *Science* **1987**, *238*, 645-650.
12. Strobel, S.A.; Moser, H.E.; Dervan, P.B. *J. Am. Chem. Soc.* **1988**, *110*, 7927-9.
13. Le Doan, T.; Perroualt, L.;Praseuth, D.;Helene, C. *Nucl.Acid Res.* **1987**, *15*, 1749-60.
14. Strobel , S.A.; Dervan, P.B. *Science* **1990**, *249*, 73-5.
15. Maher, L.J.; Wold, B.; Dervan, P.B. *Science* **1989**, *245*, 725-30.
16. Povsic, T.J.; Dervan, P.B. *J. Am. Chem. Soc.* **1990**, *112*, 9428-30.
17. Cooney, M.; Czernuszewicz, G.; Postal, E.H.; Flint, S.J.; Hogan, M.E. *Science* **1988**, *241*, 456-9.
18. Durland, R.H.; Kessler, D.J.; Gunnell, S.; Duvic, M.; Pettitt, B.M.; Hogan, M.E. *Biochemistry* **1991**, *30*, 9246-55.
19. Griffin, L.C.; Kiessling, L.L.; Beal, P.A.;Gillespie, P.; Dervan, P.B. *J. Am. Chem. Soc.* **1992**, *114*, 7976-82.
20. Griffin, L.C.; Dervan, P.B. *Science* **1989**, *245*, 967-71.
21. Koshlap, K.M.; Gillespie, P.; Dervan, P.B., Feigon, J. *J. Am. Chem. Soc.* **1993**, *115*, 7908-9.
22. Ono, A.; Ts'o, P.O.; Kan, L. *J. Am. Chem. Soc.* **1991**, *113*, 4032-3.
23. Koh, J.S.; Dervan, P.B. *J. Am. Chem. Soc.* **1992**, *114*, 1470-8.
24. Mohan, V.; Cheng, Y.K.; Marlow, G.E.; Pettitt, B.M. *Biopolymers* **1993**, *33*, 1317-25.
25. Radhakrishnan, I.; Patel, D.J. *Structure* **1994**, *2*, 17-32.
26. Burrows, A.D., Chan, C-W, Chowdhry, M.M., McGrady, J.E., Mingos, D.M.P. *Chem. Soc. Rev.* **1995**, *24*, 329-339.

27. Arnott, S.; Bond, P.J.; Selsing, E.; Smith, P.J.C. *Nucl. Acids Res.* **1976,** *10*, 2459.

28. Umemoto, K.; Sarma, M.H.; Gupta, G.; Luo, J.; Sarma, R.H. *J.Am. Chem. Soc.* **1990**, *112*, 4539-4545.

29. Howard, F.B.; Miles, H.T.;Liu, K.; Frazier, J.; Raghunathan, G.; Sasisekharan, V. *Biochemistry*, **1992**, *31*, 10671-7.

30. Macaya, R.F.; Wang, E.;Schultze, P.; Sklenar, V.; Feigon, J. *J. Mol. Biol.,* **1992**, *225*, 755-73.

31. Raghunathan, G.; Miles, H.T.; Sasisekharan, V. *Biochemistry,* **1993**, *32*, 455-62.

32. Liu, K.; Miles, H.T.; Parris, K.D.; Sasisekharan, V. *Struct. Biol,* **1994**, *1*, 11-12.

33. Laughton, C.A. *Molecular Simulation,* **1995**, *14*, 275-89.

34. Stewart, J.J.P. *J. Comp. Aided Mol. Design* **1990**, *4*, 1.

35. Frisch, M.J.; Head-Gordon, M.; Trucks, G.W.; Foresman, J.B.; Schlegel, H.B.; Ragavachari, K.; Robb, M.; Binkley, J.S.; Gonzalez, C.; Defrees, D.J.; Fox, D.J.; Whiteside, R.A.; Seeger, R.; Melius, C.F.; Baker, J.; Martin, L.R.; Kahn, L.R.; Stewart, J.J.P.; Topiol, S.; Pople, J.A. GAUSSIAN 90, **1990,** Revision J. Gaussian Inc., Pittsburgh, PA.

36. Chirlian, L.E.; Francl, M.M. *J. Comp. Chem.* **1987,** *8*, 894.

37. Pearlman, D.A.; Case, D.A.; Caldwell, J.C.; Siebel, G.L.; Singh, U.C.; Wiener, P.A.; Kollman, P.A., 1991, AMBER 4.0, University of California, San Francisco.

38. Brooks, B.R.; Bruccoleri, R.E.; Olafson, B.D.; States, D.J.; Swaminathan, S.; Karplus, M. *J. Comp. Chem.,* **1983**, *4*, 187.

J. Mol. Model. **1996**, 2, 467 – 477

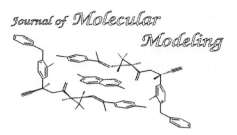

Simulation of Tumor-Specific Delivery of Radioligand Comparison of One Step, Two Step, and Genetic Transduction Systems

Tiepu Liu[†,*], Donald J. Buchsbaum, David T. Curiel, Mohamed B. Khazaeli, Ruby F. Meredith, Albert F. LoBuglio

Comprehensive Cancer Center, University of Alabama at Birmingham, Birmingham, AL 35294

† Division of Biostatistics and Epidemiology, Department of Emergency Medicine, University of Cincinnati College of Medicine, 231 Bethesda Avenue, ML-0769, Cincinnati, OH 45267-0769, USA. Phone: 513-588-5281, Fax: 513-558-5791 (Tiepu.Liu@UC.Edu)

Received: 11 June 1996 / Accepted: 23 November 1996 / Published: 5 December 1996

Abstract

A mathematical model simulation was performed to estimate the amount of radioactivity in plasma, normal tissues, and tumor tissue through three delivery approaches: one step radiolabeled monoclonal antibody (MAb) CC49 i.v. bolus injection, two step method with biotin conjugated CC49 i.v. bolus injection followed 72 hours later by i.v. bolus radiolabeled streptavidin injection, and gene therapy method to express biotin on the tumor cell surface followed by i.v. bolus radiolabeled streptavidin injection. The mathematical model was built based on a system of ordinary differential equations consisting of inputs and outputs of model components in plasma, normal tissues, and tumor tissue. Through computer modeling, we calculated concentrations of each component for plasma, tumor and normal tissues at various time points. Radioactivity ratios of tumor to plasma and tumor to normal tissues increased with time. The increase of tumor to normal tissue ratios was much faster for the gene therapy approach than for single step and two step approaches, e.g., a ratio of 24.26 vs. 2.06 and 6.24 at 72 hours after radioligand injection. Radioactivity ratios predicted by the model varied with the amount of radioactivity injected and the time interval between injections. The model could be used to evaluate different radioimmunotherapy strategies and to predict radioactivity biodistribution using other receptor-ligand systems.

Keywords: Cancer, computer simulation, gene therapy, mathematical model, MAb.

Introduction

The use of radiation therapy has improved curative treatment for many tumors. However, this technique has practical limitations in regard to limited field of therapy, normal tissue toxicity, and radioresistance mechanisms. Considerable research efforts have been directed at ways to "target" radioac-

tive isotopes to sites of malignant disease. Currently, the use of monoclonal antibodies (MAb) directed to "tumor-associated" antigens on cancer cells represents one approach to this problem which has had success in various animal model systems [1] and is the subject of considerable current phase I and II clinical trials in humans [2-10]. Such a strategy provides the ability to localize radioactive isotopes to multiple sites of disease with hopefully adequate amounts of radia-

---

* To whom correspondence should be addressed

tion to produce an antitumor effect and/or radioimmune imaging for diagnostic purposes. A second emerging strategy is to use radioactively labeled peptides able to bind to receptor positive tumor cells [11, 12]. Research efforts which provide better radioactive isotope delivery systems and/or targeting strategies will enhance our ability to apply targeted radiation therapy to the human cancer problem. Currently, the following delivery systems and targeting strategies have been actively studied in various models.

1. Single step method. This is the traditional approach with direct application of radiolabeled MAb for radioimmunodetection and radioimmunotherapy of tumors [3-10].

2. Pretargeting radioimmunotherapy strategies (two or three step pretargeting methods]. The two step approach takes advantage of the high affinity receptor-ligand systems, such as avidin-biotin, by conjugating one of the pair to a MAb for targeting and the other member of the pair in a radioligand preparation [13-15]. In this way, the conjugated MAb is administered first to bind to tumor antigen and after a certain interval to allow for plasma clearance, radiolabeled ligand with high binding affinity to the MAb conjugate is administered. Systems using either streptavidin radioligand [16-20] or biotin radioligand [16,17,21-24] have been described. The three step approach uses an additional step through an unlabeled ligand chase to clear circulating antibody conjugates before the administration of radiolabeled ligand [18,25].

3. Gene therapy approach. Gene therapy techniques have been used to induce expression of high affinity receptors/ molecules on tumor cell surfaces [26-29] and optimal radio-isotope-labeled ligands capable of being delivered to these receptors in tumors are being developed.

Successful application of radioimmunodetection and radioimmunotherapy of tumors has been affected by many

problems. While laboratory experiments and clinical trials are underway to test these strategies, optimal delivery is complicated by a complex and variable schedule of MAb conjugate infusion. Questions regarding dosing, timing, affinity constants, and rate of metabolism still need to be answered. Computer modeling and simulation has been used to gain insight into the pharmacokinetics of various schedules, including pretargeting approaches. Mathematical models applying linear and nonlinear differential equations have been developed for this purpose. As more information from experimental and clinical trials is available and new strategies become possible, mathematical modeling and computer simulation can shed new light on the complexity of the pharmacokinetics and pharmacodynamics. In this report, we present results from a mathematical model simulation to compare the amount of radioactivity in plasma, normal tissues, and tumor tissue through three delivery approaches: single step radiolabeled MAb CC49 i.v. bolus injection, two step strategy with biotin conjugated CC49 i.v. bolus injection followed 72 hours later by i.v. bolus radiolabeled streptavidin injection, and gene therapy approach to express biotin on the tumor cell surface followed by i.v. bolus radiolabeled streptavidin injection.

Methods

Conceptual models were developed for three delivery approaches as depicted in Figures 1-3. The first approach is a single step directly radiolabeled antibody CC49 (MAb*) i.v. bolus injection to target antigen (Ag) in tumor tissue. The second approach is a typical two step strategy with biotin conjugated MAb (MAb-b) i.v. bolus injection to target antigen (Ag) in tumor tissue followed 72 hours later by i.v. bolus radiolabeled streptavidin (Av*) injection. The third approach is the gene therapy approach, where the biotin gene is delivered to the tumor cells and biotin is expressed on the cell surface (b). Radiolabeled streptavidin (Av*) is injected afterward.

Monoclonal antibodies

MAb CC49 is an IgG1 which was obtained by immunizing mice with purified TAG-72 [30]. Extensive experience has been gained in using this antibody in human trials. For this study, results from a phase II trial with ^{131}I-labeled CC49 at 10 mCi/mg in prostate cancer patients were used to provide pharmacokinetic values for simulation [6]. Plasma radioactivity time series data were statistically modeled for estimation of pharmacokinetical parameters of MAb CC49 for 15 patients in the clinical trial. The phamacokinetical analysis was based on linear compartment models that can also be represented by linear differential equations. Pharmacokinetical parameters, such as $T_{1/2}$, AUC, Vd_{ss}, Cl, and MRT, were calculated using a SAS nonlinear regression program

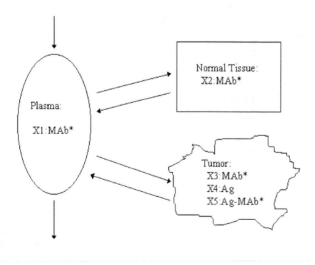

Figure 1. *Schematic representation of the one step method in which directly radiolabeled MAb (MAb*) is administered.*

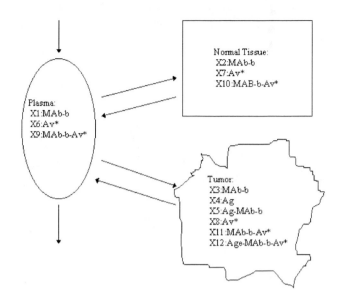

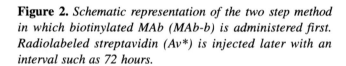

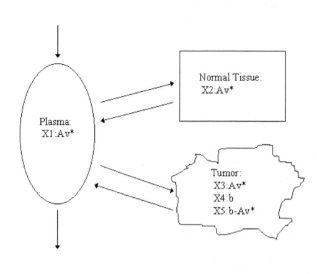

Figure 2. *Schematic representation of the two step method in which biotinylated MAb (MAb-b) is administered first. Radiolabeled streptavidin (Av*) is injected later with an interval such as 72 hours.*

Figure 3. *Schematic representation of the gene therapy method in which biotin is expressed on the tumor cell surface through genetic transduction such as direct intra-tumor injection of a viral vector. Radiolabeled streptavidin (Av*) is then injected.*

(PROC NLIN) [31-32]. Rate constants derived from these data were used in the simulation.

Avidin-biotin system

The avidin-biotin system represents a high affinity binding system for possible tissue targeting. Avidin and its analogue, streptavidin, are tetrameric proteins with a high affinity biotin-binding site on each subunit [33], which has one million-fold greater affinity than that of most antigen-antibody interactions. Since the binding is rapid and, once formed, very stable, it is expected that avidin and the vitamin biotin found in low concentrations in tissues and plasma offer possibilities for tumor imaging and therapy. Streptavidin has been reported to show less non-specific binding to tissues and is more suitable for radioiodination because it contains more tyrosine residues per subunit [14, 15]. Biotin can be chemically attached to proteins by its carboxyl end while its binding to avidin remains unaffected. Most biomolecules can be biotinylated without significant loss of biological activity [34]. Thus, the avidin-biotin system offers a universally applicable technology for cross-linking and targeting of biomolecules that has been used extensively *in vitro* [34] and *in vivo* [16-24,35-36]. In this study, we chose the avidin-biotin system with biotinylated MAb (MAb-b) CC49 to bind target cells, followed by the administration of radiolabeled streptavidin (Av*).

Gene expression

Methods to genetically induce tumor cell membrane expression of the high affinity biotin-streptavidin system are under development to allow employment of the biotin-streptavidin system for delivery of radioligands to tumor cells. This would involve the derivation of fusion genes encoding chimeric proteins derived from the RSV-G viral glycoprotein and short peptides with binding specificity for either biotin or streptavidin. In this study, we chose the approach of expressing biotin on the tumor cell surface. Radiolabeled streptavidin was administered later.

Design and production of appropriate radioactive ligands

Radiolabeling with ^{131}I and ^{125}I of MAb (MAb*) or streptavidin (Av*) uses the standard Iodogen method. Radiolabeling with other radioisotopes, such as ^{90}Y, [186]Re, and ^{111}In can be accomplished through standard procedures with commercially supplied reagents.

Localization, imaging studies and dosimetry

Localization of the various radiolabeled ligands and level of their persistence over time can be quantified experimentally using described procedures [37-42]. The absolute amount or concentration of radioactivity is proportional to the amount of ligand labeled, such as in a ratio of 10 mCi/mg. In this

Rate constants for MAb CC49 were estimated from clinical

Table 1. *Values for the Model Parameters Obtained from Literature or Clinical Trial Data. [a]*

Parameter	Value
MAb* and MAb-b: radiolabeled or biotin-conjugated CC49	
k_{12}, rate constant from plasma to tissue	5.129×10^{-6} s^{-1}
k_{21}, rate constant from tissue to plasma	1.444×10^{-5} s^{-1}
k_{el}, rate constant from plasma to environment	3.089×10^{-6} s^{-1}
Av*: radiolabeled avidin or streptavidin	
k'_{12}, rate constant from plasma to tissue	1.875×10^{-4} s^{-1}
k'_{21}, rate constant from tissue to plasma	1.870×10^{-4} s^{-1}
k'_{el}, rate constant from plasma to environment	3.710×10^{-5} s^{-1}
MAb-b-Av*	
k''_{el}, rate constant from plasma to environment	6.250×10^{-6} s^{-1}
k_f, rate constant for binding of MAb or MAb-b to antigen	10 mM^{-1} s^{-1}
k_r, rate constant for dissociation of antibody/antigen complex	10^{-5} s^{-1}
k_{met}, rate constant for internalization of antigen and expressed biotin	0
k'_f, rate constant for binding of Av to MAb-b and b	7×10^4 mM^{-1}s^{-1}
k'_r, rate constant for dissociation of Av from MAb-b and b	9×10^{-8} s^{-1}
n, valence of the antibody/antigen binding	2
n', valence of MAb-b binding to Av	1
c_0, initial free antigen or genetically expressed biotin concentration	1 mM
c_{MAb0}, initial plasma concentration of MAb* or MAb-b	3.92 mM, 39.2 mM, 392 mM
c_{Av0}, initial plasma concentration of Av*	3.92 mM, 39.2 mM, 392 mM

[a] Rate constants for MAb CC49 were estimated from clinical trial data [6]. All other model parameters were chosen from literature reports from experimental studies [14,15,18,19].

study, plasma concentration, normal tissue and tumor tissue concentrations of radioactivity will be calculated as proportional to the radiolabeled ligand concentration but without specific unit.

Mathematical modeling

The mathematical model was built based on a system of ordinary differential equations consisting of inputs and outputs of model components, such as MAb, antigen, streptavidin, and antigen, antibody, biotin, and streptavidin complexes in plasma, normal tissue, and tumor tissue. Model parameters, such as rate constants, transport coefficients, and initial val-

ues were based on previous clinical trial data or literature reports from experimental and theoretical estimates [6,19]. Through computer modeling, we calculated concentrations of each component in each tissue. Total radioactivity was calculated for each tissue by combining concentrations of free radiolabeled ligand with complexed radiolabeled ligand. Ratios of radioactivities of tumor to plasma and tumor to normal tissue were calculated for each time point.

General assumptions. Based on pharmacokinetic studies, conceptual models were developed as shown in Figures 1-3. After i.v. injection, directly labeled MAb or biotinylated MAb CC49 was distributed in the plasma, normal tissues and tumor tissue. The transport between plasma and tissues and from plasma to environment was determined by the rate constants estimated from clinical trial data, k_{12}, k_{21}, and k_{el}, listed in Table 1. The MAb binds to antigen according to a rate constant, k_f and dissociates from antigen/antibody complexes according to a rate constant k_r. When radiolabeled streptavidin

Table 2. *Predicted Radioactivity Ratios from Computer Simulation with Injected Amount of 3.92mM of Reagents. [a]*

Index	Time after radiotherapy (h)	One step method Ratio	Two step method Ratio	Two step method Relative ratio [b]	Gene therapy method Ratio	Gene therapy method Relative ratio [b]
Tumor:	4	0.08	1.39	17.38	2.38	29.75
Plasma	24	0.52	6.01	11.56	4.28	8.23
	72	0.84	– [c]	–	26.08	31.05
Tumor:	4	1.11	2.24	2.02	2.12	1.91
Normal tissue	24	1.55	4.68	3.02	3.99	2.57
	72	2.06	6.24	3.03	24.26	11.78

[a] *Same amount of MAb and streptavidin (3.92mM) was administered with an interval of 72 hr in the two step method. Radioactivity in tumor is 0.96 at 30 hr, 1.62 at 4 hr, and 1.74 at 4 hr after radiolabeled ligand injection for one step, two step and gene therapy method, respectively.*

[b] *Ratio of radioactivity ratios compared to one step method.*

[c] *Radioactivity in the plasma is completely eliminated.*

was injected into the plasma, it distributes in the body according to rate constants, k'_{12}, k'_{21}, and k'_{el}. It also binds to biotin (MAb-b and b) with high affinity, k'_f and k'_r. The MAb-b-Av* complex was eliminated from plasma at a rate constant, k''_{el}. We assume that there is no internalization of antigen or cell surface biotin, and non-specific antigen or biotin levels outside of tumor tissue are ignored. Previous experience of fitting data with pharmacokinetic models indicates that linear compartmental models fit the plasma data quite well. Therefore, it is assumed that the distribution between plasma and tissue is according to linear first-order kinetics. The linear kinetics implies that transference (input and output) is at a rate proportional to the concentration or amount of ligand.

Mathematical formulation. The conceptual model outlined above was represented by a set of differential equations - state matrix [43,44]. For a set of linear first-order differential equations, the general form is:

$$\frac{dX_i}{dt} = \sum_{j=1}^{n} k_{j-i} X_j - \sum_{j=0}^{n} k_{i-j} X_i, \quad X_i(0) = c$$

Where X_i or X_j is the state variable representing the concentration of each component, i.e., Ag, b, MAb*, MAb-b, Av*, Ag-MAb*, Ag-MAb-b, MAb-b-Av*, Ag-MAb-b-Av*,

and b-Av*, in compartment (plasma, normal tissue, and tumor tissue) i or j (i, j = 1,2,3). $X_i(0)$ is the initial concentration in compartment i. We specified that the initial concentration for antigen and genetically expressed biotin is 1 mM. The initial concentration for MAb* and MAb-b was specified as 3.92 mM with an increase of 10 times in the sensitivity analyses. The initial concentration for Av* in the two step method was specified to be equivalent to that of MAb-b. The choice of these initial values is for consistency with other investigators in mathematical modeling.

Numerical methods. A FORTRAN program, ADAPT II, was used for simulation on a DEC Alpha 3800s computer. The ADAPT II program uses the differential equation solver LSODA (Livermore Solver for Ordinary Differential equations with Automatic method switching for stiff and nonstiff problems), which uses variable order, variable step size formulations of Adam's method and Gear's method as the nonstiff and stiff equation solvers, respectively [45,46]. After the linear differential equations are defined by the state matrix and entered into the subroutine, the solution is obtained using the exponential of the matrix. This matrix exponential is approximated using an eigenvalue decomposition. The concentration of each component in each tissue is then calculated for each time point specified in the simulation.

Results

A series of computer simulations were performed based on a system of ordinary differential equations with specified rate constants for each model component, such as MAb, antigen, streptavidin, and antigen, antibody, biotin, and streptavidin complexes, as presented in Table 1. Through computer simulation, we calculated concentrations of each component in each tissue. Since our interest was in the total radioactivity in plasma, normal tissue, and tumor tissue, concentrations

Table 3. *Predicted Radioactivity Ratios from Computer Simulation with Injected Amount of 39.2mM of Reagents. [a]*

Index	Time after radiotherapy (h)	One step method Ratio	Two step method Ratio	Relative ratio [b]	Gene therapy method Ratio	Relative ratio [b]
Tumor:	4	0.08	1.17	14.63	8.70	109
Plasma	24	0.34	3.24	9.53	3634	10688
	72	0.44	– [c]	–	–	
Tumor:	4	1.04	1.00	0.96	4.95	4.76
Normal tissue	24	1.06	1.00	0.94	2180	2057
	72	1.09	1.00	0.92	–	–

[a] Same amount of MAb and streptavidin (39.2mM) was administered with an interval of 72 hr in the two step method. Radioactivity in tumor is 6.71 at 30 hr, 12.50 at 4 hr, and 25.88 at 4 hr after radiolabeled ligand injection for one step, two step and gene therapy method, respectively.
[b] Ratio of radioactivity ratios compared to one step method.
[c] Radioactivity in the plasma is completely eliminated.

from radiolabeled components, such as free radiolabeled ligand and complexed radiolabeled ligand, were combined. Ratios of radioactivities of tumor to plasma and tumor to normal tissue were calculated for each time point. The rate constants of directly labeled MAb or biotinylated MAb CC49 by i.v. injection between plasma and tissues and from plasma

to environment (k_{12}, k_{21}, and k_{el}) was estimated from clinical trial data [6]. The rate constants for MAb binding to antigen (k_f) and dissociation from antigen/antibody complexes (k_r) were from experimental data reported in the literature [19]. Radiolabeled streptavidin rate constants, k'_{12}, k'_{21}, and k'_{el} and its binding affinity to biotin, k'_f and k'_r were chosen from experimental data [14,15,18,19]. The MAb-b-Av* complex elimination rate constant, k''_{el}, was assumed as reported by Sung et al. [19].

Using phamacokinetical data from directly radiolabeled MAb CC49 in the single step approach, the ratio of tumor to plasma radioactivity increased from 0.08 at 4 hours to 0.52 at 24 hours and to 0.84 at 72 hours after radiolabeled antibody injection (Table 2). The corresponding ratios of tumor to normal tissue were 1.11, 1.55, and 2.06. By modeling the two step method, the ratio of tumor to plasma radioactivity increased from 1.39 at 4 hours to 6.01 at 24 hours after radiolabeled streptavidin injection (Table 2). The plasma radioactivity was completely eliminated after 36 hours. Ratios

Table 4. *Predicted Radioactivity Ratios from Computer Simulation with Injected Amount of 392mM of Reagents. [a]*

Index	Time after radiotherapy (h)	One step method Ratio	Two step method Ratio	Relative ratio [b]	Gene therapy method Ratio	Relative ratio [b]
Tumor:	4	0.07	0.59	8.43	1.08	15.43
Plasma	24	0.32	0.75	2.34	1.09	3.41
	72	0.41	1.94	4.73	1.25	3.05
Tumor:	4	1.00	1.00	1.00	1.01	1.01
Normal tissue	24	1.01	1.01	1.00	1.02	1.01
	72	1.01	1.01	1.00	1.17	1.16

[a] Same amount of MAb and streptavidin (392mM) was administered with an interval of 72 h in the two step method. Peak radioactivity in tumor is 64.00 at 30 hr, 93.01 at 4 hr, and 108.39 at 4 h after radiolabeled ligand

injection for one step, two step and gene therapy method, respectively.
[b] Ratio of radioactivity ratios compared to one step method.
[c] Radioactivity in the plasma is completely eliminated.

Table 5. *Predicted Radioactivity Ratios from Two Step Method with Varying Intervals.* [a]

Index	Time after radiotherapy (h)	Time Interval in Two Step Method (h)			
		24	48	72	96
Tumor:	4	0.67	1.08	1.39	1.82
Plasma	24	1.53	3.37	6.01	–
	72	13.71	– [b]	–	–
Tumor:	4	1.56	1.92	2.24	1.00
Normal tissue	24	2.40	3.49	4.68	1.00
	72	2.83	4.46	6.24	1.00

[a] *Same amount of MAb and streptavidin (3.92mM) was administered with varying intervals of 24, 48, 72, and 96 h in the two step method. Radioactivity in tumor is 1.13, 1.44, 1.62, and 1.40, respectively at 4 h after radiolabeled ligand injection.*
[b] *Radioactivity in the plasma is completely eliminated.*

of tumor to normal tissue were 2.24 at 4 hours, 4.68 at 24 hours, and 6.24 at 72 hours after radiolabeled streptavidin injection. By modeling the gene therapy approach, the ratio of tumor to plasma radioactivity increased from 2.38 at 4 hours to 4.28 at 24 hours and to 26.08 at 72 hours after radiolabeled streptavidin injection (Table 2). The corresponding ratios of tumor to normal tissue were 2.12, 3.99, and 24.26. The relative ratios of tumor to plasma radioactivity

ratios were 11.56 to 17.38 comparing the two step method to the one step method, and 8.23 to 31.05 comparing the gene therapy method to the one step method. The relative ratios of tumor to normal tissue radioactivity ratios were about 2 to 3-fold when comparing the two step method to the one step method, and about 2 to 12-fold when comparing the gene therapy method to the one step method.

Radioactivity ratios and relative ratios were also calculated from the sensitivity analysis by varying the amount of MAb and radioligand. Tables 3 and 4 present results for the amount of 39.2mM and 392mM compared to the 3.92mM results in Table 1. For the one step method, the tumor to plasma and tumor to normal tissue radioactivity ratios decreased to less than 0.5 and about 1 relative to those at the 3.92mM dose. For the two step method, both tumor to plasma and tumor to normal tissue ratios decreased markedly as the

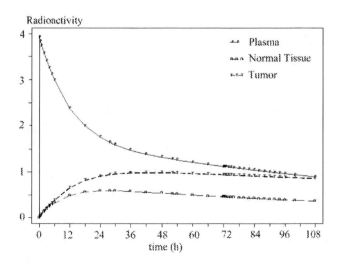

Figure 4. *Concentrations of radioactivity in plasma, normal tissues and tumor tissue in one step method following the injection of 3.92μM radiolabeled MAb.*

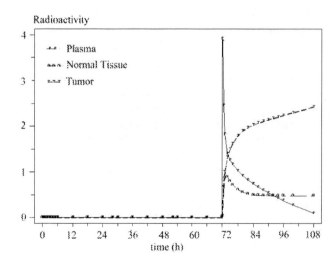

Figure 5. *Concentrations of radioactivity in plasma, normal tissues and tumor tissue in two step method with 3.92mM MAb-biotin (MAb-b) injection followed 72 hours later by injection of 3.92μM radiolabeled streptavidin.*

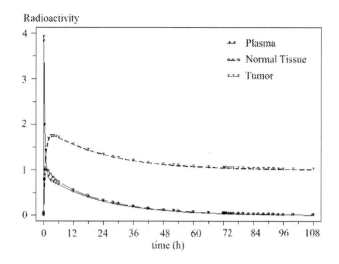

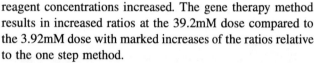

Figure 6. *Concentrations of radioactivity in plasma, normal tissues and tumor tissue in the gene therapy method following the injection of 3.92μM radiolabeled streptavidin.*

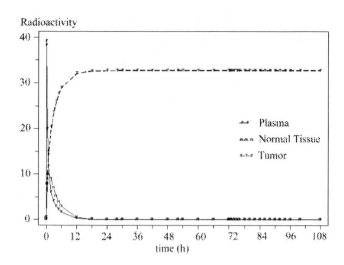

Figure 7. *Concentrations of radioactivity in plasma, normal tissues and tumor tissue in the gene therapy method following the injection of 39.2μM radiolabeled streptavidin.*

reagent concentrations increased. The gene therapy method results in increased ratios at the 39.2mM dose compared to the 3.92mM dose with marked increases of the ratios relative to the one step method.

Table 5 lists the ratios for the two step method with varying intervals between MAb-b and Av* injections. Both tumor to plasma and tumor to normal tissue ratios increased with longer intervals between 24 and 72 hours. However, the ratios decreased at the interval of 96 hours. This result indicates that an interval of about 72 hours is optimal.

Comparing the absolute concentrations of radioactivity (Figures 4-6) shows that a large amount of radioactivity remained in the plasma and the radioactivity concentration in plasma, tumor and normal tissues were not very different for a period of time in the one step method. In both the two step and gene therapy methods, plasma radioactivity disappearance was much faster due to the shorter half life of Av* compared to MAb*. The tumor concentrations were much higher than the plasma and normal tissue concentrations in both the two step method and the gene therapy method. The increase of tumor concentration was accompanied by lower levels of plasma and normal tissue concentrations in the gene therapy method than in the two step method. The increase of tumor concentration in the gene therapy method was substantial when the injected dose of radioligand increased from 3.92mM to 39.2mM (Figure 7).

Discussion

In directly radiolabeled MAb administration strategies, approaches have been taken to increase the tumor to plasma ratio by using low molecular weight fragments to achieve a lower plasma concentration than with intact antibodies. How-

ever, this is achieved at the cost of lower tumor uptake. To maximize targeting molecule (*e.g.* MAb) deposition into tumor sites while minimizing radioactive isotope exposure to the bone marrow, investigators have designed several strategies to separate these two components. One strategy has been to develop bifunctional MAb with one combining site for tumor and a second binding site for the radioactive ligand [47-49]. The bifunctional antibody is administered and allowed to circulate for several days (optimal tumor deposition) and then the radioligand is administered which has a rapid tissue distribution and short plasma half-life. This strategy allows tumor localization of the isotope to occur rapidly (matter of a few hours) with limited radiation dose to the bone marrow. The major limitation of this strategy has been the reduced affinity of the individual antigen combining sites (single rather than dual binding sites similar to Fab fragments), variable kinetics/distribution of the separate components making optimal schedules of therapy difficult to standardize. The two step pretargeting approach takes advantage of the high affinity avidin-biotin system by conjugating one of the pair to a MAb for targeting and the other member of the pair in a radioligand preparation. In this way, the conjugated MAb is administered first to bind to tumor antigen, and after a certain interval of clearance radiolabeled ligand with high binding affinity to the MAb conjugate is administered. These systems appear attractive in animal models and in radioimaging studies in humans. A major drawback of this system is that the high affinity binding of radioligand to MAb conjugates occurs with any residual MAb in the plasma or extravascular space resulting in a high level of background radiation. The use of an additional step [18,25] to clear circulating antibody conjugates improves the distribution of the radioligand but results in a complex and variable schedule of MAb conjugate infusion, plasma conjugate clearing reagent administration, and radioligand infusion. Direct intra-tumor injection of genes has been employed in a variety of anti-cancer gene therapy strategies in human trials. In this regard, Plautz *et al*. have utilized a technique of *in situ* tumor transduction in pre-clinical and clinical trials with genes encoding alloantigens to achieve anti-tumor immunization [50]. In addition, direct intra-tumor injection of viral vectors has been carried out in pre-clinical and clinical trials to achieve toxin gene delivery for therapy of gliomas [51]. Thus, the technique of direct tumor transduction is a method which would allow implementation of a variety of anti-cancer gene therapy strategies such as the avidin-biotin system. Scheduling through this approach is much simpler.

Results from experimental models have demonstrated the advantage of the two step strategy over direct radiolabeled antibody administration. Khawli *et al*. have shown that treatment of tumor-bearing nude mice with biotinylated MAb can achieve a 1.3-2.6 fold increase of tumor localization ratios at 24 hours after injection of radiolabeled streptavidin in a two step pretargeting strategy with biotinylated MAb and radiolabeled streptavidin compared to directly labeled antibodies [18]. A similar magnitude of ratios has also been reported by other investigators [13]. While neither experimental nor mathematical modeling results have been reported so far for the gene therapy approach of radioimmunotherapy of cancer, pharmacokinetic comparisons and mathematical modeling have been performed with different two step pretargeting strategies [19-22]. Sung et al. have developed a pharmacokinetic model involving two step pretargeting [19]. The model describes three compartments: an avascular tumor nodule, such as a very early primary tumor or a micrometastasis; the normal tissue; and the plasma. The results indicate that the two step protocol yields an approximately 2- to 3-fold enhancement compared with the one step direct radiolabeled antibody administration method.

In this study we have compared the two step method with a one step method based on a three compartment model similar to that developed by Sung et al [19]. However, the tumor nodule is not avascular in our model. The vascular tumor tissue represents well established rather than early tumor, which is more commonly treated with radiation therapy. Nevertheless, our results comparing the two step method with a one step method give similar ratios of tumor to normal tissue radioactivity, such as 2.02 to 3.03-fold increases from 4 to 72 hours after radioligand injection, supporting the advantage of pretargeting approaches. Moreover, we modeled the gene therapy approach in this study. Although multicompartment, numerical models of cellular events in the pharmacokinetics of gene therapies have been developed [52], our modeling of the gene therapy approach for radioimmunotherapy is new. According to the simulation results, much greater tumor to plasma and tumor to normal tissue ratios can be achieved through the gene therapy approach, such as 31.05 and 11.78-fold increases at 72 hours compared to the one step method and markedly greater ratios can be achieved through dose increment.

It is noted that all modeling results are subject to the presumed parameter values as well as the model specified. Uncertainties exist in the representation of the model to the real physiological and pathological process associated with the different approaches. Many of the parameter values used in the simulation are derived from animal model experiments or are estimates from preliminary cell culture studies. Improvement in the modeling can be made when new information comes available from experimental studies. For example, in the gene therapy method, the dramatic degree of simulated localization reflects the absence of significant amounts of biotin in the plasma or extravascular space of normal tissues and the high affinity binding of radiolabeled streptavidin to biotin expressed in the tumor. However, the degree to which this strategy will approach these dramatic values will depend on the future success of *in vivo* genetic transduction and radioligand design. Published results from our cell culture studies have indicated that the approach of genetic transduction is promising (26). Information about the amount and disappearance of genetically expressed receptors will greatly enhance the prediction through computer simulation.

Sensitivity analyses in this study showed that optimal dosing or timing schedules could be achieved with changing parameter estimates. For MAb CC49, a murine MAb with a plasma half-life of about 40 hours from one compartment modeling of phase I clinical trial data, an interval of about two half-lives between the injections of MAb-b and Av* seems most beneficial. In conclusion, this study demonstrates that the two step method can achieve a 2 to 3-fold increase in tumor to normal tissue radioactivity ratios compared to a one step method. The gene therapy method can achieve even higher ratios, e.g., 1.91 to 11.78-fold from 4 to 72 hours after radioligand injection, and the strategy appears to be easier to schedule. Concentrations of radioactivity and the relative ratios can be calculated from modeling and optimal strategies regarding dosing and timing can be evaluated from simulation. As a tool, the computer simulation is useful in examining the performance of different therapeutic strategies as long as new pharmacokinetical information is incorporated into the model from experimental and clinical studies.

Acknowledgments This study was supported in part by grant CA13148 and contract CM87215, NCI, NIH.

Reference

1. Buchsbaum, D.J.; Langmuir, V.K.; Wessels, B.W. *Med. Phys.* **1993**, *20*, 551.
2. Goldenberg, D.M. *Cancer Res* (Suppl) **1995**, *55*, 5703s.
3. Deb, N.; Goris, M.; Trisler, K.; Fowler, S.; Saal, J.; Ning, S.; Becker, M.; Marquez, C.; Knox, S. *Clin Can Res* **1996**, *2*, 1289.
4. Meredith, R.F.; Khazaeli, M.B.; Plott, W.E.; Saleh, M.N.; Liu, T; Allen, L.F.; Russell, C.D.; Orr, R.A.;

Colcher, D.; Schlom, J.; Shochat, D.; Wheeler, R.H.; LoBuglio, A.F. *J. Nucl. Med.* **1992,** *33,* 23.

5.   Meredith, R.F.; Khazaeli, M.B.; Liu, T.; Plott, G.; Wheeler, R.H.; Russell, C.; Colcher, D.; Schlom, J.; Shochat, D.; LoBuglio, A.F. *J. Nucl. Med.* **1992,** *33,* 1648.

6.   Meredith, R.F.; Bueschen, A.J.; Khazaeli, M.B.; Plott, W.E.; Grizzle, W.E.; Wheeler, R.H.; Schlom, J.; Russell, C.D.; Liu, T.; and LoBuglio, A.F. *J. Nucl. Med.* **1994,** *35,* 1017.

7.   Meredith, R.F.; Khazaeli, M.B.; Plott, W.E.; Liu, T.; Russell, C.D.; Wheeler, R.H.; LoBuglio, A.F. *Antibody, Immunoconj. and Radiopharm.* **1993,** *6,* 39.

8.   Meredith, R.F.; Khazaeli, M.B.; Grizzle, W.E.; Orr, R.A.; Plott, G.; Urist, M.M.; Liu, T.; Russell, C.D.; Wheeler, R.H.; Schlom, J.; LoBuglio, A.F. *Human Antibodies and Hybridomas* **1993,** *4,* 190.

9.   Divgi, C.R.; Scott, A.M.; Dantis, L.; Capitelli, P; Siler, K.; Hilton, S.; Finn, R.D.; Kemeny, N.; Kelsen, D.; Kostakoglu, L.; Schlom, J.; Larson, S.M. *J Nucl Med* **1995,** *36,* 586.

10. Kaminski, M.S.; Zasadny, K.R.; Francis, I.R.; Fenner, M.C.; Ross, C.W.; Milik, A.W.; Estes, J.; Tuck, M.; Regan, D.; Fisher, S.; Glenn, S.D.; Wahl, R.L. *J Clin Oncol* **1996,** *14,* 1974.

11. Kvols, L.K.; Moertel, C.G.; O'Connell, M.J.; Schut, A.J.; Rubin, J.; Hahn, R.G. *N. Engl. J. Med.* **1986,** *315,* 663.

12. Fischman, A.J.; Babich, J.W.; Strauss, H.W. *J. Nucl. Med.* **1993,** *34,* 2253.

13. Goodwin D.A. *J. Nucl. Med.* **1995,** *36,* 876.

14. Green, N.M. *Methods in Enzymology* **1990,** *184,* 51.

15. Chaiet, L.; Wolf, F.J. *Arch. Biochem. Biophys.* **1964,** *106,* 1.

16. Hnatowich, D.J.; Virzi, F.; Rusckowski, M. *J. Nucl. Med.,* **1987,** *28,* 1294.

17. Bamias, A.; Epenetos, A.A.. *Antib. Immunoconjug. and Radiopharm.* 1992, 5, 385.

18. Khawli, L.A.; Alauddin, M.M.; Miller, G.K.; Epstein, A.L. *Antib. Immunoconjug. and Radiopharm.* **1993,** *6,* 13.

19. Sung, C.; van Osdol, W.W.; Saga, T.; Neumann, R.D.; Dedrick, R.L.; Weinstein, J.N. *Cancer Res.* **1994,** *54,* 2166.

20. Saga,T.; Weinstein, J.N.; Jeong, J.M.; Heya, T.; Lee, J.T.; Le, N.; Paik, C.H.; Sung, C.; Neumann, R.D. *Cancer Res.* **1994,** *54,* 2160.

21. Sung, C.; van Osdol, W.W. *J. Nucl. Med.* 1995, 36, 867.

22. van Osdol, W.W.; Sung, C.; Dedrick, R.L.; Weinstien, J.N. *J. Nucl. Med.* **1993,** *34,* 1552.

23. del Rosario, R.B.; Wahl, R.L. *J. Nucl. Med.* **1993,** *34,* 1147.

24. Rowlinson-Busza, G.; Hnatowich, D.J.; Rusckowski, M.; Snook, D.; Epenetos, A.A. *Antib. Immunoconjug. and Radiopharm.***1993,** *6,* 97.

25. Paganelli, G.; Magnani, P.; Zito, F.; Villa, E.; Sudati, F.; Lopalco, L.; Rossetti, C.; Malcovati, M.; Chiolerio, F.; Seccamani, E.; Siccardi, A.G.; Fazio, F. *Cancer Res.* **1991,** *51,* 5960.

26. Raben, D.; Buchsbaum, D.; Khazaeli, M.B.; Rosenfeld, M.E.; Gillespie, G.Y.; Grizzle, W.E.; Liu, T.; Curiel, D.. *Gene Therapy* **1996,** *3,* 567.

27. Curiel, D.T.; Wagner, E.; Cotten, M.; Birnstiel, M.L.; Agarwal, S.; Li, C.-M.; Loechel, S.; Hu, P.-C. *Human Gene Therapy* **1992,** *3,* 147.

28. Curiel, D.T. *Nat. Immun.* 1994, 13, 141.

29. Tang, D.-C.; Johnston, S.A.; Carbone, D.P. *Cancer Gene Therapy* **1994,** *1,* 15.

30. Muraro R.; Kuroki M; Wunderlich D. *Cancer Res* **1988,** *48,* 4588.

31. SAS Institute Inc. SAS/STAT user's guide, Version 6, Fourth Ed., Volume 1 & 2, Cary, NC: SAS Institute Inc., 1989.

32. Liu T. *Controlled Clinical Trials* **1991***, 12,* 654.

33. Green, M. In: Anfunsen, C., Edsall, J., and Richards, F. Eds. Advances in Protein Chemistry, vol. 29, New York: Academic Press; 85-133, 1975.

34. Wilchek, M.; Bayer, E. *Anal Biochem.* **1988,** *171,* 1.

35. Rosebrough, S.F. *Nucl. Med. Biol.* **1993,** *20,* 663.

36. Hnatowich, D.J.; Fritz, B.; Virzi, F.; Mardirossian, G.; Rusckowski, M. *Nucl. Med. Biol.* **1993,** *20,* 189.

37. Buchsbaum, D.J.; Lawrence, T.S.; Roberson, P.L.; Heidorn, D.B.; Ten Haken, R.K.; Steplewski, Z. *Int. J. Radiat. Oncol. Biol. Phys.* **1993,** *25,* 629.

38. Buchsbaum, D.; Randall, B.; Hanna, D.; Chandler, R.; Loken, M.; Johnson, E. *Eur. J. Nucl. Med.* **1985,** *10,* 398.

39. Buchsbaum, D.J.; Hanna, D.E.; Randall, B.C.; Buchegger, F.; Mach, J-P. *Int. J. Nucl. Med. Biol.* **1985,** *12,* 79.

40. Roberson, P.L.; Buchsbaum, D.J.; Heidorn, D. B.; Ten Haken, R.K. *Int. J. Radiat. Oncol. Biol. Phys.* **1992,** *24,* 329.

41. Roberson, P.L.; Buchsbaum, D.J.; Heidorn, D.B.; Ten Haken, R.K. *Antibody, Immunoconjug., and Radiopharm.* **1992,** *5,* 397.

42. Roberson, P.L.; Heidorn, D.B.; Kessler, M.L.; Ten Haken, R.K.; Buchsbaum, D.J. *Cancer* **1994,** *73,* 912.

43. Bates, D.M.; Watts, D.G. Nonlinear regression analysis and its applications. John Wiley & Sons, New York, NY, 1988.

44. Landaw, E.M.; Distefano, J.J. *Am J Physiol.* **1984,** *246,* r665.

45. D'Argenio, D.Z.; Schumitzky, A. ADAPT II User's Guide. Biomedical Simulations Resource. University of Southern California, Los Angeles, 1992.

46. D'Argenio, D.Z.; Schumitzky, A.; Wolf, W. *Computer Methods and Programs in Biomedicine* **1988***, 27, 47.*

47. Baxter, L.T.; Yuan, F.; Jain, R.K. *Cancer Res.* **1992,** *52,* 5838.

48. Le Doussal, J.M.; Chetanneau, A.; Gruaz-Guyon, A.; Martin, M.; Gautherot, E.; Lehur, P.A.; Chatal, J.F; Delaage, M.; Barbet, J. *J. Nucl. Med.* **1993**, *34*, 1662.

49. Peltier, P.; Curtet, C.; Chatal, J.F.; Le Doussal, J.M.; Daniel, G.; Aillet, G.; Gruaz-Guyon, A.; Barbet, J.; Delaage, M. *J. Nucl. Med.* **1993**, *34*, 1267.

50. Plautz, G.E.; Yang, Z.Y.; Wu, B.Y.; Gao, X.; Huang, L.; Nabel, G.J. *Proc. Natl. Acad. Sci.* **1993**, *90*, 4645.

51. Nabel, G.J.; Chang, A.E.; Nabel, E.G.; Plautz, G.E.; Ensminger, W.; Fox, B.A.; Felgner, P.; Shu, S.; Cho, K. *Human Gene Therapy* **1994**, *5*, 57.

52. Ledley T.S.; Ledley F.D. *Human Gene Therapy* **1994**, *5*, 679.

Author index

Keyword index